ESSENTIAL MATHEMATICS
WITH APPLICATIONS

ALSO AVAILABLE FROM McGRAW-HILL

Schaum's Outline Series in Mathematics & Statistics

Most outlines include basic theory, definitions, hundreds of example problems solved in step-by-step detail, and supplementary problems with answers.

Related Titles on the Current List Include:

- Advanced Calculus
- Advanced Mathematics
- Analytic Geometry
- Basic Mathematics for Electricity & Electronics
- Basic Mathematics with Applications to Science and Technology
- Beginning Calculus
- Boolean Algebra & Switching Circuits
- Calculus
- Calculus for Business, Economics, & the Social Sciences
- College Algebra
- College Mathematics
- Complex Variables
- Descriptive Geometry
- Differential Equations
- Differential Geometry
- Discrete Mathematics
- Elementary Algebra
- Essential Computer Mathematics
- Finite Differences & Difference Equations
- Finite Mathematics
- Fourier Analysis
- General Topology
- Geometry
- Group Theory
- Laplace Transforms
- Linear Algebra
- Mathematical Handbook of Formulas & Tables
- Mathematical Methods for Business & Economics
- Mathematics for Nurses
- Matrix Operations
- Modern Abstract Algebra
- Numerical Analysis
- Partial Differential Equations
- Probability
- Probability & Statistics
- Real Variables
- Review of Elementary Mathematics
- Set Theory & Related Topics
- Statistics
- Technical Mathematics
- Tensor Calculus
- Trigonometry
- Vector Analysis

Schaum's Solved Problems Books

Each title in this series is a complete and expert source of solved problems with solutions worked out in step-by-step detail.

Related Titles on the Current List

- 3000 Solved Problems in Calculus
- 2500 Solved Problems in College Algebra and Trigonometry
- 2500 Solved Problems in Differential Equations
- 2000 Solved Problems in Discrete Mathematics
- 3000 Solved Problems in Linear Algebra
- 2000 Solved Problems in Numerical Analysis
- 3000 Solved Problems in Precalculus

Bob Miller's Math Helpers

Bob Miller's Calc I Helper
Bob Miller's Calc II Helper
Bob Miller's Precalc Helper

Available at most college bookstores, or for a complete list of titles and prices, write to: **Schaum Division**
McGraw-Hill, Inc.
Princeton Road, S-1
Hightstown, NJ 08520

ESSENTIAL MATHEMATICS
WITH APPLICATIONS

Second Edition

Lawrence A. Trivieri
DEKALB COLLEGE, CLARKSTON, GEORGIA

McGraw-Hill, Inc.
New York   St. Louis   San Francisco   Auckland   Bogotá   Caracas
Lisbon   London   Madrid   Mexico City   Milan   Montreal   New Delhi
San Juan   Singapore   Sydney   Tokyo   Toronto

ESSENTIAL MATHEMATICS WITH APPLICATIONS

Copyright © 1994, 1988 by McGraw-Hill, Inc. All rights reserved. Printed in the United States of America. Except as permitted under the United States Copyright Act of 1976, no part of this publication may be reproduced or distributed in any form or by any means, or stored in a data base or retrieval system, without the prior written permission of the publisher.

This book is printed on acid-free paper.

1 2 3 4 5 6 7 8 9 0 VNH   VNH 9 0 9 8 7 6 5 4 3

ISBN 0-07-065229-5

This book was set in Times Roman by Monotype Composition Company.
The editors were Michael Johnson, Karen M. Minette, and Margery Luhrs;
the designer was Leon Bolognese; the production supervisor was Louise Karam.
Von Hoffmann Press, Inc., was printer and binder.

Chapter opening graphic: Slide Graphics of New England, Inc.

Cover photograph: © Dr. Dennis Kunkel/PHOTOTAKE, NYC

Library of Congress Cataloging-in-Publication Data

Trivieri, Lawrence A.
    Essential mathematics with applications / Lawrence A. Trivieri.—
  2nd ed.
      p.   cm.
    Includes index.
    ISBN 0-07-065229-5 (acid-free)
    1. Mathematics.   I. Title.
QA39.2.T7325   1994
513'.122—dc20   93-19053

CONTENTS

Preface xi
Suggested Course Outlines xv
To the Student xvii

CHAPTER 1
The Whole Numbers 1

PRETEST 2

1.1 Numbers, Numerals, Counting, and Place Value 4
1.2 Ordering and Addition of Whole Numbers 7
1.3 Subtraction of Whole Numbers 14
1.4 Multiplication of Whole Numbers 22
1.5 Division of Whole Numbers 30
1.6 Rounding and Estimating 37
1.7 Mixed Operations with Whole Numbers 40
1.8 Operations Involving the Whole Number 0 47

CHAPTER SUMMARY 49
REVIEW EXERCISES 50
TEASERS 52
CHAPTER 1 TEST 53

CHAPTER 2
Fractions 55

PRETEST 56

2.1 Fractions 58
2.2 Equivalent Fractions 65
2.3 Least Common Multiple 72
2.4 Addition of Fractions 78
2.5 Ordering of Fractions; Subtraction 86
2.6 Addition and Subtraction of Mixed Numbers 91
2.7 Multiplication of Fractions and Mixed Numbers 98
2.8 Division of Fractions and Mixed Numbers 108

CHAPTER SUMMARY 113
REVIEW EXERCISES 113
TEASERS 117
CHAPTER 2 TEST 118
CUMULATIVE REVIEW, CHAPTERS 1–2 120

CHAPTER 3
Decimals 121

PRETEST 122

3.1 Decimals as Numbers 124
3.2 Addition and Subtraction of Decimals 128
3.3 Multiplication of Decimals 131
3.4 Rounding and Estimating 135
3.5 Division of Decimals 138
3.6 Decimal to Fraction and Fraction to Decimal Conversions 145
3.7 Applications Involving Decimals 150

CHAPTER SUMMARY 154
REVIEW EXERCISES 154
TEASERS 156
CHAPTER 3 TEST 157
CUMULATIVE REVIEW, CHAPTERS 1–3 158

CHAPTER 4
Percent with Applications 159

PRETEST 160

4.1 Percent 162
4.2 More on Percents 167
4.3 Problems Involving Percents 172

4.4 Percent Increase or Decrease 178
4.5 Sales and Property Tax 182
4.6 Commissions 187
4.7 Discount 192
4.8 Simple Interest 195

CHAPTER SUMMARY 201
REVIEW EXERCISES 201
TEASERS 203
CHAPTER 4 TEST 204
**CUMULATIVE REVIEW,
CHAPTERS 1–4** 205

CHAPTER 5
Measurement 207
PRETEST 208

5.1 Units of Measurement 210
5.2 Polygons and Perimeter 217
5.3 Circles and Circumference 225
5.4 Area Enclosed by Polygons and Circles; Surface Area 230
5.5 Volumes of Solids 244

CHAPTER SUMMARY 251
REVIEW EXERCISES 251
TEASERS 253
CHAPTER 5 TEST 254
**CUMULATIVE REVIEW,
CHAPTERS 1–5** 255

CHAPTER 6
Signed Numbers 257
PRETEST 258

6.1 Integers 260
6.2 Signed Numbers 264
6.3 Addition of Signed Numbers 267
6.4 Subtraction of Signed Numbers 275
6.5 Multiplication of Signed Numbers 279
6.6 Division of Signed Numbers 286
6.7 Simplest Form of a Rational Number 290
6.8 Mixed Operations with Signed Numbers 292
6.9 Applications Involving Signed Numbers 295

CHAPTER SUMMARY 299
REVIEW EXERCISES 299
TEASERS 301
CHAPTER 6 TEST 302
**CUMULATIVE REVIEW,
CHAPTERS 1–6** 303

CHAPTER 7
An Introduction to Algebra 305
PRETEST 306

7.1 Algebraic Expressions 308
7.2 Evaluating Algebraic Expressions 311
7.3 Simplifying Algebraic Expressions by Combining Like Terms 313
7.4 Simplifying Algebraic Expressions by Removing Symbols of Grouping 317
7.5 The Language of Algebra 321

CHAPTER SUMMARY 327
REVIEW EXERCISES 327
TEASERS 328
CHAPTER 7 TEST 329
**CUMULATIVE REVIEW,
CHAPTERS 1–7** 330

CHAPTER 8
Linear Equations 331
PRETEST 332

8.1 Solving Linear Equations Using Addition or Subtraction 334
8.2 Solving Linear Equations Using Multiplication or Division 340
8.3 Solving Linear Equations Using More Than One Operation 345
8.4 Simplifying First before Solving Linear Equations 349
8.5 Applications Involving Linear Equations: Part 1 354
8.6 Applications Involving Linear Equations: Part 2 360
8.7 Formulas 369
8.8 Ratio and Proportion 377

CHAPTER SUMMARY 385
REVIEW EXERCISES 386
TEASERS 387
CHAPTER 8 TEST 388

Contents

      CUMULATIVE REVIEW,
          CHAPTERS 1–8 389

CHAPTER 9
Exponents and Scientific Notation 391

PRETEST 392

- **9.1** More on Positive Integer Exponents 394
- **9.2** Powers of Products and Quotients 401
- **9.3** Zero and Negative Integer Exponents 405
- **9.4** Scientific Notation 409

      CHAPTER SUMMARY 415
      REVIEW EXERCISES 415
      TEASERS 416
      CHAPTER 9 TEST 417
      CUMULATIVE REVIEW,
          CHAPTERS 1–9 418

CHAPTER 10
Polynomials 419

PRETEST 420

- **10.1** Basic Terminology Associated with Polynomials 422
- **10.2** Addition of Polynomials 426
- **10.3** Subtraction of Polynomials 428
- **10.4** Multiplication of Polynomials 431
- **10.5** Multiplying Binomials 436
- **10.6** Division of Polynomials 441

      CHAPTER SUMMARY 448
      REVIEW EXERCISES 449
      TEASERS 450
      CHAPTER 10 TEST 451
      CUMULATIVE REVIEW,
          CHAPTERS 1–10 452

CHAPTER 11
Factoring and Special Products 453

PRETEST 454

- **11.1** Review of Prime Factorization 456
- **11.2** Common Monomial Factoring 459
- **11.3** Factoring the Difference of Two Squares 464
- **11.4** Factoring Trinomials of the Form $x^2 + bx + c$ 468
- **11.5** Factoring Quadratic Trinomials of the Form $ax^2 + bx + c$ 472
- **11.6** Factoring by Grouping 476
- **11.7** Factoring Completely 479
- **11.8** Solving Equations by Factoring 481
- **11.9** Applications Involving Equations 486

      CHAPTER SUMMARY 491
      REVIEW EXERCISES 492
      TEASERS 493
      CHAPTER 11 TEST 494
      CUMULATIVE REVIEW,
          CHAPTERS 1–11 495

CHAPTER 12
Rational Expressions 497

PRETEST 498

- **12.1** Rational Expressions and Their Simplification 500
- **12.2** Multiplication and Division of Rational Expressions 505
- **12.3** Addition and Subtraction of Like Rational Expressions 511
- **12.4** Least Common Multiple 515
- **12.5** Addition and Subtraction of Unlike Rational Expressions 518
- **12.6** Complex Rational Expressions 522
- **12.7** Equations Involving Rational Expressions 526
- **12.8** Applications Involving Rational Expressions 531

      CHAPTER SUMMARY 535
      REVIEW EXERCISES 535
      TEASERS 537
      CHAPTER 12 TEST 538
      CUMULATIVE REVIEW,
          CHAPTERS 1–12 539

CHAPTER 13
Graphing Linear Equations and Linear Inequalities 541

PRETEST 542

- **13.1** Rectangular Coordinate System 544
- **13.2** Linear Equations in Two Variables 550

13.3 Graphing Straight Lines 553
13.4 Slope of a Line 559
13.5 Linear Inequalities in One Variable and Their Graphs 565
13.6 Linear Inequalities in Two Variables 572
13.7 Graphs of Linear Inequalities in Two Variables 574

CHAPTER SUMMARY 582
REVIEW EXERCISES 582
TEASERS 584
CHAPTER 13 TEST 585
CUMULATIVE REVIEW, CHAPTERS 1–13 586

CHAPTER 14
Systems of Linear Equations 587

PRETEST 588

14.1 Graphical Method for Solving a System of Linear Equations 590
14.2 Elimination Method for Solving a System of Linear Equations 598
14.3 Substitution Method for Solving a System of Linear Equations 606
14.4 Applications Involving Systems of Equations 610

CHAPTER SUMMARY 616
REVIEW EXERCISES 616
TEASERS 618
CHAPTER 14 TEST 619
CUMULATIVE REVIEW, CHAPTERS 1–14 620

CHAPTER 15
Square Roots 621

PRETEST 622

15.1 Square Roots of Nonnegative Numbers 624
15.2 Simplifying Square Roots of Expressions without Fractions 627
15.3 Simplifying Square Roots of Expressions with Fractions 631
15.4 Addition and Subtraction of Square Roots 635
15.5 Multiplication and Division of Square Roots 639
15.6 Equations Involving Square Roots 646

CHAPTER SUMMARY 649
REVIEW EXERCISES 649
TEASERS 650
CHAPTER 15 TEST 651
CUMULATIVE REVIEW, CHAPTERS 1–15 652

CHAPTER 16
Quadratic Equations and the Pythagorean Theorem 653

PRETEST 654

16.1 The Standard Form of a Quadratic Equation 656
16.2 Solving Quadratic Equations by Factoring 658
16.3 Solving Quadratic Equations Using Square Roots 661
16.4 Solving Quadratic Equations by Completing the Square 665
16.5 The Quadratic Formula 671
16.6 Applications Involving Quadratic Equations 675
16.7 The Pythagorean Theorem 681

CHAPTER SUMMARY 690
REVIEW EXERCISES 690
TEASERS 692
CHAPTER 16 TEST 693
CUMULATIVE REVIEW, CHAPTERS 1–16 694

CHAPTER 17
Basic Concepts of Geometry 695

PRETEST 696

17.1 Terminology and Notation 698
17.2 Angles 703
17.3 Some Special Pairs of Angles 708
17.4 Parallel and Perpendicular Lines 712
17.5 Triangles 719
17.6 Similar Triangles 725

17.7 Special Right Triangles 731
17.8 Congruent Triangles 740

    CHAPTER SUMMARY 748
    REVIEW EXERCISES 748
    TEASERS 750
    CHAPTER 17 TEST 751
    CUMULATIVE REVIEW,
         CHAPTERS 1–17 752

ANSWERS TO ODD-NUMBERED EXERCISES **A-1**
ANSWERS TO CHAPTER PRETESTS **A-41**
INDEX **I-1**

PREFACE

Essential Mathematics with Applications, Second Edition, is written to provide students with the basic arithmetic and algebraic skills needed for daily life as well as further coursework in mathematics. As the title indicates, applications or word problems, chosen from real-life situations, are found throughout the text. These are included to help students learn to apply arithmetic, algebra, and geometry skills to problem-solving in everyday life. The development of topics is nonrigorous and readable. The writing style is conversational and should appeal to the student. Upon mastering the material in this text, the student should be able to succeed in an intermediate algebra course or a course designed especially for liberal arts students.

ORGANIZATION

Since the needs of a particular group of students at this level will vary, **flexibility** plays a major role in the organization of this text. For this reason a complete discussion of basic arithmetic is included, starting with whole numbers and continuing through the integers. These chapters can be tailored or omitted to fit specific needs. Chapter 5 on measurement, may be delayed without a loss of continuity. Chapters 6–12 are sequential and represent the core algebra chapters for a course at this level. Upon completion of Chapter 12, the instructor may choose materials from the remaining chapters as time and interest dictate. Suggested course outlines follow this preface.

Pretest

Each chapter begins with a pretest with each item referenced to the section where the material is covered.

Objective Statements

At the beginning of each section are objective statements. These help the student identify the important topics in the section.

Summary Sections

At the end of each chapter is a summary that includes the **key words** found in the chapter, a set of **review exercises** referenced to chapter sections, **teasers exercises** that challenge students, and **post-tests**.

Examples

Throughout the chapter there are numerous examples that illustrate the topics and procedures being introduced. These examples also reinforce definitions and rules for various arithmetic and algebraic operations.

Practice Exercises

The examples are accompanied by **practice exercises** keyed to the examples and found next to them. These practice exercises allow the student to access basic understanding and comprehension of the material being studied. For immediate feedback, the answers appear after groups of examples.

Cumulative Reviews

Starting with Chapter 2, each chapter has a set of cumulative review exercises.

Answers

Answers to all the chapter pretests as well as the odd-numbered exercises are included at the back of the book.

SUPPLEMENTS

There are a number of supplements available to assist both instructor and student.

Student's Solutions Manual

This contains solutions to all the pretests as well as solutions to the odd-numbered end-of-section exercises, end-of-chapter exercises, and cumulative reviews.

Instructor's Resource Manual

This contains solutions to all of the post-tests as well as solutions to the even-numbered end-of-section exercises, end-of-chapter exercises, and cumulative reviews. Sample tests are also included. There are multiple-choice and short answer tests for each chapter, midterm tests and final tests.

The Professor's Assistant

A computerized test generator that allows the instructor to create tests using questions generated from a standard testbank. This testing system enables the instructor to choose questions either manually or randomly by section, question type, difficulty level, and other criteria. This system is available for IBM, IBM compatible, and Macintosh computers.

Print Test Bank

This is a printed and bound copy of the questions found in the standard testbank.

Video Series

This provides students with additional instructional and visual lesson support.

Tutorial

This is a self-paced tutorial that reinforces selected topics and provides unlimited opportunities to review concepts and to practice problem solving. It requires virtually *no* computer training on the part of the student and is available for IBM, IBM-compatible, and Macintosh computers.

For further information about these supplements, please contact your local college division sales representative.

ACKNOWLEDGMENTS

I would like to express my sincere appreciation to the many people, who through their encouragement, criticisms, and suggestions, have contributed to the successful development of the First and Second Editions. I am especially grateful to the following individuals for their thoughtful reviews and suggestions: Ruth Afflack, California State University, Long Beach; Laura Cameron, University of New Mexico, Albuquerque; Mike Contino, California State University, Hayward; Terry Czerwinski, University of Illinois at Chicago; Roberta Gehrmann, California State University, Sacramento; Don Johnson, New Mexico State University; Miriam Keesey, San Diego State University; Mark Manchester, SUNY Agricultural and Technical College, Morrisville; Joan Mundy, University of New Hampshire; Susann Novales, San Francisco State University; Ellen O'Keefe, University of New Hampshire; Paul Pontius, Pan American University; Richard Semmler, Northern Virginia Community College; Doris Schoonmaker, Hudson Valley Community College; Peter Tannenbaum, California State University, Fresno; Faye Thames, Lamar University; Gerry Vidrine, Louisiana State University, Baton Rouge; and Steve Wilson, Sonoma State University.

I would like to thank Judy Knapp of the University of Wisconsin at Whitewater and Gloria Langer of the University of Colorado, Boulder, for their assistance with the accuracy check of the exercise solutions in the First Edition, Margaret Donlan of the University of Delaware, in the Second Edition, and Richard Semmler of Northern Virginia Community College, for preparing the student's solutions manual.

Special thanks go to the editorial and production staffs of McGraw-Hill, Inc., for their assistance and genuine interest in editing and producing this text. Among this dedicated group are Michael Johnson, Mathematics Editor; Karen Minette, Assistant Mathematics Editor; Margery Luhrs, Editing Supervisor; and Louise Karam, Production Supervisor. Their encouragement, guidance, and patience throughout the life of this project is sincerely appreciated.

Finally, I wish to thank my wife, Joyce B. Trivieri, for her loving support, encouragement, and patience throughout this project.

Lawrence A. Trivieri

SUGGESTED COURSE OUTLINES

BASIC COURSE: ALGEBRA WITH EXTENSIVE TREATMENT OF ARITHMETIC

Chapters 1–3, Sections 4.1–4.4, and Chapters 6–12 should be covered sequentially and completely.

The rest of Chapter 4 and all or parts of Chapter 5 may be covered, if necessary or desired.

Materials from Chapters 13–16 may also be selected depending on student preparation and instructor/class interests.

CORE COURSE: ALGEBRA WITH BRIEF REVIEW OF ARITHMETIC

Chapters 1–5 may be omitted.

Chapter 6 should be covered as quickly as the class preparation permits.

Chapters 7–12 should be covered sequentially and completely.

Materials from Chapters 13–16 may be selected depending on student preparation and instructor/class interests.

EITHER COURSE WITH A GEOMETRY COMPONENT

If geometry is to be included in the course, then either course outlined above should include Chapter 5 (which may be covered any time after Chapter 3) and all or parts of Chapter 17 (which may be covered any time after Chapter 8).

TO THE STUDENT

In order to meet with success, it is essential that you have a positive attitude and a willingness to learn mathematics. There are no spectators when it comes to learning mathematics. You learn mathematics by doing mathematics. That means *thinking* about what you are doing and then pushing the pencil across the paper in an attempt to do the exercises. It is necessary that you seek help as difficulties arise. This help can be sought from classmates, the course instructor, or the staff in the mathematics laboratory, if one exists at your school. You must do the assigned work, including homework, on a regular basis. There are no guaranteed routes to success in mathematics. However, a positive attitude, a willingness to learn, determination, patience, good study habits, and seeking additional help as needed, should help you to meet with success. Best wishes!

ESSENTIAL MATHEMATICS
WITH APPLICATIONS

CHAPTER 1
The Whole Numbers

The first type of number that you ever learned about as a child answers questions involving "How many?" such as: "How many days are in a week?" "How many fish did you catch?" "How many questions did you answer correctly?" Answers to these questions are given by numbers called "whole numbers."

In this chapter, we will review the whole numbers, together with the operations of addition, subtraction, multiplication, and division of them. Properties associated with these operations, as well as properties of the whole number 0, will also be reviewed. Ordering of the whole numbers will be examined. Rounding and estimating will also be discussed. Throughout the chapter, applications involving whole numbers will be introduced.

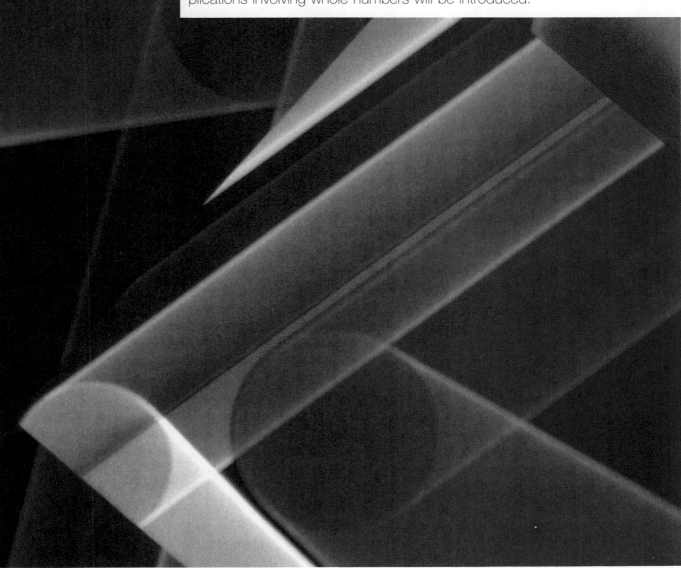

CHAPTER 1 PRETEST

This pretest covers material on whole numbers. Read each question carefully and try to answer all questions. Each question is keyed to the chapter section in which the particular topic is discussed. Check your answers with those given in the back of the text.

Questions 1–3 pertain to Section 1.1.

1. Indicate the place value of the underlined digit in 2<u>4</u>5,132.

2. Write the word name for 8,062,790.

3. Write the numeral for: Fifteen million, six hundred seven thousand, eight hundred ninety-one.

Questions 4–8 pertain to Section 1.2.

4. Circle the larger whole number: 217 or 301.

5. Circle the smaller whole number: 196 or 187.

6. Add: 23 + 57 + 9.

7. Add: 36 + 109 + 1024.

8. Add: 507
          82
    +699

Questions 9–11 pertain to Section 1.3.

9. Subtract: 302 − 87.

10. From 1608, subtract 1035.

11. How much larger is 12,035 than 9807?

Questions 12–13 pertain to Section 1.4.

12. Multiply 86 and 57.

13. Multiply: 976
            × 47

Questions 14–15 pertain to Section 1.5.

14. Divide: $576 \div 24$.

15. Divide: $54\overline{)2462}$.

Questions 16–18 pertain to Section 1.6.

16. Round 345,609 to the nearest thousand.

17. Round 4,876,002 to the nearest ten thousand.

18. Estimate the product 87×64 by rounding each factor to the nearest ten.

Questions 19–21 pertain to Section 1.7. Perform the indicated operations.

19. $5 \times 4 - 8 \div 2 + 7$

20. $56 \div (2 \times 4 - 1) - 4 \div 2$

21. $(288 \div 12) - (49 \div 7)$

Questions 22–23 pertain to Section 1.8. Perform the indicated operations. If not possible, state why not.

22. $(6 \div 2 - 3) \div 7$

23. $5 \div (0 \div 4)$

1.1 Numbers, Numerals, Counting, and Place Value

OBJECTIVES

After completing this section, you should be able to:
1. Indicate the value of each digit in a numeral.
2. Write the word name for a number.
3. Write the numeral for the word name of a number.

One of the first things that you probably learned to do with numbers was to count. Counting is used to determine how many items are in a collection. The need for counting is obvious. We may need to account for various inventories of stock, energy sources, members of a camping party, students enrolled in a class, beds in a hospital, and so on.

When we count, we use only 10 symbols or numerals (0, 1, 2, 3, 4, 5, 6, 7, 8, 9) to represent the numbers. Each one of these basic numerals is called a **digit.** When counting past 9, we go from a 1-digit numeral to a 2-digit numeral and write 10. The 2-digit numerals are 10, 11, 12, . . . , through 99. After 99, we continue with 3-digit numerals until we get to 999. The next numeral consists of 4 digits, and so forth.

As we keep counting, longer numerals are necessary to represent larger numbers. To help read the numeral, the digits are separated by commas into groups of 3 from right to left, such as 476,982,109.

In our system of writing numerals, each digit has **place value.** That is, the location of a digit in a numeral indicates the value of the digit. The first digit on the right represents the number of *ones;* the next digit to the left represents the number of *tens;* the next digit to the left represents the number of *hundreds;* and so forth as indicated in Table 1.1. The numeral shown in Table 1.1 is read as "nine hundred seventy-six thousand, three hundred two."

TABLE 1.1

Grouping	Thousands			Ones		
Numeral	9	7	6 ,	3	0	2
Place value	Hundred thousands	Ten thousands	Thousands	Hundreds	Tens	Ones

In Table 1.2, a longer numeral is given. The numeral in Table 1.2 is read "thirty-seven trillion, eight hundred ninety-six billion, five hundred forty-two million, seven hundred sixty-five thousand, seventy-four."

Notice the grouping by threes. Also notice that the singular is used for each group. For instance, we start with 37 trillion and not 37 trillions. Further, notice that the word "and" is not used between groups of digits in writing or reading the numeral. The reason for not using the word "and" will be clear when you get to the chapters on fractions and decimals.

In Table 1.2, the first 7 from the left represents the number of trillions, the next 7 to the right represents the number of hundred thousands, and the last 7 represents the number of tens.

1.1 Numbers, Numerals, Counting, and Place Value

TABLE 1.2														
Grouping	Trillions		Billions			Millions			Thousands			Ones		
Numeral	3	7 ,	8	9	6 ,	5	4	2 ,	7	6	5 ,	0	7	4
Place value	Ten trillions	Trillions	Hundred billions	Ten billions	Billions	Hundred millions	Ten millions	Millions	Hundred thousands	Ten thousands	Thousands	Hundreds	Tens	Ones

Practice Exercise 1. Indicate the place value of the digit 6 in the numeral 2634.

EXAMPLE 1 Indicate the place value of the digit 9 in the numeral 892,190.

SOLUTION: The digit 9 is in the *ten thousands* place. Its value is 90,000.

Practice Exercise 2. Indicate the value of each digit in the numeral 4596.

EXAMPLE 2 Indicate the value of each digit in the numeral 39,802.

SOLUTION:
3 is in the *ten thousands* place; it represents $3 \times 10,000 = 30,000$.
9 is in the *thousands* place; it represents $9 \times 1000 = 9000$.
8 is in the *hundreds* place; it represents $8 \times 100 = 800$.
0 is in the *tens* place; it represents $0 \times 10 = 0$.
2 is in the *ones* place; it represents $2 \times 1 = 2$.

39,802 represents $30,000 + 9000 + 800 + 2$ and is read "thirty-nine thousand, eight hundred two."

Practice Exercise 3. Write the word name for 333,222,111.

EXAMPLE 3 Write the word name for 119,692,102.

SOLUTION: 119,692,102 is written as "one hundred nineteen million, six hundred ninety-two thousand, one hundred two."

Practice Exercise 4. Write the word name for 16,598,403,172,504.

EXAMPLE 4 Write the word name for 4,912,345,765,532.

SOLUTION: 4,912,345,765,532 is written as "four trillion, nine hundred twelve billion, three hundred forty-five million, seven hundred sixty-five thousand, five hundred thirty-two."

Practice Exercise 5. Write the numeral for five thousand, six hundred eighty-seven.

EXAMPLE 5 Write the numeral for twenty-three thousand, nineteen.

SOLUTION: 23,019

Practice Exercise 6. Write the numeral for three billion, twenty-three million, seven hundred fifty thousand, nine hundred two.

EXAMPLE 6 Write the numeral for five hundred sixty million, forty-nine thousand, three hundred two.

SOLUTION: 560,049,302

ANSWERS TO PRACTICE EXERCISES

1. 600. **2.** The digit 4 represents $4 \times 1000 = 4000$; the digit 5 represents $5 \times 100 = 500$; the digit 9 represents $9 \times 10 = 90$; and the digit 6 represents $6 \times 1 = 6$. **3.** Three hundred thirty-three million, two hundred twenty-two thousand, one hundred eleven. **4.** Sixteen trillion, five hundred ninety-eight billion, four hundred three million, one hundred seventy-two thousand, five hundred four. **5.** 5687. **6.** 3,023,750,902.

EXERCISES 1.1

Indicate the place value of the underlined digit in each number given in Exercises 1–10.

1. 59,<u>8</u>41
2. 1,08<u>2</u>,796
3. 400<u>0</u>
4. 98,<u>1</u>11,472
5. <u>6</u>47,810,007
6. 164,1<u>3</u>2
7. 1,<u>0</u>00,000
8. 17,88<u>3</u>,928,143
9. 1<u>2</u>4,923,700
10. 5<u>4</u>,610,092,294,711

In Exercises 11–20, write the numeral for each of the given word names.

11. One thousand, four hundred fifty
12. Two hundred thousand, six hundred seventy-five
13. Three thousand, six
14. Four million, eight hundred seventy-three thousand, six hundred
15. Fifteen million, seven hundred eighty-five
16. Two billion, nine hundred eighteen million, eleven thousand, five hundred twenty-one
17. Three hundred twelve billion, four hundred five million, six thousand, eighty-nine
18. Five hundred billion, five hundred million, five hundred thousand
19. Six trillion, seven billion, five million, four thousand, nine hundred three
20. Two hundred thirty-seven trillion, eight hundred fifty-six billion, seven hundred nineteen million, five hundred eleven thousand, one hundred twelve

In Exercises 21–30, write the word name for each of the given numerals.

21. 968
22. 8987
23. 19,201
24. 231,692
25. 896,005
26. 2,169,726
27. 13,769,123
28. 610,526,400
29. 1,619,725,010
30. 19,169,234,999,888

31. Write the numeral in which the tens digit is 3, the units digit is 5, and the hundreds digit is 6.
32. Write the numeral in which the units digit is 9, the hundreds digit is 0, the thousands digit is 7, and the tens digit is 8.
33. If a numeral has 5 in the thousands place, then the 5 represents what number?
34. If a numeral has 7 in the hundred thousands place, then the 7 represents what number?
35. If a numeral has 0 in the hundreds place, then the 0 represents what number?
36. How many digits are in the numeral 307?
37. How many digits are in the numeral 40,506?
38. Write the numeral for the largest 3-digit number.
39. Write the numeral for the smallest 4-digit number.
40. Write the numeral for the largest 5-digit number if no digits can be used more than once.

1.2 Ordering and Addition of Whole Numbers

OBJECTIVES

After completing this section, you should be able to:
1. Order any two whole numbers.
2. Add whole numbers.
3. Find the answer to a word problem using addition of whole numbers.
4. Determine the perimeters of polygons.

The counting numbers are 1, 2, 3, 4, 5, and so forth. There is a smallest counting number, but there is no largest counting number. The counting numbers are also called the **natural numbers**. Notice that 0 is not a counting (or natural) number. However, if we include 0 with all of the natural numbers, the new set of numbers formed is called the **whole numbers**.

When we count, we get to some numbers before others. For instance, 3 comes before 8, and 7 comes before 20. The process of placing the whole numbers in their proper sequence is called **ordering**.

We use the symbol $>$ to mean "**is greater than.**" Thus, since 8 follows 3 in counting, 8 is greater than 3 and we write $8 > 3$. Similarly, $20 > 7$.

The symbol $<$ is used to mean "**is less than.**" Since 8 is greater than 3, we could also say that 3 is less than 8 and write $3 < 8$. Similarly, since $20 > 7$, we also have $7 < 20$.

In Practice Exercises 1–2, insert the symbol $>$ or $<$.

Practice Exercise 1. 18   9

EXAMPLE 1 Which of the two numbers is greater: 49 or 30?

SOLUTION: 49 is greater than 30 and is denoted by $49 > 30$ or $30 < 49$.

Practice Exercise 2. 189   209

EXAMPLE 2 Which of the two numbers is greater: 198 or 201?

SOLUTION: 201 is greater than 198 and is denoted by $201 > 198$ or $198 < 201$.

Practice Exercise 3. Order the following numbers from smallest to largest: 24, 67, 13, 32, 89, 55, 19.

EXAMPLE 3 Order the following numbers from smallest to largest:

23, 59, 12, 101, 69, 135, 9

SOLUTION: $9 < 12 < 23 < 59 < 69 < 101 < 135$

Practice Exercise 4. Order the following numbers from largest to smallest: 93, 34, 79, 64, 112, 56, 41.

EXAMPLE 4 Order the following numbers from largest to smallest:

17, 89, 303, 97, 237, 46, 3, 119

SOLUTION: $303 > 237 > 119 > 97 > 89 > 46 > 17 > 3$

ANSWERS TO PRACTICE EXERCISES
1. $>$.   2. $<$.   3. $13 < 19 < 24 < 32 < 55 < 67 < 89$.   4. $112 > 93 > 79 > 64 > 56 > 41 > 34$.

We will now consider the **addition** of whole numbers. When whole numbers are added, the digits representing the same place value are added together. That is, ones are added to ones, tens are added to tens, hundreds are added to hundreds, and so forth. To avoid errors in addition, write the numbers to be added so that the same place values are under each other: ones under ones, tens under tens, and so forth. The result in an addition problem is called the **sum.**

We note the following observations pertaining to the addition of whole numbers. Since we want to state properties that are true for all whole numbers, letters such as a, b, and c are sometimes used.

1. The order in which we add two whole numbers is not important. This is known as the **commutative property for addition** of whole numbers. In particular, we have that $2 + 3 = 3 + 2, 4 + 6 = 6 + 4, 5 + 9 = 9 + 5$, and $10 + 12 = 12 + 10$.

Instead of writing particular cases such as those just given, we could generalize the statement as follows. Let a represent one whole number and b represent a second whole number. Then, the commutative property for the addition of whole numbers would be written or symbolized as

$$a + b = b + a$$

In the above statement, the letters a and b are used to represent *any* two whole numbers. If we let $a = 3$ and $b = 13$, then $a + b = b + a$ becomes $3 + 13 = 13 + 3$. How many different whole number values can you assign to the letters a and b?

Statements such as $a + b = b + a$ come from algebra, where we use letters to represent numbers. For instance, the symbol 5 represents one and only one whole number. However, infinitely many different whole number values can be assigned to the letter a and also to the letter b.

2. To add three whole numbers, we must group them first, since only two numbers can be added at one time. However, the order in which we group the numbers is not important. That is, if we let a, b, and c represent any three whole numbers, then

$$(a + b) + c = a + (b + c)$$

This is known as the **associative property for addition** of whole numbers. Again, we are letting the letters a, b, and c represent *any* whole numbers. In particular, $(2 + 3) + 5 = 2 + (3 + 5), (5 + 7) + 2 = 5 + (7 + 2)$, and $(19 + 15) + 11 = 19 + (15 + 11)$.

3. If 0 is added to a whole number, or if a whole number is added to 0, the sum is equal to the given whole number. That is, if a is any whole number, then

$$a + 0 = a \quad \text{and} \quad 0 + a = a$$

The whole number 0 is called the **additive identity,** or the **identity element,** for the whole numbers.

Practice Exercise 5. Add: 235 + 342.

EXAMPLE 5 Add: 123 + 32 + 524.

SOLUTION:
```
  123 = (  1  hundred ) + (  2  tens) + (  3  ones)
   32 = (  0  hundreds) + (  3  tens) + (  2  ones)
 +524 = (  5  hundreds) + (  2  tens) + (  4  ones)
      = (1+0+5 hundreds) + (2+3+2 tens) + (3+2+4 ones)
  679 = (  6  hundreds) + (  7  tens) + (  9  ones)
```

or
```
  123
   32       (Note that we line up the 32 with the other
 +524       terms from the right.)
  679
```

Practice Exercise 6. Add: 369 + 27 + 1037.

EXAMPLE 6 Add: 24 + 38.

SOLUTION:
```
   24 = (  2  tens) + (  4  ones)
  +38 = (  3  tens) + (  8  ones)
      = (2+3 tens) + (4+8 ones)
      = (  5  tens) + ( 12  ones)
```

1.2 Ordering and Addition of Whole Numbers

We seem to have a problem. We added 4 ones and 8 ones and got 12 ones. The 12 ones cannot be written in the ones column since 12 is a 2-digit numeral. However, we could make use of the fact that $12 = 10 + 2$ and also that 10 ones is equal to 1 ten. Therefore,

$$
\begin{aligned}
24 &= (\ 2\ \text{tens}) + (\ 4\ \text{ones}) \\
+38 &= (\ 3\ \text{tens}) + (\ 8\ \text{ones}) \\
&= (2+3\ \text{tens}) + (4+8\ \text{ones}) \\
&= (\ 5\ \text{tens}) + (\ 12\ \text{ones}) && \text{(Note that 12 is a 2-digit number.)} \\
&= (5\ \text{tens}) + (10\ \text{ones}) + (2\ \text{ones}) && \text{(Rewriting)} \\
&= (5\ \text{tens}) + (1\ \text{ten}) + (2\ \text{ones}) && \text{(Renaming)} \\
&= (5 + 1\ \text{tens}) + (2\ \text{ones}) && \text{(Adding the tens)} \\
62 &= (6\ \text{tens}) + (2\ \text{ones}) && \text{(Simplifying)}
\end{aligned}
$$

Therefore, we have

$$
\begin{array}{r}
24 \\
+38 \\
\hline
62
\end{array}
$$

Observe that the 2 ones of the 12 were placed in the ones column and the 1 ten of the 12 was "carried" over to the tens column and added to the tens already there. We will use this procedure when adding whole numbers. However, when "renaming," we always start with units on the *right* and work to the left.

ANSWERS TO PRACTICE EXERCISES

5. 577.   6. 1433.

As we have seen, when adding two or more numbers, it is sometimes necessary to carry a number from one column in the sum over to another column, using facts such as:

- 10 ones equal 1 ten
- 10 tens equal 1 hundred
- 10 hundreds equal 1 thousand

and so forth.

The "carry" procedure is also used when adding like units such as minutes, seconds, feet, meters, pints, liters, and so forth. We will indicate this with several examples.

Practice Exercise 7. Three different recipes for punch require 3 pt; 4 qt 1 pt; and 3 gal 3 qt of soda. How much soda is required for the three recipes? (*Hint:* 2 pt = 1 qt; 4 qt = 1 gal.)

EXAMPLE 7 Andrea worked 3 days to build a playhouse. The first day she worked 4 hr and 15 min. The second day she worked 3 hr and 40 min. The third day she worked 5 hr and 12 min. What was the total time spent building the playhouse?

SOLUTION: The total time would be the *sum* of the time spent the first day, the second day, and the third day. We can arrange this as follows:

$$
\begin{aligned}
\text{First day:} &\quad 4\ \text{hr} + 15\ \text{min} && \text{(Note that 4 hr and 15 min means} \\
\text{Second day:} &\quad 3\ \text{hr} + 40\ \text{min} && \text{4 hr plus 15 min.)} \\
\text{Third day:} &\quad +5\ \text{hr} + 12\ \text{min} \\
\text{Total time:} &\quad 12\ \text{hr} + 67\ \text{min} \\
&= 12\ \text{hr} + 60\ \text{min} + 7\ \text{min} && \text{(Rewriting)} \\
&= 12\ \text{hr} + 1\ \text{hr} + 7\ \text{min} && \text{(Renaming)} \\
&= 13\ \text{hr} + 7\ \text{min} && \text{(Adding hours)}
\end{aligned}
$$

Hence, the total time spent was 13 hr and 7 min.

Practice Exercise 8. A backpacker takes 4 lb 8 oz of dehydrated food, 2 lb 7 oz of nuts and fruit, and eight 1-lb cans and an 11-oz jar of coffee. What is the total weight of food carried?

EXAMPLE 8 Michelle ordered four items from a mail-order catalog. The weights of these items were 2 lb 6 oz, 1 lb 11 oz, 4 lb 8 oz, and 9 oz. What was the total weight of these four items? (*Hint:* 16 oz = 1 lb.)

SOLUTION: The total weight of these four items could be represented as follows:

$$
\begin{aligned}
\text{First item:} &\quad 2\text{ lb} + 6\text{ oz} \\
\text{Second item:} &\quad 1\text{ lb} + 11\text{ oz} \\
\text{Third item:} &\quad 4\text{ lb} + 8\text{ oz} \\
\text{Fourth item:} &\quad \underline{9\text{ oz}} \\
\text{Total} &= 7\text{ lb} + 34\text{ oz} \\
&= 7\text{ lb} + 16\text{ oz} + 16\text{ oz} + 2\text{ oz} \quad \text{(Rewriting)} \\
&= 7\text{ lb} + 1\text{ lb} + 1\text{ lb} + 2\text{ oz} \quad \text{(Renaming)} \\
&= 9\text{ lb} + 2\text{ oz} \quad \text{(Adding the pounds)}
\end{aligned}
$$

Therefore, the total weight of the four items was 9 lb 2 oz.

Practice Exercise 9. Cindy works in a hardware store. One week she worked 4 hr 15 min on Monday, 5 hr 30 min on Tuesday, 3 hr 10 min on Wednesday, 4 hr 35 min on Thursday, and 7 hr 45 min on Saturday. How much time did she work that week?

EXAMPLE 9 Daryl has three pieces of pipe that have lengths of 5 ft 3 in., 4 ft 7 in., and 6 ft 8 in. What is the total length of the pipe? (*Hint:* 12 in. = 1 ft.)

SOLUTION: The total length of the pipe could be represented as follows:

$$
\begin{aligned}
\text{First piece:} &\quad 5\text{ ft} + 3\text{ in.} \\
\text{Second piece:} &\quad 4\text{ ft} + 7\text{ in.} \\
\text{Third piece:} &\quad \underline{6\text{ ft} + 8\text{ in.}} \\
\text{Total} &= 15\text{ ft} + 18\text{ in.} \\
&= 15\text{ ft} + 12\text{ in.} + 6\text{ in.} \quad \text{(Rewriting)} \\
&= 15\text{ ft} + 1\text{ ft} + 6\text{ in.} \quad \text{(Renaming)} \\
&= 16\text{ ft} + 6\text{ in.} \quad \text{(Adding the feet)}
\end{aligned}
$$

Therefore, the total length of pipe is 16 ft 6 in.

ANSWERS TO PRACTICE EXERCISES

7. 5 gal 1 qt.   8. 15 lb 10 oz.   9. 25 hr 15 min.

In this chapter, we introduce and discuss polygons. In Chapter 5, we will discuss them more completely. A **polygon** is defined to be a closed geometric figure consisting of at least three line segments, called its **sides**. The **perimeter of a polygon,** denoted by P, is defined to be the sum of the lengths of its sides.

Practice Exercise 10. Determine the perimeter of the triangle whose sides measure 23 ft, 34 ft, and 41 ft.

EXAMPLE 10 Determine the perimeter of the accompanying triangle indicated to the right.

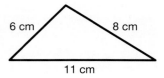

SOLUTION: A triangle is a polygon having three sides. Therefore, its perimeter is the sum of the lengths of its sides. We have

$$
\begin{aligned}
P &= 6\text{ cm} + 8\text{ cm} + 11\text{ cm} \\
&= (6 + 8 + 11)\text{ cm} \\
&= 25\text{ cm}
\end{aligned}
$$

1.2 Ordering and Addition of Whole Numbers

Practice Exercise 11. Determine the perimeter of the rectangle two of whose sides measure 31 yd each and whose other two sides measure 26 yd each.

EXAMPLE 11 Determine the perimeter of the accompanying rectangle indicated to the right.

SOLUTION: A rectangle is a polygon having four sides. Two of the sides measure 13 yd and the other two sides measure 28 yd. We have

$$P = 28 \text{ yd} + 13 \text{ yd} + 28 \text{ yd} + 13 \text{ yd}$$
$$= (28 + 13 + 28 + 13) \text{ yd}$$
$$= 82 \text{ yd}$$

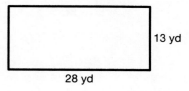

Practice Exercise 12. Determine the perimeter of the square that measures 102 in. on a side.

EXAMPLE 12 Determine the perimeter of the accompanying square indicated to the right.

SOLUTION: A square is a four-sided polygon, all of whose sides have the same length. Hence,

$$P = 17 \text{ ft} + 17 \text{ ft} + 17 \text{ ft} + 17 \text{ ft}$$
$$= (17 + 17 + 17 + 17) \text{ ft}$$
$$= 68 \text{ ft}$$

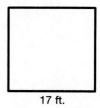

Practice Exercise 13. Determine the perimeter of a seven-sided polygon whose sides have measures of 10 cm, 12 cm, 9 cm, 16 cm, 7 cm, 11 cm, and 19 cm.

EXAMPLE 13 Determine the perimeter of the polygon indicated to the right.

SOLUTION: The perimeter of the given polygon is equal to the sum of the lengths of its sides. Hence,

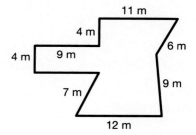

$$P = 4 \text{ m} + 9 \text{ m} + 4 \text{ m} + 11 \text{ m} + 6 \text{ m} + 9 \text{ m} + 12 \text{ m} + 7 \text{ m} + 9 \text{ m}$$
$$= (4 + 9 + 4 + 11 + 6 + 9 + 12 + 7 + 9) \text{ m}$$
$$= 71 \text{ m}$$

ANSWERS TO PRACTICE EXERCISES

10. 98 ft.   11. 114 yd.   12. 408 in.   13. 84 cm.

EXERCISES 1.2

In Exercises 1–10, insert the symbol $<$ or $>$ between the given pairs of numbers to make true statements.

1. 4   9   **2.** 9   7   **3.** 16   23   **4.** 39   42   **5.** 27   19
6. 79   93   **7.** 101   99   **8.** 209   87   **9.** 165   183   **10.** 2001   1002

Add in Exercises 11–20.

11.	173	12.	323	13.	312	14.	2012	15.	12,346
	41		451		46		451		125
	+204		+125		+531		+6124		+32,327

16.	632,405	17.	312,792	18.	4,197,605	19.	2,369,123	20.	23,446
	19,670		1,908		2,682,097		1,002,040		67,275
	279,084		23,456		19,887		4,190,756		35,649
	+ 58,697		+568,765		+ 364,903		+6,213,469		92,590
									+34,608

Add in Exercises 21–23.

21. 2037 + 12,476 + 199 + 3014
22. 4621 + 26 + 137 + 5004 + 407
23. 52,369 + 496,071 + 239 + 7098 + 504,742
24. How many hours does John spend in class each week if he has 5 hr of math, 3 hr of sociology, and 6 hr of biology?
25. A marketing research company surveys 3121 people in Mall A, 974 people in Mall B, and 2038 people in Mall C. How many people are surveyed?
26. How many kilos of strawberries are picked if 1008 kilos are picked on Friday, 2148 kilos are picked on Saturday, and 915 kilos are picked on Sunday?
27. A walk-in clinic sees the following numbers of patients over a 3-day period: 105 on Monday, 218 on Tuesday, and 98 on Wednesday. How many patients are seen during the 3 days?
28. If a library receives different donations of books numbering 91, 427, 639, and 1807, respectively, what is the total number of books donated?
29. A small store's incomes from various departments are $2139, $1802, $3447, and $962. What is the total income?
30. The numbers of gallons of fuel used over December, January, and February are 225, 207, and 186. What is the total number of gallons of fuel used?
31. Five employees received pay checks in the following amounts: $253, $311, $198, $290, and $201. Determine the total amount of the checks.
32. Four building lots are side by side. The first lot is 208 ft 4 in. wide; the second lot is 211 ft 9 in. wide; the third is 199 ft 5 in. wide; and the fourth is 189 ft 11 in. wide. What is the total width of the four lots? (*Hint:* 12 in. = 1 ft.)
33. Michael's college expenses for 1 year were as follows: $2350 for tuition, $980 for room and board, $130 for books, $85 for fees, and $416 for travel and personal expenses. What were his total expenses for the year?
34. For her work-study program in college, Karen worked as follows during 1 week: 1 hr 20 min on Monday, 2 hr 15 min on Tuesday, 55 min on Wednesday, 3 hr 10 min on Thursday, and 2 hr 35 min on Friday. How much time did she work that week?
35. Four members of a football team weigh 216 lb 7 oz, 219 lb 12 oz, 227 lb 4 oz, and 232 lb 10 oz. Find their combined weight. (*Hint:* 16 oz = 1 lb.)
36. Suppose that, during a given month, General Motors sold 1,162,207 automobiles, Ford Motor Company sold 1,097,692 automobiles, Chrysler Motor Company sold 968,895 automobiles, and Nissan Company sold 532,119 automobiles. How many automobiles were sold by the four companies during the month?
37. A janitor has four containers of different sizes. One container has a capacity of 3 gal 2 qt; one has a capacity of 2 gal 3 qt, one has a capacity of 3 qt 1 pt, and one has a capacity of 2 qt 1 pt. What is the total capacity of the four containers? (*Hint:* 1 gal = 4 qt; 1 qt = 2 pt.)
38. At a particular company, 86 salaried employees are members of a union, 169 salaried employees are not members of a union, 462 hourly employees are members of a union, and 107 hourly employees are not members of a union.
    a. How many are salaried employees?
    b. How many are hourly employees?

c. How many employees are members of a union?
d. How many employees are not members of a union?
e. What is the total number of employees in the four groups?

39. During a five-game World Series, the paid attendance was 68,312 the first day, 69,679 the second day, 71,019 the third day, 66,897 the fourth day, and 70,092 the last day. What was the combined paid attendance for the 5 days?

40. A runner passes the first timing station in a marathon after 45 min. The next section takes her 1 hr 6 min, and the final distance to the finish line takes her another 1 hr 30 min. What was her total time?

41. Three pieces of fabric measure 1 yd 6 in., 2 ft 3 in., and 18 in. What is the total measure of the fabric? (*Hint:* 12 in. = 1 ft; 3 ft = 1 yd.)

42. For a state lottery, 123,619 tickets were sold on Monday, 201,609 tickets on Tuesday, 196,918 tickets on Wednesday, 242,687 tickets on Thursday, and 251,009 tickets on Friday. How many tickets were sold for the 5 days?

43. The census figures for five states in the United States are 8,196,203; 7,097,492; 3,092,703; 1,261,079; and 4,567,832. What is the total census for the five states?

44. The numbers of traffic tickets issued in Giant City during a 7-day period were: 2069; 2814; 1905; 2005; 2709; 1895; and 1699. How many tickets were issued during the 7 days?

45. A line is painted on the road and markings are made at various points as illustrated here. What is the length of the segment indicated by the question mark?

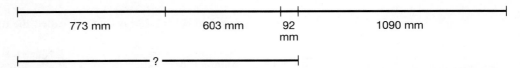

46. Determine the perimeter of a triangle whose sides have measures of 51 in., 72 in., and 87 in.

47. Determine the perimeter of a triangle two of whose sides measure 19 yd each and whose third side measures 23 yd.

48. Determine the perimeter of a triangle each of whose sides measure 67 cm.

49. A rectangle has two sides of measure 26 ft each and two sides of measure 37 ft each. Determine its perimeter.

50. Two sides of a rectangle measure 16 yd each and the other two sides measure 35 ft each. Determine the perimeter of the rectangle. (*Hint:* Note that the units of measure are not the same. Using the relationship 3 ft = 1 yd, convert yards to feet or feet to yards.)

51. A square measures 18 ft on a side. Determine its perimeter.

52. Each side of a square measures 24 yd. Determine the perimeter of the square.

53. A five-sided polygon has sides with measures 26 cm, 17 cm, 19 cm, 31 cm, and 29 cm. Determine the perimeter of the polygon.

54. A six-sided polygon is such that three of its sides measure 14 yd each, two of its sides measure 17 yd each, and the other side measures 19 yd. Determine the perimeter of the polygon.

55. A nine-sided polygon has two sides that measure 28 in. each, two sides that measure 23 in. each, and three sides that measure 34 in. each. The other two sides measure 19 in. and 31 in. Determine the perimeter of the polygon.

1.3 Subtraction of Whole Numbers

OBJECTIVES

After completing this section, you should be able to:
1. Subtract whole numbers.
2. Find the answer to a word problem using subtraction.

The process of "taking away" is called **subtraction.** It is also the process that enables us to tell how much larger one number is than another. Hence, 9 is 6 more than 3, since $9 - 3 = 6$. Also, 9 is 3 more than 6, since $9 - 6 = 3$. From the basic addition problem $6 + 3 = 9$, we get the following two subtraction problems.

Addition Problem	Corresponding Subtraction Problems
$\begin{array}{r} 6 \\ +3 \\ \hline 9 \end{array}$	$\begin{array}{r} 9 \\ -3 \\ \hline 6 \end{array}$ and $\begin{array}{r} 9 \\ -6 \\ \hline 3 \end{array}$

The answer in a subtraction problem is called the **difference.**

Practice Exercise 1. Subtract: $9 - 2$.

EXAMPLE 1 Subtract: $8 - 6$.

SOLUTION: $8 - 6 = 2$ since $6 + 2 = 8$

Practice Exercise 2. Subtract: $16 - 7$.

EXAMPLE 2 Subtract: $14 - 9$.

SOLUTION: $14 - 9 = 5$ since $9 + 5 = 14$

Practice Exercise 3. From 19, subtract 11.

EXAMPLE 3 From 18, subtract 9.

SOLUTION: From 18, subtract 9 means to subtract 9 from 18. Hence, $18 - 9 = 9$ since $9 + 9 = 18$.

Practice Exercise 4. From 13, subtract 13.

EXAMPLE 4 From 8, subtract 8.

SOLUTION: From 8, subtract 8 means to subtract 8 from 8. Hence, $8 - 8 = 0$ since $8 + 0 = 8$.

ANSWERS TO PRACTICE EXERCISES
1. 7.   2. 9.   3. 8.   4. 0.

To check the answer in subtraction, add the answer to the number being subtracted; the result should be the number you started with.

Note, then, that subtraction is the opposite (or inverse) operation of addition:

- If we start with 3 and add 5, the result is 8. However, if we start with 8 and subtract 5, the result is 3.
- If we start with 7 and add 9, the result is 16. However, if we start with 16 and subtract 9, the result is 7.

1.3 Subtraction of Whole Numbers

The order in which we add two whole numbers is not important. However, the order in which we subtract one whole number from another *is* important. For instance, to add 19 and 25, we could write

$$\begin{array}{r} 19 \\ +25 \\ \hline 44 \end{array} \quad \text{or} \quad \begin{array}{r} 25 \\ +19 \\ \hline 44 \end{array}$$

However, $25 - 19$ can be written only as

$$\begin{array}{r} 25 \\ -19 \\ \hline 6 \end{array}$$

since subtraction is *not* commutative.

If a and b are two whole numbers and a is equal to b, we write $a = b$. If c and d are two whole numbers and c is not equal to d, we write $c \neq d$ where the / through the = symbol means not.

Practice Exercise 5. Show that $12 - 9 \neq 9 - 12$.

EXAMPLE 5   Show that $9 - 2 \neq 2 - 9$.

SOLUTION: $9 - 2 = 7$ since $2 + 7 = 9$, but $2 - 9$ is not a whole number since we cannot subtract a larger whole number from a smaller whole and get a whole number. Hence,

$$9 - 2 \neq 2 - 9$$

Practice Exercise 6. Show that $7 - 0 \neq 0 - 7$.

EXAMPLE 6   Show that $13 - 8 \neq 8 - 13$.

SOLUTION: $13 - 8 = 5$ since $8 + 5 = 13$, but $8 - 13$ is not a whole number. Hence,

$$13 - 8 \neq 8 - 13$$

ANSWERS TO PRACTICE EXERCISES

5. $12 - 9 = 3$, but $9 - 12$ is not a whole number.   6. $7 - 0 = 7$, but $0 - 7$ is not a whole number.

In general, if a and b are *any* two whole numbers, then

$$a - b \neq b - a$$

Subtraction of whole numbers is *not* an associative operation, either. Note that

$$(9 - 6) - 2 = 3 - 2 = 1$$

but

$$9 - (6 - 2) = 9 - 4 = 5$$

Since $1 \neq 5$, then

$$(9 - 6) - 2 \neq 9 - (6 - 2)$$

shows that grouping *does* matter in subtraction; subtraction is *not* an associative operation. In general, if a, b, and c are *any* whole numbers, then

$$(a - b) - c \neq a - (b - c)$$

Practice Exercise 7. Show that $(9 - 4) - 3 \neq 9 - (4 - 3)$.

EXAMPLE 7 Show that $(11 - 4) - 2 \neq 11 - (4 - 2)$.

SOLUTION: $(11 - 4) - 2 = 7 - 2 = 5$, but $11 - (4 - 2) = 11 - 2 = 9$. Since $5 \neq 9$, then $(11 - 4) - 2 \neq 11 - (4 - 2)$.

Practice Exercise 8. Show that $(12 - 3) - 5 \neq 12 - (3 - 5)$.

EXAMPLE 8 Show that $(10 - 2) - 5 \neq 10 - (2 - 5)$.

SOLUTION: $(10 - 2) - 5 = 8 - 5 = 3$, but $10 - (2 - 5)$ is not a whole number since $2 - 5$ is not a whole number. Hence,

$$(10 - 2) - 5 \neq 10 - (2 - 5)$$

ANSWERS TO PRACTICE EXERCISES

7. $(9 - 4) - 3 = 5 - 3 = 2$, but $9 - (4 - 3) = 9 - 1 = 8$; $2 \neq 8$. **8.** $(12 - 3) - 5 = 9 - 5 = 4$, but $12 - (3 - 5)$ is not a whole number since $3 - 5$ is not a whole number.

To subtract a whole number from another whole number when the whole numbers have 2 or more digits, we write the smaller number under the larger number with the digits representing the same value under each other. Then, we subtract as illustrated in examples 9 through 12.

Practice Exercise 9. Subtract: $86 - 23$.

EXAMPLE 9 Subtract:

$$\begin{array}{r} 8536 \\ -5204 \\ \hline \end{array}$$

SOLUTION:

STEP 1: Subtract the ones:

$$\begin{array}{r} 8536 \\ -5204 \\ \hline 2 \end{array}$$

STEP 2: Subtract the tens:

$$\begin{array}{r} 8536 \\ -5204 \\ \hline 32 \end{array}$$

STEP 3: Subtract the hundreds:

$$\begin{array}{r} 8536 \\ -5204 \\ \hline 332 \end{array}$$

STEP 4: Subtract the thousands:

$$\begin{array}{r} 8536 \\ -5204 \\ \hline 3332 \end{array}$$

CHECK:

$$\begin{array}{r} 5204 \\ +3332 \\ \hline 8536 \end{array} \checkmark$$

1.3 Subtraction of Whole Numbers

When you are subtracting one number from another, if the digit in a place value is larger than the digit directly above it, you will need to "borrow." "Borrowing" in subtraction is very similar to "carrying" in addition. The following examples illustrate how this is done.

Practice Exercise 10. Subtract: 54 − 28.

EXAMPLE 10 Subtract:

$$\begin{array}{r} 752 \\ -324 \\ \hline \end{array}$$

SOLUTION:

STEP 1: Subtract the ones. Since 4 > 2, borrow 1 ten = 10 ones from the tens column in 752; add these 10 ones to the 2 ones already in the ones column; then subtract 4 ones from 12 ones. In this and the remaining examples, we will space the digits in the given numbers to emphasize what we are doing.

$$\begin{array}{r} \overset{41}{7\,\cancel{5}\,2} \\ -3\,2\,4 \\ \hline 8 \end{array}$$

STEP 2: Subtract the tens:

$$\begin{array}{r} \overset{41}{7\,\cancel{5}\,2} \\ -3\,2\,4 \\ \hline 2\,8 \end{array}$$

STEP 3: Subtract the hundreds:

$$\begin{array}{r} \overset{41}{7\,\cancel{5}\,2} \\ -3\,2\,4 \\ \hline 4\,2\,8 \end{array}$$

CHECK:

$$\begin{array}{r} 324 \\ +428 \\ \hline 752 \checkmark \end{array}$$

Practice Exercise 11. Subtract: 4617 − 2406.

EXAMPLE 11 Subtract:

$$\begin{array}{r} 5304 \\ -3978 \\ \hline \end{array}$$

SOLUTION:

STEP 1: Subtract the ones. Since 8 > 4, borrow 1 ten = 10 ones from the tens column in 5304; then subtract the ones:

$$\begin{array}{r} 5304 \\ -3978 \\ \hline \end{array}$$

Since there are 0 tens in 5304, we cannot borrow 1 ten. Therefore, we move to the hundreds column to borrow 1 hundred = 10 tens. The subtraction problem now becomes:

$$\begin{array}{r} \overset{21}{5\,\cancel{3}\,0\,4} \\ -3\,9\,7\,8 \\ \hline \end{array}$$

STEP 2: Borrow 1 ten = 10 ones from the tens column in $5\overset{2}{\cancel{3}}\overset{1}{0}4$, add the 10 ones to the 4 ones in the ones column, and subtract the ones:

$$\begin{array}{r} \overset{\overset{9}{}}{}\\ 5\,\overset{2}{\cancel{3}}\,\overset{\cancel{9}}{\cancel{0}}\,\overset{1}{4} \\ -3\,9\,7\,8 \\ \hline 6 \end{array}$$

STEP 3: Subtract the tens:

$$\begin{array}{r} 5\,\overset{2}{\cancel{3}}\,\overset{\overset{9}{\cancel{}}}{\cancel{0}}\,\overset{1}{4} \\ -3\,9\,7\,8 \\ \hline 2\,6 \end{array}$$

STEP 4: Subtract the hundreds. Since $9 > 2$, borrow 1 thousand = 10 hundreds from the hundreds column in $5\overset{\overset{9}{\cancel{}}}{\cancel{3}}\overset{1}{0}4$, add the 10 hundreds to the 2 hundreds, and subtract:

$$\begin{array}{r} \overset{4}{\cancel{5}}\,\overset{\overset{1}{\cancel{2}}}{\cancel{3}}\,\overset{\overset{9}{\cancel{}}}{\cancel{0}}\,\overset{1}{4} \\ -3\,9\,7\,8 \\ \hline 3\,2\,6 \end{array}$$

STEP 5: Subtract the thousands:

$$\begin{array}{r} \overset{4}{\cancel{5}}\,\overset{\overset{1}{\cancel{2}}}{\cancel{3}}\,\overset{\overset{9}{\cancel{}}}{\cancel{0}}\,\overset{1}{4} \\ -3\,9\,7\,8 \\ \hline 1\,3\,2\,6 \end{array}$$

CHECK:
$$\begin{array}{r} 3978 \\ +1326 \\ \hline 5304 \checkmark \end{array}$$

> **Practice Exercise 12.** From 4003, subtract 2917.

EXAMPLE 12 Subtract:
$$\begin{array}{r} 96003 \\ -34567 \end{array}$$

SOLUTION:

STEP 1: Borrow 10 ones from the tens column and subtract the ones:

$$\begin{array}{r} 96003 \\ -34567 \end{array}$$

Since there are 0 tens in 96003, we cannot borrow 1 ten. Therefore, we move to the hundreds column to borrow 1 hundred = 10 tens. Again, there are 0 hundreds so we move to the thousands column and borrow 1 thousand = 10 hundreds as indicated:

$$\begin{array}{r} 9\,\overset{5}{\cancel{6}}\,\overset{1}{\cancel{0}}\,0\,3 \\ -3\,4\,5\,6\,7 \end{array}$$

1.3 Subtraction of Whole Numbers

STEP 2: Borrow 10 tens from the hundreds column:

$$\begin{array}{r} \overset{9}{5\,\cancel{1}\,1} \\ 9\,6\,\cancel{0}\,0\,3 \\ -3\,4\,5\,6\,7 \\ \hline \end{array}$$

STEP 3: Borrow 10 ones from the tens column:

$$\begin{array}{r} \overset{9}{5}\,\overset{9}{\cancel{1}}\,\cancel{1} \\ 9\,6\,\cancel{0}\,\cancel{0}\,3 \\ -3\,4\,5\,6\,7 \\ \hline \end{array}$$

STEP 4: Subtract the ones:

$$\begin{array}{r} \overset{9}{5}\,\overset{9}{\cancel{1}}\,\cancel{1} \\ 9\,6\,\cancel{0}\,\cancel{0}\,3 \\ -3\,4\,5\,6\,7 \\ \hline 6 \end{array}$$

STEP 5: Subtract the tens:

$$\begin{array}{r} \overset{9}{5}\,\overset{9}{\cancel{1}}\,\cancel{1} \\ 9\,6\,\cancel{0}\,\cancel{0}\,3 \\ -3\,4\,5\,6\,7 \\ \hline 3\,6 \end{array}$$

STEP 6: Subtract the hundreds:

$$\begin{array}{r} \overset{9}{5}\,\overset{9}{\cancel{1}}\,\cancel{1} \\ 9\,6\,\cancel{0}\,\cancel{0}\,3 \\ -3\,4\,5\,6\,7 \\ \hline 4\,3\,6 \end{array}$$

STEP 7: Subtract the thousands:

$$\begin{array}{r} \overset{9}{5}\,\overset{9}{\cancel{1}}\,\cancel{1} \\ 9\,6\,\cancel{0}\,\cancel{0}\,3 \\ -3\,4\,5\,6\,7 \\ \hline 1\,4\,3\,6 \end{array}$$

STEP 8: Subtract the ten thousands:

$$\begin{array}{r} \overset{9}{5}\,\overset{9}{\cancel{1}}\,\cancel{1} \\ 9\,6\,\cancel{0}\,\cancel{0}\,3 \\ -3\,4\,5\,6\,7 \\ \hline 6\,1\,4\,3\,6 \end{array}$$

CHECK:

$$\begin{array}{r} 34567 \\ +61436 \\ \hline 96003\ \checkmark \end{array}$$

ANSWERS TO PRACTICE EXERCISES

9. 63.   **10.** 26.   **11.** 2211.   **12.** 1086.

On a particular mathematics test, Gina received a grade of 83, and Dianne received a grade of 69. How many more points did Gina receive than Dianne? Since subtraction is the process by which we can determine how much larger one whole number is than another whole number, we can use subtraction to answer the question:

$$\begin{array}{rr} \text{Gina's score:} & 83 \\ \text{Dianne's score:} & -69 \\ \hline \text{Difference:} & 14 \end{array}$$

Therefore, Gina scored 14 points higher than Dianne. Also note that Dianne scored 14 points lower than Gina.

Practice Exercise 13. If Tammy has $259 in her checking account and writes a check for $126, how much money does she have left in her account?

EXAMPLE 13 During a physical examination, a patient's systolic blood pressure was 130 mm at rest. After a brief period of active exercise, the systolic pressure was 158 mm. What was the increase between the two systolic pressure readings?

SOLUTION: The problem may be set up and solved as follows:

$$\begin{array}{rr} \text{Pressure reading after exercise:} & 158 \text{ mm} \\ \text{Pressure reading at rest:} & -130 \text{ mm} \\ \hline \text{Increase (difference):} & 28 \text{ mm} \end{array}$$

Therefore, the increase between the two systolic pressure readings was 28 mm.

Practice Exercise 14. If Gerry has $312 in his savings account and withdraws $89, how much money does he have left in his account?

EXAMPLE 14 Liz and Brian went fishing. Liz caught a northern pike that weighed 10 lb 5 oz. Brian caught a rainbow trout that weighed 6 lb 9 oz. How much heavier was Liz's fish than Brian's fish?

SOLUTION: The problem may be set up and solved as follows:

$$\begin{array}{rr} \text{Liz's fish:} & 10 \text{ lb} + 5 \text{ oz} \\ \text{Brian's fish:} & -(6 \text{ lb} + 9 \text{ oz}) \end{array}$$

Rewrite this problem by borrowing 1 lb from 10 lb, converting 1 lb to 16 oz, and adding 16 oz to 5 oz. Then, 10 lb 5 oz may be written as 9 lb 21 oz.

$$\begin{array}{rr} \text{Liz's fish:} & 9 \text{ lb} + 21 \text{ oz} \\ \text{Brian's fish:} & -(6 \text{ lb} + 9 \text{ oz}) \\ \hline \text{Difference:} & 3 \text{ lb} + 12 \text{ oz} \end{array}$$

Therefore, Liz's fish was 3 lb 12 oz heavier than Brian's fish.

ANSWERS TO PRACTICE EXERCISES

13. $133.   14. $223.

EXERCISES 1.3

Subtract in Exercises 1–15. Check your results by addition.

1. 53	2. 43	3. 63	4. 236	5. 569
−32	−17	−48	−124	− 45

1.3 Subtraction of Whole Numbers

6.	343 −197	**7.**	803 −299	**8.**	700 −359	**9.**	2000 − 587	**10.**	6509 −3405
11.	7642 −5030	**12.**	7059 −5196	**13.**	5007 −3116	**14.**	8001 −5999	**15.**	4006 − 249

Subtract in Exercises 16–21. Check your results by addition.

16. 17 − 9 **17.** 84 − 53
18. 469 − 246 **19.** 6454 − 4021
20. 86 − 59 **21.** 609 − 46
22. From 5967, subtract 2041.
23. From 6095, subtract 2082.
24. From 6239, subtract 4095.
25. Subtract 6999 from 9080.
26. Sue can type 115 words per minute, and Don can type 86 words per minute. How many more words per minute can Sue type than Don?
27. Laura jogged for 37 min 23 sec, and Marge jogged for 11 min 43 sec.
   a. How much longer did Laura jog than Marge?
   b. What was the combined jogging time for the two?
28. On a particular day, 52 U.S. senators voted in favor of a bill, 39 senators voted against the bill, and 4 senators abstained from voting.
   a. How many more senators voted in favor of the bill than voted against the bill?
   b. How many more senators voted against the bill than abstained?
   c. How many senators were present to vote that day?
   d. There are 100 U.S. senators. How many senators were not present to vote that day?
29. Easy Rider ran a horse race in 3 min 17 sec. Break-O-Day ran the same race in 2 min 29 sec. How much longer did the slower horse take to run the race?
30. A boat trailer is towed by a station wagon. The station wagon measures 17 ft 4 in. from front bumper to trailer hitch. The boat trailer measures 14 ft 8 in. from trailer hitch to rear.
   a. How much longer is the station wagon than the boat trailer?
   b. If the boat trailer is hitched to the station wagon, what is the combined length of the two?
31. Company A and Company B each employ 100 workers. Company A lost 23 work-days because of employee illness and 16 work-days because of employee accidents. Company B lost 19 work-days because of illness and 20 work-days because of accidents.
   a. Which company lost more work-days because of employee illness? How many more?
   b. Which company lost more work-days because of accidents by its employees? How many more?
   c. What is the combined number of work-days lost by Company A because of illness and accidents?
   d. What is the combined number of work-days lost by Company B because of illness and accidents?
   e. Which company lost more combined work-days because of illness and accidents? How many more?
32. The Smith Family used 129 kW of electricity and 236 ft^3 of natural gas for heating purposes during a particular month last year. For the same month, the Brown family used 146 kW of electricity and 218 ft^3 of natural gas.
   a. Which family used more electricity for the month? How much more?
   b. Which family used more natural gas for the month? How much more?
   c. What was the combined amount of electricity used by the two families for the month?
   d. What was the combined amount of natural gas used by the two families for the month?
33. Bette made three toll calls one day. The first call lasted 7 min 33 sec; the second, 23 min 8 sec; the third, 16 min 29 sec.
   a. What was the time difference between the shortest and the longest calls?
   b. What was the time difference between the two shortest calls?
   c. What was the time difference between the two longest calls?
   d. What was the combined time for the three calls?

34. Ms. Chong bought 2 lb 8 oz of ground chuck, 3 lb 2 oz of sausage, 2 lb of hot dogs, and 4 lb 1 oz of lamb chops.
    a. What was the difference in weight between the ground chuck and the sausage?
    b. What was the difference in weight between the hot dogs and the lamb chops?
    c. What was the difference in weight between the ground chuck and the lamb chops?
    d. What was the combined weight of the four meats?
35. Theresa has $468 in her savings account. If she withdraws $132, how much money will be left in her account?
36. Jay has $324 in his checking account. If he writes a check for $169, how much money will be left in his account?
37. Karole made a payment of $125 toward her charge account at the Big Bargain Store. If her account was $672, how much does she still owe?
38. John's house is assessed at $23,130; Sue's house is assessed at $26,175; Marty's house is assessed at $19,050.
    a. How much higher is the assessment for Sue's house than for John's?
    b. How much higher is the assessment for John's house than for Marty's?
    c. What is the difference between the highest and the lowest of the assessments?
    d. What is the total assessment for the three houses?

1.4 Multiplication of Whole Numbers

OBJECTIVES

After completing this section, you should be able to:
1. Multiply two whole numbers.
2. Find the answer to a word problem using multiplication.
3. Determine the area of triangular, square, and rectangular regions.

Suppose that we have 4 children and each child has 2 coins. How many coins do the 4 children have among them? We could arrange the coins as in Figure 1.1. Then, by counting, we could determine that there are 8 coins. We would then say that

$$4 \text{ times } 2 \text{ is equal to } 8$$
or
$$4 \times 2 = 8$$

FIGURE 1.1   First child's coins   Second child's coins   Third child's coins   Fourth child's coins

The symbol $\times$ is used to indicate multiplication. We could also write

$$4 \times 2 = 8$$
as $\quad (4)(2) = 8$
or $\quad 4(2) = 8$
or $\quad 4 \cdot 2 = 8$

In the expression $4 \cdot 2$, the dot $\cdot$ is written in the middle of the space between the 4 and 2. The $\cdot$ should not be confused with a decimal point, which is written on the line.

1.4 Multiplication of Whole Numbers

Multiplication may be thought of as repeated addition with all the numbers being the same. Hence,

$$4 \times 2 = 8$$

may be thought of as

$$4 \times 2 = 2 + 2 + 2 + 2 = 8$$

A multiplication problem may also be written vertically as follows:

$$\begin{array}{r} 2 \text{ (Factor)} \\ \times 4 \text{ (Factor)} \\ \hline 8 \text{ (Product)} \end{array}$$

In the preceeding problem, 2 is a number being multiplied and is called a **factor.** The number 4 indicates how many times 2 is being multiplied and is also called a factor. The result, or answer, in a multiplication problem is called the **product.**

Practice Exercise 1. Multiply: 4×6.

EXAMPLE 1 Multiply: 3×4.

SOLUTION: $3 \times 4 = 4 + 4 + 4 = 12$

Hence,
$$\begin{array}{r} 4 \text{ (Factor)} \\ \times 3 \text{ (Factor)} \\ \hline 12 \text{ (Product)} \end{array}$$

Practice Exercise 2. Multiply: 6×4.

EXAMPLE 2 Multiply: 4×3.

SOLUTION: $4 \times 3 = 3 + 3 + 3 + 3 = 12$

Hence,
$$\begin{array}{r} 3 \text{ (Factor)} \\ \times 4 \text{ (Factor)} \\ \hline 12 \text{ (Product)} \end{array}$$

Examples 1 and 2 illustrate that the order in which we multiply two whole numbers is not important. Multiplication of whole numbers is a commutative operation.

If we let *a* and *b* represent any two whole numbers, then we could state the **commutative property for multiplication** of whole numbers as follows:

$$a \times b = b \times a$$

(What other operation that you have studied so far is also commutative?)

ANSWERS TO PRACTICE EXERCISES

1. $4 \times 6 = 6 + 6 + 6 + 6 = 24$. 2. $6 \times 4 = 4 + 4 + 4 + 4 + 4 + 4 = 24$.

If a whole number is multiplied by 1, the product is that same whole number. Also, if 1 is multiplied by a whole number, the product is that same whole number. This is known as the **identity property** for multiplication of whole numbers. The whole number 1 is called the **multiplicative identity** for whole numbers. (Do you remember what the additive identity is for whole numbers?) If we let a represent any whole number, we could state the identity property for multiplication of whole numbers as follows:

$$a \times 1 = a \quad \text{and} \quad 1 \times a = a$$

We will now consider multiplying 3 whole numbers. However, just as in addition, we must group them by 2s to multiply.

Practice Exercise 3. Multiply: (a) $(4 \times 5) \times 3$; (b) $4 \times (5 \times 3)$.

EXAMPLE 3 Multiply: (a) $(2 \times 2) \times 3$; (b) $2 \times (2 \times 3)$.

SOLUTION: (a) $(2 \times 2) \times 3 = 4 \times 3 = 12$; (b) $2 \times (2 \times 3) = 2 \times 6 = 12$. Therefore, $(2 \times 2) \times 3 = 2 \times (2 \times 3)$.

Practice Exercise 4. Multiply: (a) $(7 \times 3) \times 4$; (b) $7 \times (3 \times 4)$.

EXAMPLE 4 Multiply: (a) $(2 \times 4) \times 1$; (b) $2 \times (4 \times 1)$.

SOLUTION: (a) $(2 \times 4) \times 1 = 8 \times 1 = 8$; (b) $2 \times (4 \times 1) = 2 \times 4 = 8$. Therefore, $(2 \times 4) \times 1 = 2 \times (4 \times 1)$.

> **ANSWERS TO PRACTICE EXERCISES**
>
> **3.** (a) $(4 \times 5) \times 3 = 20 \times 3 = 60$; (b) $4 \times (5 \times 3) = 4 \times 15 = 60$. **4.** (a) $(7 \times 3) \times 4 = 21 \times 4 = 84$; (b) $7 \times (3 \times 4) = 7 \times 12 = 84$.

Examples 3 and 4 illustrate that multiplication of whole numbers is an associative operation. If we let a, b, and c represent any whole numbers, then we could state the **associative property for multiplication** of whole numbers as follows:

$$(a \times b) \times c = a \times (b \times c)$$

Recall that to add numbers having more than one digit, we add the digits that have the same place value. That is, we add ones to ones, tens to tens, and so forth. To multiply numbers having more than one digit, we use a different procedure.

Rule for Multiplying Numbers

To multiply numbers having more than 1 digit, multiply the digit in each place value of one number by the digits in every place value of the other number. Then, add the products obtained.

The procedure of multiplying numbers having more than one digit is illustrated in the following examples.

Practice Exercise 5. Multiply: 9×18.

EXAMPLE 5 Multiply: 4×17.

SOLUTION:

```
     17
    × 4
     28   (The product of 4 and 7)
   + 40   (The product of 4 and 10)
     68   (The product)
```

1.4 Multiplication of Whole Numbers

An alternative method of solution for Example 5 involves the following shortcut:

Step 1:
```
   2
  17
× 4
   8
```
4 × 7 = 28

Multiply 7 ones by 4 ones, obtaining 28 ones. Write the 8 in the ones column below the 4. Carry the 2 to the tens column as indicated.

Step 2:
```
  17
× 4
  68
```
1 × 4 = 4
+2
6

Multiply the 1 ten by the 4 ones, obtaining 4 tens. Add to this the 2 tens from the previous step, obtaining 6 tens. Write the 6 in the tens column in the answer.

Practice Exercise 6. Multiply: 13 × 16.

EXAMPLE 6 Multiply the following numbers using the shortcut method:

$$\begin{array}{r} 236 \\ \times\, 342 \\ \hline \end{array}$$

SOLUTION:

Step 1:
```
   236
 × 342
   472
```
Multiply 236 by 2 ones following the same procedure as used in Example 5.

Step 2:
```
   236
 × 342
   472
   944
```
Multiply 236 by 4 tens. Notice that the answer (944) is lined up from the right under the tens column since we are multiplying by 4 *tens*.

Step 3:
```
   236
 × 342
   472
   944
   708
```
Multiply 236 by 3 hundreds. Notice that the answer (708) is lined up from the right under the hundreds column since we are multiplying by 3 *hundreds*.

Step 4:
```
    236
  × 342
    472
    944
    708
  80712
```
472 + 9440 + 70800 = 80712

It was noted earlier that we can multiply in any order. Hence, the method of checking multiplication is to interchange or reverse the factors and multiply.

Practice Exercise 7. Multiply: 23 × 12.

EXAMPLE 7 Multiply 234 and 63 and check your result.

SOLUTION:
```
    234
 ×   63
    702
   1404
  14742
```

CHECK:
$$\begin{array}{r} 63 \\ \times 234 \\ \hline 252 \\ 189 \\ 126 \\ \hline 14742 \checkmark \end{array}$$

Practice Exercise 8.
205 × 57.

EXAMPLE 8 Multiply 1037 and 205 and check your result.

SOLUTION:
$$\begin{array}{r} 1037 \\ \times 205 \\ \hline 5185 \\ 20740 \\ \hline 212585 \end{array}$$

CHECK:
$$\begin{array}{r} 205 \\ \times 1037 \\ \hline 1435 \\ 615 \\ 2050 \\ \hline 212585 \checkmark \end{array}$$

ANSWERS TO PRACTICE EXERCISES

5. 162.   6. 208.   7. 276.   8. 11,685.

Practice Exercise 9. If you make a weekly deposit of $35 to your savings account, how much money will be deposited in 52 weeks?

EXAMPLE 9 Motor oil is packed 24-qt cans to a case. How many cans of oil are in 38 cases of oil?

SOLUTION: Since there are 24 cans of oil in one case and there are 38 cases of oil, we multiply 24 and 38 to get the answer:

$$\begin{array}{rl} 24 & \text{(Cans of oil in one case)} \\ \times 38 & \text{(Number of cases of oil)} \\ \hline 192 & \\ 72 & \\ \hline 912 & \text{(Number of cans of oil in 38 cases)} \end{array}$$

Practice Exercise 10. If a salesperson travels an average of 1865 miles per week, how many miles will the person travel in 48 weeks?

EXAMPLE 10 A negotiated labor contract gave each employee of a company a $42-a-week raise in pay. If there are 217 employees at the company, what is the total weekly wage increase?

SOLUTION: Since there are 217 employees at the company and each employee received a $42 a week increase in pay, the total weekly wage increase is found by multiplying 217 and $42:

$$\begin{array}{rl} 217 & \text{(Number of employees)} \\ \times \$42 & \text{(Weekly increase for each employee)} \\ \hline 434 & \\ 868 & \\ \hline \$9114 & \text{(\textit{Total} weekly wage increase)} \end{array}$$

> **ANSWERS TO PRACTICE EXERCISES**
> 9. $1820. 10. 89,520.

In Section 1.2, we briefly introduced polygons and their perimeters. In this section, we briefly introduce the **area** of a polygon. Unlike perimeter, the area of a polygon depends upon the type of polygon we have. We will consider only triangles, rectangles, and squares.

Consider the triangle illustrated in Figure 1.2, having a base denoted by b and a height denoted by h. The area A of the triangle is given by $A = \frac{1}{2}bh$ where b and h are measured in the same unit and A is measured in the unit squared. The expression $\frac{1}{2}bh$ means $\frac{1}{2}$ times b times h.

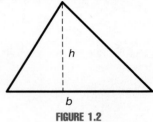

FIGURE 1.2

Practice Exercise 11. Determine the area of a triangle with base of 38 ft and height of 22 ft.

EXAMPLE 11 Determine the area of a triangle that has a base of 18 in. and a height of 10 in.

SOLUTION: We have that $b = 18$ in. and $h = 10$ in. Hence,

$$A = \frac{1}{2}bh$$

$$= \frac{1}{2}(18 \text{ in.})(10 \text{ in.})$$

$$= \frac{1}{2}(18)(10)(\text{in.})(\text{in.})$$

$$= 90(\text{in.})(\text{in.})$$

$$= 90 \text{ square inches}$$

where (inches)(inches) is written as "square inches," abbreviated "in.^{2}."

Practice Exercise 12. Determine the area of a triangle with base of 46 cm and height of 34 cm.

EXAMPLE 12 A triangle has a base of 24 yd and a height of 16 yd. Determine the area of the triangle.

SOLUTION: We have $b = 24$ yd and $h = 16$ yd. Hence,

$$A = \frac{1}{2}bh$$

$$= \frac{1}{2}(24 \text{ yd})(16 \text{ yd})$$

$$= \frac{1}{2}(24)(16)(\text{yd})(\text{yd})$$

$$= 192 \text{ yd}^2$$

Consider the rectangle illustrated in Figure 1.3, with length L and width W. The area A of the rectangle is given by $A = LW$ where L and W are given in the same unit and A is given in the unit squared.

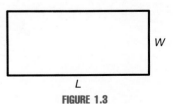

FIGURE 1.3

Practice Exercise 13. A rectangle has a length of 41 in. and a width of 31 in. Determine the area of the rectangle.

EXAMPLE 13 A rectangle has a length of 17 cm and a width of 13 cm. Determine the area of the rectangle.

SOLUTION: We have $L = 17$ cm and $W = 13$ cm. Hence,

$$\begin{aligned} A &= LW \\ &= (17 \text{ cm})(13 \text{ cm}) \\ &= (17)(13)(\text{cm})(\text{cm}) \\ &= 221 (\text{cm})(\text{cm}) \\ &= 221 \text{ cm}^2 \end{aligned}$$

Practice Exercise 14. A rectangle has a length of 6 yd and a width of 11 ft. Determine the area of the rectangle. (*Hint:* Convert yards to feet.)

EXAMPLE 14 A rectangle has a length of 29 in. and a width of 2 ft. Determine the area of the rectangle.

SOLUTION: We have $L = 29$ in. and $W = 2$ ft. However, these are different units. Converting 2 ft to inches, we have $W = 2$ ft $= 2(12$ in.$) = 24$ in. Hence,

$$\begin{aligned} A &= LW \\ &= (29 \text{ in.})(24 \text{ in.}) \\ &= (29)(24)(\text{in.})(\text{in.}) \\ &= 696 \text{ (in.)(in.)} \\ &= 696 \text{ in.}^2 \end{aligned}$$

Consider the square in Figure 1.4, with each side of measure s units. The area A of the square is given by $A = (s)(s)$ square units.

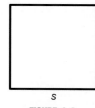

FIGURE 1.4

Practice Exercise 15. A square measures 17 in. on a side. Determine the area of the square.

EXAMPLE 15 A square measures 19 m on a side. Determine the area of the square.

SOLUTION: We have $s = 19$ m. Hence,

$$\begin{aligned} A &= (s)(s) \\ &= (19 \text{ m})(19 \text{ m}) \\ &= 361 \text{ (m)(m)} \\ &= 361 \text{ m}^2 \end{aligned}$$

Practice Exercise 16. A square measures 32 cm on a side. Determine the area of the square.

EXAMPLE 16 A square measures 23 ft on a side. Determine the area of the square.

SOLUTION: We have $s = 23$ ft. Hence,

$$\begin{aligned} A &= (s)(s) \\ &= (23 \text{ ft})(23 \text{ ft}) \\ &= (23)(23)(\text{ft})(\text{ft}) \\ &= 529 \text{ (ft)(ft)} \\ &= 529 \text{ ft}^2 \end{aligned}$$

ANSWERS TO PRACTICE EXERCISES

11. 418 ft^2.   **12.** 782 cm^2.   **13.** 1271 cm^2.   **14.** 198 ft^2.   **15.** 289 in.^{2}.   **16.** 1024 cm^2.

EXERCISES 1.4

Multiply in Exercises 1–25. Check your results.

1.	92 ×83	2.	99 ×90	3.	138 × 65	4.	176 × 94	5.	239 × 49
6.	666 × 77	7.	234 ×123	8.	399 ×268	9.	459 ×318	10.	760 ×837
11.	2367 × 191	12.	3067 × 230	13.	5699 × 936	14.	6218 ×1234	15.	8427 × 319
16.	9079 × 269	17.	8579 × 304	18.	9628 × 507	19.	6057 × 506	20.	2136 × 219
21.	9607 ×1782	22.	8792 ×3803	23.	4876 ×3456	24.	5076 ×3901	25.	7080 ×6009

26. Find the product of 62 and 23.
27. Find the product of 127 and 87.
28. Multiply 824 and 237.
29. Multiply 7209 and 906.
30. Multiply 8090 and 6027.
31. There are 24 cans of peaches in a case. How many cans of peaches are in 369 cases of peaches?
32. A particular model of car is EPA rated to get 29 mi/gal of gasoline. If the car has a gasoline tank that holds 22 gal, how many miles should the driver be able to go on a full tank of gasoline?
33. Janice's monthly salary is $1965. What is her yearly salary?
34. If it cost $62 per square foot to build a house, what will be the cost of a house that has 2140 ft^2 of floor space?
35. The average salary paid to executives of ABC Corporation is $32,605 per year. What is the total salary paid if there are 16 executives in the company? (*Hint:* Average is determined by dividing the total salaries by the number of executives.)
36. If the mortgage payment on a house is $862 per month, how much money will be paid for mortgage payments over a period of 25 yr?
37. Ben borrowed $3000 at an annual percentage interest rate of 11%, for 4 yr. His monthly payments are $78.
    a. How much money will Ben have to pay over the 4-yr period?
    b. How much more than the $3000 borrowed will he have to repay?
38. For a school picnic, it was decided that each student would have 2 hot dogs and each chaperone would have 3 hot dogs. There are 126 students and 7 chaperones.
    a. How many hot dogs must be ordered for the students?
    b. How many hot dogs must be ordered for the chaperones?
    c. How many hot dogs must be ordered for the students and the chaperones?
39. The Rotary-Reel Lawn Mower Co. manufactures a deluxe mower. Each mower requires 4 hr of skilled labor and 6 hr of unskilled labor to manufacture.
    a. How many hours of skilled labor are required to manufacture 692 deluxe mowers?
    b. How many hours of unskilled labor are required to manufacture 1168 deluxe mowers?
    c. In a particular week, 367 deluxe mowers are manufactured. How many combined hours of skilled and unskilled labor were required?
40. The Acme Oil Recycling Company collected 217 drums of oil one week, 186 drums of oil the second week, and 97 drums of oil the third week. If each drum holds 55 gal of oil, how many gallons of oil were collected during the 3-week period?
41. Determine the area of a triangle with a base of 28 in. and height of 16 in.
42. Determine the area of a triangle with a base of 4 yd and height of 8 ft.

43. A rectangle has length of 37 cm and width of 25 cm. Determine the area of the rectangle.
44. A rectangle has length of 7 yd and width of 17 ft. Determine the area of the rectangle.
45. A square measures 27 ft on a side. Determine the area of the square.
46. A square measures 108 cm on a side. Determine the area of the square.

1.5 Division of Whole Numbers

OBJECTIVES

After completing this section, you should be able to:
1. Divide whole numbers.
2. Solve a word problem using division.

We will now consider the operation of division with whole numbers. Addition, subtraction, multiplication, and division are the four basic operations of arithmetic. You have already learned that addition and subtraction are inverse (or opposite) operations of each other. In a similar manner, you will learn that multiplication and division are inverse (or opposite) operations of each other.

We considered multiplication as repeated additions, with all the terms being the same. In a similar manner, **division** may be considered as repeated subtractions. That is, we can consider how many times one number can be subtracted from another number.

Practice Exercise 1. A merchant has 54 U.S. savings bonds of equal value and wishes to divide them equally among his nine employees. How many bonds will each employee receive?

EXAMPLE 1 If a man has 24 shares of stock and he wishes to divide these equally among his 4 children, how many shares of stock will each child receive?

SOLUTION: Since 24 is greater than 4, each child will receive at least 1 share of stock. When 1 share of stock is given to each child, there would be 24 − 4, or 20, shares left. Since 20 is greater than 4, each child receives another share of stock, leaving 20 − 4, or 16, shares. Again, 16 is greater than 4, so each child receives a third share of stock, leaving 16 − 4, or 12, shares. Since 12 is also greater than 4, each child receives a fourth share of stock, leaving 12 − 4, or 8, shares. Again, 8 is greater than 4, so each child receives a fifth share of stock, leaving 8 − 4, or 4, shares. Since 4 is exactly equal to 4, each child receives a sixth share of stock, leaving 4 − 4, or 0, shares.

Therefore, the 24 shares of stock can be divided equally among 4 children with each child receiving exactly 6 shares. We may think of this problem as follows:

```
 24  shares (total)
− 4  shares (1 per child)
 20  shares left
− 4  shares (1 per child a second time)
 16  shares left
− 4  shares (1 per child a third time)
 12  shares left
− 4  shares (1 per child a fourth time)
  8  shares left
− 4  shares (1 per child a fifth time)
  4  shares left
− 4  shares (1 per child a sixth time)
  0  shares left
```

Therefore, 24 divided by 4 is 6.

1.5 Division of Whole Numbers

> **ANSWER TO PRACTICE EXERCISE**
> 1. 6.

We use the symbol ÷ to indicate division. With this symbol, we may write 24 ÷ 4 = 6 to mean 24 divided by 4 is equal to 6. The number being divided (in the preceding case, 24) is called the **dividend;** the number by which the dividend is being divided (in the preceding case, 4) is called the **divisor;** the result or answer (in the preceding case, 6) is called the **quotient.**

The above division can also be displayed as:

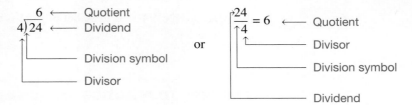

Since division is thought of as being the opposite (or inverse) operation of multiplication, we use multiplication as a check for division. (Recall how other operations have been checked using their inverse operations.) Hence,

$$24 \div 4 = 6$$

means that

$$4 \times 6 = 24$$

That is, the product of the divisor and the quotient is equal to the dividend.

$$24 \div 4 = 6$$

can be thought of as starting with 24 items and separating them into 4 groups with 6 items in each group as illustrated in Figure 1.5.

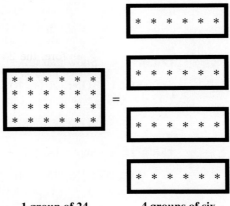

FIGURE 1.5   1 group of 24   4 groups of six

It should be noted that division is *not* a commutative operation. The order in which we divide one whole number by another whole number is important. For instance,

$$8 \div 4 = 2$$

but

$$4 \div 8$$

is not a whole number.

In general, if a and b are any two whole numbers, then

$$a \div b \neq b \div a$$

Practice Exercise 2. Show that $20 \div 5 \neq 5 \div 20$.

EXAMPLE 2 Show that $12 \div 2 \neq 2 \div 12$.

SOLUTION: $12 \div 2 = 6$, but $2 \div 12$ is not a whole number. Therefore, $12 \div 2 \neq 2 \div 12$.

Practice Exercise 3. Show that $45 \div 9 \neq 9 \div 45$.

EXAMPLE 3 Show that $24 \div 6 \neq 6 \div 24$.

SOLUTION: $24 \div 6 = 4$, but $6 \div 24$ is not a whole number. Therefore, $24 \div 6 \neq 6 \div 24$.

ANSWERS TO PRACTICE EXERCISES

2. $20 \div 5 = 4$, but $5 \div 20$ is not a whole number. Therefore, $20 \div 5 \neq 5 \div 20$. **3.** $45 \div 9 = 5$, but $9 \div 45$ is not a whole number. Therefore, $45 \div 9 \neq 9 \div 45$.

Division of whole numbers is *not* an associative operation either. Consider the following:

$$16 \div (4 \div 2) = 16 \div 2$$
$$= 8$$

but
$$(16 \div 4) \div 2 = 4 \div 2$$
$$= 2$$

Since $8 \neq 2$

then $16 \div (4 \div 2) \neq (16 \div 4) \div 2$

showing that grouping *does* matter in division.

In general, if a, b, and c are *any* whole number, then

$$(a \div b) \div c \neq a \div (b \div c)$$

Practice Exercise 4. Show that $(16 \div 8) \div 2 \neq 16 \div (8 \div 2)$.

EXAMPLE 4 Show that $(20 \div 10) \div 2 \neq 20 \div (10 \div 2)$.

SOLUTION: $(20 \div 10) \div 2 = 2 \div 2 = 1$, but $20 \div (10 \div 2) = 20 \div 5 = 4$. Since $1 \neq 4$, then $(20 \div 10) \div 2 \neq 20 \div (10 \div 2)$.

Practice Exercise 5. Show that $(27 \div 3) \div 9 \neq 27 \div (3 \div 9)$.

EXAMPLE 5 Show that $(15 \div 3) \div 5 \neq 15 \div (3 \div 5)$.

SOLUTION: $(15 \div 3) \div 5 = 5 \div 5 = 1$, but $15 \div (3 \div 5)$ is not a whole number since $3 \div 5$ is not a whole number. Therefore, $(15 \div 3) \div 5 \neq 15 \div (3 \div 5)$.

ANSWERS TO PRACTICE EXERCISES

4. $(16 \div 8) \div 2 = 2 \div 2 = 1$, but $16 \div (8 \div 2) = 16 \div 4 = 4$. Since $1 \neq 4$, then $(16 \div 8) \div 2 \neq 16 \div (8 \div 2)$. **5.** $(27 \div 3) \div 9 = 9 \div 9 = 1$, but $27 \div (3 \div 9)$ is not a whole number since $3 \div 9$ is not a whole number. Hence, $(27 \div 3) \div 9 \neq 27 \div (3 \div 9)$.

1.5 Division of Whole Numbers

Practice Exercise 6. Divide 382 by 7.

EXAMPLE 6 Divide 283 by 6.

SOLUTION: $6\overline{)283}$

STEP 1:
$$\begin{array}{r} 4 \\ 6\overline{)283} \end{array}$$

Since 2 is less than 6, we consider the first 2 digits from the left and ask how many groups of 6 are contained in 28. The answer is 4. Since 28 represents tens, we place the 4 above the 8 in the tens place in the quotient.

STEP 2:
$$\begin{array}{r} 4 \\ 6\overline{)283} \\ -240 \\ \hline 43 \end{array} \quad (40 \times 6)$$

Multiply 4 tens (in the quotient) by 6 ones (in the divisor), obtaining 240. Place the 240 under the 283 and subtract. The difference is 43.

STEP 3:
$$\begin{array}{r} 47 \\ 6\overline{)283} \\ -240 \\ \hline 43 \end{array} \quad (40 \times 6)$$

Now ask how many groups of 6 are contained in 43. The answer is 7. Since 43 represents ones, we place the 7 in the ones place in the quotient.

STEP 4:
$$\begin{array}{r} 47 \\ 6\overline{)283} \\ -240 \\ \hline 43 \\ -42 \\ \hline 1 \end{array} \quad \begin{array}{l} (40 \times 6) \\ \\ (7 \times 6) \end{array}$$

Multiply 7 ones (in the quotient) by 6 ones (in the divisor), obtaining 42. Place the 42 under the 43 already there and subtract. The difference is 1. This 1 is called the **remainder** (abbreviated R in this example) and may be written in either form, as indicated in Step 5.

STEP 5:

$$\begin{array}{r} 47 \\ 6\overline{)283} \\ -240 \\ \hline 43 \\ -42 \\ \hline 1R \end{array} \begin{array}{l} \\ (40 \times 6) \\ \\ (7 \times 6) \end{array} \quad \text{or} \quad \begin{array}{r} 47\,\text{R}1 \\ 6\overline{)283} \\ -240 \\ \hline 43 \\ -42 \\ \hline 1 \end{array} \begin{array}{l} \\ (40 \times 6) \\ \\ (7 \times 6) \end{array}$$

Hence, there are 47 groups of 6 contained in 283, with a remainder of 1.

CHECK:
$$\begin{array}{rl} 47 & \text{(Quotient)} \\ \times\ 6 & \text{(Divisor)} \\ \hline 282 & \\ +\ \ 1 & \text{(Remainder)} \\ \hline 283\ \checkmark & \text{(Dividend)} \end{array}$$

Practice Exercise 7. Divide 1435 by 7.

EXAMPLE 7 Divide 863 by 21.

SOLUTION:

STEP 1:
$$\begin{array}{r} 4 \\ 21\overline{)863} \end{array}$$

Since the divisor is a 2-digit number, we ask how many groups of 21 are contained in 86. The answer is 4; (21 × 4 = 84). Write the 4 in the quotient above the 6 in the dividend.

STEP 2:
$$\begin{array}{r} 4 \\ 21\overline{)863} \\ -840 \\ \hline 23 \end{array} \quad (40 \times 21)$$

Multiply 4 tens (in the quotient) by 21 (the divisor), obtaining 840. Place the 840 below the 863 and subtract.

STEP 3:
$$\begin{array}{r} 41 \\ 21\overline{)863} \\ -840 \quad (40 \times 21) \\ \hline 23 \end{array}$$

Now ask how many groups of 21 are contained in 23. The answer is 1. Write 1 in the quotient above the 3 in the dividend.

STEP 4:
$$\begin{array}{r} 41 \\ 21\overline{)863} \\ -840 \quad (40 \times 21) \\ \hline 23 \\ -21 \quad (1 \times 21) \\ \hline 2 \end{array}$$

Multiply 1 one (in the quotient) by 21 (the divisor), obtaining 21. Write the 21 below the 23 obtained in the previous step and subtract. The difference is 2.

STEP 5:
$$\begin{array}{r} 41 \\ 21\overline{)863} \\ -840 \quad (40 \times 21) \\ \hline 23 \\ -21 \quad (1 \times 21) \\ \hline 2\,R \end{array}$$

Now ask how many groups of 21 are contained in 2. The answer is 0. Hence, we cannot subtract any more 21s from 863. We subtracted 21 from 863 a total of 41 times and have 2 left over.

CHECK:
$$\begin{array}{r} 41 \quad \text{(Quotient)} \\ \times 21 \quad \text{(Divisor)} \\ \hline 41 \\ 82 \\ \hline 861 \\ +\;\;2 \quad \text{(Remainder)} \\ \hline 863 \quad \text{(Dividend)} \end{array}$$

We will shorten the division procedure in the following examples.

Practice Exercise 8. Divide 20306 by 21.

EXAMPLE 8 Divide 1812 by 6.

SOLUTION:
$$\begin{array}{r} 302 \\ 6\overline{)1812} \\ -18 \quad (300 \times 6) \\ \hline 01 \\ -0 \quad (0 \times 6) \\ \hline 12 \\ -12 \quad (2 \times 6) \\ \hline 0\,R \end{array}$$

CHECK:
$$\begin{array}{r} 302 \quad \text{(Quotient)} \\ \times\;\;6 \quad \text{(Divisor)} \\ \hline 1812 \\ +\;\;0 \quad \text{(Remainder)} \\ \hline 1812 \quad \text{(Dividend)} \end{array}$$

Practice Exercise 9. Divide 109072 by 39.

EXAMPLE 9 Divide 609,762 by 216.

SOLUTION:
$$\begin{array}{r} 2{,}822 \\ 216\overline{)609{,}762} \\ -432 \quad (2000 \times 216) \\ \hline 177\;7 \\ -172\;8 \quad (800 \times 216) \\ \hline 4\;96 \\ -4\;32 \quad (20 \times 216) \\ \hline 642 \\ -432 \quad (2 \times 216) \\ \hline 210\,R \end{array}$$

1.5 Division of Whole Numbers

CHECK:
$$\begin{array}{r}2822 \text{ (Quotient)}\\ \times 216 \text{ (Divisor)}\\ \hline 16932\\ 2822\\ 5644\\ \hline 609{,}552\\ +210 \text{ (Remainder)}\\ \hline 609{,}762 \checkmark \text{ (Dividend)}\end{array}$$

ANSWERS TO PRACTICE EXERCISES

6. 54 R4.   7. 205.   8. 966 R20.   9. 2796 R28.

Practice Exercise 10. An automobile dealer received a shipment of 6215 gal of oil. If the shipment arrived in drums, with each drum containing 55 gal of oil, how many drums of oil were in the shipment?

EXAMPLE 10   A certain car can travel 437 mi on 23 gal of gasoline. How far can it travel on 1 gal of gasoline?

SOLUTION: We know how many miles were traveled on 23 gal of gasoline and want to determine how many miles are traveled on 1 gal of gasoline. Therefore, we use division and divide 437 by 23:

$$\begin{array}{r}19\\ 23\overline{)437}\\ -23\quad (10\times 23)\\ \hline 207\\ -207\quad (9\times 23)\\ \hline 0\end{array}$$

Therefore, the car can travel 19 mi on 1 gal of gasoline.

Practice Exercise 11. If your annual salary is $19,812 and you are paid weekly, what is your weekly salary?

EXAMPLE 11   Martha earns $17,184 a year and is paid monthly. What is her monthly salary?

SOLUTION: To determine Martha's monthly salary, we divide her yearly salary ($17,184) by the number of months in a year (12):

$$\begin{array}{r}\$1{,}432\\ 12\overline{)\$17{,}184}\\ -12\\ \hline 5\ 1\\ -4\ 8\\ \hline 38\\ -36\\ \hline 24\\ -24\\ \hline 0\end{array}$$

Therefore, her monthly salary is $1432.

ANSWERS TO PRACTICE EXERCISES

10. 113.   11. $381.

EXERCISES 1.5

Divide in Exercises 1–20.

1. $8\overline{)120}$
2. $12\overline{)132}$
3. $24\overline{)288}$
4. $23\overline{)276}$
5. $32\overline{)391}$
6. $25\overline{)999}$
7. $17\overline{)10,825}$
8. $123\overline{)8162}$
9. $40\overline{)46,951}$
10. $469\overline{)10,231}$
11. $521\overline{)12,676}$
12. $287\overline{)23,696}$
13. $90,069 \div 207$
14. $46,907 \div 313$
15. $99,088 \div 629$
16. $102,345 \div 439$
17. $4,762,007 \div 592$
18. $169,431 \div 201$
19. $986,333 \div 812$
20. $890,096 \div 709$
21. Divide 6392 by 26.
22. Divide 8091 by 38.
23. Divide 10,213 by 101.
24. Divide 27,619 by 197.
25. Divide 313,069 by 908.
26. Divide 504,030 by 999.
27. Divide 1,360,142 by 1213.
28. Divide 1,697,001 by 2003.
29. Determine the quotient when 6897 is divided by 29.
30. Determine the quotient when 9016 is divided by 213.
31. Determine the remainder when 8672 is divided by 17.
32. Determine the remainder when 9063 is divided by 28.
33. There are 432 cans of fruit cocktail contained in 18 cases. Each case contains the same number of cans. How many cans of fruit cocktail are in each case?
34. If your annual salary is $16,432, how much do you earn per week?
35. The EPA rating for a certain car is 23 mi/gal. A man bought one of these cars, drove it for 3625 mi, and used 240 gal of gasoline. He then complained to the dealer who sold the car to him. What do you think the complaint was about? Why?
36. A state lottery prize of $259,720 was shared by 43 contestants. How much did each contestant receive, if they all received the same amount?
37. The Tasty Nut Shop received a bulk shipment of 9948 oz of cashew nuts. How many 12-oz packages of the nuts can be made up from the shipment?
38. The Tasty Nut Shop also received a bulk shipment of 600 lb of roasted peanuts. How many 12-oz packages of roasted peanuts can be made up from the shipment? (*Hint:* 1 lb = 16 oz.)
39. If 3696 work-days are lost per year due to illness at the ABC Corp., and there are 231 employees, what is the average number of work-days lost per employee?
40. If 40,880 automobile accidents occur in a particular part of the country in 1 year, how many automobile accidents occur, on the average, in a day for that part of the country? (365 days = 1 year.)
41. Sue borrowed $3000 to buy a car. The finance charges for the loan amounted to $492. If she agreed to repay the total amount of the loan and finance charges in 36 equal monthly payments, how much would each payment be?
42. Mr. Rodriguez's estate was $126,350. He left half of the estate to his wife. The other half was divided equally among his five children. How much did each child receive from the estate?
43. Community College has a budget of $10,632,690. There are 4314 students enrolled at the college. What is the average cost per student represented by the budget? (Disregard the cents, if any.)

1.6 Rounding and Estimating

OBJECTIVES

After completing this section, you should be able to:
1. Round a whole number to the nearest place value.
2. Using rounding, estimate the answer to a problem.

ROUNDING

Often, we approximate the value of a number by using a procedure known as **rounding**. For example, if 4967 students are enrolled at a college, the registrar may report that 5000 students are enrolled, rounding the actual enrollment to the nearest thousand. Or, if a large industry has an annual payroll of $36,198,261, a Chamber of Commerce spokesperson may refer to the industry's payroll as being $36,000,000, rounding the actual payroll to the nearest million dollars. Possibly, the actual payroll may be referred to as $40,000,000, rounding to the nearest ten million dollars.

Rule for Rounding Numbers

To *round a number* to a given place value, look at the digit immediately to the right of that place value:

1. If this digit is less than 5, change it and all digits to the right of it to zeros. Do not change the digit in the given place value.
2. If this digit is greater than or equal to 5, change it and all digits to its right to zeros. *Add* 1 to the digit in the given place value.

CAUTION The above rule for rounding numbers is not to be applied blindly when working with real-life problems. For instance, to mail a letter that weighs 1.1 oz, you must pay postage for 2 oz.

Practice Exercise 1. Round 45,673 to the nearest ten.

EXAMPLE 1 Round 3647 to the nearest ten.

SOLUTION: The digit in the tens place is 4. The digit to its right is 7, which is greater than 5. Part 2 of the rule applies and 3647 is rounded up to 3650 to the nearest ten.

Practice Exercise 2. Round 937,802 to the nearest thousand.

EXAMPLE 2 Round 13,269 to the nearest thousand.

SOLUTION: The digit in the thousands place is 3. The digit to its right is 2, which is less than 5. Part 1 of the rule applies and 13,269 is rounded down to 13,000 to the nearest thousand.

Practice Exercise 3. Round 123,908 to the nearest hundred.

EXAMPLE 3 Round 169,750 to the nearest hundred.

SOLUTION: The digit in the hundreds place is 7. The digit to its right is 5. Part 2 of the rule applies and 169,750 is rounded up to 169,800.

Practice Exercise 4. Round 9,876,123 to the nearest ten thousand.

EXAMPLE 4 Round 234,987 to the nearest ten thousand.

SOLUTION: The digit in the ten thousands place is 3. The digit to its right is 4. Part 1 of the rule applies and 234,987 is rounded down to 230,000.

Practice Exercise 5. Round 277,456,123 to the nearest million.

EXAMPLE 5 Round 26,567,297 to the nearest million.

SOLUTION: The digit in the millions place is 6. The digit to its right is 5. Part 2 of the rule applies and 26,567,297 is rounded up to 27,000,000.

ANSWERS TO PRACTICE EXERCISES

1. 45,670.   2. 938,000.   3. 123,900.   4. 9,880,000.   5. 277,000,000.

ESTIMATION

Sometimes, when we are doing calculations, it is only necessary to get a "ballpark" result. That is, we try **estimating** the result by rounding all of the numbers involved to the same place value. For instance, to estimate the sum of 817 and 962, we could round both numbers to the nearest ten, obtaining 820 and 960, and estimate the sum to be 1780. Or we could round the numbers to the nearest hundred, obtaining 800 and 1000, and estimate the sum to be 1800. Generally, when estimating, there is no "right" answer since the estimate depends upon how we round the given numbers.

Practice Exercise 6. Estimate the following sum by rounding each number to the nearest ten: 37 + 79 + 53.

EXAMPLE 6 Estimate the following sum by rounding each number to the nearest ten: 43 + 77 + 92.

SOLUTION: Estimate: 40 + 80 + 90 = 210. (The actual sum is 212.)

Practice Exercise 7. Estimate the following sum by rounding each number to the nearest hundred: 723 + 457 + 378 + 289.

EXAMPLE 7 Estimate the following sum by rounding each number to the nearest hundred: 237 + 587 + 912 + 168.

SOLUTION: Estimate: 200 + 600 + 900 + 200 = 1900. (The actual sum is 1904.)

Practice Exercise 8. Estimate the following product by rounding each number to the nearest ten: 53 × 87.

EXAMPLE 8 Estimate the following product by rounding each number to the nearest ten: 77 × 62.

SOLUTION: Estimate: 80 × 60 = 4800. (The actual product is 4774.)

Practice Exercise 9. Estimate the following product by rounding each number to the nearest hundred: 769 × 435.

EXAMPLE 9 Estimate the following product by rounding each number to the nearest hundred: 692 × 539.

SOLUTION: Estimate: 700 × 500 = 350,000. (The actual product is 372,988.)

Practice Exercise 10. Estimate the following difference by rounding each number to the nearest thousand: 46,045 − 22,704.

EXAMPLE 10 Estimate the following difference by rounding each number to the nearest thousand: 23,894 − 17,432.

SOLUTION: Estimate: 24,000 − 17,000 = 7000. (The actual difference is 6462.)

1.6 Rounding and Estimating

Practice Exercise 11. Estimate the following difference by rounding each number to the nearest ten thousand: 452,007 − 14,567.

EXAMPLE 11 Estimate the following difference by rounding each number to the nearest ten thousand: 156,980 − 93,789.

SOLUTION: Estimate: 160,000 − 90,000 = 70,000. (The actual difference is 63,191.)

Practice Exercise 12. Estimate the quotient when 27,934 is divided by 72.

EXAMPLE 12 Estimate the quotient when 16,129 is divided by 38.

SOLUTION: We could round 16,129 to the nearest thousand (obtaining 16,000) and 38 to the nearest ten (obtaining 40). Estimate: 16,000 ÷ 40 = 400. (The actual quotient is 424 and the remainder is 17.)

Practice Exercise 13. Estimate the quotient when 368,739 is divided by 2,935.

EXAMPLE 13 Estimate the quotient when 238,987 is divided by 1987.

SOLUTION: We could round 238,987 to the nearest ten thousand (obtaining 240,000) and 1987 to the nearest thousand (obtaining 2000). Estimate: 240,000 ÷ 2,000 = 120. (The actual quotient is 120 and the remainder is 547.)

ANSWERS TO PRACTICE EXERCISES

6. 170. **7.** 1900. **8.** 4500. **9.** 320,000. **10.** 23,000. **11.** 440,000. **12.** 400 (obtained by dividing 28,000 by 70). **13.** 123 (obtained by dividing 369,000 by 3000).

EXERCISES 1.6

In Exercises 1–20, round each whole number to the nearest place indicated.

1. 234; tens
2. 312; tens
3. 687; tens
4. 899; tens
5. 783; hundreds
6. 650; hundreds
7. 6047; hundreds
8. 8962; hundreds
9. 12,063; thousands
10. 17,910; thousands
11. 124,962; hundreds
12. 623,790; ten thousands
13. 1,234,567; thousands
14. 1,234,567; ten thousands
15. 1,234,567; millions
16. 56,023,706; hundred thousands
17. 5,500,000; millions
18. 23,692,763; ten millions
19. 18,500,000; millions
20. 986,209,706; thousands

In Exercises 21–25, estimate the following sum by rounding each of the numbers to the nearest place indicated.

21. 67 + 92 + 45 + 61; tens
22. 83 + 56 + 65 + 75 + 81; tens
23. 304 + 569 + 356 + 872; hundreds
24. 421 + 697 + 356 + 918 + 162; hundreds
25. 1234 + 5678 + 3097 + 7892; thousands

In Exercises 26–30, estimate the following differences by rounding each of the numbers to the nearest place indicated.

26. 87 − 43; tens
27. 756 − 467; hundreds
28. 1245 − 763; hundreds
29. 23,456 − 17,234; thousands
30. 367,987 − 158,003; ten thousands

In Exercises 31–35, estimate the following products by rounding each of the numbers to the nearest place indicated.

31. 67 × 79; tens
32. 456 × 38; tens
33. 279 × 356; hundreds
34. 809 × 395; hundreds
35. 1209 × 777; hundreds

In Exercises 36–40, estimate the quotients when the first number (rounded as indicated) is divided by the second number (rounded as indicated).

36. 789 (hundreds) ÷ 18 (tens)
37. 1567 (thousands) ÷ 53 (tens)
38. 33,899 (thousands) ÷ 169 (tens)
39. 121,567 (thousands) ÷ 61,001 (thousands)
40. 439,809 (ten thousands) ÷ 21,534 (thousands)

1.7 Mixed Operations with Whole Numbers

OBJECTIVE

After completing this section, you should be able to find the value of an expression using a prescribed order of operations.

We have indicated that multiplication may be thought of as being repeated additions. When performing operations involving both multiplication and addition, the multiplication is done first. The multiplications are done in the order in which they occur from left to right. Then, the additions are done.

In this section, we will introduce another property for the whole numbers. This property involves both multiplication and addition.

Perform the indicated operations, using the distributive property of multiplication over addition (or subtraction).

Practice Exercise 1.
2 × (7 + 5)

EXAMPLE 1 Determine each of the following: (a) 2 × (3 + 4); (b) (2 × 3) + (2 × 4).

SOLUTION: (a) 2 × (3 + 4) = 2 × 7 = 14; (b) (2 × 3) + (2 × 4) = 6 + 8 = 14. Therefore, 2 × (3 + 4) = (2 × 3) + (2 × 4).

Practice Exercise 2.
3 × (9 + 4)

EXAMPLE 2 Determine each of the following: (a) 3 × (2 + 5); (b) (3 × 2) + (3 × 5).

SOLUTION: (a) 3 × (2 + 5) = 3 × 7 = 21; (b) (3 × 2) + (3 × 5) = 6 + 15 = 21. Therefore, 3 × (2 + 5) = (3 × 2) + (3 × 5).

1.7 Mixed Operations with Whole Numbers

ANSWERS TO PRACTICE EXERCISES

1. $2 \times (7 + 5) = (2 \times 7) + (2 \times 5) = 14 + 10 = 24$. **2.** $3 \times (9 + 4) = (3 \times 9) + (3 \times 4) = 27 + 12 = 39$.

To compute the expression $3 \times (2 + 4)$, you could:

- First add 2 and 4, obtaining 6
- Then multiply 3 and 6, obtaining 18

or

- First multiply 3 and 2, obtaining 6
- Then multiply 3 and 4, obtaining 12
- Finally, add 6 and 12, obtaining 18

The property illustrated in Examples 1 and 2 is called the **distributive property for multiplication over addition** of whole numbers. If we let *a, b,* and *c* represent *any* whole numbers, then this property can be stated as

$$a \times (b + c) = (a \times b) + (a \times c)$$

Notice that this property involves two operations—multiplication and addition. We also have a **distributive property for multiplication over subtraction** of whole numbers. If we let *a*, *b*, and *c* represent any whole numbers, then this property can be stated as

$$a \times (b - c) = (a \times b) - (a \times c)$$

Practice Exercise 3.
$9 \times (13 - 4)$

EXAMPLE 3 Determine the value of each of the following: (a) $3 \times (6 - 2)$; (b) $(3 \times 6) - (3 \times 2)$.

SOLUTION: (a) $3 \times (6 - 2) = 3 \times 4 = 12$; (b) $(3 \times 6) - (3 \times 2) = 18 - 6 = 12$. Therefore, $3 \times (6 - 2) = (3 \times 6) - (3 \times 2)$.

Practice Exercise 4.
$12 \times (8 - 2)$

EXAMPLE 4 Determine the value of each of the following: (a) $5 \times (7 - 4)$; (b) $(5 \times 7) - (5 \times 4)$.

SOLUTION: (a) $5 \times (7 - 4) = 5 \times 3 = 15$; (b) $(5 \times 7) - (5 \times 4) = 35 - 20 = 15$. Therefore, $5 \times (7 - 4) = (5 \times 7) - (5 \times 4)$.

ANSWERS TO PRACTICE EXERCISES

3. $9 \times (13 - 4) = (9 \times 13) - (9 \times 4) = 117 - 36 = 81$. **4.** $12 \times (8 - 2) = (12 \times 8) - (12 \times 2) = 96 - 24 = 72$.

The distributive properties involve two arithmetic operations: multiplication and either addition or subtraction. We often encounter three or even all four arithmetic operations in the same problem. The order in which these operations are performed *is* important. Therefore, to perform the indicated operations, we use the following rule concerning **order of operations.**

Rule for Performing Operations on Whole Numbers

1. Do the operations within parentheses, if any.
2. Do all multiplications and divisions (whichever come first), in order from left to right.
3. Do all additions and subtractions (whichever come first), in order from left to right.

Perform the indicated operations using the preceding rule.

Practice Exercise 5.
$3 \times 4 - 2 \times 3$

EXAMPLE 5 Evaluate: $(8 - 5) \times (6 + 1) - 18 \div 6 \times 2$.

SOLUTION:

STEP 1: First, do the work within the parentheses:

$$(8 - 5) \times (6 + 1) - 18 \div 6 \times 2$$
$$= 3 \times 7 - 18 \div 6 \times 2$$

STEP 2: Next, do all multiplications and divisions (whichever come first) in order from left to right:

$$3 \times 7 - 18 \div 6 \times 2$$
$$= 21 - 18 \div 6 \times 2$$
$$= 21 - 3 \times 2$$
$$= 21 - 6$$

STEP 3: Finally, do all additions and subtractions (whichever come first) in order from left to right:

$$21 - 6 = 15$$

Therefore, $(8 - 5) \times (6 + 1) - 18 \div 6 \times 2 = 15$.

Practice Exercise 6.
$16 \div (7 - 3) + 4 \times 3$

EXAMPLE 6 Evaluate: $8 \div 2 \times 3 - 8 \div 2 + 7$.

SOLUTION:

STEP 1: There are no parentheses, so we first do all the multiplications and divisions (whichever come first) in order from left to right:

$$8 \div 2 \times 3 - 8 \div 2 + 7$$
$$= 4 \times 3 - 8 \div 2 + 7$$
$$= 12 - 8 \div 2 + 7$$
$$= 12 - 4 + 7$$

STEP 2: Next, do all additions and subtractions (whichever come first) in order from left to right:

$$12 - 4 + 7$$
$$= 8 + 7$$
$$= 15$$

Therefore, $8 \div 2 \times 3 - 8 \div 2 + 7 = 15$.

ANSWERS TO PRACTICE EXERCISES

5. 6.   6. 16.

1.7 Mixed Operations with Whole Numbers

In some problems, we encounter a whole number multiplied by itself such as 3×3. This product may also be written as 3^2, that is, the number 3 with a **superscript** 2 to the right and above the 3, as indicated. The expression 3^2 is read "3 squared" and means 3×3. The number being squared is called the **base**, and the superscript is called the **exponent**. In a similar manner, 4^3 is read "4 cubed" and means $4 \times 4 \times 4$.

- $2^2 = 2 \times 2 = 4$; the base is 2, the exponent is 2.
- $4^2 = 4 \times 4 = 16$; the base is 4, the exponent is 2.
- $2^3 = 2 \times 2 \times 2 = 8$; the base is 2, the exponent is 3.
- $3^3 = 3 \times 3 \times 3 = 27$; the base is 3, the exponent is 3.
- $2^5 = 2 \times 2 \times 2 \times 2 \times 2 = 32$; the base is 2, the exponent is 5.
- $3^4 = 3 \times 3 \times 3 \times 3 = 81$; the base is 3, the exponent is 4.

Practice Exercise 7. $6^2 - 3^3$

EXAMPLE 7 Determine the value of $7^2 + 5^2$.

SOLUTION:
$$\begin{aligned} 7^2 + 5^2 &= (7 \times 7) + (5 \times 5) \\ &= 49 + 25 \\ &= 74 \end{aligned}$$

Practice Exercise 8. $2^3 \times 3^2$

EXAMPLE 8 Determine the value of $5^3 - 3^3$.

SOLUTION:
$$\begin{aligned} 5^3 - 3^3 &= (5 \times 5 \times 5) - (3 \times 3 \times 3) \\ &= 125 - 27 \\ &= 98 \end{aligned}$$

Practice Exercise 9. $2^4 \div 4^2$

EXAMPLE 9 Determine the value of $6^2 - 3^3$.

SOLUTION:
$$\begin{aligned} 6^2 - 3^3 &= (6 \times 6) - (3 \times 3 \times 3) \\ &= 36 - 27 \\ &= 9 \end{aligned}$$

Practice Exercise 10. $3^4 + 2^5$

EXAMPLE 10 Determine the value of $3^4 \div 9^2$.

SOLUTION:
$$\begin{aligned} 3^4 \div 9^2 &= (3 \times 3 \times 3 \times 3) \div (9 \times 9) \\ &= 81 \div 81 \\ &= 1 \end{aligned}$$

ANSWERS TO PRACTICE EXERCISES
7. 9. 8. 72. 9. 1. 10. 113.

To perform mixed operations on whole numbers when exponents are also involved requires an additional step in the rule introduced earlier.

Extended Rule for Performing Operations on Whole Numbers

1. Do all operations within parentheses, if any.
2. Evaluate all expressions with exponents.
3. Do all multiplications and divisions (whichever come first) in order from left to right.
4. Do all additions and subtractions (whichever come first) in order from left to right.

P Parentheses

E Exponents

M Multiplications
   and/or   (In order from left to right)
D Divisions

A Additions
   and/or   (In order from left to right)
S Subtractions

Practice Exercise 11.
$(2^2 - 1) - 8 \div 4$

EXAMPLE 11 Evaluate: $3^2 - 2^3 \times 4 \div 8 + 3 \times (1 \times 4)$.

SOLUTION:

STEP 1: Work within parentheses:

$$3^2 - 2^3 \times 4 \div 8 + 3 \times (1 + 4)$$
$$= 3^2 - 2^3 \times 4 \div 8 + 3 \times 5$$

STEP 2: Evaluate all expressions with exponents:

$$3^2 - 2^3 \times 4 \div 8 + 3 \times 5$$
$$= 9 - 8 \times 4 \div 8 + 3 \times 5$$

STEP 3: Do all multiplications and divisions in order from left to right:

$$9 - 8 \times 4 \div 8 + 3 \times 5$$
$$= 9 - 32 \div 8 + 3 \times 5$$
$$= 9 - 4 + 3 \times 5$$
$$= 9 - 4 + 15$$

STEP 4: Do all additions and subtractions in order from left to right:

$$9 - 4 + 15$$
$$= 5 + 15$$
$$= 20$$

Therefore, $3^2 - 2^3 \times 4 \div 8 + 3 \times (1 + 4) = 20$.

Practice Exercise 12. $(2 + 3) \times (8 \div 2) \div 2^2 + (3 \times 2)^2$

EXAMPLE 12 Evaluate: $(2 + 3)^2 \div (7 - 2) + 9 - 3 \times 4 + 4^2$.

SOLUTION:

STEP 1: Work within parentheses:

$$(2 + 3)^2 \div (7 - 2) + 9 - 3 \times 4 + 4^2$$
$$= 5^2 \div 5 + 9 - 3 \times 4 + 4^2$$

1.7 Mixed Operations with Whole Numbers

STEP 2: Evaluate all expressions with exponents:

$$5^2 \div 5 + 9 - 3 \times 4 + 4^2$$
$$= 25 \div 5 + 9 - 3 \times 4 + 16$$

STEP 3: Do all multiplications and divisions in order from left to right:

$$25 \div 5 + 9 - 3 \times 4 + 16$$
$$= 5 + 9 - 3 \times 4 + 16$$
$$= 5 + 9 - 12 + 16$$

STEP 4: Do all additions and subtractions in order from left to right:

$$5 + 9 - 12 + 16$$
$$= 14 - 12 + 16$$
$$= 2 + 16$$
$$= 18$$

Therefore, $(2 + 3)^2 \div (7 - 2) + 9 - 3 \times 4 + 4^2 = 18$.

ANSWERS TO PRACTICE EXERCISES
11. 1.   12. 41.

Practice Exercise 13. Paula rented four videos at $3 each, including sales tax. She gave the salesclerk $20 to pay for her purchase. How much change should she receive?

EXAMPLE 13 Mark bought 2 shirts for $14 each, including sales tax. He gave the salesclerk $30 to pay for his purchase. How much change should he receive?

SOLUTION: To determine the amount of change that Mark should receive, we must subtract the cost for the two shirts from $30. We have

$$\$30 - (2 \times \$14)$$
$$= \$30 - \$28$$
$$= \$2$$

Hence, Mark should receive $2 in change.

Practice Exercise 14. The cube of 2 is subtracted from the square of 5. The difference is then multiplied by the sum of 3 and 8. What is the result?

EXAMPLE 14 The square of 5 is subtracted from the cube of 4. The difference is then divided by the sum of 4 and 9. What is the result?

SOLUTION: We must divide a difference $(4^3 - 5^2)$ by a sum $(4 + 9)$. We have:

$$(4^3 - 5^2) \div (4 + 9)$$
$$= (64 - 25) \div (4 + 9)$$
$$= 39 \div 13$$
$$= 3$$

In Example 14, notice that the difference, $4^3 - 5^2$, was put in parentheses, as was the sum, $4 + 9$. That is very necessary because, without the parentheses, we would have

$$4^3 - 5^2 \div 4 + 9$$

which is evaluated as

$$4^3 - 5^2 \div 4 + 9$$
$$= 64 - 25 \div 4 + 9$$

which is not a whole number since $25 \div 4$ is not a whole number.

ANSWERS TO PRACTICE EXERCISES

13. $8. **14.** 187.

CAUTION Use parentheses or other symbols of grouping when necessary.

EXERCISES 1.7

Perform the indicated operations in Exercises 1–34.

1. 4^2
2. 5^3
3. 11^2
4. 3^5
5. 5^4
6. 6^3
7. 9^4
8. 10^3
9. $4^3 - 4^2$
10. $5^3 - 5^2$
11. $7^4 - 7^3$
12. $9^3 - 9^2$
13. $8 \times (4 + 7)$
14. $13 \times (14 - 8)$
15. $19 \times (22 - 2)$
16. $23 \times (16 + 4)$
17. $4 \times 6 - 3 \times 5$
18. $3 \times (6 + 2) + 5 \times 4$
19. $3 + 1 \times 4 - 8 \div 2$
20. $9 \div (4 - 1) + 4 \div (3 + 1)$
21. $5 \times 6 - 49 \div (9 - 2)$
22. $36 \div (4 + 5) + 4 \times 2 \times 5$
23. $4^2 - 16 \div 4 \times 2 + 1$
24. $(5^2 + 1) \div (3^2 + 4) \times 1 + 3$
25. $17 - 4^3 \div 4^2 + 3^4$
26. $(21 \div 7 + 6) \div (8 - 5 \times 1) + 2$
27. $5 + 0 \div 3 - 2^3 \div 4 + 3^2$
28. $(16 \div 4 - 1) - 81 \div 9 \div 3 + 2^2$
29. $5^2 + 6 \times (5 - 2)^2 + 6^2 \div (2^3 + 1)$
30. $10 - 15 \div 3 + (6 - 3)^3 \times (7 - 2^2)$
31. $(6 - 2 \times 2)^3 + 3^2 \div (4^2 - 13) \times 1^4$
32. $(2^3 + 3 \times 4) - (8 \div 4)^2 - (5^2 - 3 \times 7)^2$
33. $3^3 - 2 \times 2^2 \times 3 + (5 \times 2 - 6)^2 \div 2^3$
34. $6 - (3^2 \times 2^2) \div (7 - 1) + 6^2 - 2 \times 3^2$
35. The Hat Shop buys hats for $14 each and sells them for $22 each. What is the total profit on the sale of 36 hats? (*Hint:* Profit is equal to the sale price minus the cost.)
36. A hardware store owner buys hammers for $4 each and sells them for $7 each. She also buys levels for $11 each and sells them for $19 each. If she sells 27 hammers and 16 levels, what is her total profit?
37. Joyce starts the month with $169 in her checking account. She writes 2 checks for $27 each, makes a deposit of $89, and then writes three checks for $41 each. What is her checking account balance at the end of these transactions if there is also a $3 service charge?
38. If 9 is multiplied by the square of the sum of 3 and 4, what is the result?
39. The cube of 4 is divided by the square of 4. The quotient is then subtracted from 10. What is the result?
40. If the sum of 1 and the square of 5 is divided by the sum of 4 and the square of 3, what is the result?

1.8 Operations Involving the Whole Number 0

OBJECTIVE

After completing this section, you should be able to perform arithmetic operations involving the whole number 0.

We have discussed the four basic operations of arithmetic and also examined the order of operations. What happens, however, when the whole number 0 is involved with one or more of these operations?

In this section, we will discuss operations involving the whole number 0. We will start with the operation of addition.

Practice Exercise 1. Add: (a) 17 + 0; (b) 0 + 9; (c) 26 + 0; (d) 0 + 43; (e) 51 + 0.

EXAMPLE 1 Add: (a) 2 + 0; (b) 3 + 0; (c) 0 + 5; (d) 15 + 0; and (e) 0 + 0.

SOLUTION: (a) 2 + 0 = 2; (b) 3 + 0 = 3; (c) 0 + 5 = 5; (d) 15 + 0 = 15; (e) 0 + 0 = 0.

ANSWERS TO PRACTICE EXERCISE

1. (a) 17; (b) 9; (c) 26; (d) 43; (e) 51.

In Example 1, we reemphasize that if 0 is added to a given whole number or if a whole number is added to 0, the sum is equal to the given whole number. We encountered this property for the whole number 0 in Section 1.2. Recall that the whole number 0 is called the *additive identity* for the whole numbers. We stated that if a represents any whole number, then:

$$a + 0 = a \quad \text{and} \quad 0 + a = a$$

What happens when the whole number 0 is involved in a subtraction problem? Clearly, we can't subtract a whole number (other than 0) from 0 since we can't subtract a larger whole number from a smaller whole number. However, we can subtract 0 from any whole number.

Practice Exercise 2. Subtract: (a) 7 − 0; (b) 19 − 0; (c) 10 − 0; (d) 87 − 0; (e) 102 − 0.

EXAMPLE 2 Subtract: (a) 2 − 0; (b) 4 − 0; (c) 9 − 0; (d) 13 − 0; and (e) 0 − 0.

SOLUTION: (a) 2 − 0 = 2; (b) 4 − 0 = 4; (c) 9 − 0 = 9; (d) 13 − 0 = 13; and (e) 0 − 0 = 0.

ANSWERS TO PRACTICE EXERCISE

2. (a) 7; (b) 19; (c) 10; (d) 87; (e) 102.

In Example 2, we note that if 0 is subtracted from a whole number, the difference is that same whole number. Hence, if a represents any whole number, then:

$$a - 0 = a$$

Remember, subtraction is *not* a commutative operation; the order in which we subtract is important.

Next, we will examine the whole number 0 involved with multiplication.

> **Practice Exercise 3.** Multiply: (a) 5 × 0; (b) 31 × 0; (c) 0 × 59; (d) 0 × 205; (e) 117 × 0.

EXAMPLE 3 Multiply: (a) 1 × 0; (b) 4 × 0; (c) 0 × 10; (d) 17 × 0; (e) 0 × 0.

SOLUTION: (a) 1 × 0 = 0; (b) 4 × 0 = 0; (c) 0 × 10 = 0; (d) 17 × 0 = 0; and (e) 0 × 0 = 0.

ANSWERS TO PRACTICE EXERCISE

3. (a) 0; (b) 0; (c) 0; (d) 0; (e) 0.

Example 3 illustrates that 0 is *not* the identity element for multiplication. (Do you remember what is?) It does, however, illustrate the following important property:

$$a \times 0 = 0 \quad \text{and} \quad 0 \times a = 0$$

where a is any whole number (including 0). That is, if 0 is multiplied by any whole number, or if a whole number is multiplied by 0, then the product is 0.

Last, we will examine the whole number 0 involved with the operation of division. As you will see in the following examples, you must exercise caution when 0 is involved in division.

> **Practice Exercise 4.** 0 ÷ 11

EXAMPLE 4 What is 0 ÷ 4?

SOLUTION: Let 0 ÷ 4 = a where a is our answer. Since our check for division is multiplication, then 0 ÷ 4 = a is the same as 0 = 4 × a. But the only whole number by which you can multiply 4 and get a product of 0 is 0. Therefore, a = 0 and 0 ÷ 4 = 0.

> **Practice Exercise 5.** 7 ÷ 0

EXAMPLE 5 What is 2 ÷ 0?

SOLUTION: Let 2 ÷ 0 = b. Then, 2 ÷ 0 = b means that 2 = 0 × b. But if b is any whole number, then 0 × b = 0. Therefore, 2 ÷ 0 *has no solution*. That is, there is *no* whole number whose product with 0 is 2. We say that 2 ÷ 0 *is not defined* since *it is impossible*.

Similarly, 8 ÷ 0 means that we are looking for a whole number p whose product with 0 is 8. There is *no* whole number that works. Also, 25 ÷ 0 means that we are looking for a whole number q whose product with 0 is 25. Again, there is *no* whole number that works. Since the product of 0 and *any* whole number is 0, there are no whole numbers that satisfy 8 ÷ 0, 25 ÷ 0, 17 ÷ 0, 39 ÷ 0, and so forth.

In Example 6, we will look at 0 ÷ 0. For the quotient 0 ÷ 0 to exist, there has to be a unique solution; that is, there has to be one and only one whole number that works.

> **Practice Exercise 6.** 0 ÷ 9

EXAMPLE 6 What is 0 ÷ 0?

SOLUTION: Let 0 ÷ 0 = c. Then, 0 ÷ 0 = c means that 0 = 0 × c. However, c can be any whole number since 0 × c = 0. We say that 0 ÷ 0 is *not defined* since there is *no unique solution*.

> **ANSWERS TO PRACTICE EXERCISES**
> 4. 0.   5. Not possible.   6. 0.

Note that in Examples 5 and 6 we saw that division by 0 is not defined. However, the division is not defined for different reasons. In Example 5, it is not defined because the division is impossible. In Example 6, the division is not defined since it is not unique.

In summary, we say that division by 0 is not defined for the reasons given above.

EXERCISES 1.8

Do the indicated operations, if possible. If not possible, indicate why not.

1. $5 + 0$
2. $0 + 2$
3. $6 - 0$
4. 7×0
5. $0 + 0$
6. 0×5
7. 0×0
8. $0 \div 3$
9. $9 - 0$
10. $0 + 9$
11. $0 \div 0$
12. $0 \div 6$
13. $6 \div 0$
14. $100 + 0$
15. $23 - 0$
16. $(7 + 0) + 0$
17. $(8 - 0) - 8$
18. $(8 - 0) - 0$
19. $(0 \times 7) \times 0$
20. $(0 \times 7) \times 7$
21. $(0 \div 9) \div 0$
22. $(0 \div 0) \div 9$
23. $2 + 3 - (4 - 0) + 5 \times 0 \div 6$
24. $(15 \div 3) - (0 \div 4) + (0 + 5)$
25. $4 \div (2 - 0) + 0 \times 7 - 6 \div 3$
26. $(0 \div 5) \times (0 \div 7) \div 9 - 0$
27. $2 - 0 \div 6 + 4 \times 0 \div 2$
28. $(16 \div 8 - 2) \div (2 \times 3 - 1)$
29. $(25 \div 5 + 3) \div (2 - 2 \times 1)$
30. $0 \times 0 + 0 \div 0 - 0$
31. $(0^2 \div 4) + (5 - 5)^2$
32. $2^3 \times 0 \div 3^2 \times (6 - 2)^2$

Chapter Summary

In this chapter, we discussed counting, numerals, and place value. Whole numbers, ordering of them, and the operations of addition, subtraction, multiplication, and division of them were reviewed. Properties associated with these operations were examined, as well as properties of 0.

The **key words** introduced in this chapter are listed below in the order in which they appeared. You should know the meaning of these words before proceeding to the next chapter.

digit (p. 4)
place value (p. 4)
natural number (p. 7)
whole number (p. 7)
ordering (p. 7)
is greater than (p. 7)
is less than (p. 7)
addition (p. 7)
sum (p. 8)
commutative property for addition (p. 8)
associative property for addition (p. 8)
additive identity (p. 8)
identity element (p. 8)
polygon (p. 10)
sides (p. 10)
perimeter of a polygon (p. 10)
subtraction (p. 14)
difference (p. 14)
multiplication (p. 23)
factor (p. 23)
product (p. 23)
commutative property for multiplication (p. 23)
identity property (p. 23)
multiplicative identity (p. 23)
associative property for multiplication (p. 24)
area (p. 27)
division (p. 30)
dividend (p. 31)
divisor (p. 31)
quotient (p. 31)
remainder (p. 33)
rounding (p. 37)
estimating (p. 38)
distributive property for multiplication over addition (p. 41)
distributive property for multiplication over subtraction (p. 41)
order of operations (p. 41)
superscript (p. 43)
base (p. 43)
exponent (p. 43)

Review Exercises

1.1 Numbers, Numerals, Counting, and Place Value

In Exercises 1–6, indicate the place value of the underlined digit.

1. 69<u>1</u>2
2. 1<u>3</u>7,645
3. 23,69<u>9</u>
4. 2,<u>6</u>17,094
5. <u>4</u>32,697,123
6. 9<u>5</u>,181,205

For each word name in Exercises 7–11, write the numeral.

7. Three hundred fifty-six
8. Two thousand, five hundred twenty-one
9. Four million, six hundred two thousand, nineteen
10. Nine hundred eighty billion, sixty-nine million, seventy-six thousand
11. Fourteen trillion, seven hundred eighty billion, nine hundred thirty-three million, seven hundred forty-five thousand, eight hundred seventy-one

For each numeral in Exercises 12–16, write the word name.

12. 2692
13. 131,069
14. 9,203,087
15. 4,695,169,007
16. 14,697,876,090,700

1.2 Ordering and Addition of Whole Numbers

In Exercises 17–20, insert the symbol $>$ or $<$ between the two given numbers to make true statements.

17. 89   96
18. 296   183
19. 1213   1123
20. 6096   5968

In Exercises 21–29, add and check your results by adding in the reverse order.

21. 123
    +462

22. 305
    +593

23. 673
    + 26

24. 36 + 25 + 12
25. 83 + 19 + 37
26. 406 + 235 + 382

27. 5964
      275
    +3094

28. 694
     705
     293
    +842

29. 2304
      597
      621
    +3008

30. Dick has the following monthly expenses: $225 for rent; $63 for utilities; $59 for car payment; $113 for food; and $109 for miscellaneous expenses. What are his total monthly expenses?
31. Ann graded four sets of mathematics tests. The first set took 3 hr 23 min to grade; the second set, 4 hr 38 min; the third set, 2 hr 55 min; the fourth set, 3 hr 47 min. What was the total time she spent grading the tests?
32. Larry caught five fish. Their weights were 6 lb 3 oz, 5 lb 9 oz, 4 lb 13 oz, 7 lb 2 oz, and 6 lb 15 oz. What was the total weight of his catch?

1.3 Subtraction of Whole Numbers

In Exercises 33–41, subtract and check your results by addition.

33. 98
    −64

34. 95
    −47

35. 80
    −59

36. 6972
    −4051

37. 7962 − 4850
38. 5807 − 3604
39. 6002 − 4555
40. 7984 − 4097
41. Subtract 3863 from 5040.
42. The price of coffee increased from $8.86 a kilo to $9.53 a kilo. How much was the increase per kilo of coffee?
43. Vin read a novel in 6 hr 23 min. Toni read the same novel in 5 hr 47 min. How much longer did it take Vin to read the novel?
44. In a recent school board election, John Dewey received 12,609 votes and James Alsoran received 10,786 votes. How many more votes did John receive than James?

1.4 Multiplication of Whole Numbers

In Exercises 45–53, multiply and check your results.

45. 23
$\times 14$

46. 57
$\times 45$

47. 6094
$\times 8723$

48. 7892
$\times 1369$

49. 89×67 **50.** 203×39 **51.** 412×374 **52.** 599×607 **53.** 6031×203

54. If 1 gal of paint will cover 450 ft^2 of wall surface, how many square feet of wall surface will 19 gal cover?

55. A grocer has 23 baskets of fruit. Each basket weighs 17 lb 6 oz. What is the total weight of the 23 baskets of fruit?

56. At Community University, there are 2163 registered freshmen and 1962 registered seniors. If each student had to complete 6 different forms during registration, how many forms were completed by all of the registered freshmen and senior students?

1.5 Division of Whole Numbers

In Exercises 57–65, divide and check your results.

57. $288 \div 12$ **58.** $960 \div 16$ **59.** $1250 \div 25$
60. $3067 \div 81$ **61.** $1231 \div 27$ **62.** $62{,}915 \div 19$
63. $739{,}021 \div 239$ **64.** $87{,}654 \div 209$ **65.** $908{,}070 \div 304$

66. A college bookstore manager received 864 books shipped in 27 boxes. If each box contains the same number of books, how many books are in each box?

67. If a particular make of car has an EPA rating of 26 mi/gal and a man drives one of these cars on a 3302-mi trip, how many gallons of gasoline should he expect to use?

68. A school district receives state aid in the form of $731 per student per year. If the district receives aid in the amount of $1,561,416, for how many students did it receive such aid?

1.6 Rounding and Estimating

In Exercises 69–72, round each whole number to the nearest place indicated.

69. 687; hundreds
70. 34,501; thousands
71. 916,258; ten thousands
72. 47,765,002; millions

73. Estimate the following sum by rounding each of the numbers to the nearest hundreds: $324 + 673 + 909 + 1289$.

74. Estimate the following difference by rounding each of the numbers to the nearest thousands: $27{,}987 - 13{,}205$.

75. Estimate the following product by rounding each of the numbers to the nearest tens: 872×89.

76. Estimate the following quotient when 65,872 (rounded to the nearest thousand) is divided by 2196 (rounded to the nearest hundred).

1.7 Mixed Operations with Whole Numbers

Perform the indicated operations.

77. $3 \times 4 - 6 \div 3 + 7$
78. $(32 \div 8 + 5) \div (8 \times 2 - 7)$
79. $12 \div 6 \times 3 - 5 + 0 \div 7$
80. $81 \div (2 \times 5 - 1) - 3 \times (1 + 4 \div 2)$
81. $3^2 - 2^3 + 4^2 \div 2^4$

82. Jay bought 2 tires for $59 each, including sales tax. He gave the clerk a $145 check to pay for the tires. How much change should he receive?

83. The cube of 3 is added to the square of 5. The sum is then divided by the product of 2 and 13. What is the result?

1.8 Operations Involving the Whole Number 0

Perform the indicated operations, if possible. If not possible, state why not.

84. 8×0 **85.** $0 \div 0$ **86.** $7 \div 0$ **87.** $9 - 0$ **88.** $[2 \times 3 - (4 + 2)] \div 5$

Teasers

1. List all possible pairs of whole numbers whose sum is 13, and whose product is less than 40.
2. Determine which is the largest number: (a) the product of 2 and 3, (b) the sum of 2 and 3, or (c) 2 raised to the power 3.
3. Determine the smallest counting number such that, when it is divided by 6, the remainder is 3 and, when divided by 7, the remainder is 2.
4. Determine the smallest whole number, greater than 25, that is a multiple of 8 and such that, when it is divided by 3, the remainder is 1 and, when it is divided by 6, the remainder is 4.
5. What is the smallest whole number that is exactly divisible by 2, 3, and 5 but, when divided by 7, leaves a remainder of 2?
6. There are six football players in a group. The smallest player weighs 190 lb and the heaviest player weighs 240 lb.
   a. Is 1025 lb a reasonable estimate for the total weight of the six players? Explain your answer.
   b. Is 1316 lb a reasonable estimate for the total weight of the six players? Explain your answer.

In your own words, but using good English and complete sentences, indicate what is meant by each statement in Exercises 7–10.

7. Multiplication of whole numbers is a commutative operation.
8. The whole number 0 is the additive identity in the set of whole numbers.
9. The whole numbers can be ordered.
10. The whole number 15 is exactly divisible by 3.

CHAPTER 1 TEST

This test covers material from Chapter 1. Read each question carefully and answer all questions. If you miss a question, review the appropriate section in the text.

1. In the number 63,792, what digit is in the thousands place?
2. Write the numeral for three million, seven hundred three thousand, eight hundred fifty.
3. Write the word name for 4239.
4. Write the word name for 3,609,714.

In Exercises 5–23, perform the indicated operations.

5. $236 + 51$
6. $23 + 46 + 32$
7.
    503
    270
    413
    + 36
8.
    2961
    327
    1470
    + 863
9.
    76
    −34
10.
    974
    −531
11. $62 - 39$
12. $5079 - 3067$
13. $6050 - 3874$
14.
    69
    ×23
15.
    307
    × 46
16. 351×148
17. 8063×205
18. $576 \div 12$
19. $2462 \div 26$
20. $726,209 \div 243$
21. $54 \div (6 \times 2 - 3) - 2 \times 3$
22. $2^3 - (3^2 \div 3) + 4 \times 5$
23. $36 \div 6 \times 3 - 7 + 0 \div 9$

In Exercises 24–25, insert the symbol $>$ or $<$ between the given number symbols to make true statements.

24. $3 \times 2 \quad 9 \div 3$
25. $14 \div 7 \quad 0 \div 7$
26. You have $319 in your checking account. After you write checks for $53 and $89, what is your new balance?
27. You owe your department store $286 and make a payment of $119. What is your new balance?
28. A manufacturer has 406 widgets in inventory. An additional 162 widgets are added to the inventory and 219 are removed. How many widgets are left in the inventory?
29. The student enrollment in a public high school decreased by 863 in 1 year. If the previous enrollment was 11,231, what is the new enrollment?
30. If canned sauerkraut is packed 24 cans to a case, how many cans of sauerkraut would be in 312 cases?
31. If 1 gal of latex paint covers 425 ft² of wall space, how many gallons of paint should be bought to paint 2680 ft² of wall space? (Assume that the paint is sold in gallon cans only.)
32. In a particular community, a special tax of $23 is levied on each adult resident. If there are 4086 residents in the community, including 1209 children, what will be the total tax levied?
33. Round 3467 to the nearest hundred.
34. Round 25,409 to the nearest thousand.
35. Estimate the following sum by rounding each term to the nearest ten: $123 + 89 + 327 + 64 + 155$.

CHAPTER 2: Fractions

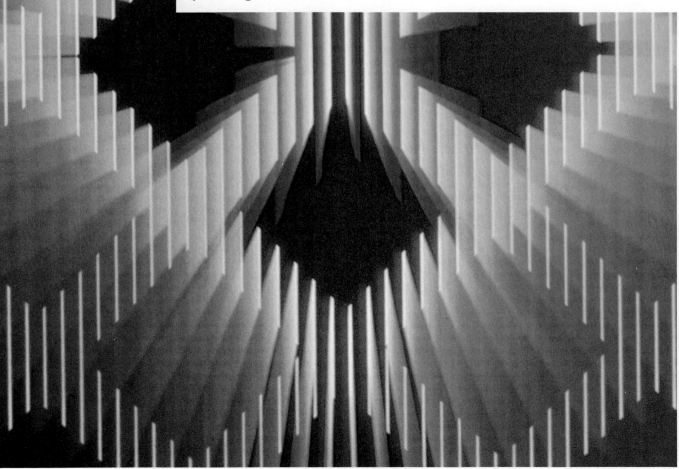

In Chapter 1, we reviewed the whole numbers and operations on them. You learned that the sum or the product of any two whole numbers is always a whole number. However, you could subtract or divide whole numbers only with certain restrictions; that is, some subtractions and divisions do not yield results that are whole numbers.

A great many of our everyday problems involving arithmetic can be solved using just the whole numbers. For others, another type of number is necessary. Measurement is one case where whole numbers alone will not suffice. For instance, if you have two unequal lengths of pipe, each between 8 and 9 meters long, how could you describe their lengths?

In this chapter, we will introduce fractions; the operations of addition, subtraction, multiplication, and division of them; and ordering of fractions. Applications involving fractions will be introduced throughout the chapter.

You will learn that fractions can be used as another way to represent division.

For instance, the fraction $\frac{3}{7}$ means $3 \div 7$. In your later study of mathematics, you will encounter the importance of fractions in statistics and probability, among other areas.

CHAPTER 2 PRETEST

This pretest covers material on fractions. Read each question carefully, and try to answer all questions. Each question is keyed to the chapter section in which the particular topic is discussed. Check your answers with those given in the back of the text.

Questions 1–5 pertain to Section 2.1.

1. Write the word name for $\frac{11}{15}$.

2. Write the fraction for eleven-sixteenths.

3. Rewrite the fraction $\frac{171}{22}$ as a mixed number.

4. Rewrite the mixed number $12\frac{2}{5}$ as an improper fraction.

5. On a 90-question test, 16 responses were incorrect. Write a fraction that represents the number of correct responses to the total number of questions.

Questions 6–8 pertain to Section 2.2.

6. The fractions $\frac{2}{5}$ and $\frac{3}{4}$ are (circle one):   Equivalent   or   Not equivalent

7. Write a fraction with a denominator 12 that is equivalent to $\frac{8}{3}$.

8. Rewrite the fraction $\frac{81}{36}$ in simplest form.

Questions 9–10 pertain to Section 2.3.

9. Rewrite the composite number 128 as a product of primes.

10. Determine the least common multiple (LCM) for 24 and 80.

Questions 11–16 pertain to Section 2.4. Write each answer in simplest form.

11. Add: $\frac{3}{18} + \frac{5}{18}$.

12. Add: $\frac{9}{29} + \frac{21}{29}$.

13. Add: $\frac{2}{7} + \frac{1}{3} + \frac{2}{5}$.

14. Add: $\frac{8}{11} + \frac{7}{29}$.

15. Add: $\frac{11}{23} + \frac{14}{31}$.

16. Add: $\frac{2}{9} + \frac{3}{11} + \frac{5}{77}$.

Questions 17–19 pertain to Section 2.5. Write each answer in simplest form.

17. Which fraction is larger? $\frac{2}{7}$ or $\frac{3}{8}$

18. Subtract: $\frac{8}{13} - \frac{3}{13}$.

19. From $\frac{5}{6}$, subtract $\frac{2}{7}$.

Questions 20–23 pertain to Section 2.6. Write your answers as mixed numbers, where appropriate.

20. Add: $3\frac{2}{7} + 4\frac{1}{6}$.

21. Add: $9\frac{2}{5} + 10\frac{3}{11}$.

22. Subtract: $19\frac{1}{2} - 13\frac{1}{7}$.

23. Subtract $10\frac{4}{9}$ from $18\frac{1}{5}$.

Questions 24–28 pertain to Section 2.7. Write each answer in simplest form.

24. Multiply: $\frac{4}{7} \times \frac{5}{12}$.

25. Multiply: $\frac{14}{27} \times \frac{9}{28}$.

26. Multiply: $4\frac{1}{3} \times 3\frac{2}{7}$.

27. Write the reciprocal for $\frac{17}{13}$.

28. Write the reciprocal, in simplest form, for $\frac{3 \times 6}{4 \times 5}$.

Questions 29–30 pertain to Section 2.8. Write each answer in simplest form.

29. Divide: $\frac{9}{14} \div \frac{3}{7}$.

30. Divide: $\frac{2}{3} \div 4\frac{2}{5}$.

2.1 Fractions

OBJECTIVES

After completing this section, you should be able to:
1. Write the word name for a fraction.
2. Classify a fraction as either a proper or an improper fraction.
3. Rewrite an improper fraction as a mixed number.
4. Rewrite a mixed number as an improper fraction.
5. Write a fraction that represents the number of equal parts chosen from the total number of parts indicated.

For most students, the word "fraction" probably means "part of." Consider the diagram in Figure 2.1 as representing a sheet of paper, or one unit. Now suppose that the sheet of paper is folded and divided into two equal parts, as indicated in Figure 2.2. Then each part is one of the two equal parts. We represent this by the symbol $\frac{1}{2}$ which is read "one-half."

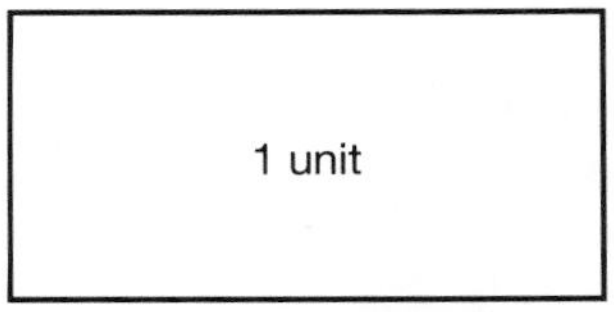

FIGURE 2.1

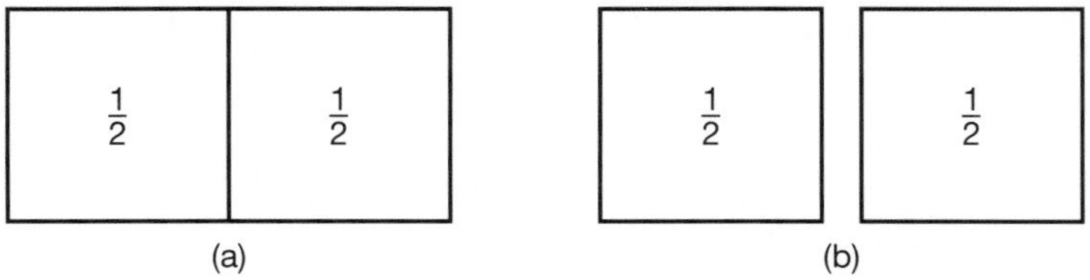

FIGURE 2.2

Similarly, the sheet of paper in Figure 2.1 could be folded and divided into three equal parts, as indicated in Figure 2.3. Each part in Figure 2.3 is one of the three equal parts. Each of these equal parts is represented by the symbol $\frac{1}{3}$, which is read "one-third."

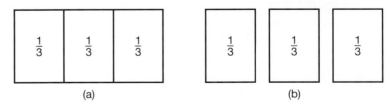

FIGURE 2.3

The sheet of paper illustrated in Figure 2.1 could also be folded and divided into four equal parts (as indicated in Figure 2.4a), with each of the equal parts being represented by the symbol $\frac{1}{4}$ (read "one-fourth"), or folded and divided into five equal parts (as indicated in Figure 2.4b), with each of the equal parts being represented by the symbol $\frac{1}{5}$ (read "one-fifth").

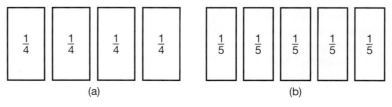

FIGURE 2.4

2.1 Fractions

The symbols $\frac{1}{2}, \frac{1}{3}, \frac{1}{4}$, and $\frac{1}{5}$ are called "fractions." Each fraction consists of two numbers. In the fraction $\frac{1}{2}$, the top number (1) is called the "numerator," and the bottom number (2) is called the "denominator." The denominator indicates the number of equal parts into which the unit is divided (or the sheet of paper is folded). The numerator indicates the number of those equal parts that are being considered.

Hence, the fraction $\frac{4}{9}$ (read "four one-ninths" or simply "four-ninths") has a denominator of 9, indicating that the unit is divided into nine equal parts. The numerator of the fraction is 4, indicating that four of those equal parts are being considered. (See Figure 2.5.)

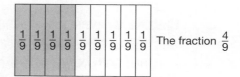

FIGURE 2.5

> **DEFINITION**
>
> A **fraction** is a number resulting from the division of one whole number a by another whole number b, with $b \neq 0$, and is written in the form $\frac{a}{b}$ or a/b or $a \div b$. The whole number a is called the **numerator** of the fraction. The non-zero whole number b is called the **denominator.**

The numerator of a fraction may be less than, equal to, or greater than its denominator. If the numerator of a fraction is less than its denominator, then the fraction is called a **proper fraction.** If the numerator of a fraction is equal to or greater than its denominator, then the fraction is called an **improper fraction.** Improper fractions will be used later in the section when discussing mixed numbers and operations with them.

Practice Exercise 1. Classify each of the following fractions as being proper or improper: (a) $\frac{5}{3}$; (b) $\frac{3}{8}$; (c) $\frac{0}{11}$; (d) $\frac{13}{13}$.

EXAMPLE 1 Classify each of the following fractions as being proper or improper: (a) $\frac{2}{5}$, (b) $\frac{7}{4}$, (c) $\frac{0}{2}$, (d) $\frac{3}{3}$.

SOLUTION: (a) $\frac{2}{5}$ (read "two-fifths") is a proper fraction. Its numerator is 2 and its denominator is 5; $2 < 5$. (b) $\frac{7}{4}$ (read "seven-fourths") is an improper fraction. Its numerator is 7 and its denominator is 4; $7 > 4$. (c) $\frac{0}{2}$ (read "zero-halves") is a proper fraction. Its numerator is 0 and its denominator is 2; $0 < 2$. (d) $\frac{3}{3}$ (read "three-thirds") is an improper fraction. Its numerator is 3 and its denominator is 3; $3 = 3$. (Can you think of another name for $\frac{3}{3}$?)

The fraction $\frac{3}{3}$ is another name for 1 since $\frac{3}{3}$ means $3 \div 3$ and $3 \div 3 = 1$. Similarly, the fractions $\frac{2}{2}, \frac{3}{3}, \frac{9}{9}$, and $\frac{13}{13}$ are also names for 1. Hence, if the numerator of a fraction is not zero and is equal to its denominator, the fraction is another name for the whole number 1.

> **ANSWERS TO PRACTICE EXERCISE**
>
> **1.** (a) Improper; (b) proper; (c) proper; (d) improper.

For Practice Exercises 2 and 3, write a fraction that represents the number of equal parts chosen from the total number of parts indicated.

Practice Exercise 2. The number of single-digit numbers among the first 19 counting numbers.

EXAMPLE 2 Write a fraction that represents 4 students chosen from a class of 23 students.

SOLUTION: To write a fraction that represents 4 students chosen from a class of 23 students, consider that (1) the numerator of the fraction would be the number of parts (students) chosen, or 4, and (2) the denominator of the fraction would be the total number of parts (students), or 23. Hence, the required fraction is $\frac{4}{23}$.

Practice Exercise 3. 323 people taking a postal examination for 36 openings.

EXAMPLE 3 Write a fraction that represents the number of minority executives selected from a group of 7 minority and 12 nonminority executives.

SOLUTION: To write a fraction that represents the number of minority executives selected from a group of 7 minority and 12 nonminority executives, consider that (1) the numerator of the fraction would be the number of parts (minority executives) selected, or 7, and (2) the denominator of the fraction would be the total number of parts (executives), or $7 + 12 = 19$. Hence, the required fraction is $\frac{7}{19}$.

ANSWERS TO PRACTICE EXERCISES

2. $\frac{9}{19}$. 3. $\frac{323}{36}$.

Recall that if the numerator of a fraction is greater than its denominator, then the fraction is an improper fraction. We will now consider rewriting such improper fractions as mixed numbers.

The fraction $\frac{5}{3}$ is an improper fraction and represents 5 one-thirds, or $5 \times \frac{1}{3}$. But $5 = 3 + 2$ and, therefore

$$\frac{5}{3} = 5 \times \frac{1}{3}$$
$$= (3 + 2) \times \frac{1}{3} \quad \text{(Substitution)}$$
$$= \left(3 \times \frac{1}{3}\right) + \left(2 \times \frac{1}{3}\right) \quad \text{(Distributive property)}$$
$$= \frac{3}{3} + \frac{2}{3} \quad \text{(Rewriting)}$$
$$= 1 + \frac{2}{3} \quad \left(\text{Since } \frac{3}{3} = 1\right)$$

Notice that $\frac{5}{3} = 1 + \frac{2}{3}$ is the *sum* of a whole number and a proper fraction. Such a number is called a "mixed number." (See Figure 2.6.)

2.1 Fractions

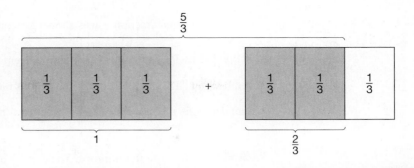

FIGURE 2.6

> **DEFINITION**
> A **mixed number** is the sum of a whole number and a proper fraction.

In a mixed number, the whole number part and the proper fraction part (or fractional part) are written side by side without the addition symbol. For instance, $\frac{5}{3} = 1 + \frac{2}{3}$ would be written as the mixed number $1\frac{2}{3}$. The mixed number $1\frac{2}{3}$ is read "1 and two-thirds." The "and" suggests addition. This also reinforces why we don't use the word "and" when reading whole numbers.

Rewrite each of the following improper fractions as a mixed number.

Practice Exercise 4. $\frac{38}{9}$

EXAMPLE 4 Show that the improper fraction $\frac{9}{7}$ represents the mixed number $1\frac{2}{7}$.

SOLUTION: The improper fraction $\frac{9}{7}$ represents the mixed number $1\frac{2}{7}$ since $\frac{9}{7}$ means $9 \times \frac{1}{7}$ and, therefore,

$$9 \times \frac{1}{7} = (7 + 2) \times \frac{1}{7}$$
$$= \left(7 \times \frac{1}{7}\right) + \left(2 \times \frac{1}{7}\right)$$
$$= \frac{7}{7} + \frac{2}{7}$$
$$= 1 + \frac{2}{7}$$

(See Figure 2.7.)

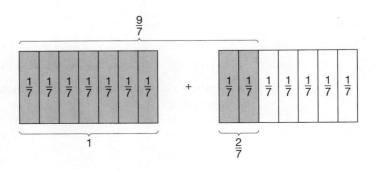

FIGURE 2.7

The improper fraction $\frac{9}{7}$ was shown to represent the mixed number $1\frac{2}{7}$. Notice that if 9 is divided by 7, the quotient is 1 and the remainder is 2, suggesting 1 $\left(\text{or } \frac{7}{7}\right)$ plus two-sevenths $\left(\frac{2}{7}\right)$. The quotient, 1, is the whole number part of the mixed number, and the remainder, 2, is the numerator of the fractional part.

Practice Exercise 5. $\frac{241}{28}$

EXAMPLE 5 Rewrite the improper fraction $\frac{37}{5}$ as a mixed number.

SOLUTION: To rewrite the improper fraction $\frac{37}{5}$ as a mixed number, divide 37 by 5. This gives a quotient of 7 and a remainder of 2. Hence, $\frac{37}{5} = 7\frac{2}{5}$.

Practice Exercise 6. $\frac{1021}{12}$

EXAMPLE 6 Rewrite the improper fraction $\frac{113}{8}$ as a mixed number.

SOLUTION: To rewrite the improper fraction $\frac{113}{8}$ as a mixed number, divide 113 by 8. This gives a quotient of 14 and a remainder of 1. Hence, $\frac{113}{8} = 14\frac{1}{8}$.

ANSWERS TO PRACTICE EXERCISES

4. $4\frac{2}{9}$. 5. $8\frac{17}{28}$. 6. $85\frac{1}{12}$.

Now, in reversing the process, note that the mixed number $3\frac{1}{2}$ means $3 + \frac{1}{2}$. But 3 can be written as $\frac{6}{2}$, or $6 \times \frac{1}{2}$. Hence,

$$3\frac{1}{2} = 3 + \frac{1}{2}$$

$$= \left(6 \times \frac{1}{2}\right) + \left(1 \times \frac{1}{2}\right) \quad \left(\text{Rewriting } 3 = \frac{6}{2} = 6 \times \frac{1}{2} \text{ and } \frac{1}{2} = 1 \times \frac{1}{2}\right)$$

$$= (6 + 1) \times \frac{1}{2} \quad \text{(Distributive property)}$$

$$= 7 \times \frac{1}{2} \quad \text{(Adding)}$$

$$= \frac{7}{2} \quad \text{(Rewriting)}$$

This suggests that to rewrite the mixed number $3\frac{1}{2}$ as an improper fraction, we simply multiply the whole number part, 3, by the denominator of the fractional part, 2, and add the numerator of the fractional part, 1. This sum, 7, is then written over the denominator of the fractional part, 2. Hence,

$$3\frac{1}{2} = \frac{(3 \times 2) + 1}{2} = \frac{6 + 1}{2} = \frac{7}{2}$$

2.1 Fractions

Rewrite each of the following mixed numbers as an improper fraction.

Practice Exercise 7. $6\frac{3}{7}$

EXAMPLE 7 Rewrite the mixed number $9\frac{1}{4}$ as an improper fraction.

SOLUTION: The mixed number $9\frac{1}{4}$ represents the improper fraction $\frac{37}{4}$ since

$$9\frac{1}{4} = \frac{(9 \times 4) + 1}{4} = \frac{36 + 1}{4} = \frac{37}{4}$$

Practice Exercise 8. $49\frac{2}{3}$

EXAMPLE 8 Rewrite the mixed number $13\frac{2}{11}$ as an improper fraction.

SOLUTION: The mixed number $13\frac{2}{11}$ represents the improper fraction $\frac{145}{11}$ since

$$13\frac{2}{11} = \frac{(13 \times 11) + 2}{11} = \frac{143 + 2}{11} = \frac{145}{11}$$

Practice Exercise 9. $63\frac{4}{5}$

EXAMPLE 9 Rewrite the mixed number $23\frac{3}{8}$ as an improper fraction.

SOLUTION:

$$23\frac{3}{8} = \frac{(23 \times 8) + 3}{8} = \frac{184 + 3}{8} = \frac{187}{8}$$

ANSWERS TO PRACTICE EXERCISES

7. $\frac{45}{7}$.   8. $\frac{149}{3}$.   9. $\frac{319}{5}$.

EXERCISES 2.1

In Exercises 1–10, write the names for each of the given fractions or mixed numbers. State whether each is a proper fraction, an improper fraction, or a mixed number.

1. $\frac{2}{9}$
2. $\frac{3}{13}$
3. $\frac{11}{20}$
4. $\frac{39}{46}$
5. $\frac{19}{51}$
6. $3\frac{3}{7}$
7. $23\frac{4}{7}$
8. $\frac{27}{15}$
9. $\frac{116}{23}$
10. $17\frac{13}{17}$

Write the fraction or mixed number for each of the word names in Exercises 11–16.

11. Seven-eighths
12. Twenty-three sevenths
13. One hundred nine sixteenths
14. One hundred and nine-sixteenths
15. Seven and twelve-twentieths
16. Two hundred sixty and three-fourths

In Exercises 17–32, rewrite each improper fraction as a mixed number.

17. $\dfrac{23}{11}$ 18. $\dfrac{31}{15}$ 19. $\dfrac{61}{11}$ 20. $\dfrac{37}{8}$

21. $\dfrac{87}{13}$ 22. $\dfrac{143}{12}$ 23. $\dfrac{105}{14}$ 24. $\dfrac{129}{17}$

25. $\dfrac{231}{19}$ 26. $\dfrac{601}{27}$ 27. $\dfrac{239}{27}$ 28. $\dfrac{909}{11}$

29. $\dfrac{1011}{18}$ 30. $\dfrac{1234}{6}$ 31. $\dfrac{2000}{17}$ 32. $\dfrac{1007}{15}$

Rewrite each mixed number in Exercises 33–48 as an improper fraction.

33. $3\dfrac{2}{3}$ 34. $4\dfrac{1}{9}$ 35. $6\dfrac{3}{7}$ 36. $7\dfrac{2}{7}$

37. $8\dfrac{4}{9}$ 38. $10\dfrac{9}{11}$ 39. $11\dfrac{1}{10}$ 40. $13\dfrac{4}{23}$

41. $17\dfrac{2}{9}$ 42. $21\dfrac{3}{8}$ 43. $101\dfrac{11}{12}$ 44. $19\dfrac{11}{12}$

45. $36\dfrac{4}{5}$ 46. $47\dfrac{9}{11}$ 47. $80\dfrac{3}{7}$ 48. $151\dfrac{1}{9}$

In Exercises 49–60, rewrite each fraction as a whole number.

49. $\dfrac{3}{3}$ 50. $\dfrac{26}{1}$ 51. $\dfrac{0}{9}$ 52. $\dfrac{17}{17}$

53. $\dfrac{0}{10}$ 54. $\dfrac{29}{1}$ 55. $\dfrac{102}{1}$ 56. $\dfrac{101}{101}$

57. $\dfrac{0}{100}$ 58. $\dfrac{0}{1000}$ 59. $\dfrac{199}{1}$ 60. $\dfrac{1000}{1000}$

In Exercises 61–75, write a fraction that represents the ratio of the number of equal parts chosen from the total number of parts indicated.

61. 3 girls chosen from a group of 17 girls
62. 13 cards from a deck of 52 cards
63. The number of vowels from the total number of letters in the word "fraction"
64. The number of @s from the collection [@, *, &, @, %, *, @, &]
65. The number of even-numbered floors in a twenty-three story building
66. The number of vowels from the collection [a, c, e, i, p, e, o, q, s]
67. The number of digits greater than 8 in the numeral 12,345,678
68. 4 buildings on a college campus containing 31 buildings
69. 23 cars on a lot containing 120 cars
70. The number of female salesclerks on a total staff of 9 male and 14 female salesclerks
71. The number of students who failed a course in a class of 60 students if 47 students passed the course and 5 students withdrew before the end of the course
72. 210 applications for a total of 17 jobs
73. The number of Republican candidates in an election with 7 Republican candidates, 9 Democrats, and 2 other candidates
74. The number of bused students in a school district with 11,203 enrolled students if 57 students are not bused
75. The number of Fords on a used car lot containing 23 GM cars, 11 Fords, and 5 Volkswagens.

2.2 Equivalent Fractions

OBJECTIVES

After completing this section, you should be able to:
1. Determine whether or not two given fractions are equivalent fractions.
2. Using the Fundamental Principle of Fractions, write different equivalent fractions for a given fraction.
3. Rewrite a fraction in its simplest form.
4. Rewrite each fraction of a pair of fractions as an equivalent fraction so that both fractions have the same denominator.

Now, consider two of the sheets of paper illustrated in Figure 2.1. They are the same size and shape as illustrated in Figure 2.8. Fold the sheet on the left into two equal parts and the sheet on the right into four equal parts, as illustrated in Figure 2.9. Shading one of the two equal parts $\left(\frac{1}{2}\right)$ on the sheet on the left and two of the four equal parts $\left(\frac{2}{4}\right)$ of the sheet on the right, we note that $\frac{1}{2}$ and $\frac{2}{4}$ represent the same amount of paper. These fractions are said to be *equivalent*.

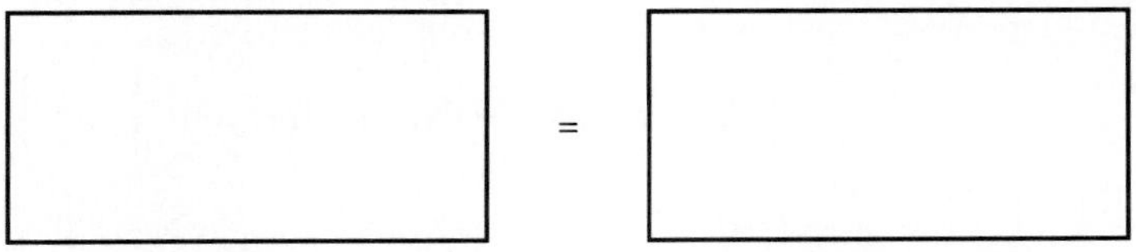

FIGURE 2.8

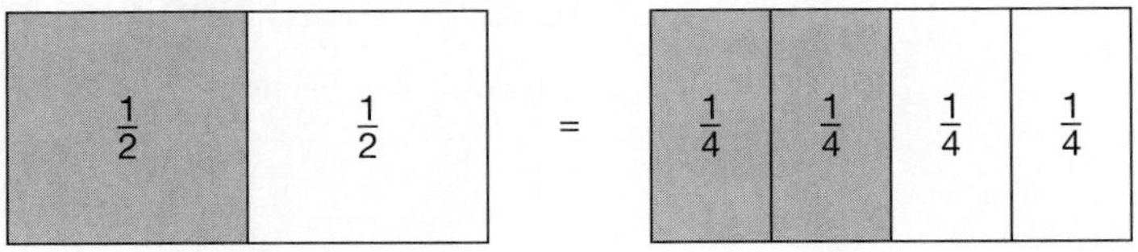

FIGURE 2.9

In Figure 2.10, we observe that $\frac{2}{3}$ and $\frac{4}{6}$ are equivalent fractions.

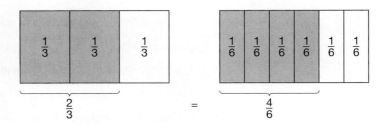

FIGURE 2.10

DEFINITION

Equivalent fractions are fractions that are different names for the same part of a unit.

We note, at this point in our discussion, that if $\frac{a}{b}$ and $\frac{c}{d}$ are equivalent fractions, then $a \times d = b \times c$. Although we will establish this fact later in this chapter, we wish to use it now in determining whether two fractions are or are not equivalent.

Determine which of the following pairs of fractions are equivalent.

Practice Exercise 1. $\frac{1}{3}$ and $\frac{2}{4}$

EXAMPLE 1 Determine whether the fractions $\frac{1}{2}$ and $\frac{3}{6}$ are equivalent.

SOLUTION: The fractions $\frac{1}{2}$ and $\frac{3}{6}$ are equivalent fractions; $1 \times 6 = 2 \times 3$.

Practice Exercise 2. $\frac{3}{5}$ and $\frac{6}{10}$

EXAMPLE 2 Determine whether the fractions $\frac{3}{3}$ and $\frac{4}{4}$ are equivalent.

SOLUTION: The fractions $\frac{3}{3}$ and $\frac{4}{4}$ are equivalent fractions; $3 \times 4 = 4 \times 3$.

Practice Exercise 3. $\frac{3}{3}$ and $\frac{7}{7}$

EXAMPLE 3 Determine whether the fractions $\frac{1}{3}$ and $\frac{2}{5}$ are equivalent.

SOLUTION: The fractions $\frac{1}{3}$ and $\frac{2}{5}$ are *not* equivalent fractions; $1 \times 5 \neq 3 \times 2$.

Practice Exercise 4. $\frac{3}{7}$ and $\frac{5}{9}$

EXAMPLE 4 Determine whether the fractions $\frac{3}{4}$ and $\frac{4}{7}$ are equivalent.

SOLUTION: The fractions $\frac{3}{4}$ and $\frac{4}{7}$ are *not* equivalent fractions: $3 \times 7 \neq 4 \times 4$.

ANSWERS TO PRACTICE EXERCISES

1. Not equivalent. 2. Equivalent. 3. Equivalent. 4. Not equivalent.

Let's consider what happens if we alter the numerator and the denominator of a fraction. Consider the fraction $\frac{4}{6}$.

- If we *multiply* both the numerator and the denominator by 2, we obtain the new fraction $\frac{8}{12}$. The fractions $\frac{4}{6}$ and $\frac{8}{12}$ are equivalent; $4 \times 12 = 6 \times 8$.
- If we *divide* both the numerator and the denominator by 2, we obtain the new fraction $\frac{2}{3}$. The fractions $\frac{4}{6}$ and $\frac{2}{3}$ are equivalent; $4 \times 3 = 6 \times 2$.
- If we *add* 2 to both the numerator and the denominator, we obtain the new fraction $\frac{6}{8}$. The fractions $\frac{4}{6}$ and $\frac{6}{8}$ are *not* equivalent; $4 \times 8 \neq 6 \times 6$.
- If we *subtract* 2 from both the numerator and the denominator, we obtain the new fraction $\frac{2}{4}$. The fractions $\frac{4}{6}$ and $\frac{2}{4}$ are *not* equivalent; $4 \times 4 \neq 6 \times 2$.

As illustrated above, we can only multiply or divide *both* the numerator and the denominator of a fraction by a whole number to obtain an equivalent fraction. Further, the whole number used must be different from 0. Remember, fractions represent divisions, and division by 0 is *not* permitted.

2.2 Equivalent Fractions

Fundamental Principle of Fractions

1. If *both* the numerator and the denominator of a fraction are *multiplied* by the same *nonzero* whole number, then an equivalent fraction is obtained.
2. If *both* the numerator and the denominator of a fraction are *divided* by the same *nonzero* whole number, then an equivalent fraction is obtained.

Using the first part of the **Fundamental Principle of Fractions,** we have:

- $\frac{2}{3} = \frac{2 \times 3}{3 \times 3} = \frac{6}{9}$; $\frac{2}{3}$ and $\frac{6}{9}$ are equivalent fractions. (*Note:* We can generate many different names for the fraction $\frac{2}{3}$ by using different multipliers. For instance, $\frac{2}{3} = \frac{2 \times 2}{3 \times 2} = \frac{4}{6}$; $\frac{2}{3} = \frac{2 \times 3}{3 \times 3} = \frac{6}{9}$; $\frac{2}{3} = \frac{2 \times 4}{3 \times 4} = \frac{8}{12}$; and so forth.)

- $\frac{4}{5} = \frac{4 \times 2}{5 \times 2} = \frac{8}{10}$; $\frac{4}{5}$ and $\frac{8}{10}$ are equivalent fractions.

- $\frac{6}{5} = \frac{6 \times 3}{5 \times 3} = \frac{18}{15}$; $\frac{6}{5}$ and $\frac{18}{15}$ are equivalent fractions.

Using the second part of the Fundamental Principle of Fractions, we have:

- $\frac{8}{12} = \frac{8 \div 4}{12 \div 4} = \frac{2}{3}$; $\frac{8}{12}$ and $\frac{2}{3}$ are equivalent fractions.

- $\frac{10}{25} = \frac{10 \div 5}{25 \div 5} = \frac{2}{5}$; $\frac{10}{25}$ and $\frac{2}{5}$ are equivalent fractions.

- $\frac{0}{6} = \frac{0 \div 3}{6 \div 3} = \frac{0}{2}$; $\frac{0}{6}$ and $\frac{0}{2}$ are equivalent fractions.

Usually, we multiply the numerator and denominator of a fraction by the same nonzero number in order to change it to an equivalent fraction having *a particular denominator*. As you will see, this is a necessary step in addition and subtraction of fractions.

Equivalent fractions play an important role in the addition and subtraction of fractions that have different denominators. We rewrite the fractions in equivalent form with the same denominators. Then, we add or subtract. This will be discussed in later sections of this chapter.

Sometimes, we may want to rewrite fractions in equivalent form by changing their numerators.

In Practice Exercises 5–7, rewrite each pair of fractions given as equivalent fractions with the denominator indicated following the semicolon.

Practice Exercise 5. $\frac{1}{3}$ and $\frac{2}{5}$; 15

EXAMPLE 5 Determine the missing number indicated by the question mark such that $\frac{2}{5}$ and $\frac{?}{15}$ are equivalent fractions.

SOLUTION: Note that $15 = 5 \times 3$. Then, using the Fundamental Principle of Fractions, we can multiply *both* the numerator and denominator of $\frac{2}{5}$ by 3. We have $\frac{2}{5} = \frac{2 \times 3}{5 \times 3} = \frac{6}{15}$. Therefore, ? = 6.

> **Practice Exercise 6.** $\frac{3}{7}$ and $\frac{5}{2}$; 14

EXAMPLE 6 Write a fraction equivalent to $\frac{2}{3}$ that has a denominator of 21.

SOLUTION: Since $21 = 3 \times 7$, we can multiply both the numerator and denominator of $\frac{2}{3}$ by 7. Hence,

$$\frac{2}{3} = \frac{2 \times 7}{3 \times 7} = \frac{14}{21}$$

and $\frac{14}{21}$ is the required fraction.

> **Practice Exercise 7.** $\frac{2}{9}$ and $\frac{3}{4}$;

EXAMPLE 7 Write a fraction equivalent to $\frac{4}{5}$ that has a numerator of 16.

SOLUTION: Since $16 = 4 \times 4$, we can multiply both the numerator and the denominator of $\frac{4}{5}$ by 4. Hence,

$$\frac{4}{5} = \frac{4 \times 4}{5 \times 4} = \frac{16}{20}$$

and $\frac{16}{20}$ is the required fraction.

> **In Practice Exercises 8 and 9, use the Fundamental Principle of Fractions to determine the missing number indicated by the question mark that makes the pairs of fractions equivalent.**
>
> **Practice Exercise 8.** $\frac{3}{4}$ and $\frac{?}{8}$

EXAMPLE 8 Write a fraction equivalent to $\frac{8}{24}$ that has a denominator of 12.

SOLUTION: Since $12 = 24 \div 2$, we can divide both the numerator and the denominator of $\frac{8}{24}$ by 2. Hence,

$$\frac{8}{24} = \frac{8 \div 2}{24 \div 2} = \frac{4}{12}$$

and $\frac{4}{12}$ is the required fraction.

> **Practice Exercise 9.** $\frac{6}{?}$ and $\frac{9}{12}$

EXAMPLE 9 Write a fraction equivalent to $\frac{9}{81}$ that has a numerator of 3.

SOLUTION: Since $3 = 9 \div 3$, we can divide both the numerator and the denominator of $\frac{9}{81}$ by 3. Hence,

$$\frac{9}{81} = \frac{9 \div 3}{81 \div 3} = \frac{3}{27}$$

and $\frac{3}{27}$ is the required fraction.

ANSWERS TO PRACTICE EXERCISES

5. $\frac{5}{15}$ and $\frac{6}{15}$. 6. $\frac{6}{14}$ and $\frac{35}{14}$. 7. $\frac{24}{108}$ and $\frac{81}{108}$. 8. 6. 9. 8.

2.2 Equivalent Fractions

When working with equivalent fractions, we generally look for a fraction that is in its "simplest" *form*.

> **DEFINITION**
> A fraction is said to be in its **simplest form** or in *lowest terms* if the numerator and denominator of the fraction cannot both be exactly divided by any whole number other than 1.

Rewrite each of the given fractions in simplest form.

Practice Exercise 10. $\frac{14}{21}$

EXAMPLE 10 Determine the simplest form of the fraction $\frac{5}{17}$.

SOLUTION: The fraction $\frac{5}{17}$ *is* in simplest form since 5 and 17 are both exactly divisible only by the whole number 1.

Practice Exercise 11. $\frac{12}{42}$

EXAMPLE 11 Determine the simplest form of the fraction $\frac{5}{10}$.

SOLUTION: The fraction $\frac{5}{10}$ is *not* in its simplest form since both the numerator and denominator are exactly divisible by 5. Hence, $\frac{5}{10} = \frac{5 \div 5}{10 \div 5} = \frac{1}{2}$. The resulting fraction, $\frac{1}{2}$, is simplified. The simplest form of the fraction $\frac{5}{10}$ is the equivalent fraction $\frac{1}{2}$.

Practice Exercise 12. $\frac{35}{120}$

EXAMPLE 12 Determine the simplest form of the fraction $\frac{16}{24}$.

SOLUTION: The fraction $\frac{16}{24}$ is *not* in simplest form since both the numerator and denominator are exactly divisible by 4. Hence, $\frac{16}{24} = \frac{16 \div 4}{24 \div 4} = \frac{4}{6}$. The resulting fraction, $\frac{4}{6}$, is still not in simplest form since both the numerator and denominator of $\frac{4}{6}$ are exactly divisible by 2. Hence, $\frac{4}{6} = \frac{4 \div 2}{6 \div 2} = \frac{2}{3}$. The fraction $\frac{2}{3}$ is in simplest form. Hence, the simplest form of the fractions $\frac{16}{24}$ and $\frac{4}{6}$ is $\frac{2}{3}$.

Practice Exercise 13. $\frac{77}{121}$

EXAMPLE 13 Determine the simplest form of the fraction $\frac{352}{847}$.

SOLUTION: The fraction $\frac{352}{847}$ is *not* in its simplest form since both 352 and 847 are exactly divisible by 11. Hence, we have $\frac{352}{847} = \frac{352 \div 11}{847 \div 11} = \frac{32}{77}$. The resulting fraction, $\frac{32}{77}$, is simplified.

ANSWERS TO PRACTICE EXERCISES

10. $\frac{2}{3}$.   11. $\frac{2}{7}$.   12. $\frac{7}{24}$.   13. $\frac{7}{11}$.

In Example 13, you may have had difficulty determining that 11 exactly divides both 352 and 847. We will consider a more systematic procedure of simplifying fractions later in the text.

We will conclude this section with an application for equivalent fractions.

Practice Exercise 14. Joyce flew for 3 hr and traveled 850 mi. Heather flew for 2 hr and traveled 590 mi. Did the two fly an equivalent number of miles per hour?

EXAMPLE 14 On a psychology test, Holly answered 16 of 20 questions correctly and Don answered 21 of 25 questions correctly. Each answered 4 questions incorrectly. Did they answer an equivalent number of questions correctly?

SOLUTION: If the fractions $\frac{16}{20}$ and $\frac{21}{25}$ are equivalent, then 16×25 must be equal to 20×21. But, $16 \times 25 = 400$ and $20 \times 21 = 420$. Since $400 \neq 420$, then the two fractions are not equivalent. Therefore, Holly and Don did not answer an equivalent number of questions correctly.

ANSWER TO PRACTICE EXERCISE

14. No, since $\frac{850}{3}$ is not equivalent to $\frac{590}{2}$.

EXERCISES 2.2

In Exercises 1–20, determine which pairs of fractions are equivalent. In Exercises 17–20, note that the division bar serves as a grouping symbol. You must simplify the numerators and the denominators first.

1. $\frac{1}{2}$ and $\frac{2}{3}$
2. $\frac{2}{5}$ and $\frac{10}{4}$
3. $\frac{2}{3}$ and $\frac{4}{6}$
4. $\frac{2}{7}$ and $\frac{4}{14}$
5. $\frac{2}{7}$ and $\frac{10}{35}$
6. $\frac{0}{3}$ and $\frac{0}{2}$
7. $\frac{1}{2}$ and $\frac{6}{12}$
8. $\frac{5}{9}$ and $\frac{4}{8}$
9. $\frac{10}{15}$ and $\frac{2}{3}$
10. $\frac{12}{16}$ and $\frac{8}{12}$
11. $\frac{6}{6}$ and $\frac{11}{11}$
12. $\frac{8}{5}$ and $\frac{7}{4}$
13. $\frac{7}{6}$ and $\frac{21}{18}$
14. $\frac{11}{3}$ and $\frac{3}{11}$
15. $\frac{9}{5}$ and $\frac{18}{10}$
16. $\frac{11}{6}$ and $\frac{12}{22}$
17. $\frac{3+2}{4}$ and $\frac{10}{6+2}$
18. $\frac{5-5}{3}$ and $\frac{6-6}{2}$
19. $\frac{8-7}{1}$ and $\frac{5}{5}$
20. $\frac{1+6}{5}$ and $\frac{4+3}{8-3}$

In Exercises 21–40, use the Fundamental Principle of Fractions to determine the missing number indicated by the question mark that makes the pair of fractions equivalent.

21. $\frac{2}{3}$ and $\frac{?}{6}$
22. $\frac{6}{8}$ and $\frac{3}{?}$
23. $\frac{8}{5}$ and $\frac{16}{?}$
24. $\frac{0}{5}$ and $\frac{?}{3}$
25. $\frac{2}{5}$ and $\frac{?}{20}$
26. $\frac{9}{12}$ and $\frac{?}{4}$
27. $\frac{8}{5}$ and $\frac{16}{?}$
28. $\frac{4}{3}$ and $\frac{4}{?}$
29. $\frac{1}{6}$ and $\frac{4}{?}$

2.2 Equivalent Fractions

30. $\frac{7}{4}$ and $\frac{?}{12}$ 31. $\frac{32}{16}$ and $\frac{8}{?}$ 32. $\frac{77}{99}$ and $\frac{7}{?}$

33. $\frac{216}{27}$ and $\frac{?}{9}$ 34. $\frac{100}{25}$ and $\frac{?}{5}$ 35. $\frac{25}{75}$ and $\frac{?}{15}$

36. $\frac{17}{19}$ and $\frac{51}{?}$ 37. $\frac{0}{116}$ and $\frac{?}{29}$ 38. $\frac{102}{42}$ and $\frac{?}{7}$

39. $\frac{200}{150}$ and $\frac{20}{?}$ 40. $\frac{100}{90}$ and $\frac{?}{18}$

Write five different fractions that are equivalent to each of the fractions in Exercises 41–50.

41. $\frac{1}{2}$ 42. $\frac{2}{3}$ 43. $\frac{3}{4}$ 44. $\frac{0}{5}$ 45. $\frac{2}{13}$

46. $\frac{4}{5}$ 47. $\frac{13}{7}$ 48. $\frac{11}{10}$ 49. $\frac{8}{5}$ 50. $\frac{11}{4}$

Rewrite each of the pairs of fractions in Exercises 51–60 as equivalent fractions with the denominator indicated following the semicolon.

51. $\frac{1}{2}$ and $\frac{3}{4}$; 8   52. $\frac{2}{3}$ and $\frac{1}{5}$; 15   53. $\frac{1}{7}$ and $\frac{3}{5}$; 35

54. $\frac{2}{9}$ and $\frac{5}{6}$; 54   55. $\frac{3}{11}$ and $\frac{1}{2}$; 66   56. $\frac{1}{5}$ and $\frac{7}{10}$; 50

57. $\frac{1}{9}$ and $\frac{3}{4}$; 72   58. $\frac{6}{5}$ and $\frac{5}{3}$; 60   59. $\frac{11}{2}$ and $\frac{4}{7}$; 14

60. $\frac{9}{11}$ and $\frac{22}{13}$; 143

In Exercises 61–95, rewrite each fraction in simplest form.

61. $\frac{6}{8}$ 62. $\frac{11}{77}$ 63. $\frac{24}{44}$ 64. $\frac{28}{91}$ 65. $\frac{40}{68}$

66. $\frac{26}{39}$ 67. $\frac{48}{72}$ 68. $\frac{44}{33}$ 69. $\frac{92}{56}$ 70. $\frac{25}{100}$

71. $\frac{121}{99}$ 72. $\frac{60}{180}$ 73. $\frac{121}{33}$ 74. $\frac{102}{42}$ 75. $\frac{25}{90}$

76. $\frac{44}{214}$ 77. $\frac{121}{88}$ 78. $\frac{206}{110}$ 79. $\frac{102}{210}$ 80. $\frac{200}{500}$

81. $\frac{308}{187}$ 82. $\frac{905}{600}$ 83. $\frac{156}{204}$ 84. $\frac{102}{234}$ 85. $\frac{162}{243}$

86. $\frac{500}{170}$ 87. $\frac{858}{286}$ 88. $\frac{513}{342}$ 89. $\frac{1386}{3276}$ 90. $\frac{6930}{3080}$

91. $\frac{2005}{905}$ 92. $\frac{1800}{5400}$ 93. $\frac{5005}{7007}$ 94. $\frac{1000}{5555}$ 95. $\frac{3150}{4158}$

96. In a baseball league, team A won 7 games out of 9 games played and team B won 6 games out of 8 games played. Did the two teams win an equivalent number of games of the games played?

97. Store A sells oranges at 3 for 50¢. Store B sells the same kind and size of oranges at 6 for $1. Are these equivalent prices?

98. On a mathematics test, Bette answered 24 out of 30 questions correctly and Luke answered 16 out of 24 questions correctly. Did the two answer an equivalent number of questions correctly?

99. Mr. Jones bought a house for $61,000 and later sold it for $64,000. Ms. Smith bought a house for $62,000 and later sold it for $65,000. Did the two make an equivalent

profit compared to purchase price? (*Hint:* Profit is the difference between the sale price and the purchase price.)

100. Jason drove his car for 3 hr and traveled 130 mi. Gretchen drove her car for 2 hr and traveled 85 mi. Did they drive an equivalent number of miles per hour?
101. Karole worked for 6 hr and received $20. Ted worked for 8 hr and received $26. Did they receive an equivalent hourly rate of pay?
102. In an automobile showroom, a new car that was originally priced to sell for $9500 was marked down to $8200. A second car that was originally priced to sell for $8700 was marked down to $7600. Does this represent an equivalent marked-down price for the two cars?
103. In the same showroom, two other cars were also marked down. One, originally priced at $8200, was marked down to $7380. The other, originally priced at $9100, was marked down to $8190. Are these equivalent marked-down prices for the two cars?
104. Joe withdrew $120 from his savings account containing $210. What fractional part of his savings was withdrawn? Write your answer in simplest form.
105. There are 35 students in a class who took a test. Thirty of the students passed the test.
    a. What fractional part of the class passed the test? Write your answer in simplest form.
    b. What fractional part of the class did not pass the test? Write your answer in simplest form.

2.3 Least Common Multiple

OBJECTIVE

After completing this section, you should be able to find the least common multiple of two or more whole numbers.

In the next section, we will consider the addition of fractions. To do so, we use the "least common denominator" of two or more denominators. To do this, we will need to know something about what are called prime and composite numbers.

DEFINITION

A **prime number** is a counting number greater than 1 that is exactly divisible only by 1 and itself. A counting number greater than 1 that is not a prime number is a **composite number.**

It should be noted from these two definitions that the counting number 1 is neither prime nor composite.

- The counting number 11 is a prime number since the only counting numbers that exactly divide 11 are 1 and 11.
- The counting number 12 is a composite number since the counting numbers 2, 3, 4, and 6 exactly divide 12, in addition to the counting numbers 1 and 12.

These Are All the Prime Numbers Less than 100

2, 3, 5, 7, 11, 13, 17, 19, 23, 29, 31, 37, 41, 43, 47, 53, 59, 61, 67, 71, 73, 79, 83, 89, 97

2.3 Least Common Multiple

Every composite number can be written as the product of prime factors. To rewrite a composite number as a product of prime factors, use the procedure given in the box.

Prime Factorization of Composite Numbers

To rewrite a composite number as a product of prime factors:

1. Write the composite number as a product of two natural (counting) numbers, other than the number itself and 1.
2. If the factors obtained in Step 1 are composite numbers, rewrite each of them as a product of two natural numbers.
3. Continue the procedure until all of the factors involved are prime.
4. The result is the **prime factorization** of the given composite number.

In Practice Exercises 1–4, rewrite each of the composite numbers as a product of primes.

Practice Exercise 1. 30

EXAMPLE 1 Determine the prime factorization of 36.

SOLUTION:

STEP 1: Write 36 as the product of two natural numbers. For instance,

$$36 = 2 \times 18$$

STEP 2: Since 18 is composite, rewrite it as a product of two natural numbers. For instance, $18 = 3 \times 6$. Therefore,

$$36 = 2 \times 3 \times 6$$

STEP 3: Since 6 is composite, rewrite it as a product of two natural numbers. For instance, $6 = 2 \times 3$. Therefore,

$$36 = 2 \times 3 \times 2 \times 3$$

STEP 4: Since all of the factors in the product in Step 3 are prime, we are done.

STEP 5: The prime factorization of 36 is:

$$36 = 2 \times 2 \times 3 \times 3$$

or, using exponents,

$$36 = 2^2 \times 3^2$$

The preceding procedure will be shortened in Examples 2–4.

Practice Exercise 2. 45

EXAMPLE 2 Determine the prime factorization for 96.

SOLUTION:
$$\begin{aligned} 96 &= 2 \times 48 \\ &= 2 \times 2 \times 24 \\ &= 2 \times 2 \times 2 \times 12 \\ &= 2 \times 2 \times 2 \times 2 \times 6 \\ &= 2 \times 2 \times 2 \times 2 \times 2 \times 3 \quad \text{(All factors are prime.)} \\ &= 2^5 \times 3 \quad \text{(Using exponents)} \end{aligned}$$

Practice Exercise 3. 82

EXAMPLE 3 Determine the prime factorization for 81.

$$
\begin{aligned}
\text{SOLUTION:} \quad 81 &= 3 \times 27 \\
&= 3 \times 3 \times 9 \\
&= 3 \times 3 \times 3 \times 3 \quad \text{(All factors are prime.)} \\
&= 3^4 \quad \text{(Using exponents)}
\end{aligned}
$$

Practice Exercise 4. 120

EXAMPLE 4 Determine the prime factorization of 220.

$$
\begin{aligned}
\text{SOLUTION:} \quad 220 &= 22 \times 10 \\
&= 2 \times 11 \times 10 \\
&= 2 \times 11 \times 2 \times 5 \quad \text{(All factors are prime.)} \\
&= 2 \times 2 \times 5 \times 11 \quad \text{(Rearranging the factors)} \\
&= 2^2 \times 5 \times 11 \quad \text{(Using exponents)}
\end{aligned}
$$

ANSWERS TO PRACTICE EXERCISES

1. $2 \times 3 \times 5$. **2.** $3^2 \times 5$. **3.** 2×41. **4.** $2^3 \times 3 \times 5$.

As the composite numbers get larger, it is sometimes difficult to rewrite them as products of two natural numbers. There are, however, basic tests of divisibility that can help you to arrive at the prime factorization of these composite numbers. A few of these are listed in the accompanying box.

Tests for Divisibility by the Prime Numbers 2, 3, and 5

- *Divisibility by 2:* A natural number is divisible by 2 if the ones digit of the number is 0, 2, 4, 6, or 8. (A number that is divisible by 2 is called an *even* number.)
- *Divisibility by 3:* A natural number is divisible by 3 if the sum of the digits of the number is divisible by 3.
- *Divisibility by 5:* A natural number is divisible by 5 if the ones digit of the number is 0 or 5.

Practice Exercise 5. Determine whether each of the given numbers is divisible by 2: (a) 27; (b) 406.

EXAMPLE 5 Determine whether each of the following numbers is divisible by 2: (a) 68; (b) 57.

SOLUTION: (a) 68 is divisible by 2 since the ones digit is 8; (b) 57 is *not* divisible by 2 since the ones digit is *not* 0, 2, 4, 6, or 8.

Practice Exercise 6. Determine whether each of the given numbers is divisible by 3: (a) 135; (b) 4120.

EXAMPLE 6 Determine whether each of the following numbers is divisible by 3: (a) 213; (b) 403.

SOLUTION: (a) 213 is divisible by 3 since $2 + 1 + 3 = 6$ is divisible by 3; (b) 403 is *not* divisible by 3 since $4 + 0 + 3 = 7$ is *not* divisible by 3.

Practice Exercise 7. Determine whether each of the given numbers is divisible by 5: (a) 2005; (b) 5552.

EXAMPLE 7 Determine whether each of the following numbers is divisible by 5: (a) 155; (b) 268.

SOLUTION: (a) 155 is divisible by 5 since the ones digit is 5; (b) 268 is *not* divisible by 5 since the ones digit is *not* 0 or 5.

2.3 Least Common Multiple

> **ANSWERS TO PRACTICE EXERCISES**
>
> **5.** (a) No; (b) yes. **6.** (a) Yes; (b) no. **7.** (a) Yes; (b) no.

Practice Exercise 8. Determine a prime factorization of 205.

EXAMPLE 8 Determine the prime factorization of 132.

SOLUTION:

STEP 1: 132 is divisible by 2. Hence,

$$132 = 2 \times 66$$

STEP 2: 66 is divisible by 2; $66 = 2 \times 33$. Hence,

$$132 = 2 \times 2 \times 33$$

STEP 3: 33 is divisible by 3; $33 = 3 \times 11$. Hence,

$$132 = 2 \times 2 \times 3 \times 11$$

STEP 4: 11 is prime. Therefore, the prime factorization of 132 is:

$$132 = 2 \times 2 \times 3 \times 11$$
or $132 = 2^2 \times 3 \times 11$ (Using exponents)

Practice Exercise 9. Determine a prime factorization of 810.

EXAMPLE 9 Determine the prime factorization of 450.

SOLUTION:

STEP 1: 450 is divisible by 2. Hence,

$$450 = 2 \times 225$$

STEP 2: 225 is divisible by 5; $225 = 5 \times 45$. Hence,

$$450 = 2 \times 5 \times 45$$

STEP 3: 45 is divisible by 5; $45 = 5 \times 9$. Hence,

$$450 = 2 \times 5 \times 5 \times 9$$

STEP 4: 9 is divisible by 3; $9 = 3 \times 3$. Hence,

$$450 = 2 \times 5 \times 5 \times 3 \times 3$$

STEP 5: 3 is prime. Therefore, the prime factorization of 450 is:

$$450 = 2 \times 3 \times 3 \times 5 \times 5$$
or $450 = 2 \times 3^2 \times 5^2$ (Using exponents)

> **ANSWERS TO PRACTICE EXERCISES**
>
> **8.** 5×41. **9.** $2 \times 3^4 \times 5$.

When working with fractions, it is often necessary to rename two or more fractions by equivalent fractions having the same denominators. This is called a "common denominator" of the fractions. To determine a common denominator, we must know how to find a common multiple of two or more whole numbers.

A "multiple" of a whole number is the product of that number and any whole number. For instance, 15 is a multiple of 5 since $15 = 5 \times 3$. Similarly, 24 is a multiple of 8 since $24 = 8 \times 3$. Every whole number has many multiples.

- Some multiples of 3 are: 0, 3, 6, 9, 12, 15, 30, 60, 90.
- Some multiples of 6 are: 0, 6, 12, 18, 24, 30, 42, 54, 72.
- Some multiples of 7 are: 0, 7, 14, 21, 28, 49, 63, 70, 105.

A whole number that is a multiple of two or more whole numbers is called a "common multiple" of those numbers. The smallest whole number, other than 0, that is a common multiple of two or more whole numbers is called the **least common multiple (LCM)** of those numbers.

Rule for Finding the LCM of Two Whole Numbers

To form the least common multiple of the numbers *a* and *b*, denoted by LCM(*a*, *b*), follow these four steps:

1. Write each number as a product of prime factors.
2. Write as a product each of the prime factors used in Step 1.
3. Write each of the prime factors in Step 2 with the *greatest* exponent that it has in any of the products in Step 1.
4. Form the LCM(*a*, *b*).

Practice Exercise 10. Find LCM(36, 90).

EXAMPLE 10 Determine LCM(6, 8).

SOLUTION:

STEP 1: Write each number as a product of prime factors:

$$6 = 2 \times 3 = 2^1 \times 3^1$$
$$8 = 2 \times 2 \times 2 = 2^3$$

STEP 2: Write as a product each of the prime factors used in Step 1:

$$2 \times 3$$

STEP 3: Write each of the prime factors in Step 2 with the *greatest* exponent that it has in any of the products in Step 1:

$$2^3 \times 3^1$$

STEP 4: Form the LCM of 6 and 8:

$$\text{LCM}(6, 8) = 2^3 \times 3^1 = 8 \times 3 = 24$$

Practice Exercise 11. Find LCM(15, 40).

EXAMPLE 11 Determine LCM(6, 27).

SOLUTION:

STEP 1: Write each number as a product of prime factors:

$$6 = 2 \times 3 = 2^1 \times 3^1$$
$$27 = 3 \times 3 \times 3 = 3^3$$

2.3 Least Common Multiple

STEP 2: Write as a product each of the prime factors used in Step 1:

$$2 \times 3$$

STEP 3: Write each of the prime factors in Step 2 with the *greatest* exponent that it has in any of the products in Step 1:

$$2^1 \times 3^3$$

STEP 4: Form the LCM of 6 and 27:

$$\text{LCM}(6, 27) = 2^1 \times 3^3 = 2 \times 27 = 54$$

This procedure can also be used to find the LCM for three or more numbers.

EXAMPLE 12 Determine LCM(8, 12, 20).

SOLUTION:

STEP 1: Write each number as a product of prime factors:

$$8 = 2 \times 2 \times 2 = 2^3$$
$$12 = 2 \times 2 \times 3 = 2^2 \times 3^1$$
$$20 = 2 \times 2 \times 5 = 2^2 \times 5^1$$

STEP 2: Write as a product each of the prime factors used in Step 1:

$$2 \times 3 \times 5$$

STEP 3: Write each of the prime factors in Step 2 with the *greatest* exponent that it has in any of the products in Step 1.

$$2^3 \times 3^1 \times 5^1$$

STEP 4: Form the LCM of 8, 12, and 20:

$$\text{LCM}(8, 12, 20) = 2^3 \times 3^1 \times 5^1 = 8 \times 3 \times 5 = 120$$

> **Practice Exercise 12.** Find LCM(10, 26, 32).

ANSWERS TO PRACTICE EXERCISES
10. 180.   **11.** 120.   **12.** 2080.

EXERCISES 2.3

In Exercises 1–20, classify each of the given numbers as being prime or composite.

1. 17	**2.** 26	**3.** 43	**4.** 78	**5.** 89
6. 97	**7.** 103	**8.** 137	**9.** 146	**10.** 151
11. 171	**12.** 173	**13.** 191	**14.** 197	**15.** 201
16. 261	**17.** 289	**18.** 401	**19.** 703	**20.** 1001

In Exercises 21–50, rewrite each of the given composite numbers as a product of primes.

21. 12　　**22.** 22　　**23.** 28　　**24.** 32　　**25.** 44
26. 48　　**27.** 56　　**28.** 60　　**29.** 66　　**30.** 78
31. 88　　**32.** 99　　**33.** 102　　**34.** 110　　**35.** 125
36. 136　　**37.** 142　　**38.** 150　　**39.** 175　　**40.** 200
41. 242　　**42.** 310　　**43.** 375　　**44.** 420　　**45.** 550
46. 605　　**47.** 790　　**48.** 844　　**49.** 965　　**50.** 1000

In Exercises 51–70, determine whether each number is divisible by (a) 2, (b) 3, (c) 5.

51. 21　　**52.** 33　　**53.** 72　　**54.** 65　　**55.** 80
56. 100　　**57.** 156　　**58.** 175　　**59.** 190　　**60.** 201
61. 245　　**62.** 269　　**63.** 310　　**64.** 321　　**65.** 507
66. 699　　**67.** 1020　　**68.** 2345　　**69.** 2678　　**70.** 3000

In Exercises 71–90, determine the least common multiple (LCM) for each group of numbers.

71. 6, 18　　**72.** 9, 12　　**73.** 8, 20　　**74.** 12, 27
75. 20, 35　　**76.** 42, 56　　**77.** 32, 88　　**78.** 50, 125
79. 62, 186　　**80.** 2, 5, 7　　**81.** 8, 12, 16　　**82.** 10, 12, 15
83. 14, 15, 21　　**84.** 20, 36, 45　　**85.** 26, 44, 55　　**86.** 32, 48, 252
87. 110, 175, 210　　**88.** 5, 7, 9, 10
89. 7, 8, 10, 18　　**90.** 24, 27, 32, 40

91. If the whole number a is a multiple of the whole number b, what is the LCM(a, b)?
92. If the whole numbers c and d are different primes, what is the LCM(c, d)?
93. What whole number, if any, is a multiple of every whole number?

2.4 Addition of Fractions

OBJECTIVE

After completing this section, you should be able to add fractions.

Now that you have learned about equivalent fractions and least common multiples, you will be introduced to the addition of fractions. Remember, so far in our discussion, you can only add "like" units such as tens to tens, ones to ones, and centimeters to centimeters. The same is true when you try to add fractions. You can only add like fractions.

DEFINITION

Like fractions are fractions that have the same denominators. Fractions that have different denominators are called **unlike fractions.**

- The fractions $\frac{2}{7}$ and $\frac{4}{7}$ are like fractions since they both have a denominator of 7.
- The fractions $\frac{1}{9}, \frac{4}{9}$, and $\frac{11}{9}$ are like fractions since they all have a denominator of 9.

- The fractions $\frac{2}{3}$ and $\frac{3}{4}$ are unlike fractions since they have different denominators.
- The fractions $\frac{2}{5}$ and $\frac{3}{8}$ are unlike fractions since they have different denominators.

If we have two like fractions, they can be added quite simply. For instance, to add $\frac{2}{7}$ and $\frac{3}{7}$, consider that

$$\frac{2}{7} + \frac{3}{7} = \left(2 \times \frac{1}{7}\right) + \left(3 \times \frac{1}{7}\right) \quad \text{(Rewriting)}$$
$$= (2 + 3) \times \frac{1}{7} \quad \text{(Distributive property)}$$
$$= 5 \times \frac{1}{7} \quad \text{(Adding)}$$
$$= \frac{5}{7} \quad \text{(Rewriting)}$$

This addition problem is illustrated in Figure 2.11.

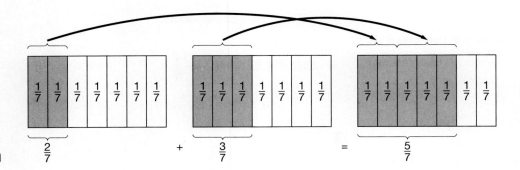

FIGURE 2.11

Therefore, to add $\frac{2}{7}$ and $\frac{3}{7}$, simply add their numerators, $2 + 3 = 5$, and write this sum over the like denominator, 7.

The rule for **addition of like fractions** is summarized in the accompanying box.

Rule for Addition of Like Fractions

To add like fractions:
1. Add the numerators and write this sum over the like denominator.
2. Simplify, if possible.

Practice Exercise 1. Add: $\frac{6}{11} + \frac{9}{11}$.

EXAMPLE 1 Add: $\frac{4}{7} + \frac{5}{7}$.

SOLUTION: $\frac{4}{7} + \frac{1}{7} = \frac{4 + 1}{7} = \frac{5}{7}$

Practice Exercise 2. Add:
$\frac{10}{21} + \frac{7}{21}$.

EXAMPLE 2 Add: $\frac{3}{11} + \frac{5}{11}$.

SOLUTION: $\frac{3}{11} + \frac{5}{11} = \frac{3+5}{11} = \frac{8}{11}$

Practice Exercise 3. Add:
$\frac{11}{17} + \frac{15}{17} + \frac{8}{17}$.

EXAMPLE 3 Add: $\frac{2}{9} + \frac{4}{9} + \frac{1}{9}$.

SOLUTION: $\frac{2}{9} + \frac{4}{9} + \frac{1}{9} = \frac{2+4+1}{9} = \frac{7}{9}$

Practice Exercise 4. Add:
$\frac{7}{35} + \frac{9}{35} + \frac{13}{35} + \frac{10}{35}$.

EXAMPLE 4 Add: $\frac{3}{13} + \frac{6}{13} + \frac{7}{13}$.

SOLUTION: $\frac{3}{13} + \frac{6}{13} + \frac{7}{13} = \frac{3+6+7}{13} = \frac{16}{13}$ or $1\frac{3}{13}$

ANSWERS TO PRACTICE EXERCISES

1. $\frac{15}{11}$ or $1\frac{4}{11}$. 2. $\frac{17}{21}$. 3. $\frac{34}{17} = 2$. 4. $\frac{39}{35}$ or $1\frac{4}{35}$.

To add the unlike fractions $\frac{2}{3}$ and $\frac{5}{6}$, we rewrite the two fractions in equivalent form with the same denominators.

Add and write your results in simplest form.

Practice Exercise 5. $\frac{3}{8} + \frac{2}{7}$

EXAMPLE 5 Add: $\frac{2}{3} + \frac{5}{6}$.

SOLUTION: To add $\frac{2}{3}$ and $\frac{5}{6}$, we rewrite

$$\frac{2}{3} = \frac{2 \times 2}{3 \times 2} = \frac{4}{6}$$

Then,

$$\frac{2}{3} + \frac{5}{6} = \frac{4}{6} + \frac{5}{6}$$
$$= \frac{4+5}{6}$$
$$= \frac{9}{6}$$

Therefore, $\frac{2}{3} + \frac{5}{6} = \frac{9}{6}$, which may be simplified to $\frac{3}{2}$ or $1\frac{1}{2}$. This addition problem is illustrated in Figure 2.12.

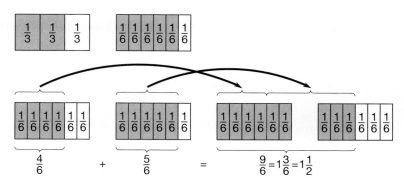

FIGURE 2.12

2.4 Addition of Fractions

In Example 5, we rewrote the fraction $\frac{2}{3}$ in equivalent form as $\frac{4}{6}$. Then $\frac{4}{6}$ and $\frac{5}{6}$ were added as like fractions. When adding two unlike fractions, it is sometimes necessary to rewrite both of them as equivalent fractions with a denominator different from either of those given. This will be illustrated in Examples 6 and 7.

Practice Exercise 6. $\frac{3}{4} + \frac{2}{5}$

EXAMPLE 6 Add: $\frac{1}{2} + \frac{2}{3}$.

SOLUTION: To add $\frac{1}{2}$ and $\frac{2}{3}$, we first rewrite the fractions as equivalent fractions with a denominator of 6. There are many other denominators that could have been chosen, but 6 is the smallest of them. Continuing, we have

$$\frac{1}{2} + \frac{2}{3} = \frac{3}{6} + \frac{4}{6}$$
$$= \frac{3 + 4}{6}$$
$$= \frac{7}{6}$$

Therefore, $\frac{1}{2} + \frac{2}{3} = \frac{7}{6}$ or $1\frac{1}{6}$. (See Figure 2.13.)

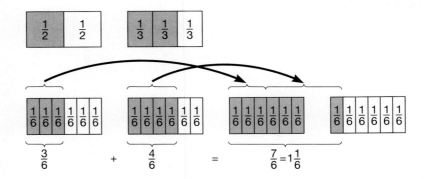

FIGURE 2.13

Practice Exercise 7. $\frac{2}{3} + \frac{3}{4}$

EXAMPLE 7 Add: $\frac{4}{7} + \frac{2}{3}$.

SOLUTION: To add $\frac{4}{7}$ and $\frac{2}{3}$, we could rewrite both fractions as equivalent fractions with a denominator of 21. We have

$$\frac{4}{7} = \frac{4 \times 3}{7 \times 3} = \frac{12}{21} \quad \text{and} \quad \frac{2}{3} = \frac{2 \times 7}{3 \times 7} = \frac{14}{21}$$

Then,

$$\frac{4}{7} + \frac{2}{3} = \frac{12}{21} + \frac{14}{21}$$
$$= \frac{12 + 14}{21}$$
$$= \frac{26}{21}$$

> **ANSWERS TO PRACTICE EXERCISES**
>
> 5. $\frac{37}{56}$.   6. $\frac{23}{20}$ or $1\frac{3}{20}$.   7. $\frac{17}{12}$ or $1\frac{5}{12}$.

In Example 7, it was not necessary to use the denominator 21. Surely other choices, such as 42, could have been made. In selecting a denominator to get like fractions, try to get the smallest whole number that will work. This will save you time and effort! In the next section, you will learn how to get the smallest or least denominator that works when you are adding fractions.

When we rename two or more fractions by equivalent fractions with a common denominator, we usually use the LCM of the denominators for the common denominator. This common denominator is called the **least** (or lowest) **common denominator** of the fractions and is denoted by **LCD**. Hence, the LCD of two or more fractions is the least common multiple of the denominators of the fractions.

We will now continue our discussion of the addition of fractions, using the least common denominator approach.

Add, using LCDs. Write your results in simplest form.

Practice Exercise 8. $\frac{2}{9} + \frac{5}{18}$

EXAMPLE 8 Add: $\frac{1}{2}$ and $\frac{3}{8}$.

SOLUTION: The LCM(2, 8) = 8. (Do you agree?) Therefore,

$$\frac{1}{2} = \frac{1 \times 4}{2 \times 4} = \frac{4}{8}$$

and

$$\frac{1}{2} + \frac{3}{8} = \frac{4}{8} + \frac{3}{8}$$

$$= \frac{4 + 3}{8} \quad \text{(Adding like fractions)}$$

$$= \frac{7}{8} \quad \text{(Simplifying)}$$

Practice Exercise 9. $\frac{4}{7} + \frac{5}{13}$

EXAMPLE 9 Add: $\frac{2}{7}$ and $\frac{3}{5}$.

SOLUTION: $\frac{2}{7} + \frac{3}{5} = \frac{2 \times 5}{7 \times 5} + \frac{3 \times 7}{5 \times 7}$ [Since LCM(5, 7) = 35]

$$= \frac{10}{35} + \frac{21}{35} \quad \text{(Simplifying)}$$

$$= \frac{10 + 21}{35} \quad \text{(Adding like fractions)}$$

$$= \frac{31}{35} \quad \text{(Simplifying)}$$

Practice Exercise 10. $\frac{3}{14} + \frac{2}{21}$

EXAMPLE 10 Add: $\frac{3}{5}$ and $\frac{3}{4}$.

SOLUTION: $\frac{3}{5} + \frac{3}{4} = \frac{3 \times 4}{5 \times 4} + \frac{3 \times 5}{4 \times 5}$ [Since LCM(4, 5) = 20]

$$= \frac{12}{20} + \frac{15}{20} \quad \text{(Simplifying)}$$

2.4 Addition of Fractions

$$= \frac{12 + 15}{20} \quad \text{(Adding like fractions)}$$

$$= \frac{27}{20} \text{ or } 1\frac{7}{20} \quad \text{(Simplifying)}$$

The steps involved in the **addition of unlike fractions** are summarized in the accompanying box.

Rule for Addition of Unlike Fractions

To add unlike fractions, do the following steps:
1. Determine the LCM of the denominators involved.
2. Rewrite all fractions involved as equivalent fractions having the LCM as denominators.
3. Add the new like fractions, using the rule for adding like fractions.
4. Simplify your result, if possible.

Practice Exercise 11. $\frac{11}{28} + \frac{4}{35}$

EXAMPLE 11 Add: $\frac{11}{18}$ and $\frac{7}{24}$.

SOLUTION: The LCM(18, 24) = 72. Therefore,

$$\frac{11}{18} = \frac{11 \times 4}{18 \times 4} = \frac{44}{72}$$

and

$$\frac{7}{24} = \frac{7 \times 3}{24 \times 3} = \frac{21}{72}$$

Adding, we now have

$$\frac{11}{18} + \frac{7}{24} = \frac{44}{72} + \frac{21}{72}$$

$$= \frac{44 + 21}{72}$$

$$= \frac{65}{72}$$

Practice Exercise 12. $\frac{3}{7} + \frac{1}{3} + \frac{3}{4}$

EXAMPLE 12 Add: $\frac{13}{15}$ and $\frac{5}{36}$.

SOLUTION: The LCM(15, 36) = 180. Therefore,

$$\frac{13}{15} = \frac{13 \times 12}{15 \times 12} = \frac{156}{180}$$

and

$$\frac{5}{36} = \frac{5 \times 5}{36 \times 5} = \frac{25}{180}$$

Adding, we now have

$$\frac{13}{15} + \frac{5}{36} = \frac{156}{180} + \frac{25}{180}$$

$$= \frac{156 + 25}{180}$$

$$= \frac{181}{180} \text{ or } 1\frac{1}{180}$$

Practice Exercise 13. $\frac{2}{5} + \frac{5}{6} + \frac{1}{2}$

EXAMPLE 13 Add: $\frac{1}{2} + \frac{1}{3} + \frac{2}{5}$.

SOLUTION: The LCM(2, 3, 5) = 30. Therefore,

$$\frac{1}{2} + \frac{1}{3} + \frac{2}{5} = \frac{1 \times 15}{2 \times 15} + \frac{1 \times 10}{3 \times 10} + \frac{2 \times 6}{5 \times 6}$$

$$= \frac{15}{30} + \frac{10}{30} + \frac{12}{30}$$

$$= \frac{15 + 10 + 12}{30}$$

$$= \frac{37}{30} \text{ or } 1\frac{7}{30}$$

Practice Exercise 14. $\frac{1}{7} + \frac{2}{11} + \frac{3}{28}$

EXAMPLE 14 Add: $\frac{1}{4} + \frac{2}{3} + \frac{3}{4} + \frac{1}{2}$.

SOLUTION: The LCM(2, 3, 4) = 12. Therefore,

$$\frac{1}{4} + \frac{2}{3} + \frac{3}{4} + \frac{1}{2} = \frac{1 \times 3}{4 \times 3} + \frac{2 \times 4}{3 \times 4} + \frac{3 \times 3}{4 \times 3} + \frac{1 \times 6}{2 \times 6}$$

$$= \frac{3}{12} + \frac{8}{12} + \frac{9}{12} + \frac{6}{12}$$

$$= \frac{3 + 8 + 9 + 6}{12}$$

$$= \frac{26}{12}$$

$$= \frac{13}{6} \text{ or } 2\frac{1}{6}$$

ANSWERS TO PRACTICE EXERCISES

8. $\frac{1}{2}$. 9. $\frac{87}{91}$. 10. $\frac{13}{42}$. 11. $\frac{71}{140}$. 12. $\frac{127}{84}$ or $1\frac{43}{84}$. 13. $\frac{26}{15}$ or $1\frac{11}{15}$. 14. $\frac{19}{44}$.

EXERCISES 2.4

Add and write your results in simplest form.

1. $\frac{2}{5} + \frac{1}{5}$
2. $\frac{3}{7} + \frac{2}{7}$
3. $\frac{4}{9} + \frac{2}{9}$
4. $\frac{3}{11} + \frac{7}{11}$
5. $\frac{9}{13} + \frac{2}{13}$
6. $\frac{15}{17} + \frac{4}{17}$
7. $\frac{2}{9} + \frac{4}{9} + \frac{1}{9}$
8. $\frac{2}{15} + \frac{7}{15} + \frac{11}{15}$
9. $\frac{11}{23} + \frac{14}{23} + \frac{17}{23}$

2.4 Addition of Fractions

10. $\dfrac{3}{37} + \dfrac{10}{37} + \dfrac{17}{37}$
11. $\dfrac{1}{2} + \dfrac{2}{5}$
12. $\dfrac{3}{7} + \dfrac{5}{4}$
13. $\dfrac{3}{8} + \dfrac{3}{4}$
14. $\dfrac{3}{7} + \dfrac{5}{8}$
15. $\dfrac{4}{13} + \dfrac{5}{9}$
16. $\dfrac{5}{11} + \dfrac{2}{7}$
17. $\dfrac{10}{19} + \dfrac{1}{3}$
18. $\dfrac{11}{13} + \dfrac{2}{11}$
19. $\dfrac{2}{9} + \dfrac{7}{15}$
20. $\dfrac{4}{7} + \dfrac{1}{13}$
21. $\dfrac{5}{18} + \dfrac{11}{12}$
22. $\dfrac{5}{24} + \dfrac{7}{32}$
23. $\dfrac{15}{26} + \dfrac{15}{36}$
24. $\dfrac{10}{21} + \dfrac{17}{70}$
25. $\dfrac{7}{16} + \dfrac{3}{20}$
26. $\dfrac{9}{16} + \dfrac{13}{24}$
27. $\dfrac{3}{22} + \dfrac{7}{33}$
28. $\dfrac{9}{40} + \dfrac{11}{12}$
29. $\dfrac{11}{19} + \dfrac{9}{13}$
30. $\dfrac{5}{21} + \dfrac{6}{17}$
31. $\dfrac{11}{19} + \dfrac{5}{11}$
32. $\dfrac{1}{2} + \dfrac{1}{3} + \dfrac{1}{4}$
33. $\dfrac{1}{7} + \dfrac{2}{9} + \dfrac{3}{5}$
34. $\dfrac{1}{2} + \dfrac{1}{3} + \dfrac{5}{6}$
35. $\dfrac{2}{5} + \dfrac{3}{7} + \dfrac{1}{9}$
36. $\dfrac{1}{2} + \dfrac{2}{3} + \dfrac{3}{4} + \dfrac{4}{5}$
37. $\dfrac{3}{5} + \dfrac{4}{7} + \dfrac{5}{6}$
38. $\dfrac{4}{7} + \dfrac{5}{9} + \dfrac{6}{11}$
39. $\dfrac{4}{11} + \dfrac{5}{9} + \dfrac{6}{7}$
40. $\dfrac{2}{15} + \dfrac{1}{16} + \dfrac{3}{4}$
41. $\dfrac{3}{8} + \dfrac{7}{10} + \dfrac{5}{12}$
42. $\dfrac{4}{15} + \dfrac{3}{16} + \dfrac{7}{20}$
43. $\dfrac{1}{10} + \dfrac{3}{20} + \dfrac{7}{30}$
44. $\dfrac{4}{5} + \dfrac{11}{20} + \dfrac{7}{25}$
45. $\dfrac{5}{6} + \dfrac{7}{12} + \dfrac{9}{10}$
46. $\dfrac{3}{4} + \dfrac{7}{10} + \dfrac{11}{30}$
47. $\dfrac{5}{6} + \dfrac{3}{28} + \dfrac{1}{42}$
48. $\dfrac{3}{14} + \dfrac{6}{35} + \dfrac{5}{21}$
49. $\dfrac{3}{8} + \dfrac{9}{14} + \dfrac{5}{36}$
50. $\dfrac{7}{15} + \dfrac{9}{21} + \dfrac{6}{35}$
51. $\dfrac{3}{4} + \dfrac{11}{36} + \dfrac{4}{81}$
52. $\dfrac{2}{7} + \dfrac{4}{9} + \dfrac{2}{3} + \dfrac{3}{4}$
53. $\dfrac{7}{9} + \dfrac{2}{3} + \dfrac{5}{6} + \dfrac{3}{4}$
54. $\dfrac{2}{9} + \dfrac{4}{7} + \dfrac{5}{9} + \dfrac{5}{7}$
55. $\dfrac{4}{11} + \dfrac{4}{9} + \dfrac{5}{8} + \dfrac{2}{9}$
56. $\dfrac{3}{5} + \dfrac{2}{3} + \dfrac{5}{7} + \dfrac{1}{8}$
57. $\dfrac{3}{7} + \dfrac{7}{10} + \dfrac{2}{9} + \dfrac{1}{8}$
58. $\dfrac{1}{2} + \dfrac{2}{3} + \dfrac{3}{4} + \dfrac{4}{5}$
59. $\dfrac{1}{7} + \dfrac{1}{8} + \dfrac{1}{9} + \dfrac{1}{10}$
60. $\dfrac{2}{5} + \dfrac{1}{6} + \dfrac{3}{7} + \dfrac{1}{8}$
61. $\dfrac{2}{3} + \dfrac{3}{5} + \dfrac{5}{7} + \dfrac{7}{9}$
62. $\dfrac{1}{10} + \dfrac{3}{50} + \dfrac{7}{100} + \dfrac{9}{1000}$
63. $\dfrac{3}{5} + \dfrac{1}{12} + \dfrac{5}{24} + \dfrac{1}{120}$

64. One piece of pipe measures $\dfrac{3}{4}$ ft and another piece of pipe measures $\dfrac{15}{16}$ ft. What is the combined length of the two pieces of pipe?

65. Brian spends $\dfrac{1}{4}$ of his monthly income for rent and $\dfrac{3}{7}$ of his monthly income for food. What fractional part of his income does he spend for rent and food?

66. Rapid Computer stock gained $\dfrac{5}{8}$ of a point during the first day of trading and $\dfrac{1}{4}$ of a point during the second day of trading. How many points did the stock gain during the two days?

67. During a furniture sale, the price of every piece of furniture was reduced by $\dfrac{1}{4}$ during the first week of the sale and an additional $\dfrac{1}{5}$ during the second week of the sale. What was the total reduction in the price of furniture during the two weeks?

68. The rainfall for one day in Our Town was $\frac{5}{8}$ in. and $\frac{3}{7}$ in. a second day. What was the total amount of rainfall during the two days?

69. Mom baked a pie and set it aside to cool. Michael came home and ate $\frac{1}{3}$ of it. Later, Karen came home and ate $\frac{1}{5}$ of the pie and gave another $\frac{1}{4}$ of it to her friend Mary Beth to eat. How much of the pie did the three of them eat?

70. Determine the total length around the triangle in the accompanying diagram. All dimensions are in feet.

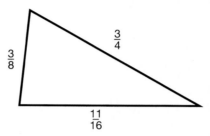

71. The price of Corporate Giant stock increased by $\frac{3}{4}$ of a point during the first hour of trading, $\frac{1}{2}$ of a point during the second hour of trading, and $\frac{7}{8}$ of a point during the third hour. What was the total increase in the price of the stock during the three hours?

72. Determine the total length around the figure in the accompanying diagram. All dimensions are in meters.

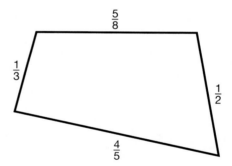

2.5 Ordering of Fractions; Subtraction

OBJECTIVES

After completing this section, you should be able to:
1. Order any two fractions.
2. Subtract one fraction from another fraction.

By now, you should be able to determine fairly easily whether one whole number is greater than or less than another. However, how would you be able to determine whether one fraction is greater than or less than another? For instance, can you readily determine that the fraction $\frac{2}{5}$ is greater than the fraction $\frac{4}{11}$?

2.5 Ordering of Fractions; Subtraction

If the fractions are like fractions, you can compare them by comparing their numerators. For instance, the fraction $\frac{2}{13}$ is less than the fraction $\frac{7}{13}$ since the fractions are like fractions and $2 < 7$.

Ordering of Fractions

In **ordering of fractions,** if two fractions have the same denominators, then the fraction with the greater numerator is the greater fraction.

Practice Exercise 1. Order the following pairs of fractions:
(a) $\frac{3}{5}$ and $\frac{7}{5}$; (b) $\frac{4}{7}$ and $\frac{9}{7}$;
(c) $\frac{5}{13}$ and $\frac{0}{13}$; (d) $\frac{12}{21}$ and $\frac{9}{21}$; (e) $\frac{15}{17}$ and $\frac{12}{17}$.

EXAMPLE 1 Order each of the following pairs of fractions: (a) $\frac{4}{7}$ and $\frac{1}{7}$; (b) $\frac{11}{6}$ and $\frac{5}{6}$; (c) $\frac{5}{9}$ and $\frac{11}{9}$; (d) $\frac{0}{3}$ and $\frac{4}{3}$.

SOLUTION: (a) $\frac{4}{7}$ and $\frac{1}{7}$ are like fractions; $\frac{4}{7} > \frac{1}{7}$ since $4 > 1$; (b) $\frac{11}{6}$ and $\frac{5}{6}$ are like fractions; $\frac{11}{6} > \frac{5}{6}$ since $11 > 5$; (c) $\frac{5}{9}$ and $\frac{11}{9}$ are like fractions; $\frac{5}{9} < \frac{11}{9}$ since $5 < 11$; (d) $\frac{0}{3}$ and $\frac{4}{3}$ are like fractions; $\frac{0}{3} < \frac{4}{3}$ since $0 < 4$.

ANSWER TO PRACTICE EXERCISE

1. (a) $\frac{3}{5} < \frac{7}{5}$; (b) $\frac{4}{7} < \frac{9}{7}$; (c) $\frac{5}{13} > \frac{0}{13}$; (d) $\frac{12}{21} > \frac{9}{21}$; (e) $\frac{15}{17} > \frac{12}{17}$.

To compare unlike fractions, we would first rewrite the fractions as equivalent fractions with the same denominators, using the LCD approach. Then, we would compare these new like fractions by comparing their numerators.

The fractions $\frac{2}{5}$ and $\frac{4}{11}$ are *not* like fractions. To compare them, we would rewrite the equivalent fraction of each, using the LCM(5, 11) = 55, and then compare their numerators. Hence,

$$\frac{2}{5} = \frac{2 \times 11}{5 \times 11} = \frac{22}{55} \text{ and } \frac{4}{11} = \frac{4 \times 5}{11 \times 5} = \frac{20}{55}$$

Since $22 > 20$, the fraction $\frac{22}{55}$ (or its equivalent $\frac{2}{5}$) is greater than the fraction $\frac{20}{55}$ (or its equivalent $\frac{4}{11}$). Therefore, $\frac{2}{5} > \frac{4}{11}$.

In Practice Exercises 2–4, insert the symbol $<$ or $>$.

Practice Exercise 2. $\frac{21}{39}$ $\frac{29}{39}$

EXAMPLE 2 Order the two fractions $\frac{2}{3}$ and $\frac{1}{4}$.

SOLUTION: $\frac{2}{3} = \frac{8}{12}$; $\frac{1}{4} = \frac{3}{12}$; $8 > 3$. Therefore, $\frac{8}{12} > \frac{3}{12}$. Hence, $\frac{2}{3} > \frac{1}{4}$.

Practice Exercise 3. $\frac{3}{9}$ $\frac{1}{2}$

EXAMPLE 3 Order the two fractions $\frac{11}{13}$ and $\frac{12}{17}$.

SOLUTION: $\frac{11}{13} = \frac{187}{221}$; $\frac{12}{17} = \frac{156}{221}$; $187 > 156$. Therefore, $\frac{187}{221} > \frac{156}{221}$. Hence, $\frac{11}{13} > \frac{12}{17}$.

Practice Exercise 4. $\dfrac{14}{23}$ $\dfrac{17}{21}$

EXAMPLE 4 Order the two fractions $\dfrac{10}{17}$ and $\dfrac{23}{29}$.

SOLUTION: $\dfrac{10}{17} = \dfrac{290}{493}; \dfrac{23}{29} = \dfrac{391}{493};$ $290 < 391$. Therefore, $\dfrac{290}{493} < \dfrac{391}{493}$. Hence, $\dfrac{10}{17} < \dfrac{23}{29}$.

> **ANSWERS TO PRACTICE EXERCISES**
> 2. <. 3. <. 4. <.

Now that we can compare fractions, we will consider subtracting one fraction from another fraction. In this section, we will consider the subtraction $\dfrac{a}{b} - \dfrac{c}{d}$ where $\dfrac{a}{b} > \dfrac{c}{d}$ or $\dfrac{a}{b} = \dfrac{c}{d}$.

The rule for **subtraction of fractions** is given in the accompanying box.

Rule for Subtraction of Fractions

To subtract fractions, do the following:

1. Determine the LCM of the denominators involved.
2. Rewrite all fractions involved as equivalent fractions having the LCM as denominator.
3. Subtract the numerators of the equivalent fractions.
4. Write the result over the LCM.
5. Simplify your results, if possible.

You are reminded that, at this point in the text, the rule for subtracting fractions always works provided you don't try to subtract a fraction from a smaller fraction. It should also be noted at this point that if two fractions have the same numerator but different denominators, then the one with the larger denominator is the smaller fraction.

Subtract and simplify your results.

Practice Exercise 5. $\dfrac{4}{7} - \dfrac{2}{9}$

EXAMPLE 5 Subtract: $\dfrac{2}{3} - \dfrac{1}{2}$.

SOLUTION: The LCM(2, 3) = 6. Therefore,

$$\dfrac{2}{3} - \dfrac{1}{2} = \dfrac{2 \times 2}{3 \times 2} - \dfrac{1 \times 3}{2 \times 3} \quad \text{(Equivalent fractions with LCD = 6)}$$

$$= \dfrac{4}{6} - \dfrac{3}{6} \quad \text{(Simplifying)}$$

$$= \dfrac{4 - 3}{6} \quad \text{(Subtract numerators; write over LCD)}$$

$$= \dfrac{1}{6} \quad \text{(Simplifying)}$$

2.5 Ordering of Fractions; Subtraction

Practice Exercise 6. $\dfrac{5}{8} - \dfrac{3}{14}$

EXAMPLE 6 Subtract: $\dfrac{31}{37} - \dfrac{9}{11}$.

SOLUTION: The LCM(11, 37) = 407. Therefore,

$$\dfrac{31}{37} - \dfrac{9}{11} = \dfrac{31 \times 11}{37 \times 11} - \dfrac{9 \times 37}{11 \times 37} \quad \text{(Equivalent fractions with LCD = 407)}$$

$$= \dfrac{341}{407} - \dfrac{333}{407} \quad \text{(Simplifying)}$$

$$= \dfrac{341 - 333}{407} \quad \text{(Subtract numerators; write over LCD)}$$

$$= \dfrac{8}{407} \quad \text{(Simplifying)}$$

Practice Exercise 7. $\dfrac{10}{17} - \dfrac{5}{11}$

EXAMPLE 7 Subtract: $\dfrac{17}{35} - \dfrac{4}{15}$.

SOLUTION: The LCM(15, 35) = 105. Therefore,

$$\dfrac{17}{35} - \dfrac{4}{15} = \dfrac{17 \times 3}{35 \times 3} - \dfrac{4 \times 7}{15 \times 7} \quad \text{(Equivalent fractions with LCD = 105)}$$

$$= \dfrac{51}{105} - \dfrac{28}{105} \quad \text{(Simplifying)}$$

$$= \dfrac{51 - 28}{105} \quad \text{(Subtract numerators; write over LCD)}$$

$$= \dfrac{23}{105} \quad \text{(Simplifying)}$$

Practice Exercise 8. If Mr. Wiley paints $\dfrac{1}{3}$ of his house one week and $\dfrac{2}{7}$ of the house the second week, how much of the house would be left unpainted?

EXAMPLE 8 If John completed $\dfrac{1}{4}$ of an assignment one day and $\dfrac{2}{9}$ of the assignment the next day, how much of the assignment was not completed?

SOLUTION: There are two parts to the solution. To determine how much of the assignment was *not* completed, we must first determine how much of the assignment was completed by adding the two parts that were completed. Next, subtract this amount from 1, which represents the total assignment. Hence, we have

$$1 - \left(\dfrac{1}{4} + \dfrac{2}{9}\right) = 1 - \left(\dfrac{9}{36} + \dfrac{8}{36}\right) \quad \text{(Since LCM = 36)}$$

$$= 1 - \left(\dfrac{9 + 8}{36}\right) \quad \text{(Adding like fractions)}$$

$$= 1 - \dfrac{17}{36} \quad \text{(Simplifying)}$$

$$= \dfrac{36}{36} - \dfrac{17}{36} \quad \left(\text{Rewriting } 1 = \dfrac{36}{36}\right)$$

$$= \dfrac{36 - 17}{36} \quad \text{(Subtracting the fractions)}$$

$$= \dfrac{19}{36}$$

Hence, $\dfrac{19}{36}$ of the assignment was *not* completed.

> **ANSWERS TO PRACTICE EXERCISES**
>
> 5. $\frac{22}{63}$. 6. $\frac{23}{56}$. 7. $\frac{25}{187}$. 8. $\frac{8}{21}$.

EXERCISES 2.5

In Exercises 1–20, insert the symbol $<$ or $>$ between each pair of fractions to make true statements.

1. $\frac{2}{3}$ $\frac{3}{4}$
2. $\frac{1}{2}$ $\frac{1}{3}$
3. $\frac{0}{2}$ $\frac{2}{3}$
4. $\frac{4}{9}$ $\frac{0}{3}$
5. $\frac{3}{4}$ $\frac{16}{17}$
6. $\frac{6}{6}$ $\frac{10}{5}$
7. $\frac{8}{9}$ $\frac{9}{8}$
8. $\frac{6}{7}$ $\frac{5}{6}$
9. $\frac{0}{5}$ $\frac{1}{7}$
10. $\frac{14}{17}$ $\frac{23}{26}$
11. $\frac{5}{13}$ $\frac{1}{27}$
12. $\frac{15}{23}$ $\frac{19}{21}$
13. $\frac{3}{4}$ $\frac{5}{9}$
14. $\frac{9}{13}$ $\frac{11}{17}$
15. $\frac{3}{7}$ $\frac{5}{14}$
16. $\frac{4}{5}$ $\frac{18}{10}$
17. $\left(\frac{1}{2}+\frac{1}{3}\right)$ $\frac{2}{5}$
18. $\left(\frac{2}{3}+\frac{1}{2}\right)$ $\frac{5}{6}$
19. $\left(\frac{2}{3}+\frac{1}{4}\right)$ $\left(\frac{3}{4}+\frac{2}{5}\right)$
20. $\left(\frac{3}{7}+\frac{1}{4}\right)$ $\left(\frac{2}{9}+\frac{4}{11}\right)$

In Exercises 21–40, subtract and write your results in simplest form.

21. $\frac{5}{8}-\frac{3}{8}$
22. $\frac{11}{16}-\frac{5}{16}$
23. $\frac{23}{37}-\frac{14}{37}$
24. $\frac{23}{38}-\frac{17}{38}$
25. $\frac{2}{3}-\frac{1}{4}$
26. $\frac{1}{2}-\frac{3}{7}$
27. $\frac{2}{9}-\frac{0}{3}$
28. $\frac{4}{5}-\frac{2}{3}$
29. $\frac{2}{3}-\frac{4}{6}$
30. $\frac{1}{5}-\frac{1}{9}$
31. $\frac{4}{11}-\frac{3}{9}$
32. $\frac{21}{13}-\frac{7}{5}$
33. $\frac{21}{17}-\frac{13}{14}$
34. $\frac{23}{11}-\frac{11}{23}$
35. $\frac{11}{13}-\frac{4}{9}$
36. $\frac{17}{19}-\frac{21}{31}$
37. $\frac{19}{23}-\frac{17}{25}$
38. $\frac{61}{13}-\frac{54}{49}$
39. $\frac{110}{9}-\frac{103}{13}$
40. $\frac{107}{15}-\frac{109}{17}$

In Exercises 41–50, perform the indicated operations and write your results in simplest form.

41. $\left(\frac{1}{2}+\frac{1}{3}\right)-\frac{1}{4}$
42. $\left(\frac{2}{3}+\frac{4}{7}\right)-\frac{3}{5}$
43. $\frac{11}{12}-\left(\frac{1}{2}+\frac{1}{6}\right)$

44. $\dfrac{23}{35} - \left(\dfrac{2}{9} + \dfrac{1}{4}\right)$ 45. $\left(\dfrac{11}{12} - \dfrac{1}{9}\right) - \dfrac{1}{3}$ 46. $\left(\dfrac{19}{20} - \dfrac{1}{3}\right) - \dfrac{2}{7}$

47. $\dfrac{8}{9} - \left(\dfrac{3}{10} - \dfrac{1}{6}\right)$ 48. $\dfrac{17}{25} - \left(\dfrac{9}{10} - \dfrac{2}{7}\right)$ 49. $\left(\dfrac{4}{7} + \dfrac{2}{9}\right) - \left(\dfrac{2}{5} + \dfrac{1}{4}\right)$

50. $\left(\dfrac{1}{2} + \dfrac{1}{3} + \dfrac{1}{4}\right) - \left(\dfrac{1}{4} + \dfrac{1}{5}\right)$

51. Joyce completed $\dfrac{10}{17}$ of her Christmas shopping before Thanksgiving. What fractional part of her shopping was not completed?

52. A certain stock gained $\dfrac{5}{8}$ of a point during the first day of trading and lost $\dfrac{1}{2}$ of a point during the second day of trading. What was the net gain (i.e., gain minus loss) during the two days of trading?

53. Fred painted $\dfrac{1}{2}$ of his house during the first week of his vacation and $\dfrac{2}{5}$ of his house during the second week of his vacation. What fractional part of his house remains to be painted?

54. A man left $\dfrac{1}{3}$ of his estate to his wife and $\dfrac{1}{8}$ of the estate to each of his three children. The rest of his estate was left to his favorite charity. What fractional part of his estate was left to his favorite charity?

55. If $\dfrac{2}{3}$ of a commercial building is occupied by a department store and $\dfrac{1}{5}$ of the building is occupied by a law office, what fractional part of the building is used for other purposes or is not occupied?

2.6 Addition and Subtraction of Mixed Numbers

OBJECTIVE

After completing this section, you should be able to add and subtract mixed numbers.

Now that you have learned how to add and subtract fractions, we will discuss the addition and subtraction of mixed numbers.

Rule for Adding Mixed Numbers

Since a mixed number is a sum of a whole number and a fraction, to add mixed numbers:

1. Add the whole number parts.
2. Add the fractional parts and simplify, if necessary.
3. If the sum of the fractional parts is an improper fraction, rewrite it as a mixed number and add the result to the sum in Step 1.

Add the following mixed numbers and simplify your results.

Practice Exercise 1. $5\frac{1}{2} + 6\frac{2}{3}$

EXAMPLE 1 Add: $3\frac{1}{2} + 2\frac{1}{3}$.

SOLUTION:

$$3\frac{1}{2} + 2\frac{1}{3} = \left(3 + \frac{1}{2}\right) + \left(2 + \frac{1}{3}\right) \quad \text{(Remember, a mixed number is a whole number plus a fraction.)}$$

$$= (3 + 2) + \left(\frac{1}{2} + \frac{1}{3}\right) \quad \text{(Regrouping)}$$

$$= (3 + 2) + \left(\frac{3}{6} + \frac{2}{6}\right) \quad \text{(Rewriting the fractions as equivalent fractions with LCD = 6)}$$

$$= 5 + \frac{5}{6} \quad \text{(Adding the whole number parts; adding the fractional parts)}$$

$$= 5\frac{5}{6} \quad \text{(Rewriting)}$$

Practice Exercise 2. $7\frac{1}{9} + 8\frac{1}{4}$

EXAMPLE 2 Add: $4\frac{2}{9} + 5\frac{5}{6}$.

SOLUTION:

$$4\frac{2}{9} + 5\frac{5}{6} = \left(4 + \frac{2}{9}\right) + \left(5 + \frac{5}{6}\right)$$

$$= (4 + 5) + \left(\frac{2}{9} + \frac{5}{6}\right)$$

$$= (4 + 5) + \left(\frac{4}{18} + \frac{15}{18}\right) \quad \text{(Equivalent fractions with LCD = 18)}$$

$$= 9 + \frac{19}{18} \quad \text{(Adding the whole number parts; adding the fractional parts)}$$

$$= 9 + \left(1 + \frac{1}{18}\right) \quad \left(\text{Since } \frac{19}{18} = 1\frac{1}{18}\right)$$

$$= (9 + 1) + \frac{1}{18} \quad \text{(Regrouping)}$$

$$= 10 + \frac{1}{18}$$

$$= 10\frac{1}{18}$$

In Example 3, we will add mixed numbers vertically.

Practice Exercise 3. $4\frac{2}{7} + 3\frac{1}{3}$

EXAMPLE 3 Add: $3\frac{1}{2} + 2\frac{1}{3} + 5\frac{3}{4}$.

SOLUTION:

$$\begin{array}{rcrcr} 3\frac{1}{2} & & 3 + \frac{1}{2} & & 3 + \frac{6}{12} \\ 2\frac{1}{3} & = & 2 + \frac{1}{3} & = & 2 + \frac{4}{12} \\ +5\frac{3}{4} & & +5 + \frac{3}{4} & & +5 + \frac{9}{12} \\ \hline \end{array} \quad \text{(Equivalent fractions with LCD = 12)}$$

2.6 Addition and Subtraction of Mixed Numbers

$$= 10 + \frac{19}{12} \quad \text{(Adding whole number parts; adding the fractional parts)}$$

$$= 10 + \left(1 + \frac{7}{12}\right)$$

$$= (10 + 1) + \frac{7}{12}$$

$$= 11 + \frac{7}{12}$$

$$= 11\frac{7}{12}$$

Practice Exercise 4.
$4\frac{1}{2} + 3\frac{1}{3} + 5\frac{1}{4}$

EXAMPLE 4 A carpenter needs pieces of wood measuring $13\frac{3}{4}$ in., $17\frac{5}{8}$ in., $15\frac{1}{2}$ in., and 16 in. If all of these pieces are to be cut from the same board, what is the shortest length of board the carpenter must use if $\frac{1}{8}$ in. is lost in each cutting?

SOLUTION: The problem could be represented as illustrated in Figure 2.14.

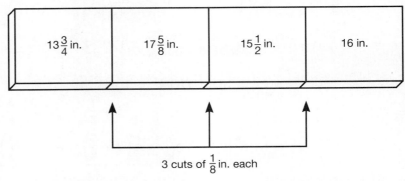

FIGURE 2.14

To determine the length of board needed, we *add* the lengths of the four pieces plus the lengths lost in cutting as follows:

$$
\begin{array}{rl}
13\frac{3}{4} & 13\frac{6}{8} \\
17\frac{5}{8} & 17\frac{5}{8} \\
15\frac{1}{2} = & 15\frac{4}{8} \\
16 & 16 \\
+ \frac{3}{8} & + \frac{3}{8} \quad \text{(Length of the 3 cuts)} \\
\hline
 & = 61\frac{18}{8} \\
 & = 63\frac{1}{4}
\end{array}
$$

Therefore, the shortest length of board must be $63\frac{1}{4}$ in. long.

ANSWERS TO PRACTICE EXERCISES

1. $12\frac{1}{6}$.   2. $15\frac{13}{36}$.   3. $7\frac{13}{21}$.   4. $13\frac{1}{12}$.

To subtract mixed numbers, we subtract the whole number parts and also subtract the fractional parts. The subtraction of mixed numbers will be illustrated in the following examples.

Subtract and write your results in simplest form.

Practice Exercise 5. $4\frac{3}{5} - 1\frac{1}{2}$

EXAMPLE 5 Subtract: $6\frac{3}{4} - 3\frac{1}{6}$.

SOLUTION:

$$6\frac{3}{4} - 3\frac{1}{6} = \left(6 + \frac{3}{4}\right) - \left(3 + \frac{1}{6}\right)$$

$$= (6 - 3) + \left(\frac{3}{4} - \frac{1}{6}\right) \quad \text{(Regrouping)}$$

$$= (6 - 3) + \left(\frac{9}{12} - \frac{2}{12}\right) \quad \text{(Equivalent fractions with LCD = 12)}$$

$$= 3 + \frac{7}{12} \quad \text{(Subtracting whole number parts; subtracting fractional parts)}$$

$$= 3\frac{7}{12} \quad \text{(Rewriting)}$$

Practice Exercise 6. $9\frac{5}{7} - 5\frac{1}{3}$

EXAMPLE 6 Subtract: $5\frac{1}{2} - 2\frac{7}{8}$.

SOLUTION:

$$\begin{array}{c} 5\frac{1}{2} \\ -2\frac{7}{8} \\ \hline \end{array} = \begin{array}{c} 5 + \frac{1}{2} \\ -\left(2 + \frac{7}{8}\right) \\ \hline \end{array} = \begin{array}{c} 5 + \frac{4}{8} \\ -\left(2 + \frac{7}{8}\right) \\ \hline \end{array} \quad \text{(Equivalent fractions with LCD = 8)}$$

But, we *cannot* subtract $\frac{7}{8}$ from $\frac{4}{8}$ since $\frac{7}{8} > \frac{4}{8}$. However, we can rewrite $5 + \frac{4}{8}$ as follows:

$$5 + \frac{4}{8} = (4 + 1) + \frac{4}{8}$$

$$= 4 + \left(1 + \frac{4}{8}\right) \quad \text{(Regrouping)}$$

$$= 4 + \left(\frac{8}{8} + \frac{4}{8}\right) \quad \left(\text{Rewriting } 1 = \frac{8}{8}\right)$$

$$= 4 + \frac{12}{8} \quad \text{(Adding the fractions)}$$

and proceed. Hence,

$$\begin{array}{c} 5 + \frac{4}{8} \\ -\left(2 + \frac{7}{8}\right) \\ \hline \end{array} = \begin{array}{c} 4 + \frac{12}{8} \\ -\left(2 + \frac{7}{8}\right) \\ \hline \end{array}$$

$$= 2 + \frac{5}{8} \quad \text{(Subtracting whole number parts; subtracting fractional parts)}$$

$$= 2\frac{5}{8} \quad \text{(Rewriting)}$$

2.6 Addition and Subtraction of Mixed Numbers

Practice Exercise 7. $8\frac{1}{3} - 3\frac{5}{7}$

EXAMPLE 7 Subtract: $9\frac{2}{7} - 4\frac{3}{4}$.

SOLUTION:

$$\begin{array}{r}9\frac{2}{7}\\-4\frac{3}{4}\\\hline\end{array} = \begin{array}{r}9 + \frac{2}{7}\\-\left(4 + \frac{3}{4}\right)\\\hline\end{array} = \begin{array}{r}9 + \frac{8}{28}\\-\left(4 + \frac{21}{28}\right)\\\hline\end{array}$$

(Equivalent fractions with LCD = 28)

But we *cannot* subtract $\frac{21}{28}$ from $\frac{8}{28}$ since $\frac{21}{28} > \frac{8}{28}$. However, we can rewrite $9 + \frac{8}{28}$ as follows:

$$9 + \frac{8}{28} = (8 + 1) + \frac{8}{28}$$
$$= 8 + \left(1 + \frac{8}{28}\right) \quad \text{(Regrouping)}$$
$$= 8 + \left(\frac{28}{28} + \frac{8}{28}\right) \quad \left(\text{Rewriting } 1 = \frac{28}{28}\right)$$
$$= 8 + \frac{36}{28} \quad \text{(Adding fractions)}$$

and proceed. Hence,

$$\begin{array}{r}9 + \frac{8}{28}\\-\left(4 + \frac{21}{28}\right)\\\hline\end{array} = \begin{array}{r}8 + \frac{36}{28}\\-\left(4 + \frac{21}{28}\right)\\\hline\\ 4 + \frac{15}{28}\end{array}$$

(Subtracting whole number parts; subtracting fractions)

$$= 4\frac{15}{28} \quad \text{(Rewriting)}$$

Practice Exercise 8. $10\frac{1}{2} - 7\frac{5}{6}$

EXAMPLE 8 Subtract: $14 - 9\frac{3}{8}$.

SOLUTION:

$$\begin{array}{r}14\\-9\frac{3}{8}\\\hline\end{array} = \begin{array}{r}13 + \frac{8}{8}\\-\left(9 + \frac{3}{8}\right)\\\hline\\ 4 + \frac{5}{8}\end{array}$$

(Rewriting)

(Subtracting whole number parts; subtracting fractional parts)

$$= 4\frac{5}{8} \quad \text{(Rewriting)}$$

ANSWERS TO PRACTICE EXERCISES

5. $3\frac{1}{10}$. 6. $4\frac{8}{21}$. 7. $4\frac{13}{21}$. 8. $2\frac{2}{3}$.

EXERCISES 2.6

In Exercises 1–30, add the mixed numbers and write your results in simplest form.

1. $3\frac{1}{4} + 2\frac{1}{4}$
2. $5\frac{1}{7} + 4\frac{3}{7}$
3. $6\frac{2}{3} + 8\frac{2}{3}$
4. $9\frac{1}{2} + 2\frac{3}{5}$
5. $7\frac{5}{6} + 4\frac{2}{5}$
6. $4\frac{3}{11} + 4\frac{3}{4}$
7. $7\frac{1}{9} + 6\frac{2}{7}$
8. $5\frac{1}{13} + 3\frac{4}{7}$
9. $6\frac{2}{3} + 7\frac{11}{13}$
10. $8\frac{1}{9} + 6\frac{1}{8}$
11. $11\frac{1}{9} + 13\frac{2}{13}$
12. $40\frac{5}{7} + 51\frac{6}{11}$
13. $68\frac{1}{4} + 73\frac{3}{17}$
14. $87\frac{4}{11} + 62\frac{5}{13}$
15. $16\frac{3}{14} + 23\frac{8}{21}$
16. $23\frac{7}{9} + 11\frac{2}{13}$
17. $37\frac{4}{17} + 20\frac{3}{13}$
18. $19\frac{6}{7} + 23$
19. $\frac{7}{8} + 19\frac{2}{15}$
20. $46\frac{2}{11} + \frac{5}{16}$
21. $2\frac{1}{2} + 3\frac{1}{3} + 4\frac{1}{4}$
22. $5\frac{1}{2} + 6\frac{2}{7} + 8\frac{1}{3}$
23. $7\frac{1}{9} + 6\frac{2}{5} + 4\frac{1}{4}$
24. $8\frac{1}{3} + 9\frac{1}{6} + 11\frac{2}{9}$
25. $5\frac{11}{12} + 9\frac{8}{15} + 11\frac{17}{20}$
26. $6\frac{7}{10} + 9\frac{13}{20} + 11\frac{17}{30}$
27. $17\frac{1}{2} + \frac{9}{11} + 9\frac{3}{10}$
28. $23 + 19\frac{7}{13} + \frac{6}{17}$
29. $3\frac{1}{2} + 4\frac{1}{3} + 5\frac{1}{4} + 3\frac{1}{6}$
30. $2\frac{1}{10} + 3\frac{1}{5} + 4\frac{1}{2} + 5\frac{1}{7}$

In Exercises 31–60, subtract the mixed numbers and write your results in simplest form.

31. $4\frac{7}{8} - 3\frac{1}{2}$
32. $9\frac{6}{7} - 5\frac{1}{3}$
33. $13\frac{1}{3} - 8\frac{4}{7}$
34. $23\frac{1}{4} - 14\frac{5}{9}$
35. $13\frac{2}{3} - 7\frac{3}{5}$
36. $19\frac{5}{8} - 7\frac{3}{14}$
37. $21\frac{19}{20} - 12\frac{7}{12}$
38. $32\frac{11}{16} - 23\frac{1}{12}$
39. $27\frac{5}{8} - 9\frac{5}{12}$
40. $15\frac{11}{13} - 8\frac{2}{7}$
41. $10\frac{11}{15} - 6\frac{3}{7}$
42. $19\frac{9}{11} - 7\frac{2}{13}$
43. $16\frac{5}{6} - 10\frac{2}{5}$
44. $24\frac{3}{11} - 15\frac{3}{4}$
45. $41\frac{2}{3} - 17\frac{11}{13}$
46. $20\frac{1}{9} - 8\frac{2}{13}$
47. $37\frac{1}{8} - 28\frac{1}{9}$
48. $49\frac{1}{4} - 29\frac{5}{17}$
49. $63\frac{9}{16} - 47\frac{2}{5}$
50. $80\frac{3}{8} - 62\frac{5}{7}$
51. $14\frac{11}{13} - 8$
52. $23\frac{9}{17} - 15$
53. $39\frac{1}{6} - \frac{5}{7}$
54. $53\frac{2}{3} - \frac{11}{13}$
55. $19 - 13\frac{2}{7}$
56. $57 - 39\frac{7}{11}$
57. $39\frac{11}{12} - 20\frac{13}{24}$
58. $56\frac{5}{6} - 17\frac{11}{25}$
59. $72\frac{5}{36} - 57\frac{11}{24}$
60. $80\frac{1}{7} - 40\frac{8}{21}$

2.6 Addition and Subtraction of Mixed Numbers

In Exercises 61–70, perform the indicated operations and write your results in simplest form.

61. $\left(2\frac{1}{2} + 3\frac{1}{3}\right) - 1\frac{7}{8}$

62. $\left(5\frac{1}{4} - 3\frac{2}{3}\right) + 4\frac{1}{7}$

63. $5\frac{1}{6} + \left(8\frac{1}{2} - 5\frac{2}{7}\right)$

64. $9\frac{1}{5} - \left(2\frac{1}{3} + 3\frac{1}{2}\right)$

65. $8\frac{1}{5} - \left(6\frac{2}{9} - 4\frac{1}{8}\right)$

66. $7\frac{2}{9} - \left(1\frac{7}{8} + 3\frac{1}{4}\right)$

67. $\left(10\frac{1}{3} - 3\frac{4}{7}\right) - 4\frac{2}{5}$

68. $\left(11\frac{1}{6} - 4\frac{2}{9}\right) - 3\frac{1}{7}$

69. $\left(4\frac{1}{6} + 5\frac{1}{5}\right) - \left(2\frac{3}{7} + 3\frac{1}{2}\right)$

70. $\left(8\frac{2}{3} - 5\frac{1}{6}\right) + \left(10\frac{2}{7} - 8\frac{5}{9}\right)$

71. Sue works part-time at Burger Queen. On Monday, she worked $3\frac{1}{2}$ hr, on Tuesday, $4\frac{1}{4}$ hr, on Wednesday, $2\frac{3}{4}$ hr, on Thursday, $6\frac{1}{5}$ hr, and on Saturday, $7\frac{4}{7}$ hr. Friday and Sunday were her days off. How many hours did Sue work that week?

72. A plane flies from City A to City B in 2 hr 12 min. It then flies from City B to City C in 1 hr 32 min. Later, it flies from City C to City D in 3 hr 18 min. What is its total flight time from City A to City D, in hours? (*Hint:* 1 hr = 60 min.)

73. Mrs. Parker bought three roasts weighing $3\frac{7}{8}$ lb, $4\frac{1}{2}$ lb, and $5\frac{2}{3}$ lb. How many pounds of roast did she buy?

74. A 12-ft-wide roll of carpet is 60 yd long. A customer wishes to buy three lengths of the carpet measuring $15\frac{1}{4}$ yd, $13\frac{7}{8}$ yd, and $16\frac{1}{5}$ yd. How many yards of the carpet does the customer wish to buy?

75. A contractor subdivided a plot of land into four building lots measuring $2\frac{2}{5}$ acres, $3\frac{1}{4}$ acres, $1\frac{7}{8}$ acres, and $3\frac{2}{3}$ acres. What is the total number of acres for the four lots?

76. Determine the total length around the triangle in the accompanying diagram. All dimensions are in centimeters.

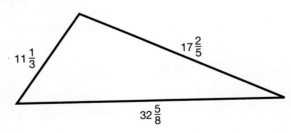

77. The accompanying diagram represents the dimensions of a room. What is the total length around the room if all dimensions are given in meters?

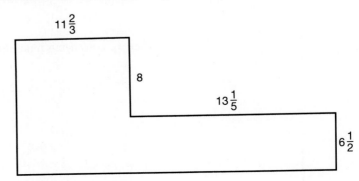

78. What number must be added to $39\frac{1}{7}$ to get a sum of $56\frac{1}{3}$?

79. A side of beef weighed $117\frac{1}{8}$ lb. A butcher sold $59\frac{3}{4}$ lb of the beef. How many pounds were left?

80. If the sum of $9\frac{1}{5}$ and $8\frac{2}{3}$ is subtracted from $29\frac{1}{4}$, what is the result?

2.7 Multiplication of Fractions and Mixed Numbers

OBJECTIVES

After completing this section, you should be able to:
1. Multiply fractions and mixed numbers and express the result in simplest form.
2. Write the reciprocal of a nonzero fraction.

Now that you have learned to add and subtract fractions, we will consider multiplying them.

Since the fractions $\frac{2}{1}, \frac{4}{1}$, and $\frac{8}{1}$ correspond, respectively, to the whole numbers 2, 4, and 8, we can define the product of the fractions $\frac{2}{1}$ and $\frac{4}{1}$ to be the fraction $\frac{8}{1}$ since $2 \times 4 = 8$. In a similar manner, since $6 \times 5 = 30$, we can define the product of the fractions $\frac{6}{1}$ and $\frac{5}{1}$ to be the fraction $\frac{30}{1}$.

What happens when we try to multiply a fraction with a denominator of 1 with any other fraction? We will give a rule for doing this multiplication. This rule will be motivated by the following three examples.

Practice Exercise 1. Multiply: $\frac{4}{1} \times \frac{5}{6}$

EXAMPLE 1 Multiply: $\frac{2}{1} \times \frac{3}{5}$.

SOLUTION: To multiply the fractions $\frac{2}{1}$ and $\frac{3}{5}$, recall that the fraction $\frac{2}{1}$ is another name for the whole number 2. Then,

$$\frac{2}{1} \times \frac{3}{5} = 2 \times \frac{3}{5}$$

But earlier we considered multiplication to be repeated addition. Hence,

$$2 \times \frac{3}{5} = \frac{3}{5} + \frac{3}{5}$$

and

$$\frac{2}{1} \times \frac{3}{5} = 2 \times \frac{3}{5}$$

$$= \frac{3}{5} + \frac{3}{5}$$

2.7 Multiplication of Fractions and Mixed Numbers

$$= \frac{3+3}{5}$$

$$= \frac{6}{5}$$

Therefore, $\frac{2}{1} \times \frac{3}{5} = \frac{6}{5}$.

Practice Exercise 2. Multiply: $\frac{3}{1} \times \frac{7}{9}$

EXAMPLE 2 Multiply $\frac{3}{1}$ and $\frac{4}{7}$.

SOLUTION:
$$\frac{3}{1} \times \frac{4}{7} = 3 \times \frac{4}{7}$$
$$= \frac{4}{7} + \frac{4}{7} + \frac{4}{7}$$
$$= \frac{4+4+4}{7}$$
$$= \frac{12}{7}$$

Therefore, $\frac{3}{1} \times \frac{4}{7} = \frac{12}{7}$.

Practice Exercise 3. Multiply: $\frac{5}{1} \times \frac{4}{7}$

EXAMPLE 3 Multiply $\frac{4}{1}$ and $\frac{6}{11}$.

SOLUTION:
$$\frac{4}{1} \times \frac{6}{11} = 4 \times \frac{6}{11}$$
$$= \frac{6}{11} + \frac{6}{11} + \frac{6}{11} + \frac{6}{11}$$
$$= \frac{6+6+6+6}{11}$$
$$= \frac{24}{11}$$

Therefore, $\frac{4}{1} \times \frac{6}{11} = \frac{24}{11}$.

ANSWERS TO PRACTICE EXERCISES

1. $\frac{10}{3}$. 2. $\frac{7}{3}$. 3. $\frac{20}{7}$.

The three preceding examples show how to take the product of any fraction and a fraction with a denominator of 1. The product is a new fraction whose numerator is the product of the numerators of the two fractions. The denominator of the new fraction is the product of the denominators of the two fractions.

To Find the Product of a Fraction with a Denominator of 1 and Any Other Fraction:

1. Multiply the numerators to get the numerator of the product.

2. Multiply the denominators to get the denominator of the product.

Multiply and write your results in simplest form.

Practice Exercise 4. $\dfrac{4}{1} \times \dfrac{2}{3}$

EXAMPLE 4 Multiply: $\dfrac{3}{1} \times \dfrac{2}{5}$.

SOLUTION: $\dfrac{3}{1} \times \dfrac{2}{5} = \dfrac{3 \times 2}{1 \times 5} = \dfrac{6}{5}$

Practice Exercise 5. $\dfrac{6}{1} \times \dfrac{1}{7}$

EXAMPLE 5 Multiply: $5 \times \dfrac{6}{7}$.

SOLUTION: $5 \times \dfrac{6}{7} = \dfrac{5}{1} \times \dfrac{6}{7} = \dfrac{5 \times 6}{1 \times 7} = \dfrac{30}{7}$

Practice Exercise 6. $5 \times \dfrac{3}{4}$

EXAMPLE 6 Multiply: $\dfrac{6}{1} \times \dfrac{11}{13}$.

SOLUTION: $\dfrac{6}{1} \times \dfrac{11}{13} = \dfrac{6 \times 11}{1 \times 13} = \dfrac{66}{13}$

ANSWERS TO PRACTICE EXERCISES

4. $\dfrac{8}{3}$. 5. $\dfrac{6}{7}$. 6. $\dfrac{15}{4}$ or $3\dfrac{3}{4}$.

Now, what happens if we try to multiply the fractions $\dfrac{3}{4}$ and $\dfrac{2}{3}$? Would we get $\dfrac{3 \times 2}{4 \times 3} = \dfrac{6}{12} = \dfrac{1}{2}$? To answer the question, consider the following. The fraction $\dfrac{3}{4}$ means "3 of 4 equal parts," and the fraction $\dfrac{2}{3}$ means "2 of 3 equal parts." Consider the rectangle, or unit, in Figure 2.15.

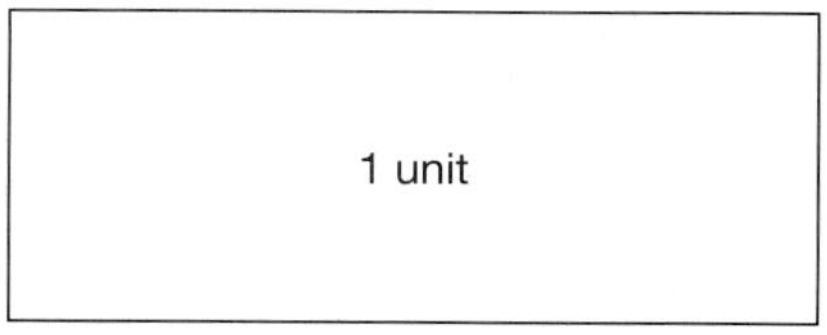

1 unit

FIGURE 2.15

Divide the rectangle *vertically* into 4 equal parts and *horizontally* into 3 equal parts, as illustrated in Figure 2.16.

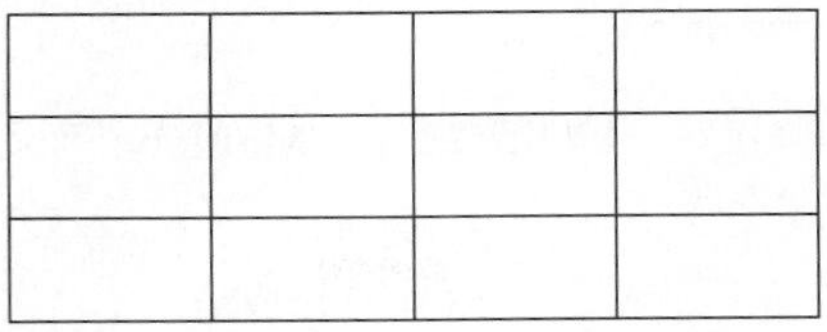

2.7 Multiplication of Fractions and Mixed Numbers

FIGURE 2.16

Next, shade the first 3 of the *vertical* parts as indicated in Figure 2.17. This shaded portion represents the fraction $\frac{3}{4}$.

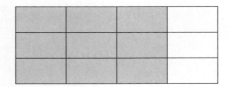

FIGURE 2.17

Now, using another color, shade the bottom 2 of the 3 *horizontal* parts of the rectangle, as in Figure 2.18. This shaded portion represents the fraction $\frac{2}{3}$.

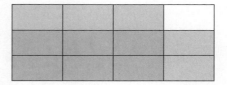

FIGURE 2.18

Finally, observe that there are 12 equal parts comprising the rectangle, or unit, and that 6 of these equal parts have been shaded twice. Hence, we have 6 of 12 equal parts of the unit, or $\frac{6}{12}$.

Observe that if we multiply the numerators of the fractions $\frac{3}{4}$ and $\frac{2}{3}$, we get 6, and the product of their denominators is 12. This suggests that the product of the fractions $\frac{3}{4}$ and $\frac{2}{3}$ is the fraction $\frac{6}{12}$, or $\frac{1}{2}$ when simplified.

We will now state the general rule for the **multiplication of fractions**.

Rule for Multiplication of Any Two Fractions

To multiply any two fractions:

1. Multiply the numerators of the two fractions to get the numerator of the product.
2. Multiply the denominators of the two fractions to get the denominator of the product.
3. Simplify the resulting fraction, if possible.

Note that, in multiplication of fractions, it is *not* necessary to get a common denominator, as is the case in addition and subtraction of fractions.

Chapter 2 Fractions

Multiply and write your results in simplest form.

Practice Exercise 7. $\frac{2}{3} \times \frac{1}{5}$

EXAMPLE 7 Multiply: $\frac{2}{3} \times \frac{5}{7}$.

SOLUTION:
$$\frac{2}{3} \times \frac{5}{7} = \frac{2 \times 5}{3 \times 7}$$
$$= \frac{10}{21}$$

Practice Exercise 8. $\frac{3}{8} \times \frac{2}{5}$

EXAMPLE 8 Multiply: $\frac{3}{4} \times \frac{6}{11}$.

SOLUTION:
$$\frac{3}{4} \times \frac{6}{11} = \frac{3 \times 6}{4 \times 11}$$
$$= \frac{18}{44}$$
$$= \frac{18 \div 2}{44 \div 2}$$
$$= \frac{9}{22}$$

Practice Exercise 9. $\frac{4}{7} \times \frac{2}{5}$

EXAMPLE 9 Multiply: $\frac{6}{11} \times \frac{0}{3}$.

SOLUTION:
$$\frac{6}{11} \times \frac{0}{3} = \frac{6 \times 0}{11 \times 3}$$
$$= \frac{0}{33}$$
$$= 0$$

Practice Exercise 10. $\frac{6}{11} \times \frac{11}{13}$

EXAMPLE 10 Multiply: $\frac{5}{7} \times \frac{2}{11}$.

SOLUTION:
$$\frac{5}{7} \times \frac{2}{11} = \frac{5 \times 2}{7 \times 11}$$
$$= \frac{10}{77}$$

When multiplying fractions, it is sometimes possible to simplify the work by first dividing the numerator and the denominator of the product by any common factors.

Practice Exercise 11. $\frac{22}{35} \times \frac{20}{66}$

EXAMPLE 11 Multiply: $\frac{3}{8} \times \frac{4}{9}$.

SOLUTION:
$$\frac{3}{8} \times \frac{4}{9} = \frac{3 \times 4}{8 \times 9}$$
$$= \frac{\overset{1}{\cancel{3}} \times 4}{8 \times \underset{3}{\cancel{9}}} \quad \text{(Dividing by 3)}$$

2.7 Multiplication of Fractions and Mixed Numbers

$$= \frac{\overset{1}{\cancel{3}} \times \overset{1}{\cancel{4}}}{\underset{2}{\cancel{8}} \times \underset{3}{\cancel{9}}} \quad \text{(Dividing by 4)}$$

$$= \frac{1 \times 1}{2 \times 3}$$

$$= \frac{1}{6}$$

This procedure could be shortened as follows:

$$\frac{\overset{1}{\cancel{3}} \times \overset{1}{\cancel{4}}}{\underset{2}{\cancel{8}} \times \underset{3}{\cancel{9}}} = \frac{1 \times 1}{2 \times 3}$$

$$= \frac{1}{6}$$

Practice Exercise 12.
$\frac{4}{9} \times \frac{16}{27} \times \frac{6}{35}$

EXAMPLE 12 Multiply: $\frac{5}{7} \times \frac{14}{25} \times \frac{21}{28}$.

SOLUTION: $\frac{5}{7} \times \frac{14}{25} \times \frac{21}{28} = \frac{5 \times 14 \times 21}{7 \times 25 \times 28}$

$$= \frac{\overset{1}{\cancel{5}} \times 14 \times 21}{7 \times \underset{5}{\cancel{25}} \times 28} \quad \text{(Dividing by 5)}$$

$$= \frac{\overset{1}{\cancel{5}} \times \overset{1}{\cancel{14}} \times 21}{7 \times \underset{5}{\cancel{25}} \times \underset{2}{\cancel{28}}} \quad \text{(Dividing by 14)}$$

$$= \frac{\overset{1}{\cancel{5}} \times \overset{1}{\cancel{14}} \times \overset{3}{\cancel{21}}}{\underset{1}{\cancel{7}} \times \underset{5}{\cancel{25}} \times \underset{2}{\cancel{28}}} \quad \text{(Dividing by 7)}$$

$$= \frac{1 \times 1 \times 3}{1 \times 5 \times 2}$$

$$= \frac{3}{10}$$

or

$$\frac{\overset{1}{\cancel{5}} \times \overset{1}{\cancel{14}} \times \overset{3}{\cancel{21}}}{\underset{1}{\cancel{7}} \times \underset{5}{\cancel{25}} \times \underset{2}{\cancel{28}}} = \frac{1 \times 1 \times 3}{1 \times 5 \times 2}$$

$$= \frac{3}{10}$$

ANSWERS TO PRACTICE EXERCISES

7. $\frac{2}{15}$. 8. $\frac{3}{20}$. 9. $\frac{8}{35}$. 10. $\frac{6}{13}$. 11. $\frac{4}{21}$. 12. $\frac{128}{2835}$.

Multiplication of Mixed Numbers

Multiplication of mixed numbers can be accomplished by rewriting the mixed numbers as improper fractions.

Multiply and write your answers in simplest form.

Practice Exercise 13. $4\frac{1}{3} \times 5\frac{1}{2}$

EXAMPLE 13 Multiply: $3\frac{1}{2} \times 4\frac{2}{3}$.

SOLUTION:
$$3\frac{1}{2} \times 4\frac{2}{3} = \frac{7}{2} \times \frac{14}{3}$$
$$= \frac{7 \times \overset{7}{\cancel{14}}}{\underset{1}{\cancel{2}} \times 3}$$
$$= \frac{7 \times 7}{1 \times 3}$$
$$= \frac{49}{3} \text{ or } 16\frac{1}{3}$$

Practice Exercise 14. $6\frac{2}{3} \times 7\frac{1}{8}$

EXAMPLE 14 Multiply: $5\frac{4}{7} \times 3\frac{2}{5}$.

SOLUTION:
$$5\frac{4}{7} \times 3\frac{2}{5} = \frac{39}{7} \times \frac{17}{5}$$
$$= \frac{39 \times 17}{7 \times 5}$$
$$= \frac{663}{35} \text{ or } 18\frac{33}{35}$$

Practice Exercise 15. $8\frac{1}{2} \times 9\frac{1}{5}$

EXAMPLE 15 Multiply: $2\frac{1}{3} \times 7\frac{1}{3}$.

SOLUTION:
$$2\frac{1}{3} \times 7\frac{1}{3} = \frac{7}{3} \times \frac{22}{3}$$
$$= \frac{7 \times 22}{3 \times 3}$$
$$= \frac{154}{9} \text{ or } 17\frac{1}{9}$$

ANSWERS TO PRACTICE EXERCISES

13. $23\frac{5}{6}$.   14. $47\frac{1}{2}$.   15. $78\frac{1}{5}$.

Practice Exercise 16. Joyce spends $\frac{5}{16}$ of her monthly income for house payments. If her monthly income is $3200, how much is her house payment?

EXAMPLE 16 Dianne works on a piece-work basis. One day, she completed 375 units of work but $\frac{1}{15}$ of the units did not pass inspection. How many units did not pass inspection?

SOLUTION: To determine how many units did not pass inspection, we note that $\frac{1}{15}$ of 375 means $\frac{1}{15}$ times 375. We have

2.7 Multiplication of Fractions and Mixed Numbers

$$\frac{1}{15} \times 375 = \frac{1}{\cancel{15}_1} \times \frac{\cancel{375}^{25}}{1}$$

$$= \frac{1 \times 25}{1 \times 1}$$

$$= \frac{25}{1}$$

$$= 25$$

Hence, 25 units did not pass inspection.

ANSWER TO PRACTICE EXERCISE

16. $1000.

Sometimes, the product of two fractions is 1. When this happens, there is a special relationship between the numerator and denominator of one fraction and the numerator and denominator of the other fraction. This will be illustrated below.

- $\frac{2}{3} \times \frac{3}{2} = \frac{2 \times 3}{3 \times 2} = \frac{6}{6} = \frac{1}{1} = 1$

- $\frac{4}{5} \times \frac{5}{4} = \frac{4 \times 5}{5 \times 4} = \frac{20}{20} = \frac{1}{1} = 1$

- $\frac{3}{7} \times \frac{7}{3} = \frac{3 \times 7}{7 \times 3} = \frac{21}{21} = \frac{1}{1} = 1$

- $\frac{7}{11} \times \frac{11}{7} = \frac{7 \times 11}{11 \times 7} = \frac{77}{77} = \frac{1}{1} = 1$

Each example above is a product of two fractions such that their product is the fraction $\frac{1}{1}$. Observe that, in each case, the numerator and the denominator of the first fraction are, respectively, the denominator and the numerator of the second fraction. Each fraction in the multiplication is called the "reciprocal" of the other. Reciprocals of fractions are used in the division of fractions. This will be discussed in the next section.

DEFINITION

Let $\frac{a}{b}$ be any *nonzero* fraction. Then the fraction $\frac{b}{a}$ is called the **reciprocal** of the fraction $\frac{a}{b}$.

Finding a Reciprocal

1. Every *nonzero* fraction has a reciprocal.
2. The *reciprocal* of a nonzero fraction is formed by interchanging the numerator and the denominator of the original fraction.

We should note that:

- The product of a *nonzero* fraction and its reciprocal is always the fraction $\frac{1}{1}$ (or simply, 1).
- The fraction $\frac{0}{1}$ does *not* have a reciprocal.

Write the reciprocal for each of the following.

Practice Exercise 17. (a) $\frac{3}{7}$; (b) $\frac{6}{11}$; (c) $2\frac{1}{6}$ (d) 3

EXAMPLE 17 Determine the reciprocal of each of the following fractions: (a) $\frac{2}{3}$; (b) $\frac{11}{4}$; (c) $\frac{5}{1}$; (d) $5\frac{1}{3}$.

SOLUTION: (a) The reciprocal of $\frac{2}{3}$ is $\frac{3}{2}$; (b) $\frac{4}{11}$ is the reciprocal of $\frac{11}{4}$; (c) $\frac{1}{5}$ is the reciprocal of $\frac{5}{1}$; (d) the reciprocal of $5\frac{1}{3} = \frac{16}{3}$ is $\frac{3}{16}$.

ANSWERS TO PRACTICE EXERCISES

17. (a) $\frac{7}{3}$; (b) $\frac{11}{6}$; (c) $\frac{6}{13}$; (d) $\frac{1}{3}$.

EXERCISES 2.7

Do the indicated operations in Exercises 1–20 and express your results in simplest form.

1. $3 \times \frac{1}{2}$
2. $5 \times \frac{2}{3}$
3. $9 \times \frac{3}{7}$
4. $6 \times \frac{2}{5}$
5. $\frac{1}{2} \times \frac{1}{6}$
6. $\frac{2}{3} \times \frac{4}{5}$
7. $\frac{2}{7} \times \frac{4}{3}$
8. $\frac{1}{5} \times \frac{2}{7}$
9. $\frac{4}{13} \times \frac{5}{9}$
10. $\frac{3}{7} \times \frac{2}{11}$
11. $\frac{3}{5} \times \frac{4}{9}$
12. $\frac{0}{6} \times \frac{11}{17}$
13. $\left(\frac{1}{2} \times \frac{1}{3}\right) \times \frac{1}{3}$
14. $\frac{2}{3} \times \left(\frac{1}{7} \times \frac{6}{5}\right)$
15. $\frac{2}{5} \times \left(\frac{1}{2} + \frac{3}{7}\right)$
16. $\frac{6}{7} \times \left(\frac{2}{9} + \frac{3}{4}\right)$
17. $\frac{3}{7} \times \left(\frac{4}{5} + \frac{3}{4}\right)$
18. $\left(\frac{2}{7} + \frac{0}{5}\right) \times \left(\frac{0}{7} + \frac{6}{5}\right)$
19. $\left(\frac{2}{9} - \frac{1}{7}\right) \times \left(\frac{1}{3} + \frac{2}{5}\right)$
20. $\left(\frac{2}{5} + \frac{3}{4}\right) \times \left(\frac{6}{7} + \frac{5}{4}\right)$

In Exercises 21–40, perform the indicated operations by first rewriting the mixed numbers as fractions.

21. $3\frac{1}{2} \times 4\frac{2}{3}$
22. $3\frac{1}{7} \times 5\frac{1}{9}$
23. $7\frac{1}{6} \times 2\frac{4}{5}$
24. $3\frac{1}{3} \times 2\frac{1}{4}$
25. $8\frac{1}{6} \times 7\frac{1}{4}$
26. $11\frac{1}{2} \times 13\frac{1}{3}$
27. $9\frac{5}{8} \times 8\frac{3}{4}$
28. $10\frac{11}{12} \times 17\frac{2}{3}$
29. $15\frac{1}{4} \times 17\frac{2}{5}$

2.7 Multiplication of Fractions and Mixed Numbers

30. $1\frac{3}{4} \times \left(4\frac{1}{2} + 3\frac{2}{3}\right)$ 31. $7\frac{1}{3} + \left(2\frac{3}{4} \times 5\frac{1}{6}\right)$ 32. $4\frac{1}{9} \times \left(2\frac{3}{4} \times 5\frac{1}{6}\right)$

33. $5\frac{1}{6} \times \left(2\frac{3}{7} + 4\frac{1}{6}\right)$ 34. $\left(5\frac{1}{7} \times 2\frac{3}{8}\right) - 6\frac{1}{5}$

35. $11\frac{1}{2} - \left(2\frac{3}{4} \times 1\frac{7}{9}\right)$ 36. $\left(4\frac{2}{9} + 3\frac{1}{4}\right) \times \left(3\frac{2}{3} + 5\frac{1}{2}\right)$

37. $\left(5\frac{2}{3} - 1\frac{1}{2}\right) \times \left(7\frac{3}{4} - 2\frac{3}{5}\right)$ 38. $\left(2\frac{9}{11} + 5\frac{7}{9}\right) \times \left(10\frac{1}{6} - 7\frac{5}{9}\right)$

39. $\left(6\frac{1}{3} \times 2\frac{1}{5}\right) - \left(3\frac{1}{4} \times 1\frac{5}{6}\right)$ 40. $\left(2\frac{1}{2} + 3\frac{2}{3} + 4\frac{1}{4}\right) \times \left(5\frac{2}{11} - 3\frac{7}{10}\right)$

Write the reciprocal for each of the fractions in Exercises 41–60. If the reciprocal does not exist, indicate this fact.

41. $\frac{2}{3}$ 42. $\frac{1}{2}$ 43. $\frac{4}{9}$ 44. $\frac{5}{6}$ 45. $\frac{9}{7}$

46. $\frac{0}{6}$ 47. $\frac{23}{11}$ 48. $\frac{17}{6}$ 49. $\frac{101}{17}$ 50. $\frac{35}{7}$

51. $\frac{3}{5}$ 52. $\frac{0}{7}$ 53. $\frac{3-2}{7}$ 54. $\frac{6}{9-7}$ 55. $\frac{5-5}{4}$

56. $\frac{2 \times 3}{7}$ 57. $\frac{6}{11+13}$ 58. $\frac{2 \times 3}{6 \times 7}$ 59. $\frac{7 \times 6}{6 \times 7}$ 60. $\frac{9-8}{3 \times 2}$

61. In a class of 36, $\frac{1}{9}$ of the students failed an English test. How many of the students failed the test?

62. The sales tax in a certain community is $4\frac{1}{2}$¢ on every dollar. What would be the sales tax on a purchase amounting to $192?

63. In a group of 120 children, $\frac{7}{60}$ were found to have cavities when checked by their dentists. How many children had cavities?

64. The circumference of a circle (the distance around the circle) is 121 cm. What is the length of $\frac{5}{11}$ of the circumference of the circle?

65. The Hazardous Duty Corporation employs 868 people. On a particular day, $\frac{1}{31}$ of the employees were absent due to injury and $\frac{1}{14}$ were absent due to illness. How many employees were absent due to injury or illness?

66. A merchant had an inventory of $15,500 in perishable stock. Due to a power blackout, $\frac{23}{25}$ of the stock was destroyed. What was the value of the stock destroyed?

67. At University College, $47\frac{1}{5}$¢ of every dollar budgeted is for the Dean of Academic Affairs' budget. If the total budget is $10,692,120, how much is in the Dean's budget?

68. Bill Wilson spends $\frac{1}{4}$ of his income for his house payment and $\frac{1}{3}$ of his income for food for himself and family. If his income is $36,000, how much is left for other purposes?

2.8 Division of Fractions and Mixed Numbers

OBJECTIVE

After completing this section, you should be able to divide fractions and mixed numbers.

Suppose that you were asked the question, What is 15 divided by $\frac{1}{3}$? Would your answer be 5? If so, you are probably thinking of the answer to the question, What is 15 divided by 3? The two questions are not the same. The answer to the first question is 45, as we will now illustrate.

In Practice Exercises 1–4, divide and write your results in simplest form.

Practice Exercise 1. $16 \div \frac{1}{4}$

EXAMPLE 1 Divide 15 by $\frac{1}{3}$.

SOLUTION:
$$15 \div \frac{1}{3} = (15 \times 1) \div \frac{1}{3}$$
$$= \left(15 \times \frac{3}{3}\right) \div \frac{1}{3} \quad \left(\text{Rewriting } 1 = \frac{3}{3}\right)$$
$$= \left(\frac{15 \times 3}{3}\right) \div \frac{1}{3}$$
$$= \frac{45}{3} \div \frac{1}{3}$$
$$= 45 \text{ thirds} \div 1 \text{ third}$$
$$= 45$$

Notice that we are dividing like units in Example 1; we are dividing one-thirds by one-thirds.

Practice Exercise 2. $\frac{1}{5} \div 6$

EXAMPLE 2 Divide $\frac{1}{3}$ by 2.

SOLUTION:
$$\frac{1}{3} \div 2 = \frac{1}{3} \div (2 \times 1)$$
$$= \frac{1}{3} \div \left(2 \times \frac{3}{3}\right)$$
$$= \frac{1}{3} \div \left(\frac{2 \times 3}{3}\right)$$
$$= \frac{1}{3} \div \frac{6}{3}$$
$$= 1 \text{ third} \div 6 \text{ thirds}$$
$$= \frac{1}{6}$$

2.8 Division of Fractions and Mixed Numbers

Practice Exercise 3. $\frac{2}{3} \div \frac{1}{2}$

EXAMPLE 3 Divide $\frac{2}{3}$ by $\frac{3}{4}$.

SOLUTION:
$$\frac{2}{3} \div \frac{3}{4} = \frac{2 \times 4}{3 \times 4} \div \frac{3 \times 3}{4 \times 3} \quad \text{(Equivalent fractions)}$$
$$= \frac{8}{12} \div \frac{9}{12}$$
$$= 8 \text{ twelfths} \div 9 \text{ twelfths}$$
$$= 8 \div 9$$
$$= \frac{8}{9}$$

Practice Exercise 4. $\frac{3}{7} \div \frac{1}{4}$

EXAMPLE 4 Divide $\frac{4}{7}$ by $\frac{2}{5}$.

SOLUTION:
$$\frac{4}{7} \div \frac{2}{5} = \frac{4 \times 5}{7 \times 5} \div \frac{2 \times 7}{5 \times 7} \quad \text{(Equivalent fractions)}$$
$$= \frac{20}{35} \div \frac{14}{35}$$
$$= 20 \text{ thirty-fifths} \div 14 \text{ thirty-fifths}$$
$$= 20 \div 14$$
$$= \frac{10}{7} \quad \text{(Simplified)}$$

ANSWERS TO PRACTICE EXERCISES

1. 64.   2. $\frac{1}{30}$.   3. $\frac{4}{3}$ or $1\frac{1}{3}$.   4. $\frac{12}{7}$ or $1\frac{5}{7}$.

The procedure illustrated in the preceding four examples for dividing a fraction by a nonzero fraction can always be used. However, as the denominators become larger, it becomes more difficult to find a common denominator for the equivalent fractions. There is another procedure that can be used. It involves converting a division problem to an appropriate multiplication problem, using the reciprocal of the divisor.

- $\qquad 15 \div \frac{1}{3} = 45 \quad$ (From Example 1)

But $\qquad 15 \times 3 = 45$

Therefore, $\qquad 15 \div \frac{1}{3} = 15 \times 3$

- $\qquad \frac{1}{3} \div 2 = \frac{1}{6} \quad$ (From Example 2)

But $\qquad \frac{1}{3} \times \frac{1}{2} = \frac{1}{6}$

Therefore, $\qquad \frac{1}{3} \div 2 = \frac{1}{3} \times \frac{1}{2}$

- $\qquad \frac{2}{3} \div \frac{3}{4} = \frac{8}{9} \quad$ (From Example 3)

But $\dfrac{2}{3} \times \dfrac{4}{3} = \dfrac{8}{9}$

Therefore, $\dfrac{2}{3} \div \dfrac{3}{4} = \dfrac{2}{3} \times \dfrac{4}{3}$

■ $\dfrac{4}{7} \div \dfrac{2}{5} = \dfrac{10}{7}$ (From Example 4)

But $\dfrac{4}{7} \times \dfrac{5}{2} = \dfrac{20}{14} = \dfrac{10}{7}$

Therefore, $\dfrac{4}{7} \div \dfrac{2}{5} = \dfrac{4}{7} \times \dfrac{5}{2}$

The preceding examples demonstrate an *operational definition* for **division of fractions** by a nonzero fraction. This is given in the following rule.

Rule for Dividing Fractions

To divide a fraction by a *nonzero* fraction:
1. Change the operation of division to multiplication.
2. Use the reciprocal of the divisor.

Using this rule, for example, to divide $\dfrac{1}{2}$ by $\dfrac{3}{4}$, we have:

$$\dfrac{1}{2} \div \dfrac{3}{4} = \dfrac{1}{2} \times \dfrac{4}{3}$$ (Changing division to multiplication; using the reciprocal of the divisor)

$$= \dfrac{1 \times 4}{2 \times 3}$$

$$= \dfrac{4}{6}$$

$$= \dfrac{2}{3}$$

Divide, using the operational definition. Write your results in simplest form.

Practice Exercise 5. $5 \div 1\dfrac{1}{2}$

EXAMPLE 5 Divide: $\dfrac{2}{3} \div \dfrac{5}{8}$.

SOLUTION:
$$\dfrac{2}{3} \div \dfrac{5}{8} = \dfrac{2}{3} \times \dfrac{8}{5}$$
$$= \dfrac{2 \times 8}{3 \times 5}$$
$$= \dfrac{16}{15} \text{ or } 1\dfrac{1}{15}$$

Practice Exercise 6. $2\dfrac{2}{3} \div 4\dfrac{1}{4}$

EXAMPLE 6 Divide: $\dfrac{4}{7} \div \dfrac{5}{6}$.

SOLUTION:
$$\dfrac{4}{7} \div \dfrac{5}{6} = \dfrac{4}{7} \times \dfrac{6}{5}$$

2.8 Division of Fractions and Mixed Numbers

$$= \frac{4 \times 6}{7 \times 5}$$

$$= \frac{24}{35}$$

Practice Exercise 7. $3\frac{2}{7} \div 6\frac{1}{3}$

EXAMPLE 7 Divide: $\frac{0}{2} \div \frac{3}{4}$.

SOLUTION:

$$\frac{0}{2} \div \frac{3}{4} = \frac{0}{2} \times \frac{4}{3}$$

$$= \frac{0 \times 4}{2 \times 3}$$

$$= \frac{0}{6}$$

$$= 0 \quad \text{(Simplified)}$$

Practice Exercise 8. How many pieces of wire $23\frac{1}{8}$ ft long can be cut from a string of wire 400 ft long?

EXAMPLE 8 How many pieces of string $15\frac{1}{2}$ in. long can be cut from a 200-in. piece of string?

SOLUTION: Since the 200-in. piece of string is to be cut into lengths of $15\frac{1}{2}$ in. each, the operation of *division* is used. We have

$$200 \div 15\frac{1}{2} = 200 \div \frac{31}{2}$$

$$= \frac{200}{1} \times \frac{2}{31}$$

$$= \frac{400}{31}$$

$$= 12\frac{28}{31}$$

Therefore, 12 pieces of string can be cut from a 200-in. piece of string. There will be a piece of string left over that measures

$$\frac{28}{31} \times 15\frac{1}{2} = \frac{28}{31} \times \frac{31}{2}$$

$$= \frac{28 \times 31}{31 \times 2}$$

$$= \frac{\overset{14}{\cancel{28}} \times \overset{1}{\cancel{31}}}{\underset{1}{\cancel{31}} \times \underset{1}{\cancel{2}}}$$

$$= 14 \text{ in.}$$

ANSWERS TO PRACTICE EXERCISES

5. $3\frac{1}{3}$.   6. $\frac{32}{51}$.   7. $\frac{69}{133}$.   8. 17.

EXERCISES 2.8

In Exercises 1–30, divide and write your results in simplest form.

1. $4 \div \dfrac{1}{3}$
2. $7 \div \dfrac{3}{8}$
3. $11 \div \dfrac{4}{9}$
4. $23 \div \dfrac{11}{17}$
5. $\dfrac{3}{5} \div \dfrac{1}{4}$
6. $\dfrac{1}{3} \div \dfrac{1}{4}$
7. $\dfrac{4}{7} \div \dfrac{5}{6}$
8. $\dfrac{5}{6} \div \dfrac{3}{7}$
9. $\dfrac{0}{6} \div \dfrac{7}{8}$
10. $\dfrac{6}{6} \div \dfrac{5}{5}$
11. $\dfrac{2}{3} \div \dfrac{5}{6}$
12. $\dfrac{3}{7} \div \dfrac{1}{2}$
13. $\dfrac{37}{23} \div \dfrac{23}{37}$
14. $\dfrac{47}{94} \div \dfrac{1}{2}$
15. $\dfrac{56}{17} \div \dfrac{2}{17}$
16. $\dfrac{6}{19} \div \dfrac{1}{38}$
17. $\dfrac{11}{23} \div \dfrac{2}{2}$
18. $\dfrac{17}{4} \div \dfrac{34}{19}$
19. $\dfrac{8}{13} \div \dfrac{94}{39}$
20. $\dfrac{17}{14} \div \dfrac{13}{28}$
21. $\dfrac{21}{30} \div \dfrac{63}{5}$
22. $\dfrac{36}{9} \div \dfrac{2}{27}$
23. $\dfrac{21}{25} \div \dfrac{3}{5}$
24. $\dfrac{11}{24} \div \dfrac{22}{45}$
25. $\dfrac{10}{27} \div \dfrac{50}{81}$
26. $\dfrac{9}{10} \div \dfrac{45}{16}$
27. $\dfrac{17}{12} \div \dfrac{25}{24}$
28. $\dfrac{14}{17} \div \dfrac{26}{85}$
29. $\dfrac{121}{135} \div \dfrac{22}{45}$
30. $\dfrac{5}{65} \div \dfrac{10}{21}$

In Exercises 31–50, divide and write your results in simplest form. Write your answers as mixed numbers whenever possible.

31. $4\dfrac{1}{2} \div \dfrac{2}{3}$
32. $\dfrac{6}{7} \div 3\dfrac{1}{9}$
33. $2\dfrac{1}{3} \div 4\dfrac{1}{4}$
34. $5\dfrac{1}{2} \div 6\dfrac{1}{4}$
35. $2\dfrac{1}{17} \div 3\dfrac{1}{2}$
36. $2\dfrac{5}{11} \div \dfrac{5}{13}$
37. $20\dfrac{1}{4} \div 4\dfrac{1}{2}$
38. $2\dfrac{3}{13} \div \dfrac{13}{19}$
39. $8\dfrac{1}{9} \div 7\dfrac{1}{6}$
40. $10\dfrac{1}{7} \div 6\dfrac{2}{3}$
41. $8\dfrac{1}{13} \div 3\dfrac{7}{20}$
42. $11\dfrac{1}{25} \div 3\dfrac{3}{5}$
43. $0 \div 17\dfrac{13}{19}$
44. $10 \div 3\dfrac{1}{10}$
45. $\dfrac{7}{8} \div 5\dfrac{7}{16}$
46. $13\dfrac{1}{10} \div \dfrac{4}{9}$
47. $23 \div 10\dfrac{7}{11}$
48. $11\dfrac{9}{11} \div 17$
49. $0 \div 59\dfrac{13}{31}$
50. $7\dfrac{4}{35} \div 9\dfrac{5}{14}$

In Exercises 51-60, do the indicated operations.

51. $2\dfrac{3}{4} \div \left(3\dfrac{1}{2} + 2\dfrac{3}{5}\right)$
52. $5\dfrac{1}{6} \div \left(6\dfrac{2}{7} - 2\dfrac{4}{9}\right)$
53. $\left(4\dfrac{2}{7} \div 3\dfrac{1}{2}\right) + \dfrac{11}{12}$
54. $\left(4\dfrac{6}{11} \div 5\dfrac{1}{2}\right) - \dfrac{3}{10}$
55. $\left(3\dfrac{2}{7} \div 4\dfrac{1}{9}\right) \div 2\dfrac{5}{7}$
56. $7\dfrac{1}{9} \div \left(\dfrac{6}{7} \div 7\dfrac{1}{6}\right)$
57. $\left(\dfrac{2}{3} \times 1\dfrac{4}{7}\right) \div \left(6\dfrac{1}{11} \times 5\dfrac{1}{3}\right)$
58. $\left(3\dfrac{11}{23} - 1\dfrac{4}{11}\right) \div \left(\dfrac{1}{10} + 5\dfrac{3}{7}\right)$
59. $\left(\dfrac{6}{11} \div 3\dfrac{2}{7}\right) \times \left(7\dfrac{2}{9} \div \dfrac{3}{7}\right)$
60. $\left(\dfrac{7}{9} \div 1\dfrac{1}{3}\right) \div \left(5\dfrac{2}{9} \div 2\dfrac{3}{7}\right)$

61. Pieces of ribbon measuring $\dfrac{1}{6}$ yd are to be cut from a ribbon measuring $10\dfrac{2}{3}$ yd long. How many such pieces can be cut?

62. A salesperson travels 30 mi in $\frac{7}{12}$ hr. What is her average speed in miles per hour? (*Hint:* To find the average speed, divide the distance traveled by the time, in hours, that it took to travel that distance.)
63. Another salesperson traveled $213\frac{1}{2}$ mi in $4\frac{1}{3}$ hr. What was his average speed in miles per hour? (See Exercise 62.)
64. An empty 50-gal tank is to be filled with water. If a $2\frac{1}{4}$ gal bucket is to be used, how many times will the bucket have to be filled before the tank is filled?
65. Which is the better buy, an 11-oz bottle of dish detergent that sells for 79¢ or a 13-oz bottle of the same detergent that sells for 95¢?
66. A senior high school class is selling peanuts to raise money for a class trip. If they have 9 sacks containing a total of $612\frac{1}{2}$ lb of peanuts and they put the peanuts into bags holding $\frac{5}{16}$ lb each, how many bags can they fill from the 9 sacks?
67. In Exercise 66, if the peanuts were put into bags holding $\frac{3}{4}$ lb each, how many bags can they fill from the 9 sacks?
68. In Exercise 66, if there were 7 sacks containing a total of $456\frac{3}{4}$ lb of peanuts, and the students were putting the peanuts into bags holding $1\frac{1}{4}$ lb each, how many bags can they fill from the 7 sacks?

Chapter Summary

In this chapter, we defined fractions as quotients of whole numbers with the denominators not equal to zero. Equivalent fractions and the Fundamental Principle of Fractions were also discussed. Rules for addition, subtraction, multiplication, and division of fractions were introduced. The simplest form of a fraction was defined, as was ordering of fractions. Mixed numbers were also introduced.

The **key words** and concepts introduced in this chapter are listed here in the order in which they appeared in the text. You should know the meaning of all of these before proceeding to the next chapter.

fractions (p. 59)
numerator (p. 59)
denominator (p. 59)
proper fraction (p. 59)
improper fraction (p. 59)
mixed number (p. 61)
equivalent fractions (p. 65)
Fundamental Principle of Fractions (p. 67)
simplest form of a fraction (p. 69)

prime number (p. 72)
composite number (p. 72)
prime factorization (p. 73)
least common multiple (LCM) (p. 76)
like fractions (p. 78)
unlike fractions (p. 78)
addition of like fractions (p. 79)
least common denominator (LCD) (p. 82)

addition of unlike fractions (p. 83)
ordering of fractions (p. 87)
subtraction of fractions (p. 88)
multiplication of fractions (p. 101)
reciprocal (p. 105)
division of fractions (p. 110)

Review Exercises

2.1 Fractions

Write the word name for each fraction in Exercises 1–5.

1. $\frac{2}{9}$
2. $\frac{3}{4}$
3. $\frac{1}{17}$
4. $\frac{14}{23}$
5. $\frac{37}{49}$

Write the fraction or mixed number for each word name in Exercises 6–11.

6. Six-elevenths.
7. Ten-sixths.
8. Nine and one-third.
9. Eleven and two-ninths.
10. One hundred three-sevenths.
11. One hundred and three-sevenths.

In Exercises 12–16, rewrite each improper fraction as a mixed number.

12. $\frac{36}{7}$ **13.** $\frac{83}{11}$ **14.** $\frac{124}{9}$ **15.** $\frac{183}{17}$ **16.** $\frac{201}{19}$

In Exercises 17–21, rewrite each mixed number as an improper fraction.

17. $5\frac{2}{7}$ **18.** $9\frac{6}{7}$ **19.** $11\frac{2}{3}$ **20.** $17\frac{4}{7}$ **21.** $23\frac{5}{11}$

For each expression in Exercises 22–24, write a fraction that represents the ratio of the number of equal parts chosen from the total number of parts indicated.

22. The number of female executives in a corporation having 7 female and 8 male executives.
23. The number of correct responses on a 94-question test if 17 responses were incorrect.
24. The number of items of production passing inspection by quality control if 163 were accepted and 46 rejected.

2.2 Equivalent Fractions

In Exercises 25–30, determine which pairs of fractions are equivalent.

25. $\frac{2}{5}$ and $\frac{6}{15}$ **26.** $\frac{3}{4}$ and $\frac{4}{5}$ **27.** $\frac{9}{5}$ and $\frac{5}{9}$

28. $\frac{7}{3}$ and $\frac{21}{9}$ **29.** $\frac{13}{2}$ and $\frac{6}{5}$ **30.** $\frac{42}{6}$ and $\frac{14}{2}$

In Exercises 31–36, use the Fundamental Principle of Fractions to determine the missing number, indicated by the question mark, so that the fractions in each pair are equivalent.

31. $\frac{3}{4}$ and $\frac{?}{12}$ **32.** $\frac{6}{11}$ and $\frac{12}{?}$ **33.** $\frac{0}{19}$ and $\frac{?}{21}$

34. $\frac{48}{9}$ and $\frac{?}{3}$ **35.** $\frac{9}{4}$ and $\frac{27}{?}$ **36.** $\frac{1}{19}$ and $\frac{6}{?}$

Rewrite each fraction in Exercises 37–41 in its simplest form.

37. $\frac{9}{15}$ **38.** $\frac{27}{81}$ **39.** $\frac{81}{36}$ **40.** $\frac{105}{21}$ **41.** $\frac{99}{143}$

42. Tammy won 6 of 8 bets. Kristin won 5 of 7 bets. Did they win an equivalent number of bets?
43. Joanne called on 24 customers and sold merchandise to 5 of them. Jill called on 26 customers and sold merchandise to 6 of them. Did they sell merchandise to an equivalent number of their customers?
44. John bought 3 shirts for $40. Tom bought 4 shirts of the same quality and paid $50. Did they pay an equivalent price per shirt?

2.3 Least Common Multiple

Classify each number in Exercises 45–49 as being prime or composite.

45. 18 **46.** 29 **47.** 88 **48.** 107 **49.** 215

In Exercises 50–54, rewrite each of the composite numbers as a product of primes.

50. 30 **51.** 69 **52.** 112 **53.** 320 **54.** 640

In Exercises 55–59, determine the least common multiple (LCM) for each group of numbers.

55. 7, 28 **56.** 15, 75 **57.** 76, 81 **58.** 9, 12, 25
59. 30, 45, 62

2.4 Addition of Fractions

In Exercises 60–70, add and write your results in simplest form.

60. $\frac{2}{7} + \frac{1}{7}$
61. $\frac{4}{11} + \frac{1}{11}$
62. $\frac{3}{10} + \frac{2}{17}$
63. $\frac{23}{3} + \frac{4}{9}$
64. $\frac{2}{5} + \frac{1}{2} + \frac{4}{9}$
65. $\frac{11}{6} + \frac{1}{9} + \frac{4}{3} + \frac{1}{4}$
66. $\frac{4}{7} + \frac{5}{11}$
67. $\frac{7}{24} + \frac{5}{32}$
68. $\frac{7}{19} + \frac{9}{11}$
69. $\frac{3}{7} + \frac{4}{9} + \frac{5}{11}$
70. $\frac{4}{15} + \frac{7}{16} + \frac{1}{4}$

71. During a computer sale, the price of every personal computer was reduced by $\frac{1}{5}$ of the regular price and then further reduced by $\frac{1}{7}$ of the regular price. What was the total reduction in the regular price for the computers?

72. Krystle sold $\frac{2}{9}$ of the total number of candy bars for a school sale. Sonya sold $\frac{3}{11}$ of the total number. What fractional part of the total number of candy bars did the two sell?

73. Determine the total length around a triangle whose sides measure $\frac{3}{7}$ cm, $\frac{4}{9}$ cm, and $\frac{5}{8}$ cm.

2.5 Ordering of Fractions; Subtraction

In Exercises 74–79, insert the symbol $<$, $=$, or $>$ between each pair of fractions to make a true statement.

74. $\frac{1}{4}$ $\frac{1}{5}$
75. $\frac{6}{7}$ $\frac{8}{9}$
76. $\frac{17}{34}$ $\frac{23}{46}$
77. $\frac{21}{11}$ $\frac{7}{6}$
78. $\frac{3}{5}$ $\left(\frac{1}{3} + \frac{2}{7}\right)$
79. $\left(\frac{2}{9} + \frac{1}{3}\right)$ $\left(\frac{3}{8} + \frac{1}{4}\right)$

In Exercises 80–85, subtract and write your results in simplest form.

80. $\frac{4}{9} - \frac{2}{5}$
81. $\frac{6}{13} - \frac{5}{19}$
82. $\frac{8}{9} - \frac{6}{7}$
83. $\frac{14}{23} - \frac{6}{11}$
84. $\frac{19}{28} - \frac{5}{14}$
85. $\frac{13}{15} - \frac{3}{4}$

86. Alyssa completed $\frac{4}{11}$ of a puzzle. What portion of the puzzle was not completed?

87. Alyssa completed $\frac{4}{11}$ of a puzzle. Byran later completed another $\frac{1}{3}$ of the puzzle. What portion of the puzzle was not completed?

88. U R Bright stock gained $\frac{7}{8}$ of a point one day, gained $\frac{3}{4}$ of a point a second day, and lost $\frac{3}{8}$ of a point on the third day. What was the net gain for the stock during the three days?

2.6 Addition and Subtraction of Mixed Numbers

In Exercises 89–93, add the mixed numbers and write your results as mixed numbers.

89. $2\frac{1}{2} + 3\frac{1}{6}$
90. $4\frac{1}{7} + 3\frac{1}{9}$
91. $6\frac{4}{5} + 7\frac{3}{4}$
92. $7\frac{1}{6} + 2\frac{1}{4} + 3\frac{3}{5}$
93. $3\frac{1}{4} + 2\frac{1}{3} + 1\frac{2}{5}$

In Exercises 94–98, subtract and write your results as mixed numbers.

94. $6\frac{2}{3} - 4\frac{1}{2}$
95. $18\frac{1}{3} - 14\frac{4}{5}$
96. $20\frac{1}{6} - 11\frac{9}{10}$
97. $17\frac{2}{9} - 9\frac{5}{7}$
98. $63\frac{4}{11} - 46\frac{5}{12}$

99. What is the total length of two pieces of carpet that measure $23\frac{1}{5}$ ft and $30\frac{5}{8}$ ft long?
100. What number must be added to $17\frac{1}{8}$ to get $40\frac{3}{7}$?
101. The price of a certain stock gained $2\frac{3}{8}$ points during the first hour of trading, gained $1\frac{7}{16}$ points during the second hour, and lost $\frac{3}{4}$ of a point during the third hour. What was the net gain for the stock during the three hours?

2.7 Multiplication of Fractions and Mixed Numbers

In Exercises 102–105, perform the indicated operations and write your results in simplest form.

102. $4 \times \frac{1}{3}$
103. $\frac{2}{3} \times \frac{3}{7}$
104. $\frac{2}{5} \times \left(\frac{1}{2} + \frac{2}{6}\right)$
105. $\left(\frac{3}{7} + \frac{4}{5}\right) \times \left(\frac{1}{10} + \frac{2}{5}\right)$

In Exercises 106–109, perform the indicated operations by first rewriting the mixed numbers as fractions.

106. $4\frac{1}{2} \times 5\frac{2}{3}$
107. $3\frac{2}{3} \times 5\frac{1}{6}$
108. $2\frac{1}{2} \times \left(4\frac{1}{3} + 5\frac{1}{4}\right)$
109. $6\frac{1}{8} \times \left(1\frac{3}{4} + 2\frac{1}{3}\right)$

In Exercises 110–113, write the reciprocal, in simplest form, for each fraction. If the reciprocal does not exist, state this fact.

110. $\frac{3}{4}$
111. $\frac{17}{11}$
112. $\frac{5}{8+9}$
113. $\frac{2 \times 4}{3 \times 5}$

114. If the sales tax is $5\frac{3}{4}$¢ on each dollar, what would be the sales tax on a purchase amounting to $240?
115. What is $\frac{11}{13}$ of $7\frac{1}{2}$?
116. If your monthly income is $1400 and your spend $\frac{1}{4}$ of your income for rent and $\frac{2}{5}$ for food and entertainment, how much of your monthly income is left for other purposes?

2.8 Division of Fractions and Mixed Numbers

In Exercises 117–122, divide and write your results in simplest form.

117. $\frac{2}{3} \div \frac{1}{2}$
118. $\frac{11}{22} \div 4$
119. $4\frac{1}{2} \div \frac{1}{7}$
120. $\frac{23}{3} \div 5\frac{1}{2}$
121. $\frac{0}{23} \div 5\frac{1}{2}$
122. $23\frac{1}{2} \div 4\frac{1}{5}$

123. A piece of string measuring $19\frac{1}{2}$ m long is to be cut into pieces measuring $\frac{1}{5}$ m long. How many such pieces of string can be cut?
124. Fifty-five gallons of a liquid detergent are contained in a drum. The detergent is to be put into $1\frac{1}{2}$-qt containers. How many such containers can be filled from the drum of detergent?
125. How many 12-oz packages of chocolate chip cookies can be obtained from 225 lb of the cookies?

Teasers

1. For a T-ball league, the coaches ordered some pizzas at the end of the season. Four meat lovers' pizzas and 6 pepperoni pizzas were ordered. Each pizza was cut into 8 pieces. After the party, there were 3 pieces of meat lovers' pizza and 5 pieces of pepperoni pizza left.
   a. What fractional part of the meat lovers' pizzas was eaten?
   b. What fractional part of the pepperoni pizzas was left?
   c. What fractional part of the total pizzas was left?

2. Joyce wallpapered $\frac{1}{4}$ of her bedroom during the first hour. During the second hour, she wallpapered $\frac{1}{3}$ of the room. She then removed $\frac{1}{2}$ of the wallpaper put up during the first 2 hours. How much of the room still needed to be wallpapered?

3. Determine a pattern that exists between the numbers in each pair: $\left(\frac{1}{2}, \frac{1}{4}\right)$, $\left(\frac{1}{16}, \frac{1}{4}\right)$, $\left(\frac{1}{5}, \frac{1}{25}\right)$, $\left(\frac{1}{81}, \frac{1}{9}\right)$, $\left(\frac{1}{7}, \frac{1}{49}\right)$, $\left(\frac{1}{121}, \text{---}\right)$, $\left(\frac{1}{10}, \text{---}\right)$ Determine the missing number in each of the two last pairs.

4. The Candy Factory produced 127 boxes of candy. Each box contains $37\frac{1}{2}$ lb of candy. How many packages, each containing $1\frac{1}{4}$ lb of candy, can be filled from this total amount of candy?

5. Let n be a prime number. Is n^2 a prime number? Explain your answer.

6. Let p be a prime number and let $q = 3p + 2$. Determine all values of p, less than 50, such that q is also prime.

In your own words, but using good English and complete sentences, indicate what is meant by each of the statements in Exercises 7–10.

7. A fraction is in its simplest form, or is simplified.
8. The whole number 0 does not have a reciprocal.
9. The fractions $\frac{a}{b}$ and $\frac{c}{d}$ are equivalent fractions.
10. Addition of fractions is an associative operation.

CHAPTER 2 TEST

This test covers material from Chapter 2. Read each question carefully and answer all questions. If you miss a question, review the appropriate section in the text.

In Exercises 1–3, write the word name for the given fraction or mixed number.

1. $\dfrac{4}{7}$

2. $\dfrac{36}{53}$

3. $5\dfrac{21}{37}$

In Exercises 4–5, write the fraction or mixed number for the given word name.

4. Nine-sevenths.

5. Seventy-three fifths.

In Exercises 6–7, rewrite the given improper fraction as a mixed number.

6. $\dfrac{13}{4}$

7. $\dfrac{261}{13}$

In Exercises 8–9, rewrite the given mixed number as an improper fraction.

8. $13\dfrac{4}{9}$

9. $31\dfrac{5}{7}$

In Exercises 10–24, perform the indicated operations and write your answers in simplest form. (Use mixed numbers where appropriate.)

10. $\dfrac{2}{3} + \dfrac{1}{6} + \dfrac{3}{4}$

11. $2\dfrac{1}{4} + 3\dfrac{1}{5}$

12. $6\dfrac{2}{7} + 5\dfrac{1}{3} + 2\dfrac{3}{7}$

13. $\dfrac{11}{13} - \dfrac{4}{7}$

14. $7\frac{2}{5} - 3\frac{1}{2}$

15. $19\frac{4}{7} - 13\frac{1}{9}$

16. $\frac{2}{3} \times \frac{4}{7}$

17. $\frac{1}{3} \times \left(\frac{1}{4} + \frac{1}{5}\right)$

18. $3\frac{2}{7} \times 5\frac{1}{6}$

19. $5\frac{1}{2} \times \left(7\frac{1}{6} - 4\frac{3}{4}\right)$

20. $5 \div \frac{2}{7}$

21. $\frac{6}{7} \div 11$

22. $2\frac{1}{7} \div 3\frac{2}{5}$

23. $\frac{2}{7} \div \left(\frac{1}{4} \times \frac{2}{3}\right)$

24. $\left(\frac{7}{9} - \frac{2}{7}\right) \div \left(\frac{3}{4} - \frac{1}{3}\right)$

In Exercises 25–26, insert the symbol $>$, $=$, or $<$ between the given fractions to make true statements.

25. $\frac{2}{7}$ $\frac{3}{8}$

26. $\frac{14}{17}$ $\frac{15}{19}$

In Exercises 27–29, write the reciprocal, in simplest form, for the indicated number.

27. 5

28. $\frac{6+2}{5+7}$

29. $\frac{3 \times 5}{1 + 3}$

30. Write a fraction that represents the ratio of the number of female police officers to the total number of officers on a 115-person police force if 87 members are male.

31. Write a fraction that represents the ratio of the number of correct responses on a test to the total number of questions, if 23 items were answered correctly, 5 items were answered incorrectly, and 3 items were left blank.

32. Determine whether the fractions $\frac{4}{3}$ and $\frac{6}{5}$ are equivalent.

33. A recipe calls for $2\frac{2}{3}$ cups of flour. A second recipe calls for $3\frac{1}{4}$ cups of flour. How much more flour is required for the second recipe?

34. Bette worked $4\frac{1}{6}$ hr doing her mathematics homework and $5\frac{1}{4}$ hr doing her chemistry homework. How many hours did she work on the two assignments?

35. A beef roast weighed $15\frac{1}{8}$ lb. The butcher cut it into two pieces. If one piece weighed $7\frac{3}{5}$ lb, how much did the other piece weigh?

CUMULATIVE REVIEW: CHAPTERS 1–2

In Exercises 1–15, perform the indicated operations. Write your results in simplest form.

1. $123 + 56 + 872$
2. $\frac{4}{7} + \frac{1}{3} + \frac{2}{5}$
3. $5\frac{1}{7} + 4\frac{1}{2} + 2\frac{1}{3}$
4. $31 + 2\frac{3}{4} + \frac{1}{3}$
5. $1024 - 809$
6. $\frac{11}{13} - \frac{2}{9}$
7. $5\frac{1}{6} - 2\frac{3}{4}$
8. $23 - 5\frac{7}{11}$
9. 207×36
10. $\frac{4}{11} \times \frac{22}{35}$
11. $4\frac{1}{9} \times 8\frac{2}{7}$
12. $\frac{9}{13} \times 3\frac{1}{6}$
13. $26{,}975 \div 17$
14. $\frac{21}{37} \div \frac{3}{2}$
15. $1\frac{4}{5} \div 3\frac{2}{3}$

In Exercises 16–18, circle the larger number.

16. $\frac{19}{17}$ or $\frac{17}{15}$
17. $4\frac{1}{2}$ or $3\frac{7}{9}$
18. $5\frac{1}{2}$ or $\frac{22}{5}$

In Exercises 19–21, perform the indicated operations. Write your results in simplest form.

19. $16 \div 2^3 + (3^2 - 4 \div 2)$
20. $5 - 0 \div 4^2 \times 7$
21. $\frac{1}{2} \times \frac{1}{3} + \frac{1}{4} \div \frac{1}{5}$

In Exercises 22–23, write the word names for the given numerals.

22. $2{,}034{,}507$
23. $16\frac{11}{15}$

24. Determine the perimeter of a rectangle whose width is 24 cm and whose length is 37 cm.

25. Determine the area of a triangular region with base of 19 yd and height of 9 yd.

CHAPTER 3 Decimals

In this chapter, we will extend our number system from the whole numbers and fractions to what are called decimals.

As was noted earlier, performing operations on fractions is not always an easy task. However, the task becomes easier if the fractions have denominators that are powers of ten, such as 10; 100; 1000; and so forth. By extending our knowledge of the place value of digits in a number that are located to the right of the ones place, we can eliminate the need for denominators in many cases. The places to the right of the ones place are called decimal places. The ones digits and the digit immediately to its right are separated by a dot, called a decimal point. These numerals are used to represent decimal numbers.

Decimals have widespread use in real-life applications. You are already familiar with dollar and cents applications. Your normal body temperature is recorded as 98.6 degrees Fahrenheit. Sobriety tests use readings generally given as decimals. Sports pages in your local newspaper include decimals such as batting averages for baseball players.

The arithmetic of decimals will be discussed in this chapter. Applications involving decimals will be given throughout the chapter.

CHAPTER 3 PRETEST

This pretest covers material on decimals. Read each question carefully, and try to answer all questions. Each question is keyed to the chapter section in which the particular topic is discussed. Check your answers with those given in the back of the text.

Questions 1–4 pertain to Section 3.1.

1. Determine the place value for the underlined digit in 2.09$\underline{7}$.

2. Write the word name for 123.469.

3. Which is the larger decimal number: 0.23 or 0.199?

4. Arrange the decimal numbers in order of size, with the smallest first: 0.62; 0.819; 0.6; 0.901

Questions 5–8 pertain to Section 3.2.

5. Add: 0.46 + 1.9 + 0.203.

6. Subtract: 56.1 − 27.964.

7. From 119.62, subtract 80.049.

8. Perform the indicated operations: 6.97 + 102.3 − 46.355.

Questions 9–10 pertain to Section 3.3.

9. Round 87.369 to the nearest hundredth.

10. Round 96.5067 to the nearest one.

Questions 11–13 pertain to Section 3.4.

11. Multiply: 2.34×1.56.

12. Multiply: $(1.37)(17.26)$.

13. Perform the indicated operations: $(1.17 + 0.7) \times (4.3 - 1.96)$

Questions 14–16 pertain to Section 3.5.

14. Divide: $25.22 \div 1.3$.

15. Divide: $\dfrac{0.0918}{0.18}$.

16. Perform the indicated operations: $633.6 \div 0.22 - 397.1$.

Questions 17–19 pertain to Section 3.6.

17. Rewrite 23.24 as a fraction in simplest form.

18. Rewrite $\dfrac{17}{125}$ as a decimal.

19. Rewrite $\dfrac{103}{17}$ as a decimal, correct to the nearest tenth.

Questions 20 pertains to Section 3.7.

20. If you drive your car for 3069 mi and use 173.7 gal of gasoline, what is the average number of miles traveled per gallon of gasoline? Give your answer to the nearest tenth of a mile.

3.1 Decimals as Numbers

OBJECTIVES

After completing this section, you should be able to:
1. Determine the place value for a particular digit in a given decimal number.
2. Read or write the name of a decimal number.
3. Compare two decimal numbers.

Recall that the digits in a whole number have place values according to their locations. For instance, in 4625:

- The digit 5 is located in the ones place and has a value of 5 since $5 \times 1 = 5$.
- The digit 2 is located in the tens place and has a value of 20 since $2 \times 10 = 20$.
- The digit 6 is located in the hundreds place and has a value of 600 since $6 \times 100 = 600$.
- The digit 4 is located in the thousands place and has a value of 4000 since $4 \times 1000 = 4000$.

We also note that as we move from the right digit to the left digit in the number, the place value of each digit is ten times the place value of the digit on its right. (See Table 3.1.)

TABLE 3.1

	4	6	2	5
	Thousands	Hundreds	Tens	Ones
Place value:	$1000 = 100 \times 10$	$100 = 10 \times 10$	$10 = 10 \times 1$	1
	←——Right to left			

If, however, we move from the left digit to the right digit in the number, the place value of each digit is $\frac{1}{10}$ of the place value of the digit on its left. (See Table 3.2.)

TABLE 3.2

	4	6	2	5
	Thousands	Hundreds	Tens	Ones
Place value:	1000	$100 = \frac{1}{10} \times 1000$	$10 = \frac{1}{10} \times 100$	$1 = \frac{1}{10} \times 10$
	Left to right——→			

If we were to continue with digits to the *right* of the ones place, the pattern indicated in Table 3.2 would continue. To do this, we place a dot, called a **decimal point,** immediately after the ones digit. The place values for the digits to the *right* of the decimal point are named as indicated in Table 3.3.

3.1 Decimals as Numbers

TABLE 3.3

	Ones	Tenths	Hundredths	Thousandths	Ten-thousandths	Hundred-thousandths	Millionths
Place value:	1	$\frac{1}{10} =$ $\frac{1}{10} \times 1$	$\frac{1}{100} =$ $\frac{1}{10} \times \frac{1}{10}$	$\frac{1}{1000} =$ $\frac{1}{10} \times \frac{1}{100}$	$\frac{1}{10,000} =$ $\frac{1}{10} \times \frac{1}{1000}$	$\frac{1}{100,000} =$ $\frac{1}{10} \times \frac{1}{10,000}$	$\frac{1}{1,000,000} =$ $\frac{1}{10} \times \frac{1}{100,000}$

In the numeral .9638:

- The digit 9 is located in the tenths place and has a value of $\frac{9}{10}$ since $9 \times \frac{1}{10} = \frac{9}{10}$.
- The digit 6 is located in the hundredths place and has a value of $\frac{6}{100}$ since $6 \times \frac{1}{100} = \frac{6}{100}$.
- The digit 3 is located in the thousandths place and has a value of $\frac{3}{1000}$ since $3 \times \frac{1}{1000} = \frac{3}{1000}$.
- The digit 8 is located in the ten-thousandths place and has a value of $\frac{8}{10,000}$ since $8 \times \frac{1}{10,000} = \frac{8}{10,000}$.

For place values to the *right* of the ones place, we write:

- $\frac{1}{10}$ as .1
- $\frac{1}{100}$ as .01
- $\frac{1}{1000}$ as .001
- $\frac{1}{10,000}$ as .0001

and so forth. Numbers such as .1, .01, .001, and .0001 are called "decimal fractions" or simply **decimal numbers.** The numbers .7, 1.3, 23.63, and 3.0059 are also examples of decimals.

To Read a Decimal Number
1. If the number is greater than 1, read the digits to the *left* of the decimal points as you would read a whole number.
2. Read the decimal point as "and."
3. Read the digits to the *right* of the decimal point as though they represent a whole number *and* state the place value of the digit on the extreme right.

Write each of the following in words.

Practice Exercise 1. (a) 0.17; (b) 2.3; (c) 19.87; (d) 0.239.

EXAMPLE 1 Write each of the following decimal numbers in words: (a) .37; (b) .623; (c) 23.78; (d) 123.45.

SOLUTION: (a) .37 is written as "thirty-seven hundredths"; (b) .623 is written as "six hundred twenty-three thousandths"; (c) 23.78 is written as "twenty-three and seventy-eight hundredths"; (d) 123.45 is written as "one hundred twenty-three and forty-five hundredths."

> **ANSWERS TO PRACTICE EXERCISE**
>
> **1.** (a) Seventeen hundredths; (b) two and three tenths; (c) nineteen and eighty-seven hundredths; (d) two hundred thirty-nine thousandths.

If the decimal number is less than 1, it may be helpful for you to write a 0 in the ones place. For instance, .6 would be written as 0.6; .139 would be written as 0.139; and so forth.

Placing zeros to the *right* of a decimal number does not change its value. For example, the decimal 0.4 can be written as 0.40 since

$$0.4 = \frac{4}{10}$$

$$0.40 = \frac{40}{100}$$

$$\frac{4}{10} = \frac{4 \times 10}{10 \times 10} = \frac{40}{100}$$

The idea of placing zeros to the right of a decimal number is helpful if we wish to compare decimals. Rewrite the decimals, if necessary, so that they have the same number of decimal places and then compare them. For instance, to compare the decimals 0.23 and 0.382, we:

1. Note that 0.23 has two decimal places, whereas 0.382 has three decimal places.
2. Rewrite 0.23 as 0.230 (with three decimal places).
3. Now compare 0.230 (two hundred thirty thousandths) and 0.382 (three hundred eighty-two thousandths).
4. Conclude that $0.382 > 0.23$.

Practice Exercise 2. (a) Which is greater, 0.619 or 0.92? (b) Which is greater, 7.01 or 7.002?

EXAMPLE 2 Which is greater: 0.78 or 0.693?

SOLUTION: Rewrite 0.78 as 0.780 and determine that 0.780 (seven hundred eighty thousandths) is greater than 0.693 (six hundred ninety-three thousandths). Hence, $0.78 > 0.693$.

Practice Exercise 3. Order the following decimals with the greatest first: 0.23; 0.196; 0.8; 0.61; 0.213.

EXAMPLE 3 Arrange the following decimals in order of size, with the smallest first: 0.17; 0.4; 0.476; 0.392.

SOLUTION: Rewrite each of the decimals with the same number (three) of decimal places:

$$0.17 = 0.170$$
$$0.4 = 0.400$$
$$0.476 = 0.476$$
$$0.392 = 0.392$$

Now, determine that $0.170 < 0.392 < 0.400 < 0.476$, and that the required order is: 0.17; 0.392; 0.4; 0.476.

ANSWERS TO PRACTICE EXERCISES

2. (a) 0.92; (b) 7.01.   **3.** 0.8; 0.61; 0.23; 0.213; 0.196.

EXERCISES 3.1

In Exercises 1–20, determine the place value for the digit 6 in each of the given decimal numbers.

1. 12.063	**2.** 0.06	**3.** 100.62	**4.** 2.1365
5. 63.174	**6.** 2.0006	**7.** 137.7546	**8.** 206.123
9. 90.00006	**10.** 0.125746	**11.** 26.39	**12.** 479.306
13. 561.02	**14.** 0.369	**15.** 650.03	**16.** 651.92
17. 36.99	**18.** 987.654	**19.** 406.05	**20.** 1639.1

In Exercises 21–30, write the name for each of the decimal numbers given.

21. 0.39	**22.** 1.3	**23.** 0.016	**24.** 14.231
25. 15.71	**26.** 2.397	**27.** 1.0276	**28.** 391.67
29. 1987.23	**30.** 10.10101		

In Exercises 31–40, write the numeral for each decimal number named.

31. One and sixteen hundredths
32. Two hundred three thousandths
33. Fifty and fifty-five hundredths
34. One thousand, seven and three tenths
35. Ninety-one and ninety-one ten thousandths
36. Three and four hundred fifty-six thousandths
37. Seventeen thousand, eighteen and twenty-two hundred thousandths
38. Four and eleven thousand, four hundred nine millionths
39. Sixty-seven thousand, seven hundred fifty and eighty-nine hundredths
40. Three million, four thousand, eight hundred thirty-seven and eighty-seven thousandths

In Excercises 41–60, insert the symbol >, =, or < between the given pairs of decimal numbers to make true statements.

41. 0.61 0.8	**42.** 0.7 0.581	**43.** 0.2 0.20
44. 8.6 8.43	**45.** 46.503 46.54	**46.** 2.1 0.21
47. 11.023 11.12	**48.** 6.702 6.75	**49.** 90.0 99.09
50. 9.001 9.0001	**51.** 27.9 27.089	**52.** 0.983 0.99
53. 10.23 10.213	**54.** 13.07 13.070	**55.** 0.080 0.08
56. 17.690 17.6090	**57.** 0.040 0.0041	**58.** 0.101 0.02
59. 2.576 2.5759	**60.** 15.023 15.2030	

In Exercises 61–70, arrange the decimal numbers in order of size, with the greatest first.

61. 0.71; 0.59; 0.9; 0.813
62. 1.01; 0.10; 10.1; 1.0101
63. 0.05; 0.2; 0.417; 0.218
64. 0.987; 0.099; 0.9; 0.0909
65. 23.1; 23.203; 22.09; 22.109
66. 57.406; 57.046; 57.44; 57.6
67. 2.453; 2.4532; 0.2567; 20.54
68. 11.13; 10.103; 1.0013; 11.131
69. 10.031; 10.31; 10.103; 10.0013
70. 9.09; 9.90; 9.99; 9.099

3.2 Addition and Subtraction of Decimals

OBJECTIVE

After completing this section, you should be able to add and subtract decimal numbers.

When we add whole numbers, we line them up with the ones under the ones, the tens under the tens, and so forth. We add decimal numbers in the same way.

Rule for Adding Decimal Numbers

To add two or more decimal numbers:

1. Arrange the decimal numbers under each other, with the decimal points lined up directly below each other.
2. Add the corresponding digits according to place value.
3. Put the decimal point in the sum directly below the other decimal points.

Practice Exercise 1. Add: 0.87 + 0.96.

EXAMPLE 1 Add: 2.39 + 16.2 + 11.867.

SOLUTION:

STEP 1: Arrange the decimal numbers under each other, lining up their decimal points:

$$
\begin{array}{r} 2.39 \\ 16.2 \\ +11.867 \end{array}
\quad \text{or} \quad
\begin{array}{r} 2.390 \\ 16.200 \\ +11.867 \end{array}
$$

STEP 2: Add the corresponding digits according to place value:

$$
\begin{array}{r} 2.390 \\ 16.200 \\ +11.867 \\ \hline 30\,457 \end{array}
$$

STEP 3: Put the decimal point in the sum directly below the other decimal points.

$$
\begin{array}{r} 2.390 \\ 16.200 \\ +11.867 \\ \hline 30.457 \end{array}
$$

This procedure for adding decimals can be shortened, as demonstrated in the next two examples.

Practice Exercise 2. Add: 0.123 + 12.21 + 1.41.

EXAMPLE 2 Add: 2.134 + 19.04 + 8.0167 + 0.073.

SOLUTION:

$$
\begin{array}{r} 2.134 \\ 19.04 \\ 8.0167 \\ +0.073 \end{array}
\quad \text{or} \quad
\begin{array}{r} 2.1340 \\ 19.0400 \\ 8.0167 \\ +0.0730 \\ \hline 29.2637 \end{array}
$$

3.2 Addition and Subtraction of Decimals

Practice Exercise 3. Add: 1.019 + 39 + 42.13 + 101.09.

EXAMPLE 3 Add 23.1 + 1.067 + 43 + 11.2034.

SOLUTION:

```
   23.1            23.1000
    1.067           1.0670
   43.         or  43.0000
 + 11.2034       + 11.2034
                  78.3704
```

Subtraction of decimal numbers is done in a similar manner.

Rule for Subtracting Decimal Numbers

To subtract decimal numbers:
1. Arrange the decimal numbers under each other, with the decimal points lined up directly below each other.
2. Subtract the corresponding digits according to place value.
3. Put the decimal point in the answer directly below the other decimal points.

Practice Exercise 4. Subtract: 47.52 − 19.78.

EXAMPLE 4 Subtract: 23.96 − 1.342.

SOLUTION:

STEP 1: Arrange the decimal numbers with their decimal points lined up under each other:

```
   23.96         23.960
 −  1.342   or  − 1.342
```

STEP 2: Subtract the corresponding digits according to place value:

```
   23.960
 −  1.342
   22 618
```

STEP 3: Put the decimal point in the answer directly below the others:

```
   23.960
 −  1.342
   22.618
```

This procedure for subtracting decimals can be shortened, as shown in the next two examples.

Practice Exercise 5. Subtract 97.56 from 609.1.

EXAMPLE 5 From 301.902, subtract 96.07.

SOLUTION:

```
   301.902         301.902
 −  96.07    or  −  96.070
                   205.832
```

Practice Exercise 6. From the sum of 19.8 and 239.046 subtract 90.31.

EXAMPLE 6 Subtract 9.0137 from the sum of 20.06 and 4.857.

SOLUTION: We first determine the sum of 20.06 and 4.857:

```
   20.060
 +  4.857
   24.917
```

Next, we subtract 9.0137 from the sum:

$$\begin{array}{r} 24.9170 \\ -9.0137 \\ \hline 15.9033 \end{array}$$

ANSWERS TO PRACTICE EXERCISES

1. 1.83.   **2.** 13.743.   **3.** 183.239.   **4.** 27.74.   **5.** 511.54.   **6.** 168.536.

EXERCISES 3.2

In Exercises 1–20, add as indicated.

1. 0.63 + 0.25
2. 0.123 + 0.345
3. 1.2 + 0.97
4. 23.96 + 1.205
5. 10.09 + 231.006
6. 2.136 + 17.37
7. 39.69 + 47.1
8. 10.63 + 2.7909
9. 53.2 + 1.3062
10. 0.02 + 0.0003
11. 29.1 + 2.009
12. 0.002 + 18.91
13. 1.6 + 23.09 + 0.172
14. 100.2 + 23 + 1.86
15. 0.5 + 5.21 + 63.051
16. 98 + 20.86 + 1.079
17. 37.698 + 203.9 + 104
18. 19.3 + 1.682 + 41.74
19. 236.1 + 10.23 + 98.765
20. 6.927 + 69.27 + 692.7 + 6927

In Exercises 21–30, add.

21.   23.1
         0.97
       +2.063

22.   0.067
          10.02
        +231.8

23.   19.23
          1.602
        +172.3

24.   1.07
           0.5
        +12.149

25.   17.2
         609.007
          14.59
         +63

26.   8.076
          19.23
         101.0069
          +0.7

27. 1236.9
           67.08
            0.0703
         +902.702

28.   16.9
          107.87
            0.049
        +4059.5

29. 23,156.73
          2,690.087
            509.1
              0.63
           +19.904

30.   3,070.7
         14,986.003
            607.96
              0.8007
           +19.146

In Exercises 31–50, subtract.

31. 69.3 − 37.07
32. 23.27 − 2.136
33. 39.69 − 27.1
34. 38.134 − 17.097
35. 96 − 59.076
36. 0.967 − 0.8998

37. 769.02 − 490.87
38. 209.201 − 58.03
39. 76.002 − 7.6002
40. 200.003 − 79.8764
41. 30.7 − 8.092
42. 1.23 − 0.3456
43. 86.46 − 53
44. 64 − 46.79
45. 39 − 1.872
46. 0.98 − 0.592
47. 1.2 − 0.3456
48. 2.0304 − 0.999
49. 592.31 − 87.099
50. 62.072 − 9.0909

In Exercises 51–60, subtract.

51. 123.2
    − 87.19

52. 0.69
    − 0.3742

53. 309.231
    − 126.94

54. 19.0069
    − 4.968

55. 0.0869
    − 0.009

56. 256.02
    − 196.8764

57. 999.9
    − 876.99

58. 706.05
    − 409.0807

59. 2360.01
    − 407.9056

60. 0.008006
    − 0.00509009

In Exercises 61–65, perform the indicated operations.

61. 63.2 + 9.076 − 47.89
62. 59.23 − 1.962 − 30.7
63. 103.4 − (23.96 + 51.003)
64. (9.03 + 102.976) − (63.007 + 0.90061)
65. (87.01 − 23.904) + (0.087 − 0.0099)
66. Subtract 17.123 from the sum of 63.1 and 29.07.
67. Subtract 72.9 from the sum of 123.96 and 52.069.
68. Subtract the sum of 68.05 and 47.207 from 203.96.
69. From 869.2, subtract the sum of 203.97 and 470.909.
70. From the sum of 69.3 and 102.73, subtract the sum of 37.6 and 40.96.

3.3 Multiplication of Decimals

OBJECTIVES

After completing this section, you should be able to:
1. Multiply decimal numbers.
2. Multiply a decimal number by a power of 10.

Multiplication of decimal numbers is performed in the same manner as multiplication of whole numbers. The only difference is that, in the multiplication of decimal numbers, we must decide where to place the decimal point in the product.

To formulate a rule for determining the number of decimal places in the product of two decimals, consider multiplying 0.6 and 0.23. Since

$$0.6 = \frac{6}{10}$$

and

$$0.23 = \frac{23}{100}$$

we have

$$0.6 \times 0.23 = \frac{6}{10} \times \frac{23}{100}$$
$$= \frac{6 \times 23}{10 \times 100}$$
$$= \frac{138}{1000}$$
$$= 0.138$$

There is one decimal place in the factor 0.6, and there are two decimal places in the factor 0.23; the product has three (1 + 2) decimal places.

Now, consider multiplying 0.17 and 0.32. Since

$$0.17 = \frac{17}{100}$$

and

$$0.32 = \frac{32}{100}$$

we have

$$0.17 \times 0.32 = \frac{17}{100} \times \frac{32}{100}$$
$$= \frac{17 \times 32}{100 \times 100}$$
$$= \frac{544}{10,000}$$
$$= 0.0544$$

There are two decimal places in the factor 0.17 and two decimal places in the factor 0.32; the product has four (2 + 2) decimal places. Note that in order to get the four decimal places in the product, it was necessary to place a 0 between the decimal point and the 5. Place value is important here. Without the 0, we would have five hundred forty-four *thousandths* instead of five hundred forty-four *ten thousandths*.

To Multiply Decimal Numbers

1. Multiply the decimal numbers as though they were whole numbers.
2. The number of decimal places in the product is the *sum* of the number of decimal places in each factor.

Practice Exercise 1. Multiply: 2.34 × 1.7.

EXAMPLE 1 Multiply 0.175 and 0.34.

SOLUTION:

```
   0.175    (A three-decimal place number)
 × 0.34     (A two-decimal place number)
   700      (Multiplying as though the decimal
   525       numbers were whole numbers)
 0.05950    [A five-decimal (3 + 2) place number]
```

3.3 Multiplication of Decimals

Practice Exercise 2. Multiply: 0.345 × 18.2.

EXAMPLE 2 Multiply 24.13 and 7.94.

SOLUTION:
```
     24.13      (A two-decimal place number)
    ×7.94       (A two-decimal place number)
     9652       (Multiplying as though the decimal
    21 717      numbers were whole numbers)
   168 91
   191.5922    [A four-decimal (2 + 2) place number]
```

Practice Exercise 3. Multiply: 0.069 × 11.2.

EXAMPLE 3 Multiply 0.00123 and 17.4.

SOLUTION:
```
    0.00123
    × 17.4
       492
       861
       123
    0.021402
```

ANSWERS TO PRACTICE EXERCISES

1. 3.978.   2. 6.279.   3. 0.7728.

Frequently, we multiply decimals by powers of 10 such as 10; 100; 1000; and so forth. We will now examine such multiplications.

Practice Exercise 4. Multiply: 1.78 × 100.

EXAMPLE 4 Multiply 46.926 × 10.

SOLUTION:
```
    46.926
    × 10
   469.260 or 469.26
```

Practice Exercise 5. Multiply: 0.45 × 10.

EXAMPLE 5 Multiply 46.926 by 100.

SOLUTION:
```
    46.926
    × 100
   4692.600 or 4692.6
```

Practice Exercise 6. Multiply: 14.167 × 1000.

EXAMPLE 6 Multiply 46.926 by 1000.

SOLUTION:
```
     46.926
    × 1000
   46926.000 or 46,926
```

In the preceding examples, observe that:

- Multiplying a decimal number by 10 has the effect of moving the decimal point in the decimal number *one* place to the *right*.
- Multiplying a decimal number by 100 has the effect of moving the decimal point in the decimal number *two* places to the *right*.
- Multiplying a decimal number by 1000 has the effect of moving the decimal point in the decimal number *three* places to the *right*.

We note that 10; 100; 1000; 10,000; and so forth are called **powers of 10**. We have:

$$10^1 = 10$$
$$10^2 = 10 \times 10 = 100$$
$$10^3 = 10 \times 10 \times 10 = 1000$$
$$10^4 = 10 \times 10 \times 10 \times 10 = 10,000$$

and so forth.
Also, note that

- 10^1 is the *first* power of 10 and has *1* zero following the 1.
- 10^2 is the *second* power of 10 and has *2* zeros following the 1.
- 10^3 is the *third* power of 10 and has *3* zeros following the 1.
- 10^4 is the *fourth* power of ten and has *4* zeros following the 1.

And so forth.
To multiply a decimal number by a power of 10, we use the following rule.

Rule for Multiplying a Decimal Number by a Power of 10

To multiply a decimal number by 10^n (where n is a counting number), move the decimal point in the decimal number n places to the *right*.

Practice Exercise 7. Multiply: 27.69×10^2.

EXAMPLE 7

- $2.169 \times 10 = 2.169 \times 10^1 = 21.69$
- $3.029 \times 100 = 3.029 \times 10^2 = 302.9$
- $0.06 \times 100 = 0.06 \times 10^2 = 6$
- $1.2 \times 1000 = 1.2 \times 10^3 = 1200$ (Note that two zeros had to be placed following the 2.)
- $26.060974 \times 10^4 = 260,609.74$.

ANSWERS TO PRACTICE EXERCISES

4. 178.   5. 4.5.   6. 14,167.   7. 2769.

EXERCISES 3.3

Multiply in Exercises 1–30.

1. 19.6×1.23
2. 0.69×23.9
3. 23.4×30.7
4. 58.17×19.002
5. 2.312×87.9
6. 1.231×3.04

7. (92.063)(14.7)
8. (7.06)(101.003)
9. (83.902)(0.067)
10. (0.004)(0.00003)
11. (0.12)(0.17)(0.19)
12. (1.2)(2.3)(4.67)
13. $0.02 \times 0.03 \times 0.04$
14. $0.06 \times 9.06 \times 21.7$
15. $6.107 \times 11.09 \times 16$
16. $20.02 \times 30.07 \times 40$
17. $38.391 \times 19.1 \times 0.06$
18. $0.812 \times 0.16 \times 1.7$
19. $3.005 \times 2.06 \times 1000$
20. $60.903 \times 5.007 \times 100$
21. $43.4 \times 1.9 \times 2.03$
22. (23.69)(123.4)
23. (3.967)(20.04)
24. (78.999)(1.504)
25. (89.697)(3.72)
26. (3.967)(23.23)
27. (5.006)(234.72)
28. (7.609)(9.056)
29. (63.9)(736.42)
30. (109.02)(506.023)

In Exercises 31–50, multiply using the rule for multiplying decimals by powers of 10.

31. 23.1×10
32. 0.69×10
33. 1.789×10^1
34. 2.0697×10^1
35. 59.2×100
36. 0.769×100
37. 73.479×10^2
38. 0.0896×10^2
39. 14.1658×1000
40. 2.09876×1000
41. 0.789×10^3
42. 12.010203×10^3
43. $57.1 \times 10,000$
44. $6.79041 \times 10,000$
45. 8.79547×10^4
46. 0.008734×10^4
47. 21.4798×10^5
48. 0.0003456789×10^6
49. $(2.3 \times 10^2) \times 10^2$
50. $(0.07946 \times 10^3) \times 10^2$

In Exercises 51–56, perform the indicated operations.

51. $(2.16 + 10.3) \times 1.2$
52. $(42.06 - 18.9) + 7.9$
53. $39.81 + 2.93 \times 9.01$
54. $569.1 - 12.03 \times 11.9$
55. $(19.6 + 4.39) \times (56.2 - 30.93)$
56. $(76.92 - 59.7) \times (14.02 + 9.9)$

3.4 Rounding and Estimating

OBJECTIVES

After completing this section, you should be able to:
1. Round a decimal number to a nearest place value.
2. Using rounding, estimate the answer to a problem.

ROUNDING

In Section 1.6, we introduced and discussed rounding of whole numbers and estimating the answer to a problem. In this section, the rule for **rounding** numbers will now be extended to include decimal numbers.

Rule for Rounding Numbers

To round a number to a given place value, look at the digit immediately to the *right* of that place value:

1a. If the number is a whole number and this digit is less than 5, change it and all digits to the right of it to zeros. Do not change the digit in the given place value.

b. If the number is a decimal number and this digit is less than 5 and is to the right of the decimal point, drop it and all digits to its right.

2a. If the number is a whole number and this digit is greater than or equal to 5, change it and all digits to its right to zero. *Add* 1 to the digit in the given place value.

b. If the number is a decimal number and this digit is greater than or equal to 5 and is to the right of the decimal point, drop it and all digits to its right. *Add* 1 to the digit in the given place value.

Practice Exercise 1. Round 57,658 to the nearest thousand.

EXAMPLE 1 Round 5847 to the nearest ten.

SOLUTION: The digit in the tens place is 4. The digit to its right is 7, which is greater than 5. Part 2*a* of the rule applies and 5847 is rounded up to 5850 to the nearest ten.

Practice Exercise 2. Round 4.7478 to the nearest thousandth.

EXAMPLE 2 Round 73,247 to the nearest thousand.

SOLUTION: The digit in the thousands place is 3. The digit to its right is 2, which is less than 5. Part 1*a* of the rule applies and 73,247 is rounded down to 73,000 to the nearest thousand.

Practice Exercise 3. Round 106.91 to the nearest one.

EXAMPLE 3 Round 314,750 to the nearest hundred.

SOLUTION: The digit in the hundreds place is 7. The digit to its right is 5. Part 2*a* of the rule applies and 314,750 is rounded up to 314,800.

Practice Exercise 4. Round 28.687 to the nearest tenth.

EXAMPLE 4 Round 7.2687 to the nearest hundredth.

SOLUTION: The digit in the hundredths place is 6. The digit to its right is 8, which is greater than 5. Part 2*b* of the rule applies and 7.2687 is rounded up to 7.27.

Practice Exercise 5. Round 2098.34 to the nearest hundred.

EXAMPLE 5 Round 20.8524 to nearest thousandth.

SOLUTION: The digit in the thousandths place is 2. The digit to its right is 4, which is less than 5. Part 1*b* of the rule applies and 20.8524 is rounded down to 20.852.

Practice Exercise 6. Round 20.9834 to the nearest hundredth.

EXAMPLE 6 Round 0.987 to the nearest tenth.

SOLUTION: The digit in the tenths place is 9. The digit to its right is 8, which is greater than 5. Part 2*b* of the rule applies and 0.987 is rounded up to 1.0.

ANSWERS TO PRACTICE EXERCISES

1. 58,000.   2. 4.748.   3. 107.   4. 28.7.   5. 2100.   6. 20.98.

ESTIMATION

In Section 1.6, we discussed **estimation** of answers using rounding. We continue that discussion with decimal numbers.

Practice Exercise 7. Estimate the following sum by rounding each number to the nearest tenth: 0.37 + 0.79 + 0.53.

EXAMPLE 7 Estimate the following sum by rounding each number to the nearest tenth: 0.43 + 0.77 + 0.92.

SOLUTION: Estimate: 0.4 + 0.8 + 0.9 = 2.1. (The actual sum is 2.12.)

3.4 Rounding and Estimating

Practice Exercise 8. Estimate the following sum by rounding each number to the nearest hundredth: 0.723 + 0.457 + 0.378 + 0.289.

EXAMPLE 8 Estimate the following sum by rounding each number to the nearest hundredth: 0.237 + 0.587 + 0.912 + 0.168.

SOLUTION: Estimate: 0.24 + 0.59 + 0.91 + 0.17 = 1.91. (The actual sum is 1.904.)

Practice Exercise 9. Estimate the following product by rounding each number to the nearest tenth: (0.53)(0.87).

EXAMPLE 9 Estimate the following product by rounding each number to the nearest tenth: (0.77)(0.62).

SOLUTION: Estimate: (0.8)(0.6) = 0.48. (The actual product is 0.4774.)

Practice Exercise 10. Estimate the perimeter, P, of a rectangle with width of 12.098 yd and length of 18.345 yd by rounding each measurement to the nearest tenth.

EXAMPLE 10 The sides of a triangle measure 2.68 ft, 3.59 ft, and 5.43 ft. Estimate the perimeter, P, of the triangle by rounding each of the measurements to the nearest tenth.

SOLUTION:
$$P = 2.68 \text{ ft} + 3.59 \text{ ft} + 5.47 \text{ ft}$$
$$= (2.68 + 3.59 + 5.47) \text{ ft}$$

Estimate:
$$P = (2.7 + 3.6 + 5.5) \text{ ft}$$
$$= 11.8 \text{ ft}$$

(The actual value of P is 11.74 ft.)

Practice Exercise 11. Estimate the area, A, of a square whose sides each measure 10.65 in. by rounding the measurement to the nearest tenth.

EXAMPLE 11 A rectangle has a width of 20.169 cm and length of 17.894 cm. Estimate the area, A, of the rectangle by rounding each of the measurements to the nearest tenth.

SOLUTION:
$$A = (20.169 \text{ cm})(17.894 \text{ cm})$$
$$= (20.169)(17.894) \text{ cm}^2$$

Estimate:
$$A = (20.2)(17.9) \text{ cm}^2$$
$$= 361.58 \text{ cm}^2$$

(The actual value of A is 360.904086 cm^2.)

ANSWERS TO PRACTICE EXERCISES

7. 1.7. **8.** 1.85. **9.** 0.45. **10.** 60.8 yd. **11.** 114.49 in.^{2}.

EXERCISES 3.4

In Exercises 1–40, round each decimal number to the nearest place indicated.

1. 2.34; tenths
2. 0.312; tenths
3. 6.87; tenths
4. 0.899; tenths
5. 0.783; hundredths
6. 6.509; hundredths
7. 6.047; hundredths
8. 2.8962; hundredths
9. 1.2063; thousandths
10. 0.17910; thousandths
11. 124.962; hundredths
12. 6.23790; ten thousandths
13. 123.4567; thousandths
14. 1.234567; ten thousandths
15. 0.1234567; millionths
16. 56.023706; hundred thousandths
17. 5.5500; tenths
18. 23,692.763; tenths
19. 0.018500; hundredths
20. 986.209706; thousandths

21. 0.123; tenths
22. 0.666; tenths
23. 1.049; tenths
24. 3.962; tenths
25. 0.6041; hundredths
26. 0.9999; hundredths
27. 1.0050; hundredths
28. 12.0632; hundredths
29. 0.12394; thousandths
30. 2.00902; thousandths
31. 40.1234; thousandths
32. 59.9065; thousandths
33. 0.12345; ten thousandths
34. 0.24689; ten thousandths
35. 9.909090; ten thousandths
36. 9.090909; hundred thousandths
37. 1.2345678; millionths
38. 2.4824683; millionths
39. 10.0009999; millionths
40. 9.9999999; millionths
41. Multiply 23.697 and 2.469 and round the product to the nearest thousandth.
42. Multiply 46.29 and 372.096 and round the product to the nearest hundredth.
43. Multiply (3.09)(12.12)(9.6) and round the product to the nearest tenth.
44. Multiply 88.888 and 9.999 and round the product to the nearest ten thousandth.
45. Estimate the following sum by rounding each of the terms to the nearest tenth: 23.56 + 6.977 + 0.986 + 13.42.
46. Estimate the following sum by rounding each of the terms to the nearest hundredth: 4.097 + 0.345 + 6.938 + 0.098 + 5.9999.
47. Estimate the following product by rounding each of the factors to the nearest tenth: (19.87)(24.29).
48. A triangle has sides of measures 43.25 ft, 37.89 ft, and 49.34 ft. Estimate the perimeter of the triangle by rounding each of the measures to the nearest tenth.
49. A rectangle has a width of 17.86 in. and a length of 27.54 in. Estimate the perimeter of the rectangle by rounding its width and length to the nearest tenth.
50. A square has sides each of measure 15.67 cm. Estimate the area of the square by rounding the length of each side to the nearest tenth.

3.5 Division of Decimals

OBJECTIVES

After completing this section, you should be able to:
1. Divide decimal numbers.
2. Divide a decimal number by a power of 10.

We divide decimals in the same manner as we divide whole numbers. The only difference is that, in the division of decimal numbers, we must decide where to place the decimal point in the quotient. For instance, to divide 184.24 by 7, we proceed as follows:

$$\begin{array}{r} 26 \\ 7{\overline{\smash{\big)}\,184.24}} \\ \underline{14} \\ 44 \\ \underline{42} \\ 2 \end{array}$$

The division of 184 by 7 has a quotient of 26 and a remainder of 2. But we still have 0.24 to divide by 7. Therefore, we have 2 + 0.24 = 2.24 to divide by 7. We note that 2.2 is 22 tenths. If we divide 22 tenths by 7, we get a quotient of 3 tenths. This suggests that

3.5 Division of Decimals

we could continue the division by placing a decimal point after the 6 in the quotient and keep dividing as we would with whole numbers. We have:

$$
\begin{array}{r}
26.32 \\
7\overline{)184.24} \\
\underline{14} \\
44 \\
\underline{42} \\
2\,2 \\
\underline{2\,1} \\
14 \\
\underline{14} \\
0
\end{array}
$$

We can check this division by multiplying the quotient (26.32) by the divisor (7) and see if we obtain the dividend (184.24). You are encouraged to do this check and determine that $26.32 \times 7 = 184.24$.

Let us now divide 0.0786 by 1.31. The divisor is a decimal. We know that we may multiply both the numerator and denominator of a fraction by the same nonzero number without changing the value of the fraction. In particular, if we multiply both the dividend (0.0786) and the divisor (1.31) by 100, we obtain the following:

$$\frac{0.0786}{1.31} = \frac{0.0786 \times 100}{1.31 \times 100}$$

$$= \frac{7.86}{131}$$

Therefore, $1.31\overline{)0.0786}$ is equivalent to $131\overline{)7.86}$. We now proceed as before:

$$
\begin{array}{r}
0.06 \\
131\overline{)7.86} \\
\underline{7\,86} \\
0
\end{array}
$$

Hence, $0.0786 \div 1.31 = 0.06$.

CHECK:
$$
\begin{array}{r}
1.31 \quad \text{(Divisor)} \\
\times 0.06 \quad \text{(Quotient)} \\
\hline
0.0786 \checkmark \quad \text{(Dividend)}
\end{array}
$$

In the preceding division, note that we moved the decimal point the same number of places to the *right* in *both* the divisor and the dividend. When the divisor is a decimal number, we determine where the decimal point will be in the quotient by multiplying both the dividend and divisor by 10, 100, 1000, etc., to make the divisor a whole number.

Rule for Division of Decimal Numbers

1. If the divisor is not a whole number, move the decimal point as many places as necessary until the decimal point is to the right of all the digits.
2. Move the decimal point in the dividend the same number of places to the right.
3. Place the decimal point in the quotient directly above the decimal point in the dividend (as determined in Step 2).
4. Divide as you would with whole numbers.

Practice Exercise 1. Divide: 23.112 ÷ 43.2.

EXAMPLE 1 Divide 38.391 by 19.1.

SOLUTION: $19.1\overline{)38.391}$ or $191\overline{)383.91}$. (For the latter both the dividend and the divisor are multiplied by 10.) Hence, $19.1\overline{)38.391}$ becomes:

$$
\begin{array}{r}
2.01 \\
191\overline{)383.91} \\
\underline{382} \\
1\,91 \\
\underline{1\,91} \\
0
\end{array}
$$

Therefore, 38.391 ÷ 19.1 = 2.01.

CHECK:

$$
\begin{array}{r}
19.1 \quad \text{(Divisor)} \\
\times 2.01 \quad \text{(Quotient)} \\
\hline
191 \\
38\,20 \\
\hline
38.391 \checkmark \quad \text{(Dividend)}
\end{array}
$$

Practice Exercise 2. Divide: 316.8 ÷ 0.22.

EXAMPLE 2 Divide 197.442 by 0.4701.

SOLUTION: $0.4701\overline{)197.442}$ or $4701\overline{)1974420}$. (For the latter both the dividend and the divisor are multiplied by 10,000.) Hence, $0.4701\overline{)197.422}$ becomes:

$$
\begin{array}{r}
420 \\
4701\overline{)1974420.} \\
\underline{18804} \\
9402 \\
\underline{9402} \\
0
\end{array}
$$

Therefore, 197.442 ÷ 0.4701 = 420.

CHECK:

$$
\begin{array}{r}
0.4701 \quad \text{(Divisor)} \\
\times 420 \quad \text{(Quotient)} \\
\hline
9\,4020 \\
188\,04 \\
\hline
197.4420 \checkmark \quad \text{(Dividend)}
\end{array}
$$

Practice Exercise 3. Divide: $6.9\overline{)8.28}$.

EXAMPLE 3 Divide 29.92 by 0.022.

SOLUTION: $0.022\overline{)29.92}$ or $22\overline{)29920}$. (For the latter both the dividend and the divisor are multiplied by 1000.) Hence, $0.022\overline{)29.92}$ becomes:

$$
\begin{array}{r}
1360. \\
22\overline{)29920.} \\
\underline{22} \\
79 \\
\underline{66} \\
132 \\
\underline{132} \\
0
\end{array}
$$

3.5 Division of Decimals

Therefore, $29.92 \div 0.022 = 1360$.

$$
\begin{array}{r}
\text{CHECK:} \quad 0.022 \quad \text{(Divisor)} \\
\times 1360 \quad \text{(Quotient)} \\
\hline
1\;320 \\
6\;6 \\
22 \\
\hline
29.920 \checkmark \quad \text{(Dividend)}
\end{array}
$$

ANSWERS TO PRACTICE EXERCISES

1. 0.535. **2.** 1440. **3.** 1.2.

Just as we frequently multiply decimals by powers of 10, we also frequently divide decimals by powers of 10. We will now examine such divisions.

Practice Exercise 4. Divide: $67.9 \div 100$.

EXAMPLE 4   Divide 67.9 by 10.

SOLUTION:

$$
\begin{array}{r}
6.79 \\
10\overline{)67.90} \\
\underline{60} \\
7\;9 \\
\underline{7\;0} \\
90 \\
\underline{90} \\
0
\end{array}
$$

Practice Exercise 5. Divide: $0.769 \div 10$.

EXAMPLE 5   Divide 56.8 by 100.

SOLUTION:

$$
\begin{array}{r}
0.568 \\
100\overline{)56.800} \\
\underline{50\;0} \\
6\;80 \\
\underline{6\;00} \\
800 \\
\underline{800} \\
0
\end{array}
$$

Practice Exercise 6. Divide: $698.7 \div 10^4$.

EXAMPLE 6   Divide 23.12 by 1000.

SOLUTION:

$$
\begin{array}{r}
0.02312 \\
1000\overline{)23.12000} \\
\underline{20\;00} \\
3\;120 \\
\underline{3\;000} \\
1200 \\
\underline{1000} \\
2000 \\
\underline{2000} \\
0
\end{array}
$$

In the preceding examples, observe that:

- Dividing a decimal number by 10 has the effect of moving the decimal point in the decimal number *one* place to the *left*.

- Dividing a decimal number by 100 has the effect of moving the decimal point in the decimal number *two* places to the *left*.
- Dividing a decimal number by 1000 has the effect of moving the decimal point in the decimal number *three* places to the *left*.

Rule for Dividing a Decimal Number by a Power of 10

To divide a decimal number by 10^n (where n is a counting number), move the decimal point in the decimal number n places to the *left*.

Practice Exercise 7. Divide: $23.97 \div 10^3$.

EXAMPLE 7 Divide: $2.169 \div 10$.

SOLUTION: $2.169 \div 10 = 2.169 \div 10^1 = 0.2169$

Practice Exercise 8. Divide: $678.4 \div 1000$.

EXAMPLE 8 Divide: $13.029 \div 100$.

SOLUTION: $13.029 \div 100 = 13.029 \div 10^2 = 0.13029$

Practice Exercise 9. Divide: $3.2 \div 1000$.

EXAMPLE 9 Divide: $1.2 \div 1000$.

SOLUTION: $1.2 \div 1000 = 1.2 \div 10^3 = 0.0012$

Practice Exercise 10. Divide: $1234.5 \div 10^5$.

EXAMPLE 10 Divide: $2679.1 \div 10^4$.

SOLUTION: $2679.1 \div 10^4 = 0.26791$

ANSWERS TO PRACTICE EXERCISES

4. 0.679. **5.** 0.0769. **6.** 0.06987. **7.** 0.02397. **8.** 0.6784. **9.** 0.0032. **10.** 0.012345.

When dividing whole numbers, we sometimes have a remainder other than zero. When 25 is divided by 6, the remainder is 1; we write $\frac{25}{6} = 4\frac{1}{6}$. When dividing decimals, there may also be a nonzero remainder. If there is, we place zeros to the right of the last digit in the dividend and keep dividing, as in the following examples.

Divide and write the quotient to the nearest decimal place indicated.

Practice Exercise 11. $87.9 \div 0.6$; tenths.

EXAMPLE 11 Divide 41.2 by 5.

SOLUTION:

```
      8.2                8.24
   5)41.2       or    5)41.20
     40                  40
     ──                  ──
     1 2                 1 2
     1 0                 1 0
     ──                  ──
       2                   20
                           20
                           ──
                            0
```

CHECK: 8.24   (Quotient)
                  ×5   (Divisor)
                ─────
                41.20 ✓   (Dividend)

3.5 Division of Decimals

Sometimes the division does not terminate no matter how many zeros are placed to the right of the last digit in the dividend. The quotient, however, can be carried out to as many decimal places as desired.

Practice Exercise 12.
12.69 ÷ 0.14; thousandths.

EXAMPLE 12 Determine the quotient, to the nearest tenth, when 23.69 is divided by 1.4.

SOLUTION: To round the quotient to the nearest tenth, it is necessary to carry the division to the hundredths place. Hence,

$$1.4 \overline{)23.69}$$

becomes

```
        16.92
   14 )236.90
       14
       ---
        96
        84
        ---
        12 9
        12 6
        ----
           30
           28
           --
            2
```

Therefore, to the nearest tenth, 23.69 ÷ 1.4 is 16.9. The quotient is only approximate, due to rounding. We will indicate this by the use of the symbol ≈. Hence, 23.69 ÷ 1.4 ≈ 16.9, which is read "23.69 divided by 1.4 *is approximately equal to* 16.9."

Practice Exercise 13.
56.792 ÷ 1.8; hundredths.

EXAMPLE 13 Determine the quotient, to the nearest hundredth, when 609.14 is divided by 23.9.

SOLUTION: To round the quotient to the nearest hundredth, it is necessary to carry the division to thousandths. Hence,

$$23.9 \overline{)609.14}$$

becomes

```
          25.487
   239 )6091.400
        478
        ---
        1311
        1195
        ----
         116 4
          95 6
         -----
          20 80
          19 12
          -----
           1 680
           1 673
           -----
               7
```

Therefore, 609.14 ÷ 23.9 ≈ 25.49.

ANSWERS TO PRACTICE EXERCISES

11. 146.5.   12. 90.643.   13. 31.55.

EXERCISES 3.5

Divide in Exercises 1–30.

1. $1.07\overline{)2.4824}$
2. $2.32\overline{)0.1624}$
3. $51.7\overline{)2186.91}$
4. $86.4\overline{)46.224}$
5. $0.44\overline{)633.6}$
6. $470.1\overline{)394.884}$
7. $3.72\overline{)86.27424}$
8. $7.5\overline{)10.8525}$
9. $0.694\overline{)0.85362}$
10. $10\overline{)1.23}$
11. $100\overline{)246.9}$
12. $1000\overline{)47.97}$
13. $0.39\overline{)0.0975}$
14. $20\overline{)87.12}$
15. $40\overline{)96.7}$
16. $5.7 \div 2.5$
17. $12.312 \div 34.2$
18. $18.71712 \div 2.68$
19. $28.125 \div 0.25$
20. $104.252 \div 2.68$
21. $34.38 \div 7.2$
22. $16.821 \div 6.23$
23. $29.904 \div 0.48$
24. $981.12 \div 11.2$
25. $1.60011 \div 0.069$
26. $30.6936 \div 6.09$
27. $24.3756 \div 3.33$
28. $197.508 \div 90.6$
29. $18.564 \div 4.76$
30. $7.9992 \div 9.09$

In Exercises 31–50, divide using the rule for dividing decimals by powers of 10.

31. $23.1 \div 10$
32. $0.69 \div 10$
33. $17.89 \div 10^1$
34. $206.97 \div 10^1$
35. $59.2 \div 100$
36. $0.769 \div 100$
37. $7347.9 \div 10^2$
38. $0.0896 \div 10^2$
39. $1416.58 \div 1000$
40. $2098.76 \div 1000$
41. $7891.2 \div 10^3$
42. $12010.203 \div 10^3$
43. $5719 \div 10,000$
44. $67904.1 \div 10,000$
45. $87.9547 \div 10^4$
46. $8.734 \div 10^4$
47. $21479.8 \div 10^5$
48. $3456.789 \div 10^6$
49. $(234.5 \div 10^2) \div 10^2$
50. $(79.46 \div 10^3) \div 10^2$

In Exercises 51–60, divide and write the quotient to the nearest decimal place indicated.

51. $12.3 \div 0.4$; tenths
52. $239.4 \div 0.19$; tenths
53. $0.036 \div 0.34$; hundredths
54. $123.69 \div 1.23$; hundredths
55. $368.123 \div 2.69$; hundredths
56. $0.069 \div 0.39$; thousandths
57. $1.23 \div 0.024$; thousandths
58. $3269.7 \div 46.78$; hundredths
59. $4678.9 \div 57.04$; hundredths
60. $99.999 \div 8.08$; thousandths

In Exercises 61–66, perform the indicated operations.

61. $(23.171 + 15.22) \div 19.1$
62. $(63.4 - 21.82) \div 0.9$
63. $97.7184 \div 3.6 \div 1.17$
64. $(7.9 + 8.921) \div (8.03 - 1.8)$
65. $(1962.24 \div 11.2) \times 0.006$
66. $204.5 - 74.866 \div 0.82$

3.6 Decimal to Fraction and Fraction to Decimal Conversions

OBJECTIVES

After completing this section, you should be able to:
1. Convert a decimal number to a fraction.
2. Convert a fraction to a decimal number.

Converting a decimal number to a fraction is not difficult. Let's examine the decimal number 0.7 and the fraction $\frac{7}{10}$. Both are read "seven tenths." Similarly, both the decimal number 0.93 and the fraction $\frac{93}{100}$ are read "ninety-three hundredths." Hence, to convert a decimal number to a fraction, read the decimal and write the corresponding fraction that has the same name. Simplify the fraction, if necessary.

In Practice Exercises 1–4, write each of the following decimal numbers as a fraction in simplest form.

Practice Exercise 1.

EXAMPLE 1 Convert 0.32 to a fraction.

SOLUTION:

STEP 1: The decimal number 0.32 is read "thirty-two hundredths."

STEP 2: The corresponding fraction is $\frac{32}{100}$.

STEP 3: $$0.32 = \frac{32}{100}$$
$$= \frac{8}{25} \quad \text{(Simplified)}$$

Practice Exercise 2.

EXAMPLE 2 Convert 0.684 to a fraction.

SOLUTION:

STEP 1: The decimal number 0.684 is read "six hundred eighty-four thousandths."

STEP 2: The corresponding fraction is $\frac{684}{1000}$.

STEP 3: $$0.684 = \frac{684}{1000}$$
$$= \frac{171}{250} \quad \text{(Simplified)}$$

To write 2.39 as a fraction, consider:

$$2.39 = 2 + 0.39$$
$$= \frac{200}{100} + \frac{39}{100} \quad \text{(Like fractions)}$$
$$= \frac{239}{100}$$

These examples illustrate the following rule for converting a decimal number to a fraction.

Rule for Converting a Decimal Number to a Fraction

1. The *numerator* of the fraction is the decimal number without the decimal point.
2. The *denominator* of the fraction is a 1 followed by as many zeros as there are decimal places in the decimal number.
3. Simplify the fraction, if necessary.

Practice Exercise 3. 1.25

EXAMPLE 3 Write 0.425 as a fraction.

SOLUTION:

- The numerator is 425.
- The denominator is 1000 since there are three decimal places in the decimal number.
- The fraction is $\frac{425}{1000} = \frac{17}{40}$ (simplified).

Practice Exercise 4. 36.08

EXAMPLE 4 Write 23.12 as a fraction.

SOLUTION:

- The numerator is 2312.
- The denominator is 100 since there are two decimal places in the decimal number.
- The fraction is $\frac{2312}{100} = \frac{578}{25}$ (simplified).

ANSWERS TO PRACTICE EXERCISES

1. $\frac{19}{50}$.   2. $\frac{109}{500}$.   3. $\frac{5}{4}$.   4. $\frac{902}{25}$.

To convert a fraction to a decimal number, we make use of the fact that a fraction involves division. For example, $\frac{3}{6}$ means $3 \div 6$; $\frac{7}{8}$ means $7 \div 8$; and $\frac{16}{5}$ means $16 \div 5$.

In Practice Exercises 5–9, convert each of the following fractions to decimal numbers.

Practice Exercise 5. $\frac{1}{16}$

EXAMPLE 5 Convert the fraction $\frac{3}{6}$ to a decimal number.

SOLUTION:

$$3 \div 6 = 0.5 \text{ since } 6 \overline{)3.0}$$
$$\phantom{3 \div 6 = 0.5 \text{ since }}\underline{3\,0}$$
$$\phantom{3 \div 6 = 0.5 \text{ since 6)}}0$$

Practice Exercise 6. $\frac{13}{20}$

EXAMPLE 6 Convert the fraction $\frac{7}{8}$ to a decimal number.

SOLUTION:

$$7 \div 8 = 0.875 \text{ since } 8 \overline{)7.000}$$
$$\underline{6\,4}$$
$$60$$
$$\underline{56}$$
$$40$$
$$\underline{40}$$
$$0$$

3.6 Decimal to Fraction and Fraction to Decimal Conversions

Practice Exercise 7. $\frac{45}{32}$

EXAMPLE 7 Convert the fraction $\frac{16}{5}$ to a decimal number.

SOLUTION:

$$16 \div 5 = 3.2 \text{ since } 5 \overline{\smash{)}16.0}$$

$$\begin{array}{r} 3.2 \\ 5\overline{)16.0} \\ \underline{15} \\ 1\,0 \\ \underline{1\,0} \\ 0 \end{array}$$

Note that these three fractions were converted to decimals by dividing the numerator of the fraction by the denominator of the fraction.

Rule for Converting a Fraction to a Decimal Number

To convert a fraction to a decimal number, divide the numerator of the fraction by the denominator of the fraction.

Practice Exercise 8. $\frac{163}{25}$

EXAMPLE 8 Convert $\frac{17}{4}$ to a decimal number.

SOLUTION: Divide 17 by 4:

$$\begin{array}{r} 4.25 \\ 4\overline{)17.00} \\ \underline{16} \\ 1\,0 \\ \underline{8} \\ 20 \\ \underline{20} \\ 0 \end{array}$$

Therefore, $\frac{17}{4} = 4.25$.

Practice Exercise 9. $\frac{96}{400}$

EXAMPLE 9 Convert $\frac{63}{25}$ to a decimal number.

SOLUTION: Divide 63 by 25:

$$\begin{array}{r} 2.52 \\ 25\overline{)63.00} \\ \underline{50} \\ 13\,0 \\ \underline{12\,5} \\ 50 \\ \underline{50} \\ 0 \end{array}$$

Therefore, $\frac{63}{25} = 2.52$.

ANSWERS TO PRACTICE EXERCISES

5. 0.0625.   6. 0.65.   7. 1.40625.   8. 6.52.   9. 0.24.

So far in our discussion of converting a fraction to a decimal number, the division had a zero remainder. In such cases, the corresponding decimal numbers are called **terminating decimals.** In some cases, however, the division will never have a zero remainder no matter how far we carry out the division. Consider the fraction $\frac{1}{3}$. If we divide the numerator, 1, by the denominator, 3, the division does not terminate:

$$\begin{array}{r} 0.333 \\ 3\overline{)1.000} \\ \underline{9} \\ 10 \\ \underline{9} \\ 10 \\ \underline{9} \\ 1 \end{array}$$

For each step in the division process, a remainder of 1 was obtained and the division continued. Since the division will never have a remainder of 0, we can approximate $\frac{1}{3}$ by a decimal such as:

$$\frac{1}{3} \approx 0.3 \quad \text{(To the nearest tenth)}$$
$$\frac{1}{3} \approx 0.33 \quad \text{(To the nearest hundredth)}$$
$$\frac{1}{3} \approx 0.333 \quad \text{(To the nearest thousandth)}$$

We also represent the decimal number for $\frac{1}{3}$ as:

$$\frac{1}{3} = 0.333\ldots$$

where the three dots, . . . , indicate the repeating nature of the decimal number. Such decimal numbers are called **repeating decimals.**

In a similar manner, the fraction $\frac{3}{11}$ can be written as the repeating decimal

$$\frac{3}{11} = 0.272727\ldots$$

with the two digits, 27, repeating.

A repeating decimal may also be written with a bar over the digit or digits that repeat, instead of using the three dots. For example, we write

$$\frac{1}{3} = 0.\overline{3} \quad \text{(Since the digit 3 repeats)}$$

and

$$\frac{3}{11} = 0.\overline{27} \quad \text{(Since the digits 27 repeat)}$$

When we convert a fraction to a decimal number and the quotient is a repeating decimal, we may want to write the quotient correct to the nearest hundredth, for instance. To do so, we would have to carry the division out to the thousandths place and round the quotient to the nearest hundredth.

3.6 Decimal to Fraction and Fraction to Decimal Conversions

EXAMPLE 10 Write $\frac{7}{11}$ as a decimal number correct to the nearest hundredth.

SOLUTION: Since we want the decimal number to the nearest hundredth, we carry the division out three decimal places and round the quotient to the nearest hundredth:

$$\begin{array}{r} 0.636 \\ 11\overline{)7.000} \\ \underline{6\;6} \\ 40 \\ \underline{33} \\ 70 \\ \underline{66} \\ 4 \end{array}$$

Therefore, $\frac{7}{11} = 0.64$ to the nearest hundredth.

Practice Exercise 10. (a) Write $\frac{1}{7}$ as a decimal number, correct to the nearest hundredth. (b) Write $\frac{11}{13}$ as a decimal number, correct to the nearest thousandth. (c) Write $\frac{2}{11}$ as a decimal number, correct to the nearest hundredth. (d) Write $\frac{3}{17}$ as a decimal number, correct to the nearest thousandth.

ANSWERS TO PRACTICE EXERCISE

10. (a) 0.14. (b) 0.846. (c) 0.18. (d) 0.176.

EXERCISES 3.6

In Exercises 1–25, convert each of the decimal numbers to a fraction in simplest form.

1. 0.3
2. 3.2
3. 0.16
4. 0.05
5. 0.125
6. 8.4
7. 16.6
8. 2.036
9. 0.0234
10. 0.00125
11. 9.85
12. 7.562
13. 0.24375
14. 5.068
15. 19.02
16. 7.025
17. 20.204
18. 3.1255
19. 1.10015
20. 22.222
21. 0.00525
22. 100.25
23. 17.284
24. 69.035
25. 100.0565

In Exercises 26–40, convert each of the fractions to a decimal number. Carry out the division until the remainder is zero.

26. $\frac{3}{4}$
27. $\frac{5}{8}$
28. $\frac{3}{5}$
29. $\frac{7}{16}$
30. $\frac{19}{16}$
31. $\frac{21}{8}$
32. $\frac{6}{25}$
33. $\frac{8}{5}$
34. $\frac{17}{16}$
35. $\frac{101}{200}$
36. $\frac{53}{64}$
37. $\frac{59}{20}$
38. $\frac{89}{125}$
39. $\frac{219}{400}$
40. $\frac{1111}{800}$

In Exercises 41–50, convert each of the fractions to a decimal number correct to the nearest hundredth.

41. $\frac{7}{6}$
42. $\frac{8}{15}$
43. $\frac{5}{9}$
44. $\frac{4}{7}$

45. $\dfrac{23}{13}$ **46.** $\dfrac{29}{15}$ **47.** $\dfrac{63}{13}$ **48.** $\dfrac{7}{31}$

49. $\dfrac{12}{43}$ **50.** $\dfrac{15}{51}$

In Exercises 51–60, convert each of the fractions to a decimal number correct to the nearest thousandth.

51. $\dfrac{7}{15}$ **52.** $\dfrac{11}{13}$ **53.** $\dfrac{22}{15}$ **54.** $\dfrac{17}{31}$

55. $\dfrac{19}{41}$ **56.** $\dfrac{37}{51}$ **57.** $\dfrac{23}{12}$ **58.** $\dfrac{107}{6}$

59. $\dfrac{41}{17}$ **60.** $\dfrac{53}{37}$

3.7 Applications Involving Decimals

OBJECTIVE

After completing this section, you should be able to:
1. Solve simple word problems involving decimals.
2. Determine the circumference of a circle.

In this section, we will examine different types of word problems that involve decimal numbers. A basic problem involving decimals is that of cost. Cost is generally given in dollars and cents. However, 1 cent is equal to $\dfrac{1}{100}$ of a dollar. Using $ to represent dollars, we would write:

$$1 \text{ cent} = \$.01$$

Hence,

$$93 \text{ cents} = \$.93$$

and

$$2 \text{ dollars and } 71 \text{ cents} = \$2.71$$

Sometimes, the cost is given in tenths of a cent. For instance, 23.6 cents would be written as

$$23.6 \text{ cents} = \$.236$$

Practice Exercise 1. You plan to vacation and estimate that you will spend $38.50 per day. How many days of vacation can you take if you have $500.50?

EXAMPLE 1 Carol went on a shopping spree and spent $29.99 for a dress, $19.65 for a sweater, $8.35 for a hat, and $13.06 for cosmetics. How much was the total cost of these items?

SOLUTION: The total cost of the four items would be determined by addition as follows:

3.7 Applications Involving Decimals

$29.99 (Cost of dress)
19.65 (Cost of sweater)
8.35 (Cost of hat)
+ 13.06 (Cost of cosmetics)
$71.05 (Total cost)

Practice Exercise 2. Mike has a temporary job that pays $5.85 per hour. If he works 265 hr, how much will he earn?

EXAMPLE 2 If no-lead gasoline sells for 92.9¢ per gallon, how much will 23.4 gal of gasoline cost (to the nearest cent)?

SOLUTION: This problem involves multiplication since we are given the cost of 1 gal of gasoline and wish to determine the cost of 23.4 gal of gasoline. We also note that 92.9¢ can be written as $.929. Hence,

$.929 (Price for 1 gal of gasoline)
× 23.4 (Number of gallons of gasoline)
3716
2 787
18 58
$21.7386 (Total cost)

To the nearest cent, the total cost of the gasoline would be $21.74.

Practice Exercise 3. Maria made three purchases that amounted to $15.63; $13.57; and $6.99, plus sales tax. If the sales tax is 5 cents per dollar, what was the total amount of Maria's purchasing including sales tax?

EXAMPLE 3 A man left 0.6 of his estate to his favorite charity. If the charity received $50,190, what was the value of the total estate?

SOLUTION: This problem involves division since we want to determine

$$0.6 \times ? = \$50{,}190$$

To determine the value of "?", we divide $50,190 by 0.6 as follows:

$$0.6\overline{)\$50190}$$

or

$$6\overline{)\$501900} \quad \$83650$$
48
21
18
39
36
30
30
0

Therefore, the total estate was valued at $83,650.

CHECK: $83,650
  × 0.6
  $50,190.0 ✔

Practice Exercise 4. If a new car loses 0.26 of its value during the first year, how much would a 1-year-old car be worth if it cost $10,655 when it was new?

EXAMPLE 4 Liz had $139.02 in her checking account. She wrote a check for $60.84. What was the new balance?

SOLUTION: This problem involves subtraction since Liz starts with $139.02 and "takes away" $60.84. Hence, we have:

$139.02 (Starting balance)
− 60.84 (Amount of check)
$78.18 (New balance)

Practice Exercise 5. If 1 in. is equal to 2.54 cm, how many centimeters are in 27.8 in.?

EXAMPLE 5 If fresh tomatoes sell for $2.75 a basket and there are 14 lb of tomatoes in a basket, how much will 2380 lb of tomatoes cost?

SOLUTION: There are two parts to this problem. Since we are given the price for 14 lb of tomatoes, in order to arrive at the price for 2380 lb of tomatoes, we must determine how many groups of 14 are contained in 2380. This can be done by division:

$$\begin{array}{r} 170 \\ 14\overline{)2380} \\ \underline{14} \\ 98 \\ \underline{90} \\ 0 \end{array}$$

Hence, the 2380 lb of tomatoes would be equivalent to 170 baskets of tomatoes. We know the price for 1 basket of tomatoes. To determine the total price, we now multiply:

$$\begin{array}{rl} \$2.75 & \text{(Cost for 1 basket of tomatoes)} \\ \times 170 & \text{(Number of baskets of tomatoes)} \\ \hline 192\ 50 & \\ 275\ \ \ & \\ \hline \$467.50 & \text{(Total cost)} \end{array}$$

Hence, at the rate of $2.75 per basket, 2380 lb of tomatoes would cost $467.50.

Now we turn to using decimals to determine the **circumference of a circle.**

Practice Exercise 6. Determine the radius, r, of a circle (to the nearest tenth of an inch) whose circumference, C, is 356.8 in. (Use 3.14 for π.)

EXAMPLE 6 The circumference, C, of a circle is determined using the formula $C = 2\pi r$. Using 3.14 for the value of π, determine the circumference of a circle whose radius, r, is 12.6 cm.

SOLUTION: We have:

$$\begin{aligned} C &= 2\pi r \\ &= 2(3.14)(12.6 \text{ cm}) \\ &= 2(3.14)(12.6) \text{ cm} \\ &= 79.128 \text{ cm} \end{aligned}$$

Therefore, the circumference of the circle is 79.128 cm.

Practice Exercise 7. The circumference, C, of a circle is 137.8 cm. Using 3.14 for the value of π, determine its radius, r, to the nearest tenth of a centimeter.

EXAMPLE 7 The circumference, C, of a circle is 239.6 in. Using 3.14 for the value of π, determine its radius, r, to the nearest tenth of an inch.

SOLUTION: We have:

$$\begin{aligned} C &= 2\pi r \\ 239.6 \text{ in.} &= 2(3.14)r \\ 239.6 \text{ in.} &= 6.28r \\ \frac{239.6 \text{ in.}}{6.28} &= r \\ 38.15 \text{ in.} &= r \quad \text{(Division carried out to the hundredths place)} \\ 38.2 \text{ in.} &= r \quad \text{(To the nearest tenth of an inch)} \end{aligned}$$

ANSWERS TO PRACTICE EXERCISES

1. 13. **2.** $1550.25. **3.** $38.00. **4.** $7884.70. **5.** 70.612 cm. **6.** 56.8 in. **7.** 21.9 cm.

EXERCISES 3.7

1. If 3 large grapefruit sell for 89¢, how much will $1\frac{1}{2}$ dozen grapefruit cost?
2. For a particular movie, a theater was selling tickets at $2.75 for adults and $1.50 for children. During one performance, 868 adult tickets and 671 child tickets were sold. What was the amount received for the tickets sold?
3. Gina made three purchases that amounted to $16.82; $11.25; and $8.63 plus sales tax. If the sales tax is 4¢ per dollar, what was the total amount of Gina's purchases including sales tax?
4. A man drove his car for 1368 mi and used 81.6 gal of gasoline. What was the average number of miles traveled per gallon of gasoline? Give your answer to the nearest tenth of a mile.
5. In a local community, real estate taxes are $79.12 per $1000 of assessed valuation. If a house is assessed at $39,500, how much is the real estate tax?
6. In an urban school district, the annual budget is $3,469,123. If transportation costs represent 0.13 of the budget, how much is the transportation part of the budget?
7. The budget for a rural community during a particular year was $72,000,000. If 0.16 of the budget is for health services, 0.22 is for police and fire protection, 0.09 is for education, 0.38 is for welfare, and 0.15 is for other expenses, how much is budgeted for:
    a. Health services?
    b. Police and fire protection?
    c. Education?
    d. Welfare?
    e. Other expenses?
8. Ms. Jackson died and left a will designating that 0.25 of her estate be given to her husband, 0.35 to her daughter, 0.3 to her son, and 0.1 to her confidential secretary. How much would each receive if her estate was valued at $83,950?
9. If a new car loses 0.27 of its value during the first year, how much would a 1-year-old car be worth if it cost $6800 when it was new?
10. Mom baked a pie and set it aside to cool. Karen came home and ate 0.2 of the pie. Later, Michael came home and ate 0.25 of what was left. After the two had eaten their pieces of pie, how much of the pie was left?
11. If whole-kernel corn sells at 3 cans for 89¢ and if there are 24 cans of corn in a case, how much would a case of corn cost?
12. If large cans of cut green beans sell at 2 cans for 79¢, and if there are 36 cans of beans in a case, how much would a case of beans cost?
13. If excise tax on your car is levied at the rate of $66 per thousand, how much tax do you have to pay on a car valued at $4980?
14. On December 1, my bank balance was $1742.25. During the month of December, I wrote checks for $11.75; $13.99; $53.62; $475; and $102.28. I also deposited two checks of $494.77 each. What was my balance at the end of the month, if these were the only transactions?
15. If gasoline sells for $1.29 per gallon, how much will 17.4 gal of gasoline cost?
16. The Smiths spend an evening out and pay $3.25 for parking, $8.75 for a babysitter, $45 for theater tickets, and $63.98 for dinner. How much does this cost?
17. If heating oil sells for $1.09 per gallon, how much will a delivery of 245.7 gal cost?
18. A kg is approximately 2.2 lb. How much does a 7-lb baby weigh in kilograms?
19. A chemist runs several reactions to accumulate a certain compound. On successive experiments, her yields are 11.782 g, 19.892 g, and 8.007 g. What is the total amount of this compound?
20. At a weight clinic, the instructor calculates the total weight loss of the class. The following losses were recorded; 1.5 kg, 2 kg, 0.7 kg, 1.1 kg, 2.2 kg, 0.4 kg, 1.6 kg, 1.7 kg, 0.9 kg, 1.2 kg, and 0.8 kg. What is the total weight loss?
21. An average weight loss can be calculated by dividing the total weight loss by the number of people weighed. What is the average weight loss of the class in Exercise 20?

22. Calculate, to the nearest tenth of a gallon, the miles per gallon (number of miles/gallons) performance of a car using 15.4 gal of gasoline for a 367.4-mi trip.
23. Three friends agreed to split the cost of a trip equally among them. The cost were: gas, $54.98; meals, $117.24; turnpike tolls, $3.78; and entertainment, $16.75. How much did each person pay?
24. The following numbers of tickets to a charity concert were sold:

    128 tickets @ $2.25 per ticket
    109 tickets @ $1.75 per ticket
     43 tickets @ $3.40 per ticket
    213 tickets @ $1.00 per ticket

    What was the total amount from the sale of the tickets?
25. In a circus act, a troupe forms a five-man "totem pole," with one man holding four others on his shoulders. If the weights of the other four members of the troupe are 63.7 kg, 57.5 kg, 67.4 kg, and 59.9 kg, what is the total weight the bottom man must support?

In Exercises 26–30, use 3.14 for π and determine the circumference, C, of a circle with radius, r, indicated.

26. $r = 1.7$ yd
27. $r = 10.6$ ft
28. $r = 13.67$ cm
29. $r = 26.56$ in.
30. $r = 106.1$ in.

In Exercises 31–35, use 3.14 for π and determine the radius, r, of a circle with circumference, C, indicated. Round your answer to the nearest tenth.

31. $C = 100.8$ ft
32. $C = 216.78$ in.
33. $C = 489.68$ cm
34. $C = 87.93$ yd
35. $C = 1000.9$ in.

Chapter Summary

In this chapter, you were introduced to decimal numbers. Operations involving decimals were discussed. Relationships between fractions and decimals were also examined.

The **key words** introduced in this chapter are listed here in the order in which they appeared in the chapter. You should know what they mean before proceeding to the next chapter.

decimal point (p. 124)
decimal number (p. 125)
powers of 10 (p. 134)
rounding (p. 135)
estimation (p. 136)
terminating decimals (p. 148)
repeating decimals (p. 148)
circumference of a circle (p. 152)

Review Exercises

3.1 Decimals as Numbers

In Exercises 1–5, determine the place value for the underlined digit in each of the given decimal numbers.

1. 31.0<u>4</u>6
2. 17.<u>9</u>07
3. 1.02<u>5</u>
4. 9<u>6</u>2.1
5. 24.<u>0</u>778

In Exercises 6–10, write the name of each of the decimal numbers given.

6. 0.27
7. 3.9
8. 19.057
9. 101.01
10. 327.689

Review Exercises

In Exercises 11–15, insert the symbol <, =, or > between the numbers in each pair to make true statements.

11. 0.59   0.9   **12.** 0.39   0.117   **13** 0.8   0.80
14. 2.3   2.297   **15.** 10.01   10.001

In Exercises 16–18, arrange the decimal numbers in order of size, with the greatest first.

16. 0.72, 0.891, 0.6, 0.593   **17.** 0.09, 0.99, 0.909, 0.099
18. 19.305, 19.035, 19.33, 19.2

3.2 Addition and Subtraction of Decimals

Add in Exercises 19–21.

19. 0.37 + 1.092 + 14.3   **20.** 46.19 + 207.5 + 0.694
21. 0.07 + 10.006 + 201.0009

Subtract in Exercises 22–25.

22. 92.1 − 37.069   **23.** 9.603 − 0.9603
24. 121.234 − 17.06 − 47.9
25. From 69.21, subtract 46.093.

In Exercises 26–27, perform the indicated operations.

26. 76.5 + 2.09 − 39.67   **27.** 468.732 − (0.97 + 276.9)

3.3 Multiplication of Decimals

Multiply in Exercises 28–30.

28. 17.6 × 0.06   **29.** (1.23)(19.62)
30. 0.07 × 1.1 × 29.06

In Exercises 31–32, perform the indicated operations.

31. (1.19 + 30.002) × 3.1   **32.** 2.39 + 1.97 × 0.069

3.4 Rounding and Estimating

In Exercises 33–37, round each number to the nearest place indicated.

33. 69.99; tenths   **34.** 7649.987; hundreds
35. 0.02976; thousandths   **36.** 8065.0; tens
37. 0.209067; ten thousandths

3.5 Division of Decimals

Divide in Exercises 38–40.

38. 2.6)25.22   **39.** 0.09)0.0459
40. 437.382 ÷ 51.7

In Exercises 41–42, perform the indicated operations.

41. (462.24 ÷ 86.4) + 0.096   **42.** 633.6 ÷ 0.44 ÷ 24

In Exercises 43–44, divide and write your quotient to the nearest decimal place indicated.

43. 37.9 ÷ 0.5; tenths   **44.** 1267 ÷ 0.29; hundredths

3.6 Decimal to Fraction and Fraction to Decimal Conversions

In Exercises 45–48, convert each of the decimal numbers to a fraction in simplest form.

45. 0.26   **46.** 1.096   **47.** 30.46   **48.** 69.0355

In Exercises 49–52, convert each of the fractions to a terminating decimal.

49. $\dfrac{7}{4}$ **50.** $\dfrac{19}{20}$ **51.** $\dfrac{6}{125}$ **52.** $\dfrac{87}{20}$

In Exercises 53–56, convert each of the fractions to a decimal number correct to the nearest place indicated.

53. $\dfrac{11}{6}$; tenths **54.** $\dfrac{28}{15}$; hundredths

55. $\dfrac{43}{31}$; thousandths **56.** $\dfrac{27}{56}$; thousandths

3.7 Applications Involving Decimals

57. What is 0.8 of 153.6?
58. Ms. Traveler drove her car for 2169 mi and used 121.8 gal of gasoline. What was the average number of miles traveled per gallon of gasoline? Give your answer to the nearest tenth of a mile.
59. If a gallon of gasoline cost $1.23, how many gallons of gasoline can you buy for $15? Give your answer correct to the nearest tenth of a gallon.
60. If you bought two shirts at $11.95 each, a tie for $7.99, and a belt for $8.49, how much change should you get back if you pay for the purchases with a $45.65 check and the sales tax on your purchases amounts to $1.61?
61. If a VCR that regularly sells for $339.95 is marked down $53.85, what is its sale price?
62. Which is larger, 0.13 of 187.4 or 0.27 of 267.9? How much larger?
63. Cynthia starts the month with $362.36 in her checking account. During the month, she writes checks for $43.15, $50.92, $103.75, and $87.69. She also makes a deposit of $263.80. What is her balance at the end of the month if there is also a service charge of $2.65?
64. The Royal order of the Friendly Irish sold green carnations for St. Patrick's Day. If the carnations cost $7.98 a dozen and were sold for $1 each, how much profit was made from the sale of 25 dozen carnations?
65. A welder earns $14.75 an hour and is paid time-and-a-half (1.5 times the hourly rate) for every hour over 40 worked in a single week. If the welder worked 47.5 hours in one week, what would be his pay? (Round your answer to the nearest cent.)

1. A secretarial service offers the following two types of service. For a flat monthly rate of $200, you can make an unlimited number of copies of documents. Or you can pay $50 per month plus $0.06 for each copy in excess of 600 copies. How many copies would you need to make each month in order for the flat rate to be the better buy?
2. The width of a rectangle is 16.2 cm and the area is 333.72 cm². Using good English and complete sentences, give a step-by-step set of instructions to indicate how one could determine the perimeter of the rectangle. What is the perimeter?
3. Is it possible for the sum of two decimal numbers to be a whole number? Explain your answer.
4. A student adds 13.045 and 1.2613 and obtains a sum of 25.658. The correct answer is 14.3063. What do you think that the student did wrong?
5. Another student inadvertently adds 23.12 to a number and gets a sum of 92.71. However, he should have subtracted 23.12 from the number. What is the number? What should have been the correct answer?
6. Using good English and complete sentences, give a step-by-step set of instructions to determine how one could convert the decimal number 2.305 to an equivalent fraction simplified.

In your own words, but using good English and complete sentences, indicate what is meant by each of the statements given in Exercises 7–10.

7. The decimal number $21.1\overline{37}$ is a repeating decimal.
8. The decimal numbers 1.9 and 1.90 are equivalent decimals and represent the same fraction.
9. If a, b, and c represent any decimal numbers, then $a(b + c) = ab + ac$.
10. Every whole number can be written as a decimal number.

CHAPTER 3 TEST

This test covers materials from Chapter 3. Read each question carefully, and answer all questions. If you miss a question, review the appropriate section of the text.

1. Rewrite $\frac{71}{100}$ as a decimal number.

2. Rewrite $\frac{139}{10,000}$ as a decimal number.

In Exercises 3–16, perform the indicated operations.

3. 143.02
    8.961
    +23.4

4. 906.793
    27.1
    +4.0967

5. $23.12 + 6.971 + 107$

6. $106.37 + 12.9 + 1.796$

7. 23.4
    −9.76

8. 207.972
    −89.69

9. $126.7 - 47.51$

10. $80.907 - 16.09$

11. $(23.2)(1.36)$

12. $0.76(1.67 + 0.5)$

13. $50.44 \div 2.6$

14. $\frac{0.2916}{0.18} = ?$

15. $(23.2 + 6.76) \times 2.9$

16. $(231.12 \div 43.2) - 0.096$

17. Round 2.03546 to the nearest thousandth.

18. Round 0.0976 to the nearest tenth.

19. Jim ran the 300-m shuttle run in 59.23 sec. Jack ran the same run in 61.18 sec. How much faster was Jim's time?

20. If 237.4 gal of gasoline were used during a 4791.8-mi trip, what was the average number of miles per gallon for the trip? (Give your answer to the nearest tenth.)

CUMULATIVE REVIEW: CHAPTERS 1–3

In Exercises 1–3, circle the larger number.

1. 231.42 or 238

2. $\frac{1}{2}$ or 0.469

3. $\frac{21}{4}$ or 5.025

In Exercises 4–18, perform the indicated operations.

4. $36 + 157$

5. $\frac{14}{5} + \frac{4}{15}$

6. $2.09 + 1.309$

7. $\frac{3}{4} + 23.9$

8. $9\frac{2}{7} + 6\frac{1}{8}$

9. $1980 - 908$

10. $13\frac{9}{10} - 5\frac{7}{9}$

11. $43 - 21.07$

12. $5 - \frac{11}{12}$

13. $83 \times 43\frac{1}{2}$

14. 3.08×4.56

15. $\frac{18}{5} \times \frac{35}{54}$

16. $7\frac{2}{9} \times 8\frac{1}{4}$

17. $4\frac{3}{7} \div 5\frac{1}{3}$

18. $462.24 \div 43.2$

19. Round 2304 to the nearest ten.

20. Round 37.069 to the nearest tenth.

21. Round 56.10839 to the nearest thousandth.

22. The sides of a triangle measure 3.79 ft, 4.68 ft, and 6.54 ft. Estimate the perimeter, P, of the triangle by rounding each of the measurements to the nearest tenth.

23. A rectangle has a width of 30.279 cm and length of 27.905 cm. Estimate the area, A, of the rectangle by rounding each of the measurements to the nearest tenth.

24. Rewrite the decimal 32.462 as a simplified fraction.

25. Rewrite the fraction $\frac{17}{3}$ as a decimal, correct to the nearest hundredth.

CHAPTER 4 Percent with Applications

Decimal numbers are frequently associated with what is called "percent." We often read or hear references to the percent of the number of people in a local community who are unemployed, the percent of increase in the cost of home heating oil, the percent of college students who make the dean's list, the percent of precipitation in the weather forecast, or the percent of employees from among certain minority groups.

In this chapter, we will introduce the concept of percent and the relationship between percent and decimals. Several applications of percent will be discussed. These will include sales and property tax, commissions, discount, and simple interest problems.

CHAPTER 4 PRETEST

This pretest covers material on percent. Read each question carefully and try to answer all questions. Each question is keyed to the chapter section in which the particular topic is discussed. Check your answers with those given in the back of the text.

Questions 1–2 pertain to Section 4.1.

1. Rewrite $\frac{7}{4}$ as a percent.

2. Rewrite 205% as a fraction in simplest form.

Questions 3–7 pertain to Section 4.2.

3. Change 13.8% to a fraction in lowest terms.

4. Change 0.37 to a percent.

5. Change 7 to a percent.

6. Change 1.96% to a decimal.

7. Change 1.256 to a percent.

Questions 8–11 pertain to Section 4.3.

8. What is 14% of 161.2?

9. 16 is what percent of 64?

10. What percent of 80 is 60?

11. 120 is what percent of 40?

Questions 12–13 pertain to Section 4.4.

12. The temperature increased from 65° F to 78° F during a 6-hr period. What was the percent of increase in temperature during that period?

13. The price of a VCR was decreased by 18% for a 3-day sale. If the regular price of the appliance was $369, what is the sale price?

Questions 14–15 pertain to Section 4.5.

14. Determine the amount of sales tax on a pair of shoes that sells for $37.50 if the sales tax rate is 6%.

15. Determine the total price for a sailboat that sells for $1090 with a sales tax rate of 4.5%.

16. Sue works in a retail store. She receives a weekly salary of $60 and a commission of 7% of net sales. If her net sales for a week were $1350, what was her total pay for the week? (Section 4.6)

17. A small copying machine has a marked price of $699 and a sales price of $487.95. What is the discount? (Section 4.7).

18. Determine the interest on $850 at 8% per year for 9 months. (Section 4.8)

4.1 Percent

OBJECTIVES

After completing this section, you should be able to:
1. Write fractions as percents.
2. Write percents as fractions in simplest form.

DEFINITION
A **percent** is a fraction with a denominator of 100.

- The fraction $\frac{25}{100}$ can also be written as 25 percent.
- The fraction $\frac{137}{100}$ can also be written as 137 percent.

The word "percent" means "per hundred" or "so many parts per 100." Instead of writing the word "percent," it is common to use the symbol %. Hence, in the examples just given, we could rewrite 25 percent as 25% and 137 percent as 137%.

Fractions that have denominators other than 100 can also be written as percents, as we will illustrate in the following examples.

Rewrite each of the following numbers as percents.

Practice Exercise 1. $\frac{1}{4}$

EXAMPLE 1 Rewrite the fraction $\frac{1}{2}$ as a percent.

SOLUTION:
$$\frac{1}{2} = \frac{1 \times 50}{2 \times 50}$$
$$= \frac{50}{100}$$
$$= 50\%$$

Practice Exercise 2. $\frac{3}{8}$

EXAMPLE 2 Rewrite the fraction $\frac{3}{4}$ as a percent.

SOLUTION:
$$\frac{3}{4} = \frac{3 \times 25}{4 \times 25}$$
$$= \frac{75}{100}$$
$$= 75\%$$

Practice Exercise 3. $\frac{2}{3}$

EXAMPLE 3 Rewrite the fraction $\frac{1}{3}$ as a percent.

SOLUTION:
$$\frac{1}{3} = \frac{1 \times 100}{3 \times 100}$$
$$= \frac{100 \times 1}{3 \times 100}$$
$$= \left(\frac{100}{3}\right)\left(\frac{1}{100}\right)$$

4.1 Percent

$$= \left(33\frac{1}{3}\right)\left(\frac{1}{100}\right)$$

$$= \frac{33\frac{1}{3}}{100}$$

$$= 33\frac{1}{3}\%$$

Notice, in Examples 1–3, that the fractions given represent numbers that are less than 1. These have been converted to percents less than 100 percent. In Example 4, we observe that the whole number 1 is written as 100 percent.

Practice Exercise 4. $\frac{5}{6}$

EXAMPLE 4 Rewrite 1 as a percent.

SOLUTION:
$$1 = \frac{1}{1}$$
$$= \frac{1 \times 100}{1 \times 100}$$
$$= \frac{100}{100}$$
$$= 100\%$$

In Examples 5–7, you will note that numbers greater than 1 are written as percents greater than 100 percent.

Practice Exercise 5. $\frac{9}{4}$

EXAMPLE 5 Rewrite $\frac{7}{4}$ as a percent.

SOLUTION:
$$\frac{7}{4} = \frac{7 \times 25}{4 \times 25}$$
$$= \frac{175}{100}$$
$$= 175\%$$

Practice Exercise 6. 6

EXAMPLE 6 Rewrite 4 as a percent.

SOLUTION:
$$4 = \frac{4}{1}$$
$$= \frac{4 \times 100}{1 \times 100}$$
$$= \frac{400}{100}$$
$$= 400\%$$

Practice Exercise 7. 7.8

EXAMPLE 7 Rewrite $3\frac{1}{4}$ as a percent.

SOLUTION:
$$3\frac{1}{4} = \frac{13}{4}$$
$$= \frac{13 \times 25}{4 \times 25}$$
$$= \frac{325}{100}$$
$$= 325\%$$

> **ANSWERS TO PRACTICE EXERCISES**
>
> 1. 25%.   2. 37.5%.   3. $66\frac{2}{3}$%.   4. $83\frac{1}{3}$%.   5. 225%.   6. 600%.   7. 780%.

Examples 1–7 illustrate the procedure for rewriting a fraction as a percent. Note that the same results are obtained using the following rule.

Rule for Rewriting a Fraction as a Percent

To rewrite a fraction as a percent:

1. Divide the numerator of the fraction by its denominator.
2. Multiply the quotient obtained in Step 1 by 100.
3. Add the % symbol after the result obtained in Step 2.

Rewrite each number as a percent.

Practice Exercise 8. $\frac{2}{10}$

EXAMPLE 8 Rewrite the fraction $\frac{3}{8}$ as a percent.

SOLUTION:

Step 1: Divide 3 (the numerator) by 8 (the denominator), obtaining 0.375.
Step 2: Multiply 0.375 by 100, obtaining 37.5.
Step 3: Add the % symbol after 37.5, obtaining 37.5%.

Therefore, $\frac{3}{8} = 37.5\%$.

Practice Exercise 9. 9

EXAMPLE 9 Rewrite $\frac{12}{5}$ as a percent.

SOLUTION:

Step 1: Divide 12 by 5, obtaining 2.4.
Step 2: Multiply 2.4 by 100, obtaining 240.
Step 3: Add the % symbol after 240, obtaining 240%.

Therefore, $\frac{12}{5} = 240\%$.

Practice Exercise 10. $\frac{9}{5}$

EXAMPLE 10 Rewrite 7 as a percent.

SOLUTION:

Step 1: There is no division required here since the denominator is understood to be 1.
Step 2: Multiply 7 by 100, obtaining 700.
Step 3: Add the % symbol after 700, obtaining 700%.

Therefore, $7 = 700\%$.

> **ANSWERS TO PRACTICE EXERCISES**
>
> 8. 20%.   9. 900%.   10. 180%.

4.1 Percent

We will now examine the conversion of percents to fractions.

Rewrite each of the following percents as a fraction.

Practice Exercise 11. 65%

EXAMPLE 11 Rewrite 75% as a fraction.

SOLUTION:
$$75\% = \frac{75}{100}$$
$$= \frac{75 \div 25}{100 \div 25}$$
$$= \frac{3}{4}$$

Practice Exercise 12. 238%

EXAMPLE 12 Rewrite 126% as a fraction.

SOLUTION:
$$126\% = \frac{126}{100}$$
$$= \frac{126 \div 2}{100 \div 2}$$
$$= \frac{63}{50}$$

Practice Exercise 13. $16\frac{2}{3}\%$

EXAMPLE 13 Rewrite $33\frac{1}{3}\%$ as a fraction.

SOLUTION:
$$33\frac{1}{3}\% = \frac{33\frac{1}{3}}{100}$$
$$= \left(33\frac{1}{3}\right)\left(\frac{1}{100}\right)$$
$$= \left(\frac{\cancel{100}^{1}}{3}\right)\left(\frac{1}{\cancel{100}_{1}}\right)$$
$$= \frac{1}{3}$$

ANSWERS TO PRACTICE EXERCISES

11. $\frac{13}{20}$. 12. $\frac{119}{50}$. 13. $\frac{1}{6}$.

Examples 11–13 illustrate the procedure for rewriting a percent as a fraction. Note that the same results are obtained using the following rule.

Rule for Rewriting a Percent as a Fraction

To rewrite a percent as a fraction:
1. Delete the % symbol after the number.
2. Write the number over 100.
3. Simplify, if necessary.

Rewrite the given percents as equivalent fractions in simplest form.

Practice Exercise 14. 54%

EXAMPLE 14 Rewrite 40% as a simplified fraction.

SOLUTION:

STEP 1: Delete the % symbol from 40%, obtaining 40.

STEP 2: Write 40 over 100, obtaining $\frac{40}{100}$.

STEP 3: Simplify $\frac{40}{100}$, obtaining $\frac{40}{100} = \frac{4}{10} = \frac{2}{5}$.

Therefore, $40\% = \frac{2}{5}$.

Practice Exercise 15. 155%

EXAMPLE 15 Rewrite 165% as a simplified fraction.

SOLUTION:

STEP 1: Delete the % symbol from 165%, obtaining 165.

STEP 2: Write 165 over 100, obtaining $\frac{165}{100}$.

STEP 3: Simplify $\frac{165}{100}$, obtaining $\frac{165}{100} = \frac{33}{20}$.

Therefore, $165\% = \frac{33}{20}$.

ANSWERS TO PRACTICE EXERCISES

14. $\frac{27}{50}$.   15. $\frac{31}{20}$.

EXERCISES 4.1

In Exercises 1–25, rewrite each number as a percent.

1. $\frac{1}{5}$ 2. $\frac{1}{20}$ 3. $\frac{3}{4}$ 4. $\frac{2}{3}$

5. $\frac{8}{5}$ 6. $\frac{11}{20}$ 7. $\frac{3}{2}$ 8. $\frac{1}{9}$

9. $\frac{1}{50}$ 10. $\frac{3}{25}$ 11. $\frac{1}{7}$ 12. $\frac{5}{3}$

13. $\frac{7}{4}$ 14. $\frac{11}{50}$ 15. 4   16. 6

17. 11   18. 27   19. 63   20. 100

21. $2\frac{1}{2}$ 22. $3\frac{1}{4}$ 23. $7\frac{1}{5}$ 24. $6\frac{1}{10}$

25. $15\frac{1}{6}$

In Exercises 26–50, rewrite the given percents as fractions in simplest form.

26. 18%   27. $22\frac{2}{9}$%   28. $10\frac{1}{2}$%   29. 85%
30. 68%   31. 55%   32. 73%   33. 98%
34. $56\frac{1}{2}$%   35. 92%   36. $70\frac{1}{4}$%   37. 100%
38. 102%   39. 125%   40. 170%   41. 185%
42. 325%   43. 450%   44. 500%   45. 610%
46. 725%   47. 805%   48. 900%   49. 936%
50. 1000%

51. Jan got 20% of her psychology exam questions wrong. What fractional part of the number of exam questions did she get wrong?
52. If 87.5% of the employees of the Acme Construction Company are hourly employees, what fractional part of the number of employees are hourly employees?
53. Joe spends 9.5% of his time each day watching TV. What fractional part of the day does he watch TV?
54. Labor costs accounted for 64% of the cost of a car repair. What fractional part of the cost was for labor?
55. Sarah's annual salary accounts for 92.6% of her total income. What fractional part of her total income is salary?
56. One-fourth of John's income is spent for rent. What percent of his income is spent for rent?
57. Three-eighths of Susan's income is spent for clothes. What percent of her income is spent for clothes?
58. From a survey of employees in a mall, it was determined that $\frac{5}{8}$ of the employees did not smoke. What percent of the employees did smoke?
59. Krystle answered 66 out of 80 questions on a multiple-choice exam correctly. What percent of the questions did she not answer correctly?
60. If 12 of 20 models in a fashion agency are over 70 in. tall, what percent of the models are not over 70 in. tall?

4.2 More on Percents

OBJECTIVES

After completing this section, you should be able to:
1. Convert a given percent to a decimal.
2. Convert a given decimal to a percent.

In this section, we will examine the relationships among percents, decimals, and fractions. In Section 4.1, you learned to convert percents to fractions. In Chapter 3, you learned to convert fractions to decimals. We will now combine the two conversions and learn how to convert percents to decimals.

Rewrite each of the following percents as decimals.

Practice Exercise 1. 35%

EXAMPLE 1 Rewrite 25% as a decimal.

SOLUTION: 25% means $\frac{25}{100}$. But

$$\frac{25}{100} = 0.25$$

Therefore, 25% = 0.25.

Practice Exercise 2. 167%

EXAMPLE 2 Rewrite 125% as a decimal.

SOLUTION: 125% means $\frac{125}{100}$. But

$$\frac{125}{100} = 1.25$$

Therefore, 125% = 1.25.

Practice Exercise 3. 470%

EXAMPLE 3 Rewrite 350% as a decimal.

SOLUTION: 350% means $\frac{350}{100}$. But

$$\frac{350}{100} = 3.50$$

Therefore, 350% = 3.50.

Practice Exercise 4. 2.3%

EXAMPLE 4 Rewrite 1.2% as a decimal.

SOLUTION: 1.2% means $\frac{1.2}{100}$. But

$$\frac{1.2}{100} = \frac{1.2 \times 10}{100 \times 10}$$ (Multiplying both numerator and denominator by 10 to remove decimals)

$$= \frac{12}{1000}$$

$$= 0.012$$

Therefore, 1.2% = 0.012.

Practice Exercise 5. 0.92%

EXAMPLE 5 Rewrite 0.6% as a decimal.

SOLUTION: 0.6% means $\frac{0.6}{100}$. But

$$\frac{0.6}{100} = \frac{0.6 \times 10}{100 \times 10}$$

$$= \frac{6}{1000}$$

$$= 0.006$$

Therefore, 0.6% = 0.006.

4.2 More on Percents

> **ANSWERS TO PRACTICE EXERCISES**
> 1. 0.35. 2. 1.67. 3. 4.7. 4. 0.023. 5. 0.0092.

Examples 1–5 illustrate the following rule for rewriting a percent as an equivalent decimal.

Rule for Rewriting a Percent as a Decimal

To rewrite a percent as a decimal:
1. Move the decimal point *two* places to the *left*.
2. Remove the % symbol.

Rewrite the given percents as equivalent decimals.

Practice Exercise 6. 14%

EXAMPLE 6 Rewrite 18% as a decimal.

SOLUTION: 18% = 0.18. (Note that the decimal point in 18% is understood to be immediately after the 8.)

Practice Exercise 7. 129%

EXAMPLE 7 Rewrite 150% as a decimal.

SOLUTION: 150% = 1.5

Practice Exercise 8. 4.6%

EXAMPLE 8 Rewrite 1.4% as a decimal.

SOLUTION: 1.4% = 01.4%
= 0.014

Practice Exercise 9. 0.08%

EXAMPLE 9 Rewrite 0.03% as a decimal.

SOLUTION: 0.03% = 00.03% = 0.0003

Practice Exercise 10. 0.007%

EXAMPLE 10 Rewrite 0.002% as a percent.

SOLUTION: 0.002% = 00.002%
= 0.00002

> **ANSWERS TO PRACTICE EXERCISES**
> 6. 0.14. 7. 1.29. 8. 0.046. 9. 0.0008. 10. 0.00007.

Having learned how to convert a percent to a decimal, you will now learn how to convert a decimal to a percent. The rule that you will learn is based on first converting a decimal to a fraction and then converting the fraction to a percent.

Rewrite each of the following decimal numbers as a percent.

Practice Exercise 11. 0.32

EXAMPLE 11 Rewrite 0.17 as a percent.

SOLUTION: $0.17 = \dfrac{17}{100}$. But $\dfrac{17}{100}$ means 17%. Therefore, 0.17 = 17%.

Practice Exercise 12. 2.37

EXAMPLE 12 Rewrite 1.23 as a percent.

SOLUTION: $1.23 = \frac{123}{100}$. But $\frac{123}{100}$ means 123%. Therefore, $1.23 = 123\%$.

Practice Exercise 13. 8.76

EXAMPLE 13 Rewrite 9.05 as a percent.

SOLUTION: $9.05 = \frac{905}{100}$. But $\frac{905}{100}$ means 905%. Therefore, $9.05 = 905\%$.

Practice Exercise 14. 0.03

EXAMPLE 14 Rewrite 0.06 as a percent.

SOLUTION: $0.06 = \frac{6}{100}$. But $\frac{6}{100}$ means 6%. Therefore, $0.06 = 6\%$.

Practice Exercise 15. 0.053

EXAMPLE 15 Rewrite 0.085 as a percent.

SOLUTION: $0.085 = \frac{85}{1000}$. But

$$\frac{85}{1000} = \frac{85 \div 10}{1000 \div 10}$$
$$= \frac{8.5}{100}$$
$$= 8.5\%$$

Therefore, $0.085 = 8.5\%$

ANSWERS TO PRACTICE EXERCISES

11. 32%. **12.** 237%. **13.** 876%. **14.** 3%. **15.** 5.3%.

Examples 11–15 illustrate the following rule for converting a decimal to a percent.

Rule for Rewriting a Decimal as a Percent

To rewrite a decimal as a percent:

1. Move the decimal point *two* places to the *right*.
2. Place the % symbol after the resulting number.

Rewrite the given decimals as equivalent percents.

Practice Exercise 16. 0.19

EXAMPLE 16 Rewrite 0.25 as a percent.

SOLUTION: $0.25 = 25\%$

Practice Exercise 17. 0.7

EXAMPLE 17 Rewrite 0.4 as a percent.

SOLUTION:

$0.4 = 0.40$
$ = 40\%$

(Note that a 0 was placed after the 4 so that the decimal point could be moved 2 places to the right.)

4.2 More on Percents

Practice Exercise 18. 8.7

EXAMPLE 18 Rewrite 1.6 as a percent.

SOLUTION: 1.6 = 1.60
$\qquad$ = 160%

Practice Exercise 19. 16.234

EXAMPLE 19 Rewrite 23.1 as a percent.

SOLUTION: 23.1 = 23.10 = 2310%

Practice Exercise 20. 0.073

EXAMPLE 20 Rewrite 0.065 as a percent.

SOLUTION: 0.065 = 6.5%

ANSWERS TO PRACTICE EXERCISES

16. 19%. **17.** 70%. **18.** 870%. **19.** 1623.4%. **20.** 7.3%.

EXERCISES 4.2

In Exercises 1–20, rewrite each percent as an equivalent fraction in simplest form.

1. 12%	**2.** 23%	**3.** 36%	**4.** 48%
5. 50%	**6.** 57%	**7.** 62%	**8.** 70%
9. 75%	**10.** 89%	**11.** 110%	**12.** 112%
13. 150%	**14.** 168%	**15.** 200%	**16.** 215%
17. 270%	**18.** 286%	**19.** 405%	**20.** 500%

In Exercises 21–40, rewrite each percent as an equivalent fraction in simplest form.

21. 8.2%	**22.** 12.5%	**23.** 23.68%	**24.** 39.45%
25. 62.9%	**26.** 77.5%	**27.** 81.65%	**28.** 120.36%
29. 239.4%	**30.** 250.5%	**31.** 280.06%	**32.** 297.18%
33. 362.2%	**34.** 385.5%	**35.** 390.05%	**36.** 416.24%
37. 500.8%	**38.** 600.5%	**39.** 700.25%	**40.** 925.88%

In Exercises 41–60, rewrite each percent as an equivalent decimal.

41. 36%	**42.** 47%	**43.** 81%	**44.** 99%
45. 127%	**46.** 200%	**47.** 256%	**48.** 400%
49. 496%	**50.** 602%	**51.** 783%	**52.** 999%
53. 6.9%	**54.** 17.2%	**55.** 42.69%	**56.** 53.73%
57. 0.06%	**58.** 0.059%	**59.** 0.087%	**60.** 0.0016%

In Exercises 61–80, rewrite each decimal as an equivalent percent.

61. 0.17	**62.** 0.93	**63.** 0.123	**64.** 0.367
65. 0.4359	**66.** 0.6709	**67.** 0.8215	**68.** 0.9999
69. 2.38	**70.** 14.39	**71.** 63.75	**72.** 86.23
73. 123.02	**74.** 237.39	**75.** 286.96	**76.** 301.01
77. 12.962	**78.** 0.0023	**79.** 0.0107	**80.** 0.00037

In Exercises 81–100, rewrite each fraction as an equivalent percent.

81. $\dfrac{1}{5}$ 82. $\dfrac{4}{5}$ 83. $\dfrac{7}{5}$ 84. $\dfrac{9}{5}$

85. $\dfrac{1}{8}$ 86. $\dfrac{3}{8}$ 87. $\dfrac{15}{8}$ 88. $\dfrac{17}{8}$

89. $\dfrac{1}{10}$ 90. $\dfrac{6}{10}$ 91. $\dfrac{12}{10}$ 92. $\dfrac{23}{10}$

93. $\dfrac{3}{16}$ 94. $\dfrac{1}{32}$ 95. $\dfrac{5}{64}$ 96. $\dfrac{4}{25}$

97. $\dfrac{13}{80}$ 98. $\dfrac{5}{320}$ 99. $\dfrac{123}{640}$ 100. $\dfrac{239}{1000}$

101. Express 12% as a
 a. Decimal.
 b. Simplified fraction.
102. Express 67.5% as a
 a. Decimal.
 b. Simplified fraction.
103. Express 81.6% as a
 a. Decimal.
 b. Simplified fraction.
104. Express 204% as a
 a. Decimal.
 b. Simplified fraction.
105. Express 0.36 as a
 a. Percent.
 b. Simplified fraction.
106. Express 0.82 as a
 a. Percent.
 b. Simplified fraction.
107. Express 2.34 as a
 a. Percent.
 b. Simplified fraction.
108. Express 39.4 as a
 a. Percent.
 b. Simplified fraction.
109. Express $\dfrac{3}{40}$ as a
 a. Decimal.
 b. Percent.
110. Express $\dfrac{110}{80}$ as a
 a. Decimal.
 b. Percent.

4.3 Problems Involving Percents

OBJECTIVES

After completing this section, you should be able to:
1. Find a percent of a given number.
2. Find what percent one number is of another number.
3. Find a number when a percent of it is given.

In this section, we will solve problems that involve percents. Let's consider the following three questions:

1. If 80% of 20 questions are answered correctly, how many questions are answered correctly?
2. If 16 of 20 questions are answered correctly, what percent of the questions is answered correctly?
3. If 16 questions represent 80% of the total number of questions answered correctly, how many questions were there?

4.3 Problems Involving Percents

Basically, the three preceding questions can be restated as the following three questions, respectively:

1. What is 80% of 20?
2. 16 is what percent of 20?
3. 16 is 80% of what number?

Note that in all three questions, there are three quantities:

- A *total* number, called the *base,* which we will denote by *B*.
- An *amount,* or part of the base, which we will denote by *A*.
- A *rate,* given as a percent, which we will denote by *R*.

The basic relationship among these three quantities is given by

$$A = R \times B$$

where *R* is given as a percent, *B* is the number that follows the word "of," and *A* is the remaining number. (*Reminder:* The word "of" means multiplication.)

Practice Exercise 1. What is 11.6% of 86?

EXAMPLE 1 What is 80% of 20?

SOLUTION: *A*, *R*, and *B* are determined as follows:

$$\text{What is } 80\% \text{ of } 20?$$
$$A = 80\% \times 20$$

or

$$A = 0.80 \times 20 \quad \text{(Converting } 80\% = 0.80\text{)}$$
$$= 16$$

Therefore, 16 is 80% of 20.

Practice Exercise 2. 6 is 7.5% of what number?

EXAMPLE 2 16 is what percent of 20?

SOLUTION: *A*, *R*, and *B* are determined as follows:

$$16 \text{ is what percent of } 20?$$
$$16 = R \times 20$$

or

$$R = \frac{16}{20}$$
$$= 0.80$$
$$= 80\%$$

Therefore, 16 is 80% of 20.

Practice Exercise 3. What percent of 40 is 16?

EXAMPLE 3 16 is 80% of what number?

SOLUTION: A, R, and B are determined as follows:

$$16 \text{ is } 80\% \text{ of what number?}$$
$$16 = 80\% \times B$$

or

$$16 = 0.80 \times B$$

Hence,

$$B = \frac{16}{0.80}$$
$$= 20$$

Therefore, 16 is 80% of 20.

Practice Exercise 4. 160 is what percent of 40?

EXAMPLE 4 What is 18.5% of 62.4?

SOLUTION: A, R, and B are determined as follows:

$$\text{What is } 18.5\% \text{ of } 62.4?$$
$$A = 18.5\% \times 62.4$$

or

$$A = 0.185 \times 62.4$$
$$= 11.544$$

Therefore, 11.544 is 18.5% of 62.4.

Practice Exercise 5. 25% of what number is 12?

EXAMPLE 5 28 is 35% of what number?

SOLUTION: A, R, and B are determined as follows:

$$28 \text{ is } 35\% \text{ of what number?}$$

or

$$28 = 0.35 \times B$$

Hence,

$$B = \frac{28}{0.35}$$
$$= 80$$

Therefore, 28 is 35% of 80.

Practice Exercise 6. 150 is what percent of 30?

EXAMPLE 6 40 is what percent of 20?

SOLUTION: A, R, and B are determined as follows:

$$40 \text{ is what percent of } 20?$$

4.3 Problems Involving Percents

or

$$40 = R \times 20$$

Hence,

$$R = \frac{40}{20}$$
$$= 2$$
$$= 200\%$$

Therefore, 40 is 200% of 20.

ANSWERS TO PRACTICE EXERCISES
1. 9.976. 2. 80. 3. 40%. 4. 400%. 5. 48. 6. 500%.

Practice Exercise 7. A videotape is on sale for $9. This is 60% of the regular price for the tape. What is the regular price?

EXAMPLE 7 During a summer clearance sale, the Beauty Boutique offered all its merchandise at 60% of the regular price (40% off). If an item regularly sold for $39.95, what would be the price?

SOLUTION: Since

$$\text{Sales price is 60\% of regular price}$$

we have

$$\text{Sales price} = 0.60 \times \$39.95$$
$$= \$23.97$$

Therefore, the sale price would be $23.97.

Practice Exercise 8. In a survey of 200 people, 165 said that they voted in the last general election. What percent of those surveyed voted in the last general election?

EXAMPLE 8 Dawn took a mathematics examination consisting of 20 questions. She answered 17 questions correctly. What percent of the total number of questions did Dawn answer correctly?

SOLUTION: We are asking

$$\text{What percent of 20 is 17?}$$

or

$$R \times 20 = 17$$

Hence,

$$R = \frac{17}{20}$$
$$= 0.85$$
$$= 85\%$$

Therefore, Dawn answered 85% of the examination questions correctly.

> **Practice Exercise 9.** Jay has a monthly car payment of $270. This is 30% of his monthly salary. What is his monthly salary?

EXAMPLE 9 A family pays a house rent of $350 a month. That is 25% of the family's monthly income. What is the family's monthly income?

SOLUTION: We are asking

$$\$350 \text{ is } 25\% \text{ of what?}$$

or

$$\$350 = 0.25 \times B$$

Hence,

$$B = \frac{\$350}{0.25}$$
$$= \$1400$$

Therefore, the family's monthly income is $1400.

CHECK:

$$\begin{array}{r} \$1400 \\ \times 0.25 \\ \hline 70\ 00 \\ 280\ 00 \\ \hline \$350.00 \end{array}$$

(Monthly income)
(25%)

(Monthly rent) ✓

ANSWERS TO PRACTICE EXERCISES

7. $15.   8. 82.5%.   9. $900.

All of the percent problems given in this section can be solved by using the equation

$$A = R \times B$$

where A is the *amount* or part of the total, R is the *rate,* given as a percent, and B is the *base,* or the total number.

If values for any two of the variables A, R, or B are known, then the value for the third variable can be determined, as was illustrated in the examples of this section.

EXERCISES 4.3

In Exercises 1–20, find the value of each expression.

1. 18% of 23
2. 23% of 18
3. 11.3% of 40
4. 23.9% of 15
5. 40.2% of 56
6. 8% of 1260

4.3 Problems Involving Percents

7. 9.6% of 1300
8. 0.75% of 120
9. 0.03% of 1000
10. 0.99% of 1100
11. 82.4% of 231
12. 96.5% of 1156
13. 0.04% of 11,215
14. 1.06% of 500
15. 2.63% of 210
16. 3.43% of 750
17. 5.19% of 2000
18. 19.8% of 1396
19. 107.6% of 19
20. 125.4% of 36
21. 9 is what percent of 36?
22. 16 is what percent of 480?
23. 50 is what percent of 100?
24. 170 is what percent of 85?
25. 1000 is what percent of 100?
26. 17 is what percent of 25?
27. 20 is what percent of 1?
28. 100 is what percent of 1?
29. 4 is what percent of 3?
30. 8 is what percent of 1?
31. 60 is what percent of 30?
32. 60 is what percent of 120?
33. 50 is what percent of 50?
34. 34 is what percent of 17?
35. 34 is what percent of 68?
36. 25 is what percent of 36? (Give your answer to the nearest tenth of a percent.)
37. 39 is what percent of 80? (Give your answer to the nearest tenth of a percent.)
38. 126 is what percent of 214? (Give your answer to the nearest hundredth of a percent.)
39. 209 is what percent of 317? (Give your answer to the nearest hundredth of a percent.)
40. 381 is what percent of 125? (Give your answer to the nearest hundredth of a percent.)
41. What percent of 20 is 10?
42. What percent of 60 is 20?
43. What percent of 40 is 80?
44. What percent of 85 is 170?
45. What percent of 120 is 50?
46. What percent of 6.9 is 2.3?
47. What percent of 72.4 is 18.1?
48. What percent of 19.8 is 39.6?
49. What percent of 0.05 is 0.03?
50. What percent of 19 is 76?
51. 18 is 25% of what number?
52. 100 is 50% of what number?
53. 100 is 1000% of what number?
54. 2 is 20% of what number?
55. 60 is 200% of what number?
56. 60 is 60% of what number?
57. 6.9 is 300% of what number?
58. 1.9 is 25% of what number?
59. 0.02 is 40% of what number?
60. 444 is 22.2% of what number?
61. 27 is 3% of what number?
62. 16 is 0.4% of what number?
63. 18.4 is 0.5% of what number?
64. 1 is 100% of what number?
65. 3 is 200% of what number?
66. 70 is 500% of what number?
67. 4.52 is 40% of what number?
68. 0.3 is 0.03% of what number?
69. 103.8 is 5.19% of what number?
70. 21.52 is 107.6% of what number?
71. If 4 students are absent out of a class of 24, what percent of the class is absent?
72. If a state sales tax rate is 4%, what is the tax on an appliance priced at $459?
73. There are 16 males and 20 females on a Charter Reform Committee. What percent of the committee is female? What percent of the committee is male?
74. If 56 members of a college faculty are females and they represent 40% of the faculty, how many faculty members are there?
75. A skirt sells for $12 on sale. This is 80% of the regular price for the skirt. What is the regular price?
76. Voting on a major defense issue, 52 U.S. senators voted in favor of the issue. If this was 65% of those present, how many senators were present to vote?
77. A baseball team played 24 games and won 15 of them. What percent of the games played did the team win?
78. Forty-six percent of people surveyed said that they exercised on a fairly regular basis. If 11,200 people were surveyed, how many of them exercise?
79. If the value of a new car depreciates 28% the first year, what would be the value of a 1-year-old car that originally cost $9750?
80. One hundred sixty-nine radio listeners indicated that they were very satisfied with the programming of radio station WBAD. If this was 26% of those surveyed, how many listeners were surveyed?

4.4 Percent Increase or Decrease

OBJECTIVES

After completing this section, you should be able to:
1. Find a new number when an old number is increased or decreased by a certain percent.
2. Find the percent of increase or decrease of a number.

Frequently, we encounter applications involving percents in terms of an increase or a decrease of some amount. You probably have read about tax increases of 15% in 1 year, the price of coffee increasing 60% due to a drought, or the price of gasoline decreasing by 40% due to an oil surplus on the market.

To calculate the amount of **increase** or **decrease,** we multiply the rate by the original amount.

Amount of increase or decrease = rate × original amount

Practice Exercise 1. If the price of an appliance was increased by 20%, determine the amount of increase in the price of an appliance that originally sold for $439.

EXAMPLE 1 The price of coffee increased by 60%. If coffee sold for $1.95 per pound before the increase, how much was the increase for a pound of coffee?

SOLUTION: Since the original price was $1.95 and the rate of increase is 60%, we have:

$$\begin{aligned}\text{Amount of increase} &= \text{rate} \times \text{original price} \\ &= 60\% \times \$1.95 \\ &= (0.60)(\$1.95) \\ &= \$1.170\end{aligned}$$

Therefore, the *increase* for a pound of coffee was $1.17.

Practice Exercise 2. The price of a liter of cola was decreased by 35% during a weekend special. If the regular price of a liter of the cola was $1.19, what was the amount of the decrease?

EXAMPLE 2 The price of gasoline decreased by 40%. If a gallon of gasoline sold for $1.29 before the decrease, what was the amount of decrease?

SOLUTION: Since the original price was $1.29 per gallon and the rate of decrease is 40%, we have:

$$\begin{aligned}\text{Amount of decrease} &= \text{rate} \times \text{original price} \\ &= 40\% \times \$1.29 \\ &= (0.40)(\$1.29) \\ &= \$.516\end{aligned}$$

Therefore, the *decrease* for a gallon of gasoline was 51.6¢.

Practice Exercise 3. If the number 512 is decreased by 30% of itself, what is the new number?

EXAMPLE 3 In 1970, the population of Little City was 180. During the period 1970–1975, the population of Little City increased by 20%. What was the population of Little City in 1975?

SOLUTION: To determine the population of Little City in 1975, we must do *two* things.

4.4 Percent Increase or Decrease

STEP 1: First, determine the *amount of increase* (or gain) in the population as follows:

$$\text{Amount of increase} = \text{rate} \times \text{original population}$$
$$= 20\% \times 180$$
$$= (0.20)(180)$$
$$= 36$$

STEP 2: Then *add* the amount of increase to the population in 1970 to obtain the new population:

```
 180   (Population in 1970)
+ 36   (Amount of increase)
 216   (Population in 1975)
```

Therefore, the population of Little City in 1975 was 216.

EXAMPLE 4 Bill weighed 220 lb and his doctor advised him to lose 15% of his weight. If Bill followed his doctor's advice and lost 15% of his weight, what would be his new weight?

SOLUTION: To determine Bill's new weight, we must do *two* things:

STEP 1: First, determine the *amount of decrease* (or loss) in Bill's weight as follows:

$$\text{Amount of decrease} = \text{rate} \times \text{original weight}$$
$$= 15\% \times 220 \text{ lb}$$
$$= (0.15)(220 \text{ lb})$$
$$= 33 \text{ lb}$$

STEP 2: Then *subtract* the amount of loss from the original weight.

```
 220   (Original weight in pounds)
- 33   (Amount of loss in pounds)
 187   (New weight in pounds)
```

Therefore, Bill's new weight would be 187 lb.

> **Practice Exercise 4.** Fern's salary increased by 6.5% in 1 yr. If her old salary was $17,900, what is Fern's new salary?

ANSWERS TO PRACTICE EXERCISES

1. $87.80.   2. 41.65¢.   3. 358.4.   4. $19,063.50.

Sometimes we know the original amount and the new amount but wish to determine the rate of increase or decrease. To do so, we use:

$$\text{Amount of increase or decrease} = \text{rate} \times \text{original amount}$$

and *divide* the amount of increase (or decrease) by the original amount.

$$\text{Rate of increase or decrease} = \frac{\text{amount of increase or decrease}}{\text{original amount}}$$

> **Practice Exercise 5.** If a number is increased from 20 to 25, determine the percent of increase.

EXAMPLE 5 The price of motor oil increased from 90¢ per liter to 99¢ per liter. What was the percent of increase?

SOLUTION: To solve this problem, we must do *two* things.

STEP 1: First, determine the *amount of increase* as follows:

$$\begin{aligned} \$.99 \quad &\text{(New price per liter)} \\ -.90 \quad &\text{(Old price per liter)} \\ \hline \$.09 \quad &\text{(\textit{Amount} of increase per liter)} \end{aligned}$$

STEP 2: Then determine the *rate of increase* as follows:

$$\begin{aligned} \text{Rate of increase} &= \frac{\text{amount of increase}}{\text{original price}} \\ &= \frac{\$.09}{\$.90} \\ &= \frac{1}{10} \\ &= 10\% \end{aligned}$$

Therefore, the percent of increase was 10%.

> **Practice Exercise 6.** If a number is decreased from 350 to 280, determine the percent of decrease.

EXAMPLE 6 A new car was bought for $6800. One year later, its value was $5100. What was the percent of decrease in the value of the car for the one year?

SOLUTION:

STEP 1: Determine the *amount of decrease* as follows:

$$\begin{aligned} \$6800 \quad &\text{(New car value)} \\ -5100 \quad &\text{(Value of car one year later)} \\ \hline \$1700 \quad &\text{(\textit{Amount} of decrease in value)} \end{aligned}$$

STEP 2: Determine the *rate of decrease* as follows:

$$\begin{aligned} \text{Rate of decrease} &= \frac{\text{amount of decrease}}{\text{original price}} \\ &= \frac{\$1700}{\$6800} \\ &= \frac{1}{4} \\ &= 25\% \end{aligned}$$

Therefore, there was a 25% decrease in value of the car for one year.

ANSWERS TO PRACTICE EXERCISES

5. 25%.   6. 20%.

EXERCISES 4.4

1. A worker's salary is increased by 7%. If the old salary was $9200 per year, what is the worker's new salary?
2. Another worker's salary is increased by 11%. If the old salary was $8700 per year, what is the worker's new salary?
3. The population is a small village decreased from 980 to 878. What was the percent of decrease?
4. Due to inflation, the price of an item was increased from $18 to $18.90. What was the percent of increase?
5. During a sale, the price of a washing machine was decreased by 8%. If the machine originally sold for $459, what was the sale price?
6. During the same sale, the price of a clothes dryer was decreased by 15%. If the dryer originally sold for $405, what was the sale price?
7. To conserve energy, a thermostat was reset from 72° F to 66° F. What was the percent of decrease in the thermostat readings?
8. If the thermostat in Exercise 7 was reset from 72° F to 62° F, what was the percent of decrease in the thermostat reading?
9. Due to enrollment problems, the size of the faculty at State College was decreased from 218 to 198. What was the percent of decrease in the size of the faculty? (Give your answer to the nearest tenth of a percent.)
10. Dawn had $317 in her savings account. She then made a deposit that increased her savings by 12%. What was her new balance?
11. Penny had $868 in her savings account. She then withdrew some money, leaving a balance of $720.44. What was the percent of decrease in her account?
12. The number of unemployed people in one state increased by 2.6% during a 3-month period. If 67,300 people were unemployed at the start of the 3-month period, how many people were unemployed at the end of the period? (Give your answer to the nearest whole number.)
13. A compact car was averaging 18 mi to a gallon of gasoline. After an engine tune-up, the car was averaging 21.5 mi to a gallon of gasoline. What was the percent increase in miles per gallon after the tune-up? (Give your answer to the nearest tenth of a percent.)
14. A certain stock listed at $70 per share at the start of a trading day. At the close of the day, the same stock listed for $79 per share. What was the percent of increase in the price of a share of the stock during that day? (Give your answer to the nearest tenth of a percent.)
15. A stock listed at $89 at the start of a trading day. At the close of the day, the stock listed for $81 per share. What was the percent of decrease in the price of a share of the stock during that day? (Give your answer to the nearest tenth of a percent.)
16. Karen bought a new dress for 15% less than the original price. If the dress originally sold for $49, how much did Karen pay for the dress?
17. In 1977, Don paid $1128 in federal income taxes. In 1978, he paid $1246. What was the percent increase in income taxes paid from 1977 to 1978? (Give your answer to the nearest tenth of a percent.)
18. During a 1-yr period, the number of worker strikes increased from 312 to 370. What was the percent increase of strikes during the period? (Give your answer to the nearest tenth of a percent.)
19. At a particular theater, the number of moviegoers decreased by 20% from the beginning of March through the end of April. If there were 1120 moviegoers at the beginning of March, how many were there at the end of April?
20. Paul's weight increased from 173 lb to 192 lb. What was the percent of increase in Paul's weight? (Give your answer to the nearest tenth of a percent.)
21. Federal grants to Big City were reduced from $3,619,200 to $2,896,000. What was the percent of decrease in the amount of the grants? (Give your answer to the nearest percent.)

22. Slugger's batting average increased by 15%. If his old batting average was 0.232, what is his new batting average?
23. Due to an aggressive affirmative action program at United Supply Company, the number of minority workers increased by 100% over a given period of time. If there were 117 minority workers at the beginning of the period, how many minority workers were there at the end of the period?
24. The price of a textbook was $7.95. Two years later, the price of the same textbook was $11.40. What was the percent of increase in the price of the book? (Give your answer to the nearest percent.)
25. In 3 years, the price of a textbook increased from $7.95 to $14.95. What was the percent of increase in the price of the book? (Give your answer to the nearest percent.)

4.5 Sales and Property Tax

OBJECTIVES

After completing this section, you should be able to:
1. Find the sales tax on the purchase of an item.
2. Find the sales tax rate, given the purchase price of an item and the amount of sales tax.
3. Find the total price of an item, including sales tax.
4. Find the property tax on a piece of property (house or land), given the assessed value and the tax rate.
5. Find the property tax rate, given the total budget and the total assessed value of property.

Most of the states in the United States have a tax that is called a "sales" tax. This sales tax is usually charged on most retail sales as well as on heating oil, gasoline, entertainment, hotel and motel rooms, and even on meals eaten in restaurants. In addition to state sales tax, there are also city, town, and county sales taxes.

DEFINITION

A **sales tax** is an amount to be *added* to the price of an item or service and is defined as follows:

$$\text{Sales tax} = \text{sales tax rate} \times \text{purchase price}$$

Practice Exercise 1. If the sales tax rate is 6%, what is the sales tax on a purchase of $56.80?

EXAMPLE 1 Harold bought a car for $5650. He lives in a state where there is a sales tax of 4%. How much sales tax must Harold pay on his car?

SOLUTION: Since

$$\text{Sales tax} = \text{sales tax rate} \times \text{purchase price}$$

4.5 Sales and Property Tax

we have

$$\text{Sales tax} = 4\% \times \$5650$$
$$= (0.04)(\$5650)$$
$$= \$226$$

Therefore, Harold must pay a $226 sales tax on his car.

Practice Exercise 2. Determine the total amount of an item that sells for $795 if the sales tax rate is 4.5%.

EXAMPLE 2 Laura bought a dress for $39.95. She lives in Higher Tax City where the state sales tax rate is 3%, the city sales tax rate is 1.5%, and the county sales tax rate is 1%. How much did Laura pay for the dress, including sales tax?

SOLUTION:

STEP 1: We must first determine the total sales tax rate:

State:	3.0%
City:	1.5%
County:	1.0%
Total sales tax rate:	5.5%

STEP 2: Next, we compute the amount of sales tax:

$$\text{Sales tax} = \text{sales tax rate} \times \text{purchase price}$$
$$= 5.5\% \times \$39.95$$
$$= (0.055)(\$39.95)$$
$$= \$2.19725$$

Hence, the amount of sales tax is $2.20 (since sales tax is generally computed to the *nearest* cent).

STEP 3: Finally, we determine the total amount of the purchase:

$39.95	(Purchase price)
+2.20	(Sales tax)
$42.15	(Total amount)

Therefore, the total price, including sales tax, is $42.15.

Sales Tax and Total Price

1. To find the *amount* of sales tax on a purchase, convert the sales tax rate (percent) to a decimal number and multiply the purchase price by the decimal number.
2. To find the *total price* for an item, *add* the amount of sales tax to the purchase price.

Sometimes we know the price of an item and the amount of sales tax. To determine the sales **tax rate**, we use:

$$\text{Sales tax} = \text{sales tax rate} \times \text{purchase price}$$

and *divide* the amount of sales tax by the purchase price.

$$\text{Sales tax rate} = \frac{\text{amount of sales tax}}{\text{purchase price}}$$

Practice Exercise 3. Determine the sales tax rate if an item sells for $630 and the sales tax is $25.20.

EXAMPLE 3 Connie bought a new TV for $600 and paid a sales tax of $42. The total purchase price was $642. What was the sales tax rate?

SOLUTION: Since

$$\text{Sales tax rate} = \frac{\text{amount of sales tax}}{\text{purchase price}}$$

we have

$$\text{Sales tax rate} = \frac{\$42}{\$600}$$

$$= \frac{7}{100}$$

$$= 7\%$$

Therefore, the sales tax rate was 7%.

Practice Exercise 4. Determine the sales tax rate if an item sells for $895 and the sales tax is $53.70.

EXAMPLE 4 The sales tax on the purchase of an item for $400 is $12. What is the sales tax rate?

SOLUTION: Since

$$\text{Sales tax rate} = \frac{\text{amount of sales tax}}{\text{purchase price}}$$

we have

$$\text{Sales tax rate} = \frac{\$12}{\$400}$$

$$= \frac{3}{100}$$

$$= 3\%$$

Therefore, the sales tax rate is 3%.

ANSWERS TO PRACTICE EXERCISES

1. $3.41. 2. $830.78. 3. 4%. 4. 6%.

A property tax is another form of tax used in most states in the United States; it is a tax on property. There are two kinds of property taxes: *real* property tax and *personal* property tax. **Real property tax** is tax paid on lands and buildings owned. Personal property tax is tax paid on household furnishings, clothing, cars, jewelry, and other personal belongings. We will consider only real property tax in this chapter.

4.5 Sales and Property Tax

To compute a real property tax, you must know the value given to the property and the tax rate. The value given to the property is called the **assessed value.**

Property tax = property tax rate × assessed value of property

---

The property tax rate can be stated as a percent of the assessed value, such as 2.634% of the assessed value. It may also be given in dollars per $1000 of assessed value, such as $26.34 per $1000 of assessed value.

Practice Exercise 5. If a house is assessed at $44,800 and the property tax rate is 2.87%, determine the amount of property tax on the house.

EXAMPLE 5 If the tax rate in a given county is 2.692% of the assessed value, what is the property tax for a piece of property that is assessed at $23,540?

SOLUTION: Since

$$\text{Property tax} = \text{property tax rate} \times \text{assessed value}$$

we have

$$\begin{aligned}\text{Property tax} &= 2.692\% \times \$23{,}540 \\ &= (0.02692)(\$23{,}540) \\ &= \$633.69680 \quad \text{(Verify!)}\end{aligned}$$

Therefore, the property tax is $633.70 (to the *nearest* cent).

Practice Exercise 6. If a house is assessed at $59,900 and the property tax rate is $20.96 per $1000, determine the amount of property tax on the house.

EXAMPLE 6 The property tax rate in Lo Valley is $18.65 per $1000 of assessed value. What is the amount of property tax a homeowner in Lo Valley must pay if his property is assessed at $36,400?

SOLUTION: Since the tax rate is given per $1000, we first determine how many groups of 1000 are in 36,400 as follows:

$$36{,}400 \div 1000 = 36.4$$

We then multiply 36.4 by the tax rate per $1000:

$$\begin{aligned}\text{Property tax} &= \text{property tax rate} \times \text{assessed value} \\ &= (\$18.65 \text{ per } \$1000) \times (\text{number of } \$1000) \\ &= (\$18.35)(36.4) \\ &= \$678.860\end{aligned}$$

Therefore, the amount of tax is $678.86.

If the total budget amount and the total assessed value of property in a community are known, then the property tax rate can also be determined.

$$\text{Property tax rate} = \frac{\text{amount of total budget}}{\text{amount of total assessed value}}$$

Practice Exercise 7. If the total budget for a county is $20,697,200 and the total assessed value in the county is $967,345,900, what is the property tax rate, to the nearest thousandth of a percent, if the taxes cover the budget?

EXAMPLE 7 The total budget this year for Histown is $6,545,695, and the total assessed value of the property in Histown is $545,100,200. Find the tax rate, to the nearest thousandth of a percent, if taxes cover the budget.

SOLUTION: Since

$$\text{Property tax rate} = \frac{\text{amount of total budget}}{\text{amount of total assessed value}}$$

we have

$$\text{Property tax rate} = \frac{\$6,545,695}{\$545,100,200}$$
$$\approx 0.012008 \quad \text{(Verify!)}$$
$$= 1.2008\%$$

Therefore, the tax rate, to the nearest thousandth of a percent, would be 1.201% of assessed value. (This is also equal to $12.01 per $1000 assessed value.)

Practice Exercise 8. If the total budget for a city is $31,269,500 and the total assessed value in the city is $1,501,670,100, what is the property tax rate, to the nearest thousandth of a percent, if the taxes cover the budget?

EXAMPLE 8 If the total budget for a village is $1,901,400, and the total assessed value is $10,241,000, what is the proeprty tax rate, to the nearest thousandth of a percent, if the taxes cover the budget?

SOLUTION: Since

$$\text{Property tax rate} = \frac{\text{total amount of budget}}{\text{total amount of assessed value}}$$

we have

$$\text{Property tax rate} = \frac{\$1,901,400}{\$10,241,000}$$
$$\approx 0.185665 \quad \text{(Verify!)}$$
$$\approx 18.567\%$$

Therefore, to the nearest thousandth of a percent, the tax rate is 18.567%.

ANSWERS TO PRACTICE EXERCISES

5. $1285.76.   6. $1255.50.   7. 2.140%.   8. 2.082%.

EXERCISES 4.5

In Exercises 1–10, determine the *amount* of sales tax for the given item and the given sales tax rate.

	Price of Item	*Tax Rate*		*Price of Item*	*Tax Rate*
1.	$18.00	4%	2.	$37.50	3%
3.	$79.95	2%	4.	$120.00	4.5%
5.	$397.69	3%	6.	$465.50	5%
7.	$1260.00	4%	8.	$868.00	2.5%
9.	$713.00	4.5%	10.	$905.70	3.5%

In Exercises 11–20, determine the *total price* for the given purchase.

	Price of Item	Tax Rate		Price of Item	Tax Rate
11.	$17.00	4%	12.	$46.50	3%
13.	$87.95	2%	14.	$150.00	4.5%
15.	$345.20	3%	16.	$487.75	5%
17.	$969.00	4%	18.	$1020.00	2.5%
19.	$1120.00	4.5%	20.	$1350.00	3.5%

In Exercises 21–30, determine the *sales tax rate* for the given sales tax amount and purchase price.

	Sales Tax	Price of Item		Sales Tax	Price of Item
21.	$10.00	$200.00	22.	$10.50	$350.00
23.	$21.00	$420.00	24.	$9.80	$140.00
25.	$2.54	$63.50	26.	$28.35	$810.00
27.	$18.36	$459.00	28.	$78.40	$1120.00
29.	$98.00	$1960.00	30.	$70.50	$2350.00

In Exercises 31–40, determine the *amount* of property taxes to be paid, given the assessed value and the property tax rate.

	Assessed Value	Tax Rate		Assessed Value	Tax Rate
31.	$7650	2.63%	32.	$8500	2.59%
33.	$22,400	2.83%	34.	$34,600	1.94%
35.	$47,800	2.04%	36.	$55,500	2.39%
37.	$34,500	$16.50 per $1000	38.	$43,800	$17.95 per $1000
39.	$57,100	$23.43 per $1000	40.	$59,700	$26.97 per $1000

In Exercises 41–50, determine the *property tax rate*, given the total budget and the total assessed value. (Give your answer to the nearest tenth of a percent.)

	Budget	Assessed Value		Budget	Assessed Value
41.	$316,000	$1,450,000	42.	$726,500	$3,150,800
43.	$950,700	$5,120,500	44.	$1,290,000	$23,800,000
45.	$1,540,690	$37,000,800	46.	$2,670,820	$55,600,000
47.	$3,716,490	$89,750,100	48.	$4,629,723	$103,896,750
49.	$5,500,000	$121,600,000	50.	$6,000,890	$301,209,750

4.6 Commissions

OBJECTIVES

After completing this section, you should be able to:
1. Determine the amount of commission, given the net sales and the commission rate.
2. Determine a worker's total pay, given the worker's salary and commission.
3. Determine the amount of net sales, given the amount of gross sales and returns.

Some people work at jobs where they are paid the same amount of money each week or each month. This amount is called a "salary." However, most people who work at jobs that involve selling goods in retail stores, selling real estate, or selling automobiles are paid what is called a **commission.** If a person receives all of his or her pay from commission, this is called **straight commission.** Some people receive a combination of a salary and a commission. To find the commission earned, we multiply the amount of sales by a certain percent of the sales called the **commission rate.**

Commission = commission rate × amount of net sales

> **Practice Exercise 1.** Determine the commission on $1750 of sales if the commission rate is 8%.

EXAMPLE 1 Nancy works for a straight commission and receives a commission rate of 5% on the total amount of sales. If her total sales for a week amounted to $3150, what was her commission?

SOLUTION: Since

$$\text{Commission} = \text{commission rate} \times \text{amount of net sales}$$

we have

$$\begin{aligned}\text{Commission} &= 5\% \times \$3150 \\ &= (0.05)(\$3150) \\ &= \$157.50\end{aligned}$$

Therefore, her commission was $157.50.

> **Practice Exercise 2.** Determine the commission for the sale of a house that sells for $129,900 if the commission rate is 7%.

EXAMPLE 2 Mary sells real estate and receives a commission which is 7% of the sale price. If Mary sells a house for $63,500, what is her commission?

SOLUTION: Since

$$\text{Commission} = \text{commission rate} \times \text{amount of net sales}$$

we have

$$\begin{aligned}\text{Commission} &= 7\% \times \$63{,}500 \\ &= (0.07)(\$63{,}500) \\ &= \$4445.00\end{aligned}$$

Therefore, Mary's commission is $4445.

Sometimes, returns are made after the sale of merchandise. The amount of the returns (in dollars and cents) must be subtracted from the total sales (in dollars and cents) before the commission is computed. The total amount of sales is called **gross sales.** The amount of the sales after the amount of the returns has been subtracted is called the **net sales.** Commissions are based on net sales.

> **Practice Exercise 3.** If the gross sales are $21,692 and returns amount to $987, determine the commission on net sales if the commission rate is 4%.

EXAMPLE 3 During one month, Brian had gross sales of $19,625 and his sales returns amounted to $873. He received 8% commission on his net sales. What was the amount of Brian's commission for the month?

4.6 Commissions

SOLUTION: There are *two* things to do.

STEP 1: First, we must find the *net* sales:

$19,625 (Gross sales)
− 873   (Returns)
$18,752 (Net sales)

STEP 2: Then we must find the commission:

$18,752   (Net sales)
× 0.08   (8% commission rate)
$1500.16 (Commission)

Hence, Brian's commission was $1500.16 for the month.

EXAMPLE 4 Dan works in an automotive supply store. He receives a salary of $90 per week plus a commission of 5% on his net sales. If Dan has gross sales of $1372 and returns of $109, what is his total pay for the week?

SOLUTION: We must do *three* things.

STEP 1: Determine the net sales:

$1372   (Gross sales)
− 109   (Returns)
$1263   (Net sales)

STEP 2: Determine the commission:

$1263   (Net sales)
× 0.05   (5% commission rate)
$63.15   (Commission)

STEP 3: Determine the total pay:

$90.00   (Salary)
+ 63.15   (Commission)
$153.15   (Total pay)

Therefore, Dan's total pay for the week is $153.15.

> **Practice Exercise 4.** Michelle receives a weekly salary of $73.60 plus a commission of 8% of net sales. If her net sales for the week were $2012, what was her total pay for the week?

ANSWERS TO PRACTICE EXERCISES

1. $140.   2. $9093.   3. $828.20.   4. $234.56.

Since

Commission = commission rate × amount of net sales

we can also determine the amount of net sales if we are given the commission and the commission rate.

$$\text{Amount of net sales} = \frac{\text{amount of commission}}{\text{commission rate}}$$

Practice Exercise 5. Larry works on a straight commission of 7% of his net sales. If his commission for a given week was $149.80, what was the amount of his net sales?

EXAMPLE 5 Bette works on a straight commission of 6% of her net sales. If her commission for a given week was $144, what was the amount of her net sales?

SOLUTION: Since

$$\text{Amount of net sales} = \frac{\text{amount of commission}}{\text{commission rate}}$$

we have

$$\text{Amount of net sales} = \frac{\$144}{6\%}$$

$$= \frac{\$144}{0.06}$$

$$= \$2400.00$$

Therefore, Bette's net sales amounted to $2400 for the week.

Practice Exercise 6. Evangelos sells Greek olives. He earns a salary of $1200 per month plus a commission. If his total pay for a month was $3630 and his commission rate is 15%, what was the amount of his net sales for the month?

EXAMPLE 6 Cindy works in a hardware store and receives a salary of $60 per week plus a commission of 7% of net sales. If her total pay for the week amounted to $104.10, what was the amount of her net sales?

SOLUTION: We must do *two* things.

STEP 1: Determine the amount of commission:

$$\begin{array}{rl} \$104.10 & \text{(Total pay)} \\ -60.00 & \text{(Salary)} \\ \hline \$44.10 & \text{(Commission)} \end{array}$$

STEP 2: Determine the amount of net sales:

$$\text{Amount of net sales} = \frac{\text{amount of commission}}{\text{commission rate}}$$

$$= \frac{\$44.10}{7\%}$$

$$= \frac{\$44.10}{0.07}$$

$$= \$630$$

Therefore, Cindy's net sales were $630 for the week.

ANSWERS TO PRACTICE EXERCISES

5. $2140.   6. $16,200.

EXERCISES 4.6

1. Michael sells TVs on commission at 12% of net sales. If his net sales are $1130, what is his commission?
2. In Exercise 1, what would Michael's commission be if he received only an 11% commission?
3. Laura sells cosmetics on commission at 18% of net sales. If her net sales are $324, what is her commission?
4. Daryl sells art supplies on commission at 15% of net sales. If his net sales are $965, what is his commission?
5. Karen sells women's wear on commission at 13% of net sales. If her net sales are $862, what is her commission?
6. Larry sells tires on commission at 9.5% of net sales. If his net sales are $1340, what is his commission?
7. Gina sells office supplies. She earns a salary of $40 per week plus a commission of 8% of net sales. If her net sales are $962, what is her total pay for the week?
8. In Exercise 7, if Gina's net sales are $1074, what is her total pay for the week?
9. Dick sells duplicating machines. He earns a salary of $500 per month plus a commission of 17% of net sales. If his net sales are $13,624, what is his total pay for the month?
10. Dianne sells magazines at a newsstand. She receives a commission of 15% of net sales. If her gross sales are $1380 and returns amount to $137, what is her commission?
11. George sells real estate and receives a commission of 9% of the sale price. If he sold a house for $57,900 and another house for $61,200, what was his commission for the two sales?
12. Liz sells undeveloped land and receives a commission of 11% of the sale price. If she sold three tracts of land for $11,600, $13,900, and $37,450, what was the amount of her commission for the total sales?
13. Brian sells real estate for a commission of 9% and undeveloped land for a commission of 11%. If he sells two houses for $37,900 and $72,500 and a tract of land for $162,000, what will be his total commission for the three sales?
14. Andrea sells real estate. She receives a salary of $650 per month plus a commission rate of 7% of the sale price. If her sales for the month amounted to $237,650, what was her total amount of pay?
15. In Exercise 14, if Andrea's total pay for one month was $1210, what was the total amount of her sales for the month?
16. Bill works for a commission of sales. His commission is 4% of the first $5000 in sales, 5% for the next $4000 in sales, and 6% for all sales over $9000. If his total net sales amounted to $13,500, what was his commission?
17. In Exercise 16, what would be Bill's total commission if the commission on sales over $9000 was raised to 10%?
18. Michelle works for a commission. Her commission is 3% of the first $3500 in sales, 5% on the next $5000 in sales, and 7% on all sales over $8500. If her total net sales amounted to $9750, what was her commission?
19. Cindy receives 7% commission on net sales. If her commission was $268.80, what was the amount of her net sales?
20. Ivan receives 8% commission on net sales. If his commission was $1052.96, what was the amount of his net sales?
21. Dawn works in a store and receives a salary of $60 per week plus a commission of 6.5% of net sales. If her total pay for a week was $147.10, what was the amount of her net sales?
22. Jim sells real estate and receives 8% commission on the sale price. If his commission on the sale of a house was $5464, what was the sale price for the house?
23. Penny sells real estate and receives 9.5% commission on the sale price. If her commission on the sale of a house was $6830.50, what was the sale price for the house?

24. Theron sells textbooks and receives 7% commission on net sales. For a particular period, he had returns of $2350 and received a commission of $6111.35. What were his gross sales for the period?
25. In Exercise 24, if Theron had returns of $3169 and received a commission of $5333.30, what were his gross sales for the period?

4.7 Discount

OBJECTIVES

After completing this section, you should be able to:
1. Determine the amount of discount, given the rate of discount and the marked price.
2. Determine the sale price, given the marked price and the discount.
3. Determine the rate of discount, given the discount and the marked price.

When a particular store advertises a sale, it is usually indicated by a certain percent off the regular price. For instance, a TV set may be on sale. Its regular price is $300; it is on sale at 40% off the regular price. The sale price is $180, which is determined as follows:

STEP 1: $300   (Regular price)
         ×0.40   (40% off)
         $120.00   (Savings)

STEP 2: $300   (Regular price)
         −120   (Savings)
         $180   (Sale price)

The regular price of an item is called the **marked price**. The percent off the marked price is called the **rate of discount**. The amount off is called the **discount**. The discounted price is called the **sale price**.

In the case of the TV set, the marked price is $300, the rate of discount is 40%, the discount is $120, and the sale price is $180.

Discount = rate of discount × marked price

Sale price = marked price − discount

Practice Exercise 1. An item is on sale at 30% off the marked price of $189. What is the discount?

EXAMPLE 1 Tires are advertised at 25% off the regular price. If a tire is marked at $79.80, determine (1) the discount, and (2) the sale price.

SOLUTION:

STEP 1: Discount = rate of discount × marked price
                 =   25%   × $79.80
                 = (0.25)($79.80)
                 = $19.95

Therefore, the discount is $19.95.

4.7 Discount

STEP 2: Sale price = marked price − discount
= $79.80 − $19.95
= $59.85

Therefore, the sale price of the tire is $59.85.

Practice Exercise 2. An item is on sale at 40% off the marked price of $560. What is the sale price?

EXAMPLE 2 During a cash liquidation sale, men's clothing is offered at 35% off the marked price. If the marked price of a particular suit is $160, what is the sale price?

SOLUTION: We must do *two* things.

STEP 1: First, determine the discount:

Discount = rate of discount × marked price
= 35% × $160
= (0.35)($160)
= $56.00

STEP 2: Then determine the sale price:

Sale price = marked price − discount
= $160 − $56
= $104

Therefore, the sale price for the suit is $104.

ANSWERS TO PRACTICE EXERCISES
1. $56.70. 2. $336.00.

Sometimes, we know the marked price and the discount. To determine the rate of discount, we use:

Discount = rate of discount × marked price

and *divide* the discount by the marked price.

$$\text{Rate of discount} = \frac{\text{discount}}{\text{marked price}}$$

Practice Exercise 3. The marked price of an item is $600. The discount is $240. What is the rate of discount?

EXAMPLE 3 Jerry's Bookstore has a best-seller with a marked price of $13.95 on sale for $11.16. What is the rate of discount?

SOLUTION: We must do *two* things.

STEP 1: Determine the discount:

Discount = marked price − sales price
= $13.95 − $11.16
= $2.79

Step 2: Determine the rate of discount:

$$\text{Rate of discount} = \frac{\text{discount}}{\text{marked price}}$$
$$= \frac{\$2.79}{\$13.95}$$
$$= 0.20 \quad \text{(Verify!)}$$
$$= 20\%$$

Therefore, the rate of discount is 20%.

Practice Exercise 4. The marked price of an item is $280. The discount is $42. What is the rate of discount?

EXAMPLE 4 Connie's Fashions is having an after-Christmas sale. The marked price of a coat is $140. The sale price is $84. What is the rate of discount?

SOLUTION:

Step 1: Determine the discount:

$$\begin{aligned}\text{Discount} &= \text{marked price} - \text{sale price} \\ &= \$140 - \$84 \\ &= \$56\end{aligned}$$

Step 2: Determine the rate of discount:

$$\text{Rate of discount} = \frac{\text{discount}}{\text{marked price}}$$
$$= \frac{\$56}{\$140}$$
$$= 0.40$$
$$= 40\%$$

Therefore, the rate of discount is 40%.

ANSWERS TO PRACTICE EXERCISES

3. 40%. 4. 15%.

EXERCISES 4.7

1. A VCR has a marked price of $319 and a sale price of $162.50. Determine the discount.
2. A water bed has a discount of $42.50 and a sale price of $409.25. Determine the marked price.
3. A microwave oven has a marked price of $519 and a discount of $113.75. Determine the sale price.
4. A lamp has a marked price of $75 and a discount of $15. Determine the rate of discount.
5. An automobile tire is advertised at 25% off the marked price of $82. What is the discount?
6. A yard tractor is advertised at $\frac{1}{4}$ off the marked price of $520. What is the discount?
7. A refrigerator is advertised at 30% off the marked price of $560. What is the sale price?
8. A color TV is advertised at $\frac{1}{3}$ off the marked price of $750.90. What is the sale price?

9. For a summer clearance sale, a picnic table is advertised at 60% off the marked price of $115.50. What is the sale price?
10. The marked price of a racing bike is $300. The discount is $120. What is the rate of discount?
11. The marked price of a home computer is $900. The discount is $360. What is the rate of discount?
12. The marked price of a piece of luggage is $56. The rate of discount is 15%. What is the sale price?
13. The marked price of a self-propelled lawn mower is $339. The rate of discount is 40%. What is the sale price?
14. A furniture store advertised furniture on sale up to 40% off the marked price. A chair is on sale for $169. The marked price is $250. Is this an honest sale? Why?
15. Another furniture store advertised furniture on sale for at least 20% off the marked price. A sofa is on sale for $350. The marked price is $419. Is this an honest sale? Why?
16. If the sale price on a dress is $89 and the store advertises a 25% rate of discount on all items, what was the original price?
17. Suppose a firm gets a 5% discount if it pays its supplier's bill within 10 days. How much would the firm pay for a $1172.53 bill if paid within 10 days?
18. In Exercise 17, suppose the terms are a 3% discount if the bill is paid within 10–20 days. The firm pays on the fifteenth day. How much does it have to pay?
19. I can take a 12% discount on my purchases from a company for which I work. How much should I pay for a $112 item?
20. A senior citizen can get an 8% discount on medicine at a certain pharmacy. How much does she pay for a prescription normally costing $13.75?

4.8 Simple Interest

OBJECTIVES

After completing this section, you should be able to:
1. Find the simple interest given the principal, rate, and time.
2. Find the rate of interest given the principal, interest, and time.
3. Find the principal given the interest, rate, and time.
4. Find the amount given the principal and interest.

In the previous section, we saw that money saved on the purchase of an item is called a "discount." However, when people borrow money, they have to pay for the use of the money. The price paid for the use of the borrowed money is called **interest.** If you borrow $100 for a period of 1 year, then the interest is paid at the *end* of the year. The amount of money that is borrowed is called the **principal.**

Interest is also the money that you earn on your deposit in a bank or other financial institution. You deposit an amount of money and at the end of a period of time, such as a month or a year, you receive interest for the use of your money.

Whether you borrow money and pay interest or save money and earn interest, the interest is calculated using three factors which are:

- The *principal* or the amount borrowed (or saved), denoted by P.
- The *rate of interest*, given as a percent per period of time (such as month, year, etc.), denoted by r.
- The *time*, given in years (or part of a year), denoted by t.

Interest, denoted by *I*, is the product of the principal, the rate, and the time.

$$\text{Interest} = \text{principal} \times \text{rate} \times \text{time}$$

or

$$I = P \times r \times t$$

The formula $I = P \times r \times t$ is referred to as the formula for **simple interest,** with the interest calculated once. However, the interest on money borrowed is generally calculated frequently, such as monthly, and the borrower pays interest on the accumulated interest. This is called "compound interest." The same is true with interest earned on your savings. It is generally calculated monthly or quarterly, and you earn interest on the accumulated interest.

In this text, we will use simple interest only. Henceforth, we will refer to simple interest as just interest.

Practice Exercise 1. Determine the interest when $700 is borrowed (or saved) at 9% per year for 1 year.

EXAMPLE 1 Determine the interest when $600 is borrowed (or saved) at 8% per year for 1 year.

SOLUTION:
$$\begin{aligned}\text{Interest} &= \text{principal} \times \text{rate} \times \text{time}\\ &= \$600 \times 8\% \times 1\\ &= (\$600)(0.08)(1)\\ &= \$48\end{aligned}$$

Therefore, the interest is $48.

Practice Exercise 2. Determine the interest when $800 is borrowed (or saved) at 7% per year for 18 months.

EXAMPLE 2 Determine the interest when $900 is borrowed (or saved) at 7% per year for 9 months.

SOLUTION: In this problem, the time = 9 months = $\frac{9}{12} = \frac{3}{4}$ year. Hence,

$$\begin{aligned}\text{Interest} &= \text{principal} \times \text{rate} \times \text{time}\\ &= \$900 \times 7\% \times \frac{3}{4}\\ &= \$47.25\end{aligned}$$

Therefore, the interest is $47.25.

Practice Exercise 3. Determine the interest when $1500 is borrowed (or saved) at 8% per year for 42 months.

EXAMPLE 3 Determine the interest when $1200 is borrowed (or saved) at 7.5% per year for 30 months.

SOLUTION: The time = 30 months = $\frac{30}{12} = \frac{5}{2}$ years. Hence,

$$\begin{aligned}\text{Interest} &= \text{principal} \times \text{rate} \times \text{time}\\ &= \$1200 \times 7.5\% \times \frac{5}{2}\\ &= \$1200 \times 0.075 \times \frac{5}{2}\\ &= \$225\end{aligned}$$

Therefore, the interest is $225.

4.8 Simple Interest

Practice Exercise 4. Determine the interest when $3600 is borrowed (or saved) at 9.5% per year for 90 days.

EXAMPLE 4 Determine the interest when $500 is borrowed (or saved) at 9% per year for 60 days.

SOLUTION: The time = 60 days = $\frac{60}{360} = \frac{1}{6}$ year. Hence,

$$\begin{aligned}\text{Interest} &= \text{principal} \times \text{rate} \times \text{time} \\ &= \$500 \times 9\% \times \frac{1}{6} \\ &= \$500 \times 0.09 \times \frac{1}{6} \\ &= \$7.50\end{aligned}$$

Therefore, the interest is $7.50.

In Example 4, we used 360 days in 1 year for what is called *ordinary interest*. In some cases, bankers use 365 days for 1 year.

ANSWERS TO PRACTICE EXERCISES

1. $63.   2. $84.   3. $420.   4. $85.50.

Sometimes we know the principal, the time, and the interest. We can determine the rate by *dividing* the interest by the product of the principal and the time (in years).

$$\text{Rate} = \frac{\text{interest}}{\text{principal} \times \text{time}}$$

or

$$r = \frac{I}{P \times t}$$

Practice Exercise 5. Determine the yearly rate of interest if $600 is borrowed (or saved) for 1 year with an interest of $48.

EXAMPLE 5 Determine the yearly rate of interest if $500 is borrowed (or saved) for 8 months with an interest of $30.

SOLUTION: Here, time = 8 months = $\frac{8}{12} = \frac{2}{3}$ year. Hence,

$$\begin{aligned}\text{Rate} &= \frac{\text{interest}}{\text{principal} \times \text{time}} \\ &= \frac{\$30}{\$500 \times (2 \div 3)} \\ &= \frac{\$30}{\$1000 \div 3} \\ &= \frac{30}{1} \times \frac{3}{1000} \\ &= \frac{90}{1000}\end{aligned}$$

$$= \frac{9}{100}$$
$$= 9\%$$

Therefore, the rate is 9% per year.

Practice Exercise 6. Determine the yearly rate of interest if $1800 is borrowed (or saved) for 30 months with an interest of $540.

EXAMPLE 6 Determine the yearly rate of interest if $800 is borrowed (or saved) for 6 months with an interest of $26.

SOLUTION: Here, time = 6 months = $\frac{6}{12} = \frac{1}{2}$ year. Hence,

$$\text{Rate} = \frac{\text{interest}}{\text{principal} \times \text{time}}$$
$$= \frac{\$26}{\$800 \times (1 \div 2)}$$
$$= \frac{\$26}{\$400}$$
$$= \frac{6.5}{100}$$
$$= 6.5\%$$

Therefore, the rate is 6.5% per year.

ANSWERS TO PRACTICE EXERCISES

5. 8% per year. **6.** 12% per year.

Sometimes we know the interest, the rate, and the time. We can determine the principal by *dividing* the interest by the product of the rate and the time (in years).

$$\text{Principal} = \frac{\text{interest}}{\text{rate} \times \text{time}}$$

or

$$P = \frac{I}{r \times t}$$

Practice Exercise 7. Determine the principal if $300 interest is paid on a loan (or deposit) at 10% per year for 1 year.

EXAMPLE 7 Determine the principal if $45 interest is paid on a loan at 8% per year for 9 months.

SOLUTION: Here, time = 9 months = $\frac{9}{12} = \frac{3}{4}$ year. Hence,

$$\text{Principal} = \frac{\text{interest}}{\text{rate} \times \text{time}}$$
$$= \frac{\$45}{0.08 \times (3 \div 4)}$$

4.8 Simple Interest

$$= \frac{\$45}{0.06}$$
$$= \$750$$

Therefore, the principal is $750.

Practice Exercise 8. Determine the principal if $150 interest is paid on a loan (or deposit) at 12% per year for 6 months.

EXAMPLE 8 Determine the principal if $131.25 interest is earned on a deposit at 7% per year for 18 months.

SOLUTION: Here, time = 18 months = $\frac{18}{12} = \frac{3}{2}$ years. Hence,

$$\text{Principal} = \frac{\text{interest}}{\text{rate} \times \text{time}}$$
$$= \frac{\$131.25}{0.07 \times (3 \div 2)}$$
$$= \frac{\$131.25}{0.07 \times 1.5}$$
$$= \frac{\$131.25}{0.105}$$
$$= \$1250$$

Therefore, the principal is $1250.

The sum of the principal and the interest is called the "amount." To find the amount of a loan (or accumulated deposit), *add* the interest to the principal.

Amount = principal + interest

Practice Exercise 9. Dick saved $1800 at 8.5% per year for 18 months. At the end of 18 months, how much does he have in his account?

EXAMPLE 9 Sharon borrowed $500 at 7% per year for 6 months. At the end of 6 months, how much does she owe?

SOLUTION: We must do *two* things.

STEP 1: First, determine the *interest*:

$$\text{Interest} = \text{principal} \times \text{rate} \times \text{time}$$
$$= \$500 \times 0.07 \times \frac{1}{2}$$
$$= \$17.50$$

STEP 2: Then, determine the *amount*:

$$\text{Amount} = \text{principal} + \text{interest}$$
$$= \$500 + \$17.50$$
$$= \$517.50$$

Therefore, the total amount owed at the end of the 6 months is $517.50.

ANSWERS TO PRACTICE EXERCISES

7. $3000.   8. $2500.   9. $2029.50.

EXERCISES 4.8

For each of the following, P = principal, r = rate (per year), t = time (in years), I = interest, and A = amount. (Use 360 days = 1 year.)

In Exercises 1–10, determine I.

1. $P = \$550$, $r = 9\%$, and $t = 1$ year
2. $P = \$750$, $r = 8\%$, and $t = 6$ months
3. $P = \$1200$, $r = 9\%$, and $t = 8$ months
4. $P = \$3600$, $r = 6\%$, and $t = 9$ months
5. $P = \$400$, $r = 8\%$, and $t = 90$ days
6. $P = \$800$, $r = 7.5\%$, and $t = 60$ days
7. $P = \$720$, $r = 8.5\%$, and $t = 30$ days
8. $P = \$1800$, $r = 11\%$, and $t = 120$ days
9. $P = \$960$, $r = 9.5\%$, and $t = 6$ months
10. $P = \$2300$, $r = 10.5\%$, and $t = 1$ year

In Exercises 11–20, determine r.

11. $P = \$1200$, $I = \$36$, and $t = 6$ months
12. $P = \$900$, $I = \$54$, and $t = 9$ months
13. $P = \$800$, $I = \$56$, and $t = 1$ year
14. $P = \$300$, $I = \$2.25$, and $t = 30$ days
15. $P = \$450$, $I = \$20.25$, and $t = 6$ months
16. $P = \$2000$, $I = \$300$, and $t = 18$ months
17. $P = \$3500$, $I = \$210$, and $t = 6$ months
18. $P = \$1000$, $I = \$325$, and $t = 30$ months
19. $P = \$1200$, $I = \$16$, and $t = 60$ days
20. $P = \$900$, $I = \$243$, and $t = 3$ years

In Exercises 21–30, determine P.

21. $r = 7\%$, $I = \$70$, and $t = 1$ year
22. $r = 8\%$, $I = \$48$, and $t = 6$ months
23. $r = 9\%$, $I = \$27$, and $t = 60$ days
24. $r = 6.5\%$, $I = \$58.50$, and $t = 1$ year
25. $r = 10\%$, $I = \$37.50$, and $t = 9$ months
26. $r = 11\%$, $I = \$550$, and $t = 30$ months
27. $r = 8.5\%$, $I = \$59.50$, and $t = 6$ months
28. $r = 12\%$, $I = \$336$, and $t = 42$ months
29. $r = 10.5\%$, $I = \$2100$, and $t = 4$ years
30. $r = 7\%$, $I = \$1260$, and $t = 18$ months
31. Determine A if $P = \$600$ and $I = \$30$.
32. Determine A if $P = \$750$ and $I = \$33.50$.
33. Determine P if $A = \$108$ and $I = \$9.50$.
34. Determine P if $A = \$1600$ and $I = \$135$.
35. Determine I if $A = \$860$ and $P = \$805.50$.
36. Determine I if $A = \$725$ and $P = \$690$.
37. Determine the total amount of a loan of $600 at 9% for 6 months.
38. Determine the total amount of a loan of $1120 at 8.5% for 1 year.
39. Determine the total amount of a loan of $3600 at 10% for 90 days.
40. Determine the total amount of a loan of $1080 at 9% for 8 months.

Chapter Summary

In this chapter, you were introduced to the relationships between percents, decimals, and fractions. You learned to rewrite a percent as a decimal and a decimal as a percent. You also were introduced to the basic types of problems involving percents. Applications examined included percent increase or decrease, sales and property taxes, commissions, discount, and simple interest.

The **key words** introduced in this chapter are listed here in the order in which they appeared in the chapter.

percent (p. 162)	commission (p. 188)	discount (p. 192)
increase (p. 178)	straight commission (p. 188)	sale price (p. 192)
decrease (p. 178)	commission rate (p. 188)	interest (p. 195)
sales tax (p. 182)	gross sales (p. 188)	principal (p. 195)
tax rate (p. 183)	net sales (p. 188)	rate of interest (p. 195)
real property tax (p. 184)	marked price (p. 192)	simple interest (p. 196)
assessed value (p. 185)	rate of discount (p. 192)	

Review Exercises

4.1 Percent

Rewrite each of the indicated numbers as a percent.

1. $\dfrac{2}{5}$ 2. $\dfrac{9}{4}$ 3. 31   4. $5\dfrac{1}{2}$

Rewrite each of the indicated percents as equivalent fractions in simplest form.

5. 6%   6. 45%   7. 215%   8. 630%

4.2 More on Percents

9. Change 35% to a fraction in lowest terms.
10. Change 12.6% to a fraction in lowest terms.
11. Change 23.65 to a fraction in lowest terms.
12. Change 96% to a decimal.
13. Change 8.9% to a decimal.
14. Change 0.29 to a percent.
15. Change 2.39 to a percent.
16. Change 5 to a percent.
17. Change $\dfrac{2}{5}$ to a percent.
18. Change $\dfrac{9}{4}$ to a percent.

4.3 Problems Involving Percents

19. What is 11% of 172.6?
20. What is 0.5% of 16.48?
21. 8 is what percent of 32?
22. 41 is what percent of 25?
23. What percent of 60 is 40?
24. What percent of 0.05 is 0.02?
25. 9 is 25% of what number?
26. 60 is 30% of what number?
27. 0.02 is 400% of what number?
28. 3 is 100% of what number?

4.4 Percent Increase or Decrease

29. A worker's salary is increased by 8.5%. If the old salary was $8600 per year, what is the worker's new salary?
30. The price of ground beef increased $.25 per pound in 1 week. If the old price was $1.40 per pound, what was the percent of increase during the week?

31. Mary had $875 in her savings account. She then withdrew some money, leaving a balance of $750. What was the percent of decrease in her account? (Give your answer to the nearest tenth of a percent.)

4.5 Sales and Property Tax

32. Determine the amount of sales tax on a toaster that sells for $45.60 with a sales tax of 5%.
33. Determine the total price for a used car that sells for $1200 with a sales tax of 4.5%.
34. Determine the sales tax rate for a TV that sells for $300 and has a sales tax of $21.
35. Determine the amount of property tax on a house with assessed value of $32,600 and rate of 3.5%.
36. Determine the property tax rate for a budget of $695,000 and total assessed value of $3,169,200. (Give your answer to the nearest hundredth of a percent.)

4.6 Commissions

37. Debbie works in a hardware store. She receives $40 per week plus a 10% commission of net sales. If her net sales for the week are $865, what is her total salary?
38. John sells real estate. He receives a salary of $450 per month plus a commission of 8% of the sales price. If his sales for the month amounted to $16,500, what was his total amount of pay?
39. Walt works in a record store. He receives a weekly salary of $65 and a commission of 6% of net sales. If his gross sales for a week were $1250 and returns amounted to $63, what was his total pay for the week?

4.7 Discount

40. A garden tractor has a marked price of $969.95 and a sales price of $637.50. Determine the discount.
41. A 10-speed bicycle has a marked price of $90 and a discount of $22.50. Determine the rate of discount.
42. A 5-horsepower snow thrower is on sale at 40% off the marked price of $650. What is the sale price?

4.8 Simple Interest

43. Determine the interest on $560 at 8% for 1 year.
44. Determine the interest on $980 at 8% for 9 months.
45. Determine the principal if $r = 7\%$, $I = \$3.50$, and $t = 60$ days.
46. Determine the rate of interest if $P = \$1800$, $I = \$58.50$, and $t = 6$ months.

Teasers

1. At Corporation A, $\frac{1}{4}$ of the executive positions are filled by females. At Corporation B, 8 of every 25 executive positions are filled by females. Which corporation has the higher percentage of executive positions filled by females?
2. An employee was paid a salary of $10,000. Due to poor market conditions, he agreed to take a 10% reduction in salary. Six months later, due to improved market conditions, his employer gave him a 10% increase in salary. What is the employee's new salary? (Assume that no other changes in salary were made.)
3. Judy works at Matrix Industries and receives a salary of $24,000 per year. Fringe benefits paid by the employer on her behalf amount to 23.6% of her salary. A renegotiated contract called for a 6.8% increase in salary plus additional fringe benefits, bringing the total fringe benefit package to 27.2% of the salary. How much is now paid to and on behalf of Judy by the employer?
4. An unsuspecting hourly employee approached his employer and asked for a 5% raise in pay. Without much hestitation, the employer proposed to increase the employee's hours by 25% and his gross pay by 30%. Did the employee, in fact, receive a net 5% hourly raise? Explain your answer.
5. Andrea invested $12,500 at 8.5% per year and $11,750 at 9.2% per year. At the end of the year, she gave one-half of the larger amount of interest earned to charity. How much did she give to charity?
6. John gave Jim a promissory note for a loan of $6000 at 8% per year for 9 months. Seven months later, Jim, in need of money, offered to give back the note to John in exchange of $6200 payment in full. Assuming that John had the money, should he pay back the loan in order to save in interest? If so, how much savings would be realized?
7. The Smart Shoppe sells designer dresses at discount prices. Immediately upon arrival at the store, the price of a dress is discounted 15%. Fifteen days later, the discounted price of the dress is further discounted 10%. Thirty days later, the new discounted price is reduced an additional 25%. Stacey buys a dress 48 days after the dress arrives at the store. If the arrival price of the dress was $875, how much should Stacey have to pay for the dress, if the sales tax rate is 5%.
8. A couple dine out and receive a senior citizen discount of 10%. Their check amounts to $42.00. They wish to leave a 15% gratuity. Should they add the gratuity to the check amount and then take a 10% discount? Or should they take a 10% discount on the check amount and then add 15% to the discounted amount? Is there a difference in the final figure either way? Explain your answer.
9. A new car can be bought with no downpayment or trade-in value at 2.9% per year for 48 months. Or the customer can receive a $1500 rebate from the manufacturer which can be applied to the purchase price, if the balance due is financed at 9.5% per year for 48 months. You buy a $15,000 car. What is the better deal for you if the rebate is applied to the purchase price and you finance the balance over 48 months? (Assume there is no other downpayment.)
10. Joshua works in a retail store and receives a salary of $80 per week plus 6.5% commission on net sales. Using good English and complete sentences, give a step-by-step procedure to determine what Joshua's weekly pay would be if he had gross sales of $3187 and returns of $512 for the week.

CHAPTER 4 TEST

This test covers material from Chapter 4. Read each question carefully and answer all questions. If you miss a question, review the appropriate section of the text.

1. Rewrite $\frac{4}{5}$ as a percent.

2. Rewrite $3\frac{1}{4}$ as a percent.

3. Rewrite 7% as a fraction in simplest form.

4. Rewrite 135% as a fraction in simplest form.

5. Rewrite 17.65% as a fraction in simplest form.

6. Rewrite 87% as a decimal.

7. Rewrite 14.6% as a decimal.

8. What is 13% of 57.3?

9. 9 is what percent of 36?

10. 0.03 is 60% of what number?

11. A worker's salary is increased by 9.5%. If the old salary was $7800 per year, what is the worker's new salary?

12. After an engine tune-up, the average number of miles per gallon for a particular car increased from 24 to 28. What was the percent increase?

13. Determine the amount of sales tax on an item that sells for $57.60 if the sales tax rate is 5%.

14. Dawn works in a retail store and earns $45 per week plus a 12% commission of net sales. If the net sales for a week are $750, what is her total pay?

15. A 100-W rack stereo with compact disc player has a marked price of $865.95 and a sale price of $599.99. Determine the discount.

16. A sofa bed is on sale at 30% off the marked price of $1150. What is the sale price?

17. Determine the simple interest on $650 at 8% per year for 1 year.

18. Determine the yearly rate of simple interest if the principal is $2000, interest is $90, and time is 6 months.

CUMULATIVE REVIEW: CHAPTERS 1–4

In Exercises 1–6, perform the indicated operations. Write your results in simplest form.

1. $\dfrac{5}{7} - \dfrac{9}{14}$

2. $0.04 + 3.2 + 5.007 + 19$

3. $23.1\% \times 126$

4. $962.64 \div 0.6$

5. $8\dfrac{1}{3} \times 10\dfrac{1}{5}$

6. $63 \div 10\%$

In Exercises 7–10, round each number as indicated.

7. 407.396; hundredths

8. $5\dfrac{3}{7}$; ones

9. $\dfrac{11}{13}$; tenths

10. 12.3% of 36; tenths

In Exercises 11–15, change each number as indicated.

11. 0.52 to a fraction in simplest form

12. 6.39% to a decimal

13. 0.023 to a percent

14. 68% to a fraction in simplest form

15. $\dfrac{4}{9}$ to an equivalent fraction with a numerator of 20

In Exercises 16–18, circle the larger number.

16. 23.1%   or   2.34

17. 0.06   or   5.6%

18. 37% of $\dfrac{1}{2}$ or   0.21

19. What is 12% of 213.4?

20. What percent of 120 is 40?

21. 70 is 40% of what number?

22. Determine the amount of sales tax on a coffee maker that sells for $24.98 with a sales tax of 5%.

23. Joshua works in a record store. He receives $55 per week plus a 10% commission of net sales. If his sales for the week are $995 and returns amount to $120, what is his total weekly salary?

24. Determine the perimeter of a triangular region whose sides measure 7 ft, 9 ft, and 4 yd.

25. Determine the area of a rectangular region with width 17 m and length 26 m.

CHAPTER 5 Measurement

Measurement is used to describe many quantities in everyday situations. We measure length, area, volume, liquid capacity, weight, time, temperature, and speed, to mention just a few. Each of these measurements has various units assigned to it.

In this chapter, we will introduce the concept of measurement, conversions within and between systems of measurement, and applications of measurement to geometric figures.

CHAPTER 5 PRETEST

This pretest covers material on measurement. Read each question carefully, and try to answer all questions. Each question is keyed to the chapter section in which the particular topic is discussed. Check your answers with those given in the back of the text.

Questions 1–5 pertain to Section 5.1.

1. Convert 28 ft to yards.

2. Convert 3 lb to grams.

3. Convert 3 hr, 12 min to seconds.

4. Convert 27 cm to millimeters.

5. Convert 1080 g to kilograms.

Questions 6–8 pertain to Section 5.2.

6. Determine the perimeter of a triangle with sides of lengths 6.3 cm, 5.6 cm, and 6.1 cm.

7. Determine the perimeter of an equilateral triangle with sides of length 10.3 ft.

8. Determine the perimeter of a rectangle with length 8.1 dm and width 6.8 dm.

Questions 9–10 pertain to Section 5.3. (Use 3.14 for π.)

9. Determine the circumference of a circle whose radius is 3.9 yd.

10. Determine the diameter of a circle whose circumference is 168.2 cm. (Give your answer to the nearest tenth of a centimeter.)

Questions 11–16 pertain to Section 5.4.

$\left(\text{Use } \dfrac{22}{7} \text{ for } \pi.\right)$

11. Determine the area of a rectangular region that is 12.3 ft long and 2.1 yd wide.

12. Determine the area of a circular region whose diameter is 27 cm.

13. Determine the area of a square region whose sides measure 13.9 dm each.

14. A plane region is enclosed by a parallelogram with height of 4 ft and base of 7 ft. Determine the area of the region.

15. A triangular region has a height of 7 ft and a base of 5 ft. Determine the area of the region.

16. Determine the surface area of a rectangular solid that is 8 cm long, 5 cm wide, and 4 cm high.

Questions 17–20 pertain to Section 5.5. (Leave your answers in terms of π.)

17. Determine the volume of a rectangular solid that is 1.7 cm high, 2.8 cm wide, and 2.7 cm long.

18. Determine the volume of a cube with edge of length 1.9 yd.

19. Determine the volume of a circular cylinder whose height is 12 cm and whose base has a diameter of 8 cm.

20. Determine the volume of a sphere with radius 7 in.

5.1 Units of Measurement

OBJECTIVE

After completing this section, you should be able to convert various measures of length, weight, liquid capacity, or time from one system to another or within the same system.

From the time of your birth, various units of measurement have been associated with you. At birth, you were probably measured and declared to be so many inches long. Your weight was recorded in pounds and ounces. Your temperature was given as so many degrees Fahrenheit. Your pulse was stated as so many beats per minute, and so forth. As you grew older, you learned that the distance from one city to another was so many miles or so many kilometers, that the floor space of your room was so many square feet or so many square meters, and that your weight, in pounds, changed from year to year.

Measurements are stated in generally agreed upon terms called "units."

The following are examples of measurements.

- 26 feet
- 2.3 meters
- 14.9 centimeters
- 17 ounces
- 219 grams
- 6 quarts
- 3.2 liters

There are two major systems of units in use throughout the world today. These are the *English system*, used in some countries, including the United States, and the *metric system*, used in other countries and the scientific community. The United States is currently changing from the English system to the metric system. In many ways, the metric system is more convenient since, as we shall see, it involves only the use of powers of 10 for converting within the system.

In Table 5.1, some of the more common units of measurement from both the English and metric systems are given. A few of the basic relationships between the two systems are also given. Abbreviations for some units of measure have already been given in the text; those and others are shown in the footnote to Table 5.1.

In Examples 1–5, we will illustrate conversions within the English system of measurement.

Practice Exercise 1. Convert 17 yd to feet.

EXAMPLE 1 Convert 27 in. to feet.

SOLUTION: From Table 5.1, we note that

$$12 \text{ in.} = 1 \text{ ft}$$

Hence,

$$1 \text{ in.} = \frac{1}{12} \text{ ft} \qquad \text{(Dividing both sides of the equation by 12)}$$

$$27 \text{ in.} = 27\left(\frac{1}{12} \text{ ft}\right) \qquad \text{(Multiplying both sides of the equation by 27)}$$

$$= 2.25 \text{ ft}$$

Therefore, 27 in. = 2.25 ft.

5.1 Units of Measurement

TABLE 5.1

Measure	English	Metric	Conversion
Length	1 ft = 12 in. 1 yd = 3 ft 1 mi = 5280 ft	1 cm = 10 mm 1 dm = 10 cm 1 m = 10 dm 1 km = 1000 m	1 in. = 2.54 cm 1 m = 39.4 in. 1 mi = 1.61 km 1 km = 0.6 mi
Weight	1 lb = 16 oz 1 ton = 2000 lb	1 g = 1000 mg 1 kg = 1000 g	1 lb = 454 g 1 kg = 2.20 lb
Liquid capacity	1 qt = 2 pt 1 gal = 4 qt 1 qt = 32 oz	1 L = 1000 mL 1 kL = 1000 L	1 L = 1.06 qt 1 qt = 946 mL
Time	1 min = 60 sec 1 hr = 60 min 1 day = 24 hr	1 min = 60 sec 1 hr = 60 min 1 day = 24 hr	The units are the same in both systems.

Abbreviations: ft (feet); in. (inches); yd (yards); mi (miles); lb (pounds); oz (ounces); cm (centimeters); mm (millimeters); dm (decimeters); m (meters); km (kilometers); g (grams); mg (milligrams); kg (kilograms); qt (quarts); pt (pints); L (liters); mL (milliliters); kL (kiloliters); min (minutes); sec (seconds); hr (hours)

Practice Exercise 2. Convert 86 oz to pounds.

EXAMPLE 2 Convert 4 tons to pounds.

SOLUTION: From Table 5.1, we note that

$$1 \text{ ton} = 2000 \text{ lb}$$

Hence,

$$4 \text{ tons} = 4(2000 \text{ lb}) \quad \text{(Multiplying both sides of the equation by 4)}$$
$$= 8000 \text{ lb}$$

Therefore, 4 tons = 8000 lb.

In Examples 1 and 2, each of the conversions involved only two units of measure. Further, these units of measure are directly related. For instance, to convert inches to feet, we used the relationship 12 in. = 1 ft, and to convert tons to pounds, we used 1 ton = 2000 lb.

When the conversions involve more than two units, a different procedure is used to help keep track of all the units of measurement. This procedure will be illustrated in the following examples.

Practice Exercise 3. Convert 180 sec to hours.

EXAMPLE 3 Convert 3 gal to pints.

SOLUTION: From Table 5.1, we note that 1 gal = 4 qt and that 1 qt = 2 pt. Hence,

$$3 \text{ gal} = \frac{3 \text{ gal}}{1} \times \frac{4 \text{ qt}}{1 \text{ gal}} \times \frac{2 \text{ pt}}{1 \text{ qt}}$$

Changing gallons to quarts ⟶ (1 gal = 4 qt)
Changing quarts to pints ⟶ (1 qt = 2 pt)

$$= \frac{3 \cancel{\text{gal}}}{1} \times \frac{4 \cancel{\text{qt}}}{1 \cancel{\text{gal}}} \times \frac{2 \text{ pt}}{1 \cancel{\text{qt}}}$$

$$= \frac{3 \times 4 \times 2 \text{ pt}}{1 \times 1 \times 1}$$

$$= \frac{24 \text{ pt}}{1}$$

$$= 24 \text{ pt}$$

Therefore, 3 gal = 24 pt.

Practice Exercise 4. Convert 60 mph to yards per second.

EXAMPLE 4 A car travels 40 mph. What is the rate of speed in feet per second?

SOLUTION: From Table 5.1, we note that 1 mi = 5280 ft, 1 hr = 60 min, and 1 min = 60 sec. Hence,

$$40 \text{ mph} = \frac{40 \text{ mi}}{1 \text{ hr}} \times \frac{5280 \text{ ft}}{1 \text{ mi}} \times \frac{1 \text{ hr}}{60 \text{ min}} \times \frac{1 \text{ min}}{60 \text{ sec}}$$

Changing miles to feet
Changing hours to minutes
Changing minutes to seconds

(1 mi = 5280 ft)
(1 hr = 60 min)
(1 min = 60 sec)

$$= \frac{\overset{2}{\cancel{40}} \cancel{\text{mi}}}{1 \cancel{\text{hr}}} \times \frac{\overset{88}{\cancel{5280}} \text{ ft}}{1 \cancel{\text{mi}}} \times \frac{1 \cancel{\text{hr}}}{\underset{3}{\cancel{60}} \cancel{\text{min}}} \times \frac{1 \cancel{\text{min}}}{\underset{1}{\cancel{60}} \text{ sec}}$$

$$= \frac{2 \times 88 \times 1 \times 1 \text{ ft}}{1 \times 1 \times 3 \times 1 \text{ sec}}$$

$$= \frac{176 \text{ ft}}{3 \text{ sec}}$$

$$= 58\frac{2}{3} \text{ feet per second}$$

Therefore, 40 mph = $58\frac{2}{3}$ feet per second.

ANSWERS TO PRACTICE EXERCISES

1. 51 ft. **2.** $5\frac{3}{8}$ lb. **3.** $\frac{1}{20}$ hr. **4.** $29\frac{1}{3}$ yards per second.

Units in the metric system are converted by using powers of 10. There is a basic unit for each measure of length, volume, and mass. All of the other units are given in terms of the basic units, using prefixes as given in Table 5.2.

Note that the symbol m in Table 5.2 means milli. We also have the symbol m in Table 5.1 which means meter. When the letter m is used alone, it always represents meters; hence, 6 m means 6 meters. If a symbol has two letters and the first one is m, then it represents the prefix milli; hence 5 mm means 5 millimeters and 10 mL means 10 milliliters. You may refer to the footnote in Table 5.1 and to Table 5.2 for abbreviations for units of measure.

If the basic unit is grams (g), then:

- 3 g means 3 grams.
- 4 mg means 4 milligrams.
- 5 kg means 5 kilograms.
- 7 hg means 7 hectograms.

In Examples 5–7, we will illustrate conversions within the metric system of measurement.

5.1 Units of Measurement

TABLE 5.2

Prefix	Symbol	Value
milli	m	$\frac{1}{1000}$ times basic unit
centi	c	$\frac{1}{100}$ times basic unit
deci	d	$\frac{1}{10}$ times basic unit
Basic unit		
deka	dk	10 times basic unit
hecto	h	100 times basic unit
kilo	k	1000 times basic unit

> **Practice Exercise 5.** Convert 4 L to kiloliters.

EXAMPLE 5 Convert 7 m to centimeters.

SOLUTION: Since 1 m = 100 cm, we have

$$7\text{ m} = 7(100\text{ cm}) \quad \text{(Multiplying both sides of the equation by 7)}$$
$$= 700\text{ cm}$$

Therefore, 7 m = 700 cm.

> **Practice Exercise 6.** Convert 77 cm to decimeters.

EXAMPLE 6 Convert 900 g to kilograms.

SOLUTION: Since 1 kg = 1000 g, we have

$$1\text{ g} = \frac{1}{1000}\text{ kg}$$

Hence,

$$900\text{ g} = 900\left(\frac{1}{1000}\text{ kg}\right) \quad \text{(Multiplying both sides of the equation by 900)}$$
$$= 0.9\text{ kg}$$

Therefore, 900 g = 0.9 kg.

> **Practice Exercise 7.** Convert 4000 mL to dekaliters.

EXAMPLE 7 Convert 6 dkL to milliliters.

SOLUTION: Since 1 dkL = 10L, we have

$$6\text{ dkL} = \frac{6\text{ dkL}}{1} \times \frac{10\text{ L}}{1\text{ dkL}}$$

Also, 1 L = 1000 mL
Hence,

$$6\text{ dkL} = \frac{6\text{ dkL}}{1} \times \frac{10\text{ L}}{1\text{ dkL}} \times \frac{1000\text{ mL}}{1\text{ L}}$$

Changing dekaliters to liters ⎤
Changing liters to milliliters ⎦

(1 dkL = 10 L)
(1 L = 1000 mL)

$$= \frac{6 \text{ dkL}}{1} \times \frac{10 \text{ L}}{1 \text{ dkL}} \times \frac{1000 \text{ mL}}{1 \text{ L}}$$

$$= \frac{60{,}000 \text{ mL}}{1}$$

$$= 60{,}000 \text{ mL}$$

Therefore, 6 dkL = 60,000 mL.

ANSWERS TO PRACTICE EXERCISES

5. 0.004 kL.   **6.** 7.7 dm.   **7.** 0.4 dkL.

As you travel along highways, you will note that some road signs now indicate the distance to various locations in terms of kilometers instead of the more familiar mile designation. Also, packages of cereals now have the weight given in terms of grams instead of ounces. Your favorite soda comes in bottles with liquid capacity such as 2 L instead of the more familiar quart designation.

Table 5.1 lists conversions between the English and metric systems. In Examples 8–10, we will illustrate some of these conversions.

Practice Exercise 8. Convert 90 km to miles.

EXAMPLE 8 Convert 3 pt to liters.

SOLUTION: From Table 5.1, we note that 1 qt = 2 pt and that 1 L = 1.06 qt. Hence,

$$3 \text{ pt} = \frac{3 \text{ pt}}{1} \times \frac{1 \text{ qt}}{2 \text{ pt}} \times \frac{1 \text{ L}}{1.06 \text{ qt}}$$

Changing pints to quarts ⎯⎯⎯⎯⎯⎯⎯⎯↑   (1 qt = 2 pt)
Changing quarts to liters ⎯⎯⎯⎯⎯⎯⎯⎯⎯⎯↑   (1 L = 1.06 qt)

$$= \frac{3 \text{ pt}}{1} \times \frac{1 \text{ qt}}{2 \text{ pt}} \times \frac{1 \text{ L}}{1.06 \text{ qt}}$$

$$= \frac{3 \text{ L}}{2 \times 1.06}$$

$$= 1.4 \text{ L} \qquad \text{(Rounded to the nearest tenth)}$$

Therefore, 3 pt equal 1.4 L, rounded to the nearest tenth.

Practice Exercise 9. Convert 212 qt to liters.

EXAMPLE 9 Convert 0.8 ton to kilograms.

SOLUTION: From Table 5.1, we note that 1 ton = 2000 lb and that 1 kg = 2.20 lb. Hence,

$$0.8 \text{ ton} = \frac{0.8 \text{ ton}}{1} \times \frac{2000 \text{ lb}}{1 \text{ ton}} \times \frac{1 \text{ kg}}{2.20 \text{ lb}}$$

Changing tons to pounds ⎯⎯⎯⎯⎯⎯⎯↑   (1 ton = 2000 lb)
Changing pounds to kilograms ⎯⎯⎯⎯⎯⎯⎯⎯↑   (1 kg = 2.20 lb)

$$= \frac{0.8 \text{ ton}}{1} \times \frac{2000 \text{ lb}}{1 \text{ ton}} \times \frac{1 \text{ kg}}{2.20 \text{ lb}}$$

$$= \frac{(0.8)(2000) \text{ kg}}{2.20}$$

$$= 727.3 \text{ kg} \qquad \text{(Rounded to the nearest tenth)}$$

Therefore, 0.8 ton equals 727.3 kilograms, rounded to the nearest tenth.

5.1 Units of Measurement

Practice Exercise 10. Convert 40 km to inches.

EXAMPLE 10 Convert 126 qt to dekaliters.

SOLUTION: From Table 5.1, we note that 1 qt = 946 mL and that 100 mL = 1 L. From Table 5.2, we note that 1 dkL = 10 L. Hence,

$$126 \text{ qt} = \frac{126 \text{ qt}}{1} \times \frac{946 \text{ mL}}{1 \text{ qt}} \times \frac{1 \text{ L}}{1000 \text{ mL}} \times \frac{1 \text{ dkL}}{10 \text{ L}}$$

Changing quarts to milliliters
Changing milliliters to liters
Changing liters to dekaliters

(1 qt = 946 mL)
(1 L = 1000 mL)
(1 dkL = 10 L)

$$= \frac{126 \text{ qt}}{1} \times \frac{946 \text{ mL}}{1 \text{ qt}} \times \frac{1 \text{ L}}{1000 \text{ mL}} \times \frac{1 \text{ dkL}}{10 \text{ L}}$$

$$= \frac{126 \times 946 \times 1 \times 1 \text{ dkL}}{1 \times 1 \times 1000 \times 10}$$

$$= \frac{119{,}196 \text{ dkL}}{10{,}000}$$

$$= 11.9196 \text{ dkL}$$

Therefore, 126 qt equal 11.9196 dkL.

Practice Exercise 11. If the speed limit is posted as 55 mph, what is the speed limit in kilometers per hour?

EXAMPLE 11 If the speed limit is posted as 100 km per hour, what is the speed limit in miles per hour?

SOLUTION: From Table 5.1, we note that

$$1 \text{ km} = 0.6 \text{ mi}$$

Therefore,

$$100 \text{ km} = 100(0.6 \text{ mi}) \quad \text{(Multiplying both sides of the equation by 100)}$$
$$= 60 \text{ mi}$$

Therefore, 100 km per hour is equal to 60 mph.

Practice Exercise 12. How many seconds are there in 5 weeks?

EXAMPLE 12 How many minutes are there in the month of June?

SOLUTION: Since there are 30 days in the month of June and each day has 24 hr, we have

$$1 \text{ day} = 24 \text{ hr}$$
$$30 \text{ days} = 30(24 \text{ hr}) \quad \text{(Multiplying both sides of the equation by 30)}$$
$$= 720 \text{ hr}$$

Since

$$1 \text{ hr} = 60 \text{ min}$$
$$720 \text{ hr} = 720(60 \text{ min}) \quad \text{(Multiplying both sides of the equation by 720)}$$
$$= 43{,}200 \text{ min}$$

Therefore, the month of June (30 days) has 43,200 min.

ANSWERS TO PRACTICE EXERCISES

8. 54 mi. **9.** 200 L. **10.** 1,576,000 in. **11.** 88.55 km per hr. **12.** 3,024,000 sec.

EXERCISES 5.1

In Exercises 1–6, convert the given measurements to feet.

1. 4 yd
2. 19 in.
3. 2 mi
4. 1 in.
5. 15 yd
6. 0.5 mi

In Exercises 7–10, convert the given measurements to pounds.

7. 64 oz
8. 16 oz
9. 2 tons
10. 0.25 ton

In Exercises 11–16, convert the given measurements to centimeters.

11. 3 dm
12. 4 m
13. 3 dkm
14. 30 mm
15. 0.5 hm
16. 0.6 m

In Exercises 17–22, convert the given measurements to kilograms.

17. 256 g
18. 1250 mg
19. 17 hg
20. 675 cg
21. 1770 dkg
22. 60 dg

In Exercises 23–28, convert the given measurements to pints.

23. 5 qt
24. 64 oz
25. 3 gal
26. 0.25 qt
27. 48 oz
28. 0.5 gal

In Exercises 29–34, convert the given measurements to minutes.

29. 3 hr
30. 90 sec
31. 0.5 hr
32. 150 sec
33. 1 hr, 23 min
34. 2 hr, 15 sec

In Exercises 35–39, convert the given measurements as indicated.

35. 4 mi to centimeters
36. 2.5 mi to kilometers
37. 70 km to feet
38. 5 m to feet
39. 0.3 km to feet

In Exercises 40–43, convert the given measurements as indicated.

40. 3 lb to grams
41. 13 lb to kilograms
42. 4 kg to pounds
43. 2 kg to ounces

In Exercises 44–47, convert the given measurements as indicated.

44. 3 L to quarts
45. 5 qt to liters
46. 0.4 qt to milliliters
47. 1.5 L to ounces
48. The distance between two cities is 53 mi. What is the distance in kilometers?
49. A car traveled at 55 mph. How fast did the car travel in kilometers per minute?
50. A box contains 420 g of cereal. What is the weight of the cereal in ounces? (Round, if necessary, to the nearest ounce.)
51. A container holds 34 oz of liquid soap detergent. What is this liquid capacity in liters?
52. How many seconds are there in 4 hr, 23 min, and 12 sec?
53. A young athlete recorded a standing broad jump of 7 ft 10 in. What was her accomplishment in decimeters?
54. A sheet of paper measures 8.75 in. by 11.5 in. Determine the dimensions of the sheet of paper in centimeters. (Round, if necessary, to the nearest hundredth of a centimeter.)

5.2 Polygons and Perimeter

OBJECTIVES

After completing this section, you should be able to:
1. Classify polygons according to the number of sides.
2. Classify triangles as being equilateral, isosceles, or scalene.
3. Determine the perimeter of certain polygons.

In this section, we will consider some basic geometric figures known as polygons. We will also consider some of the measurements associated with these figures.

One of the most basic geometric figures is a **line segment.** It is part of a line between two distinct (different) points and includes the two points, called its "endpoints." A line segment is named by identifying its endpoints. In Figure 5.1, some line segments are illustrated and named.

In Figure 5.2, we illustrate some geometric figures consisting of line segments. Observe that the shapes in Figures 5.2b, 5.2c, and 5.2d are closed. That is, you can start at any point on the figure, trace completely around the figure, and return to the starting point without retracing any part. That is not true of Figures 5.2a and 5.2e.

> **DEFINITION**
> A **polygon** is a closed geometric figure that consists of line segments. The line segments are called its "sides."

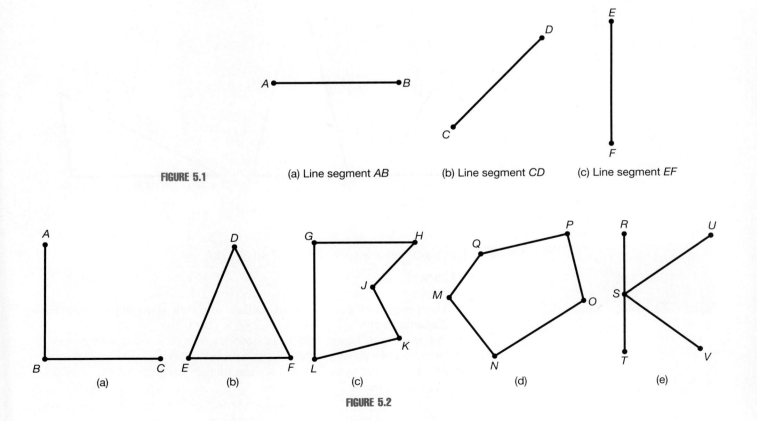

FIGURE 5.1 (a) Line segment AB   (b) Line segment CD   (c) Line segment EF

FIGURE 5.2

Polygons are named according to the number of sides that they have. A **triangle** is a polygon with three sides; a **quadrilateral** is a polygon with four sides; a **pentagon** is a polygon with five sides; a **hexagon** is a polygon with six sides; and so forth. (See Figure 5.3.)

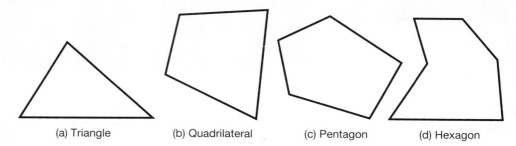

FIGURE 5.3 (a) Triangle (b) Quadrilateral (c) Pentagon (d) Hexagon

> **DEFINITION**
> A triangle can be classified according to the lengths of its sides.
> 1. If all three sides of a triangle have the same length, the triangle is an **equilateral triangle**.
> 2. If exactly two of the sides of a triangle have the same length, the triangle is an **isosceles triangle**.
> 3. If no two sides of a triangle have the same length, the triangle is a **scalene triangle**.

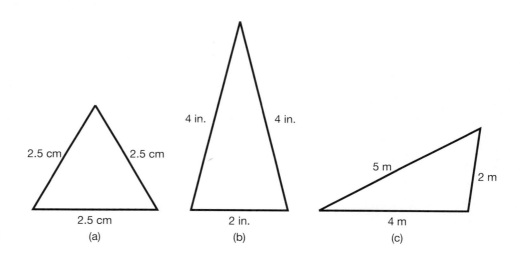

FIGURE 5.4 (a) (b) (c)

Practice Exercise 1. A triangle having all three sides of the same length is called a(n) _____ triangle.

EXAMPLE 1 Classify the triangles in Figure 5.4.

SOLUTION:

- The triangle in Figure 5.4a is an equilateral triangle since all three sides have the same length.
- The triangle in Figure 5.4b is an isosceles triangle since exactly two of its sides have the same length.
- The triangle in Figure 5.4c is a scalene triangle since the lengths of its three sides are different.

There are also special quadrilaterals.

5.2 Polygons and Perimeters

DEFINITION

1. If both pairs of the opposite sides of a quadrilateral are parallel (that is, they do not meet if extended), then the quadrilateral is called a **parallelogram**. (The opposite sides are also equal in length.)
2. A **rectangle** is a parallelogram with four right angles (that is, square corners).
3. A **square** is a rectangle all of whose sides have the same length.

Observe that squares and rectangles are special cases of parallelograms and that a square is a special case of a rectangle. (See Figure 5.5)

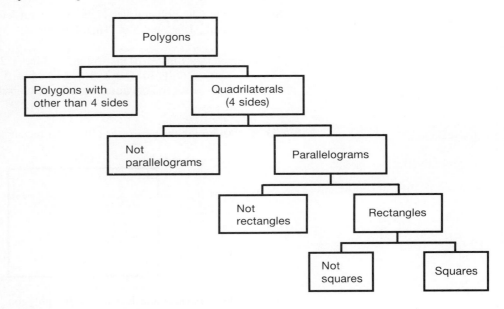

FIGURE 5.5

Practice Exercise 2. A polygon having four sides is called a _____ .

EXAMPLE 2 Classify each of the quadrilaterals given in Figure 5.6.

SOLUTION:

- The quadrilateral in Figure 5.6a is a rectangle.
- The quadrilateral in Figure 5.6b is a parallelogram.
- The quadrilateral in Figure 5.6c is a square.

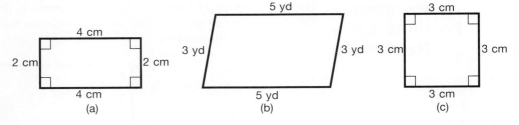

FIGURE 5.6

We will now examine the concept of perimeter of a polygon. The distance around a geometric figure is called its "perimeter."

DEFINITION

The **perimeter of a polygon** is the sum of the lengths of its sides.

Practice Exercise 3. The sum of the lengths of the sides of a polygon is called its _____.

EXAMPLE 3 Determine the perimeter of the following polygon.

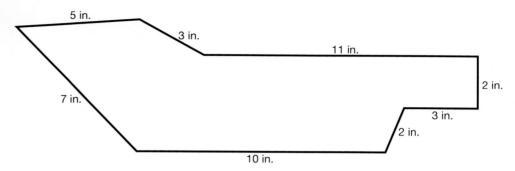

SOLUTION: The polygon has eight sides with lengths of 7 in., 10 in., 2 in., 3 in., 2 in., 11 in., 3 in., and 5 in. Therefore its

Perimeter = 7 in. + 10 in. + 2 in. + 3 in. + 2 in. + 11 in. + 3 in. + 5 in.
= (7 + 10 + 2 + 3 + 2 + 11 + 3 + 5) in.
= 43 in.

Practice Exercise 4. Determine the perimeter (P) of a rectangle with sides of lengths 2 yd and 4 ft.

EXAMPLE 4 Determine the perimeter (P) of the rectangle illustrated here.

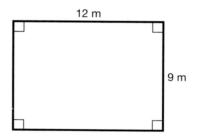

SOLUTION: A rectangle is a four-sided polygon. Its opposite sides have the same lengths. Hence,

Perimeter = 9 m + 12 m + 9 m + 12 m
= (9 + 12 + 9 + 12) m
= 42 m

Perimeter Formula for a Rectangle

The *perimeter of a rectangle* is equal to 2 times its length plus 2 times its width. If the length is given as L and the width as W, then the perimeter, P, of the rectangle is given by the formula

$$P = 2L + 2W$$

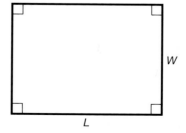

where P, L, and W are all measured in the *same* unit.

5.2 Polygons and Perimeter

Practice Exercise 5. Determine the perimeter of a rectangle with a length of 20.3 in. and a width of 18.1 in.

EXAMPLE 5 Determine the perimeter (P) of a rectangle with length of 17.6 cm and width of 10.3 cm.

SOLUTION: The rectangle has length (L) of 17.6 cm and width (W) of 10.3 cm. Hence,

$$P = 2L + 2W$$
$$= 2(17.6 \text{ cm}) + 2(10.3 \text{ cm})$$
$$= 35.2 \text{ cm} + 20.6 \text{ cm}$$
$$= 55.8 \text{ cm}$$

Practice Exercise 6. Determine the perimeter of a square that measures 17.16 cm on a side.

EXAMPLE 6 A square has a side of length 2.5 feet. What is its perimeter (P)?

SOLUTION: A square is a polygon having four sides. Therefore, its perimeter is the sum of the lengths of the four sides. However, all four sides of a square have the same length. Hence,

$$P = 2.5 \text{ ft} + 2.5 \text{ ft} + 2.5 \text{ ft} + 2.5 \text{ ft}$$
$$= (2.5 + 2.5 + 2.5 + 2.5) \text{ ft}$$
$$= 10 \text{ ft}$$

Perimeter Formula for a Square

The *perimeter of a square* is 4 times the length of a side of the square. If the square has a side of length s units, then the perimeter, P, of the square is given by the formula

$$P = 4s$$

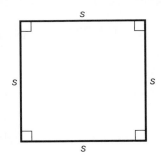

where P and s are measured in the *same* unit.

Practice Exercise 7. If the perimeter of a square is 76.68 yd, what is the measure of each side of the square?

EXAMPLE 7 Determine the perimeter (P) of a square whose sides measure 9.9 yd each.

SOLUTION: The square has a side (s) of 9.9 yards. Hence,

$$P = 4s$$
$$= 4(9.9 \text{ yd})$$
$$= 39.6 \text{ yd}$$

Practice Exercise 8. Determine the perimeter of a parallelogram whose dimensions are 11.2 cm by 26.3 cm.

EXAMPLE 8 Determine the perimeter (P) of the following parallelogram.

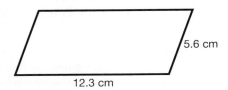

SOLUTION: Since the opposite sides of a parallelogram have equal lengths, the given parallelogram would have *two* sides of length 12.3 cm and *two* sides of length 5.6 cm. Hence,

$$P = 12.3 \text{ cm} + 5.6 \text{ cm} + 12.3 \text{ cm} + 5.6 \text{ cm}$$
$$= (12.3 + 5.6 + 12.3 + 5.6) \text{ cm}$$
$$= 35.8 \text{ cm}$$

Perimeter Formula for a Parallelogram

If a parallelogram has one pair of opposite sides of length *a* units each and another pair of opposite sides of length *b* units, then the *perimeter, P,* of the parallelogram is given by the formula

$$P = 2a + 2b$$

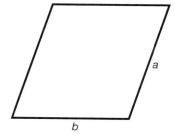

where *P*, *a*, and *b* are all measured in the *same* unit.

Practice Exercise 9. A parallelogram has a perimeter of 74 ft. If one dimension of the parallelogram is 17.4 ft, determine the other dimension.

EXAMPLE 9 Determine the perimeter of the parallelogram illustrated here.

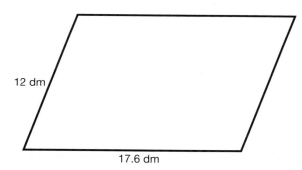

SOLUTION: For this parallelogram, $a = 12$ dm and $b = 17.6$ dm. Hence,

$$P = 2a + 2b$$
$$= 2(12 \text{ dm}) + 2(17.6 \text{ dm})$$
$$= 24 \text{ dm} + 35.2 \text{ dm}$$
$$= 59.2 \text{ dm}$$

Practice Exercise 10. The length of a rectangle is twice its width. If the perimeter of the rectangle is 75.6 m, determine its dimensions.

EXAMPLE 10 Part of a backyard is to be enclosed by a fence. If the enclosed land is in the shape of a rectangle, how much fence would be needed if the rectangle measures 23 ft by 6 yd?

SOLUTION: Since the enclosed land is in the form of a rectangle, there would be *two* sides 23 ft long and *two* sides 6 yd long. Hence,

$$P = 2(23 \text{ ft}) + 2(6 \text{ yd})$$

However, we observe that the units of measurement are not the same; hence, we must convert one to the other. If we convert yards to feet, then 6 yd become 18 ft and

5.2 Polygons and Perimeters

$$P = 2(23 \text{ ft}) + 2(6 \text{ yd})$$
$$= 2(23 \text{ ft}) + 2(6 \text{ yd})\left(\frac{3 \text{ ft}}{1 \text{ yd}}\right)$$
$$= 2(23 \text{ ft}) + 2(18 \text{ ft})$$
$$= 46 \text{ ft} + 36 \text{ ft}$$
$$= 82 \text{ ft}$$

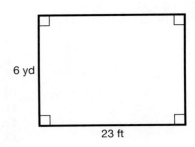

Therefore, 82 ft of fencing would be required.

ANSWERS TO PRACTICE EXERCISES

1. Equilateral. 2. Quadrilateral. 3. Perimeter. 4. 20 ft or $6\frac{2}{3}$ yd. 5. 76.8 in. 6. 68.64 cm.
7. 19.17 yd. 8. 75 cm. 9. 19.6 ft. 10. 12.6 m by 25.2 m.

EXERCISES 5.2

In Exercises 1–6, classify each of the given triangles as being equilateral, isosceles, or scalene.

1. A triangle with sides of lengths 2 in., 5 in., and 5 in.
2. A triangle with sides of lengths 4 cm, 7 cm, and 8 cm.
3.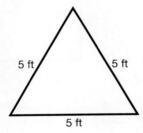
4.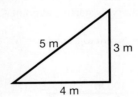
5.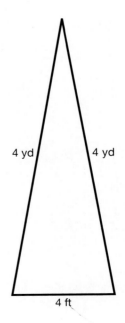
6.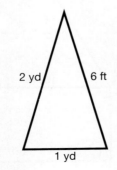

In Exercises 7–15, determine the perimeter of the indicated polygon.

7. A triangle with sides of lengths 3 in., 5 in., and 6 in.
8. An equilateral triangle with sides of length 4.3 cm.
9. A square with sides of length 3.8 ft.
10. A rectangle with length 7 cm and width 4 cm.
11. A hexagon, all of whose sides have length 1.7 cm.
12. A triangle with sides of lengths 2.3 yd, 3.1 yd, and 7.2 ft.
13. An isosceles triangle whose equal sides each have a length of 3.4 dm and whose other side has a length of 23 cm.
14. A pentagon, three of whose sides have lengths of 11.2 in., 10.9 in., and 2.7 in., and whose other sides are both of length 9.6 in.
15. An octagon (an eight-sided polygon) having three sides each of length 10.2 cm, two sides each of length 7.6 cm, two sides each of length 9.4 cm, and one side of length 4.7 cm.
16. A pentagon is formed by placing an equilateral triangle upon a rectangle as indicated in the accompanying figure. Determine the perimeter of the pentagon.

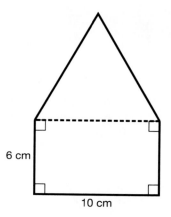

17. Markers are to be erected along the perimeter of the accompanying figure. If the markers are to be placed at 5-m intervals starting at point A, how many markers would be required?

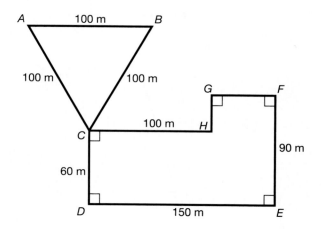

18. Determine the perimeter of the accompanying figure.

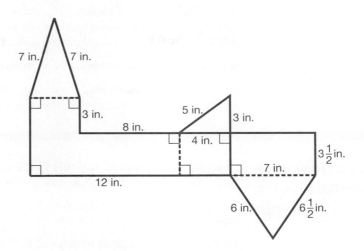

19. A strip of tape is to be placed on a gym floor in the shape of a rectangle 42 ft by 11 yd. If the tape sells for $0.11 per foot, what will be the cost for the required strip of tape? (Disregard the width of the tape.)
20. A man wants to weatherstrip the four sides of all the windows in his house. There are five windows that measure 5.25 ft long and 3.5 ft wide, and seven windows that measure 4.75 ft long and 2.75 ft wide. If the weatherstrip is sold in rolls of 100 ft for $15.75 a roll, how much will the required weatherstripping cost?

5.3 Circles and Circumference

OBJECTIVES

After completing this section, you should be able to:
1. Determine the circumference of a given circle if its radius or diameter is given.
2. Determine the radius of a given circle if its diameter or circumference is given.
3. Determine the diameter of a given circle if its radius or circumference is given.

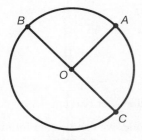

FIGURE 5.7

Another basic geometric figure is a **circle**. In Figure 5.7, we illustrate a circle with its center at the point O and a radius r. Every point on the circle is the *same* distance from the center; this distance is the **radius**.

The line segment from point B to point C in Figure 5.7 joins two points on the circle and passes through the center of the circle. The distance between any two such points is called the **diameter** of the circle. Observe that the diameter of a circle is equal to twice the radius of the circle. If we let d represent the diameter of a circle and r represent its radius, we have the relationship

$$d = 2r$$

The perimeter of a circle is called its "circumference." Unlike a polygon, we cannot find the perimeter of a circle by adding the lengths of its line segments, since there are none. To determine the circumference of a circle, we introduce a number, π, called pi. This number results from dividing the circumference of any circle by its diameter and is approximately equal to 3.14 (to two decimal places) or 3.1416 (to four decimal places). Another approximation for π is $\frac{22}{7}$. The number π is an irrational number that does not have an exact decimal representation. It is customary to use the approximations given for π for convenience's sake in approximating distance, area, and volume.

Circumference Formula for a Circle

The **circumference of a circle** is the product of π and twice its radius. If a circle has radius of r units, then its circumference, C, is given by the formula

$$C = 2\pi r$$

where C and r are both measured in the same unit.

The circumference of a circle can also be given by the formula

$$C = \pi d$$

where d is the diameter of the circle. (This follows from the fact that $d = 2r$.) Both C and d are measured in the same unit.

Practice Exercise 1. Determine the circumference of a circle if its radius is 14 cm. (Leave answer in terms of π.)

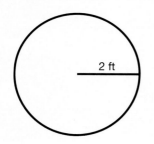

EXAMPLE 1 Determine the circumference of a circle whose radius is 2 ft.

SOLUTION: Using the formula

$$C = 2\pi r$$

with $r = 2$ ft, we have

$$C = 2\pi(2 \text{ ft})$$
$$= 4\pi \text{ ft}$$

Using the approximation $\frac{22}{7}$ for π, we know that the approximate value of the circumference is

$$C \approx \left(4 \times \frac{22}{7}\right) \text{ ft}$$
$$= \frac{88}{7} \text{ ft}$$
$$= 12\frac{4}{7} \text{ ft}$$

Practice Exercise 2. Determine the circumference of a circle if its diameter is 21 ft. (Leave answer in terms of π.)

EXAMPLE 2 Determine the circumference of a circle whose diameter is 6 cm.

SOLUTION: Since we are given the diameter of the circle, we can use the formula

$$C = \pi d$$

5.3 Circles and Circumference

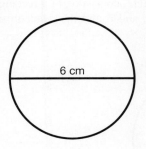

with $d = 6$ cm. Hence,

$$C = \pi(6 \text{ cm})$$
$$= 6\pi \text{ cm}$$

Using the approximation 3.14 for π, we know the approximate value of the circumference to be

$$C \approx 6 \times 3.14 \text{ cm}$$
$$= 18.84 \text{ cm}$$

Practice Exercise 3. Determine the radius of a circle if its circumference is 24 ft. $\left(\text{Use } \dfrac{22}{7} \text{ for } \pi.\right)$

EXAMPLE 3 If the circumference of a circle is 27 cm, determine its diameter, to the nearest tenth of a centimeter. (Use 3.14 for π.)

SOLUTION: Using the formula

$$C = \pi d$$

with $C = 27$ cm and $\pi = 3.14$, we know that

$$27 \text{ cm} \approx 3.14 \times d$$

or that

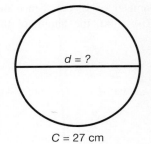

$$d \approx \dfrac{27 \text{ cm}}{3.14} \quad \text{(Dividing both sides of the equation by 3.14)}$$
$$= 8.59 \text{ cm}$$
$$= 8.6 \text{ cm} \quad \text{(To the nearest tenth of a centimeter)}$$

Practice Exercise 4. Determine the diameter of a circle if its circumference is 0.6 yd. $\left(\text{Use } \dfrac{22}{7} \text{ for } \pi.\right)$

EXAMPLE 4 If the circumference of a circle is 18 m, determine its radius, to the nearest tenth of a centimeter. (Use 3.14 for π.)

SOLUTION: Using the formula

$$C = 2\pi r$$

with $C = 18$ m and $\pi = 3.14$, we have

$$18 \text{ m} \approx 2 \times 3.14 \times r$$

or that

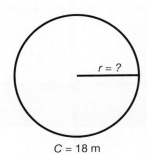

$$r \approx \dfrac{18 \text{ m}}{2 \times 3.14} \quad \text{(Dividing both sides of the equation by } 2 \times 3.14\text{)}$$
$$= \dfrac{18 \text{ m}}{6.28}$$
$$= 2.9 \text{ m} \quad \text{(To the nearest tenth of a meter)}$$

ANSWERS TO PRACTICE EXERCISES

1. 28π cm.   2. 21π ft.   3. $3\dfrac{9}{11}$ ft.   4. $\dfrac{21}{110}$ yd.

Practice Exercise 5. A rectangle has width of 9 yd and length of 14 yd. A semicircle is on the length of the rectangle. Determine the total length around the figure.

EXAMPLE 5 Determine the perimeter of the accompanying figure, consisting of a rectangle and a semicircle (i.e., half of a circle). Use 3.14 for π, and round the result to the nearest tenth.

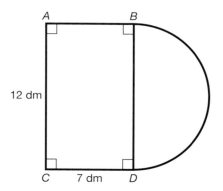

SOLUTION: The perimeter of the figure is the sum of the distance from A to B, the curved distance from B to D, and the distances from D to C and C to A. The distance from A to B is 7 dm (since the opposite sides of a rectangle have the same length); the distance from D to C is 7 dm; the distance from C to A is 12 dm. The distance from B to D is the length of the semicircle with diameter of 12 dm. Since the circumference of a circle is given by the formula $C = \pi d$, the length of the semicircle is

$$\frac{1}{2}C = \frac{1}{2}\pi d$$
$$\approx \frac{1}{2} \times 3.14 \times 12 \text{ dm}$$
$$= 18.84 \text{ dm}$$
$$= 18.8 \text{ dm} \qquad \text{(To the nearest tenth)}$$

Therefore, the perimeter of the given figure is

$$P \approx 7 \text{ dm} + 12 \text{ dm} + 18.8 \text{ dm} + 7 \text{ dm}$$
$$= 44.8 \text{ dm}$$

ANSWER TO PRACTICE EXERCISE

5. 53.98 yd.

EXERCISES 5.3

In Exercises 1–8, determine the circumference for each circle, given its radius or diameter. Use 3.14 for π. Round your results as indicated.

1. $r = 12$ cm (tenths)
2. $r = 3.5$ ft (hundredths)
3. $r = 10.2$ in. (hundredths)
4. $r = 23$ mm (tenths)
5. $d = 26.92$ yd (hundredths)
6. $d = 9.6$ ft (tenths)
7. $d = 10.7$ m (hundredths)
8. $d = 262.1$ dm (ones)

In Exercises 9–16, determine the radius for each circle, given its diameter or circumference. Use $\frac{22}{7}$ for π. Round your results as indicated.

9. $d = 123.64$ cm (tenths)
10. $d = 24.92$ ft (tenths)
11. $d = 607.1$ dm (ones)
12. $d = 0.69$ yd (tenths)
13. $C = 117.23$ m (tenths)
14. $C = 86.562$ dm (hundredths)
15. $C = 960.02$ mm (ones)
16. $C = 137$ yd (ones)

17. Determine the perimeter of the figure at left below, which consists of a rectangle and a semicircle. (A semicircle is half a circle.) Use 3.14 for π, and round your result to the nearest tenth.

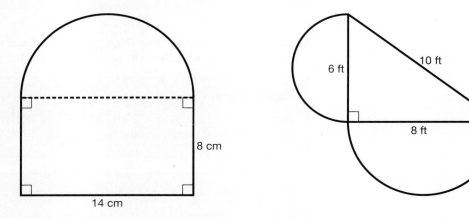

18. Determine the perimeter of the figure at right above, which consists of a right triangle and two semicircles. Use $\frac{22}{7}$ for π.

19. Determine the perimeter of the figure at left below, which consists of a rectangle, an equilateral triangle, and two semicircles. Use 3.14 for π, and round your results to the nearest tenth.

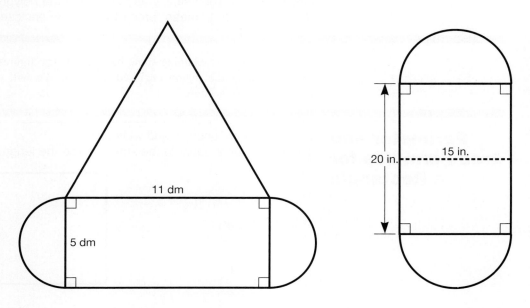

20. A mirror has the shape of a rectangle with two semicircles at each end, as shown in the figure at right above. Calculate the length of molding needed to frame it. (Disregard the width of the molding.)

5.4 Area Enclosed by Polygons and Circles; Surface Area

OBJECTIVES

After completing this section, you should be able to:
1. Find the area of triangular regions.
2. Find the area of regions enclosed by polygons.
3. Find the area of circular regions.
4. Find the surface area of certain solids.

In the previous section of this chapter, we considered the perimeter of polygons and the circumference of circles. We will now consider another measurement called "area."

DEFINITION

Area is the measure of the region that lies completely within a closed curve. Area is given in square units.

DEFINITION

A *square unit* is a square whose sides have measures of one unit.

It follows, then, that the area of a plane region is the number of square units contained in it.

For the rest of this section, we will refer to the area of a region of the plane enclosed by a polygon simply as the area of the polygon. For instance, we will write "area of square" to mean "area of the region enclosed by a square."

We will now examine some basic geometric figures and the corresponding formulas for the area of the region enclosed by them. We will also review the formulas for the perimeter of each figure.

Perimeter and Area Formulas for a Rectangle

Rectangle with length L and width W.
The *perimeter* is equal to the sum of twice the length and twice the width:

$P = 2L + 2W$

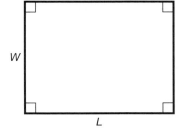

The *area* is equal to the product of the length and width:

$$A = LW$$

5.4 Area Enclosed by Polygons and Circles; Surface Area

Practice Exercise 1. A rectangle has a length of 12 yd and a width of 11 yd. Determine its perimeter and area of the region it encloses.

EXAMPLE 1 A rectangle has a length (L) of 8 ft and a width (W) of 6 ft. Determine its perimeter and the area of the rectangular region it encloses.

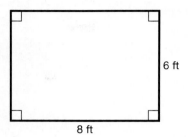

SOLUTION: Since the length is 8 ft and the width is 6 ft, we have

$$P = (2 \times L) + (2 \times W)$$
$$= (2 \times 8 \text{ ft}) + (2 \times 6 \text{ ft})$$
$$= 16 \text{ ft} + 12 \text{ ft}$$
$$= 28 \text{ ft}$$

and, using the formula for the area of rectangle, we have

$$A = L \times W$$
$$= (8 \text{ ft}) \times (6 \text{ ft})$$
$$= 48 \text{ ft}^2$$

where ft^2 is read "square feet" and results from "feet $\times$ feet."

Suppose that the rectangle in Example 1 is divided into squares as shown in Figure 5.8, each measuring 1 ft by 1 ft. Since the rectangle is 8 ft long and 6 ft wide, there will be 48 squares in the rectangular region. The area of each square is 1 ft^2. Hence, the area of the rectangular region is 48 ft^2. This illustrates why area is measured in square units.

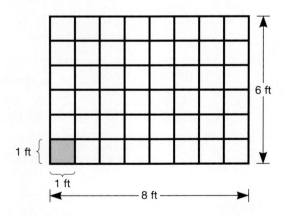

FIGURE 5.8

Practice Exercise 2. Determine the width of a rectangle if its length is 23.1 cm and its area is 277.2 cm².

EXAMPLE 2 Determine the length of a rectangle if its width is 9 cm and its area is 144 cm².

SOLUTION: Since $W = 9$ cm and $A = 144$ cm², we have

$$A = LW$$
$$144 \text{ cm}^2 = (L)(9 \text{ cm})$$
$$L = \frac{144 \text{ cm}^2}{9 \text{ cm}} \quad \text{(Dividing both sides of the equation by 9 cm)}$$
$$= 16 \text{ cm}$$

Perimeter and Area Formulas for a Square

Square with side *s*.

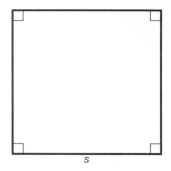

The *perimeter* is equal to 4 times the length of its side:

$$P = 4s$$

The *area* is equal to the square of the length of its side:

$$A = s^2$$

Note that finding the area of a square is a special case of finding the area of a rectangle with $L = W = s$.

Practice Exercise 3. Determine the area of the square whose sides each measure 2.9 in.

EXAMPLE 3 Determine the area of a square with side 3.2 yd long.

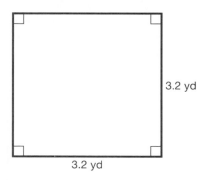

SOLUTION: Since $s = 3.2$ yd, we have

$$\begin{aligned} A &= s^2 \\ &= (3.2 \text{ yd})^2 \\ &= 10.24 \text{ yd}^2 \end{aligned}$$

Practice Exercise 4. Determine the length of a rectangle with width 4.2 yd and area 78.12 yd².

EXAMPLE 4 Determine the area (*A*) of the accompanying figure (all units in feet).

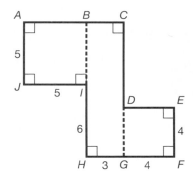

SOLUTION: The figure consists of the square *ABIJ*, the rectangle *BHGC*, and the square *GFED*. Therefore, we have

$$A = \text{area}(ABIJ) + \text{area}(BHGC) + \text{area}(GFED)$$
$$= s^2 + LW + s^2$$
$$= (5 \text{ ft})^2 + (11 \text{ ft})(3 \text{ ft}) + (4 \text{ ft})^2$$
$$= 25 \text{ ft}^2 + 33 \text{ ft}^2 + 16 \text{ ft}^2$$
$$= 74 \text{ ft}^2$$

Perimeter and Area Formulas for a Parallelogram

Parallelogram with base *b*, width *a*, and height *h*.

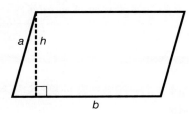

The *perimeter* is equal to the sum of twice the base and twice the width:

$$P = 2a + 2b$$

The *area* is equal to the product of the base and the height:

$$A = bh$$

Practice Exercise 5. Determine the area of a parallelogram with height 8.1 ft and whose base is 15.2 ft.

EXAMPLE 5 Determine the area of a parallelogram with height 9 cm and whose base has measure 14 cm.

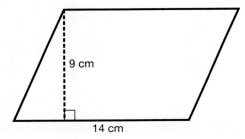

SOLUTION: Since $b = 14$ cm and $h = 9$ cm, we have

$$A = bh$$
$$= (14 \text{ cm})(9 \text{ cm})$$
$$= 126 \text{ cm}^2$$

Practice Exercise 6. Determine the measure of the base of a parallelogram with height 11.2 cm, if its area is 324.8 cm².

EXAMPLE 6 Determine the height of a parallelogram if its base measures 27 yd and its area is 337.5 yd².

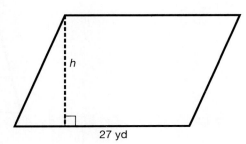

SOLUTION: Since $A = 337.5$ yd² and $b = 27$ yd, we have

$$A = bh$$
$$337.5 \text{ yd}^2 = (27 \text{ yd})(h)$$
$$h = \frac{337.5 \text{ yd}^2}{27 \text{ yd}} \quad \text{(Dividing both sides of the equation by 27 yd)}$$
$$= 12.5 \text{ yd}$$

ANSWERS TO PRACTICE EXERCISES

1. $P = 46$ yd; $A = 132$ yd². 2. 12 cm. 3. 8.41 in². 4. 18.6 yd. 5. 123.12 ft². 6. 29 cm.

Perimeter and Area Formulas for a Triangle

Triangle with sides *a*, *b*, and *c*, and height *h*.

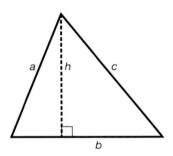

The *perimeter* is equal to the sum of the length of its sides:

$$P = a + b + c$$

The *area* is equal to one-half the product of its base and height:

$$A = \frac{1}{2}bh$$

Finding the area of a triangle follows from finding the area of a parallelogram. Note, in Figure 5.9, that the line segment from D to B divides the parallelogram into two triangles, triangle ABD and triangle BCD. The base of triangle ABD has the same measure as the base of triangle BCD. The heights of the triangles are also equal. Since the area of the parallelogram is given by $A = bh$, it follows that the area of each triangle is given by $A = \frac{1}{2}bh$.

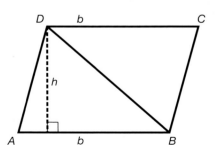

FIGURE 5.9

5.4 Area Enclosed by Polygons and Circles; Surface Area

Practice Exercise 7. The base of a triangle is 16 in. The other two sides measure 17 in. each. The height is 15 in. Determine the perimeter of the triangle and the area of the triangular region enclosed.

EXAMPLE 7 Determine the perimeter of the accompanying triangle and the area of the triangular region.

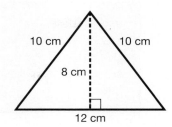

SOLUTION: The three sides of the triangle have lengths of 12 cm, 10 cm, and 10 cm. Since all of the measurements are in the same unit, we have

$$P = a + b + c$$
$$= 12 \text{ cm} + 10 \text{ cm} + 10 \text{ cm}$$
$$= 32 \text{ cm}$$

In the triangle, the base is 12 cm and the height is 8 cm. Hence,

$$A = \frac{1}{2} \times b \times h$$
$$= \frac{1}{2} \times (12 \text{ cm}) \times (8 \text{ cm})$$
$$= 48 \text{ cm}^2$$

Practice Exercise 8. Determine the area of the triangle with height 12 yd and whose base measures 17 yd.

EXAMPLE 8 Determine the area of a triangle whose base is 14 dm and whose height is 17 dm.

SOLUTION: Since $b = 14$ dm and $h = 17$ dm, we have

$$A = \frac{1}{2}bh$$
$$= \frac{1}{2}(14 \text{ dm})(17 \text{ dm})$$
$$= 119 \text{ dm}^2$$

Practice Exercise 9. Determine the measure of the base of a triangle with area 42 cm² if its height is 12 cm.

EXAMPLE 9 Determine the height of a triangle whose area is 35 in.² and whose base is 17.5 in.

SOLUTION: Since $A = 35$ in.² and $b = 17.5$ in., we have

$$A = \frac{1}{2}bh$$
$$35 \text{ in.}^2 = \frac{1}{2}(17.5 \text{ in.})(h)$$
$$70 \text{ in.}^2 = (17.5 \text{ in.})h$$
$$h = \frac{70 \text{ in.}^2}{17.5 \text{ in.}} \quad \text{(Dividing both sides of the equation by 17.5 in.)}$$
$$= 4 \text{ in.}$$

We defined a parallelogram to be a quadrilateral (a four-sided polygon) with two pairs of parallel sides. There also exist quadrilaterals with only one pair of parallel sides.

> **DEFINITION**
> A **trapezoid** is a quadrilateral with exactly one pair of parallel sides. The parallel sides of the trapezoid are called its *bases.* If the two nonparallel sides of the trapezoid have the same length, then the trapezoid is an *isosceles trapezoid.*

Area Formula for a Trapezoid

The *area of a trapezoid* is found by taking one-half the product of the height with the sum of the measures of its bases.

$$A = \frac{1}{2}(b + B)h$$

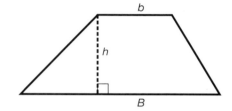

Practice Exercise 10. Determine the area of a trapezoid with height 16 ft if its bases have measures 8 ft and 21 ft.

EXAMPLE 10 Determine the area of the trapezoid whose bases measure 5 yd and 8 yd and with height 6 yd.

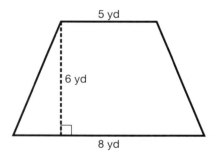

SOLUTION: Since $b = 5$ yd, $B = 8$ yd, and $h = 6$ yd, we have

$$A = \frac{1}{2}(b + B)h$$
$$= \frac{1}{2}(5 \text{ yd} + 8 \text{ yd})(6 \text{ yd})$$
$$= \frac{1}{2}(13 \text{ yd})(6 \text{ yd})$$
$$= 39 \text{ yd}^2$$

5.4 Area Enclosed by Polygons and Circles; Surface Area

Practice Exercise 11. Determine the measure of the shorter base of a trapezoid with area 289 m² if its height is 17 m and its longer base measures 22 m.

EXAMPLE 11 Determine the height of a trapezoid that has an area of 112.5 cm² and bases with measures 12 cm and 18 cm.

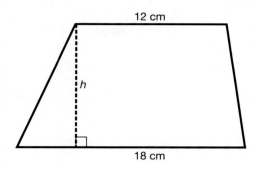

SOLUTION: Since $A = 112.5$ cm², $b = 12$ cm, and $B = 18$ cm, we have

$$A = \frac{1}{2}(b + B)h$$

$$112.5 \text{ cm}^2 = \frac{1}{2}(12 \text{ cm} + 18 \text{ cm})h$$

$$= \frac{1}{2}(30 \text{ cm})h$$

$$= (15 \text{ cm})h$$

$$h = \frac{112.5 \text{ cm}^2}{15 \text{ cm}} \quad \text{(Dividing both sides of the equation by 15 cm)}$$

$$= 7.5 \text{ cm}$$

ANSWERS TO PRACTICE EXERCISES

7. $P = 50$ in.; $A = 120$ in.²   8. 102 yd².   9. 7 cm.   10. 232 ft².   11. 12 m.

Circumference and Area Formulas for a Circle

Circle with radius *r* and diameter *d*.

The *circumference* is equal to the product of twice its radius and π:

$$C = 2\pi r \quad (\text{or } C = \pi d)$$

The *area* is equal to the product of π and the square of its radius:

$$A = \pi r^2$$

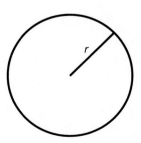

Practice Exercise 12. Determine the area of the triangular region whose base measures 8 ft and whose height is 2 yd.

EXAMPLE 12 A circle has a diameter of 14 cm. Determine its circumference and the area of the circular region it encloses. (Leave answers in terms of π.)

SOLUTION: Since $d = 14$ cm, we have

$$C = \pi d$$
$$= \pi(14 \text{ cm})$$
$$= 14\pi \text{ cm}$$

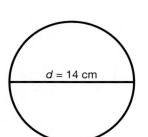

To determine the area of the circular region, we first determine the radius:

$$r = \frac{1}{2}d$$
$$= \frac{1}{4}(14 \text{ cm})$$
$$= 7 \text{ cm}$$

Then, using the formula for the area of a circle, we have

$$A = \pi r^2$$
$$= \pi(7 \text{ cm})^2$$
$$= \pi(49 \text{ cm}^2)$$
$$= 49\pi \text{ cm}^2$$

Practice Exercise 13. Determine the area of the circular region whose diameter is 9 m. Use $\frac{22}{7}$ for π, and round your answer to the nearest tenth.

EXAMPLE 13 Given the accompanying figure, determine the perimeter and the area of the enclosed region.

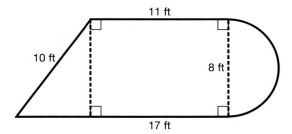

SOLUTION: To determine the perimeter, we observe that the figure consists of a quadrilateral and a semicircle. The circumference of the semicircle is equal to

$$\frac{1}{2}\pi d = \frac{1}{2}\pi(8 \text{ ft})$$
$$= 4\pi \text{ ft}$$

Therefore,

$$P = 17 \text{ ft} + 4\pi \text{ ft} + 11 \text{ ft} + 10 \text{ ft}$$
$$= (38 + 4\pi) \text{ ft}$$
$$\approx 50.56 \text{ ft (using 3.14 for } \pi)$$

To determine the area, A, of the region, we consider the given figure in three parts, as shown here.

5.4 Area Enclosed by Polygons and Circles; Surface Area

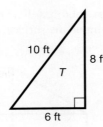
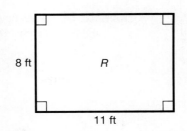
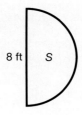

Let $A = T + R + S$, where T is the area of the triangular region with $b = 6$ ft and $h = 8$ ft; R is the area of the rectangular region with $L = 11$ ft and $W = 8$ ft; and S is a semicircular region with $r = 4$ ft. Hence,

$$T = \frac{1}{2}bh = \frac{1}{2}(6 \text{ ft})(8 \text{ ft}) = 24 \text{ ft}^2$$

$$R = LW = (11 \text{ ft})(8 \text{ ft}) = 88 \text{ ft}^2$$

$$S = \frac{1}{2}\pi r^2 = \frac{1}{2}\pi(4 \text{ ft})^2 = \frac{1}{2}\pi(16 \text{ ft}^2) = 8\pi \text{ ft}^2$$

Therefore,

$$A = 24 \text{ ft}^2 + 88 \text{ ft}^2 + 8\pi \text{ ft}^2$$
$$= (112 + 8\pi) \text{ ft}^2$$
$$\approx 137.12 \text{ ft}^2 \text{ (using 3.14 for } \pi\text{)}$$

ANSWERS TO PRACTICE EXERCISES

12. 24 ft².   13. 63.6 m².

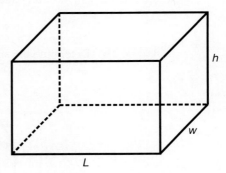

FIGURE 5.10

We will now consider the **surface area** of certain geometric figures. Consider the rectangular solid illustrated in Figure 5.10 that has length $= L$, width $= W$, and height $= h$. This solid has six sides, or *faces,* all of which are rectangular:

- There are two sides that are rectangular and have the dimensions L and W. Hence, the area of these two sides would be $2LW$.
- There are two sides that are rectangular and have dimensions L and h. Hence, the area of these two sides would be $2Lh$.
- There are two sides that are rectangular and have dimensions W and h. Hence, the area of these two sides would be $2Wh$.

The *surface area* of the solid in Figure 5.10 is the sum of the areas of the faces. If we denote the surface area of this solid by S, we have

$$S = 2LW + 2Lh + 2Wh$$

where S is measured in *square units*.

Practice Exercise 14. Determine the surface area of the rectangular solid whose dimensions are 6 cm × 8 cm × 2.5 cm.

EXAMPLE 14 A rectangular solid has dimensions as follows: $L = 18$ cm, $W = 14$ cm, and $h = 11$ cm. Determine the surface area of the solid.

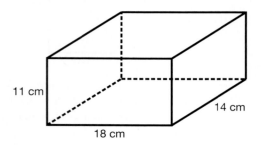

SOLUTION: To determine the surface area of the solid, we use the formula

$$S = 2LW + 2Lh + 2Wh$$

with $L = 18$ cm, $W = 14$ cm, and $h = 11$ cm. Hence,

$$S = 2(18 \text{ cm})(14 \text{ cm}) + 2(18 \text{ cm})(11 \text{ cm}) + 2(14 \text{ cm})(11 \text{ cm})$$
$$= 504 \text{ cm}^2 + 396 \text{ cm}^2 + 308 \text{ cm}^2$$
$$= 1208 \text{ cm}^2$$

If all three dimensions of a rectangular solid are equal, then the solid is called a "cube." The surface area of a cube is given by the formula

$$S = 6e^2$$

where e is the length of an edge of the cube.

Practice Exercise 15. Determine the surface area of a cube whose edges have length 2.7 ft.

EXAMPLE 15 Determine the surface area of a cube whose edges measure 4.1 ft.

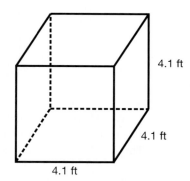

SOLUTION: Since $e = 4.1$ ft, we have

$$S = 6e^2$$
$$= 6(4.1 \text{ ft})^2$$
$$= 6(16.81 \text{ ft}^2)$$
$$= 100.86 \text{ ft}^2$$

ANSWERS TO PRACTICE EXERCISES

14. 166 cm². **15.** 43.74 ft².

EXERCISES 5.4

1. Determine the perimeter and the area of a rectangle whose width is 6 ft and whose length is 8 ft.
2. Determine the perimeter and the area of a rectangle whose width is 16 cm and whose length is 23 cm.
3. Determine the perimeter and the area of a rectangle whose width is 3 yd and whose length is 15 ft.
4. Determine the perimeter and the area of a square whose sides measure 2.7 dm.
5. Determine the perimeter and the area of a square whose sides measure 19.6 in.
6. If the perimeter of a square is 64.8 ft, determine its area.
7. If the area of a square is 256 m², determine its perimeter.
8. If the area of a rectangle is 323 ft² and its width is 17 ft, determine its perimeter.
9. If the perimeter of a rectangle is 69.2 cm and its length is 20.1 cm, determine its area.
10. Determine the area of a parallelogram whose height is 9.6 cm and whose base measures 12.7 cm.
11. Determine the height of a parallelogram whose area is 290 in.² and whose base measures 20 in.
12. Determine the area of a triangle whose height is 32.6 ft and whose base measures 38.4 ft.
13. Determine the area of a triangle whose height is 1.9 dm and whose base measures 18 cm.
14. Determine the height of a triangle whose area is 204.6 yd² and whose base measures 22 yd.
15. Determine the measure of the base of a triangle whose area is 1508 cm² and whose height is 10.4 dm.

In Exercises 16–21, determine the area of each of the triangular regions.

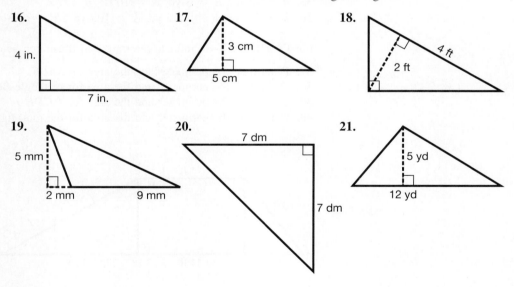

In Exercises 22–27, determine the area of the regions enclosed by each polygon.

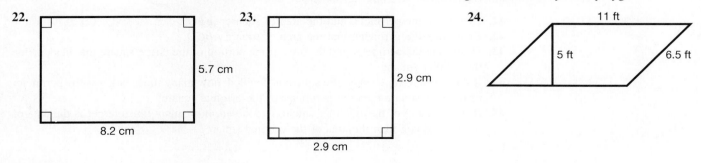

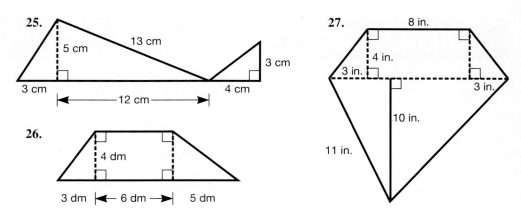

For Exercises 28–32, determine the area of the trapezoid if the bases are given by the first two measurements and the height by the third measurement.

28. 8 ft, 11 ft; 7 ft
29. 9.2 yd, 11.8 yd; 7.6 yd
30. 11 ft, 5 yd; 72 in.
31. 10.2 cm, 18.6 cm; 9 cm
32. 2.7 dm, 36.2 cm; 182 mm

For Exercises 33–36, use the formula for finding the area of a trapezoid, and determine the value of the indicated part.

33. $b = 9$ ft, $B = 11$ ft, $h = 3.6$ ft; $A = ?$
34. $b = 12$ cm, $h = 8.6$ cm, $A = 129$ cm^2; $B = ?$
35. $h = 24$ in., $B = 3$ yd, $A = 16$ ft^2; $b = ?$
36. $b = 9.6$ yd, $B = 12.2$ yd, $A = 109$ yd^2; $h = ?$

For Exercises 37–40, refer to the trapezoid illustrated below.

37. Determine the perimeter and the area of $\triangle ABF$.
38. Determine the perimeter and the area for rectangle $BCEF$.
39. Determine the perimeter and the area for $\triangle CDE$.
40. Determine the perimeter and the area for the total figure.

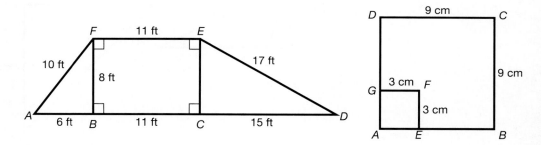

For Exercises 41–43, refer to the square above.

41. Determine the perimeter and the area of square $AEFG$.
42. Determine the perimeter and the area of square $ABCD$.
43. Determine the perimeter and the area of the portion of the larger square that lies outside the smaller square.
44. If the measure of the side of a square is doubled, how many times larger is the perimeter of the new square than the perimeter of the original square?
45. If the measure of the side of a square is doubled, how many times larger is the area of the new square than the area of the original square?

46. If the width of a rectangle is doubled and the length is multiplied by 3, how many times larger is the area of the new rectangle than the area of the original rectangle?
47. Determine the perimeter of the total figure illustrated here. (All measurements are in centimeters.)

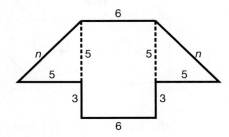

In Exercises 48–55, the radius or diameter of a circular region is given. Determine the area. Use 3.14 for π and round results to the nearest tenth.

48. $r = 2.3$ cm
49. $r = 1.9$ ft
50. $r = 2.01$ m
51. $r = 27.1$ mm
52. $d = 10.56$ yd
53. $d = 8.92$ dm
54. $d = 4.66$ in.
55. $d = 20.04$ cm

56. A square has a side of length 23 cm.
    a. Determine the perimeter of the square.
    b. Determine the area of the square region.
57. A rectangular room measures 11.5 ft by 14.2 ft. How many square yards of rug will be necessary to completely cover the floor of this room? (*Hint:* 1 yd^2 = 9 ft^2)

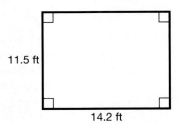

58. A room measures 14 ft long, 13 ft wide, and 8 ft high. One wall of the room has a window that measures 3.25 ft by 5 ft and a door that measures 3.5 ft by 6.5 ft. Another wall has a window that measures 3 ft by 4.5 ft. There are no other windows or doors in the room. If the walls of the room were to be painted, how many gallons of paint would be required if one gallon of paint covers 450 ft^2 of wall space and paint is sold in gallon containers only?
59. A walkway is to be constructed around a circular pool, as indicated in the figure. What is the area of the walkway? (*Hint:* The area of the walkway is the area of the large circular region minus the area of the pool.)

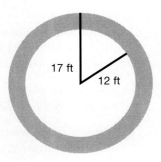

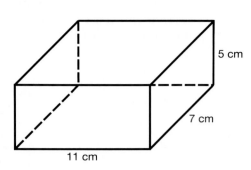
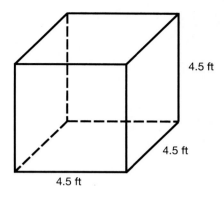

60. A rectangular solid (above left) has length 11 cm, width 7 cm, and height 5 cm.
    a. What is the surface area of the solid?
    b. What is the total length of the edges of the solid?
61. A cube (below right) has an edge of 4.5 ft.
    a. What is the surface area of the cube?
    b. What is the total length of the edges of the cube?

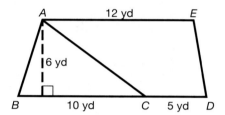
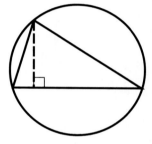

62. Referring to the trapezoid pictured at left above, determine the area of the region that lies outside of the triangle *ABC* but inside the trapezoid *ABDE*.
63. Referring to the figure at right above, determine the area of the region that lies outside the triangle but inside the circle. The circle has a radius of 5 ft. The triangle has a height of 6 ft and a base of 9 ft. $\left(\text{Use } \frac{22}{7} \text{ for } \pi.\right)$

5.5 Volumes of Solids

OBJECTIVE

After completing this section, you should be able to determine the volumes of rectangular solids, prisms, cylinders, and spheres.

In Section 5.4, we considered the area enclosed by polygons and circles. Polygons and circles are two-dimensional figures. We will now consider the measure of some three dimensional figures. Such a measure is called "volume."

DEFINITION

Volume is the measure of the space enclosed by a three-dimensional solid. Volume is given in cubic units.

5.5 Volumes of Solids

> **DEFINITION**
>
> A **cubic unit** is a cube whose sides each have measure of 1 unit. (See Figure 5.11.)

It follows, then, that the volume of a solid is the number of cubic units contained in it. A rectangular solid, each of whose sides has a measure of 5 units, has a volume of 125 cubic units. (See Figure 5.12.) A rectangular solid that is 6 units long, 5 units wide, and 4 units high has a volume of 120 cubic units. (See Figure 5.13.)

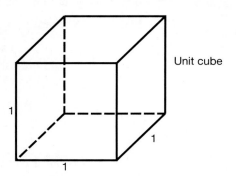

FIGURE 5.11

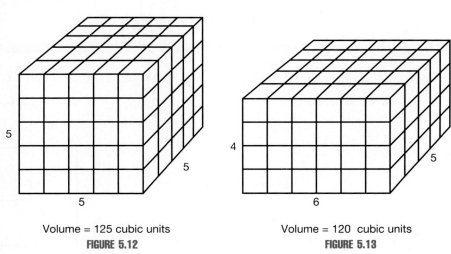

Volume = 125 cubic units
FIGURE 5.12

Volume = 120 cubic units
FIGURE 5.13

A **rectangular solid** has rectangles for its faces (sides). The volume, V, of a rectangular solid is given by the product of the measures of its length, L, width, W, and height, h. All measures are in the same units and the volume is in cubic units.

Volume Formula for a Rectangular Solid

The *volume*, V, of a *rectangular solid* with length, L, width, W, and height, h, is given by the formula

$$V = LWh$$

where L, W, and h are in the same units and V is in cubic units.

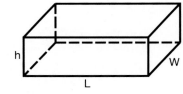

Practice Exercise 1. Determine the volume of a rectangular solid whose dimensions are 6 cm by 8 cm by 2.5 cm.

EXAMPLE 1 Determine the volume of a rectangular solid with length 5 ft, width 4 ft, and height 18 in.

SOLUTION: We first note that the units given are not all the same. Converting 18 in. to 1.5 ft, we have $L = 5$ ft, $W = 4$ ft, and $h = 1.5$ ft. Therefore,

$$V = LWh$$
$$= (5 \text{ ft})(4 \text{ ft})(1.5 \text{ ft})$$
$$= 30 \text{ ft}^3 \quad (\text{ft}^3 \text{ is read "cubic feet"})$$

Practice Exercise 2. Determine the length of a rectangular solid with width 3.9 ft and height 4.2 ft, if its volume is 98.28 ft³.

EXAMPLE 2 Determine the height of a rectangular solid with length 3.5 yd and width 2.6 yd, if its volume is 11.83 yd³.

SOLUTION: Since $L = 3.5$ yd, $W = 2.6$ yd, and $V = 11.83$ yd³, we have

$$V = LWh$$
$$11.83 \text{ yd}^3 = (3.5 \text{ yd})(2.6 \text{ yd})h$$
$$11.83 \text{ yd}^3 = (9.1 \text{ yd}^2)h$$
$$h = \frac{11.83 \text{ yd}^3}{9.1 \text{ yd}^2} \quad \text{(Dividing both sides of the equation by 9.1 yd}^2\text{)}$$
$$= 1.3 \text{ yd}$$

A special case of a rectangular solid is a cube. A *cube* is a rectangular solid with $L = W = h$. If we denote the length of the edge of a cube by e, then the volume, V, of the cube is given by the formula $V = e^3$.

Volume Formula for a Cube

The *volume V* of a **cube** with edge of length e is given by the formula

$V = e^3$

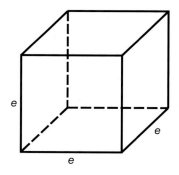

where e is measured in units and V is measured in cubic units.

Practice Exercise 3. Determine the volume of a cube whose edges have length 2.7 ft.

EXAMPLE 3 Determine the volume of the cube whose edges each have length 4.1 ft.

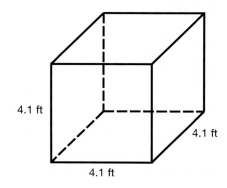

SOLUTION: To determine the volume of the cube, we use the formula

$$V = e^3$$

with $e = 4.1$ ft. Hence,

$$V = (4.1 \text{ ft})^3$$
$$= 68.921 \text{ ft}^3$$

ANSWERS TO PRACTICE EXERCISES

1. 120 cm³. 2. 6 ft. 3. 19.683 ft³.

In the formula $V = LWh$, we note that LW gives the area of the base of the rectangular solid. (Do you agree?) If we denote the area of the base by B, we have

$$V = LWh$$
$$= (LW)h \quad \text{(Grouping)}$$
$$= Bh \quad \text{(Substitution)}$$

The formula $V = Bh$ for the volume of a rectangular solid can be used to obtain the formulas for finding the volumes of other solids.

We can now extend our discussion of determining the volume of rectangular solids to determining the volume of what is called a "prism."

> **DEFINITION**
> **Congruent polygons** have the same shape and the same size.

> **DEFINITION**
> A **prism** is a solid with two bases that are congruent polygons lying in parallel planes. A prism is named according to its base.

In Figure 5.14, several named prisms are illustrated. If we denote the area of the base of a prism by B and its height by h, then the volume, V, of the prism is given by the formula $V = Bh$.

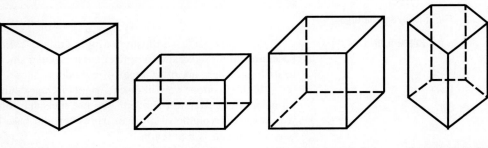

Triangular prism   Rectangular prism   Square prism (cube)   Pentagonal prism

Volume Formula for a Prism

The volume, V, of a *prism* is found by multiplying the area of its base, B, by its height, h. We have

$$V = Bh$$

EXAMPLE 4 Determine the volume of a prism whose base measures 17 cm² and whose height is 8 cm.

SOLUTION: Since $B = 17$ cm² and $h = 8$ cm, we have

$$\begin{aligned} V &= Bh \\ &= (17 \text{ cm}^2)(8 \text{ cm}) \\ &= 136 \text{ cm}^3 \end{aligned}$$

> **Practice Exercise 4.** Determine the volume of a prism whose base measures 23.5 ft² and whose height is 8.2 ft.

EXAMPLE 5 A prism has a square base each of whose sides is 11 yd long and a height of 9 ft. Determine the volume of the prism.

SOLUTION: Since the base of the prism is a square 11 yd on a side, we have

$$\begin{aligned} B &= s^2 \\ &= (11 \text{ yd})^2 \\ &= 121 \text{ yd}^2 \end{aligned}$$

Since $h = 9$ ft $= 3$ yd, we now have

$$\begin{aligned} V &= Bh \\ &= (121 \text{ yd}^2)(3 \text{ yd}) \\ &= 363 \text{ yd}^3 \end{aligned}$$

> **Practice Exercise 5.** Determine the volume of a prism with square base measuring 14 yd on a side and whose height is 15 ft.

ANSWERS TO PRACTICE EXERCISES

4. 192.7 ft³   5. 980 yd³.

Another solid that we now consider is a right circular cylinder.

> **DEFINITION**
>
> A **circular cylinder** is a solid with two bases that are congruent circles in parallel planes.

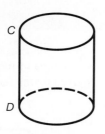

Right circular cylinder

FIGURE 5.15

In Figure 5.15, we illustrate a right circular cylinder. It is a circular cylinder since its bases are circles. It is called a right circular cylinder since the line segment $\overline{CD}$ is perpendicular to its base. There are other types of cylinders.

In this section, we will discuss right circular cylinders only. We will refer to "cylinder" to mean "right circular cylinder." We use the same formula for the volume of a cylinder as we do for the volume of a prism. However, the base of a cylinder is a circle whereas the base of a prism is a polygon.

5.5 Volumes of Solids

In Section 5.3, we noted that the area of a circular region with radius r is given by the formula

$$A = \pi r^2$$

We now have the following formula for the volume of a cylinder.

Volume Formula for a Cylinder

The *volume of a cylinder* is found by multiplying the area of its base, $B = \pi r^2$, by its height, h:

$$V = Bh$$

or

$$V = \pi r^2 h$$

Practice Exercise 6. Determine the volume of a cylinder with height 12 ft and radius 4 ft.

EXAMPLE 6 Determine the volume of a cylinder whose radius is 2 yd and whose height is 5 yd.

SOLUTION: Since $r = 2$ yd and $h = 5$ yd, we have

$$\begin{aligned} V &= \pi r^2 h \\ &= \pi (2 \text{ yd})^2 (5 \text{ yd}) \\ &= \pi (4 \text{ yd}^2)(5 \text{ yd}) \\ &= 20\pi \text{ yd}^3 \end{aligned}$$

In Example 6, we could use 3.14 for the approximate value of π and obtain an approximate numerical value for the volume of the cylinder. For the rest of this section, however, we will leave all answers in terms of π.

Practice Exercise 7. Determine the height of a cylinder with radius 5 cm if its volume is 625π cm³.

EXAMPLE 7 Determine the height of a cylinder if its volume is 63π cm³ and the diameter of its base is 6 cm.

SOLUTION: Since the diameter of the base is 6 cm, the radius is 3 cm. Since $V = 63\pi$ cm³ and $r = 3$ cm, we have

$$\begin{aligned} V &= \pi r^2 h \\ 63\pi \text{ cm}^3 &= \pi (3 \text{ cm})^2 h \\ &= (9\pi \text{ cm}^2) h \\ h &= \frac{63\pi \text{ cm}^3}{9\pi \text{ cm}^2} \quad \text{(Dividing both sides of the equation by } 9\pi \text{ cm}^2\text{)} \\ &= 7 \text{ cm} \end{aligned}$$

The last solid we will consider is a sphere.

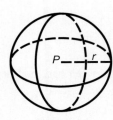

FIGURE 5.16

DEFINITION

A **sphere** is a set of points in space that are at a given distance from a given point. The given point is called the "center of the sphere," and the given distance is called the "radius of the sphere."

In Figure 5.16, we illustrate a sphere with center at P and radius r.

Volume Formula for a Sphere

The volume, V, of a *sphere* with a radius r is given by the formula

$$V = \frac{4}{3}\pi r^3$$

Practice Exercise 8. Determine the diameter of a sphere whose volume is 36π m³.

EXAMPLE 8 Determine the volume of a sphere with radius 9 in.

SOLUTION: Since $r = 9$ in., we have

$$V = \frac{4}{3}\pi r^3$$
$$= \frac{4}{3}\pi(9 \text{ in.})^3$$
$$= \frac{4}{3}\pi(729 \text{ in.}^3)$$
$$= 4\pi(243 \text{ in.}^3)$$
$$= 972\pi \text{ in.}^3$$

ANSWERS TO PRACTICE EXERCISES

6. 192π ft³.   7. 25 cm.   8. 6 m.

EXERCISES 5.5

For Exercises 1–8, use the formula $V = LWh$ for the volume of a rectangular solid to determine the value of the indicated variable.

1. $L = 12$ ft, $W = 10$ ft, $h = 7$ ft; $V = ?$
2. $L = 18$ ft, $W = 5$ yd, $h = 36$ in.; $V = ?$
3. $V = 494.5$ in.³, $L = 23$ in., $W = 8.6$ in.; $h = ?$
4. $V = 1632$ m³, $L = 17$ m, $h = 8$m; $W = ?$
5. $V = 21.16$ cm³, $W = 2.3$ cm, $h = 2.3$ cm; $L = ?$
6. $L = W = h = 19$ ft; $V = ?$
7. $L = W = 9$ cm, $V = 615.6$ cm³; $h = ?$
8. $V = 722.5$ dm³; $h = 2.5$ dm, $L = W$; $L = ?$
9. Determine the volume of a cube all of whose sides measure 12 in.
10. Determine the volume of a cube all of whose sides measure 17 cm.

For Exercises 11–16, determine the volume of each prism with the given base and height.

11. Base is a square 4 yd on a side; $h = 11$ ft.
12. Base is triangular with area 17.6 m³; $h = 9$ m.
13. Base is a pentagon (five-sided polygon) with area 28 cm²; $h = 7.9$ cm.
14. Base is a hexagon (six-sided polygon) with area 78.4 dm²; $h = 210$ cm.

5.5 Volumes of Solids

15. Base is a rectangle with $L = 23$ ft and $W = 12.1$ ft; $h = 1.9$ ft.
16. Base is a trapezoid with bases 6 ft and 8 ft and height 5 ft; $h = 4.6$ ft.

In Exercises 17–20, determine the height of the prism with the given base area, B, and volume, V.

17. $B = 23$ cm^2; $V = 276$ cm^3
18. $B = 39.4$ ft^2; $V = 669.8$ ft^3
19. $B = 64$ in.^{2}; $V = 2496$ in.^{3}
20. $B = 117$ yd^2; $V = 596.7$ yd^3

In Exercises 21–25, determine the volume of the cylinder with the given radius and height.

21. $r = 2$ ft; $h = 4$ ft
22. $r = 2$ yd; $h = 7$ ft
23. $r = 13$ cm; $h = 8.6$ cm
24. $r = 5.5$ cm; $h = 9.4$ cm
25. $r = 17$ in., $h = 0.7$ ft

In Exercises 26–30, the volume of a cylinder is given. Either the radius or the height of the cylinder is also given. Determine the other part.

26. $V = 175\pi$ yd^3, $r = 5$ yd; $h = ?$
27. $V = 400\pi$ ft^3, $r = 10$ ft; $h = ?$
28. $V = 692.9\pi$ cm^3, $r = 13$ cm; $h = ?$
29. $V = 108\pi$ cm^3, $h = 3$ cm; $r = ?$
30. $V = 614.4\pi$ m^3, $h = 2.4$ m; $r = ?$

In Exercises 31–35, determine the volume of the sphere with given radius or diameter. Leave all answers in terms of π.

31. $r = 8$ cm
32. $r = 10$ in.
33. $r = 6.1$ yd
34. $d = 14$ dm
35. $d = 21.4$ ft

Chapter Summary

In this chapter, you learned about the concept of measurement, and conversions of units of measure. You also considered applications of measurement to some basic geometric figures. Among the types of measure encountered were perimeter, circumference, area, surface area, and volume.

The **key words** introduced in the chapter are listed below in the order in which they appeared in the chapter.

line segment (p. 217)
polygon (p. 217)
triangle (p. 218)
quadrilateral (p. 218)
pentagon (p. 218)
hexagon (p. 218)
equilateral triangle (p. 218)
isosceles triangle (p. 218)
scalene triangle (p. 218)
parallelogram (p. 219)

rectangle (p. 219)
square (p. 219)
perimeter of a polygon
 (p. 219)
circle (p. 225)
radius (p. 225)
diameter (p. 225)
circumference of a circle
 (p. 226
area (p. 230)

trapezoid (p. 236)
surface area (p. 239)
volume (p. 245)
cubic unit (p. 245)
rectangular solid (p. 245)
cube (p. 246)
congruent polygons (p. 246)
prism (p. 246)
circular cylinder (p. 248)
sphere (p. 249)

Review Exercises

5.1 Units of Measurement

In Exercises 1–20, convert the measurements as indicated.

1. 17 ft to yards
2. 21 dm to centimeters
3. 41 mi to feet
4. 18 hm to meters
5. 3.6 mi to kilometers
6. 0.9 km to feet
7. 137 yd to kilometers
8. 14.6 km to yards
9. 5 L to quarts
10. 0.7 qt to liters
11. 17.5 L to pints
12. 47 pt to liters
13. 4 lb to grams
14. 2 kg to ounces
15. 420 g to pounds
16. 1000 lb to kilograms
17. 4.5 hr to minutes
18. 4550 sec to hours

19. Convert 55 mph to kilometers per second.
20. Convert 80 km per hour to feet per minute.

5.2 Polygons and Perimeter

In Exercises 21–30, determine the perimeter of the indicated polygon.

21. A triangle with sides of lengths 4.7 in., 5.2 in., and 6.6 in.
22. An equilateral triangle with sides of length 11.9 cm
23. An isosceles triangle whose equal sides have length of 1.9 yd and whose other side is 4.7 ft long
24. A square whose sides are 23.1 cm long
25. A square whose sides are 4 ft 7 in. long
26. A rectangle with length 9.6 dm and width 63 cm
27. A rectangle with length 17 ft and width 3.5 yd
28. A parallelogram whose base is 123 cm and whose width is 34 dm
29. A pentagon having two sides measuring 4.6 in., two sides measuring 6.1 in., and the other side measuring 3.9 in.
30. A hexagon with three sides each measuring 4.8 dm, two sides each measuring 6.1 dm, and the other side measuring 53 cm

5.3 Circles and Circumference

In Exercises 31–36, use 3.14 for π and round your results as indicated.

31. Determine the circumference of a circle whose radius is 4.7 ft; ones.
32. Determine the circumference of a circle whose diameter is 13.57 cm; tenths.
33. Determine the radius of a circle whose diameter is 123.69 dm; ones.
34. Determine the radius of a circle whose circumference is 229.67 yd; tenths.
35. Determine the diameter of a circle whose radius is 35.89 ft; tenths.
36. Determine the diameter of a circle whose circumference is 439.45 cm; tenths.

In Exercises 37–40, use $\frac{22}{7}$ for π and round your results as indicated.

37. Determine the circumference of a circle whose radius is 16.8 yd; ones.
38. Determine the circumference of a circle whose diameter is 56.79 in.; tenths.
39. Determine the radius of a circle whose circumference is 40 m; tenths.
40. Determine the diameter of a circle whose circumference is 239.4 yd; ones.

5.4 Area Enclosed by Polygons and Circles; Surface Area

41. Determine the area of a square region that is 34.5 cm on a side.
42. Determine the area of a square region whose perimeter is 68 yd.
43. Determine the area of a rectangular region that is 18 in. long and 14 in. wide.
44. Determine the area of a rectangular region that is 39 cm long and 2.1 dm wide.
45. Determine the area of a circular region whose radius is 34.6 ft. Use 3.14 for π, and round your results to the nearest tenth.
46. Determine the area of a circular region whose diameter is 1.3 m. Use $\frac{22}{7}$ for π, and round your results to the nearest one.
47. Determine the height of a parallelogram whose area is 580 yd^2 and whose base measures 40 yd.
48. The bases of a trapezoid are 2.3 yd and 4.1 ft. Determine the area of the trapezoid if its height is 6.8 ft.
49. A rectangular solid is 2.3 ft long, 1.9 ft wide, and 0.7 ft high. Determine the surface area of the solid.

5.5 Volumes of Solids

50. Determine the volume of the rectangular solid given in Exercise 49.
51. A cube has edges that measure 3.9 yd. Determine the surface area of the cube.
52. Determine the volume of the cube in Exercise 51.

For Exercises 53–55, leave your answers in terms of π.

53. Determine the volume of a triangular prism whose base has area 26.4 in.^{2} and whose height is 17 in.
54. Determine the volume of a cylinder whose base has diameter 14 m and whose height is 7 m.
55. Determine the volume of a sphere whose radius is 6 cm.

Review Exercises

Teasers

1. How many pints of water are contained in seventeen 55-gal barrels of water, if each barrel is 80% full of water.
2. If $1 billion, in $1 bills, were counted out on the floor of the U.S. Senate at the rate of three bills per second, how long would it take, nonstop, to count out the $1 billion?
3. Bette drove 53 mi in 70 min. How fast did she travel in kilometers per second?
4. The width of one rectangle is 12 cm and its length is 26 cm. The width of a second rectangle is one-half of the length of the first rectangle. The length of the second rectangle is 3 times the width of the first rectangle. Determine the combined area of the two rectangular regions.
5. A rectangle has dimensions of 18 ft by 8 yd. The sides of a square each measure 3 times the average of the dimensions of the rectangle. Determine the perimeter of the square.
6. One square region has an area of 49 yd². Another square region has an area of 144 yd². Determine the area of a rectangular region whose dimensions are the measures of the sides of the two square regions.
7. Is it possible for a triangle to have sides that measure 5.5 dm, 20.6 cm, and 344 mm? Explain your answer.
8. Is it possible for the perimeter of a rectangular region to be equal to the area of a square region? Explain your answer.
9. The height of a trapezoid is 7.5 in. and the bases measure 12 in. and 18 in. If the height is doubled, the shorter base is increased by 3 in., and the larger base is decreased by 2 in., what is the area of the region formed by the new trapezoid?
10. Determine the surface area of a cube whose edges each measure the same as the diameter of a circle whose circumference is 66 ft. $\left(\text{Use } \frac{22}{7} \text{ for } \pi.\right)$

CHAPTER 5 TEST

This test covers material on measurement. Read each question carefully and answer all questions. If you miss a question, review the appropriate section of the text.

In Exercises 1–10, convert each measurement as indicated.

1. 3 yd to inches
2. 2.4 mi to kilometers
3. 173 cm to decimeters
4. 3 kg to ounces
5. 6 L to quarts
6. 72.4 hr to minutes
7. 8000 sec to hours
8. 5 lb to grams
9. Determine the perimeter of a triangle whose sides measure 4.6 cm, 6.2 cm, and 7.3 cm.
10. Determine the perimeter of a pentagon having three sides measuring 8.2 in. each and the other two sides measuring 6.9 in. and 7.6 in.
11. Determine the circumference of a circle whose radius is 5.2 cm. Use 3.14 for π and give your answer to the nearest one.
12. Determine the diameter of a circle whose circumference is 234.6 yd. Use 3.14 for π, and give your answer to the nearest tenth.
13. Determine the area of a rectangular region that is 23 cm long and 16 cm wide.
14. Determine the area of a circular region whose diameter is 16.5 ft. Use 3.14 for π, and give your answer to the nearest hundredth.
15. Determine the volume of a cube that has edges 7.6 cm long.
16. Determine the surface area of a rectangular solid that is 4.2 ft long, 3.6 ft wide, and 1.2 ft high.
17. Determine the area of a triangle whose base measures 12.4 cm and whose height is 6 cm.
18. Determine the area of a trapezoid whose bases measure 10 ft and 15 ft and whose height is 9 ft.
19. Determine the volume of a circular cylinder with radius 4 in. and height 1 ft.
20. Determine the volume of a sphere with radius 13 cm.

CUMULATIVE REVIEW: CHAPTERS 1–5

In Exercises 1–6, perform the indicated operations. Write your results in simplest form.

1. $23\% \div \dfrac{1}{3}$
2. $12.4\% \times 109$
3. $18\dfrac{1}{2} - 11\dfrac{7}{8}$
4. $\dfrac{9}{11} + \dfrac{2}{3} + \dfrac{3}{4}$
5. $3^2 - 4^2 \div 8 + 0.5$
6. $1.09 + 23 + 0.691$
7. Round $18\dfrac{4}{7}$ to the nearest whole number.
8. Round 20.906 to the nearest hundredth.
9. Round 307.99 to the nearest ten.
10. Rewrite 63.4% as a fraction in simplest form.
11. Rewrite 0.03 as a percent.
12. Rewrite $\dfrac{11}{8}$ as a percent.
13. Rewrite 25% as a fraction with a numerator of 5.

In Exercises 14–16, circle the smaller number.

14. 1.19 or 11.8%
15. $\dfrac{2}{3}$ or 67%
16. $\dfrac{1}{3}$ of 4 or 27% of 3

17. Determine the perimeter of a square whose sides measure 23.4 m each.
18. Determine the area of a circular region whose diameter is 12 ft. (Leave your answer in terms of π.)
19. Determine the perimeter of a pentagon with sides of 16.2 cm, 10.9 cm, 8.8 cm, 9.4 cm, and 11.5 cm.
20. Determine the volume of a rectangular solid that is 2.3 ft high, 6.9 ft wide, and 11.2 ft long.
21. Rewrite 12.3% as a decimal.
22. Rewrite 30.6% as a simplified fraction.
23. Estimate the product of 872.9 and 68.83 by rounding each number to the nearest ten.
24. You and your guest go out for dinner. The check is in the amount of $56.45. You decide to tip your server 15% of that amount. What is the amount of the tip, to the nearest dime?
25. Ann sells textbooks and is paid a commission of 2.7% on net sales. If her gross sales for a particular period are $63,465 and returns are $8068, what is her commission for the period?

CHAPTER 6: Signed Numbers

So far in this text, we have been discussing our number system from the whole numbers to the fractions and decimals. Although we have considered several applications using these numbers, we have no numbers that would enable us to distinguish a 5-yard gain from a 5-yard loss, 20 degrees above zero from 20 degrees below zero, or 50 miles traveled east from 50 miles traveled west.

In this chapter, we will introduce signed numbers. These include integers and rational numbers. As you will see, integers are really extensions of whole numbers. Whole numbers answer the question "How many?" Integers answer the question "How many and in which direction?" Similarly, rational numbers are extensions of fractions.

We will discuss the arithmetic of signed numbers and examine applications involving them.

CHAPTER 6 PRETEST

This pretest covers material on signed numbers. Read each question carefully and try to answer all questions. Each question is keyed to the chapter section in which the particular topic is discussed. Check your answers against those given in the back of the text.

Questions 1–3 pertain to Section 6.1.

1. Write the opposite of -16.

2. Determine the value of: $|-5| - |+3|$.

3. Circle the number with the larger absolute value: $+8$ or -9.

Questions 4–5 pertain to Section 6.2

4. Write the opposite of $\dfrac{23}{-7}$.

5. Determine the absolute value of $\dfrac{-17}{-36}$.

Questions 6–8 pertain to Section 6.3. Add.

6. $(+12) + (-8) + (-5)$

7. $\left(-2\dfrac{1}{3}\right) + \left(-4\dfrac{1}{2}\right) + \left(+5\dfrac{1}{4}\right)$

8. $(-11.02) + (+8.9) + (-0.402)$

Questions 9–11 pertain to Section 6.4. Subtract.

9. $(-31) - (-23)$

10. $(+8.7) - (-2.31) - (+5.06)$

11. $\left(-23\dfrac{1}{8}\right) - \left(-37\dfrac{2}{5}\right)$

Questions 12–13 pertain to Section 6.5. Multiply.

12. $(-11) \times (-16)$

13. $(-1.2) \times (+0.6) \times (-14)$

Questions 14–15 pertain to Section 6.6. Divide.

14. $(-39) \div (-13)$

15. $\left(-7\frac{1}{6}\right) \div \left(+3\frac{2}{5}\right)$

16. Write the simplest form for $\dfrac{16}{-6}$. (Section 6.7)

Questions 17–19 pertain to Section 6.8. Perform the indicated operations.

17. $(-9) \times (+3) - (-15) \div (-5)$

18. $(-2)^3 + (-3)^3 \div (-9) + (-5)$

19. $\dfrac{-2}{5} \times \left(\dfrac{-1}{2} + \dfrac{-3}{7}\right)$

Question 20 pertains to Section 6.9.

20. Joan bought a new car. The total cost was $9650, and she received a trade-in allowance of $3700 for her old car. She financed the balance for 48 months, with equal monthly payments of $143. What was the total finance charge?

6.1 Integers

OBJECTIVES

After completing this section, you should be able to:
1. Write the opposite of an integer.
2. Classify an integer as being positive, negative, or zero.
3. Order any two integers.
4. Determine the absolute value of an integer.

Consider the temperature readings on a particular day. Suppose that at 9:00 A.M. the temperature reading in a certain locality was 0° C (degrees Celsius) and that during the next 2 hr the temperature dropped 5° C. Suppose that during the next 4 hours the temperature rose 5° C. What would the temperature reading be at 3:00 P.M. that day in the given locality? Did you get 0° C for an answer?

We could represent the drop in temperature by the symbol -5 (in degrees Celsius). The rise in temperature could be represented by the symbol $+5$ (in degrees Celsius). Then, the *change* in temperature would be

$$(-5) + (+5) = 0 \text{ (in degrees Celsius)}$$

Notice that the rise in temperature is the opposite of the drop in temperature. Hence, $+5$ may be thought of as the opposite of -5.

- Drop in temperature is the opposite of rise in temperature.
- Loss of yardage in a football game is the opposite of gain of yardage.
- Down is the opposite of up.
- Traveling south is the opposite of traveling north.

So far in this text, we have been discussing whole numbers or parts of whole numbers (decimals and fractions). Whole numbers can be illustrated by equally spaced points on a line called a "number line," as indicated in Figure 6.1. The arrow at the right end of the line indicates that the number line continues to the right indefinitely.

FIGURE 6.1

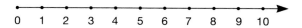

Whole numbers can be used to represent the number of yards *gained* in a football game, the number of degrees of temperature *rise*, the number of dollars *deposited* in a checking account, and so forth. However, whole numbers cannot be used to represent the opposite of these numbers; instead, integers are used.

An integer may be thought of in terms of its location on a number line relative to zero. If we extend the number line in Figure 6.1 to the left of 0 and continue to mark off equally spaced points to the left of 0, these new points would be identified with the numbers -1, -2, -3, -4, -5, and so forth, as illustrated in Figure 6.2. The arrow at the left end of the line indicates that the number line continues to the left indefinitely.

FIGURE 6.2

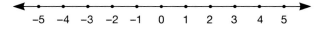

DEFINITION

The **integers** are the numbers

$$\ldots, -6, -5, -4, -3, -2, -1, 0, 1, 2, 3, 4, 5, 6, 7, \ldots$$

6.1 Integers

As we move from left to right on the number line, we move in a *positive direction*. As we move from right to left on the number line, we move in a *negative direction*. (See Figure 6.3.) We call the integers graphed to the right of 0 the **positive integers.** These were also called the "counting" (or "natural") numbers in Chapter 1.

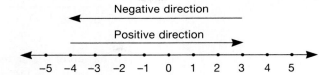

FIGURE 6.3

The integers graphed to the left of 0 are called **negative integers.** The integer 0 is neither positive nor negative. (See Figure 6.4.) Positive integers are usually written without any sign before them, although they can be written with a + sign. For instance, both 5 and +5 represent the integer positive five. Note, however, that negative integers are *always* written with a − sign in front of them.

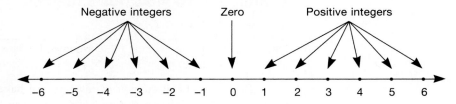

FIGURE 6.4

Note that the integers 2 and −2 are both located two units away from 0 on the number line but in opposite directions. Similarly, −4 and 4 are both located four units from 0 on the number line but in opposite directions. In fact, for every positive integer, there is a negative integer such that the two integers are the same distance from 0 on the number line but in opposite directions. Such pairs of integers are called opposites and each integer is the *opposite* of the other.

Practice Exercise 1. Write the opposite of each of the given integers: (a) +4; (b) −21; (c) −299; (d) 27; (e) −39.

EXAMPLE 1 Write the opposite for each of the following: (a) positive ten (+10); (b) negative seven (−7); (c) fifteen (15); (d) negative twenty-three (−23); (e) zero (0).

SOLUTION: (a) Negative ten (−10) is the opposite of positive ten (+10); (b) seven (7) is the opposite of negative seven (−7); (c) negative fifteen (−15) is the opposite of fifteen (15); (d) positive twenty-three (+23) is the opposite of negative twenty-three (−23); (e) zero (0) is the opposite of zero (0).

ANSWERS TO PRACTICE EXERCISE

1. (a) −4; (b) 21 (or +21); (c) 299 (or +299); (d) −27; (e) 39 (or +39).

In Chapter 1, we indicated that the whole number 3 is less than the whole number 8 since 3 comes before 8 in counting. Similarly, 20 > 7 since 20 comes after 7 in counting. On the number line, note that 3 is to the left of 8 and that 20 is to the right of 7. We order integers by comparing their locations on the number line.

DEFINITIONS

1. If the integer a is to the right of the integer b on the number line, then a is greater than b, denoted by $a > b$.
2. If the integer c is to the left of the integer d on the number line, then c is less than d, denoted by $c < d$.

Practice Exercise 2. Insert the symbol > or < between each pair of integers to make true statements:
(a) −9 __ +7;   (b) −5 __ −7;
(c) −5 __ 0;   (d) +6 __ 0.

EXAMPLE 2 Insert the symbol > or < between each pair of integers to make true statements:
(a) −3 __ −2; (b) 0 __ −4; (c) −5 __ −7; (d) −6 __ −3.

SOLUTION: Using the number line given below, we determine that

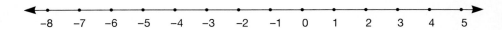

(a) −3 is to the left of −2 on the number line; hence, −3 is less than −2, denoted by −3 < −2; (b) 0 is to the right of −4 on the number line; hence, 0 is greater than −4, denoted by 0 > −4; (c) −5 is to the right of −7 on the number line; hence, −5 is greater than −7, denoted by −5 > −7; (d) −6 is to the left of −3 on the number line; hence, −6 is less than −3, denoted by −6 < −3.

ANSWERS TO PRACTICE EXERCISE

2. (a) <; (b) >; (c) <; (d) >.

So far in our discussion of integers, we have been concerned with the sign of the integer. However, when performing arithmetic operations with integers, we are interested in the numerical value of an integer, or the integer without its sign. This value is called the "absolute value" of an integer.

DEFINITION
The **absolute value** of an integer is the distance that the number is from 0 on the number line. The symbol $|n|$ is used to denote absolute value of n.

- $|5| = 5$ since 5 is 5 units from 0 on the number line in Figure 6.5.

FIGURE 6.5

- $|-5| = 5$ since −5 is 5 units from 0 on the number line in Figure 6.6.

FIGURE 6.6

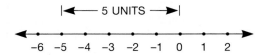

- +3 is 3 units from 0 on the number line; $|+3| = 3$.
- −2 is 2 units from 0 on the number line; $|-2| = 2$.
- −7 is 7 units from 0 on the number line; $|-7| = 7$.
- 9 is 9 units from 0 on the number line; $|9| = 9$.

6.1 Integers

Practice Exercise 3. Insert the symbol $>$, $=$, or $<$ between each pair of numbers to make true statements:
(a) $-9 \quad +7$
(b) $|-7| - |-5| \quad |-3| \times 0$
(c) $-3 \quad -|4 - 6|$
(d) $|-8| \div |-2| \quad |-5|$

EXAMPLE 3 Insert the symbol $>$, $=$, or $<$ between the two given numbers in each of the following to make true statements: (a) $|-3| \quad |+2|$. (b) $|-2| \times |-4| \quad |-(2 \times 4)|$. (c) $|5| - |3| \quad |-(5-3)|$. (d) $-2 \quad |-(7-4)|$.

SOLUTION: (a) $|-3| = 3$ and $|+2| = 2$. Since $3 > 2$, then $|-3| > |+2|$. (b) $|-2| = 2$ and $|-4| = 4$. Therefore, $|-2| \times |-4| = 2 \times 4 = 8$. Also, $|-(2 \times 4)| = |-8| = 8$. Since $8 = 8$, then $|-2| \times |-4| = |-(2 \times 4)|$. (c) $|5| = 5$ and $|3| = 3$. Therefore, $|5| - |3| = 5 - 3 = 2$. Also, $|-(5-3)| = |-2| = 2$. Since $2 = 2$, then $|5| - |3| = |-(5-3)|$. (d) $|-(7-4)| = |-3| = 3$. Since $-2 < 3$, then $-2 < |-(7-4)|$.

ANSWERS TO PRACTICE EXERCISE

3. (a) $<$; (b) $>$; (c) $<$; (d) $<$.

EXERCISES 6.1

In Exercises 1–15, write the opposite for the given integers.

1. $+5$
2. $+7$
3. -6
4. -7
5. $+10$
6. $+17$
7. -19
8. -23
9. 0
10. -81
11. $+27$
12. $+36$
13. -93
14. $+213$
15. -319

In Exercises 16–30, classify the indicated integers as being positive, negative, or neither.

16. A gain of 7 yd
17. A loss of $3
18. A temperature rise of 13 degrees
19. 15 mi south
20. 12 km north
21. A deposit of $71
22. A temperature drop of 19 degrees
23. A withdrawal of $92
24. An increase of 111 units
25. A decrease of 470 units
26. A price increase of $361
27. A loss of 569 oz
28. A wage reduction of $999
29. A profit of $333
30. A salary increase of $1019

In Exercises 31–54, evaluate the given expressions.

31. $|+3|$
32. $|-6|$
33. $|9|$
34. $|-17|$
35. $|0|$
36. $|-47|$
37. $|-101|$
38. $|96|$
39. $|+97|$
40. $|+123|$
41. $|-981|$
42. $|-602|$
43. $|3 \times 2|$
44. $|8 - 6|$
45. $|0 \div 6|$
46. $|-(1 + 3)|$
47. $|-(4 \times 4)|$
48. $|-(18 \div 9)|$
49. $|+3| + |-4|$
50. $|-5| + |-6|$
51. $|+7| + |-7|$
52. $|+8| - |-8|$
53. $|-9| \div |+9|$
54. $|-2| \times |-3|$

In Exercises 55–69, insert the symbol, $<$, $=$, or $>$ between the given pairs of numbers to make true statements.

55. $-3 \quad +4$
56. $-7 \quad -6$
57. $+8 \quad 0$
58. $-7 \quad 0$
59. $+8 \quad +11$
60. $-19 \quad -3$
61. $|-3| \quad 0$
62. $0 \quad |-4|$
63. $|-3| \quad |+4|$
64. $|-7| \quad |-8|$
65. $|9| \quad |6 + 3|$
66. $|8| \quad |10 - 2|$
67. $-3 \quad |1 + 3|$
68. $|6| - |3| \quad |6 - 3|$
69. $|-2| \times |-3| \quad |2 \times 3|$

6.2 Signed Numbers

OBJECTIVES

After completing this section, you should be able to:
1. Write the opposite of a rational number or a decimal.
2. Classify a rational number or decimal as being positive, negative, or zero.
3. Order any two rational numbers or decimals.
4. Determine the absolute value of rational numbers and decimals.

In Section 6.1, it was noted that for every positive integer (or counting number) there is a negative integer (its opposite) such that the two integers are the same distance from 0 on the number line but in opposite directions. Such pairs of integers are called "opposites," and each integer is the opposite of the other.

In Chapter 2, we noted that a fraction is a quotient of whole numbers with a nonzero denominator. Just as the whole numbers were extended to the integers (by introducing the negative integers), the fractions will be extended to what are called "rational numbers."

DEFINITION

A **rational number** is a quotient of two integers with a nonzero denominator.

- $\dfrac{-2}{+3}$ is a rational number since it is the quotient of the integer -2 divided by the nonzero integer $+3$.

- $\dfrac{-4}{-7}$ is a rational number since it is the quotient of the integer -4 divided by the nonzero integer -7.

- $\dfrac{0}{-5}$ is a rational number since it is the quotient of the integer 0 divided by the nonzero integer -5.

- $\dfrac{+4}{0}$ is *not* a rational number since the denominator is 0.

The rational numbers can be classified as being positive, negative, or zero as follows:

1. If the numerator and the denominator of a rational number are both positive or both negative, then the rational number is a positive rational number. For example, $\dfrac{4}{7}$ and $\dfrac{-3}{-5}$ are positive rational numbers.

2. If the numerator and the denominator of a rational number have different signs, then the rational number is a negative rational number. $\dfrac{-2}{9}$ and $\dfrac{11}{-13}$ are negative rational numbers.

3. If the numerator of a rational number is 0 and the denominator is not zero, then the rational number is zero. For example, $\dfrac{0}{-6}$ and $\dfrac{0}{7}$ are different names for 0.

The opposite of a positive integer is negative and the opposite of a negative integer is positive. Similarly, we can determine the opposite of a rational number.

6.2 Signed Numbers

Determine the opposite of each of the following rational numbers.

Practice Exercise 1. $\dfrac{-4}{+3}$

EXAMPLE 1 Determine the opposite of $\dfrac{-2}{+3}$.

SOLUTION: The opposite of $\dfrac{-2}{+3}$ is $\dfrac{+2}{+3}$ or $\dfrac{-2}{-3}$. $\dfrac{-2}{+3}$ is a negative number. Its opposite is a positive number which may be written either as $\dfrac{+2}{+3}$ or $\dfrac{-2}{-3}$. Note that $\dfrac{+2}{+3}$ may also be written as $\dfrac{2}{3}$.

Practice Exercise 2. $\dfrac{0}{-7}$

EXAMPLE 2 Determine the opposite of $\dfrac{3}{7}$.

SOLUTION: The opposite of $\dfrac{3}{7}$ is $\dfrac{-3}{7}$ or $\dfrac{3}{-7}$. $\dfrac{3}{7}$ is a positive number. Its opposite is a negative number which may be written either as $\dfrac{-3}{7}$ or $\dfrac{3}{-7}$.

Practice Exercise 3. $\dfrac{-5}{-6}$

EXAMPLE 3 Determine the opposite of $\dfrac{-4}{-9}$.

SOLUTION: The opposite of $\dfrac{-4}{-9}$ is $\dfrac{+4}{-9}$ or $\dfrac{-4}{+9}$.

Practice Exercise 4. $\dfrac{7}{9}$

EXAMPLE 4 Determine the opposite of $\dfrac{0}{+5}$.

SOLUTION: The opposite of $\dfrac{0}{+5}$ is $\dfrac{0}{+5}$ or $\dfrac{0}{-5}$. Note that both $\dfrac{0}{+5}$ and $\dfrac{0}{-5}$ are equal to 0. That is, the opposite of 0 is 0.

ANSWERS TO PRACTICE EXERCISES

1. $\dfrac{4}{3}$ or $\dfrac{-4}{-3}$. 2. $\dfrac{0}{-7}$ or $\dfrac{0}{+7}$. 3. $\dfrac{5}{-6}$ or $\dfrac{-5}{6}$. 4. $\dfrac{-7}{9}$ or $\dfrac{7}{-9}$.

In Section 6.1, we ordered the integers according to their locations on the number line. Rational numbers can also be ordered in a similar manner. For instance, in Figure 6.7, the rational number $\dfrac{-3}{2}$ is to the left of the rational number $\dfrac{1}{2}$. Hence, $\dfrac{-3}{2} < \dfrac{1}{2}$. Note that $\dfrac{1}{2} > \dfrac{-3}{2}$. In Figure 6.7, we note that the rational numbers $\dfrac{-3}{2}$ and $\dfrac{3}{2}$ are the same distance from 0 but are on opposite sides of 0. Therefore, their absolute values are equal. We have

$$\left|\dfrac{-3}{2}\right| = \left|\dfrac{3}{2}\right| = \dfrac{3}{2}$$

We also recall from Chapter 3 that some fractions can be rewritten as equivalent decimal numbers. For instance, $\dfrac{1}{2} = 0.5$ and $\dfrac{9}{4} = 2.25$. Similarly, the rational numbers can be written as decimals.

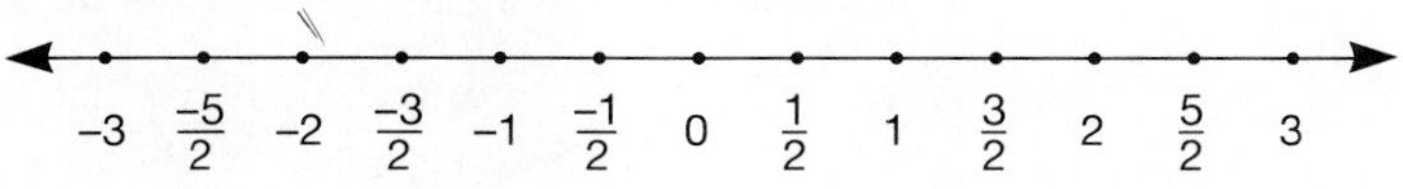

FIGURE 6.7 $-3, \dfrac{-5}{2}, -2, \dfrac{-3}{2}, -1, \dfrac{-1}{2}, 0, \dfrac{1}{2}, 1, \dfrac{3}{2}, 2, \dfrac{5}{2}, 3$

Practice Exercise 5. Rewrite each of the given rational numbers as decimals: (a) $\frac{7}{10}$; (b) $\frac{-9}{20}$; (c) $\frac{15}{40}$; (d) $\frac{-12}{18}$.

EXAMPLE 5 Rewrite each of the given rational numbers as decimals: (a) $\frac{2}{10}$; (b) $\frac{-3}{10}$; (c) $\frac{4}{-5}$; (d) $\frac{10}{15}$.

SOLUTION: (a) $\frac{2}{10} = 0.2$; (b) $\frac{-3}{10} = -\left(\frac{3}{10}\right) = -(0.3) = -0.3$; (c) $\frac{4}{-5} = -\left(\frac{4}{5}\right) = -(0.8) = -0.8$; (d) $\frac{10}{15} = \frac{2}{3} = 0.6\overline{6}$.

ANSWERS TO PRACTICE EXERCISE

6. (a) 0.7; (b) -0.45; (c) 0.375; (d) $-0.6\overline{6}$.

For the remainder of this chapter, we shall refer to integers and rational numbers as **signed numbers.** Signed numbers will also include decimals.

EXERCISES 6.2

In Exercises 1–15, write the opposite for each of the given rational numbers.

1. $\frac{-5}{7}$ 2. $\frac{-4}{-5}$ 3. $\frac{3}{2}$ 4. $\frac{-6}{1}$ 5. $\frac{-7}{2}$

6. $\frac{3}{-8}$ 7. $\frac{6}{5}$ 8. $\frac{-2}{-7}$ 9. $\frac{10}{3}$ 10. $\frac{-11}{-13}$

11. $\frac{14}{-9}$ 12. $\frac{-7}{13}$ 13. $\frac{0}{3}$ 14. $\frac{-9}{17}$ 15. $\frac{-0}{13}$

In Exercises 16–30, classify the indicated rational numbers as being positive, negative, or zero.

16. $\frac{-2}{3}$ 17. $\frac{-9}{-5}$ 18. $\frac{0}{-4}$ 19. $\frac{14}{19}$ 20. $\frac{-23}{121}$

21. $\frac{27}{-5}$ 22. $\frac{-101}{-203}$ 23. $\frac{-0}{-9}$ 24. $\frac{14}{-3}$ 25. $\frac{-9}{39}$

26. $\frac{-81}{-92}$ 27. $\frac{171}{-20}$ 28. $\frac{-38}{-41}$ 29. $\frac{0}{10}$ 30. $\frac{181}{-96}$

In Exercises 31–45, determine the absolute value for each of the following expressions.

31. $\frac{-3}{7}$ 32. $\frac{0}{-9}$ 33. -2.31 34. $\frac{-11}{-19}$ 35. $\frac{-23}{101}$

36. $\frac{-0}{11}$ 37. 4.03 38. $\frac{-19}{23}$ 39. -37.1 40. $\frac{-129}{-387}$

41. $\frac{401}{-38}$ 42. $\frac{-89}{209}$ 43. -0.81 44. $\frac{-107}{-810}$ 45. -1.007

In Exercises 46–55, insert the symbol $>$, $=$, or $<$ between the two given number symbols to make true statements. (*Hint:* Rewrite all rational numbers in simplest form.)

46. $\frac{-2}{3}$ $\frac{3}{-5}$ 47. $\frac{-1}{2}$ -0.4 48. $\frac{-2}{3}$ -0.73

49. $\dfrac{0}{-9}$ $\dfrac{0}{7}$ 50. $\dfrac{-11}{-22}$ $\dfrac{4}{5}$ 51. $\dfrac{-0}{11}$ -2.34

52. -3.1 -4.96 53. -1.09 -3.1 54. $\dfrac{-4}{7}$ $\dfrac{8}{-14}$

55. -4.09 $\dfrac{-400}{100}$

6.3 Addition of Signed Numbers

OBJECTIVE

After completing this section, you should be able to add any two signed numbers, using the appropriate rules for addition of signed numbers.

To discuss the addition of two signed numbers, we must consider the following three cases:

1. Both signed numbers have the same sign.
2. The two signed numbers have different signs.
3. One of the signed numbers is 0.

We can consider the last case very quickly. In Section 1.7, we noted that if 0 is added to a whole number or a whole number is added to 0, the sum is that whole number. The number 0 is called the "identity element for addition." This rule extends to signed numbers.

If a is any signed number, then:

$$a + 0 = a \quad \text{and} \quad 0 + a = a$$

- $(+4) + 0 = +4$
- $0 + (-7) = -7$
- $0 + 0 = 0$
- $(-19) + 0 = -19$

To add signed numbers, we can make use of the number line.

Add, using the appropriate rule.

Practice Exercise 1.
$(+7) + (+11)$

EXAMPLE 1 Add: $(+4) + (+2)$.

SOLUTION: On the number line, begin at 0 and move 4 units to the *right* (to $+4$). Then move 2 units *more* to the *right* (to $+6$). Therefore,

$$(+4) + (+2) = 6$$

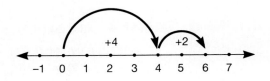

Note that we started at 0 and, after two moves, ended at +6. Since +4 and +2 are alternative ways of writing 4 and 2, respectively, we would expect that (+4) + (+2) = +6 since 4 + 2 = 6.

Practice Exercise 2.
(−4.1) + (−0.71)

EXAMPLE 2 Add: (−3) + (−4).

SOLUTION: On the number line, begin at 0 and move 3 units to the *left* (to −3). Then move 4 units *more* to the *left* (to −7). Therefore,

$$(-3) + (-4) = -7$$

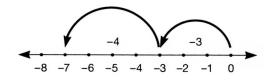

Note that we started at 0 and, after two moves, ended at −7.

Practice Exercise 3.
$\left(+\frac{1}{2}\right) + \left(+\frac{3}{7}\right)$

EXAMPLE 3 Add: (−2.5) + (−3.5).

SOLUTION: On the number line, begin at 0 and move 2.5 units to the *left* (to −2.5). Then move 3.5 units *more* to the *left* (to −6.0). Therefore,

$$(-2.5) + (-3.5) = -6.0$$

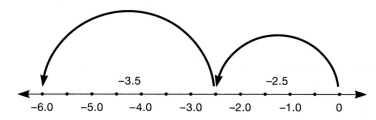

Note that we started at 0 and, after two moves, ended at −6.0.

When using the number line to add signed numbers with the same signs, we moved in the *same* direction for each number in the sum. The result had the same sign as the numbers added.

Rule for Adding Signed Numbers with the Same Sign

To add two signed numbers with the *same* sign:

1. Add their absolute values.
2. Keep the same sign for the sum.

Practice Exercise 4.
(−19) + (−7.3)

EXAMPLE 4 Add: (a) (+5) + (+8); (b) (−7) + (−9).

SOLUTION: (a) (+5) + (+8) = +(5 + 8) = +13; (b) (−7) + (−9) = −(7 + 9) = −16.

6.3 Addition of Signed Numbers

Practice Exercise 5.
$\left(\frac{+1}{5}\right) + (8.03)$

EXAMPLE 5 Add: (a) $(+2.3) + (+4.56)$; (b) $(-0.005) + (-7.09)$.

SOLUTION: (a) $(+2.3) + (+4.56) = +(2.3 + 4.56) = +6.86$; (b) $(-0.005) + (-7.09) = -(0.005 + 7.09) = -7.095$.

Practice Exercise 6.
$\left(\frac{-4}{7}\right) + \left(\frac{-5}{9}\right)$

EXAMPLE 6 Add: (a) $\left(\frac{+2}{3}\right) + \left(\frac{+1}{2}\right)$; (b) $\left(\frac{-1}{4}\right) + \left(\frac{-1}{3}\right)$.

SOLUTION: (a) $\left(\frac{+2}{3}\right) + \left(\frac{+1}{2}\right) = +\left(\frac{2}{3} + \frac{1}{2}\right) = +\frac{7}{6}$ or $+1\frac{1}{6}$; (b) $\left(\frac{-1}{4}\right) + \left(\frac{-1}{3}\right) = -\left(\frac{1}{4} + \frac{1}{3}\right) = \frac{-7}{12}$.

ANSWERS TO PRACTICE EXERCISES

1. $+18$. 2. -4.81. 3. $\frac{+13}{14}$. 4. -26.3. 5. $+8.23$. 6. $-1\frac{8}{63}$.

Using the number line, let's now add two signed numbers with different signs.

Add, using the appropriate rule.

Practice Exercise 7.
$(-9) + (+6)$

EXAMPLE 7 Add: $(+8) + (-5)$.

SOLUTION: On the number line, begin at 0 and move 8 units to the *right* (to $+8$). Then move 5 units to the *left* (to $+3$). Therefore,

$$(+8) + (-5) = +3$$

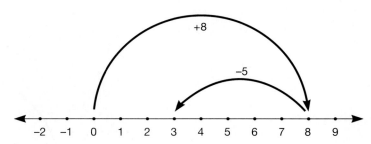

Note that we started at 0 and, after two moves, ended at $+3$.

Practice Exercise 8.
$(-11) + (+23)$

EXAMPLE 8 Add: $(+3) + (-6)$.

SOLUTION: On the number line, begin at 0 and move 3 units to the *right* (to $+3$). Then move 6 units to the *left* (to -3). Therefore,

$$(+3) + (-6) = -3$$

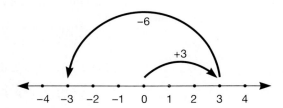

Note that we started at 0 and, after two moves, ended at -3.

Practice Exercise 9.

$\left(\dfrac{+8}{3}\right) + \left(\dfrac{-17}{3}\right)$

EXAMPLE 9 Add: $\left(\dfrac{-3}{2}\right) + \left(\dfrac{+5}{2}\right)$.

SOLUTION: On the number line, begin at 0 and move $\dfrac{3}{2}$ units to the *left* $\left(\text{to } \dfrac{-3}{2}\right)$. Then move $\dfrac{5}{2}$ units to the *right* (to $+1$). Therefore,

$$\left(\dfrac{-3}{2}\right) + \left(\dfrac{+5}{2}\right) = +1$$

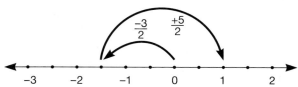

Note that we started at 0 and, after two moves, ended at $+1$.

Practice Exercise 10.

$(-6) + \left(\dfrac{+9}{4}\right)$

EXAMPLE 10 Add: $\left(\dfrac{-9}{2}\right) + (+2)$.

SOLUTION: On the number line, begin at 0 and move $\dfrac{9}{2}$ units to the *left* $\left(\text{to } \dfrac{-9}{2}\right)$. Then move 2 units to the *right* $\left(\text{to } \dfrac{-5}{2}\right)$. Therefore,

$$\left(\dfrac{-9}{2}\right) + (+2) = \dfrac{-5}{2}$$

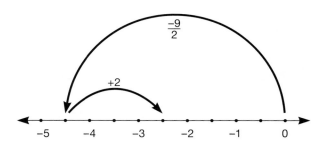

Note that we started at 0 and, after two moves, ended at $\dfrac{-5}{2}$.

We can always use the number line to find the sum of two signed numbers. However, that is time consuming and not an efficient way to add the numbers when their numerical values are large.

Reexamining Examples 7–10, we note:

- $(+8) + (-5) = +3$, or $(+8) + (-5) = +(8 - 5) = +3$
- $(+3) + (-6) = -3$, or $(+3) + (-6) = -(6 - 3) = -3$
- $\left(\dfrac{-3}{2}\right) + \left(\dfrac{+5}{2}\right) = +1$, or $\left(\dfrac{-3}{2}\right) + \left(\dfrac{+5}{2}\right) = +\left(\dfrac{5}{2} - \dfrac{3}{2}\right) = +\left(\dfrac{2}{2}\right) = +1$

6.3 Addition of Signed Numbers

- $\left(\dfrac{-9}{2}\right) + \left(\dfrac{+5}{2}\right) = -2$, or $\left(\dfrac{-9}{2}\right) + \left(\dfrac{+5}{2}\right) = -\left(\dfrac{9}{2} - \dfrac{5}{2}\right) = -\left(\dfrac{4}{2}\right) = -2$

These examples illustrate the following rule.

Rule for Adding Signed Numbers with Different Signs

To add signed numbers with different signs:
1. Take the absolute value of each number.
2. Subtract the smaller absolute value from the larger absolute value.
3. For the sum, take the sign of the number with the larger absolute value.

Practice Exercise 11.
$(-17) + (+23)$

EXAMPLE 11 Add: $(+4) + (-7)$.

SOLUTION: $(+4) + (-7) = -(7 - 4) = -3$ (Since $|-7| > |+4|$, take $-$ sign)

Practice Exercise 12.
$(+19) + (-23)$

EXAMPLE 12 Add: $(-5) + (+9)$.

SOLUTION: $(-5) + (+9) = +(9 - 5) = +4$ (Since $|+9| > |-5|$, take $+$ sign)

Practice Exercise 13.
$(-5.3) + (+8.1)$

EXAMPLE 13 Add: $(+3.7) + (-1.6)$.

SOLUTION: $(+3.7) + (-1.6) = +(3.7 - 1.6) = +2.1$ (Since $|+3.7| > |-1.6|$, take $+$ sign)

Practice Exercise 14.
$(+81.3) + (-53.49)$

EXAMPLE 14 Add: $(-13.09) + (+9.5)$.

SOLUTION: $(-13.09) + (+9.5) = -(13.09 - 9.5) = -3.59$ (Since $|-13.09| > |+9.5|$, take $-$ sign)

Practice Exercise 15.
$\left(\dfrac{-13}{7}\right) + \left(\dfrac{+3}{8}\right)$

EXAMPLE 15 Add: $\left(\dfrac{+4}{5}\right) + \left(\dfrac{-1}{2}\right)$.

SOLUTION: $\left(\dfrac{+4}{5}\right) + \left(\dfrac{-1}{2}\right) = +\left(\dfrac{4}{5} - \dfrac{1}{2}\right) = \dfrac{+3}{10}$ (Since $\left|\dfrac{+4}{5}\right| > \left|\dfrac{-1}{2}\right|$, take $+$ sign)

Practice Exercise 16.
$\left(\dfrac{+25}{3}\right) + \left(\dfrac{-17}{4}\right)$

EXAMPLE 16 Add: $\left(\dfrac{-11}{12}\right) + \left(\dfrac{+4}{5}\right)$.

SOLUTION: $\left(\dfrac{-11}{12}\right) + \left(\dfrac{+4}{5}\right) = -\left(\dfrac{11}{12} - \dfrac{4}{5}\right) = \dfrac{-7}{60}$ (Since $\left|\dfrac{-11}{12}\right| > \left|\dfrac{4}{5}\right|$, take $-$ sign)

ANSWERS TO PRACTICE EXERCISES

7. -3. 8. $+12$. 9. -3. 10. $\dfrac{-15}{4}$. 11. $+6$. 12. -4. 13. $+2.8$. 14. $+27.81$.
15. $-1\dfrac{27}{56}$. 16. $\dfrac{+49}{12}$.

To add two signed numbers, you must first determine if the numbers have the same signs or different signs and then apply the appropriate rule. Also note that the addition of signed numbers is both commutative and associative.

Add, using the appropriate rule.

Practice Exercise 17.
$(+8) + (+6)$

EXAMPLE 17 Find the sum of $(+8)$ and $(+7)$.

SOLUTION:

STEP 1: Examine the signs of the numbers. They are both the same (positive).

STEP 2: Apply the appropriate rule. Since the signs are the same, we *add* the absolute values of the numbers and place the common sign in front of the sum. Therefore,

$$(+8) + (+7) = +(8 + 7) = +15$$

Practice Exercise 18.
$(-11) + (-13)$

EXAMPLE 18 Find the sum of $(+6.1)$ and (-9.7).

SOLUTION:

STEP 1: Examine the signs of the numbers. They are different.

STEP 2: Apply the appropriate rule. Since the signs are different, we *subtract* the absolute values and place the sign of the number that has the larger absolute value in front of this difference. Therefore,

$$(+6.1) + (-9.7) = -(9.7 - 6.1) = -3.6$$

Practice Exercise 19.
$(-19.1) + (+13.62)$

EXAMPLE 19 Add: $(+2) + (-4) + (+3)$.

SOLUTION: We shall solve this problem by adding the numbers, two at a time, from left to right:

$$(+2) + (-4) + (+3)$$
$$= -(4 - 2) + (+3)$$
$$= (-2) + (+3)$$
$$= +(3 - 2)$$
$$= +1$$

Practice Exercise 20.
$\left(-3\frac{1}{2}\right) + \left(+4\frac{3}{7}\right)$

EXAMPLE 20 Add: $(-4) + (+5) + (-1) + (+7)$.

SOLUTION: We shall solve this problem by adding two pairs of integers together:

$$(-4) + (+5) + (-1) + (+7)$$
$$= +(5 - 4) + [+(7 - 1)]$$
$$= (+1) + (+6)$$
$$= +(1 + 6)$$
$$= +7$$

ANSWERS TO PRACTICE EXERCISES

17. $+14$. 18. -24. 19. -5.48. 20. $\frac{13}{14}$.

EXERCISES 6.3

In Exercises 1–35, add the indicated integers, using appropriate rules.

1. $(+9) + (+7)$
2. $(+13) + (+6)$
3. $(-7) + (-5)$
4. $(-6) + (-11)$
5. $(+9) + (-9)$
6. $(-13) + 0$
7. $0 + (+17)$
8. $(-19) + (+19)$
9. $(+7) + (-11)$
10. $(-13) + (+19)$
11. $(+14) + (-21)$
12. $(-63) + (-32)$
13. $(-54) + (+101)$
14. $(+127) + (-127)$
15. $(+561) + (+219)$
16. $(+586) + (-239)$
17. $(-304) + (+259)$
18. $(+492) + (-639)$
19. $(-807) + (+569)$
20. $(+762) + (-809)$
21. $(+6) + (+3) + (+5)$
22. $(-7) + (-9) + (-3)$
23. $(+8) + (+9) + (-11)$
24. $(-13) + (-9) + (+21)$
25. $(-23) + (+12) + (-36)$
26. $(-57) + (-12) + (+49)$
27. $(+47) + (-53) + (-32)$
28. $(+85) + (-101) + (-85)$
29. $(-109) + (-116) + (+319)$
30. $(-407) + (+509) + (-101)$
31. $(+4) + (-5) + (+8) + (-13) + (+7)$
32. $(-6) + (-8) + (+9) + (-17) + (-10)$
33. $(-23) + (-35) + (+51) + (-31) + (+19)$
34. $(+18) + (-27) + (+29) + (-46) + (+59) + (-63)$
35. $(-57) + (-61) + (-82) + (+101) + (+62) + (+93)$

In Exercises 36–60, add the indicated decimals, using appropriate rules.

36. $(+2.31) + (+5.05)$
37. $(+13.07) + (+6.9)$
38. $(-6.9) + (-0.79)$
39. $(-80.3) + (-11.99)$
40. $(+4.87) + (-4.87)$
41. $(-102.7) + 0$
42. $0 + (-86.2)$
43. $(-17.62) + (+1.3)$
44. $(+0.87) + (-3.7)$
45. $(-17.46) + (+10.009)$
46. $(+41.76) + (-27.5)$
47. $(-59.462) + (-9.71)$
48. $(-8.734) + (+40.52)$
49. $(+0.092) + (-0.39)$
50. $(+29.7) + (+3.992)$
51. $(+73.1) + (-90.021)$
52. $(-49.54) + (+121.7)$
53. $(+90.09) + (-123.812)$
54. $(-7.692) + (+12.09)$
55. $(+192.1) + (-67.809)$
56. $(+13.1) + (+0.97) + (+6.001)$
57. $(-7.6) + (-9.47) + (-3.201)$
58. $(+8.03) + (+1.7) + (-11.023)$
59. $(-5.7) + (-12.62) + (+47.9)$
60. $(-1.09) + (-11.6) + (+0.319)$

In Exercises 61–80, add the rational numbers, using appropriate rules.

61. $\left(\frac{+3}{7}\right) + \left(\frac{-4}{9}\right)$
62. $\left(\frac{-4}{5}\right) + \left(\frac{-6}{7}\right)$
63. $\left(\frac{+2}{3}\right) + \left(\frac{-5}{4}\right)$
64. $\left(\frac{-1}{7}\right) + \left(\frac{-2}{9}\right)$
65. $\left(\frac{-15}{11}\right) + \left(\frac{-2}{7}\right)$
66. $\left(\frac{+10}{19}\right) + \left(\frac{-1}{3}\right)$
67. $\left(\frac{+1}{2}\right) + \left(\frac{+1}{6}\right)$
68. $\left(\frac{+2}{3}\right) + \left(\frac{-4}{5}\right)$
69. $\left(\frac{-2}{7}\right) + \left(\frac{-4}{3}\right)$
70. $\left(\frac{-1}{5}\right) + \left(\frac{-2}{7}\right)$

71. $\left(\frac{+4}{13}\right) + \left(\frac{-5}{9}\right)$ 72. $\left(\frac{-3}{7}\right) + \left(\frac{-2}{11}\right)$

73. $\left(\frac{-3}{7}\right) + \left(\frac{-4}{9}\right)$ 74. $\left(\frac{0}{6}\right) + \left(\frac{-11}{17}\right)$

75. $\left(\frac{+1}{2}\right) + \left(\frac{+1}{3}\right) + \left(\frac{+1}{4}\right)$ 76. $\left(\frac{-2}{3}\right) + \left(\frac{-1}{7}\right) + \left(\frac{-6}{5}\right)$

77. $\left(\frac{-1}{2}\right) + \left(\frac{+2}{3}\right) + \left(\frac{-1}{4}\right)$ 78. $\left(\frac{-4}{7}\right) + \left(\frac{-2}{9}\right) + \left(\frac{+3}{5}\right)$

79. $\left(\frac{-2}{3}\right) + \left(\frac{+1}{4}\right) + \left(\frac{+1}{5}\right)$ 80. $\left(\frac{-4}{11}\right) + \left(\frac{-2}{3}\right) + \left(\frac{+4}{7}\right)$

In Exercises 81–90, add the indicated mixed numbers by first rewriting the mixed numbers as rational numbers (i.e., signed improper fractions).

81. $\left(+3\frac{1}{2}\right) + \left(+4\frac{2}{3}\right)$ 82. $\left(-3\frac{1}{7}\right) + \left(+5\frac{1}{9}\right)$

83. $\left(+7\frac{1}{6}\right) + \left(-2\frac{4}{5}\right)$ 84. $\left(-3\frac{1}{3}\right) + \left(-2\frac{1}{4}\right)$

85. $\left(-1\frac{3}{4}\right) + \left(-4\frac{1}{2}\right)$ 86. $\left(+3\frac{2}{3}\right) + \left(-7\frac{1}{4}\right)$

87. $\left(+2\frac{1}{3}\right) + \left(-3\frac{1}{2}\right) + \left(-5\frac{1}{4}\right)$ 88. $\left(-7\frac{4}{5}\right) + \left(-6\frac{2}{3}\right) + \left(-4\frac{1}{7}\right)$

89. $\left(+1\frac{2}{9}\right) + \left(-8\frac{3}{5}\right) + \left(-6\frac{3}{4}\right)$ 90. $\left(-5\frac{3}{7}\right) + \left(-9\frac{3}{4}\right) + \left(+2\frac{3}{5}\right)$

91. What is the sum of any signed number and 0?
92. What is the sum of any signed number and its opposite?
93. At the start of a week, the opening price of a share of stock was $43. The following changes were listed during the week: up 2, down 5, down 1, up 4, up 3. What was the closing price (in dollars) for a share of the stock at the end of the week?
94. Nancy had $368 in her checking account. She wrote checks for $117, $80, and $109, made a deposit of $423, and wrote additional checks for $172 and $96. What was her checking account balance after these transactions?
95. Holly weighed 135 pounds when she started to record her weight changes (in pounds) for the next 6 months. If the changes were: gained 3, lost 4, lost 2, gained 5, lost 1, and gained 3, what was her weight after all these changes?
96. Barb had $267.83 in her checking account. She wrote checks for $23.67, $123.45, and $51.32, made a deposit of $372.79, and wrote additional checks for $102.36 and $64.45. What was her checking balance after these transactions?
97. A builder owned $3\frac{3}{4}$ acres of land. He than bought $5\frac{1}{3}$ acres adjacent to the other land he owned. He then divided up the combined acreage and sold two portions that were $\frac{7}{8}$ of an acre and $1\frac{2}{5}$ acres. How many acres did he have left after the two sales?
98. The temperature at 6:00 A.M. was 26.2° C. During the next 2 hours, the temperature dropped 2.6 degrees. During the next 3 hours, the temperature rose 4.8 degrees. During the next hour, the temperature rose 1.8 degrees. During the next 4 hours, the temperature dropped 3.4 degrees. What was the temperature at 4:00 P.M.?
99. My stock opened at $38\frac{1}{4}$ points. During the next 2 hours, the stock gained $1\frac{1}{8}$ points. It gained another $2\frac{3}{4}$ points during the next 3 hours and then lost $5\frac{7}{8}$ points during the balance of the trading day. What was the closing price of my stock that day?

100. Your stock opened at $42\frac{5}{8}$ points. During the next 4 hours, the stock lost $3\frac{1}{4}$ points. It lost another $2\frac{3}{8}$ points during the next 2 hours. During the balance of the trading day, the stock gained $1\frac{3}{4}$ points. What was the closing price of your stock that day?

6.4 Subtraction of Signed Numbers

OBJECTIVE

After completing this section, you should be able to subtract one signed number from another signed number, using the appropriate rules for subtraction of signed numbers.

To subtract the whole number 3 from the whole number 8, we ask "What whole number must be added to 3 to obtain 8?" We have $8 - 3 = 5$ since $3 + 5 = 8$. In a similar manner, to subtract the integer -6 from the integer $+2$, we ask, "What integer must be added to -6 to obtain $+2$?" We have $(+2) - (-6) = +8$ since $(-6) + (+8) = +2$.

For each of the following, subtract.

Practice Exercise 1.
$(+15) - (+27)$

EXAMPLE 1 Subtract: $(+4) - (+6)$.

SOLUTION: $(+4) - (+6) = -2$ since $(+6) + (-2) = +4$

Practice Exercise 2.
$(-32) - (-21)$

EXAMPLE 2 Subtract: $(-7) - (+5)$.

SOLUTION: $(-7) - (+5) = -12$ since $(+5) + (-12) = -7$

Practice Exercise 3.
$(+4.32) - (-11.9)$

EXAMPLE 3 Subtract: $(-5.1) - (-4.6)$.

SOLUTION: $(-5.1) - (-4.6) = -0.5$ since $(-4.6) + (-0.5) = -5.1$

Practice Exercise 4.
$\left(\frac{+4}{5}\right) - \left(\frac{+6}{7}\right)$

EXAMPLE 4 Subtract: $0 - \left(-6\frac{1}{2}\right)$.

SOLUTION: $0 - \left(-6\frac{1}{2}\right) = 6\frac{1}{2}$ since $\left(-6\frac{1}{2}\right) + \left(6\frac{1}{2}\right) = 0$

Let's look carefully at the following:

- $(+4) - (+6) = -2$ (from Example 1); also, $(+4) + (-6) = -2$
- $(-7) - (+5) = -12$ (from Example 2); also, $(-7) + (-5) = -12$
- $(-5.1) - (-4.6) = -0.5$ (from Example 3); also, $(-5.1) + (+4.6) = -0.5$
- $0 - \left(-6\frac{1}{2}\right) = 6\frac{1}{2}$ (from Example 4); also, $0 + \left(+6\frac{1}{2}\right) = 6\frac{1}{2}$

From these examples, we see that subtracting a signed number is the same as adding its opposite.

Rule for Subtracting Signed Numbers

To subtract one signed number from another signed number:
1. Replace the number to be subtracted with its opposite.
2. Add.

[That is, $a - b = a + (-b)$].

We will now apply this rule in the following examples.

Practice Exercise 5.
$\left(\frac{-4}{5}\right) - \left(\frac{-2}{11}\right)$

EXAMPLE 5 Subtract: $\left(+6\frac{1}{3}\right) - \left(+3\frac{1}{2}\right)$.

SOLUTION:

$$\left(+6\frac{1}{3}\right) - \left(+3\frac{1}{2}\right) = \left(+6\frac{1}{3}\right) + \left(-3\frac{1}{2}\right)$$ Instead of subtracting $+3\frac{1}{2}$, we add its opposite, $-3\frac{1}{2}$.

$$= +\left(6\frac{1}{3} - 3\frac{1}{2}\right)$$

$$= +2\frac{5}{6}$$

CHECK: $\left(+3\frac{1}{2}\right) + \left(+2\frac{5}{6}\right) = +\left(3\frac{1}{2} + 2\frac{5}{6}\right) = +6\frac{1}{3}$ ✓

Practice Exercise 6.
$\left(-4\frac{2}{7}\right) - \left(+5\frac{3}{8}\right)$

EXAMPLE 6 Subtract: $(-9) - (-3)$.

SOLUTION:

$(-9) - (-3) = (-9) + (+3)$ Instead of subtracting -3, we add its opposite, $+3$.

$= -(9 - 3)$
$= -6$

CHECK: $(-3) + (-6) = -(3 + 6) = -9$ ✓

Practice Exercise 7.
$\left(+5\frac{1}{2}\right) - (-1.7)$

EXAMPLE 7 Subtract: $(-5.9) - (+3.12)$.

SOLUTION:

$(-5.9) - (+3.12) = (-5.9) + (-3.12)$ Instead of subtracting $+3.12$, we add its opposite, -3.12.

$= -(5.9 + 3.12)$
$= -9.02$

CHECK: $(+3.12) + (-9.12) = -(9.02 - 3.12) = -5.9$ ✓

ANSWERS TO PRACTICE EXERCISES

1. -12. 2. -11. 3. 16.22. 4. $\frac{-2}{35}$. 5. $\frac{-34}{55}$. 6. $-9\frac{37}{56}$. 7. 7.2.

6.4 Subtraction of Signed Numbers

There are times when we encounter problems involving both addition and subtraction of signed numbers. In such cases, we rewrite the problems to involve only additions, as illustrated in the following examples.

Perform the indicated operations.

Practice Exercise 8.
$(2.91) + (-3.4) - (-0.59)$

EXAMPLE 8 Perform the indicated operations:
$$(+6) + (-3) - (-2) + (-5) - (+3)$$

SOLUTION:

$$\begin{aligned}
&(+6) + (-3) - (-2) + (-5) - (+3) &&\text{(Rewriting the subtractions} \\
&= (+6) + (-3) + (+2) + (-5) + (-3) &&\text{as equivalent additions)} \\
&= [(+6) + (+2)] + [(-3) + (-5) + (-3)] &&\text{(Regrouping)} \\
&= +(6 + 2) + [-(3 + 5 + 3)] \\
&= (+8) + (-11) \\
&= -(11 - 8) \\
&= -3
\end{aligned}$$

In the preceding example, we regrouped the terms in the sum so that all the positive numbers were grouped together and all the negative numbers were grouped together. This simplifies the addtion.

Practice Exercise 9.
$\left(-2\frac{1}{2}\right) + \left(+3\frac{1}{4}\right) - \left(+4\frac{1}{5}\right)$

EXAMPLE 9 Perform the indicated operations:
$$(-5) - (-4) - (+2) - (-7) + (-6)$$

SOLUTION:
$$\begin{aligned}
&(-5) - (-4) - (+2) - (-7) + (-6) \\
&= (-5) + (+4) + (-2) + (+7) + (-6) \\
&= [(-5) + (-2) + (-6)] + [(+4) + (+7)] \\
&= -(5 + 2 + 6) + [+(4 + 7)] \\
&= (-13) + (+11) \\
&= -(13 - 11) \\
&= -2
\end{aligned}$$

ANSWERS TO PRACTICE EXERCISES

8. 0.1.   9. $-3\frac{9}{20}$.

EXERCISES 6.4

In Exercises 1–20, perform the indicated subtractions.

1. $(+6) - (+4)$
2. $(-5) - (-3)$
3. $(+7) - (-3)$
4. $(-9) - (+7)$
5. $(-3) - (+6)$
6. $(-1) - (-6)$
7. $(+5) - 0$
8. $0 - (+2)$
9. $0 - (-3)$
10. $(-11) - (-9)$
11. $(+17) - (-13)$
12. $(+21) - (-29)$

13. $(-41) - (-39)$
14. $(+57) - (-81)$
15. $(-103) - (+86)$
16. $(+210) - (-26)$
17. $(-113) - (+71)$
18. $(+609) - (+492)$
19. $(-517) - (+621)$
20. $(-831) - (-611)$

In Exercises 21–30, subtract the given decimals.

21. $(+2.31) - (+4.09)$
22. $(-12.06) - (-9.6)$
23. $(-8.5) - (-0.58)$
24. $(-60.9) - (-23.33)$
25. $(+7.89) - (-7.89)$
26. $(-345.3) - 0$
27. $0 - (-45.78)$
28. $(-23.54) - (+4.9)$
29. $(+0.93) - (-5.6)$
30. $(-23.56) - (+14.008)$

In Exercises 31–40, subtract the given rational numbers.

31. $\left(\frac{+3}{7}\right) - \left(\frac{-4}{9}\right)$
32. $\left(\frac{-4}{5}\right) - \left(\frac{-6}{11}\right)$
33. $\left(\frac{-4}{7}\right) - \left(\frac{-5}{4}\right)$
34. $\left(\frac{-3}{7}\right) - \left(\frac{-2}{9}\right)$
35. $\left(\frac{-7}{11}\right) - \left(\frac{-5}{7}\right)$
36. $\left(\frac{+10}{17}\right) - \left(\frac{-1}{3}\right)$
37. $\left(\frac{-2}{9}\right) - \left(\frac{-4}{5}\right)$
38. $\left(\frac{-3}{14}\right) - \left(\frac{-1}{5}\right)$
39. $\left(\frac{-4}{13}\right) - \left(\frac{-5}{7}\right)$
40. $\left(\frac{0}{6}\right) - \left(\frac{-13}{19}\right)$

In Exercises 41–50, subtract the indicated mixed numbers.

41. $\left(-4\frac{2}{3}\right) - \left(-3\frac{1}{2}\right)$
42. $\left(+5\frac{1}{7}\right) - \left(-3\frac{2}{9}\right)$
43. $\left(-2\frac{4}{9}\right) - \left(+7\frac{2}{5}\right)$
44. $\left(-2\frac{1}{3}\right) - \left(-3\frac{3}{4}\right)$
45. $\left(-4\frac{3}{5}\right) - \left(-1\frac{3}{4}\right)$
46. $\left(-7\frac{1}{6}\right) - \left(+3\frac{3}{7}\right)$
47. $\left(+12\frac{1}{9}\right) - \left(-23\frac{9}{11}\right)$
48. $\left(-30\frac{10}{15}\right) - \left(-8\frac{2}{7}\right)$
49. $\left(-56\frac{11}{13}\right) - \left(-46\frac{2}{5}\right)$
50. $\left(+87\frac{3}{11}\right) - \left(-64\frac{5}{9}\right)$

In Exercises 51–70, perform the indicated operations.

51. $(-1.1) + (+1.3) - (-1.7)$
52. $(+22) - (-37) + (-16)$
53. $(-30) - (-10) - (-40)$
54. $(+6.9) + (-3.2) - (-1.9)$
55. $(-0.57) - (+3.9) - (-23)$
56. $(-10.1) + (+2.6) - (-1.96)$
57. $(-2.03) - (+30.4) - (-40.5)$
58. $(-6.37) - (-19.6) + (-2.97)$
59. $(+11) + (-13) - (-21) + (-64)$
60. $(+99) - (+88) - (+77) - (+66)$
61. $(-23.1) - (+17.98) - (-3.009)$
62. $(+0.987) - (-9.65) + (-23.1)$
63. $(+2.67) + (-34.1) - (-4.057)$
64. $(-9.09) - (+8.08) - (-7.07)$
65. $\left(\frac{-9}{11}\right) + \left(\frac{-7}{10}\right) - \left(\frac{-6}{7}\right)$
66. $\left(\frac{+7}{8}\right) - \left(\frac{-9}{13}\right) - \left(\frac{+3}{4}\right)$
67. $\left(\frac{-2}{7}\right) - \left(\frac{-5}{8}\right) + \left(\frac{-7}{9}\right) - \left(\frac{-9}{16}\right)$
68. $\left(+4\frac{2}{3}\right) - \left(-5\frac{7}{9}\right) - \left(+7\frac{4}{7}\right)$

69. $\left(-6\frac{1}{2}\right) - \left(-8\frac{2}{3}\right) + \left(+7\frac{3}{4}\right) - \left(-5\frac{4}{5}\right)$

70. $\left(-5\frac{2}{7}\right) - \left(+8\frac{4}{11}\right) + \left(-9\frac{1}{2}\right) - \left(+3\frac{13}{12}\right)$

71. Subtract $+23$ from -19.
72. Subtract -4.2 from -3.7.
73. Subtract -28 from 49.
74. Subtract -2.3 from -0.36.
75. From $\frac{-4}{7}$, subtract $\frac{-5}{11}$.
76. From $\frac{12}{13}$, subtract $\frac{-7}{8}$.

77. What is the distance between -13 and 16 on the number line?
78. If -26 is subtracted from the sum of 39 and -56, what is the result?
79. The top of a high-rise building is 267 ft above ground level. The floor of a sub-basement of the building is 35 ft below ground level. What is the distance between the top of the building and the floor of the sub-basement?
80. A helicopter is hovering directly above a submarine. If the helicopter is 335 m above the water level and the submarine is 82 m below the water level, what is the distance between the two?
81. Liz starts with $5\frac{2}{3}$ c of flour for a particular recipe. She adds to it $4\frac{1}{2}$ c of flour. Later, she uses $3\frac{1}{4}$ c of the flour for part of the recipe. How much of the flour does she have left?
82. During a gasoline price war, a dealer starts with 2098.5 gal of gasoline, sells 1876.4 gal, receives a shipment of 1530.8 gal, and sells an additional 1673.2 gal of gasoline. How much gasoline does she have left to sell?
83. A new car sold for $11,267.85. The buyer of the car received a trade-in allowance of $4895. Dealer preparation costs to the buyer were $324.60. Registration of the new car cost $43.85. Sales tax amounted to $411.37. If the buyer also received a manufacturer's rebate of $700 and paid cash for the balance due, how much did the buyer pay?
84. The sum of 34.56, -67.8, and -19.09 is subtracted from the sum of -87.06, 46.7, and -32.19. What is the result?
85. The sum of $-9\frac{4}{7}$, $-7\frac{3}{5}$, and $10\frac{2}{3}$ is subtracted from the sum of $-45\frac{2}{9}$ and $39\frac{3}{4}$. What is the result?

6.5 Multiplication of Signed Numbers

OBJECTIVE

After completing this section, you should be able to multiply any two signed numbers, using the appropriate rules for multiplication of signed numbers.

In examining the multiplication of two signed numbers, we must consider the following cases:

1. At least one of the numbers is 0.
2. The two numbers are both positive.
3. One number is positive and the other number is negative.
4. The two numbers are both negative.

We can consider the first case very quickly. In Chapter 1, we noted that if a whole number is multiplied by 0 or 0 is multiplied by a whole number, the product is 0. This is the multiplication property of 0. The rule extends to signed numbers as well.

Rule for Multiplication of Signed Numbers Involving 0

If a is any signed number, then

$$a \times 0 = 0 \quad \text{and} \quad 0 \times a = 0$$

- $(+3) \times 0 = 0$
- $(-2.31) \times 0 = 0$
- $0 \times \left(\dfrac{-4}{5}\right) = 0$
- $0 \times \left(-3\dfrac{1}{7}\right) = 0$

Since a positive integer is also a counting (or natural) number, the product of two positive integers is always positive. That is, we multiply two positive integers as counting numbers. This rule also extends to other signed numbers. That is, the product of any two positive numbers is always positive.

Multiply.

Practice Exercise 1.
$(+7) \times (+9)$

EXAMPLE 1 Multiply: $(+2) \times (+3)$.

SOLUTION: $(+2) \times (+3) = 2 \times 3 = 6 = +6$

Practice Exercise 2.
$(+11) \times (+1.2)$

EXAMPLE 2 Multiply: $(+2.3) \times (+3.1)$.

SOLUTION: $(+2.3) \times (+3.1) = 2.3 \times 3.1 = 7.13 = +7.13$

Practice Exercise 3.
$\left(\dfrac{+2}{7}\right) \times \left(\dfrac{+4}{9}\right)$

EXAMPLE 3 Multiply: $\left(\dfrac{+2}{3}\right) \times \left(\dfrac{+4}{7}\right)$.

SOLUTION: $\left(\dfrac{+2}{3}\right) \times \left(\dfrac{+4}{7}\right) = \dfrac{2}{3} \times \dfrac{4}{7} = \dfrac{8}{21} = \dfrac{+8}{21}$

Practice Exercise 4.
$\left(+1\dfrac{2}{3}\right) \times \left(+2\dfrac{1}{6}\right)$

EXAMPLE 4 Multiply: $\left(+3\dfrac{1}{2}\right) \times \left(+4\dfrac{1}{3}\right)$.

SOLUTION: $\left(+3\dfrac{1}{2}\right) \times \left(+4\dfrac{1}{3}\right) = 3\dfrac{1}{2} \times 4\dfrac{1}{3} = 15\dfrac{1}{6} = +15\dfrac{1}{6}$

To obtain a rule for multiplying two signed numbers with different signs, recall that multiplication can be thought of as repeated addition. For instance, 4×3 means the sum of four groups of 3, or

$$4 \times 3 = 3 + 3 + 3 + 3 = 12$$

In a similar manner, $(+4) \times (-3)$ can be thought of as being the sum of four (-3)s, or

$$(+4) \times (-3) = (-3) + (-3) + (-3) + (-3) = -12$$

6.5 Multiplication of Signed Numbers

To find the product of $(-4) \times (+3)$, we note that the order in which we multiply two numbers is not important; the product will be the same. Hence:

$$(-4) \times (+3) = (+3) \times (-4)$$
$$= (-4) + (-4) + (-4)$$
$$= -12$$

We note, then, that the product of a positive integer and a negative integer is always negative. That is, we multiply a positive integer and a negative integer as though they were counting numbers but make their product negative.

Practice Exercise 5.
$(-23) \times (+12)$

EXAMPLE 5 Multiply: $(+5) \times (-6)$.

SOLUTION: $(+5) \times (-6) = -(5 \times 6) = -30$

Practice Exercise 6.
$(+23) \times (-14)$

EXAMPLE 6 Multiply: $(+8) \times (-11)$.

SOLUTION: $(+8) \times (-11) = -(8 \times 11) = -88$

Practice Exercise 7.
$(-39) \times (+12)$

EXAMPLE 7 Multiply: $(-9) \times (+12)$.

SOLUTION: $(-9) \times (+12) = -(9 \times 12) = -108$

Practice Exercise 8.
$(+43) \times (-17)$

EXAMPLE 8 Multiply: $(-11) \times (+50)$.

SOLUTION: $(-11) \times (+50) = -(11 \times 50) = -550$

This rule for multiplying integers with different signs also extends to the other signed numbers.

Practice Exercise 9.
$(+16.1) \times (-0.87)$

EXAMPLE 9 Multiply: $(+1.2) \times (-1.3)$.

SOLUTION: $(+1.2) \times (-1.3) = -(1.2 \times 1.3) = -1.56$

Practice Exercise 10.
$\left(\dfrac{-4}{9}\right) \times \left(\dfrac{+5}{7}\right)$

EXAMPLE 10 Multiply: $\left(\dfrac{-2}{7}\right) \times \left(\dfrac{+4}{9}\right)$.

SOLUTION: $\left(\dfrac{-2}{7}\right) \times \left(\dfrac{+4}{9}\right) = -\left(\dfrac{2}{7} \times \dfrac{4}{9}\right) = \dfrac{-8}{63}$

Practice Exercise 11.
$\left(+10\dfrac{1}{2}\right) \times \left(-7\dfrac{1}{3}\right)$

EXAMPLE 11 Multiply: $\left(+5\dfrac{1}{3}\right) \times \left(-6\dfrac{3}{4}\right)$.

SOLUTION: $\left(+5\dfrac{1}{3}\right) \times \left(-6\dfrac{3}{4}\right) = -\left(5\dfrac{1}{3} \times 6\dfrac{3}{4}\right) = -36$

Practice Exercise 12.
$(-13) \times \left(\dfrac{+7}{9}\right)$

EXAMPLE 12 Multiply: $\left(-\dfrac{3}{8}\right) \times \left(+4\dfrac{1}{7}\right)$.

SOLUTION: $\left(\dfrac{-3}{8}\right) \times \left(+4\dfrac{1}{7}\right) = -\left(\dfrac{3}{8} \times 4\dfrac{1}{7}\right) = -1\dfrac{31}{36}$

ANSWERS TO PRACTICE EXERCISES

1. $+63$. 2. $+13.2$. 3. $\frac{+8}{63}$. 4. $+3\frac{11}{18}$. 5. -276. 6. -322. 7. -468. 8. -731.
9. -14.007. 10. $\frac{-20}{63}$. 11. -77. 12. $-10\frac{1}{9}$.

What is the product of two negative integers? To determine the product of (-2) and (-3), consider the following problem:

$$(-2)[(-3) + (+3)] \quad \text{(The sum of an integer and its}$$
$$= (-2)(0) \quad \text{opposite is 0.)}$$

$$= 0 \quad \text{(Multiplication of integers involving 0)}$$

We will now solve the same problem using the distributive property of multiplication over addition. We have:

$$(-2)[(-3) + (+3)]$$
$$= [(-2)(-3)] + [(-2)(+3)] \quad \text{(Distributive property)}$$
$$= [(-2)(-3)] + (-6) \quad \text{(Product of integers with different signs)}$$

Now, we already know (from above) that the answer to this problem is 0. Hence,

$$[(-2)(-3)] + (-6) = 0$$
or
$$? \quad + (-6) = 0$$

But the only integer that can be substituted for the question mark is $+6$. (Do you agree?) Therefore,

$$(-2)(-3) = +6$$

Multiply.

Practice Exercise 13.
$(-11)(-23)$

EXAMPLE 13 Multiply: $(-4)(-5)$.

SOLUTION:
$$(-4)[(-5) + (+5)] = (-4)(0)$$
$$= 0$$

Also, $(-4)[(-5) + (+5)] = [(-4)(-5)] + [(-4)(+5)]$
$$= [(-4)(-5)] + (-20)$$
$$= ? + (-20)$$

But since $? + (-20) = 0$, we have that $? = (+20)$. Therefore, $(-4)(-5) = +20$.

Practice Exercise 14.
$(-32)(-17)$

EXAMPLE 14 Multiply: $(-7)(-8)$.

SOLUTION:
$$(-7)[(-8) + (+8)] = (-7)(0)$$
$$= 0$$

Also, $(-7)[(-8) + (+8)] = [(-7)(-8)] + [(-7)(+8)]$
$$= [(-7)(-8)] + (-56)$$
$$= ? + (-56)$$

But since $? + (-56) = 0$, we have that $? = (+56)$. Therefore, $(-7)(-8) = +56$.

ANSWERS TO PRACTICE EXERCISES

13. $+253$. 14. $+544$.

6.5 Multiplication of Signed Numbers

The preceding examples illustrate the rule for multiplying two negative signed numbers. That is, the product of two negative signed numbers is *always* positive. We will now summarize our rules for multiplying signed numbers.

Rules for Multiplying Two Signed Numbers

1. If either number is 0, the product is 0.
2. If the two numbers have the *same* sign, multiply their absolute values and make the product *positive*.
3. If the two numbers have *different* signs, multiply their absolute values and make the product *negative*.

Multiply, using the appropriate rules.

Practice Exercise 15.
$(+23) \times (-11)$

EXAMPLE 15 Multiply: $(+4) \times (+6)$

SOLUTION: $(+4) \times (+6) = +(4 \times 6) = +24$ (Same signs)

Practice Exercise 16.
$(-3.5) \times (-2.7)$

EXAMPLE 16 Multiply: $(-5) \times (+0.8)$.

SOLUTION: $(-5) \times (+0.8) = -(5 \times 0.8) = -4$ (Different signs)

Practice Exercise 17.
$(-43) \times (+1.5)$

EXAMPLE 17 Multiply: $(+7.03) \times 0$.

SOLUTION: $(+7.03) \times 0 = 0$ (One factor is 0)

Practice Exercise 18.
$\left(\frac{-8}{9}\right) \times \left(\frac{-3}{4}\right)$

EXAMPLE 18 Multiply: $(-0.6) \times (-9)$.

SOLUTION: $(-0.6) \times (-9) = +(0.6 \times 9) = +5.4$ (Same signs)

Practice Exercise 19.
$(-17.1) \times (-21.6)$

EXAMPLE 19 Multiply: $(-5.1) \times (-7.2)$.

SOLUTION: $(-5.1) \times (-7.2) = +(5.1 \times 7.2) = +36.72$

Practice Exercise 20.
$\left(-2\frac{3}{8}\right) \times \left(+3\frac{1}{3}\right)$

EXAMPLE 20 Multiply: $\left(\frac{-4}{5}\right) \times \left(\frac{+3}{8}\right)$.

SOLUTION: $\left(\frac{-4}{5}\right) \times \left(\frac{+3}{8}\right) = -\left(\frac{4}{5} \times \frac{3}{8}\right) = \frac{-3}{10}$

Practice Exercise 21.
$(-2.36) \times \left(-4\frac{1}{4}\right)$

EXAMPLE 21 Multiply: $\left(-5\frac{1}{2}\right) \times \left(-6\frac{2}{3}\right)$.

SOLUTION: $\left(-5\frac{1}{2}\right) \times \left(-6\frac{2}{3}\right) = +\left(5\frac{1}{2} \times 6\frac{2}{3}\right) = +36\frac{2}{3}$

ANSWERS TO PRACTICE EXERCISES

15. -253.   16. $+9.45$.   17. -64.5.   18. $\frac{+2}{3}$.   19. 369.36.   20. $-7\frac{11}{12}$.   21. -10.03.

In the following examples, we will consider multiplying three or more signed numbers.

Multiply, using the appropriate rules.

Practice Exercise 22.
$(+4) \times (-3) \times (-9)$

EXAMPLE 22 Multiply: $(+8) \times (-6) \times (+5)$.

SOLUTION:

$$
\begin{aligned}
&(+8) \times (-6) \times (+5) \\
&= -(8 \times 6) \times (+5) \quad \text{(Different signs)} \\
&= (-48) \times (+5) \\
&= -(48 \times 5) \quad \text{(Different signs)} \\
&= -240
\end{aligned}
$$

Practice Exercise 23.
$(-17) \times 0 \times (+43)$

EXAMPLE 23 Multiply: $(-3) \times (+4) \times (-5) \times (-6) \times (-2)$.

SOLUTION:

$$
\begin{aligned}
&(-3) \times (+4) \times (-5) \times (-6) \times (-2) \\
&= -(3 \times 4) \times (-5) \times (-6) \times (-2) \quad \text{(Different signs)} \\
&= (-12) \times (-5) \times (-6) \times (-2) \\
&= +(12 \times 5) \times (-6) \times (-2) \quad \text{(Same signs)} \\
&= (+60) \times (-6) \times (-2) \\
&= -(60 \times 6) \times (-2) \quad \text{(Different signs)} \\
&= (-360) \times (-2) \\
&= +(360 \times 2) \quad \text{(Same signs)} \\
&= +720
\end{aligned}
$$

An *even* number (such as 0, 2, 6, or 32) is a whole number that is divisible by 2. An *odd* number (such as 1, 7, 11, or 29) is a whole number that is not even. Note, in the preceding examples, that if there is an *odd* number of negative factors in a product, the product is *negative* (unless, of course, there is a factor of 0 in the product).

When multiplying signed numbers, we can ignore the signs and multiply their absolute values.

1. The product will be positive if there is an even number of negative factors (and no 0 factor).
2. The product will be negative if there is an odd number of negative factors (and no 0 factor).
3. The product will be 0 if at least one of the factors is 0.

Practice Exercise 24. $(-2.9) \times (-13.1) \times (-7) \times (+2)$

EXAMPLE 24 Multiply: $(-2) \times (-3) \times (+2) \times (-4) \times (+5)$.

SOLUTION: There are 3 (which is odd) negative factors in the product and no 0 factor; the product will be negative. We have

$$
\begin{aligned}
&(-2) + (-3) \times (+2) \times (-4) \times (+5) \\
&= -(2 \times 3 \times 2 \times 4 \times 5) \\
&= -240
\end{aligned}
$$

Practice Exercise 25.
$\left(\dfrac{-2}{3}\right) \times \left(\dfrac{+6}{7}\right) \times \left(\dfrac{+7}{15}\right)$

EXAMPLE 25 Multiply: $(-0.2) \times (+1.1) \times (+2.0) \times (-3.2)$.

SOLUTION: There are 2 (which is even) negative factors in the product and no 0 factor; the product will be positive. We have

6.5 Multiplication of Signed Numbers

$$(-0.2) \times (+1.1) \times (+2.0) \times (-3.2)$$
$$= +(0.2 \times 1.1 \times 2.0 \times 3.2)$$
$$= +1.408$$

> **ANSWERS TO PRACTICE EXERCISES**
>
> 22. 108.   23. 0.   24. -531.86.   25. $\dfrac{-4}{15}$.

EXERCISES 6.5

In Exercises 1–70, perform the indicated operations using the appropriate rules. Square brackets are used as additional symbols for grouping. Simplify within each square bracket first.

1. $(+6) \times (+4)$
2. $(+11) \times (+9)$
3. $(-5) \times (-9)$
4. $(-8) \times (-7)$
5. $(+7) \times (-8)$
6. $(-6) \times (+9)$
7. $(-1.1) \times (+13)$
8. $(+22) \times (+17)$
9. $(-39) \times 0$
10. $0 \times (-0.43)$
11. $(-29) \times (-12)$
12. $(+57) \times (+23)$
13. $(-87) \times (-16)$
14. $(-96) \times (+17)$
15. $(-0.7) \times (-0.6) \times (+8)$
16. $(+4) \times (-11) \times (+17)$
17. $(+11) \times (+12) \times (+10)$
18. $(-17) \times (-9) \times (-11)$
19. $(-9) \times (-19) \times (+29)$
20. $(-15) \times (+25) \times (+3.5)$
21. $(-10) \times (-10) \times (-10)$
22. $(-22) \times (-33) \times (-23)$
23. $(+69) \times (-11) \times (+40)$
24. $(-80) \times (-75) \times (-60)$
25. $(+7) \times (-5) \times (-4) \times (+3)$
26. $(-6) \times (-8) \times (-9) \times (+11)$
27. $(+23) \times [(-3.9) - (-5.8)]$
28. $[(+17) \times (-23)] \times (-11)$
29. $[(+12) \times (-13)] \times [(+23) \times (-36)]$
30. $[(+47) - (+35)] \times [(+16) - (+21)]$
31. $(-2.3) \times (+3.7)$
32. $(+5.01) \times (-3.2)$
33. $(-0.03) \times (-0.02)$
34. $(+9.1) \times (0.56)$
35. $(-19.6) \times (+1.23)$
36. $(+0.69) \times (+23.9)$
37. $(+1.53) \times (-23.2)$
38. $(-17.6) \times (+0.134)$
39. $(-49.7) \times (-3.64)$
40. $(+0.187) \times (+100.3)$
41. $(-0.02) \times (-0.02) \times (-0.04)$
42. $(-1.2) \times (+2.3) \times (-4.67)$
43. $(-27.61) \times (-1.793) \times (-1000)$
44. $(+0.812) \times (-0.16) \times (+1.7)$
45. $(-1.7) \times (26.01) \times (-0.05)$
46. $(-1.07) \times (-3.9) \times (-45.09)$
47. $[(-2.1) - (3.6)] \times (-9.1)$
48. $[(-0.6) \times (-7.5)] \times (-1.03)$
49. $(-8.01) \times [(-7.1) - (-6.21)]$
50. $[(-0.91) - (-1.8)] \times [(+3.9) \times (-7.1)]$
51. $\left(\dfrac{+1}{2}\right) \times \left(\dfrac{-1}{6}\right)$
52. $\left(\dfrac{-2}{3}\right) \times \left(\dfrac{-4}{5}\right)$
53. $\left(\dfrac{-2}{7}\right) \times \left(\dfrac{+4}{3}\right)$
54. $\left(\dfrac{-1}{5}\right) \times \left(\dfrac{-2}{7}\right)$
55. $\left(\dfrac{+4}{13}\right) \times \left(\dfrac{+5}{9}\right)$
56. $\left(\dfrac{+3}{7}\right) \times \left(\dfrac{-2}{11}\right)$
57. $\left(\dfrac{0}{6}\right) \times \left(\dfrac{-11}{17}\right)$
58. $\left(\dfrac{-14}{15}\right) \times \left(\dfrac{-10}{21}\right)$

59. $\left(\frac{-1}{2}\right) \times \left(\frac{-1}{3}\right) \times \left(\frac{-1}{4}\right)$ 60. $\left(\frac{-2}{3}\right) \times \left(\frac{-1}{7}\right) \times \left(\frac{+6}{11}\right)$

61. $\left[\left(\frac{-2}{3}\right) + \left(\frac{+4}{5}\right)\right] \times \left(\frac{-3}{4}\right)$ 62. $\left(\frac{+1}{4}\right) \times \left[\left(\frac{-1}{7}\right) - \left(\frac{+2}{9}\right)\right]$

63. $\left(+3\frac{1}{2}\right) \times \left(-4\frac{2}{3}\right)$ 64. $\left(-3\frac{1}{7}\right) \times \left(-5\frac{1}{9}\right)$

65. $\left(+7\frac{1}{6}\right) \times \left(-2\frac{4}{5}\right)$ 66. $\left(+3\frac{1}{3}\right) \times \left(+2\frac{1}{4}\right)$

67. $\left(-6\frac{1}{2}\right) \times \left(+4\frac{2}{7}\right)$ 68. $\left(-5\frac{1}{3}\right) \times \left(-7\frac{1}{2}\right)$

69. $\left(-1\frac{3}{4}\right) \times \left[\left(+4\frac{1}{2}\right) + \left(+3\frac{2}{3}\right)\right]$ 70. $\left(-7\frac{1}{3}\right) \times \left[\left(-2\frac{3}{4}\right) - \left(-5\frac{1}{6}\right)\right]$

71. The sum of -26 and $+12$ is multiplied by -15. What is the result?
72. The product of -17 and $+23$ is subtracted from -16. What is the result?
73. The sum of -9 and $+2.1$ is multiplied by the sum of -7 and 3.6. What is the result?
74. The product of $-4\frac{1}{5}$ and $+6\frac{2}{3}$ is subtracted from -28. What is the result?
75. The sum of $+2\frac{1}{3}$ and $-4\frac{1}{4}$ is multiplied by the difference, $\left(-5\frac{1}{2}\right) - (7)$. What is the result?

6.6 Division of Signed Numbers

OBJECTIVE

After completing this section, you should be able to divide a signed number by a nonzero number, using the appropriate rules for dividing signed numbers.

You have already learned to add, subtract, and multiply signed numbers, using the appropriate rules for doing so. We will now examine the division of signed numbers. We must consider the following cases:

1. The numerator is 0 and the denominator is nonzero.
2. Both the numerator and denominator have the same sign.
3. The numerator and denominator have different signs.

Since division and multiplication are inverse operations, any division problem can be rewritten as an equivalent multiplication problem, as we have done in previous chapters. Therefore, the rules for dividing signed numbers are similar to those for multiplying signed numbers.

6.6 Division of Signed Numbers

Rules for Dividing Signed Numbers

1. If the numerator is 0 and the denominator is nonzero, the quotient is 0.
2. If the numerator and denominator have the *same* sign, divide their absolute values and make the quotient *positive*.
3. If the numerator and denominator have *different* signs, divide their absolute values and make the quotient *negative*.
4. Division by 0 is not defined.

Divide, using the appropriate rules.

Practice Exercise 1.
$(-12) \div (+4)$

EXAMPLE 1 Divide: $(+12) \div (+3)$.

SOLUTION: $(+12) \div (+3) = +(12 \div 3) = +4$ (Same signs)

Practice Exercise 2.
$0 \div (-0.23)$

EXAMPLE 2 Divide: $(-10) \div (-2)$.

SOLUTION: $(-10) \div (-2) = +(10 \div 2) = +5$ (Same signs)

Practice Exercise 3.
$(-36) \div (-9)$

EXAMPLE 3 Divide: $0 \div (-17)$.

SOLUTION: $0 \div (-17) = 0$ (Numerator = 0, denominator ≠ 0)

Practice Exercise 4.
$(-1.8) \div (+0.04)$

EXAMPLE 4 Divide: $(-18) \div (+3)$.

SOLUTION: $(-18) \div (+3) = -(18 \div 3) = -6$ (Different signs)

Practice Exercise 5.
$(-25) \div (-8)$

EXAMPLE 5 Divide: $(+28) \div (-4)$.

SOLUTION: $(+28) \div (-4) = -(28 \div 4) = -7$ (Different signs)

So far in our discussion of integers, we have noted that the sum, difference, or product of any two integers is always an integer. That is not true with the quotient of integers. In Examples 1–5, the quotient was an integer. In Examples 6–9, the quotient is not an integer.

Practice Exercise 6.
$(+0.26) \div (+1.6)$

EXAMPLE 6 Divide: $(-18) \div (+5)$.

SOLUTION: $(-18) \div (+5) = -(18 \div 5) = -3.6$ (Different signs)

Practice Exercise 7.
$(-81) \div (-162)$

EXAMPLE 7 Divide: $(+38) \div (-10)$.

SOLUTION: $(+38) \div (-10) = -(38 \div 10) = -3.8$ (Different signs)

Practice Exercise 8.
$(+14) \div (-4)$

EXAMPLE 8 Divide: $(-2) \div (-4)$.

SOLUTION: $(-2) \div (-4) = +(2 \div 4) = +0.5$ (Same signs)

Practice Exercise 9.
$(+242) \div (+16)$

EXAMPLE 9 Divide; $(+19) \div (+8)$.

SOLUTION: $(+19) \div (+8) = +(19 \div 8) = +2.375$ (Same signs)

ANSWERS TO PRACTICE EXERCISES

1. -3. 2. 0. 3. 4. 4. -45. 5. 3.125. 6. 0.1625. 7. 0.5. 8. -3.5. 9. 15.125.

Except for determining the sign of the quotient, we divide rational numbers as we do fractions. That is, we change the operation of division to multiplication and use the reciprocal of the divisor.

Divide, using the appropriate rules.

Practice Exercise 10.
$\left(\dfrac{-2}{5}\right) \div \left(\dfrac{-3}{7}\right)$

EXAMPLE 10 Divide: $\left(\dfrac{+2}{3}\right) \div \left(\dfrac{+5}{8}\right)$.

SOLUTION:
$$\left(\dfrac{+2}{3}\right) \div \left(\dfrac{+5}{8}\right) = \left(\dfrac{+2}{3}\right) \times \left(\dfrac{+8}{5}\right)$$
$$= \dfrac{(+2) \times (+8)}{3 \times 5}$$
$$= \dfrac{+(2 \times 8)}{15}$$
$$= \dfrac{+16}{15} \left(\text{or } +1\dfrac{1}{15}\right)$$

Practice Exercise 11.
$\left(\dfrac{0}{-6}\right) \div \left(\dfrac{-3}{4}\right)$

EXAMPLE 11 Divide: $\left(\dfrac{-4}{7}\right) \div \left(\dfrac{-5}{6}\right)$.

SOLUTION:
$$\left(\dfrac{-4}{7}\right) \div \left(\dfrac{-5}{6}\right) = \left(\dfrac{-4}{7}\right) \times \left(\dfrac{-6}{5}\right)$$
$$= \dfrac{(-4) \times (-6)}{7 \times 5}$$
$$= \dfrac{+(4 \times 6)}{35}$$
$$= \dfrac{+24}{35}$$

Practice Exercise 12.
$\left(\dfrac{-4}{7}\right) \div \left(\dfrac{+5}{9}\right)$

EXAMPLE 12 Divide; $\left(\dfrac{0}{-2}\right) \div \left(\dfrac{-3}{4}\right)$.

SOLUTION:
$$\left(\dfrac{0}{-2}\right) \div \left(\dfrac{-3}{4}\right) = \left(\dfrac{0}{-2}\right) \times \left(\dfrac{-4}{3}\right)$$
$$= \dfrac{0 \times (-4)}{(-2) \times 3}$$
$$= \dfrac{0}{-(2 \times 3)}$$
$$= \dfrac{0}{-6}$$
$$= 0$$

We divide signed mixed numbers by first rewriting the numbers as rational numbers.

Practice Exercise 13.
$\left(-4\dfrac{1}{5}\right) \div \left(-2\dfrac{3}{4}\right)$

EXAMPLE 13 Divide; $\left(-10\dfrac{1}{2}\right) \div \left(+2\dfrac{2}{3}\right)$.

SOLUTION:
$$\left(-10\dfrac{1}{2}\right) \div \left(+2\dfrac{2}{3}\right) = \left(\dfrac{-21}{2}\right) \div \left(\dfrac{+8}{3}\right)$$

$$= \left(\frac{-21}{2}\right) \times \left(\frac{+3}{8}\right)$$

$$= \frac{(-21) \times (+3)}{2 \times 8}$$

$$= \frac{-(21 \times 3)}{16}$$

$$= \frac{-63}{16}$$

$$= -3\frac{15}{16}$$

ANSWERS TO PRACTICE EXERCISES

10. $\frac{+14}{15}$. 11. 0 12. $\frac{-36}{35}$ or $-1\frac{1}{35}$. 13. $+1\frac{29}{55}$.

EXERCISES 6.6

In Exercises 1–56, perform the indicated operations, using the appropriate rules.

1. $(-6) \div (+2)$
2. $(-9) \div (-3)$
3. $(+1.6) \div (-4)$
4. $(-25) \div (+0.5)$
5. $0 \div (-3)$
6. $(+81) \div (-9)$
7. $(-0.12) \div (+5)$
8. $(-63) \div (-0.08)$
9. $(-42) \div (-3)$
10. $(+57) \div (-3)$
11. $(-100) \div (+16)$
12. $(+121) \div (-8)$
13. $(+24.9) \div (+0.3)$
14. $0 \div (-16)$
15. $(-57) \div (-16)$
16. $(-0.361) \div (-3.2)$
17. $(-162) \div (+27)$
18. $0 \div (-51)$
19. $(-9.5) \div (+20)$
20. $(-0.21) \div (-40)$
21. $[(+2) + (+10)] \div [(-8) + (+2)]$
22. $[(-5) - (+1)] \div [(-4) - (-2)]$
23. $[(+4) \times (-6)] \div [(-16) \div (-4)]$
24. $[(+10) - (-20)] \div [(-0.2) \times (+0.3)]$
25. $[(+6.4) \div (-8)] \div [(-16) \div (-0.4)]$
26. $[(-32) \div (+8)] \div [(-17) - (-9)]$
27. $\left(\frac{1}{3}\right) \div \left(\frac{-1}{4}\right)$
28. $\left(\frac{-4}{7}\right) \div \left(\frac{+5}{6}\right)$
29. $\left(\frac{-5}{6}\right) \div \left(\frac{-3}{7}\right)$
30. $\left(\frac{0}{7}\right) \div \left(\frac{-7}{8}\right)$
31. $\left(\frac{-6}{6}\right) \div \left(\frac{-5}{5}\right)$
32. $\left(\frac{2}{3}\right) \div \left(\frac{-5}{6}\right)$
33. $\left(\frac{-3}{7}\right) \div \left(\frac{-1}{2}\right)$
34. $\left(\frac{-37}{23}\right) \div \left(\frac{-23}{37}\right)$
35. $\left(\frac{47}{94}\right) \div \left(\frac{-1}{2}\right)$
36. $\left(\frac{-5}{9}\right) \div \left(\frac{+4}{5}\right)$

37. $\left(\frac{2}{9}\right) \div \left(\frac{3}{4}\right)$

38. $\left(\frac{0}{7}\right) \div \left(\frac{-11}{13}\right)$

39. $\left(2\frac{1}{3}\right) \div \left(-4\frac{1}{4}\right)$

40. $\left(-5\frac{1}{2}\right) \div \left(6\frac{1}{4}\right)$

41. $\left(-8\frac{1}{9}\right) \div \left(-7\frac{1}{6}\right)$

42. $\left(-10\frac{1}{7}\right) \div \left(-6\frac{2}{3}\right)$

43. $\left(-6\frac{1}{5}\right) \div \left(7\frac{1}{3}\right)$

44. $\left(8\frac{2}{3}\right) \div \left(7\frac{1}{9}\right)$

45. $(-14) \div \left(-3\frac{1}{2}\right)$

46. $\left(5\frac{1}{8}\right) \div (-11)$

47. $\left(\frac{-2}{3}\right) \div \left(4\frac{1}{5}\right)$

48. $\left(-7\frac{1}{2}\right) \div \left(\frac{-4}{5}\right)$

49. $\left(\frac{-2}{7}\right) \div \left[\left(\frac{-1}{3}\right) + \left(\frac{2}{7}\right)\right]$

50. $\left[\left(\frac{1}{5}\right) - \left(\frac{-2}{3}\right)\right] \div \left(\frac{-1}{4}\right)$

51. $\left[\left(-3\frac{1}{2}\right) - \left(4\frac{1}{5}\right)\right] \div \left(\frac{-1}{3}\right)$

52. $\left(\frac{1}{6}\right) \div \left[\left(-1\frac{1}{2}\right) + \left(5\frac{2}{3}\right)\right]$

53. $[(-21) + (36)] \div \left(-7\frac{1}{3}\right)$

54. $\left[\left(2\frac{1}{2}\right) - (-12)\right] \div \left(\frac{-4}{7}\right)$

55. $[12 \div (-8)] \div [(-25) \div (-4)]$

56. $[(-13) \div 16] \div [(-17) + (-8)]$

57. The product of two integers is -2106. One of the integers is -26. What is the other?

58. The product of two integers is 2520. One of the integers is -35. What is the other?

59. The product of two numbers is -12. One number is $-4\frac{1}{2}$. What is the other?

60. The sum of -23 and -16 is divided by $\frac{-2}{3}$. What is the result?

61. The quotient when $-7\frac{1}{2}$ is divided by a number is $2\frac{1}{3}$. What is the number?

62. What is the quotient when the product of $\frac{-2}{3}$ and $4\frac{1}{2}$ is divided by -32?

63. What is the quotient when the sum of $2\frac{1}{3}$ and $-5\frac{1}{2}$ is divided by $\frac{-4}{7}$?

64. What is the quotient when the sum of -17.1 and 23.4 is divided by the sum of 23 and -15?

6.7 Simplest Form of a Rational Number

OBJECTIVE

After completing this section, you should be able to write a rational number in its simplest form.

Most of the rules that you learned about fractions apply to rational numbers as well. Recall that if *both* the numerator and the denominator of a fraction are *multiplied* or *divided* by the same *nonzero* whole number, the resulting fraction is equivalent to the original fraction.

6.7 Simplest Form of a Rational Number

Similarly, if both the numerator and the denominator of a rational number are multiplied or divided by a *nonzero integer*, the resulting rational number is equivalent to the original rational number.

- $\dfrac{2}{-3} = \dfrac{2 \times 2}{-3 \times 2} = \dfrac{4}{-6}$

- $\dfrac{6}{7} = \dfrac{6 \times (-4)}{7 \times (-4)} = \dfrac{-24}{-28}$

- $\dfrac{0}{6} = \dfrac{0 \times (-5)}{6 \times (-5)} = \dfrac{0}{-30}$

- $\dfrac{8}{-24} = \dfrac{8 \div 4}{-24 \div 4} = \dfrac{2}{-6}$

- $\dfrac{-9}{-27} = \dfrac{-9 \div (-3)}{-27 \div (-3)} = \dfrac{3}{9}$

- $\dfrac{3}{-4} = \dfrac{3 \times (-1)}{-4 \times (-1)} = \dfrac{-3}{4}$

In Example 1 of Section 6.2, we noted that the opposite of $\dfrac{-2}{3}$ is $\dfrac{+2}{3}$ or $\dfrac{-2}{-3}$. Is one form preferred over the other? The answer is given in the following definition.

DEFINITION

A rational number is said to be in its **simplest form,** or **simplified,** if:

1. The denominator is *positive*.
2. The numerator and denominator can only be divided by 1 and -1.

The simplest form of $\dfrac{2}{-3}$ is $\dfrac{-2}{3}$. Also, the simplest form of $-\left(\dfrac{2}{3}\right)$ is $\dfrac{-2}{3}$. It should be noted, then, that $\dfrac{2}{-3}$, $\dfrac{-2}{3}$, and $-\left(\dfrac{2}{3}\right)$ all name the same number. That is, regardless of whether a $-$ sign appears only in front of the numerator, or only in front of the denominator, or only in front of the entire fraction, it means the same thing.

Determine the simplest form of the given rational numbers.

Practice Exercise 1. $\dfrac{-16}{24}$

EXAMPLE 1 Determine the simplest form of the rational number $\dfrac{5}{-7}$.

SOLUTION: The simplest form of the rational number $\dfrac{5}{-7}$ is $\dfrac{-5}{7}$.

Practice Exercise 2. $\dfrac{0}{-3}$

EXAMPLE 2 Determine the simplest form of the rational number $\dfrac{-5}{17}$.

SOLUTION: The rational number $\dfrac{-5}{17}$ is in its simplest form.

Practice Exercise 3. $\dfrac{105}{-20}$

EXAMPLE 3 Determine the simplest form of the rational number $\dfrac{-12}{-16}$.

SOLUTION: The simplest form of the rational number $\dfrac{-12}{-16}$ is $\dfrac{3}{4}$.

Practice Exercise 4. $\dfrac{122}{-80}$

EXAMPLE 4 Determine the simplest form of the rational number $\dfrac{0}{-3}$.

SOLUTION: The simplest form of the rational number $\dfrac{0}{-3}$ is $\dfrac{0}{1} = 0$.

ANSWERS TO PRACTICE EXERCISES

1. $\dfrac{-2}{3}$. 2. 0. 3. $\dfrac{-21}{4}$. 4. $\dfrac{-61}{40}$.

EXERCISES 6.7

Determine the simplest form for each of the given rational numbers.

1. $\dfrac{4}{-9}$
2. $\dfrac{0}{-3}$
3. $\dfrac{-4}{12}$
4. $\dfrac{-16}{32}$
5. $\dfrac{56}{-23}$
6. $\dfrac{86}{-40}$
7. $\dfrac{-23}{-69}$
8. $\dfrac{0}{7}$
9. $\dfrac{-405}{-85}$
10. $\dfrac{200}{-60}$
11. $\dfrac{72}{-40}$
12. $\dfrac{-120}{-316}$
13. $\dfrac{0}{-87}$
14. $\dfrac{-10}{200}$
15. $\dfrac{105}{-35}$
16. $\dfrac{75}{-90}$
17. $\dfrac{0}{-31}$
18. $\dfrac{-160}{25}$
19. $\dfrac{70}{-110}$
20. $\dfrac{-68}{-138}$
21. $\dfrac{-17}{68}$
22. $\dfrac{-19}{-152}$
23. $\dfrac{-500}{600}$
24. $\dfrac{-478}{-336}$
25. $\dfrac{-256}{200}$
26. $\dfrac{-618}{-584}$
27. $\dfrac{1098}{-385}$
28. $\dfrac{-3639}{-180}$
29. $\dfrac{-56}{4200}$
30. $\dfrac{8800}{-2900}$

6.8 Mixed Operations with Signed Numbers

OBJECTIVE

After completing this section, you should be able to perform mixed operations with signed numbers, using rules for doing certain operations before others.

To do mixed operations on signed numbers, we use the same rule (Please Excuse My Dear Aunt Sally) as we did with whole numbers (see Section 1.7). However, keep in mind that subtraction of signed numbers can be converted to a corresponding addition problem.

6.8 Mixed Operations with Signed Numbers

Perform the indicated operations.

Practice Exercise 1.
$(-2) \times (-3) + (-4)$

EXAMPLE 1 Perform the indicated operations:

$$(+2) \times (-3)^2 - [(-5) + (-3)] \times [(+4) - (-6)]$$

SOLUTION:

STEP 1: Work within the square brackets first:

$$(+2) \times (-3)^2 - [(-5) + (-3)] \times [(+4) - (-6)]$$
$$= (+2) \times (-3)^2 - [-(5 + 3)] \times [(+4) + (+6)]$$
$$= (+2) \times (-3)^2 - (-8) \times [+(4 + 6)]$$
$$= (+2) \times (-3)^2 - (-8) \times (+10)$$

STEP 2: Evaluate all expressions with exponents:

$$(+2) \times (-3)^2 - (-8) \times (+10)$$
$$= (+2) \times (+9) - (-8) \times (+10)$$

STEP 3: Perform all multiplications and divisions (whichever come first) in order, from left to right:

$$(+2) \times (+9) - (-8) \times (+10)$$
$$= +(2 \times 9) - (-8) \times (+10)$$
$$= (+18) - (-8) \times (+10)$$
$$= (+18) - [-(8 \times 10)]$$
$$= (+18) - (-80)$$

STEP 4: Perform all additions and subtractions (whichever come first) in order, from left to right:

$$(+18) - (-80)$$
$$= (+18) + (+80)$$
$$= +(18 + 80)$$
$$= +98$$

Practice Exercise 2.
$(3.6)^2 - 0 \div (-2.5)^2$

EXAMPLE 2 (a) $(-2)^2 = (-2) \times (-2) = +4$
(b) $-2^2 = -(2 \times 2) = -4$

[Note that $-2^2 \neq (-2)^2$. In the expression $(-2)^2$, we squared (-2) and obtained $+4$. In the expression -2^2, we first squared 2, obtaining $+4$, and then multiplied $+4$ by -1, obtaining -4. Do not confuse the two expressions because they are not equal.]

Practice Exercise 3.
$[-6^2 - (-6)] \div$
$[(-18) \div (+3)]$

EXAMPLE 3 Perform the indicated operations:

$$-3^3 \div (-3)^2 + (2.1)^2 - [(-4.0) \times (-0.8)]^2$$

SOLUTION:

STEP 1: Work within the square brackets first:

$$-3^3 \div (-3)^2 + (2.1)^2 - [(-4.0) \times (-0.8)]^2$$
$$= -3^3 \div (-3)^2 + (2.1)^2 - [+(4.0 \times 0.8)]^2$$
$$= -3^2 \div (-3)^2 + (2.1)^2 - (+3.2)^2$$

STEP 2: Evaluate all expressions with exponents:

$$-3^3 \div (-3)^2 + (2.1)^2 - (3.2)^2$$
$$= -27 \div (+9) + (4.41) - (10.24)$$

STEP 3: Perform all multiplications and divisions (whichever come first) in order, from left to right:

$$-27 \div (+9) + (4.41) - (10.24)$$
$$= -(27 \div 9) + (4.41) - (10.24)$$
$$= (-3) + (4.41) - (10.24)$$

STEP 4: Perform all additions and subtractions (whichever come first) in order, from left to right:

$$(-3) + (4.41) - (10.24)$$
$$= +(4.41 - 3) - (10.24)$$
$$= (+1.41) - (10.24)$$
$$= (+1.41) + (-10.24)$$
$$= -(10.24 - 1.41)$$
$$= -8.83$$

ANSWERS TO PRACTICE EXERCISES

1. +2.   2. 12.96.   3. +5.

EXERCISES 6.8

Perform the indicated operations.

1. $(-3) + (+2) \times (-4)$
2. $(+8) \div (-4) + (-1) \times (+5)$
3. $[(-7) + (+2)] \div [(-2.5) \div (-5)]$
4. $-2^2 - (+3)^3 + (-2)^3$
5. $(-3)^2 \times (-2^3) \times (+4)^2$
6. $(-5) \times (-1)^3 + [(-2) \times (-3)] \div (-6)$
7. $[(+2) \times (-1)]^2 + (-3)$
8. $[(+3) + (-3)] \div (-19) + (-2)$
9. $(-12) \div (-2)^2 + (-3) \times (-1)^3$
10. $[(-3) + (+4)]^3 \times [(+7) - (+9)]^2$
11. $(-0.2) \times (-3) \times (+4) - (-3) \times [(-2) + (+0.4)]$
12. $[(-0.07) + (-0.3)] \div [(-5) + (+3)] \div [(-2.0) \div (-5)]$
13. $[(+3.1) + (-2.9)]^2 + [(+2.4) + (-3.7)]^3 - (-6.3)$
14. $[(+17) - (+3)] \div [(-9) + (+2)] - [0 \div (+23.39)]$
15. $(-3.6) \div [(+5.4) \div (-18)] \times [(-2.0) \div (+5)]$
16. $[(+50) \div (-5)] \div [(-48.6) \div (+24.3)] + (5.9)$
17. $-2^3 \div (-4) + (-1)^3 \times (-5) - (-6)$
18. $[(+0.3)^2 - (+9)] \div (-0.4)^3 + (-7) \times (-0.3)^2$
19. $(-2)^2 \times (+3) \times (-1)^3 \div [(-2) \times (+3)] - (-9)$
20. $[(-60) \div (-2)] \div (-15) - (-2)^2 \times (+3.7) + (-5.9)$
21. $\left(\dfrac{-1}{3}\right) + \left(\dfrac{+1}{2}\right) \times \left(\dfrac{-1}{4}\right)$
22. $\left(\dfrac{+3}{8}\right) \div \left(\dfrac{-2}{5}\right) + \left(\dfrac{-1}{4}\right)$
23. $\left[\left(\dfrac{-3}{7}\right) + \left(\dfrac{+2}{5}\right)\right] \div \left(\dfrac{-3}{4}\right)$
24. $\left(\dfrac{-1}{3}\right)^2 - \left(\dfrac{-1}{2}\right)^3 + \left(\dfrac{-2}{3}\right)^2$
25. $\left(\dfrac{-2}{7}\right)^2 \times \left(\dfrac{-1}{4}\right) + \left(\dfrac{+1}{3}\right)^2$
26. $\left(\dfrac{-2}{9}\right) \times \left(\dfrac{-1}{4}\right)^3 + \left[\left(\dfrac{2}{3}\right)^2 \div \left(\dfrac{-1}{4}\right)\right]$
27. $\left[\left(\dfrac{-3}{7}\right) \times \left(\dfrac{-1}{2}\right)\right]^2$
28. $\left[\left(\dfrac{-2}{5}\right)^2 - \left(\dfrac{-3}{4}\right)\right]^3$

29. $\left(\dfrac{2}{3} - \dfrac{3}{4}\right)^3 \div \dfrac{-2^2}{5}$

30. $\dfrac{-1^3}{4} \times \left(\dfrac{-1}{3}\right)^2 \div \dfrac{-1^5}{5}$

31. $1 \div \left[\left(\dfrac{-3}{7}\right) \times \left(\dfrac{2}{3}\right)\right]^3$

32. $-3^2 - \left[\dfrac{-2^3}{9} \div \dfrac{4}{(-3)^3}\right]$

33. $\left[\dfrac{2}{3} - \left(\dfrac{-3}{4}\right)\right]^3 \div \left[\left(\dfrac{-1}{2}\right) - \left(\dfrac{-3}{8}\right)\right]^2$

34. $\left[\dfrac{1}{2} \div \left(\dfrac{-3}{7}\right)\right]^3 - \left[\left(\dfrac{-2}{9}\right) \times \left(\dfrac{-3^2}{11}\right)\right]$

35. $\left(-3\dfrac{1}{2}\right)^2 \times \left(2\dfrac{1}{3}\right) - \left(-4\dfrac{1}{4}\right)$

36. $\left(4\dfrac{1}{5}\right) - \left(-2\dfrac{1}{2}\right)^2 \div \left(-3\dfrac{1}{3}\right)$

37. $\left(-1\dfrac{1}{4}\right)^3 \div \left(-2\dfrac{1}{3}\right)^2 - \left(\dfrac{-4}{7}\right)$

38. $\left[\left(-4\dfrac{1}{6}\right) - \left(-5\dfrac{1}{3}\right)\right]^2 \div \left(-3\dfrac{1}{2}\right)$

39. $\left(-7\dfrac{1}{2}\right) + \left(-3\dfrac{1}{4}\right)^2 \times \left(2\dfrac{1}{3}\right)^2 \div \left(\dfrac{-2}{5}\right)$

40. $\left[\left(-5\dfrac{1}{2}\right) \div \left(-3\dfrac{3}{4}\right)\right]^2 + \left(2\dfrac{1}{3}\right)^3 \div \left(-1\dfrac{1}{2}\right)$

6.9 Applications Involving Signed Numbers

OBJECTIVE

After completing this section, you should be able to solve simple word problems involving signed numbers.

In this section, we will continue to examine the solution of verbal, or word, problems. These, however, will involve signed numbers. The same strategies used in previous chapters for solving word problems will also be used here.

Practice Exercise 1. For three horse races, a woman lost $35, won $40, and lost $25. What were her net earnings for the three races?

EXAMPLE 1 Suppose that you have $258 in your savings account and that you make two deposits: one for $73 and the other for $119. Three weeks later, you withdraw $185. How much money is left in your savings account? (Disregard interest earned.)

SOLUTION: Note that you start with money in your savings account. You then make two deposits, which means that you *add* these amounts to the amount already in your account. Finally, the withdrawal is *subtracted* from the sum of the original amount and the two deposits. Hence, we have

$258 (Original amount)
  73 (Two deposits)
+119
─────
$450 (New balance after deposits)
−185 (Withdrawal)
─────
$265 (Net balance after withdrawal)

Hence, there will be $265 left in the savings account.

Practice Exercise 2. You have $1100 in your checking account. For how many weeks can you withdraw $35 each week if no additional amount is added to your account?

EXAMPLE 2 At 8:00 A.M., the temperature was 45° F at the New International Airport. At 4:00 P.M. the same day, the temperature at the airport was 73° F. What was the amount of change in the temperature (in degrees Fahrenheit)?

SOLUTION: We are looking for a *difference* in temperature readings. Hence, we must subtract as follows:

$$
\begin{array}{ll}
73° \text{ F} & \text{(Temperature reading at 4:00 P.M.)} \\
-45° \text{ F} & \text{(Temperature reading at 8:00 A.M.)} \\
\hline
28° \text{ F} & \text{(Change in temperature readings)}
\end{array}
$$

Hence, the temperature rose by 28° F during the designated period.

Practice Exercise 3. In Exercise 2, how much money will be left in your account after the last $35 withdrawal?

EXAMPLE 3 A new car lists for $14,637. The dealer preparation charges are $365; taxes amount to $656; and registration fees are $35. The manufacturer of the car gives a $1500 rebate. Joe Smith buys the car and receives a trade-in allowance of $1150 for his old car. How much will Joe actually spend for his new car? (Disregard finance charges.)

SOLUTION: The amount that Joe will actually pay for the new car will be the list price *plus* the dealer preparation charges *plus* the taxes *plus* the registration fees *minus* the manufacturer's rebate *minus* the trade-in allowance for his old car. Hence, we have

$$
\begin{array}{ll}
\$14{,}637 & \text{(List price)} \\
365 & \text{(Dealer preparation charges)} \\
656 & \text{(Taxes)} \\
+\ 35 & \text{(Registration fees)} \\
\hline
\$15{,}693 & \text{(Total costs)} \\
-\ 1{,}500 & \text{(Manufacturer's rebate)} \\
\hline
\$14{,}193 & \\
-\ 1{,}150 & \text{(Trade-in allowance)} \\
\hline
\$13{,}043 & \text{(Joe's actual cost)}
\end{array}
$$

Hence, Joe actually spent $13,043 for his new car.

Practice Exercise 4. Ms. Way's credit card balance was $1236.43. She made a payment of $213.75, charged an additional $416.34, and made another payment of $709.85. What was the new balance on the credit card after these transactions? (Disregard interest charges.)

EXAMPLE 4 Mr. and Mrs. Thomas both work. He earns $27,500 per year and she earns $8400 per year working part-time. They file a joint income tax return claiming $6123 in itemized deductions. They have three dependent children, and each exemption on their tax return is worth $2000. What is their "net" taxable income? (Assume no other deductions.)

SOLUTION: The net taxable income will be Mr. Thomas's income *plus* Mrs. Thomas's income *minus* the itemized deductions *minus* $2000 for each exemption. Notice that $2000 is subtracted for *each* exemption. An exemption is claimed for both Mr. Thomas and Mrs. Thomas, as well as for each of their three dependent children. Hence, we have

$$
\begin{array}{ll}
\$2000 & \text{(For each exemption)} \\
\times\ 5 & \text{(The number of exemptions)} \\
\hline
\$10{,}000 & \text{(Total amount for exemptions)}
\end{array}
$$

Now we determine

$$
\begin{array}{ll}
\$27{,}500 & \text{(Mr. Thomas's income)} \\
+\ 8{,}400 & \text{(Mrs. Thomas's income)} \\
\hline
\$35{,}900 & \text{(Combined incomes)} \\
-\ 6{,}123 & \text{(Itemized deductions)} \\
\hline
\$29{,}777 & \\
-\ 10{,}000 & \text{(Total for exemptions)} \\
\hline
\$19{,}777 & \text{(Net taxable income)}
\end{array}
$$

Hence, the Thomas's net taxable income is $19,777.

6.9 Applications Involving Signed Numbers

Practice Exercise 5. If -39 is subtracted from the sum of $+47$ and a third integer, the difference is -79. What is the third integer?

EXAMPLE 5 The product of two numbers is -9.62. If one number is 2.6, what is the other number?

SOLUTION: Since we are given one of the two numbers and also their product, we must *divide* to determine the other number. We have

$$(\text{One number}) \times (\text{other number}) = \text{product}$$

or

$$(2.6) \times (\text{other number}) = -9.62$$

Therefore,

$$\text{Other number} = -9.62 \div 2.6$$
$$= -3.7$$

ANSWERS TO PRACTICE EXERCISES

1. $-\$20$. 2. 31 weeks. 3. \$15. 4. \$729.17. 5. -165.

EXERCISES 6.9

1. What is the positive difference, in degrees Celsius, between 11 degrees below zero and 23 degrees above zero?
2. During a football game, two plays yielded a loss of 6 yd and a gain of 4 yd. What was the net gain, in yards, for the two plays?
3. Andrea bought a skirt for \$18.95, a blouse for \$13.98, and a purse for \$9.50. She received a discount of \$5 for the purchases. How much did she actually pay for the three items? (Assume no other costs such as taxes.)
4. For three horse races, a man won \$55, lost \$35, and lost \$40. What were his net earnings for the three races?
5. A student opened a savings account with \$100. For each of the next 15 weeks, the student deposited \$25 in the account. During the period, there were three withdrawals for \$12.63, \$37.11, and \$46.82. How much money was in the account at the end of the 15 weeks? (Disregard interest earned.)
6. A person saved \$1349 over a period of 38 weeks. If the person saved the same amount for each of the 38 weeks, how much was saved per week?
7. Michelle started college with \$500 in her non-interest-bearing checking account. If she withdraws \$15 per week, how many such withdrawals can she make and how much will be left in the checking account after the last \$15 withdrawal?
8. A major appliance retails for \$675.79. The dealer grants a discount of \$125 and the manufacturer of the appliance awards a \$50 rebate. If the finance charges amount to \$86.40, what will be the total cost for the appliance, if the purchase is financed?
9. Mr. and Mrs. Hill both work full-time. She earns \$14,900 per year and he earns \$11,200 per year. They file a joint income tax return claiming \$4962 in itemized deductions. They have two dependent children, and each exemption claimed on their tax return is worth \$2000. What is their net taxable income? (Assume no other deductions or exemptions.)
10. A gasoline storage depot has tanks containing 1,000,000 gal of gasoline. From these tanks each week are shipped 2500 gal of gasoline to Dealer A, 3000 gal of gasoline to Dealer B, and 3500 gal of gasoline to Dealer C. For how many weeks can the three

dealers be supplied with gasoline from the depot if no additional gasoline is brought into the depot?

11. A certain share of stock lost, on the average, $5.35 a week for 19 weeks. For each of the next 7 weeks, the stock gained $6.45 per share. What was the net change in the price of the stock during the 26 weeks?

12. The product of two integers is -1053. If one of the integers is $+39$, what is the other integer?

13. If -17 is subtracted from the sum of $+23$ and a third integer, the difference is -11. What is the third integer?

14. If the product of $+56$ and -39 is divided by a third integer, then the quotient is -78. What is the third integer?

15. The Acme Construction Company employs 42 unskilled laborers who are paid $7 per hour and 36 skilled laborers who are paid $11 per hour. If each employee works 40 hr per week, what is the total amount of wages paid the 78 employees per week?

16. The Boiler Works employs 26 people. Each worker is paid $8 per hour with time-and-a-half for each hour in excess of 40 hours per week. If each employee works 50 hr per week, what is the total amount of wages paid the 26 employees per week?

17. The inventory for Part A decreased 12 units per day for 8 days and then increased 7 units per day for 7 days. During the same period of time, the inventory for Part B increased 9 units per day for 6 days and then decreased 5 units per day for 9 days. During this 15-day period, what was the net change in the combined inventory for the two parts?

18. A thief is being pursued by a police officer in a 45-story office building. The police officer enters an elevator on the ground floor at the time that the thief is on the 23rd floor. The thief simultaneously enters an elevator on the 23rd floor, rides it to the 32nd floor, changes elevators, and descends 27 floors. Meanwhile, the police officer gets off on the 22nd floor. How many floors apart are the thief and the police officer?

19. A ball is fired 67.38 m into the air from the top of a mountain and descends onto a ledge that is 85.72 m below the top of the mountain. What is its distance of descent to the ledge?

20. On the average, a secretary in a particular office types 23 letters per week, answers 59 telephone calls per week, and handles 63 pieces of incoming mail per week. During a year, what is the combined number of these three activities performed by the secretary?

21. Mary had a temperature of 98.8° F at the start of one day. During the day, it rose another 3.8° F before falling 2.6° F by the end of the day. What was her temperature at the end of that day?

22. During each of 6 consecutive hours, the temperature in my home town dropped 6.8° F. How much of a change in temperature was there during the 6 hours?

23. The Lynch family owed their credit union $2136.40. They borrowed an additional $1180, repaid $1425, and borrowed another $700. What was the amount owed the credit union after these transactions? (Disregard interest charges.)

24. The cost-of-living index had the following changes during 6 months: a loss of 0.6%, a loss of 0.2%, a gain of 0.5%, a gain of 0.2%, a loss of 0.4%, and a loss of 0.3%. What was the net change in the index over these six months?

25. A pilot is landing an airplane, and the plane descends 500 ft/min for 6 min and then descends 400 ft/min for 2 min. Due to problems on the ground, the pilot is then instructed not to land and ascends 800 ft/min for 11 min. What was the altitude of the plane at this time, relative to its altitude 19 min earlier?

26. An object is dropped from a tower. To determine its position 6 sec after it begins to drop, we must evaluate the expression $\left(\frac{1}{2}\right)(-32)(6)^2$. If distance is measured in feet, determine the position of the object after it has dropped for 6 sec.

27. The square of -3.1 is subtracted from -1.92. What is the result?

28. The product of two numbers is $-27\frac{5}{8}$. If one number is $3\frac{1}{4}$, what is the other number?

29. If a number is divided by 0.8, the quotient is -1.875. What is the number?

30. If a number is multiplied by the square of $2\frac{1}{3}$, the product is $-6\frac{29}{36}$. What is the number?

Chapter Summary

In this chapter, we introduced and discussed signed numbers. The arithmetic operations of addition, subtraction, multiplication, and division of signed numbers were examined and the various rules for performing these operations were discussed. Verbal problems involving signed numbers were also considered.

The **key words** introduced in this chapter are listed here in the order in which they appeared in the chapter. You should know what these words mean before proceeding to the next chapter.

integers (p. 260)
positive integer (p. 261)
negative integer (p. 261)
absolute value (p. 262)
rational number (p. 264)
signed numbers (p. 266)
simplest form of a rational number (p. 290)

Review Exercises

6.1 Integers

In Exercises 1–9, write the opposite of each integer.

1. $+6$
2. -8
3. $+12$
4. -96
5. $+127$
6. -312
7. 0
8. -1
9. -1010

In Exercises 10–18, classify the given integer as being positive, negative, or zero.

10. $+6$
11. -3
12. 0
13. -13
14. 27
15. $+56$
16. -396
17. -999
18. $+501$

In Exercises 19–24, determine the value for each expression.

19. $|+7|$
20. $|-3|$
21. $|0|$
22. $|+5| + |-3|$
23. $|+5| - |-3|$
24. $|-3| \times |+2|$

In Exercises 25–30, insert the symbol $<$, $=$, or $>$ between the two given numbers to make true statements.

25. $+8 \quad +6$
26. $-8 \quad -6$
27. $-9 \quad +8$
28. $0 \quad -3$
29. $|-4| \quad |-5|$
30. $|-8| \quad |+6|$

6.2 Signed Numbers

In Exercises 31–36, write the opposite for each of the given rational numbers.

31. $\dfrac{-4}{9}$
32. $\dfrac{+4}{7}$
33. $\dfrac{2}{11}$
34. $\dfrac{+3}{-7}$
35. $\dfrac{-4}{-13}$
36. $\dfrac{0}{9}$

In Exercises 37–42, classify the indicated rational numbers as being positive, negative, or zero.

37. $\dfrac{-4}{5}$
38. $\dfrac{0}{-5}$
39. $\dfrac{-7}{-19}$
40. $\dfrac{23}{-6}$
41. $\dfrac{+13}{+9}$
42. $\dfrac{-0}{-7}$

In Exercises 43–48, determine the absolute value for each of the following expressions.

43. $\dfrac{-5}{8}$
44. -4.09
45. $\dfrac{16}{-9}$
46. -0.09
47. $\dfrac{-23}{-37}$
48. $\dfrac{0}{-8}$

In Exercises 49–54, insert the symbol, $<$, $=$, or $>$ between the two given number symbols to make true statements.

49. $\dfrac{-1}{2} \quad \dfrac{3}{-4}$
50. $-0.6 \quad \dfrac{-3}{8}$
51. $\dfrac{-4}{7} \quad \dfrac{0}{-6}$
52. $-3.91 \quad -4.56$
53. $-1.09 \quad \dfrac{-109}{100}$
54. $\dfrac{-11}{22} \quad \dfrac{+33}{-66}$

6.3 Addition of Signed Numbers

In Exercises 55–64, add.

55. $(+8) + (+6)$
56. $(-7.2) + (-5.6)$
57. $\left(\dfrac{-3}{7}\right) + \left(\dfrac{2}{5}\right)$
58. $\left(-11\dfrac{1}{3}\right) + \left(+8\dfrac{1}{5}\right)$
59. $(-23.06) + (+19.9)$
60. $(+11) + (-3) + (+7)$
61. $\left(-58\dfrac{2}{7}\right) + \left(+39\dfrac{3}{8}\right)$
62. $\left(\dfrac{-5}{11}\right) + \left(\dfrac{-3}{7}\right)$
63. $(-12.3) + (+2.36) + (+12.3)$
64. $(+50.1) + (-39.9) + (-2.69)$

6.4 Subtraction of Signed Numbers

In Exercises 65–74, subtract.

65. $(+17) - (+9)$
66. $(+5.6) - (-3.2)$
67. $\left(-23\dfrac{1}{2}\right) - \left(-6\dfrac{1}{5}\right)$
68. $\left(\dfrac{-11}{13}\right) - \left(\dfrac{-2}{+7}\right)$
69. $(+87) - (96)$
70. $(-10.1) - (-20.9)$
71. $\left(+96\dfrac{2}{9}\right) - \left(+101\dfrac{3}{8}\right)$
72. $\left(-59\dfrac{4}{7}\right) - \left(-69\dfrac{2}{5}\right)$
73. $(-103.1) - (+69.2)$
74. $(+20.01) - (-197.7)$

In Exercises 75–88, perform the indicated operations.

75. $(+7) + (-6) - (+5)$
76. $(-13) - (+16) + (-23)$
77. $(-2.3) - (-3.7) - (-5.9)$
78. $(+0.67) - (5.9) - (+0.72)$
79. $(+101) - (+190) + (-203)$
80. $\left(-5\dfrac{1}{2}\right) + \left(-6\dfrac{1}{5}\right) - \left(+8\dfrac{1}{3}\right)$

6.5 Multiplication of Signed Numbers

In Exercises 81–90, multiply.

81. $(+6) \times (+9)$
82. $(-7) \times (-8)$
83. $(+11) \times (-12)$
84. $(-23) \times (+17)$
85. $\left(+4\dfrac{1}{2}\right) \times \left(-3\dfrac{1}{3}\right) \times \left(-2\dfrac{1}{5}\right)$
86. $\left(+4\dfrac{1}{5}\right) \times \left(+3\dfrac{1}{2}\right) \times \left(-2\dfrac{3}{4}\right)$
87. $(-13) \times (+7) \times (-19)$
88. $(-3.2) \times (-0.9) \times (-12)$
89. $(-10) \times (-20) \times (-123)$
90. $(+87) \times (-17) + (-11)$

6.6 Division of Signed Numbers

In Exercises 91–100, divide.

91. $(+27) \div (+3)$
92. $(-6.4) \div (-0.4)$
93. $(+81) \div (-27)$
94. $(-13.2) \div (+11)$
95. $0 \div \left(-12\dfrac{1}{9}\right)$
96. $\left(-3\dfrac{1}{7}\right) \div \left(+4\dfrac{1}{5}\right)$
97. $(-96) \div (-12)$
98. $0 \div (+117)$
99. $\left(-20\dfrac{1}{3}\right) \div \left(-12\dfrac{1}{2}\right)$
100. $(-420) \div (-20)$

6.7 Simplest Form of a Rational Number

In Exercises 101–106, determine the simplest form for each of the given rational numbers.

101. $\dfrac{-6}{10}$ **102.** $\dfrac{5}{-3}$ **103.** $\dfrac{-7}{-14}$ **104.** $\dfrac{0}{-9}$ **105.** $\dfrac{25}{-15}$ **106.** $\dfrac{-27}{81}$

6.8 Mixed Operations with Signed Numbers

In Exercises 107–116, perform the indicated operations.

107. $(+8) \times (-2) - (-12) \div (-3)$
108. $(-2)^2 + (-3)^3 \div (-9) + (+4)$
109. $[(+6) \times (-4)] \div [(+8) - (+4)]$
110. $[(-4)^2 \div (-8)]^3 \div [(-64) \div (+8)]$
111. $[(+4)^2 - (-2)^4] \div [(+5) - (-8)]^2$
112. $\dfrac{-1}{2} + \left(\dfrac{-2}{3} - \dfrac{1}{4}\right)$
113. $\dfrac{2}{5} \times \left(\dfrac{-1}{2} + \dfrac{-2}{5}\right)$
114. $\left(\dfrac{-3}{7} - \dfrac{-4}{9}\right) \times \left(\dfrac{-1}{10} + \dfrac{-2}{5}\right)$
115. $-2\dfrac{1}{2} \times \left(-4\dfrac{1}{3} + 5\dfrac{1}{4}\right)$
116. $6\dfrac{1}{8} \times \left(-1\dfrac{3}{4} - 2\dfrac{1}{2}\right)$

6.9 Applications Involving Signed Numbers

117. What is the difference, in feet, between $73\dfrac{1}{2}$ ft above ground level and $22\dfrac{3}{4}$ ft below ground level?

118. If you have $316.39 in your savings account and make two withdrawals of $85.19 and $96.07 and one deposit of $113.85, how much money will be in your savings account? (Disregard interest earned.)

119. If an item that regularly sells for $465 is on sale for $395 plus $30 for shipping charges and $6 for tax, how much would be the total cost of the item if purchased for cash?

120. Two opposing armies are positioned facing each other 300.5 km apart. One army advances 275.8 km, forcing the other army to retreat 100 km. How far apart are the two armies after these maneuvers?

121. Tom bought a new car. The total cost was $14,850, and he received a trade-in allowance of $2100 for his old car. He financed the balance for 36 months, with equal monthly payments of $428. What was the total finance charge?

122. Tom, Dick, and Mary each invested $5000 in a business. During the first year, their total income from the business was $136,500, and their net profits, after taxes, amounted to $52,500. If the initial investment was subtracted from the net profits and the balance divided equally among the three, how much would each receive?

Teasers

1. Indicate how you could determine the sum of the first 1000 positive integers without actually adding them.
2. A rectangle has a length of 14 yd and a width of 9 yd. Does there exist a square region, with integer dimensions, that has the same area as the area of the rectangular region? Explain your answer.
3. Which of the operations, on the set of integers, is commutative?
4. Indicate, in your own words, what is meant by the statement that multiplication of integers is an associative operation.
5. Explain, in your own words, what is meant by the symbol $-|-7|$.
6. Explain, in your own words, the difference between the symbols -6^2 and $(-6)^2$.
7. If a and b represent integers and $ab = 0$, must both a and b be zero? Explain your answer.
8. What property does the set of integers have that the set of whole numbers does not have?
9. a. What, if any, is the smallest integer?
   b. What, if any, is the largest negative integer?
   c. What, if any, is the smallest nonnegative integer?
10. Does there exist an integer that is both nonpositive and nonnegative? If so, identify all such integers.

CHAPTER 6 TEST

This test covers material from Chapter 6. Read each question carefully and answer all questions. If you miss a question, review the appropriate section of the text.

1. Write the opposite of $+113$.
2. Write the opposite of $-23\frac{1}{2}$.

In Exercises 3–5, determine the value of the indicated expressions.

3. $|-5| - |-3|$
4. $|-3| \times |+2|$
5. $|-1.8| \div |-9|$

In Exercises 6–23, perform the indicated operations.

6. $(+8) + (-5)$
7. $\frac{-2}{7} + \frac{3}{5}$
8. $(-7) + (-4)$
9. $0 - (-19)$
10. $\left(\frac{-4}{7}\right) - \left(\frac{-7}{8}\right)$
11. $\left(-23\frac{2}{5}\right) - \left(-56\frac{1}{3}\right)$
12. $(5.43) - (-89.6)$
13. $(-3.6) \times (-4.1)$
14. $-2\frac{1}{3} + 4\frac{2}{7}$
15. $(+16) \times (-23)$
16. $\frac{-4}{13} \div \frac{-12}{25}$
17. $(-36) \div (-4)$
18. $-5\frac{4}{11} \times \frac{-5}{-8}$
19. $-3\frac{1}{2} \div -4\frac{2}{9}$
20. $(+7) - (-6) + (-5)$
21. $(-2.3) + (-3.1) + (+0.19)$
22. $[(-12) + (-23)] \times [0 \div (-7)]$
23. $(-3)^2 \times (-1)^3 \div (-3)$

24. Determine whether the rational numbers $\frac{-3}{11}$ and $\frac{6}{-22}$ are or are not equivalent.
25. Determine the simplest form for the rational number $\frac{-2}{14}$.
26. Determine the simplest form for the rational number $\frac{3}{-8}$.

In Exercises 27–28, insert the symbol $<$ or $>$ between the pairs of numbers to make true statements.

27. $\frac{-7}{8} \quad \frac{6}{-9}$
28. $\frac{4}{7} \quad \frac{-3}{-5}$

29. If you have $393 in your savings account, and you make two withdrawals of $87 and $79 and one deposit of $96, what will be your new balance? (Disregard interest earned.)
30. Sonya bought a new car. The total cost was $6450, less a trade-in allowance of $2700. She financed the balance for 36 months, with equal monthly payments of $138. What was the total finance charge?

CUMULATIVE REVIEW: CHAPTERS 1-6

Perform the indicated operations in Exercises 1–8.

1. $6.29 + 0.008 + 13.1$
2. $(-7) + 14 - (-13) + (-8)$
3. $\dfrac{14}{23} - \dfrac{2}{5}$
4. $(-14)(+23)$
5. $(-80) \div (-20) \times (-3)$
6. $68.7\% \times 23$
7. $-2^3 \div (-2)^2$
8. $\left(-3\dfrac{1}{4}\right)\left(2\dfrac{3}{5}\right)$

In Exercises 9–12, circle the larger number.

9. $(1.2)(-8)$ or $(-3)(4.1)$
10. 1.3% of 6 or $|-8|$
11. $|-19|$ or -23
12. $\dfrac{17}{-13}$ or $\dfrac{-12}{17}$

In Exercises 13–15, round as indicated.

13. $6\dfrac{3}{5}$; ones
14. 4.069; tenths
15. $\dfrac{4}{11}$; hundredths

16. Determine the perimeter of a rectangular region 17.3 cm long and 11.6 cm wide.
17. Determine the area of a triangular region with base 4.1 yd and height 3 ft.
18. Determine the volume of a cube with edge of length 1.7 dm.
19. At the start of a month, a checking account had an opening balance of $69.31. The following transactions were made during the month: deposit of $55.40, check written for $39.16, check written for $40.10, deposit of $120, check written for $73.83, and a deposit of $35. What was the balance at the end of the month, if these were the only transactions and there was no service charge?
20. Write a fraction that represents the ratio of the number of female professors to the total number of faculty on a 267-member faculty if 182 are male.
21. Determine whether the fractions $\dfrac{11}{19}$ and $\dfrac{19}{27}$ are equivalent.
22. Write the word name for $29\dfrac{5}{8}$.
23. Simplify: $\dfrac{1}{3} \times \dfrac{2}{5} + \dfrac{3}{4} \div \dfrac{1}{6}$.
24. An auto mechanic earns $17.50 an hour and is paid time-and-a-half (1.5 times the hourly rate) for every hour over 40 worked in a single week. If the mechanic worked $48\dfrac{3}{4}$ hours in 1 week, what would be his pay?
25. Determine the simple interest on $1300 at 7.5% for 18 months.

CHAPTER 7: An Introduction to Algebra

Algebra is a generalization of arithmetic in which we use letters to represent numbers. For instance, the formula used for finding the area of a rectangular region is $A = LW$. In this formula, L and W are just place holders for the length and width, respectively, of *any* rectangle. They will be different numbers for different rectangles. Similarly, in the equation $x + 2 = 5$, the variable x can be any number. However, our usual objective is to find the one number that makes a *true* statement.

In this chapter, we will discuss algebraic expressions. You will learn to evaluate and combine algebraic expressions, including removing symbols of grouping. We will conclude the chapter with an introduction to the language of algebra.

CHAPTER 7 PRETEST

This pretest covers material on an introduction to algebra. Read each question carefully and try to answer all questions. Each question is keyed to the chapter section in which the particular topic is discussed. Check your answers with those given in the back of the text.

Questions 1–4 pertain to Section 7.1. Fill in the blanks to make true statements.

1. The algebraic expression $4u + 5v$ is a sum of two _____.

2. The algebraic expression $5abc$ is a _____ of four factors.

3. The numerical coefficient of the second term in the expression $3x - 2y + 4z$ is _____.

4. The second factor in the expression $-3xyz$ is _____.

Questions 5–6 pertain to Section 7.2.

5. Evaluate the expression $3x - 2y$ if $x = -2$ and $y = 3$.

6. Evaluate the expression $-2u^2v - u$ if $u = -2$ and $v = 1.6$.

Question 7 pertains to Section 7.3.

7. Simplify the expression $2x - 3y + x + 2y$ by combining like terms.

Questions 8–9 pertain to Section 7.4. Simplify each of the given expressions by combining like terms.

8. $3(a + 2b) - 2(3a - b)$

9. $2[a - 3(a + 2b) + 4b]$

Questions 10–15 pertain to Section 7.5. Write an algebraic expression for each of the following word expressions.

10. The sum of 7 and three times p.

11. One-half the sum of q and 9.

12. Four less than twice x.

13. The sum of x and the square of y.

14. 7 is added to three times y, and the result is diminished by 10.

15. The square of the sum of x and 2 increased by the quotient of the cube of x and 3.

7.1 Algebraic Expressions

OBJECTIVES

After completing this section, you should be able to:
1. Classify algebraic expressions as sums or products.
2. Determine the number of terms or factors in an algebraic expression.
3. Determine coefficients of terms in an algebraic expression.

Algebra is a generalization of arithmetic in which we use letters to represent numbers. Such letters are called **variables**. Sometimes, a variable can represent only one number. In such a case, the variable is called a **constant**.

When introducing the commutative property for the addition of whole numbers, we wrote

$$a + b = b + a$$

where a and b represent whole numbers. The symbols a and b represent *any* whole numbers, unlike the numerals 3 and 7 that represent *specific* whole numbers. The expressions $a + b$ and $b + a$ are called "algebraic expressions."

> **DEFINITION**
>
> An **algebraic expression** is an expression that consists only of numbers, letters, symbols of operations, and symbols of grouping.

Note that *not all parts* of the definition need be present in an algebraic expression. The following are algebraic expressions:

- $a + b + c$
- $\dfrac{1}{u - v}$
- $\sqrt{4y - 1}$
- $3x + y$
- $p^2 - 3q^4$
- $2u^3 + 5u^2 - 7u + \dfrac{1}{2}$

In arithmetic, we have the four basic operations of addition, subtraction, multiplication, and division. However, in algebra, we have two basic operations which are addition and multiplication. A difference can be treated as a sum and a quotient can be treated as a product.

> **DEFINITION**
>
> If an algebraic expression is a *sum*, then each part of the sum is called a "term." If an algebraic expression is a *product*, then each part of the product is called a "factor."

It is emphasized that terms are part of an algebraic expression that are separated by + or − signs. Each + or − sign is taken with the expression that follows it. Hence,

7.1 Algebraic Expressions

the expression $3x \div 4y$ is a single term. The expression $a - b \div c + d$ is a sum of three terms which are a, $-b \div c$, and d.

> **Practice Exercise 1.** Identify the terms in the expression $3ab - 5cd$.

EXAMPLE 1 The expression $8x - 9y$ is a sum. Identify its terms.

SOLUTION: The expression $8x - 9y$ can be rewritten as $8x + (-9y)$, a *sum of two terms*. The terms are $8x$ and $-9y$.

> **Practice Exercise 2.** Identify the terms in the expression $xy - 3x^2y - 7xy^3$.

EXAMPLE 2 The expression $a^2b - c + 4d$ is a sum. Identify its terms.

SOLUTION: The expression $a^2b - c + 4d$ can be rewritten as $a^2b + (-c) + 4d$, a *sum of three terms*. The terms are a^2b, $-c$, and $4d$. (Notice that the first term is a product of the factors a^2 and b.)

> **Practice Exercise 3.** Identify the factors in the expression $9a^2b^3c$.

EXAMPLE 3 The expression $3x^2y$ is a product. Identify its factors.

SOLUTION: The expression $3x^2y$ is a *product of three factors*, which are 3, x^2, and y.

> **Practice Exercise 4.** Identify the factors in the expression $-5u^2vw^5z$.

EXAMPLE 4 The expression $2x^2(3x + 5)$ is a product. Identify its factors.

SOLUTION: The expression $2x^2(3x + 5)$ is a *product of three factors*, which are 2, x^2, and $3x + 5$. Notice that the third factor is a *sum* of the terms $3x$ and 5. However, the given expression is considered to be *one term*.

ANSWERS TO PRACTICE EXERCISES

1. $3ab$ and $-5cd$. 2. xy, $-3x^2y$, and $-7xy^3$. 3. 9, a^2, b^3, and c. 4. -5, u^2, v, w^5, and z.

In the product $2xy$, the first factor, 2, is a number factor. The other two factors are variable factors.

> **DEFINITION**
> In a term, the number factor is called the **numerical coefficient**, or, simply, **coefficient**.

> **Practice Exercise 5.** Determine the coefficient in the expression $10a^2bc$.

EXAMPLE 5 Determine the coefficient in the expression $3x^2y^3$.

SOLUTION: The expression $3x^2y^3$ is a product of the factors 3, x^2, and y^3. The coefficient is 3.

> **Practice Exercise 6.** Determine the coefficient in the expression $-7x^6y$.

EXAMPLE 6 Determine the coefficient in the expression $-8rs^4t$.

SOLUTION: The expression $-8rs^4t$ is a product of the factors -8, r, s^4, and t. The coefficient is -8.

> **Practice Exercise 7.** Determine the coefficient of each term in the sum $-3r + 4r - 5s$.

EXAMPLE 7 The expression $5x - 6y - 9z$ is a sum of three terms. Determine the coefficient of each of the terms.

SOLUTION: The coefficients are 5, -6, and -9.

> **ANSWERS TO PRACTICE EXERCISES**
> 5. 10.   6. -7.   7. $-3, 4,$ and -5.

The coefficient of *xy* is 1 even though there is no number written. The expression *xy* can be written as 1(*xy*). Similarly, the coefficient of $-xy$ is -1 since $-xy = (-1)(xy)$.

CAUTION The numerical coefficient in an algebraic expression is multiplied by the product of all the other factors. Thus, $2xy$ means $2(xy)$. It does *not* mean $2(x)2(y)$.

EXERCISES 7.1

In Exercises 1–10, list (a) all of the different constants and (b) all of the different variables in each expression.

1. $3x - 5y$
2. $4ab - 3c$
3. $u^2 - 3uv$
4. $\dfrac{7z}{5x - 8y^2}$
5. $(2x - 3y) + (4x^2 - y)$
6. $R^2 - 3S + 5T$
7. $2p(3p - 4q)^5$
8. $(r - st)^3 - 5rs^2t$
9. $A - \dfrac{1}{2}bh$
10. $F - \dfrac{9}{5}C - 32$

In Exercises 11–20, determine the number of terms in each of the given expressions and identify each term.

11. $3x^2y$
12. $2a - 3b + c$
13. $x^2 - y^2$
14. $x - y + x^2 + y^2$
15. $(a + b)(a - b)$
16. $(2xy + x) - 3(x - y)$
17. $3S^2 - 4T(3S + R^3)$
18. $(3a - 4b^2) \div 5c$
19. $2 - 3[x - 5(6 - 5y)]$
20. $(a \div 2) + (2b - c) \div 3 - 4d \div 5$

In Exercises 21–26, determine the numerical coefficient for each of the given expressions.

21. $2ab^2$
22. $-3xy^2z^4$
23. $-\dfrac{1}{6}rs^3t$
24. $-2.7(a + b)c$
25. $81(x + y)^2(x - y)$
26. $-53a^2b^3c^4d^5$

In Exercises 27–34, determine the numerical coefficient for *each term* in the given expression.

27. $2a - 3b + 4c$
28. $-4(a + b)^2 - 5b^3$
29. $-2(a^2 - 2ab + b^2)$
30. $a^3 - 3a^2b + 3ab^2 - b^3$
31. $xy - yz + xyz$
32. $x^2 - 4xy - y^3$
33. $2p^2q - 3pq^2 + 4p^2q$
34. $2u - (3 - v - w) + x$

7.2 Evaluating Algebraic Expressions

OBJECTIVE: After completing this section, you should be able to evaluate algebraic expressions using given values for the indicated variables.

The operations of addition, subtraction, and division in algebra are indicated by the same symbols that are used in arithmetic. However, for multiplication in algebra, we usually do *not* use the × symbol between two factors in a product, as is the case for multiplication in arithmetic. For example, the symbol ab represents the product of the two factors a and b; no multiplication symbol is written between the two factors. However, if number factors are involved, we must be careful in writing them. The product of 2, 3, x, and y would be written as $(2)(3)xy$ or $2 \cdot 3 \cdot xy$ and *not* as $23xy$.

Rule for Evaluating Algebraic Expressions

To **evaluate** an algebraic expression:
1. Replace each variable in the expression by the given numerical value.
2. Enclose the numerical values in parentheses.
3. Perform the arithmetic operations, according to the rules for the order of operations.

Practice Exercise 1. Evaluate the expression $p^2 \div 3q + 4r$ when $p = -3$, $q = -1$, and $r = 4$.

EXAMPLE 1 Evaluate the algebraic expression $2a + b$ when $a = 2$ and $b = 3$.

SOLUTION: Substituting the value 2 for a and the value 3 for b in the given expression, we have

$$2a + b$$
$$= 2(2) + (3)$$
$$= 4 + 3$$
$$= 7$$

Practice Exercise 2. Evaluate the expression $2ab^2 - 3a$ when $a = 2$ and $b = -2$.

EXAMPLE 2 Evaluate $2xy$ when $x = \frac{1}{3}$ and $y = \frac{-1}{4}$.

SOLUTION: Substituting, we have

$$2xy$$
$$= 2\left(\frac{1}{3}\right)\left(\frac{-1}{4}\right)$$
$$= \left(\frac{2}{3}\right)\left(\frac{-1}{4}\right)$$
$$= \frac{-2}{12}$$
$$= \frac{-1}{6}$$

(Note that the coefficient 2 was multiplied by the product of x and y. We did *not* multiply x by 2 and y by 2.)

Notice that in the preceding examples, after we substituted the given values for the indicated variables, we evaluated the expression using the rule for order of operations.

Practice Exercise 3. Evaluate the expression $(2x + 3y)^2$ when $x = 2$ and $y = -3$.

EXAMPLE 3 Evaluate the algebraic expression $xy + 3x^2 - 4y$ when $x = -3$ and $y = -4$.

SOLUTION: Substituting, we have

$$xy + 3x^2 - 4y$$
$$= (-3)(-4) + 3(-3)^2 - 4(-4)$$
$$= (-3)(-4) + 3(+9) - 4(-4)$$
$$= 12 + 27 - (-16)$$
$$= 39 - (-16)$$
$$= 55$$

Practice Exercise 4. Evaluate the expression $4x - 5y$ when $x = -0.1$ and $y = -0.3$.

EXAMPLE 4 Evaluate the algebraic expression $2a - (b - 3a) + 4b$ when $a = 0.2$ and $b = -0.3$.

SOLUTION: Substituting, we have

$$2a - (b - 3a) + 4b$$
$$= 2(0.2) - [(-0.3) - 3(0.2)] + 4(-0.3)$$
$$= 2(0.2) - [(-0.3) - (0.6)] + 4(-0.3)$$
$$= 2(0.2) - (-0.9) + 4(-0.3)$$
$$= 0.4 - (-0.9) + (-1.2)$$
$$= 1.3 + (-1.2)$$
$$= 0.1$$

ANSWERS TO PRACTICE EXERCISES

1. 13. 2. 10. 3. 25. 4. 1.1.

EXERCISES 7.2

Evaluate each of the following expressions, using the given values for the indicated variables.

1. $2x + 3y$ when $x = 3$ and $y = 4$
2. $3x - 5y$ when $x = -2$ and $y = -3$
3. $2a - 3b + 4c$ when $a = 0.2$, $b = -0.1$, and $c = -0.2$
4. $3a^2 + b^3$ when $a = \frac{1}{3}$ and $b = \frac{1}{2}$
5. $(3a + b)^3$ when $a = 2$ and $b = -3$
6. $3cd^2 - 2c^2d$ when $c = -2$ and $d = 3$
7. $p^2q - 3pq + 4q^2$ when $p = -0.1$ and $q = -0.2$
8. $u + 3v - uv + v^3$ when $u = \frac{1}{3}$ and $v = \frac{-1}{2}$
9. $2x^2 + 4xy$ when $x = 3$ and $y = 4$
10. $-3r^3s^2t$ when $r = 2$, $s = \frac{-1}{2}$, and $t = \frac{1}{4}$
11. $-3(x^2 + y)$ when $x = 2.1$ and $y = -5.3$
12. $(x + y) - (x - y)$ when $x = -3.3$ and $y = -4.5$

13. $(abc)^2$ when $a = -2$, $b = -3$, and $c = -4$
14. $a^2 + b^2 + c^2$ when $a = 3$, $b = -2$, and $c = 1$
15. $(a + b + c)^2$ when $a = 3$, $b = -2$, and $c = 1$
16. $(x + y)(x - y)$ when $x = \frac{2}{3}$ and $y = \frac{-1}{4}$
17. $a(b + c)^2$ when $a = -2$, $b = -0.3$, and $c = -0.2$
18. $(x - y)(x^2 + y^2)$ when $x = 1.5$ and $y = -7.2$
19. $(a^3 - b^3) \div (a - b)$ when $a = -2$ and $b = -3$
20. $(u^3 + v^3) \div (u + v)$ when $u = \frac{-1}{2}$ and $v = \frac{-1}{3}$
21. $(2x + y)(x - 3y)$ when $x = -2$ and $y = 3$
22. $u^2 - 3uv + 4v^2$ when $u = -1$ and $v = -2$
23. $2p^3 + 4p^2 - 5p - 1$ when $p = -0.1$
24. $-3(m - 2n)^2$ when $m = \frac{-1}{3}$ and $n = \frac{1}{4}$
25. $\frac{2r - 3s^2}{s + t}$ when $r = -3$, $s = -2$, and $t = 5$
26. $\frac{t^2 - 4}{t - 2}$ when $t = -0.3$
27. $w^2(3w - 2)(w + 1)$ when $w = -2$
28. $(3r - 2s)^3(2r + 3s)^2$ when $r = \frac{-1}{2}$ and $s = \frac{-1}{3}$
29. $(2a - 3b^2 + 4c)^4$ when $a = -5$, $b = -2$, and $c = 6$
30. $-5s^3tu^2w$ when $s = 2$, $t = \frac{-1}{3}$, $u = -3$, and $w = \frac{-1}{2}$

7.3 Simplifying Algebraic Expressions by Combining Like Terms

OBJECTIVE

After completing this section, you should be able to simplify algebraic expressions by combining like terms.

Suppose that Michael starts the day with 4 dollars. He later mows the lawn for a neighbor and receives 3 dollars. He then receives 6 dollars for helping Mrs. Smith with a chore. At the end of the day, he pays back to his brother the 5 dollars that he borrowed the week before. How many dollars does Michael have at the end of the day?

To determine the answer to this question, we note that Michael starts the day with 4 dollars, receives 3 dollars, again receives 6 dollars, and then pays back 5 dollars. Hence, we have

$$4 \text{ dollars} + 3 \text{ dollars} + 6 \text{ dollars} - 5 \text{ dollars}$$
$$= 7 \text{ dollars} + 6 \text{ dollars} - 5 \text{ dollars}$$
$$= 13 \text{ dollars} - 5 \text{ dollars}$$
$$= 8 \text{ dollars}$$

Therefore, at the end of the day, Michael would have 8 dollars.

In this example, we were working with "like" or "similar terms." That is, we were combining dollars.

> **DEFINITION**
> In an algebraic expression, terms that contain the same variables with the same exponents are called **like terms**. All constant terms are like terms.

In this definition, it should be noted that for like terms, only the numerical coefficients may differ. The following expressions are like terms:

- $2a, -4a, 5a, \frac{2}{3}a$ (These are a terms.)

- $2pq, -3pq, \frac{1}{2}pq$ (These are pq terms.)

- $3y^2, y^2, -4y^2, \frac{-1}{3}y^2$ (These are y^2 terms.)

- $0, -3, 0.7, \frac{2}{7}$ (These are constant terms.)

Terms that are not like terms are called **unlike terms**. The following expressions are unlike terms:

- $2a, 3b$ (The variables are different.)
- $x^2, 2x^3$ (The exponents are different.)
- p^2q, pq^2 (The exponents are different.)
- $y^3, 4$ (y^3 term and constant term)

Suppose that you were now asked to add: $2x + 3x + 7x$. How would you do it? You would probably recognize that $2x$, $3x$, and $7x$ are all like terms. If x represented dollars, you would probably think in terms of adding 2 dollars, 3 dollars, and 7 dollars; the sum would be $(2 + 3 + 7)$ dollars, or 12 dollars. Hence, you would probably conclude that

$$2x + 3x + 7x = (2 + 3 + 7)x = 12x$$

which is correct.

Now, suppose that you were asked to add: $2x + 3y$. How would you do it? Did you recognize that $2x$ and $3y$ are *not* like terms since the variable parts are not the same? If so, good! If not, look again. We *cannot* get a single term for the sum of $2x + 3y$.

Rule for Combining Like Terms

To *combine like terms* means to add them. To add like terms, add their numerical coefficients and multiply the sum by the common variable part.

In the examples that follow, note the use of the Distributive Property for Multiplication over Addition when combining like terms.

7.3 Simplifying Algebraic Expressions by Combining Like Terms

Combine like terms in each of the following expressions.

Practice Exercise 1.
$-6a + 11a$

EXAMPLE 1 Combine: $10c + 5c$.

SOLUTION: $10c + 5c = (10 + 5)c = 15c$

Practice Exercise 2. $5b + b$

EXAMPLE 2 Combine: $a + 3a$.

SOLUTION: $a + 3a = (1 + 3)a = 4a$ [Remember that a means $(1)a$.]

Practice Exercise 3.
$-0.6y^4 + 0.4y^4$

EXAMPLE 3 Combine: $3x^2 + 5x^2$.

SOLUTION: $3x^2 + 5x^2 = (3 + 5)x^2 = 8x^2$

Practice Exercise 4. $t - 5t - t$

EXAMPLE 4 Combine: $2y + 3y - y$.

SOLUTION: $2y + 3y - y = (2 + 3 - 1)y = 4y$ [Remember: $-y = (-1)y$.]

Practice Exercise 5.
$5r^2s^3 - 2r^2s^3 - r^2s^3$

EXAMPLE 5 Combine: $4x^2y + 2x^2y - x^2y$.

SOLUTION: $4x^2y + 2x^2y - x^2y = (4 + 2 - 1)x^2y = 5x^2y$

Practice Exercise 6.
$\frac{1}{4}s + \frac{2}{5}s - 2s$

EXAMPLE 6 Combine: $0.2u - 1.9u + 3.7u - 2.8u$.

SOLUTION: $0.2u - 1.9u + 3.7u - 2.8u = (0.2 - 1.9 + 3.7 - 2.8)u = -0.8u$

Practice Exercise 7.
$0.7t - 1.5t - 2.3t + 4.6t$

EXAMPLE 7 Combine: $\frac{8}{9}u - \frac{1}{9}u + \frac{4}{9}u$.

SOLUTION: $\frac{8}{9}u - \frac{1}{9}u + \frac{4}{9}u = \left(\frac{8}{9} - \frac{1}{9} + \frac{4}{9}\right)u = \frac{11}{9}u$

ANSWERS TO PRACTICE EXERCISES

1. $5a$. 2. $6b$. 3. $-0.2y^4$. 4. $-5t$. 5. $2r^2s^3$. 6. $\frac{-27}{20}s$. 7. $1.5t$.

Unlike terms cannot be combined. However, if an algebraic expression is a sum containing more than two terms and if some of the terms are like terms, we may combine them, using the commutative and associative properties for addition.

Combine like terms in each of the following expressions.

Practice Exercise 8.
$4y^2 + y^2 - 2y^2$

EXAMPLE 8 Combine the like terms in $2a + 3b + 4a - 5b$.

SOLUTION: The given expression is a sum of four terms. The first and third terms are like terms and may be grouped together. The second and fourth terms are like terms and may also be grouped together. We have

$2a + 3b + 4a - 5b$
$= (2a + 4a) + (3b - 5b)$ (Regrouping the terms)
$= (2 + 4)a + (3 - 5)b$ (Adding like terms in each group)
$= 6a + (-2)b$ (Simplifying)
$= 6a - 2b$ (Rewriting, using the definition of subtraction)

Since $6a$ and $-2b$ have different variable factors, they are not like terms and cannot be combined.

Practice Exercise 9.

EXAMPLE 9 Simplify: $2x^2y - 3xy^2 + 4x^2y$.

SOLUTION: The given expression is a sum of three terms. The first and third terms are like terms and may be combined. We have

$$2x^2y - 3xy^2 + 4x^2y$$
$$= (2x^2y + 4x^2y) - 3xy^2 \quad \text{(Regrouping)}$$
$$= (2 + 4)x^2y - 3xy^2 \quad \text{(Adding like terms)}$$
$$= 6x^2y - 3xy^2 \quad \text{(Simplifying)}$$

Since $6x^2y$ and $-3xy^2$ are unlike terms (the exponents are different), they cannot be combined.

Practice Exercise 10.

EXAMPLE 10 Combine like terms: $3pq^2 - p^2q + 4p^2q - pq^2 + pq$.

SOLUTION: The given expression is a sum of five terms. The first and fourth terms are like terms and may be combined. The second and third terms are like terms and may also be combined. We have

$$3pq^2 - p^2q + 4p^2q - pq^2 + pq$$
$$= (3pq^2 - pq^2) + (-p^2q + 4p^2q) + pq \quad \text{(Regrouping)}$$
$$= (3 - 1)pq^2 + (-1 + 4)p^2q + pq \quad \text{(Adding like terms)}$$
$$= 2pq^2 + 3p^2q + pq \quad \text{(Simplifying)}$$

The result is a sum of three unlike terms.

ANSWERS TO PRACTICE EXERCISES

8. $3y^2$. 9. $2pq^2 + 3p^2q$. 10. $2r^2s^3 + 4r^3s^2 - r^2s^2$.

EXERCISES 7.3

Simplify each of the given expressions by combining like terms.

1. $x - y + 2x + 3y$
2. $3a - 4b + b - a$
3. $x - 2y + 3x - 4y + 2x$
4. $0.2a + 0.3b + 1 + 0.2b - 0.4a$
5. $\frac{1}{2}x^2y - \frac{1}{3}xy^2 + \frac{1}{4}x^2y$
6. $4rs - 5r^2s + 2rs^2 - 3rs - r^2s$
7. $2a + 4b + c + 3a - 2b - 4c$
8. $3.1x - 1.7y + 4.6z + 3.8y - 2.5z + x$
9. $a^2b - ab^2 + 2a^2b^2 + 3ab^2 - 4a^2b^2$
10. $\frac{1}{3}r + \frac{1}{2}s - \frac{2}{3}t + r - \frac{3}{4}t + \frac{4}{5}r - \frac{1}{5}s$
11. $1 - 2y + 3x - 4 + 5y - x + 2$
12. $3a - b + c - 2a - b + 3c - 5b$
13. $10.2u + 2.39v - 3.9 - 4.07 + 5.17u - 3.2v$
14. $3a^2b^3 - 2a^2b + 4ab^3 - 4a^2b - ab^3 - 2a^3b^3$
15. $2a - 3b + 3c - a + 2b - b - 2a - 4c + c$

16. $4r - s - t - 3s - r + 2t + 3t - 2r - 5s$
17. $x - 2.09y - 3.7z - 2.87x - y + z - y + 4.01z - x$
18. $x^2 - 2y + 3xy + 3x + y - xy - 2x^2 + 4xy + y^2$
19. $2ab - 3bc + a - 2b + ab - 2bc - 3a + 4b + 5bc$
20. $\frac{2}{3}a - \frac{4}{5} + \frac{1}{3} - \frac{3}{4}a + 1 + \frac{2}{5}a - \frac{2}{5}$
21. $x - y + 1 - 3x - 2y + 4 + 4x - 7y - 9$
22. $m^2n - 3mn^2 - m^2n^2 - mn^2 + 4m^2n + 2m^2n^2$
23. $\frac{1}{2}p^2q + \frac{1}{3}pq^2 - \frac{1}{5}p^2q - 2pq^2$
24. $\frac{3}{2}uv - \frac{7}{3}u^2v - \frac{1}{4}uv + \frac{1}{3}uv^2 - \frac{1}{7}uv^2 + u^2v$
25. $0.3x^2 - 0.1x + 0.9 - 0.7x^2 - x^3 + 1.7x^3 - 0.9x$
26. $2st + s^2t - 3st^2 - 3st + 0.2st^2 - 1.3s^2t$
27. $\frac{2}{3}x^2y^3 - \frac{1}{2}xy^2 + \frac{1}{4}xy^2 - \frac{2}{5}x^2y^3 + \frac{1}{9}x^2y - 2x^2y^3$
28. $2(x + y) - 3(y - z) - 4(x + y) + 6(y - z) - 5(x + y)$ (Hint: There are $(x + y)$ terms and $(y - z)$ terms in this exercise.)
29. $-0.3(p^2 - 3q) + 0.4(p - q^2) - 0.5(p - q^2) + 1.9(p^2 - 3q) + 2.1(p - q^2)$
30. $\frac{1}{2}(x^2 - y^2) - \frac{1}{3}(x + y) - \frac{1}{3}(x^2 - y^2) + \frac{1}{5}(x - y) + 2(x + y) - 3(x - y)$

7.4 Simplifying Algebraic Expressions by Removing Symbols of Grouping

OBJECTIVE

After completing this section, you should be able to simplify algebraic expressions by removing symbols of grouping before combining like terms.

In Section 7.3, you learned to simplify algebraic expressions by combining like terms. Sometimes, algebraic expressions contain symbols of grouping which must be removed before simplifying.

Rules for Removing Symbols of Grouping

1. When removing parentheses (or other symbols of groupings) preceded by a + sign, drop the parentheses and the + sign, and leave the sign of each term within the grouping symbol as is. (This is equivalent to multiplying whatever is in the parentheses by 1.)
2. When removing parentheses (or other symbols of grouping) preceded by a − sign, drop the parentheses and the − sign, *but change* the sign of *each term* within the grouping symbol. (This is equivalent to multiplying whatever is in the parentheses by −1.)
3. When more than one set of grouping symbols are involved, remove the innermost symbols first and work outwards.

It should be noted that if no sign precedes a symbol of grouping, it is understood to be a + sign. Similarly, if no sign precedes a term within a grouping symbol, it is also understood to be a + sign.

Remove the symbols of grouping from each of the following.

Practice Exercise 1.
$2a - (b - c)$

EXAMPLE 1 Remove the parentheses: $x + (y - z)$.

SOLUTION:

$$x + (y - z)$$
$$= x + (+y - z) \quad \text{(Sign preceding } y \text{ is understood to be +.)}$$
$$\quad\quad\quad\quad\quad\quad\quad \text{(Drop these: Rule 1.)}$$
$$= x \quad + y - z \quad \text{(Leave the sign of each term within the parentheses as is.)}$$
$$= x + y - z \quad \text{(Writing more compactly)}$$

Practice Exercise 2.
$x + (y - z) - w$

EXAMPLE 2 Remove the parentheses: $5x - (2y - z)$.

SOLUTION:

$$5x - (2y - z)$$
$$= 5x - (+2y - z) \quad \text{(Sign preceding } 2y \text{ is understood to be +.)}$$
$$\quad\quad\quad\quad\quad\quad\quad \text{(Drop these: Rule 2.)}$$
$$= 5x \quad - 2y + z \quad \text{(Change the sign of each term within the parentheses.)}$$
$$= 5x - 2y + z \quad \text{(Writing more compactly)}$$

Practice Exercise 3.
$t - [u + (v - y) - (x - s)]$

EXAMPLE 3 Remove the symbols of grouping: $2x - [y + (z - 3w)]$.

SOLUTION:

$$2x - [y + (z - 3w)] \quad \text{(Start with innermost parentheses.)}$$
$$= 2x - [y + (+z - 3w)] \quad \text{(Sign preceding } z \text{ is understood to be +.)}$$
$$= 2x - [y \quad + z - 3w] \quad \text{(Drop parentheses and + sign preceding;}$$
$$\quad\quad\quad\quad\quad\quad\quad\quad\quad\quad \text{leave sign for each term within parentheses as is.)}$$
$$= 2x - [+y + z - 3w] \quad \text{(Sign preceding } y \text{ is understood to be +.)}$$
$$= 2x \quad - y - z + 3w \quad \text{(Drop brackets and } - \text{ sign preceding;}$$
$$\quad\quad\quad\quad\quad\quad\quad\quad\quad\quad \text{change sign of each term within brackets.)}$$
$$= 2x - y - z + 3w \quad \text{(Writing more compactly)}$$

ANSWERS TO PRACTICE EXERCISES

1. $2a - b + c$. 2. $x + y - z - w$. 3. $t - u - v + y + x - s$.

To combine like terms in an algebraic expression that involves symbols of grouping, we first remove the grouping symbols and then proceed as in the previous section.

Simplify each of the following by combining like terms.

Practice Exercise 4.
$x + (y - x) - (2x - 3y)$

EXAMPLE 4 Simplify by combining like terms:

$$(2x - 3y) - (3z - 4x) - (y - 2z)$$

7.4 Simplifying Algebraic Expressions by Removing Symbols of Grouping

SOLUTION:

$$
\begin{aligned}
&(2x - 3y) - (3z - 4x) - (y - 2z) \\
&= 2x - 3y - (3z - 4x) - (y - 2z) && \text{(Removing parentheses preceded by a + sign understood)} \\
&= 2x - 3y - 3z + 4x - (y - 2z) && \text{(Removing parentheses preceded by a − sign)} \\
&= 2x - 3y - 3z + 4x - y + 2z && \text{(Removing parentheses preceded by a − sign)} \\
&= 6x - 3y - 3z - y + 2z && \text{(Adding the } x \text{ terms)} \\
&= 6x - 4y - 3z + 2z && \text{(Adding the } y \text{ terms)} \\
&= 6x - 4y - z && \text{(Adding the } z \text{ terms)}
\end{aligned}
$$

Practice Exercise 5.
$p - (q + r) - (p - q)$

EXAMPLE 5 Simplify by combining like terms:

$$-(3a - 2y) + (3a + 2y)$$

SOLUTION:

$$
\begin{aligned}
&-(3a - 2y) + (3a + 2y) \\
&= -3a + 2y + (3a + 2y) && \text{(Removing parentheses preceded by a − sign)} \\
&= -3a + 2y + 3a + 2y && \text{(Removing parentheses preceded by a + sign)} \\
&= (0)a + 2y + 2y && \text{(Adding the } a \text{ terms)} \\
&= (0)a + 4y && \text{(Adding the } y \text{ terms)} \\
&= 0 + 4y && \text{(Multiplying)} \\
&= 4y && \text{(Adding)}
\end{aligned}
$$

Practice Exercise 6.
$x - [y + (x - 2y) - (y - 3x)]$

EXAMPLE 6 Simplify by combining like terms:

$$2a - [3b - (2a - 3b)] - 5a$$

SOLUTION:

$$
\begin{aligned}
&2a - [3b - (2a - 3b)] - 5a \\
&= 2a - [3b - 2a + 3b] - 5a && \text{(Removing parentheses preceded by a − sign)} \\
&= 2a - [6b - 2a] - 5a && \text{(Adding the } b \text{ terms)} \\
&= 2a - 6b + 2a - 5a && \text{(Removing parentheses preceded by a − sign)} \\
&= -a - 6b && \text{(Adding the } a \text{ terms)}
\end{aligned}
$$

Practice Exercise 7.
$-(3a + 2b) - (3a - 2b)$

EXAMPLE 7 Simplify by combining like terms:

$$x - \{2 + y + [3 - 2x - (y - x) + 1] - 2\} + 2x$$

SOLUTION:

$$
\begin{aligned}
&x - \{2 + y + [3 - 2x - (y - x) + 1] - 2\} + 2x \\
&= x - \{2 + y + [3 - 2x - y + x + 1] - 2\} + 2x && \text{(Removing parentheses preceded by a − sign)} \\
&= x - \{2 + y + [4 - x - y] - 2\} + 2x && \text{(Adding like terms)} \\
&= x - \{2 + y + 4 - x - y - 2\} + 2x && \text{(Removing parentheses preceded by a + sign)} \\
&= x - \{4 - x\} + 2x && \text{(Adding like terms)} \\
&= x - 4 + x + 2x && \text{(Removing parentheses preceded by a − sign)} \\
&= 4x - 4 && \text{(Adding } x \text{ terms)}
\end{aligned}
$$

ANSWERS TO PRACTICE EXERCISES

4. $-2x + 4y$. 5. $-r$. 6. $-3x + 2y$. 7. $-6a$.

In the previous chapters on arithmetic, we encountered the distributive property for multiplication over addition which we symbolized as

$$a(b + c) = ab + ac$$

This property is also used in algebra. For instance, if an algebraic sum is enclosed within parentheses and the sum is to be multiplied by a factor, then *every* term of the sum *must* be multiplied by that factor.

Practice Exercise 8. Multiply: $5(2x - 3y)$.

EXAMPLE 8 Multiply: $2(a + b)$.

SOLUTION:

$2(a + b)$
$= (2)(a) + (2)(b)$ (*Each* term within the parentheses is multiplied by the factor 2.)
$= 2a + 2b$ (Simplifying)

Practice Exercise 9. Multiply: $-6(r - 2s + 4t)$.

EXAMPLE 9 Multiply: $-3(2x - 3y + z)$.

SOLUTION:

$-3(2x - 3y + z)$
$= (-3)(2x) + (-3)(-3y) + (-3)(z)$ (*Each* term within the parentheses is multiplied by -3.)
$= -6x + 9y - 3z$ (Simplifying)

Practice Exercise 10. Multiply: $(a - 2b - 3c)(-5)$.

EXAMPLE 10 Multiply: $(2p - 3q)(-4)$.

SOLUTION:

$(2p - 3q)(-4)$
$= (2p)(-4) + (-3q)(-4)$ (*Each* term within the parentheses is multiplied by the factor -4.)
$= -8p + 12q$ (Simplifying)

Simplify the following expression by combining like terms:

Practice Exercise 11.
$-4(p - 2q) - 5(3q - 4p)$.

EXAMPLE 11 Simplify the following expression by combining like terms:

$$2(x - y) - 3(2y - 4x)$$

SOLUTION:

$2(x - y) - 3(2y - 4x)$
$= (2)(x) - (2)(y) - 3(2y - 4x)$ (Using the distributive property)
$= 2x - 2y - 3(2y - 4x)$ (Simplifying)
$= 2x - 2y + (-3)(2y - 4x)$ (Rewriting, using the definition of subtraction)
$= 2x - 2y + (-3)(2y) + (-3)(-4x)$ (Distributive property)
$= 2x - 2y + (-6y) + (12x)$ (Simplifying)
$= 14x - 2y + (-6y)$ (Adding x terms)
$= 14x - 8y$ (Adding y terms)

ANSWERS TO PRACTICE EXERCISES

8. $10x - 15y$. **9.** $-6r + 12s - 24t$. **10.** $-5a + 10b + 15c$. **11.** $16p - 7q$.

7.5 The Language of Algebra

EXERCISES 7.4

For each of the following, remove the grouping symbols and, wherever possible, combine like terms.

1. $c + (a - b)$
2. $4 - (b + 2)$
3. $(p - 2q) - 3$
4. $(T + S) - (R - V)$
5. $(3a + 2b + 1) - 4a$
6. $x + (y + 3x) - (4y + 2x)$
7. $(2x^2y - 3xy^2) - (4x^2y + 5xy^2 + 3x^2y^2)$
8. $4rs - (5r^2s + 2rs^2) - (3rs - r^2s)$
9. $(a^2b - ab^2 + 2a^2b^2) + (3ab^2 - 4a^2b^2)$
10. $(0.1r + 0.2s - 0.3t) + (0.4r - 0.6t) + (0.5r - 0.4s)$
11. $\left[1 - \frac{1}{2}y\right] - \left[\frac{1}{3}x - \frac{3}{4}\right] + \left[\frac{1}{5}y - x + \frac{1}{2}\right]$
12. $(3a^2b^3 - 2a^2b^2 + 4ab^3) - (4a^2b - ab^3 - 2a^2b^3)$
13. $6 - 2(c - d)$
14. $-3(x - 2y) - 3y$
15. $-2(x + 3y) + 3(y - x)$
16. $x - 2(y - 3x) + 4(2x - y)$
17. $3(2a - 3b + c) - 4(3c - a + 2b)$
18. $-4(3s + r + 2t) - 5(3t - 2r - 5s)$
19. $\frac{1}{2}(4r - s - t) - \frac{1}{3}(3s - r + 2t)$
20. $\frac{-1}{4}(2x + y - 3z) - \frac{1}{5}(x - 2y - 4z)$
21. $7(x^2 - 2y + 3xy) + 2(3x + y - xy) - 4(2x^2 - 4xy - y^2)$
22. $-3(2ab - 3bc + a) - \frac{1}{2}(2b - ab + 2bc) - \frac{1}{3}(3a - 4b - 6bc)$
23. $2a - [4 - (3 - 3a)] + 1$
24. $x - [y - (1 + 3x) - (2y - 4)]$
25. $2.7 - (3.9 - x) - [1 - (2.7x - 3.9) + (3.1x + 1.7)]$
26. $a - 2(3a - 2b) - [-3(b - 2a) - (4a - b)]$
27. $3[-2a - 4(b - 2a)] - 4[2b - 5(a - 3b) + 3a]$
28. $-\{x - [x - (x - y) + (2y - x)] - 2x\}$
29. $a - \{(3a - 2b) - [2(b - 2a) - 3(4a - b)]\}$
30. $-2[a - 3(b - 2a)] - \{2b - [4b - 2(a - 5b)] + 3a\}$

7.5 The Language of Algebra

OBJECTIVES

After completing this section, you should be able to:
1. Identify the parts of an equation.
2. Translate verbal expressions into algebraic expressions.

There are many real-life problems that can be solved by using algebra. However, most of these problems are expressed in words. To use algebra to solve these problems, you must understand not only the English language but the language of algebra as well. For the rest of this text, you will be learning how to use algebra to help solve some of these real-life problems.

When using algebra to solve problems stated in words, you first must be able to change the word expressions into algebraic expressions, that is, into the "language of algebra."

To help you change word expressions into algebraic expressions, a partial list of commonly used word expressions is given in Table 7.1, together with the corresponding algebraic expressions.

For each of the following, write the corresponding algebraic expression.

Practice Exercise 1. The product of 2 and m.

EXAMPLE 1 Write an algebraic expression for "the sum of x and y."

SOLUTION: The word expression "the sum of x and y" is written as $x + y$.

Practice Exercise 2. 6 less than y.

EXAMPLE 2 Write an algebraic expression for "p increased by 7."

SOLUTION: The word expression "p increased by 7" is written as $p + 7$.

TABLE 7.1

	Word Expression	Algebraic Expression
Addition	Add; added to; the sum of; more than; increased by; the total of; plus	$+$
	Add x and y	$x + y$
	y added to 7	$7 + y$
	The sum of a and b	$a + b$
	m more than n	$n + m$
	p increased by 10	$p + 10$
	The total of q and 10	$q + 10$
	9 plus m	$9 + m$
Subtraction	Subtract; subtract from; difference between; less; less than; decreased by; diminished by; take away; reduced by; exceeds; minus	$-$
	Subtract x from y	$y - x$
	From x, subtract y	$x - y$
	The difference between x and 7	$x - 7$
	10 less m	$10 - m$
	10 less than m	$m - 10$
	p decreased by 11	$p - 11$
	8 diminished by w	$8 - w$
	y take away z	$y - z$
	p reduced by 6	$p - 6$
	x exceeds y	$x - y$
	r minus s	$r - s$
Multiplication	Multiply; times; the product of; multiplied by; times as much; of	$\times$
	Multiply x and y	xy
	7 times y	$7y$
	The product of x and y	xy
	5 multiplied by y	$5y$
	$\frac{1}{5}$ of p	$\frac{1}{5}p$
Division	Divide; divides; divided by; the quotient of; the ratio of; equal amounts of; per	$\div$

(continued)

7.5 The Language of Algebra

Practice Exercise 3. The square of x, diminished by 5.

Practice Exercise 4. Twice the sum of a and b.

Practice Exercise 5. The quotient of twice y and three times x.

EXAMPLE 3 Write an algebraic expression for "the total of $2y$ and $7y$."

SOLUTION: The word expression "the total of $2y$ and $7y$" is written as $2y + 7y$.

EXAMPLE 4 Write an algebraic expression for "y decreased by x."

SOLUTION: The word expression "y decreased by x" is written as $y - x$.

EXAMPLE 5 Write an algebraic expression for "one-third of m."

SOLUTION: The word expression "one-third of m" is written as $\frac{1}{3}m$.

ANSWERS TO PRACTICE EXERCISES

1. $2m$. 2. $y - 6$. 3. $x^2 - 5$. 4. $2(a + b)$. 5. $\frac{2y}{3x}$.

TABLE 7.1 (continued)

	Word Expression	Algebraic Expression
	Divide x by 6	$\frac{x}{6}$ or $x \div 6$
	7 divides x	$\frac{x}{7}$ or $x \div 7$
	7 divided by x	$\frac{7}{x}$ or $7 \div x$
	The quotient of y and 5	$\frac{y}{5}$ or $y \div 5$
	The ratio of u to v	$\frac{u}{v}$ or $u \div v$
	u separated into 4 equal parts	$\frac{u}{4}$ or $u \div 4$
	5 parts per 100 parts	$\frac{5}{100}$
Power	The square of y	y^2
	The cube of k	k^3
	t raised to the fourth power	t^4
Equals	Is equal to; the same as; is; are; the result of; will be; was	$=$
	x is equal to y	$x = y$
	p is the same as q	$p = q$
Multiplication by 2	Two, two times; twice; twice as much as; double	2
	Twice z	$2z$
	y doubled	$2y$
Multiplication by $\frac{1}{2}$	Half of; one-half of; half as much as; one-half times	$\frac{1}{2}$
	Half of u	$\frac{1}{2}u$
	One-half times m	$\frac{1}{2}m$

In the previous sections of this chapter, you learned about algebraic expressions. We will now introduce the algebraic equation.

> **DEFINITION**
>
> If two algebraic expressions are set equal to each other, then the resulting statement is called an **algebraic equation.**

In this definition, note that an equation consists of three parts:

1. The equal sign (=).
2. The algebraic expression to the left of the = sign (called the left-hand side or member of the equation).
3. The algebraic expression to the right of the = sign (called the right-hand side or member of the equation.)

The following are examples of algebraic equations.

- $a + b = b + a$
- $2x - 3 = 5$
- $\dfrac{2p - 7}{3} = 4 - 5p$
- $-3(4y + 3) - 2y = 2(1 - 5y)$
- $4u^2 - 5u + 7 = 0$

The = sign in an equation means that the two algebraic expressions on either side of the = sign have the same value. In the second equation just above, this means that the algebraic expression $2x - 3$ has the same value as 5. However, if we substitute different values for x, the expression $2x - 3$ also assumes different values. And, for most of these values of x, the expression $2x - 3$ will *not* have the value 5.

> **DEFINITION**
>
> A **solution of an algebraic equation** involving a single variable is a value which, when substituted for the variable involved, results in a true statement. *To solve an equation* involving a single variable means to determine its solutions.

Practice Exercise 6. Determine if -4 is a solution for the equation $2y + 1 = -7$.

EXAMPLE 6 Determine if 2 is a solution for the equation $x + 3 = 5$.

SOLUTION: 2 is a solution for the equation $x + 3 = 5$ since $2 + 3 = 5$ is a true statement.

Practice Exercise 7. Determine if 0.5 is a solution for the equation $3 - 2u = 5$.

EXAMPLE 7 Determine if 3 is a solution for the equation $2y = 7$.

SOLUTION: 3 is *not* a solution for the equation $2y = 7$ since $(2)(3) = 7$ is *not* a true statement.

7.5 The Language of Algebra

Practice Exercise 8. Determine if $\frac{4}{3}$ is a solution for the equation $2x - 1 = 3 - x$.

EXAMPLE 8 Determine if $\frac{2}{3}$ is a solution for the equation $2u - \frac{1}{3} = 1$.

SOLUTION: $\frac{2}{3}$ is a solution for the equation $2u - \frac{1}{3} = 1$ since

$$(2)\left(\frac{2}{3}\right) - \left(\frac{1}{3}\right) = \left(\frac{4}{3}\right) - \left(\frac{1}{3}\right)$$
$$= \frac{3}{3}$$
$$= 1$$

is a true statement.

Practice Exercise 9. Determine if $1\frac{1}{2}$ is a solution for the equation $2y^2 = 3y - 2$.

EXAMPLE 9 Determine if $\frac{1}{2}$ is a solution for the equation $5x - 2 = 2 - 3x$.

SOLUTION: $\frac{1}{2}$ is a solution for the equation $5x - 2 = 2 - 3x$. When $x = \frac{1}{2}$, the left-hand side is

$$5x - 2 = 5\left(\frac{1}{2}\right) - 2$$
$$= \frac{5}{2} - 2$$
$$= \frac{1}{2}$$

and the right-hand side is

$$2 - 3x = 2 - 3\left(\frac{1}{2}\right)$$
$$= 2 - \frac{3}{2}$$
$$= \frac{1}{2}$$

The left-hand side of the equation is equal to the right-hand side.

ANSWERS TO PRACTICE EXERCISES

6. Yes. 7. No. 8. Yes. 9. No.

- The equation $y^2 = -2$ has *no* solution. Whatever value is substituted for y in the equation, y^2 will never be negative. (Do you agree?)
- The equation $2(u + 1) = 2u + 2$ has infinitely many solutions. No matter what value is substituted for u, the left-hand side of the equation will be equal to the right-hand side.

EXERCISES 7.5

In Exercises 1–30, write the corresponding algebraic expression.

1. The sum of a and b
2. The sum of c and five times d
3. Twice the sum of e, f, and g
4. Six times the cube of h
5. Seven more than twice the cube of i
6. The sum of j and k minus their product
7. The product of the squares of m and n
8. The square of the product of m and n
9. The quotient of p and q
10. The cube of r, increased by 7
11. Seven more than the cube of r
12. The sum of the squares of s, t, and u
13. The square of the sum of s, t, and u
14. The sum of v and w, divided by their product
15. The product of v and w, divided by their sum
16. The product of the sum and difference of x and y
17. The square of the sum of x and y, minus the cube of their difference
18. The product of x and the sum of y and z
19. The sum of three consecutive integers, if z represents the smallest of the three integers
20. The product of the squares of two consecutive integers, if z represents the larger of the two integers
21. The sum of twice x and five times y
22. The product of y and 4 more than y
23. One-third y diminished by one-ninth w
24. One-half of u diminished by three times v
25. The square of twice the sum of w and 5
26. The sum of 7 and a number, diminished by the number (let n represent the number)
27. The difference of a number and 4 divided by the product of 9 and the number (let p represent the number)
28. One-tenth of the sum of 5 and two-thirds of y
29. Twice the sum of m and 7 divided by 3 times the difference of 4 and n
30. Negative 3 times the difference of four times t and 7, divided by the product of t and 9

In Exercises 31–50, determine if the given value is a solution for the equation shown.

31. $-2y = 6$; -3
32. $3w - 1 = 4$; $\dfrac{5}{3}$
33. $3 - 4p = 0$; $\dfrac{4}{3}$
34. $w^2 = 0.9$; 0.3
35. $\dfrac{1}{2}u = -5$; -10
36. $2(w - 3) = w - 6$; 0
37. $0.2(t - 1) = 1.8$; 10
38. $2(y - 3) = -2(2 - 3y)$; -4
39. $\dfrac{2}{3} - \dfrac{1}{2}u = \dfrac{3}{5}u$; -1
40. $x^2 - 1 = (x - 1)^2$; -1
41. $x + 7 = 2$; -5
42. $2y - 3 = 4$; 1
43. $2 - 3u = 5$; -1
44. $\dfrac{p + 7}{4} = 1$; -3
45. $\dfrac{2q}{3} - 5 = 1$; 9
46. $\dfrac{2y - 1}{13} = 0$; $\dfrac{1}{2}$

47. $3(y - 1) = 2(4 - y); \dfrac{11}{5}$

48. $-5(2u - 1) = 4(3 - 2u); \dfrac{-7}{2}$

49. $2(x - 3) - \dfrac{3}{4}(x + 2) = \dfrac{7}{10}; -3$

50. $0.2(2t - 1) = 0.3(4 - 3t); \dfrac{-1}{8}$

Chapter Summary

In this chapter, you were introduced to algebraic expressions and their evaluation. You also learned to simplify expressions by combining like terms. Sometimes it is necessary to remove symbols of grouping before combining like terms. You were also introduced to the language of algebra, how to change word expressions to algebraic expressions, and the meaning of the solution of an equation in a single variable.

The **key words** introduced in this chapter are listed here in the order in which they appeared in the chapter. You should know what these words mean before proceeding to the next chapter.

algebra (p. 308)	numerical coefficient (p. 309)	unlike terms (p. 314)
variables (p. 308)	evaluate (p. 311)	algebraic equation (p. 324)
constant (p. 308)	like terms (p. 314)	solution of an algebraic equation (p. 324)
algebraic expression (p. 308)		

Review Exercises

7.1 Algebraic Expressions

In Exercises 1–4, determine the number of terms in the given expression and identify *each* term.

1. $(2a^2b - c) + b^3c^4$
2. $(2x - 3y)(x^2 - 3y) - x^3y + 4(x + y)$
3. $-3(a^2 - 2ab + 3b^3)$
4. $2(p - q) + 3(q - p) - 4p - 5q$

In Exercises 5–8, determine the numerical coefficient for the given expressions.

5. $a^2b^3c^4$
6. $-4x^3y^4z$
7. $-3(a + b)^2c$
8. $17(p - q)(p + q)(p^2q^3)$

7.2 Evaluating Algebraic Expressions

In Exercises 9–13, evaluate the given algebraic expressions using the values for the indicated variables.

9. $3u - 2uv; u = 2$ and $v = -3$
10. $2a^2 - 3b^3; a = 3$ and $b = -2$
11. $(2x - 3y)^2; x = -2$ and $y = -3$
12. $-2(x^2 - y) \div (x + y); x = -1$ and $y = -2$
13. $3(a^2 - b^2) \div (2a - 3b); a = 2$ and $b = -2$

7.3 Simplifying Algebraic Expressions by Combining Like Terms

In Exercises 14–17, simplify the given expressions by combining like terms.

14. $3x + 4y - 1 - 5x + y$
15. $a^2b - 2ab^2 - 3a^2b + 4ab^2 - 2a^2b^2$
16. $-3x^2y - 2xy^2 + x^2y^2 + x^2y - 3x^2y^2$
17. $2p - 4q + 3p + 5q - p - 7q$

7.4 Simplifying Algebraic Expressions by Removing Symbols of Grouping

In Exercises 18–21, simplify the given expressions by removing grouping symbols and combining like terms.

18. $3(2m - 3n + 4p) + (2p - m - 7n)$
19. $2(1 - 2x) - 3(x - 3y) - 4(4y - 2)$
20. $3a - [1 - 5(2a - 3)] - 2a$
21. $b - \{2a - [4b + 2(a - 3b) - 6(2a + 3b)]\}$

7.5 The Language of Algebra

In Exercises 22–26, write the corresponding algebraic expression for the given word expression.

22. The sum of p, q, and r
23. Four times the square of k
24. Three times the cube of the sum of a and b
25. Six less than three times t
26. The sum of x and the product of y and z

1. Write an algebraic expression that is the sum of three terms.
2. Write an algebraic expression that is the product of three factors.
3. Write an algebraic expression that is a sum of three terms and such that the sum of the coefficients of the terms is -7.
4. Write an algebraic expression that is the product of two factors and such that each factor is the sum of two terms.
5. Write an algebraic expression that is the sum of two terms and such that each term is the product of two factors.
6. Write an algebraic expression that is a product of a constant factor and two variable factors.
7. Write an algebraic expression that is a product of three factors with each factor written with exponents and such that the sum of the exponents is 0.
8. Let q be three less than twice p and let r be four more than one-half of p. Rewrite the expression $p + 2q - 3r$ in terms of p only and simplify your result.
9. Determine the value of the square of the sum of x, y, and 7, if $x = -2$ and $y = -3$.
10. Determine the value of the cube of the sum of y and 1, multiplied by the square of the sum of y and 1, if $y = -0.5$.

CHAPTER 7 TEST

This test covers materials from Chapter 7. Read each question carefully and answer all questions. If you miss a question, review the appropriate section of the text.

In Exercises 1–5, fill in the blanks to make true statements.

1. The algebraic expression $3x - 5y$ is a __(a)__ of two __(b)__.
2. The algebraic expression $4xyu$ is a __(a)__ of __(b)__ factors.
3. The numerical coefficient in the expression $-2a^2b^3c$ is _____.
4. The numerical coefficient of the second term in the expression $2x - y + 4z$ is _____.
5. The first term in the expression $2x(4 - x) - 3x^2y$ is a __(a)__ of __(b)__ __(c)__.

In Exercises 6–8, evaluate the given expressions for the indicated values of the variables.

6. $2a - 3b$ if $a = -3$ and $b = 2$
7. $4x^2 - (x + y)$ if $x = -2$ and $y = 4$
8. $2p^3q - p$ if $p = -1$ and $q = -3$

In Exercises 9–13, simplify the given expressions by combining like terms.

9. $a + 2b - 2a - b$
10. $2(x + 2y) - 3(2x - y)$
11. $-3(u + 2v + w) - 2(4v - 5w) + 5u$
12. $2p - [3q - 2(p - q) + 4p]$
13. $u - 3(u - v) + 2[v - (2u - 3v)]$

In Exercises 14–20, write an algebraic expression for each of the word expressions.

14. The sum of twice n and 5.
15. Twice the sum of n and 5.
16. Three more than 4 times the square of w.
17. The cube of the sum of w and 7.
18. The square of the sum of x, y, and 7.
19. 6 is diminished by x, and the result diminished by twice the sum of x and 5.
20. The cube of the sum of y and 1, multiplied by the square of the sum of y and 1.

CUMULATIVE REVIEW: CHAPTERS 1–7

Perform the indicated operations in Exercises 1–6.

1. $32.7\% \times 12.1$
2. $\left(-2\dfrac{7}{8}\right)\left(\dfrac{-11}{12}\right)$
3. $-3^2 \div (-3)^3$
4. $\dfrac{23}{25} - \dfrac{11}{13}$
5. $(8 - 13) \times (10 - 17)$
6. $|-8|^2 \div (-2)^3$

In Exercises 7–10, circle the larger number.

7. $\dfrac{-12}{17}$ or $\dfrac{16}{-13}$
8. $|-17|$ or 13
9. 1.6% of 11 or $|-3|$
10. -3^2 or $(-2)^3$
11. Round 2.3×1.7 to the nearest tenth.
12. Round 13.6% of 2.03 to the nearest hundredth.
13. Determine the perimeter of a triangular region whose sides measure 16.29 cm, 11.3 cm, and 11.04 cm.
14. Determine the volume of a cube with edge of length 3.7 ft.
15. Evaluate the expression $2a - 4b$ if $a = -2$ and $b = 3$.
16. Evaluate the expression $-3x^2y + 2z$ if $x = 3$, $y = -2$, and $z = -3$.
17. Simplify the expression $2(3a - b) - 3(a - 2b)$ by combining like terms.
18. Simplify the expression $-3[p - 2(q + 2p) - 3q] + 4p$ by combining like terms.
19. Write an algebraic expression for the sum of x and 10 divided by the product of twice x and 5.
20. Write an algebraic expression for three times n diminished by the quotient of 11 and n.
21. Rewrite 26.8% as a simplified fraction.
22. Determine the simple interest on $1750 at 6.5% for 42 months.
23. The width of a rectangle is 23.7 cm and its length is 36.9 cm. Estimate the area, A, of the rectangle by rounding each dimension to the nearest ten.
24. Maria works on commission and receives 4.5% of net sales. Determine her commission for a week when her gross sales were $3785 and returns amounted to $423.
25. Determine whether the fractions $\dfrac{16}{21}$ and $\dfrac{17}{22}$ are equivalent.

CHAPTER 8
Linear Equations

In the previous chapter, you were introduced to the study of algebra. Algebraic expressions were examined and various algebraic operations were discussed. Applications were introduced together with their solutions. The solutions sometimes involved setting two expressions (either arithmetic or algebraic) equal to each other. The equality of two such expressions is called an equation. Algebra is one of the basic tools necessary for problem solving and many problems are solved by the use of equations.

In this chapter, we will introduce linear equations in a single variable and their solutions. Methods for solving more difficult equations will be given later in the text. Word problems, of the type that can be solved by using linear equations, will also be examined in this chapter.

CHAPTER 8 PRETEST

This pretest covers material on linear equations. Read each question carefully and try to answer all questions. Each question is keyed to the chapter section in which the particular topic is discussed. Check your answers with those given in the back of the text.

Questions 1–4 pertain to Section 8.1. Solve each equation for the given variable. Check your results.

1. $x + 5 = -2$

2. $x - (-2.6) = 1.9$

3. $y - \dfrac{1}{2} = \dfrac{-1}{3}$

4. $w + (-3.09) = 0.5$

Questions 5–8 pertain to Section 8.2. Solve each equation for the given variable. Check your results.

5. $-3p = 17$

6. $\dfrac{w}{-0.3} = 1.6$

7. $y \div \dfrac{2}{3} = \dfrac{-1}{4}$

8. $\dfrac{3}{4}u = \dfrac{-1}{7}$

Questions 9–10 pertain to Section 8.3. Solve each equation for the given variable. Check your results.

9. $3(1 - 2x) = 7$

10. $\dfrac{3u - 1}{7} = 2.6$

Questions 11–12 pertain to Section 8.4. Solve each equation for the given variable. Check your results.

11. $4(1 - 2y) = -5(y + 1)$

12. $3.2(w + 1) = -2.3(w - 3)$

Questions 13–14 pertain to Section 8.5.

13. The sum of three consecutive integers is 54. What are the integers?

14. The perimeter of a rectangle is 300 m. If the length of the rectangle is 3 m less than twice its width, what are the dimensions of the rectangle?

Questions 15–16 pertain to Section 8.6.

15. Two cyclists travel toward each other, in parallel lanes, from points 200 mi apart, at speeds of 45 mph and 55 mph, respectively. When will they pass each other?

16. Jane is one-half her mother's age. In 18 yr, her mother will be one and one-half times as old as Jane is then. How old is each now?

Questions 17–18 pertain to Section 8.7.

17. Solve the formula $P = 2L + 2W$ for L in terms of P and W. Then evaluate L when $P = 54$ m and $W = 11$ m.

18. Solve the formula $T = 2R - 3S^2W$ for W in terms of T, R, and S. Then evaluate W when $T = -13$, $R = -2$, and $S = -1$.

Questions 19–20 pertain to Section 8.8.

19. Write as a ratio: The number of face cards in an ordinary deck of playing cards to the total number of cards in the deck.

20. Solve for n: $\dfrac{2n}{7} = \dfrac{14}{35}$.

8.1 Solving Linear Equations Using Addition or Subtraction

OBJECTIVE

After completing this section, you should be able to solve a simple linear equation, in one variable, using either addition or subtraction.

In Chapter 7, we defined an **equation** to be an equality between two arithmetic or algebraic expressions. An equation may or may not contain **variables.** If an equation contains variables, we are interested in determining what value or values of the variables will make the equation a true statement. Such values are called "solutions" of the equation. Finding all solutions of the equation is called "solving the equation."

In this chapter, we will study the solution of algebraic equations that are called "linear equations in a single variable."

DEFINITION

A **linear equation in a single variable** is an algebraic equation that involves only one variable, with an exponent of 1, and such that there are no fractions with the variable in the denominator.

The following are examples of linear equations in a single variable.

- $x + 4 = 7$ is a linear equation in the variable x.
- $0.2y - 1.3 = 0.5$ is a linear equation in the variable y.
- $\frac{-3}{4}(t + 1) = \frac{2}{3}(5 - 3t)$ is a linear equation in the variable t.
- $4s - 5(s + 1) = 19 - 7(2s - 3)$ is a linear equation in the variable s.

The following are examples of equations that are *not* linear equations in a single variable.

- $x^2 + 3 = 0$ (The variable x has an exponent 2.)
- $xy = 3$ (There are two variables.)
- $2y^{1/2} - y + 5 = 0$ (One term involving the variable y has exponent $\frac{1}{2}$.)
- $3 - \frac{1}{u} = 1$ (The second term has a variable in the denominator.)

If an equation is true for *all* values of the variable involved, then the equation is called an "identity." The following equations are identities:

- $x + 3 = 3 + x$
- $2(y - 1) = 2y - 2$
- $4u - 1 - 3u = u - 1$

It should be noted that some equations in a single variable have *no solutions.* For instance, the equation $x + 2 = x + 1$ has no solutions. No matter what value is substituted for x, 2 more than that value is *not* equal to 1 more than that value.

8.1 Solving Linear Equations Using Addition or Subtraction

In this chapter, most of the linear equations in a single variable will have only one solution. Such equations are called "conditional equations."

Some linear equations are so simple that the solution can be obtained simply by inspection, that is, without doing any calculations. Other linear equations require doing some calculations to determine their solutions.

The basic strategy used to solve a linear equation in one variable is to write an equivalent equation in which the variable is all by itself, with a coefficient of 1, on one side of the equation.

If two linear equations in the same variable have the same solutions, then the equations are called **equivalent equations.** Using this basic strategy to solve a linear equation, we keep changing the given equation into equivalent equations until the variable is all by itself, with a coefficient of 1, on one side of the equation. We will illustrate this with the several examples that follow.

Practice Exercise 1. $x - 7 = 2$

EXAMPLE 1 Solve the equation $x - 4 = 2$ for x.

SOLUTION: To solve the equation, we must isolate x; that is, we must get x all by itself on one side of the equation. If we start with x and subtract 4, to get back to x, what can we do? We can add 4 since adding 4 is the inverse or opposite of subtracting 4. But, if we add 4 to $x - 4$, we must also add 4 to 2 in order to maintain the equality. Hence,

$$x - 4 = 2$$

becomes

$$\begin{aligned}(x - 4) + 4 &= 2 + 4 \quad \text{(Adding 4 to \textit{both} sides)}\\ [x + (-4)] + 4 &= 2 + 4 \quad \text{(Rewriting, using the definition of subtraction)}\\ x + (-4 + 4) &= 2 + 4 \quad \text{(Regrouping)}\\ x + 0 &= 6 \quad \text{(Simplifying)}\\ x &= 6 \quad \text{(Simplifying)}\end{aligned}$$

Therefore, 6 is the required solution. We can check this by substituting 6 for x in the *original* equation.

CHECK: If $x = 6$, does $x - 4 = 2$?

$$\begin{aligned} x - 4 &= 2 \\ 6 - 4 &\overset{?}{=} 2 \\ 2 &= 2 \checkmark \end{aligned}$$

Practice Exercise 2. $y - 0.91 = -1.6$

EXAMPLE 2 Solve the equation $y - 3 = 6$ for y.

SOLUTION: To solve this equation, we add 3 to both sides of the equation (since adding 3 is the inverse or opposite of subtracting 3). Hence:

$$y - 3 = 6$$

becomes

$$\begin{aligned}(y - 3) + 3 &= 6 + 3 \quad \text{(Adding 3 to \textit{both} sides)}\\ [y + (-3)] + 3 &= 6 + 3 \quad \text{(Rewriting, using the definition of subtraction)}\end{aligned}$$

$$y + (-3 + 3) = 6 + 3 \quad \text{(Regrouping)}$$
$$y + 0 = 9 \quad \text{(Simplifying)}$$
$$y = 9 \quad \text{(Simplifying)}$$

Therefore, 9 is the required solution.

CHECK: If $y = 9$, does $y - 3 = 6$?

$$y - 3 = 6$$
$$9 - 3 \; ? \; 6$$
$$6 = 6 \; \checkmark$$

In the following examples, some of the steps in the solutions will be combined.

Practice Exercise 3. $z - \dfrac{3}{4} = \dfrac{4}{3}$

EXAMPLE 3 Solve the equation $w - \dfrac{2}{3} = \dfrac{4}{7}$ for w.

SOLUTION:
$$w - \frac{2}{3} = \frac{4}{7} \quad \left(\frac{2}{3} \text{ is being } \textit{subtracted} \text{ from } w.\right)$$
$$w = \frac{4}{7} + \frac{2}{3} \quad \left(\text{Adding } \frac{2}{3} \text{ to } \textit{both} \text{ sides}\right)$$
$$w = \frac{26}{21} \quad \text{(Adding fractions)}$$

CHECK:
$$w - \frac{2}{3} = \frac{4}{7}$$
$$\frac{26}{21} - \frac{2}{3} \; ? \; \frac{4}{7}$$
$$\frac{26}{21} - \frac{14}{21} \; ? \; \frac{4}{7}$$
$$\frac{12}{21} \; ? \; \frac{4}{7}$$
$$\frac{4}{7} = \frac{4}{7} \; \checkmark$$

Practice Exercise 4.
$t - 2\dfrac{1}{2} = -1\dfrac{2}{5}$

EXAMPLE 4 Solve the equation $r - 1.46 = -2.69$ for r.

SOLUTION:

$$r - 1.46 = -2.69 \quad \text{(1.46 is being } \textit{subtracted} \text{ from } r)$$
$$r = -2.69 + 1.46 \quad \text{(Adding 1.46 to } \textit{both} \text{ sides)}$$
$$= -1.23 \quad \text{(Adding decimals)}$$

CHECK:
$$r - 1.46 = -2.69$$
$$-1.23 - 1.46 \; ? \; -2.69$$
$$-1.23 + (-1.46) \; ? \; -2.69$$
$$-2.69 = -2.69 \; \checkmark$$

ANSWERS TO PRACTICE EXERCISES

1. $x = 9$. 2. $y = -0.69$. 3. $z = \dfrac{25}{12} \left(\text{or } 2\dfrac{1}{12}\right)$. 4. $t = \dfrac{11}{10} \left(\text{or } 1\dfrac{1}{10}\right)$.

8.1 Solving Linear Equations Using Addition or Subtraction

Examples 1–4 illustrate the following rule for solving linear equations by using addition.

Addition Rule for Solving Linear Equations

The *same* quantity may be *added to both sides* of a linear equation to obtain an equivalent equation. That is, if

$$x - a = b$$

then

$$(x - a) + a = b + a$$

or

$$x = b + a$$

Practice Exercise 5.
$w + \dfrac{1}{2} = \dfrac{-1}{3}$

EXAMPLE 5 Solve the equation $u + 5 = 2$ for u.

SOLUTION: To solve the equation, we must isolate u; that is, we must get u all by itself on one side of the equation. If we start with u and add 5, to get back to u, what can we do? We can subtract 5 since subtracting 5 is the inverse or opposite of adding 5. But, if we subtract 5 from $u + 5$, we must also subtract 5 from 2 in order to maintain the equality. Hence,

$$u + 5 = 2$$

becomes
$(u + 5) - 5 = 2 - 5$ (Subtracting 5 from *both* sides)
$u + (5 - 5) = 2 - 5$ (Regrouping)
$u + 0 = -3$ (Simplifying)
$u = -3$ (Simplifying)

Therefore, -3 is the required solution.

CHECK:
$u + 5 = 2$
$-3 + 5 \;?\; 2$
$2 = 2 \;\checkmark$

Practice Exercise 6.
$t + 2.3 = 1.41$

EXAMPLE 6 Solve the equation $y + 5 = 9$ for y.

SOLUTION:
$y + 5 = 9$ (5 is being *added* to y.)
$y = 9 - 5$ (*Subtracting* 5 from *both* sides)
$y = 4$ (Simplifying)

CHECK:
$y + 5 = 9$
$4 + 5 \;?\; 9$
$9 = 9 \;\checkmark$

Practice Exercise 7.
$v + 17 = -3$

EXAMPLE 7 Solve the equation $x + 7 = 1$ for x.

SOLUTION:
$x + 7 = 1$ (7 is being *added* to x.)
$x = 1 - 7$ (*Subtracting* 7 from *both* sides)
$x = -6$ (Simplifying)

CHECK:
$x + 7 = 1$
$-6 + 7 \;?\; 1$
$1 = 1 \;\checkmark$

Practice Exercise 8.
$x + (-8) = -2$

EXAMPLE 8 Solve the equation $s + \frac{1}{2} = \frac{1}{3}$ for s.

SOLUTION:

$s + \frac{1}{2} = \frac{1}{3}$ ($\frac{1}{2}$ is being *added* to s.)

$s = \frac{1}{3} - \frac{1}{2}$ (*Subtracting* $\frac{1}{2}$ from *both* sides)

$s = \frac{-1}{6}$ (Simplifying)

CHECK:

$s + \frac{1}{2} = \frac{1}{3}$

$\frac{-1}{6} + \frac{1}{2} \;?\; \frac{1}{3}$

$\frac{-1}{6} + \frac{3}{6} \;?\; \frac{1}{3}$

$\frac{2}{6} \;?\; \frac{1}{3}$

$\frac{1}{3} = \frac{1}{3}$ ✓

Practice Exercise 9.
$y + \left(-1\frac{1}{3}\right) = 3\frac{2}{7}$

EXAMPLE 9 Solve the equation $t + 0.4 = 0.23$ for t.

SOLUTION:

$t + 0.4 = 0.23$ (0.4 is being *added* to t.)
$t = 0.23 - 0.4$ (*Subtracting* 0.4 from *both* sides)
$t = -0.17$ (Simplifying)

CHECK:

$t + 0.4 = 0.23$
$-0.17 + 0.4 \;?\; 0.23$
$0.23 = 0.23$ ✓

ANSWERS TO PRACTICE EXERCISES

5. $w = \frac{-5}{6}$.   6. $t = -0.89$.   7. $v = -20$.   8. $x = 6$.   9. $y = 4\frac{13}{21}$.

Examples 5–9 illustrate the following rule for solving linear equations using subtraction.

Subtraction Rule for Solving Linear Equations

The *same* quantity may be *subtracted* from *both* sides of a linear equation to obtain an equivalent equation. That is, if

$$x + a = b$$

then

$$(x + a) - a = b - a$$

or

$$x = b - a$$

EXERCISES 8.1

Solve each of the following equations for the given variable. Check your results.

1. $x + 5 = 4$
2. $u + 5 = -2$
3. $y + 7 = 1$
4. $z + 19 = -2$
5. $w + 6 = -4$
6. $y + \frac{1}{5} = \frac{-2}{7}$
7. $p + (-2) = -3$
8. $q + (-11) = 17$
9. $y - 7 = 2$
10. $x - 9 = 0$
11. $x - \frac{1}{4} = \frac{2}{3}$
12. $z - \frac{5}{7} = \frac{-1}{9}$
13. $z - \frac{2}{3} = \frac{1}{2}$
14. $u - \frac{1}{4} = \frac{1}{5}$
15. $x + \frac{1}{2} = \frac{2}{3}$
16. $y + \frac{1}{5} = \frac{-2}{9}$
17. $p + (-2.9) = -5.2$
18. $m - (-1.03) = 2.6$
19. $u + 1.2 = 3.1$
20. $w + 6.1 = -2.34$
21. $z - \frac{4}{3} = \frac{1}{2}$
22. $u - \frac{1}{4} = \frac{4}{5}$
23. $s + \frac{2}{3} = \frac{3}{11}$
24. $t - \frac{4}{5} = \frac{-1}{7}$
25. $r - 1.6 = 2.7$
26. $s - 0.2 = -1.93$
27. $t - 19 = -26$
28. $w - \frac{6}{7} = \frac{1}{9}$
29. $s + (-3) = 4$
30. $y - (-1.2) = 6.01$
31. $m + 7 = \frac{1}{2}$
32. $n - 2\frac{1}{2} = 3\frac{1}{4}$
33. $q - \left(\frac{-1}{2}\right) = \frac{2}{3}$
34. $p + \frac{1}{2} = 0.3$
35. $x - 11.7 = 9.83$
36. $y - 23.31 = -12.7$
37. $z + \frac{11}{19} = 2\frac{2}{3}$
38. $p - 4\frac{2}{7} = -5\frac{1}{3}$
39. $q + \left(-3\frac{5}{9}\right) = -2\frac{1}{6}$
40. $t - 101.7 = -69.06$

8.2 Solving Linear Equations Using Multiplication or Division

OBJECTIVE

After completing this section, you should be able to solve a simple linear equation, in one variable, using either multiplication or division.

In Section 8.1, we solved linear equations using either addition or subtraction. In the following examples, we will solve linear equations using multiplication or division.

Practice Exercise 1.
$\frac{y}{-3} = -4$

EXAMPLE 1 Solve the equation $\frac{t}{4} = 5$ for t.

SOLUTION: In the given equation, observe that t is being divided by 4 (which is *nonzero*) and the quotient is 5. To solve the equation, then, we multiply both sides of the equation by 4 (since multiplying by 4 is the inverse or opposite of dividing by 4). Hence,

$$\frac{t}{4} = 5$$

becomes

$$(4)\left(\frac{t}{4}\right) = (4)(5) \quad \text{(Multiplying \textit{both} sides by 4)}$$

or

$$t = 20 \quad \text{(Simplifying)}$$

Therefore, 20 is the required solution.

CHECK:
$$\frac{t}{4} = 5$$
$$\frac{20}{4} \ ? \ 5$$
$$5 = 5 \ \checkmark$$

Practice Exercise 2.
$\frac{x}{6.1} = 2.4$

EXAMPLE 2 Solve the equation $\frac{x}{5} = 3.2$ for x.

SOLUTION:
$$\frac{x}{5} = 3.2 \quad (x \text{ is being \textit{divided} by 5.})$$
$$(5)\left(\frac{x}{5}\right) = (5)(3.2) \quad (\textit{Multiplying both} \text{ sides by 5})$$
$$x = 16 \quad \text{(Solution)}$$

CHECK:
$$x \div 5 = 3.2$$
$$16 \div 5 \ ? \ 3.2$$
$$3.2 = 3.2 \ \checkmark$$

8.2 Solving Linear Equations Using Multiplication or Division

Practice Exercise 3. $w \div \frac{2}{7} = \frac{1}{4}$

EXAMPLE 3 Solve the equation $\dfrac{y}{-2} = \dfrac{1}{3}$ for y.

SOLUTION:

$$\dfrac{y}{-2} = \dfrac{1}{3} \qquad \text{(y is being } divided \text{ by } -2.)$$

$$(-2)\left(\dfrac{y}{-2}\right) = (-2)\left(\dfrac{1}{3}\right) \qquad \text{(\emph{Multiplying both} sides by } -2)$$

$$y = \dfrac{-2}{3} \qquad \text{(Solution)}$$

Check:

$$\dfrac{y}{-2} = \dfrac{1}{3}$$

$$\left(\dfrac{-2}{3}\right) \div (-2) \;?\; \dfrac{1}{3}$$

$$\dfrac{1}{3} = \dfrac{1}{3} \checkmark$$

Practice Exercise 4. $t \div 2\dfrac{1}{4} = -3\dfrac{1}{2}$

EXAMPLE 4 Solve the equation $\dfrac{u}{1.2} = 3.1$ for u.

SOLUTION:

$$\dfrac{u}{1.2} = 3.1 \qquad \text{(u is being } divided \text{ by } 1.2.)$$

$$(1.2)\left(\dfrac{u}{1.2}\right) = (1.2)(3.1) \qquad \text{(\emph{Multiplying both} sides by } 1.2)$$

$$u = 3.72 \qquad \text{(Solution)}$$

Check:

$$\dfrac{u}{1.2} = 3.1$$

$$\dfrac{3.72}{1.2} \;?\; 3.1$$

$$3.1 = 3.1 \checkmark$$

ANSWERS TO PRACTICE EXERCISES

1. $y = 12.$ 2. $x = 14.64.$ 3. $w = \dfrac{1}{14}.$ 4. $t = -7\dfrac{7}{8}.$

Examples 1–4 illustrate the following rule for solving linear equations using multiplication.

Multiplication Rule for Solving Linear Equations

Both sides of a linear equation may be *multiplied* by the same *nonzero* quantity to obtain an equivalent equation. That is, if

$$\dfrac{x}{a} = b \qquad (a \neq 0)$$

then
$$a\left(\frac{x}{a}\right) = (a)(b)$$
or
$$x = ab$$

Practice Exercise 5. $-4x = 6$

EXAMPLE 5 Solve the equation $2u = 5$ for u.

SOLUTION: To solve the equation, we divide both sides of the equation by 2 (since dividing by 2 is the inverse or opposite of multiplying by 2). Hence,

$$2u = 5$$

becomes

$$\frac{2u}{2} = \frac{5}{2} \quad \text{(Dividing } both \text{ sides by 2)}$$

or

$$u = \frac{5}{2} \quad \text{(Simplifying)}$$

Therefore, $\frac{5}{2}$ is the required solution.

CHECK:
$$2u = 5$$
$$2\left(\frac{5}{2}\right) \; ? \; 5$$
$$5 = 5 \checkmark$$

Practice Exercise 6. $0.13y = 2.6$

EXAMPLE 6 Solve the equation $3x = 23$ for x.

SOLUTION:
$$3x = 23 \quad (x \text{ is being } multiplied \text{ by 3.})$$
$$\frac{3x}{3} = \frac{23}{3} \quad \text{(Dividing both sides by 3)}$$
$$x = \frac{23}{3} \left(\text{or } 7\frac{2}{3}\right) \quad \text{(Solution)}$$

CHECK:
$$3x = 23$$
$$3\left(\frac{23}{3}\right) \; ? \; 23$$
$$23 = 23 \checkmark$$

Practice Exercise 7. $\frac{3}{5}w = \frac{-1}{2}$

EXAMPLE 7 Solve the equation $2.3y = 0.69$ for y.

SOLUTION:
$$2.3y = 0.69 \quad (y \text{ is being } multiplied \text{ by 2.3.})$$
$$\frac{2.3y}{2.3} = \frac{0.69}{2.3} \quad \text{(Dividing both sides by 2.3)}$$
$$y = 0.3 \quad \text{(Solution)}$$

8.2 Solving Linear Equations Using Multiplication or Division

$$\text{CHECK:} \quad 2.3y = 0.69$$
$$(2.3)(0.3) \; ? \; 0.69$$
$$0.69 = 0.69 \; ✓$$

Practice Exercise 8.

$$\left(2\frac{3}{8}\right)u = -4\frac{2}{5}$$

EXAMPLE 8 Solve the equation $\frac{3}{4}t = \frac{1}{3}$ for t.

SOLUTION:
$$\frac{3}{4}t = \frac{1}{3} \qquad \left(t \text{ is being } multiplied \text{ by } \frac{3}{4}.\right)$$

$$\frac{3}{4}t \div \frac{3}{4} = \frac{1}{3} \div \frac{3}{4} \qquad \left(\text{Dividing both sides by } \frac{3}{4}\right)$$

$$\frac{3}{4}t \times \frac{4}{3} = \frac{1}{3} \times \frac{4}{3} \qquad \text{(Multiplying by reciprocal)}$$

$$\left(\frac{3}{4} \times \frac{4}{3}\right)t = \frac{1}{3} \times \frac{4}{3} \qquad \text{(Rewriting)}$$

$$t = \frac{4}{9} \qquad \text{(Solution)}$$

CHECK:
$$\frac{3}{4}t = \frac{1}{3}$$
$$\left(\frac{3}{4}\right)\left(\frac{4}{9}\right) \; ? \; \frac{1}{3}$$
$$\frac{1}{3} = \frac{1}{3} \; ✓$$

ANSWERS TO PRACTICE EXERCISES

5. $x = \frac{-3}{2}$ (or $-1\frac{1}{2}$).   6. $y = 20$.   7. $w = \frac{-5}{6}$.   8. $u = -1\frac{81}{95}$.

Examples 5–8 illustrate the following rule for solving linear equations using division.

Division Rule for Solving Linear Equations

Both sides of a linear equation may be *divided* by the same *nonzero* quantity to obtain an equivalent equation. That is, if

$$ax = b \qquad (a \neq 0)$$

then

$$\frac{ax}{a} = \frac{b}{a}$$

or

$$x = \frac{b}{a}$$

EXERCISES 8.2

Solve each of the following equations for the given variable. Check your results.

1. $x \div 3 = 6$
2. $y \div 4 = -2$
3. $t \div 5 = 1.2$
4. $p \div (-4) = 6.5$
5. $z \div (-2) = 4.2$
6. $u \div (-1.2) = 7$
7. $3u = 9.6$
8. $5.2u = 1.56$
9. $2x = 7$
10. $3x = 4$
11. $-4y = 9$
12. $-5z = -1.2$
13. $\frac{2}{3}x = \frac{1}{7}$
14. $\frac{-1}{7}y = \frac{-3}{4}$
15. $r \div 3 = \frac{-1}{5}$
16. $s \div (-4) = \frac{-2}{3}$
17. $w \div (-0.5) = -1.1$
18. $t \div (-1.9) = 3$
19. $\frac{2}{3}u = \frac{1}{2}$
20. $\frac{1}{7}t = \frac{2}{3}$
21. $\frac{-4}{11}t = \frac{-2}{3}$
22. $\frac{5}{7}q = \frac{-7}{5}$
23. $9.2w = -2.76$
24. $-1.7s = -0.085$
25. $\frac{m}{-6} = \frac{13}{27}$
26. $\frac{n}{6} = \frac{-11}{17}$
27. $-23t = 470$
28. $56u = -1247$
29. $y \div 2.3 = 6.1$
30. $1.9t = 0.38$
31. $\frac{4}{7}p = \frac{-2}{3}$
32. $w \div 0.9 = 37$
33. $\left(3\frac{1}{3}\right)x = 2\frac{1}{4}$
34. $\left(-5\frac{1}{7}\right)y = -3\frac{1}{4}$
35. $p \div (-0.2) = -3.19$
36. $q \div (-11.2) = 0.03$
37. $x \div \frac{4}{9} = \frac{-1}{2}$
38. $y \div \left(-6\frac{5}{7}\right) = -4\frac{1}{9}$
39. $t \div (-12.7) = \frac{1}{4}$
40. $w \div (-0.13) = -7\frac{1}{2}$

8.3 Solving Linear Equations Using More than One Operation

OBJECTIVE

After completing this section, you should be able to solve linear equations in one variable that involve more than one operation.

It is often necessary to use more than one operation to solve an equation. The basic procedure involved is to write an equivalent equation in which all the terms containing the variable are on one side of the equation and all other terms are on the other side of the equation.

Practice Exercise 1.
$3y + 4 = -2$

EXAMPLE 1 Solve the equation $2x - 3 = 4$ for x.

SOLUTION: In the equation $2x - 3 = 4$, observe that x is multiplied by 2. Then, 3 is subtracted from the product, $2x$, to produce the difference 4. We will look at this example in two parts.

PART 1: Forming the equation. (Here, we examine how the equation was formed.)

STEP 1: x is multiplied by 2 to get $2x$.

STEP 2: 3 is subtracted from $2x$ to get $2x - 3$.

Result: The result is 4.

PART 2: Solving the equation. To solve the equation, we perform the inverse or opposite operation in each step in Part 1 *but in reverse order*.

STEP 1: *Add* 3 to $2x - 3$ (and also *add* 3 to 4). Therefore,

$$2x - 3 = 4$$

becomes

$$(2x - 3) + 3 = 4 + 3$$

or

$$2x = 7$$

STEP 2: *Divide* $2x$ by 2 (and also *divide* 7 by 2). Therefore,

$$2x = 7$$

becomes

$$\frac{2x}{2} = \frac{7}{2}$$

or

$$x = \frac{7}{2}$$

Therefore, $\frac{7}{2}$ is the required solution.

CHECK:
$$2x - 3 = 4$$
$$2\left(\frac{7}{2}\right) - 3 \; ? \; 4$$
$$7 - 3 \; ? \; 4$$
$$4 = 4 \checkmark$$

> **Practice Exercise 2.**
> $(s - 3) \div 2 = \frac{1}{5}$

EXAMPLE 2 Solve the equation $3y + 1 = 7$ for y.

SOLUTION: In the equation $3y + 1 = 7$, observe that y is multiplied by 3. Then 1 is added to the product, $3y$, to produce the sum of 7.

PART 1: Forming the equation.
  STEP 1: y is multiplied by 3 to get $3y$.
  STEP 2: 1 is added to $3y$ to get $3y + 1$.
  Result: The result is 7.

PART 2: Solving the equation. To solve the equation, we perform the inverse or opposite operation in each step in Part 1 *but in reverse order*.
  STEP 1: *Subtract* 1 from $3y + 1$ (and also *subtract* 1 from 7). Therefore,
  $$3y + 1 = 7$$
  becomes
  $$(3y + 1) - 1 = 7 - 1$$
  or
  $$3y = 6$$
  STEP 2: *Divide* $3y$ by 3 (and also *divide* 6 by 3). Therefore,
  $$3y = 6$$
  becomes
  $$\frac{3y}{3} = \frac{6}{3}$$
  or
  $$y = 2$$

Therefore, 2 is the required solution.

CHECK:
$$3y + 1 = 7$$
$$3(2) + 1 \; ? \; 7$$
$$6 + 1 \; ? \; 7$$
$$7 = 7 \checkmark$$

> **Practice Exercise 3.**
> $4(3 - 2t) = 5$

EXAMPLE 3 Solve the equation $\frac{u - 3}{4} = 2$ for u.

SOLUTION: In the given equation, observe that 3 is subtracted from u. Then, the result, $u - 3$, is divided by 4, to produce the quotient 2.

PART 1: Forming the equation.
  STEP 1: 3 is subtracted from u to get $u - 3$.
  STEP 2: $u - 3$ is divided by 4 to get $\frac{u - 3}{4}$.
  Result: The result is 2.

8.3 Solving Linear Equations Using More than One Operation

PART 2: Solving the equation.

STEP 1: *Multiply* $\frac{u-3}{4}$ by 4 (and also *multiply* 2 by 4). Therefore,

$$\frac{u-3}{4} = 2$$

becomes

$$4\left(\frac{u-3}{4}\right) = 4(2)$$

or

$$u - 3 = 8$$

STEP 2: *Add* 3 to $u - 3$ (and also *add* 3 to 8). Therefore,

$$u - 3 = 8$$

becomes

$$u = 11$$

Therefore, 11 is the required solution.

CHECK:
$$\frac{u-3}{4} = 2$$
$$\frac{11-3}{4} \,?\, 2$$
$$\frac{8}{4} \,?\, 2$$
$$2 = 2 \checkmark$$

Practice Exercise 4.
$0.3s - 0.19 = 2.0$

EXAMPLE 4 Solve the equation $2(y - 3) = 7$ for y.

SOLUTION: Using the distributive property, we rewrite $2(y - 3) = 2y - 6$. Therefore, the equation $2(y - 3) = 7$ becomes $2y - 6 = 7$. In the later equation, observe that we start with y, multiply by 2, subtract 6 from the product, and the result is 7.

PART 1: Forming the equation.

STEP 1: y is multiplied by 2 to get $2y$.

STEP 2: 6 is subtracted from $2y$ to get $2y - 6$.

Result: The result is 7.

PART 2: Solving the equation.

STEP 1: *Add* 6 to $2y - 6$ (and also *add* 6 to 7). Therefore,

$$2y - 6 = 7$$

becomes

$$(2y - 6) + 6 = 7 + 6$$

or

$$2y = 13$$

Step 2: *Divide* 2y by 2 (and also *divide* 13 by 2). Therefore,

$$2y = 13$$

becomes

$$\frac{2y}{2} = \frac{13}{2}$$

or

$$y = \frac{13}{2}$$

Therefore, $\frac{13}{2}$ is the required solution. The check is left to you.

ANSWERS TO PRACTICE EXERCISES

1. $y = -2$. 2. $s = \frac{17}{5}$ (or $3\frac{2}{5}$). 3. $t = \frac{7}{8}$. 4. $s = 7.3$.

EXERCISES 8.3

Solve each of the following equations for the given variable. Check your results in the given equations.

1. $2x - 1 = 3$
2. $3y + \frac{1}{2} = \frac{2}{3}$
3. $3m + 1 = 2$
4. $\frac{t}{-2} + 3 = 4$
5. $(x + 3) \div 2 = 4$
6. $(y - 4) \div 3 = 0$
7. $(t - 5) \div 2 = \frac{1}{3}$
8. $(7 - u) \div 5 = 6$
9. $4y - 3 = \frac{1}{2}$
10. $6 - 2t = 7$
11. $0.2p - 0.3 = 1.7$
12. $1.2q + 2.3 = -0.9$
13. $3(x + 4) = 7$
14. $4(5 - 3y) = 6$
15. $-2(u + 6) = 5$
16. $-3(t + 1) = 6$
17. $2.3(w - 1) = 4.9$
18. $-0.7(t + 0.2) = 1.17$
19. $2(1 + 3r) = 6$
20. $(9 - q) \div 7 = 6$
21. $(2x + 5) \div 3 = 4$
22. $(5 - 2y) \div 4 = 1$
23. $\frac{4}{7}t - 5 = 2$
24. $5 - \frac{2}{3}t = 7$
25. $5.1w + 0.19 = -1.6$
26. $23.5 - 1.7v = -3.91$
27. $(7 - u) \div 5 = \frac{2}{3}$
28. $-4(1 - w) = 3$
29. $-5y + 7 = \frac{1}{2}$
30. $(2s + 5) \div 3 = 0$
31. $\frac{-3}{4}x + 5 = \frac{2}{3}$
32. $\frac{-2}{3}(y + 5) = 1$

33. $(3 - 2w) \div (-4) = \dfrac{1}{6}$

34. $4(3 - 7r) = -11$

35. $(11t - 3) \div (-2) = 5$

36. $\dfrac{2}{7}(5x + 2) = \dfrac{-1}{3}$

37. $\dfrac{2}{7}u - 3\dfrac{1}{2} = 6\dfrac{2}{3}$

38. $\left(p - 4\dfrac{2}{9}\right) \div 3 = 6\dfrac{1}{3}$

39. $5 - \left(q \div \dfrac{2}{3}\right) = 5\dfrac{1}{2}$

40. $(2.7 - 3.2t) \div 4 = 1.9$

8.4 Simplifying First before Solving Linear Equations

OBJECTIVE

After completing this section, you should be able to solve linear equations containing algebraic expressions that can be simplified by combining like terms.

Sometimes a linear equation in a single variable contains more than one term with the variable. Whenever possible, combine like terms first and rewrite the equation so that all of the variable terms are on one side of the equation. This procedure will be illustrated in the following examples.

Practice Exercise 1.
$2q + 3 = 1 + q$

EXAMPLE 1 Solve the equation $4p - 3 = 5 - p$ for p.

SOLUTION: In the equation $4p - 3 = 5 - p$, the variable p is found on both sides. We must try to write an equivalent equation in which all terms containing p are on one side of the equation and all the remaining terms on the other side. If we add p to both sides of the equation, then only the left side of the equation will contain variable terms. Hence,

$$4p - 3 = 5 - p$$

becomes

$$(4p - 3) + p = (5 - p) + p \quad \text{(Adding } p \text{ to } both \text{ sides)}$$
$$5p - 3 = 5 \quad \text{(Simplifying)}$$
$$(5p - 3) + 3 = 5 + 3 \quad \text{(Adding 3 to } both \text{ sides)}$$
$$5p = 8 \quad \text{(Simplifying)}$$
$$\dfrac{5p}{5} = \dfrac{8}{5} \quad \text{(Dividing } both \text{ sides by 5)}$$
$$p = \dfrac{8}{5} \quad \text{(Simplifying)}$$

Therefore, the required solution is $\dfrac{8}{5}$.

CHECK:
$$4p - 3 = 5 - p$$
$$4\left(\frac{8}{5}\right) - 3 \stackrel{?}{=} 5 - \frac{8}{5}$$
$$\frac{32}{5} - 3 \stackrel{?}{=} \frac{17}{5}$$
$$\frac{17}{5} = \frac{17}{5} \checkmark$$

Practice Exercise 2. $-2(r + 1) - 5 = 5 + 2(r - 2)$

EXAMPLE 2 Solve the equation $2(t - 3) + 4 = 3 - 4(t + 1)$ for t.

SOLUTION:

$$2(t - 3) + 4 = 3 - 4(t + 1)$$
$$2t - 6 + 4 = 3 - 4t - 4 \quad \text{(Removing parentheses)}$$
$$2t - 2 = -4t - 1 \quad \text{(Combining like terms)}$$
$$6t - 2 = -1 \quad \text{(Adding } 4t \text{ to } both \text{ sides)}$$
$$6t = 1 \quad \text{(Adding 2 to } both \text{ sides)}$$
$$t = \frac{1}{6} \quad \text{(Dividing } both \text{ sides by 6)}$$

Therefore, the required solution is $\frac{1}{6}$. The check is left to you. Remember, however, to check in the *original* equation.

Practice Exercise 3.
$\frac{1}{2}(t - 4) = 7 - \frac{1}{3}(t + 6)$

EXAMPLE 3 Solve the equation $3\left(m + \frac{1}{2}\right) = 6 - 2\left(m - \frac{1}{3}\right)$ for m.

SOLUTION:

$$3\left(m + \frac{1}{2}\right) = 6 - 2\left(m - \frac{1}{3}\right)$$
$$3m + \frac{3}{2} = 6 - 2m + \frac{2}{3} \quad \text{(Removing parentheses)}$$
$$3m + \frac{3}{2} = \frac{20}{3} - 2m \quad \text{(Combining like terms)}$$
$$5m + \frac{3}{2} = \frac{20}{3} \quad \text{(Adding } 2m \text{ to } both \text{ sides)}$$
$$5m = \frac{31}{6} \quad \left(\text{Subtracting } \frac{3}{2} \text{ from } both \text{ sides}\right)$$
$$m = \frac{31}{30} \quad \text{(Dividing } both \text{ sides by 5)}$$

Therefore, the required solution is $m = \frac{31}{30}$. You should check this result.

Practice Exercise 4.
$3(x + 1) = \frac{2x - 3}{3}$

EXAMPLE 4 Solve the equation $\frac{p - 3}{4} = 6(p + 1)$ for p.

SOLUTION:

$$\frac{p - 3}{4} = 6(p + 1)$$
$$p - 3 = 24(p + 1) \quad \text{(Multiplying } both \text{ sides by 4 to remove the fraction)}$$

8.4 Simplifying First before Solving Linear Equations

$$p - 3 = 24p + 24 \quad \text{(Removing parentheses)}$$
$$p = 24p + 27 \quad \text{(Adding 3 to both sides)}$$
$$-23p = 27 \quad \text{(Subtracting 24p from both sides)}$$
$$p = \frac{-27}{23} \quad \text{(Dividing both sides by } -23\text{)}$$

Therefore, the required solution is $\frac{-27}{23}$. Again, you should check this result in the *original* equation.

Practice Exercise 5. $3(x - 2) - 5x + 6 = -(x + 3)$

EXAMPLE 5 Solve the equation $4y - 2(y + 3) - 7 = \frac{1}{2}(2y - 3)$ for y.

SOLUTION:

$$4y - 2(y + 3) - 7 = \frac{1}{2}(2y - 3)$$
$$8y - 4(y + 3) - 14 = 2y - 3 \quad \text{(Multiplying both sides by 2 to remove the fraction)}$$
$$8y - 4y - 12 - 14 = 2y - 3 \quad \text{(Removing parentheses)}$$
$$4y - 26 = 2y - 3 \quad \text{(Combining like terms)}$$
$$2y - 26 = -3 \quad \text{(Subtracting 2y from both sides)}$$
$$2y = 23 \quad \text{(Adding 26 to both sides)}$$
$$y = \frac{23}{2} \quad \text{(Dividing both sides by 2)}$$

Therefore, the required solution is $\frac{23}{2}$.

Practice Exercise 6.
$6s - 5(s - 2) = 4 + 2(s - 3)$

EXAMPLE 6 Solve the equation $0.2(r + 3) - 0.7r = -0.4(1 - 2r)$ for r.

SOLUTION:

$$0.2(r + 3) - 0.7r = -0.4(1 - 2r)$$
$$10[0.2(r + 3) - 0.7r] = 10[-0.4(1 - 2r)] \quad \text{(Multiplying both sides by 10 to remove decimals)}$$
$$2(r + 3) - 7r = -4(1 - 2r) \quad \text{(Simplifying)}$$
$$2r + 6 - 7r = -4 + 8r \quad \text{(Removing parentheses)}$$
$$6 - 5r = -4 + 8r \quad \text{(Combining like terms)}$$
$$6 = -4 + 13r \quad \text{(Adding 5r to both sides)}$$
$$10 = 13r \quad \text{(Adding 4 to both sides)}$$
$$\frac{10}{13} = r \quad \text{(Dividing both sides by 13)}$$

Therefore, the required solution is $\frac{10}{13}$.

CHECK:
$$0.2(r + 3) - 0.7r = -0.4(1 - 2r)$$
$$0.2\left(\frac{10}{13} + 3\right) - 0.7\left(\frac{10}{13}\right) \; ? \; -0.4\left[1 - 2\left(\frac{10}{13}\right)\right]$$
$$0.2\left(\frac{49}{13}\right) - \left(\frac{7}{13}\right) \; ? \; -0.4\left[1 - \left(\frac{20}{13}\right)\right]$$
$$\left(\frac{9.8}{13}\right) - \left(\frac{7}{13}\right) \; ? \; -0.4\left(\frac{-7}{13}\right)$$
$$\frac{2.8}{13} = \frac{2.8}{13} \checkmark$$

Practice Exercise 7.
$$\frac{u-3}{2} = \frac{3u-2}{3}$$

EXAMPLE 7 Solve the equation $\frac{x+1}{4} = \frac{2x-1}{3}$ for x.

SOLUTION:

$$\frac{x+1}{4} = \frac{2x-1}{3}$$

$$12\left(\frac{x+1}{4}\right) = 12\left(\frac{2x-1}{3}\right) \quad \text{(Multiplying both sides by 12 to remove denominators)}$$

$$3(x+1) = 4(2x-1) \quad \text{(Simplifying)}$$
$$3x + 3 = 8x - 4 \quad \text{(Removing parentheses)}$$
$$3 = 5x - 4 \quad \text{(Subtracting } 3x \text{ from both sides)}$$
$$7 = 5x \quad \text{(Adding 4 to both sides)}$$
$$\frac{7}{5} = x \quad \text{(Dividing both sides by 5)}$$

Therefore, the required solution is $\frac{7}{5}$. The check is left to you.

ANSWERS TO PRACTICE EXERCISES

1. $q = -2$. 2. $r = -2$. 3. $t = \frac{42}{5}$. 4. $x = \frac{-12}{7}$. 5. $x = 3$. 6. $s = 12$. 7. $u = \frac{-5}{3}$.

We will now summarize the procedure that we have developed for solving linear equations in one variable.

Procedure for Solving Linear Equations in One Variable

1. If possible, multiply both sides of the equation by the same nonzero quantity to get an equivalent equation without fractions or decimals.
2. Remove all symbols of grouping, if any.
3. Combine like terms on each side of the equation, if appropriate.
4. Add (or subtract) the same quantity to (or from) *both* sides of the equation to isolate the variable term.
5. Multiply (or divide) *both* sides of the equation by the same *nonzero* number to get a coefficient of 1.
6. Check your results in the *original equation*.

EXERCISES 8.4

Solve each of the following equations for the given variable. Check your results in the given equations.

1. $2(x - 1) = x$
2. $3x - 2 = 2(x - 1)$
3. $5(x - 3) = 7 - 2x$
4. $3y - 7 = 2(4 - 3y)$

8.4 Simplifying First before Solving Linear Equations

5. $2(w + 7) = 3w - 1$
6. $-4(u + 1) = 5u - 3$
7. $-3(t - 4) = 5(2 - 3t)$
8. $2p - 3(p + 1) = 5(1 - p)$
9. $\dfrac{x - 3}{4} = \dfrac{2x + 1}{5}$
10. $\dfrac{2 - y}{5} = \dfrac{2y - 1}{3}$
11. $\dfrac{2r - 1}{-3} = \dfrac{3 - 2r}{-4}$
12. $2.1(w - 3) = -0.6(2w + 5)$
13. $4(x + 1) = -2(x + 3)$
14. $3(y - 2) = -5(y + 1)$
15. $\dfrac{1}{2}(v - 3) = \dfrac{-2}{3}(2 - 5v)$
16. $-2.09(1 - w) = 3.2(3 + w)$
17. $-2(3 - 2p) = 5(1 - 4p)$
18. $\dfrac{1}{2}(q - 3) = \dfrac{2}{3}(q + 1)$
19. $\dfrac{1}{4}(t + 2) = \dfrac{2}{5}(3t - 1)$
20. $\dfrac{1}{4}(2 - w) = \dfrac{2}{7}(3w + 1)$
21. $2(x - 1) + x = 3(1 - x)$
22. $4y - 3(y + 1) = 2y + 4(1 - y)$
23. $3(x + 4) - 4x = 2 + 3(x - 3)$
24. $\dfrac{y}{2} - 2(y + 4) = 5(y - 1) - 6$
25. $6t - (4 - 5t) = \dfrac{t + 1}{2}$
26. $\dfrac{2(p - 3)}{4} = \dfrac{3(p - 2)}{5}$
27. $2(x - 3) - 3(1 - 2x) = 4(x - 5)$
28. $2 - 3(p + 1) = 4(p - 3) - 5(2p + 3)$
29. $0.3(y - 1) = 1.1$
30. $2t - 0.31 = 3(t - 0.5)$
31. $3(w - 0.1) = 0.7 - 2w$
32. $3s - 0.17 = 2(0.4 - 3s)$
33. $\dfrac{p - 0.3}{4} = \dfrac{2p + 1.3}{5}$
34. $\dfrac{1.2 - q}{7} = \dfrac{2q - 2.3}{3}$
35. $\dfrac{0.6 - 2u}{-3} = \dfrac{3u - 0.7}{-4}$
36. $2.72(t - 1) = -0.6(2 - 3t)$
37. $0.5(w - 2) = 0.7(3 - 2w)$
38. $0.3(w - 5) - 0.4w = 2 + 0.3(w + 4)$
39. $0.7w - (0.1 - 2w) = \dfrac{w + 1.2}{3}$
40. $0.3(x - 1) = 0.5(x - 2) - 0.3(2x + 5)$
41. $3(2 - 5y) - 17 = 9y - 4(4 + 12y)$
42. $3(2 - 3y) + 10 = 8y - 6(3 - 4y)$
43. $0.2(2q - 5) + 2 = 0.7q + 0.3(4 - q)$
44. $18(3 + 4r) - 28r = 34 + 4(6 + 11r)$
45. $-3[4t - 2(6 + 2t)] = 36$
46. $5[3(t - 5) + 8t] = 35$
47. $\dfrac{1}{4}[4 - 2(5 - s)] = \dfrac{2}{3}(3s - 12)$
48. $\dfrac{1}{4}[4(6 - s) - 2] = \dfrac{1}{3}(9 - 5s)$
49. $3[2u - 3(u - 5)] = 4[3u + 2(3 - 4u)]$
50. $1.2[u + 5(3 - 2u)] = 2.1[3(u - 4) - 5u]$

8.5 Applications Involving Linear Equations: Part 1

OBJECTIVES

After completing this section, you should be able to:
1. Translate a word problem into a linear equation in a single variable.
2. Solve application problems.

In this section, we will use linear equations in one variable to solve some application problems. The following procedure should be used.

Procedure for Solving a Word Problem

1. Read (and, if necessary, re-read) the word problem very carefully.
2. Determine what it is that you are asked to find.
3. Represent each unknown quantity by one variable.
4. Translate (rewrite) the verbal statement(s) into an equation. (It is not always necessary to use all of the given data.)
   a. Is there enough information to solve the problem?
   b. Is there more information than is needed to solve the problem?
5. Solve the equation.
6. Check your results.
   a. Are the results reasonable?
   b. Do the results satisfy all of the conditions of the problem?

Practice Exercise 1. If 9 is subtracted from a number, the difference is -16. What is the number?

EXAMPLE 1 If 6 is added to a number, the sum is 14. What is the number?

SOLUTION: This simple example illustrates the procedure used to solve a word problem. The unknown in this problem is "the number." We could start by writing:

Let $y =$ the number.   (*Note:* This statement should be written on your paper.)

Then, reading the statement carefully, we see that the sum of the number and 6 is 14. The word statement, then, is:

The number plus 6 is 14.

Substituting y for "the number," $+$ for "plus," and $=$ for "is," we can translate (rewrite) the word statement as the mathematical statement (equation):

$$y + 6 = 14$$

Solving the equation, we get

$$y = 14 - 6$$
$$= 8$$

Finally, we substitute $y = 8$ in the equation $y + 6 = 14$ and get $8 + 6 = 14$, which is a true statement. This only checks the solution for the equation. As a final check, go to the original problem: If 6 is added to 8, the sum is 14. This is true. Therefore, 8 is the solution to the problem.

8.5 Applications Involving Linear Equations: Part 1

Practice Exercise 2. The perimeter of a triangle is 58 cm. One side of the triangle is twice as long as a second side. The third side is 25 cm long. Determine the lengths of the other two sides.

EXAMPLE 2 A piece of string 24 in. long is cut into two pieces such that one piece is 4 in. less than 3 times the length of the other piece. Determine the length of the two pieces of string.

SOLUTION: There are two pieces of string and the combined length is 24 in. Since we don't know the length of either piece, we

Let t = length of one piece (in inches).

Then, $24 - t$ = length of the other piece (in inches).

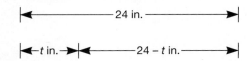

The verbal statement is:

The length of one piece is 4 in. less than 3 times the length of the other piece.

Rewriting this verbal statement, we have:

The length of one piece is 3 times the length of the other piece minus 4.

Substituting t for "the length of one piece," $24 - t$ for "the length of the other piece," $=$ for "is," we have

$$t = 3 \text{ times } (24 - t) \text{ minus } 4$$

or

$$t = 3(24 - t) - 4$$

Solving this equation, we have

$$t = 3(24 - t) - 4$$
$$t = 72 - 3t - 4$$
$$t = 68 - 3t$$
$$4t = 68$$
$$t = 17$$

Therefore,

$t = 17$ in.   (The length of one piece)
$24 - t = 7$ in.   (The length of the other piece)

The two pieces of string, then, are 7 in. and 17 in. long. Check to be sure that these lengths satisfy all conditions of the original problem:

1. The two pieces together have a length of $(17 + 7)$ in. or 24 in.
2. The length of one piece is 4 in. less than 3 times the length of the other piece.

Practice Exercise 3. In an election, 35%, or 1673, of the voters voted for Candidate A. How many people voted in the election?

EXAMPLE 3 In a particular school district, 11%, or 2167, of the students ride a bus to school. How many students are there in the school district?

SOLUTION: In this problem, we have to determine the number of students in the school district.

Let n = the number of students.

The verbal problem is:

> 11% of the students in the school district is 2167.

Substituting 0.11 for "11%," the operation of multiplication for "of," n for "students," and $=$ for "is," we have the mathematical statement (equation):

$$0.11n = 2167$$

Solving this equation, we have:

$$0.11n = 2167$$
$$n = \frac{2167}{0.11}$$
$$= 19{,}700$$

Hence, there are $n = 19{,}700$ students in the school district.

CHECK:
$$0.11n = 2167$$
$$(0.11)(19{,}700) \; ? \; 2167$$
$$2167 = 2167 \; \checkmark$$

EXAMPLE 4 The sum of four consecutive integers is 350. What are the numbers?

SOLUTION: Consecutive integers are integers that follow one right after the other as you count. Since there are four integers involved in the problem, we:

	Let $y = $ the first integer	(The smallest one)
Then,	$y + 1 = $ the second integer	(That is, add 1 to the first integer)
	$y + 2 = $ the third integer	
and	$y + 3 = $ the fourth integer	(The largest one)

The verbal statement for this problem is:

> The sum of the four consecutive integers is 350.

The equation, then, becomes:

$$y + (y + 1) + (y + 2) + (y + 3) = 350$$

Solving the equation for y, we have:

$$y + (y + 1) + (y + 2) + (y + 3) = 350$$
$$y + y + 1 + y + 2 + y + 3 = 350$$
$$4y + 6 = 350$$
$$4y = 344$$
$$y = 86$$

Hence, the required solution is:

$$y = 86$$
$$y + 1 = 87$$
$$y + 2 = 88$$
$$y + 3 = 89$$

indicating that the four consecutive integers are 86, 87, 88, and 89. Check to determine that their sum is 350.

> **Practice Exercise 4.** The sum of two positive numbers is 81. If one-ninth of the smaller number is 5 less than one-fifth of the larger number, what are the numbers?

8.5 Applications Involving Linear Equations: Part 1

Practice Exercise 5. The perimeter of a rectangle is 78 yd. If the length of the rectangle is 3 yd less than twice the width, what are the dimensions of the rectangle?

EXAMPLE 5 A rectangular parking lot is to be constructed so that its length is 3 times as long as its width. What are the dimensions of the lot if the perimeter is to be 640 ft?

SOLUTION: Since we are to determine the dimensions of the lot, we

Let W = its width   (In feet)
and $3W$ = its length   (In feet)

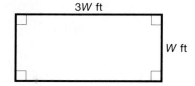

The perimeter of the rectangular lot is given by:

Perimeter (640) is equal to twice the width (W) plus twice the length ($3W$).

Substituting, we obtain the equation

$$640 = 2W + 2(3W)$$

Solving the equation, we have

$$640 = 2W + 2(3W)$$
$$640 = 2W + 6W$$
$$640 = 8W$$
$$80 = W$$

Hence, the required solution is

$$W = 80 \text{ ft} \quad \text{(Width)}$$
$$3W = 240 \text{ ft} \quad \text{(Length)}$$

Check to determine that the perimeter is 640 ft.

Practice Exercise 6. Three times a number increased by 4 is the same as the number increased by 22. What is the number?

EXAMPLE 6 The logo for the Ace Construction Company is to be triangular in shape, with a perimeter of 200 in. If the longest side is 10 in. less than twice the shortest side, and the third side is 10 in. more than the shortest side, what are the lengths of the sides of the triangle?

SOLUTION: Since we are to determine the lengths of the sides of the triangle, we

Let s = the length of the shortest side   (In inches)
Then, $2s - 10$ = the length of the longest side   (In inches)
and $s + 10$ = the length of the third side   (In inches)

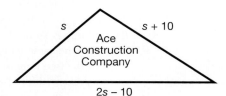

The perimeter of the triangle is the sum of the lengths of its sides. Substituting, we obtain the equation

$$s + (2s - 10) + (s + 10) = 200$$

Solving the equation, we have

$$s + (2s - 10) + (s + 10) = 200$$
$$s + 2s - 10 + s + 10 = 200$$
$$4s = 200$$
$$s = 50$$

Hence, the required solution is

$$s = 50 \text{ in.} \quad \text{(Length of shortest side)}$$
$$2s - 10 = 90 \text{ in.} \quad \text{(Length of longest side)}$$
$$s + 10 = 60 \text{ in.} \quad \text{(Length of third side)}$$

Check to determine that the perimeter is 200 in.

ANSWERS TO PRACTICE EXERCISES

1. −7. **2.** 11 cm and 22 cm. **3.** 4780. **4.** 36 and 45. **5.** 14 yd by 25 yd. **6.** 9.

EXERCISES 8.5

1. Four times a number increased by 15 is the same as the number decreased by 6. Determine the number.
2. Five less than twice a number is equal to the number increased by 4. Determine the number.
3. The sum of two numbers is 165. If one number is twice as large as the other, determine the two numbers.
4. The sum of two numbers is 27. If one-fourth of the larger number is 7 less than the smaller number, determine the two numbers.
5. One number is two-fifths of another number. Their sum is 105. Determine the numbers.
6. A garden plot is in the form of a rectangle. The perimeter is 400 m. The length of the plot is 4 times its width. Determine the dimensions of the plot.
7. The perimeter of a triangle is 30 cm. The shortest side is two-thirds the length of the longest side. The other side is five-sixths the length of the longest side. Determine the dimensions of the triangle.
8. The perimeter of a four-sided polygon is 143 yd. Three sides have the same length. The length of each of those sides is 6 yd less than twice the length of the fourth side. What are the lengths of the sides of the polygon?
9. A building is in the shape of a pentagon (a five-sided polygon). The perimeter around the building is 91 m. The two longest sides have the same length. The two shortest sides also have the same lengths. Each of the two shortest sides is 4 m less than the fifth side. Each of the two longest sides is 3 m less than twice the fifth side. Determine the lengths of the sides of the polygon.
10. The sum of the three angles of a triangle is 180 degrees. In the triangle pictured at the top of the next page, if the largest angle is twice as large as the smallest angle, and the third angle is 20 degrees more than the smallest angle, how many degrees are in each of the three angles? (*Hint:* Let x = the number of degrees in the smallest angle.)

8.5 Applications Involving Linear Equations: Part 1

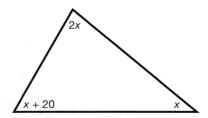

11. A ladder leans against a building and forms a right triangle with the ground and the building. (See accompanying figure.) If angle C has a measure that is 15 degrees more than twice the measure of angle A, determine the measure for each of the angles A and C.

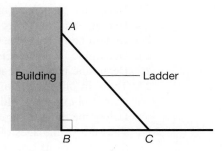

12. The area of a rectangular region is given by the equation $A = LW$ where L represents the length and W represents the width. If the area of a rectangular region is 736 cm^2 and the length is 32 cm, what is its width?

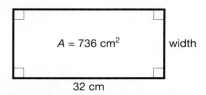

13. A parking lot for a medical building is in the form of a rectangle. Determine the length of the rectangular lot if its width is 23.6 ft and its area is 710.36 ft^2.
14. A picture frame is in the form of a rectangle. The length of the frame is 12 in. more than its width. The perimeter around the frame is 144 in. Determine the length and width of the frame.
15. Determine the selling price for a sweater if the cost of the sweater is $12.10 and the profit is 20% of the cost.
16. Determine the selling price for an electric toaster that costs $10.22 if the profit is 30% of the selling price.
17. An automobile repair bill came to $288.90. This included $179 for parts, $18.90 for state sales tax, and $26 for each hour of labor. Determine the number of hours of labor.
18. A meal consisting of a cheeseburger, french fries, and a milkshake contains 863 calories. If the cheeseburger contains 119 more calories than the french fries and the french fries contain 54 more calories than the milkshake, how many calories are in each of the items?
19. In a college general biology course, there are freshmen, sophomores, and juniors enrolled. There are 50% more sophomores than freshmen. The junior enrollment is 13 more than half of the freshmen enrollment. Additionally, there are 4 senior citizens enrolled. If the total enrollment in the course is 2717, how many sophomores are enrolled?
20. Michael wants to fence in his garden behind his house. He wants it to be rectangular in shape but needs to fence only three sides since the back of the house will serve as one side. (See figure, top of next page.) He has 64 m of fence and wants the length to be 8 m less than twice the width. What will be the dimensions of the garden?

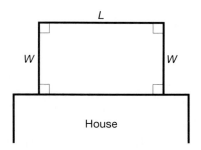

21. For an American Heart Run, Sarah received pledges that were $2 more than 3 times the amount of pledges received by Norayne. If the total of their pledges was $94, how much did each of the two runners receive in pledges?
22. Determine two consecutive whole numbers whose sum is 39.
23. Determine three consecutive whole numbers whose sum is 114.
24. Determine three consecutive integers whose sum is -48.
25. Determine four consecutive whole numbers whose sum is 90.
26. Determine two consecutive odd integers whose sum is 28.
27. The sum of the ages of three children is 22. The middle child is 7 years old. Twelve years from now, the oldest child will be twice as old as the youngest child will be then. Determine the present ages of the oldest and youngest children.
28. The sum of the ages of Bill and Lois is 20 years. One year from now, Bill will be 9 times the age of Lois 1 year ago. What is the present age of each?
29. John is older than Nancy. The sum of their ages is 72. Six years from now, John will be twice as old as Nancy will be then. Determine their present ages.
30. A wooden bench is 42 years older than an antique vase. Twenty years ago, the bench was three times as old as the vase was then. Determine the present ages of the two items.

8.6 Applications Involving Linear Equations: Part 2

OBJECTIVE After completing this section, you should be able to use linear equations to solve other types of word problems.

In this section, we will examine additional word problems that can be solved by using simple linear equations.

Practice Exercise 1. A merchant has two types of party snacks, one that sells for $1.59 per pound and another that sells for $1.89 per pound. How much of each should the merchant mix together in order to have a 100-lb mixture that can be sold on special for $1.77 per pound?

EXAMPLE 1 A merchant has two grades of grass seed, one that sells for 89¢ a pound and another that sells for 69¢ a pound. How much of each should the merchant mix together in order to have a 100-lb mixture of grass seed that can be sold during a weekend sale at 77¢ a pound?

SOLUTION: We do not know how many pounds of each grade of grass seed are to be mixed to form the mixture. However, we

Let g = the number of pounds of grass seed @ 89¢ per pound
Then, $100 - g$ = the number of pounds of grass seed @ 69¢ per pound

8.6 Applications Involving Linear Equations: Part 2

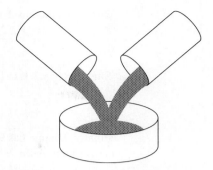

The verbal statement associated with this problem is:

The *value* of the 89¢ per pound grass seed plus the *value* of the 69¢ per pound grass seed is equal to the *value* of the 77¢ per pound mixture.

The value of the 89¢ per pound grass seed is equal to the cost per pound ($.89) times the number of pounds (g). Similarly, the value of the 69¢ per pound grass seed is equal to the cost per pound ($.69) times the number of pounds ($100 - g$). The value of the 77¢ per pound mixture is equal to the cost per pound ($.77) times the number of pounds (100). The equation associated with the verbal statement is

$$(\$.89)(g) + (\$.69)(100 - g) = (\$.77)(100)$$

which becomes

$$89g + 69(100 - g) = 77(100)$$

by multiplying both sides of the equation by 100 cents per dollar. Solving for g, we have

$$89g + 69(100 - g) = 77(100)$$
$$20g + 6900 = 7700$$
$$20g = 800$$
$$g = 40$$

Therefore, the merchant should mix

with $\quad g = 40$ lb of 89¢ per pound grass seed
$\quad 100 - g = 60$ lb of 69¢ per pound grass seed

to form the 77¢ per pound mixture.

CHECK:

$.89	$.69	$.77
×40	×60	×100
$35.60	× $41.40	= $77.00
(Value of $.89/lb seed)	(Value of $.69/lb seed)	(Value of $.77/lb mixture)

Example 1 represents what is known as a mixture problem. Instead of grass seed, we could have different grades or types of ground meat, candy, gasoline, and so forth.

EXAMPLE 2 Dianne has a collection of coins consisting of pennies, nickels, dimes, and quarters. She has 1 more dime than 3 times the number of nickels; she has 3 less quarters than twice the number of nickels; and she has 9 more pennies than she has quarters. If the total collection is worth $6.37, how many of each kind of coin does Dianne have?

SOLUTION: The number of pennies, dimes, and quarters is given in terms of the number of nickels. Therefore, we

Let $n = $ the number of nickels

Practice Exercise 2. Joshua has a collection of coins consisting of nickels, dimes, and quarters. He has twice as many dimes as nickels and 3 times as many quarters as nickels. If his total collection is worth $14, how many of each kind of coin does he have?

Since Dianne has 1 more dime than 3 times the number of nickels, we have

$$3n + 1 = \text{the number of dimes}$$

She also has 3 less quarters than twice the number of nickels. Hence,

$$2n - 3 = \text{the number of quarters}$$

Finally, Dianne has 9 more pennies than she has quarters. Since the number of quarters is given as $2n - 3$, the number of pennies would be $(2n - 3) + 9$ or

$$2n + 6 = \text{the number of pennies}$$

The verbal statement for this problem is:

The *value* of the pennies plus the *value* of the nickels plus the *value* of the dimes plus the *value* of the quarters is equal to *total value* of the collection.

The corresponding equation, then, is

$$\$.01(2n + 6) + \$.05n + \$.10(3n + 1) + \$.25(2n - 3) = \$6.37$$

which is equivalent to

$$(2n + 6) + 5n + 10(3n + 1) + 25(2n - 3) = 637$$

Solving the equation for n, we have

$$(2n + 6) + 5n + 10(3n + 1) + 25(2n - 3) = 637$$
$$2n + 6 + 5n + 30n + 10 + 50n - 75 = 637$$
$$87n - 59 = 637$$
$$87n = 696$$
$$n = 8$$

Hence, Dianne has

$$2n + 6 = 22 \text{ pennies}$$
$$n = 8 \text{ nickels}$$
$$3n + 1 = 25 \text{ dimes}$$
$$2n - 3 = 13 \text{ quarters}$$

Check to determine that the collection is worth $6.37.

Another type of problem that can be solved by simple linear equations is a uniform motion and distance problem. Uniform motion involves an object that travels at a constant rate of speed along a particular path. The speed of the object refers to the distance traveled along the path during a particular period of time and is given as so many feet per second, kilometers per hour, and so forth. If the rate of speed is denoted by r, the time by t, and the distance by d, then the relationship existing among the three variables is given by the equation

$$d = rt$$

Practice Exercise 3. Determine the time that it takes an airplane to fly 1575 mi at 350 mph.

EXAMPLE 3 Determine the speed of a snowmobile that travels 150 mi in 4 hr.

SOLUTION: In this problem, we are given the distance and the time. Hence, to determine the rate of speed, we use the equation

8.6 Applications Involving Linear Equations: Part 2

with
$$d = rt$$
$$d = 150 \text{ mi}$$
and
$$t = 4 \text{ hr}$$

|←———— 150 mi ————→|

Hence,
$$150 \text{ mi} = r \times (4 \text{ hr})$$

To solve for r, divide both sides of the equation by 4 hr. Hence,
$$r = (150 \text{ mi}) \div (4 \text{ hr})$$
or
$$r = 37.5 \text{ mph}$$

Practice Exercise 4. Two runners start at points 20 miles apart and run toward each other at speeds of 3 mph and 3.5 mph. If they start at the same time, how long will it take for them to meet?

EXAMPLE 4 Two runners start from the same place and travel in opposite directions at speeds of 3 mph and 4 mph, respectively. How long after they start will they be 15 mi apart?

SOLUTION: In this problem, we have two runners traveling at different rates of speed. We must determine the time, t, when they will be 15 mi apart. First, we diagram the problem as shown. Obviously, the faster runner will travel further than the slower runner during the same period of time. If we

Let t = time traveled (in hours)

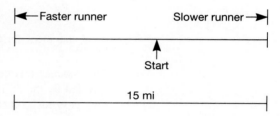

then we could determine the distance, rate, and time for each runner using the equation $d = rt$ as follows:

Slower runner:
$$d = rt$$
$$= (3 \text{ mph})(t \text{ hr})$$
$$= 3t \text{ mi}$$

Faster runner:
$$d = rt$$
$$= (4 \text{ mph})(t \text{ hr})$$
$$= 4t \text{ hr}$$

The verbal statement for this problem is:

The *distance* traveled by the slower runner plus the *distance* traveled by the faster runner is equal to the *total distance* traveled.

The corresponding equation, then, is
$$3t + 4t = 15$$
or
$$7t = 15$$
$$t = 2\frac{1}{7}$$

Therefore, it would be $2\frac{1}{7}$ hr for the two runners to be 15 mi apart.

> **Practice Exercise 5.** Krystle traveled a distance of 320 mi. She traveled 4 hr on interstate highways and 2 hr on country roads. If she drove 20 mph faster on the interstate highways than she did on the county roads, determine how fast she traveled on each part of the trip.

EXAMPLE 5 Two cyclists travel toward each other, in parallel lanes, from points 500 mi apart at speeds of 40 mph and 50 mph, respectively. When will they pass each other if they start at the same instant?

SOLUTION: In this problem, we have two cyclists traveling toward each other and at different rates of speed. When they pass each other, the problem is over. Hence, we must determine the time, t. We can diagram the problem as shown. If we

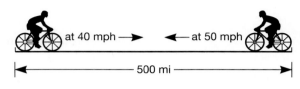

Let $t =$ time traveled (in hours)

then the distance traveled by each cyclist can be determined as follows:

For 40 mph cyclist:
$$d = rt$$
$$= (40 \text{ mph})(t \text{ hours})$$
$$= 40t \text{ mi}$$

For 50 mph cyclist:
$$d = rt$$
$$= (50 \text{ mph})(t \text{ hours})$$
$$= 50t \text{ mi}$$

The verbal statement for this problem is:

The *distance* traveled by the 40 mph cyclist plus the *distance* traveled by the 50 mph cyclist is equal to 500 mi (at the time they pass each other).

The corresponding equation, then, is

$$40t + 50t = 500$$
or
$$90t = 500$$
$$t = 5\frac{5}{9}$$

Therefore, it would be $5\frac{5}{9}$ hr before the two cyclists pass each other. You should check this result.

From physics, we get another type of problem that can be solved using simple linear equations. This type of problem is the lever problem. In physics, a "lever" is defined as a rigid rod supported at one point called the "fulcrum." A familiar example of a lever is a teeterboard or seesaw. If a mass of x pounds is placed on a lever at a distance of d units from its fulcrum, then the product of x and d is called the "moment" of x about the fulcrum. That is

$$\text{Moment} = xd$$

The lever principle states that when the lever is in a position of equilibrium, the sum of the moments of all the masses on one side of the fulcrum is equal to the sum of the moments of all the masses on the other side.

8.6 Applications Involving Linear Equations: Part 2

Practice Exercise 6. Two children weighing 85 lb and 60 lb sit on opposite ends of a teeterboard, 9 ft and 12 ft, respectively, away from the fulcrum. Is the board in a position of equilibrium?

EXAMPLE 6 If two children weighing 120 lb and 70 lb, respectively, sit on opposite ends of a teeterboard which is 19 ft long, where should the board be supported to maintain a position of equilibrium?

SOLUTION: The problem is diagrammed here, with the fulcrum being closer to the heavier child.

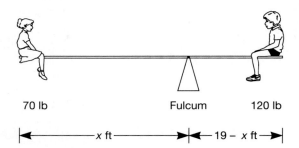

Let x = distance of the fulcrum from the 70-lb child (in feet)
Then, $19 - x$ = distance of the fulcrum from the 120-lb child (in feet)

since the total length of the teeterboard is 19 ft. Therefore, the moment for the 70-lb child is $70x$ and the moment for the 120-lb child is $120(19 - x)$. Using the lever principle, the corresponding equation is

$$70x = 120(19 - x)$$
or
$$70x = 2280 - 120x$$
$$190x = 2280$$
$$x = 12$$

Therefore, the fulcrum should be placed 12 ft from the 70-lb child.

Practice Exercise 7. Two children weighing 80 lb and 95 lb sit on the same side of a teeterboard, 4.5 ft and 6.5 ft, respectively, from the fulcrum. Two other children weighing 65 lb and 80 lb sit on the other side of the teeterboard, 6 ft and 7.5 ft, respectively, from the fulcrum. Is the teeterboard in a position of equilibrium?

EXAMPLE 7 Two children weighing 60 lb and 80 lb sit on the same side of a teeterboard 5 ft and 9 ft, respectively, from the fulcrum which is in the center of the board. A third child weighing 90 lb sits 7 ft from the fulcrum on the other side. Where should a fourth child weighing 100 lb sit to balance the board?

SOLUTION: The problem is diagrammed here. The moment for the 80-lb child is $(80)(9) = 720$. The moment for the 60-lb child is $(60)(5) = 300$. The sum of all the moments on the *left* side of the fulcrum is $720 + 300 = 1020$.

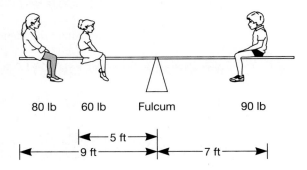

The moment for the 90-lb child is $(90)(7) = 630$

We note that $630 < 1020$. Therefore, according to the lever principle, the teeterboard will be in equilibrium if the fourth child is placed on the board on the same side of the fulcrum as the 90-lb child. How far from the fulcrum?

Let y = distance of the fourth child from the fulcrum (in feet)

Then the moment for the 100-lb child is $100y$. We now have the following:

The sum of *all* the moments on the *left* $= (80)(9) + (60)(5)$
The sum of *all* the moments on the *right* $= (100)(y) + (90)(7)$

Since the sum of all the moments on the left must equal the sum of all the moments on the right, we have

$$(80)(9) + (60)(5) = (100)(y) + (90)(7)$$

or
$$720 + 300 = 100y + 630$$
$$1020 = 100y + 630$$
$$390 = 100y$$
$$3.9y = y$$

Hence, the fourth child should be placed $y = 3.9$ ft from the fulcrum and on the same side of the board as the 90-lb child.

In Section 4.8, we considered simple interest problems. Recall that interest is the price paid for the use of money. Interest, denoted by I, is the product of the principal (the amount borrowed or saved), the rate (given as a percent per period of time), and time (given in years or part of a year). Using P for principal, r for rate, and t for time, we have

$$I = Prt$$

We will now extend our discussion of simple interest by considering an application involving more than one interest rate. We will use the procedures introduced in this section to solve such problems.

Practice Exercise 8. Barbara invested $13,300, part at 6% annual interest and part at 8.5% annual interest. How much did she invest at each rate if the total interest earned was $993?

EXAMPLE 8 Linda had $10,000 to invest. She invested part of the money in a savings account at 8% per year and the rest in a certificate of deposit that pays 9.5% per year. If the total interest for the year was $890, how much did Linda invest at each rate?

SOLUTION: We are to determine the amount of money that Linda invested at each rate.

Let $p =$ the amount invested at 8% (in dollars)

Then

$10,000 - p =$ the amount invested at 9.5% (in dollars)

We now determine the interest received on each investment as follows:

At 8%:
$$I = Prt$$
$$= p \times 8\% \times 1 \text{ yr}$$
$$= p \times 0.08 \times 1$$
$$= 0.08p \text{ dollars}$$

At 9.5%:
$$I = Prt$$
$$= (10,000 - p) \times 9.5\% \times 1 \text{ yr}$$
$$= (10,000 - p) \times 0.095 \times 1$$
$$= 0.095(10,000 - p) \text{ dollars}$$

The verbal statement for this problem is:

The interest received from the 8% investment plus the interest received from the 9.5% investment is equal to the total interest received.

8.6 Applications Involving Linear Equations: Part 2

The corresponding equation, then, is:

$$0.08p + 0.095(10{,}000 - p) = 890$$
$$80p + 95(10{,}000 - p) = 890{,}000 \quad \text{(Multiplying both sides of the equation by 1000 to remove all decimals)}$$
$$80p + 950{,}000 - 95p = 890{,}000$$
$$950{,}000 - 15p = 890{,}000$$
$$-15p = -60{,}000$$
$$p = 4000$$

and $\quad 10{,}000 - p = 6000$

Therefore, Linda invested $4000 at 8% and $6000 at 9.5%.

ANSWERS TO PRACTICE EXERCISES

1. 40 lb @ $1.59 and 60 lb @ $1.89. 2. 14 nickels, 28 dimes, and 42 quarters. 3. 4.5 hr. 4. $3\frac{1}{13}$ hr. 5. 60 mph on interstate highways and 40 mph on county roads. 6. No. 7. No. 8. $5500 @ 6% and $7800 @ 8.5%

EXERCISES 8.6

1. Michael has a collection of coins, consisting of pennies, nickels, dimes, and quarters. He has half as many quarters as dimes and two more nickels than dimes. The number of pennies in his collection is 3 less than twice the number of dimes. If the total collection is worth $3.02, how many of each kind of coin does Michael have?

2. Liz has a collection of coins consisting of pennies, dimes, and quarters. She has 7 more pennies than dimes, and the number of quarters is 3 less than one-half the number of dimes. If the total collection is worth $2.14, how many of each kind of coin does she have?

3. Gina has a collection of coins consisting of nickels, dimes, and quarters. She has twice as many dimes as quarters and 8 more nickels than dimes. If her total collection of coins is worth $5.90, how many of each kind of coin does she have?

4. A butcher has two grades of ground meat, one that sells for $1.49 per pound and another that sells for $1.79 per pound. How much of each should the butcher mix together in order to have a 100-lb mixture that can be sold during a weekend special as $1.61 per pound?

5. A collection of stamps consists of 22¢ stamps, 39¢ stamps, and 17¢ stamps. The number of 17¢ stamps is 3 less than twice the number of 39¢ stamps. There are half as many 17¢ stamps as there are 22¢ stamps. If the total collection of stamps is worth $11.05, how many of each kind of stamps are there in the collection?

6. A grocery shopper redeemed manufacturer coupons worth $5.85. The coupons were worth 25¢, 30¢, and 40¢ each. If the number of 30¢ coupons redeemed was 3 less than twice the number of 25¢ coupons, and the number of 40¢ coupons redeemed was 5 less than the number of 25¢ coupons, how many of each kind of coupon did the shopper redeem?

7. A football game was attended by 4730 people, some of whom paid $1.50 each for reserved seats and the rest of whom paid $.90 for general admission tickets. If the total receipts for the game were $5172, how many tickets of each kind were sold?

8. Two people in automobiles travel toward each other, in parallel lanes, from points 600 mi apart at speeds of 45 mph and 60 mph, respectively. When will they pass each other if they start at the same instant?

9. One boy can run 100 yd in 17 sec, and a second boy can run the same distance in 16 sec. In a race between them, how long will it take the faster boy to gain a lead of 2 yd over the other boy?

10. A train leaves the station traveling at the rate of 50 mph. One hour later, another train leaves the station, traveling in the same direction but along parallel tracks, at the rate of 70 mph. How long will it take for the second train to catch up with the first train?

11. A motorcyclist starts from a certain location and travels along a roadway at the rate of 15 mph. A motorist starts at the same location 3 hours later and travels along the same path at the rate of 40 mph. Assuming that the cyclist is still traveling, when will the motorist catch up to her?

12. Nick and Larry leave work, at the same location, in their cars and travel home in opposite directions on a straight road. Nick travels 10 mph faster than Larry. After 30 min, they are 35 mi apart. How fast is each man driving?

13. Brandon works 360 mi away from his home and returns home on weekends. For the trip home, he travels 6 hr on interstate hghways and 1 hour on two-lane roads. If he drives 25 mph faster on the interstate highways than he does on the two-lane roads, determine how fast he travels on each part of the trip home.

14. At 8:00 A.M., a boat leaves Marina A and travels to Marina B, a distance of 180 mi, at 25 mph. Two hours later, a second boat leaves Marina B and travels to Marina A at 35 mph. At what time will they pass each other?

15. Larry can run 4 mph. His friend, Kathy, can run 6 mph. If they start running together in the same direction, how long would it take Kathy to get 5 mi ahead of Larry?

16. Sarah left her office at 8:30 A.M. to go on a service call to the Downstate Computing Company. The service call took 45 min. She then returned directly back to her office along the same route, arriving at 10:25 A.M. If she traveled out at the rate of 40 mph and returned at the rate of 30 mph, how far is the company located from her office?

17. A child weighing 85 lb sits 6 ft from the fulcrum on a teeterboard. How far from the other side of the fulcrum should a 70-lb child sit in order to balance the other child?

18. A girl weighing 65 lb sits on a teeterboard 4 ft from the fulcrum. How far from the other side of the fulcrum should a 75-lb boy sit in order to balance the girl?

19. Two children weighing 60 lb and 80 lb sit on the same side of a teeterboard, 4 ft and 7 ft, respectively, from the fulcrum. Where should a third child weighing 100 lb sit to balance the board?

20. One child weighing 50 lb sits on a teeterboard 5 ft from the fulcrum. A second child weighing 70 lb sits on the teeterboard 4 ft from the other side of the fulcrum. Where should a third child weighing 60 lb sit on the teeterboard to balance the other two children?

21. Two children weighing 70 lb and 85 lb sit on the same side of a teeterboard, 5 ft and 7 ft, respectively, from the fulcrum. Two other children weighing 75 lb and 80 lb sit on the other side of the teeterboard, 4 ft and $7\frac{1}{2}$ ft, respectively, from the fulcrum. Determine whether the teeterboard is balanced.

22. If two children, weighing 120 lb and 70 lb, respectively, sit at opposite ends of a teeterboard which is 19 ft long, where should the board be supported to maintain a position of equilibrium?

23. Johnny weighs 80 lb and sits 1 ft from the end of a teeterboard which is 20 ft long. Mary weighs 70 lb and sits 2 ft from the other end of the teeterboard. Where should the teeterboard be supported to maintain a position of equilibrium?

24. Two children weighing 50 lb and 60 lb sit on the same side of a teeterboard, 1 ft and 2 ft, respectively, from the end of the teeterboard. A third child, weighing 90 lb, sits at the other end of the teeterboard. If the teeterboard is 16 ft long, where should it be supported to maintain a position of equilibrium?

25. How many gallons of a solution of water with 45% antifreeze should be added to 22 gal of a solution with 25% antifreeze to obtain a solution with 38% antifreeze?

26. How many liters of 12% salt solution should a nurse add to 18 L of a 20% salt solution to obtain a 15% salt solution?
27. Joyce has a savings account that earns 6.5% annual interest and a certificate of deposit that earns 11.5% annual interest. If the total amount invested is $19,000 and the total annual interest is $1835, how much does she have invested in each account?
28. Ann invested $17,000, part at 7% interest and the balance at 8.5% interest per year. If the total amount of interest earned at the end of the year is $1310, how much was invested at each rate?
29. Kevin invested $8000 in a certificate of deposit that pays 9% interest per year. He also invested an amount of money in a stock option that pays 8% interest at the end of the year. If the interest from the $8000 investment was $160 more than the interest from the stock investment, how much did Kevin invest in the stock option?
30. Patti invested $7200 with an annual simple interest rate of 10%. How much should she invest in a second account at 12% simple interest so that the interest earned by each account is the same?
31. Julia has $15,000 to invest. She invests $9000 in a certificate of deposit at a rate of 12% per year. At what rate should she invest the remaining amount if she wants a total annual interest of $1650 from the two investments?
32. Daryl invested $13,400 at 11.5% per year. How much additional money must he invest at 10% per year so that the total annual interest from both investments will be $2000?
33. Alexa received a cash payment of $65,000 from the sale of her house. She invested $25,000 of this amount at 12% and $30,000 at 13.5%. At what rate must she invest the remaining $10,000 so that the total annual interest from the three investments will be $8000?
34. In an isosceles triangle, two of the angles have the same measure. If the third angle has a measure that is 30 degrees less than twice the measure of one of the equal angles, determine the measure of the equal angles. (*Hint:* The angle sum of any triangle is 180 degrees.)
35. The pilot of a light airplane flew from City A to City B at an average speed of 150 mph. On the return flight from City B to City A, the pilot flew at an average speed of 120 mph. If the total flying time was 6 hr, how far apart are the two cities?

8.7 Formulas

OBJECTIVES

After completing this section, you should be able to:
1. Evaluate a formula.
2. Use a formula to solve an applied problem.

Many problems in mathematics, the physical sciences, and business involve the use of formulas. A mathematical **formula** is a rule that expresses a relationship between two or more quantities. For example, we have already used the formula $C = \pi d$ to express the relationship between the circumference, C, of a circle and its diameter, d. Also, the formula $I = Prt$ expresses the relationship between interest (I), and the principal (P), rate (r), and time (t).

We will now consider the evaluation of a formula.

Evaluating a Formula

To evaluate a formula:
1. Substitute the given (or known) values for the variables in the formula.
2. Simplify the resulting expression, using the order of operations rules.

> **Practice Exercise 1.** How long will it take Brandan to travel 308 mi at 55 mph?

EXAMPLE 1 The formula $d = rt$ expresses the relationship between the distance (d) traveled by an object in uniform motion, and the rate (r) and time (t). What distance does Joanne travel if she drives at a rate of 55 mph for $2\frac{1}{2}$ hr?

SOLUTION: We must evaluate d (in miles) given that $r = 55$ mph and $t = 2\frac{1}{2}$ hr. Using the given formula, we have:

$$d = rt$$
$$= (55 \text{ mph})\left(2\frac{1}{2} \text{ hr}\right)$$
$$= \left(\frac{55 \text{ mi}}{1 \text{ hr}}\right)\left(\frac{5}{2} \text{ hr}\right)$$
$$= \frac{(55 \text{ mi})(5 \text{ hr})}{(1 \text{ hr})(2)}$$
$$= \frac{275 \text{ mi}}{2}$$
$$= 137.5 \text{ mi}$$

Therefore, Joanne will travel 137.5 mi in $2\frac{1}{2}$ hr at 55 mph.

> **Practice Exercise 2.** Determine a Celsius temperature if the Fahrenheit temperature is 86°.

EXAMPLE 2 In the United States, temperature readings are usually given in terms of the Fahrenheit scale. However, a movement is underway to give temperature readings using the Celsius (or centigrade) scale. Using C for Celsius and F for Fahrenheit, we have the conversion formula $C = \frac{5}{9}(F - 32°)$. Determine a Celsius temperature if the Fahrenheit temperature is 77°.

SOLUTION: Using the given formula with F = 77°, we have

$$C = \frac{5}{9}(F - 32°)$$
$$= \frac{5}{9}(77° - 32°)$$
$$= \frac{5}{9}(45°)$$
$$= \frac{(5)(\overset{5°}{\cancel{45°}})}{\underset{1}{\cancel{9}}}$$
$$= (5)(5°)$$
$$= 25°$$

Therefore, a Fahrenheit temperature of 77° is equivalent to a Celsius temperature of 25°.

8.7 Formulas

Practice Exercise 3. Determine the mental age of 14-year-old child who has an IQ of 120.

EXAMPLE 3 IQ (or intelligence quotient) is the relationship between mental age (M) and chronological age (C). For children, this relationship is given by the formula $IQ = \frac{100M}{C}$. Determine the IQ of a 12-year-old child who is found to have a mental age of 15 years.

SOLUTION: We are required to solve for IQ given that $M = 15$ yr and $C = 12$ yr. We have

$$IQ = \frac{100M}{C}$$

$$= \frac{(100)(15 \text{ yr})}{12 \text{ yr}}$$

$$= \frac{\overset{125}{\cancel{1500} \text{ yr}}}{\underset{1}{\cancel{12} \text{ yr}}}$$

$$= 125$$

Therefore, the child has an IQ of 125.

Almost all students are familiar with averages. We make references to batting averages, average speed traveled, average test grades, average amount of rainfall, average temperature readings, and so forth. Frequently, when we have a set of numbers, the average of that set of numbers is used to represent or describe the entire set. To calculate an average of a set of numbers, add all of the numbers in the set, and then divide that sum by the number of numbers that were added.

Practice Exercise 4. The average of three test scores is 87. Two of the scores are 79 and 91. Determine the third score.

EXAMPLE 4 The average for four test scores is given by the formula $A = \frac{a + b + c + d}{4}$ where $a = $ the first test score, $b = $ the second test score, $c = $ the third test score, and $d = $ the fourth test score. Evaluate A if $a = 83$, $b = 79$, $c = 70$, and $d = 78$.

SOLUTION: Substituting the test scores in the given formula, we have:

$$A = \frac{a + b + c + d}{4}$$

$$= \frac{83 + 79 + 70 + 78}{4}$$

$$= \frac{310}{4}$$

$$= 77.5$$

Therefore, the average of the four test scores is 77.5.

Practice Exercise 5. Using the formula $s = \frac{1}{2}gt^2$ with $g = 32$ ft/sec², determine s if $t = 5.4$ sec.

EXAMPLE 5 From physics, we use the formula $s = \frac{1}{2}gt^2$ to determine the distance (s) that a free-falling object will travel in a given amount of time (t) under the effects of gravity (g). Determine s when $t = 4.5$ sec. (We will use $g = 32$ ft/sec².)

SOLUTION: Substituting the given values in the formula, we have

$$s = \frac{1}{2}gt^2$$

$$= \left(\frac{1}{2}\right)\left(\frac{32 \text{ ft}}{1 \text{ sec}^2}\right)\left(4\frac{1}{2} \text{ sec}\right)^2$$

$$= \left(\frac{1}{2}\right)\left(\frac{32 \text{ ft}}{1 \text{ sec}^2}\right)\left(\frac{9 \text{ sec}}{2}\right)^2$$

$$= \left(\frac{1}{2}\right)\left(\frac{32 \text{ ft}}{1 \text{ sec}^2}\right)\left(\frac{81 \text{ sec}^2}{4}\right)$$

$$= \frac{(1)(32 \text{ ft})(81 \text{ sec}^2)}{(2)(1 \text{ sec}^2)(4)}$$

$$= \frac{(1)(\overset{8}{\cancel{32}} \text{ ft})(81 \cancel{\text{ sec}^2})}{(2)(1 \cancel{\text{ sec}^2})(\underset{1}{\cancel{4}})}$$

$$= \frac{(1)(\overset{4}{\cancel{8}} \text{ ft})(81)}{\underset{1}{\cancel{2}}}$$

$$= (1)(4 \text{ ft})(81)$$

$$= 324 \text{ ft}$$

Therefore, the object will fall 324 ft in 4.5 sec.

The formula $E = mc^2$ is familiar to most people. It represents Einstein's theory of relativity and expresses the equivalence between energy (E) and the mass (m) of an object and the speed of light (c). The formula, as written, is easy to use to solve for E, if we are given m. (c is a constant.) If we wanted to determine the mass, given the energy, we would solve the given formula for m as follows:

$$E = mc^2$$

$$\frac{E}{c^2} = m \qquad \text{(Dividing both sides of the equation by } c^2\text{)}$$

Therefore, we have $m = \frac{E}{c^2}$.

> **Practice Exercise 6.** Using the formula $E = IR$, determine the value of R when $E = 110$ V and $I = 30$ A.

EXAMPLE 6 Ohm's law is used in the study of electrical circuits. It is usually written in the form $E = IR$, where E represents the number of volts of an electromotive force of a circuit, I represents the number of amperes (amps) of current flowing through the circuit, and R represents the number of ohms (Ω) of resistance present in the circuit. Solve the formula for R in terms of E and I, and evaluate R given that $E = 220$ V and $I = 60$ A.

SOLUTION: Starting with the formula $E = IR$, we solve for R as follows:

$$E = IR$$

$$\frac{E}{I} = R \qquad \text{(Dividing both sides of the quation by } I\text{)}$$

Therefore, $R = \frac{E}{I}$.

8.7 Formulas

We now evaluate R given $E = 220$ V and $I = 60$ A:

$$R = \frac{E}{I}$$
$$= \frac{220 \text{ V}}{60 \text{ A}}$$
$$= \frac{\overset{11}{\cancel{220}} \text{ V}}{\underset{3}{\cancel{60}} \text{ A}}$$
$$= \frac{11}{3} \text{ V/A}$$
$$= \frac{11}{3} \Omega \quad (\text{V/A} = \Omega)$$

Therefore, $R = \frac{11}{3} \Omega$ when $E = 220$ V and $I = 60$ A.

> **Practice Exercise 7.** Using the formula $V = \frac{4}{3}\pi r^3$, determine V when $r = 4$ cm. (Use $\frac{22}{7}$ for π.)

EXAMPLE 7 The average (A) of the three test scores a, b, and c is given by the formula $A = \frac{a + b + c}{3}$. If $a = 66$ and $b = 76$, what score should a student receive on a third test to give the student an average of 80?

SOLUTION: There are two parts to this problem.

PART 1: Starting with the given formula, we could solve for the third test score (c) as follows:

$$A = \frac{a + b + c}{3}$$

$3A = a + b + c$ (Multiplying both sides of the equation by 3)
$3A = (a + b) + c$ (Grouping)
$3A - (a + b) = c$ (Subtracting $a + b$ from both sides of the equation)

Therefore, $c = 3A - (a + b)$.

PART 2: Using the result of the first part, we can now evaluate c for the given values of A, a, and b:

$$c = 3A - (a + b)$$
$$= 3(80) - (66 + 76)$$
$$= 240 - 142$$
$$= 98$$

Therefore, the student must receive a score of 98 on the third test to get an average of 80. (Given the first two test scores, is this very likely to happen?)

> **Practice Exercise 8.** Using the formula given in Example 8, determine the value for m, if $C = \$89.60$ and $d = 8$.

EXAMPLE 8 The Economy Motor Agency rents cars on a short-term basis. It determines the rental cost by using the formula $C = \frac{400d + 9m}{100}$ where C represents the rental cost (in dollars), d represents the number of rental days, and m represents the number of miles traveled. If you have \$65 for a car rental and expect to travel approximately 500 mi, for how many days can you rent a car from the agency?

SOLUTION: Again, there are two parts to this problem.

PART 1: First, solve the given formula for d:

$$C = \frac{400d + 9m}{100}$$

$100C = 400d + 9m$ (Multiplying both sides of the equation by 100)

$100C - 9m = 400d$ (Subtracting 9m from both sides of the equation)

$\dfrac{100C - 9m}{400} = d$ (Dividing both sides of the quation by 400)

Therefore, $d = \dfrac{100C - 9m}{400}$.

PART 2: We now evaluate d for the given values of C and m:

$$d = \frac{100C - 9m}{400}$$

$$= \frac{(100)(65) - (9)(500)}{400}$$

$$= \frac{6500 - 4500}{400}$$

$$= \frac{2000}{400}$$

$$= 5$$

Therefore, you would be able to rent the car for 5 days.

ANSWERS TO PRACTICE EXERCISES

1. 5.6 hr. 2. 30° C. 3. 16.8 yr. 4. 91. 5. 466.56 ft. 6. $\dfrac{11}{3}$ ohms. 7. $268\dfrac{4}{21}$ cm³. 8. 640 mi.

EXERCISES 8.7

In Exercises 1–5, use the formula $d = rt$.

1. Evaluate d if $r = 45$ mph and $t = 2\dfrac{1}{4}$ hr.
2. Evaluate d if $r = 50$ m/sec and $t = 17$ sec.
3. Solve for r in terms of d and t and then evaluate r when $d = 117$ mi and $t = 3$ hr.
4. Solve for t in terms of d and r and then evaluate t when $d = 221$ ft and $r = 34$ ft/sec.
5. How long does it take a snowmobiler to travel 600 ft if she travels at a speed of 40 ft/sec?

In Exercises 6–10, use the formulas $F = \dfrac{9}{5}C + 32°$ and $C = \dfrac{5}{9}(F - 32°)$ for converting temperature readings between the Fahrenheit and Celsius scales.

6. A thermometer registers the temperature at 20° C. What is the equivalent temperature in degrees Fahrenheit?

8.7 Formulas

7. If the temperature outside your dorm is 15° C, is it freezing? (*Hint:* Convert to degrees Fahrenheit.)
8. What is the equivalent, in degrees Celsius, of a temperature reading of 65° F?
9. Start with the formula $F = \frac{9}{5}C + 32°$ and solve for C in terms of F.
10. Start with the formula $C = \frac{5}{9}(F - 32°)$ and solve for F in terms of C.

In Exercises 11–13, use the formula $IQ = \frac{100M}{C}$ for intelligence quotient (IQ), where M represents the mental age and C is the chronological age of a child.

11. Determine the IQ of a 10-year-old child who has a mental age of 9 years.
12. Solve for M in terms of IQ and C and evaluate M if $IQ = 140$ and $C = 15$ years.
13. Solve for C in terms of IQ and M and evaluate C if $IQ = 80$ and $M = 8$ years.

In Exercises 14–19, use the formula $A = \frac{a + b + c + d}{4}$ which represents the average of the four numbers a, b, c, and d.

14. Determine A if $a = 61$, $b = 72$, $c = 59$, and $d = 88$.
15. Determine A if $a = -23$, $b = 17$, $c = 18$, and $d = -42$.
16. A price comparison of four stores was made for the price of a compact radio. In the first store, the price was $38.75; in the second store, $41.45; in the third store, $39.25; and in the fourth store, $39.95. The radios in all four stores were identical. What was the average price for the radio for the four stores?
17. A test was repeated four times to determine the braking distance of a new car. The test results were 218 ft, 211 ft, 223 ft, and 207 ft. Determine the average braking distance for the car.
18. For the first three weeks of February, Liz received commissions of $291, $276, and $227 working in a hardware store. What must her commission be for the fourth week to have an average weekly commission of $260 for the four weeks.?
19. If you have three exam scores of 88, 79, and 80, what score do you need on the fourth exam to have an average of 84?

In Exercises 20–22, use the formula $s = \frac{1}{2}gt^2$ which represents the distance (s) that a free-falling object travels in a given amount of time (t) under the effects of gravity ($g = 32$ ft/sec^2).

20. A student's ID card is dropped from the top of a 600-ft dormitory building. How far will the card fall in 5.2 sec?
21. A book is dropped from the top of a 300-ft building. How close to the ground will the book be after 3.5 sec?
22. A ball is dropped from a window of a building. The window is 200 ft above ground level. One second later, another ball is dropped from another window of the building. The second window, 120 ft above ground level, is directly below the first window. How far apart will the two balls be 3 sec after the first ball is dropped?

In Exercises 23–25, use the formula $E = IR$ for Ohm's law where E represents the electromotive force of a circuit (in volts), I represents the number of amperes of current flowing through the circuit, and R represents the number of ohms of resistance present in the circuit.

23. Determine the number of volts in a 15-A circuit with 8 Ω of resistance.
24. Solve for I in terms of E and R and determine the number of amps in a 220-V circuit with resistance of 14 Ω.
25. Solve for R in terms of E and I and determine the number of ohms in a 220-V circuit with 30 A of current flowing through it.

In Exercises 26–29, use the formula $C = \dfrac{400d + 9m}{100}$ where C represents the rental cost (in dollars) of an automobile, d represents the number of days of travel, and m represents the number of miles traveled.

26. Karole rented a car to drive home from college during the spring break. If she traveled 420 mi and rented the car for 3 days, what was the rental cost?
27. Brenda rented a car to drive to various locations for job interviews. If she traveled 1100 mi and rented the car for 15 days, what was the rental cost?
28. Joanne rented a car for 4 days. If the rental cost was $44.80, how many miles did she travel?
29. Jim rented a car and traveled 480 mi. If the rental cost was $63.20, for how many days did he rent the car?

In Exercises 30–32, use the formula $C = \dfrac{aD}{a + 12}$ which relates a child's dose of drugs (C) to an adult's dosage (D). (a = age of the child.)

30. Determine the correct dose of an antibiotic for 6-year-old Jimmy if the adult dose is 30 mg.
31. Solve the given formula for D in terms of C and a.
32. Using the results obtained in Exercise 31, determine an adult's dose of penicillin if the dose for an 8-year-old child is 12 mg.

For Exercises 33–36, use the formula $V = C(1 - rt)$ which represents the present value (V) of an item that is depreciated each year. C is the original cost, r is the rate of depreciation, and t is time.

33. An automobile depreciates at the rate of 15% per year. Determine the value of an automobile 2 yr after it was purchased if the purchase price was $11,500.
34. A microwave oven depreciates at the rate of 8% per year. Determine the value of a microwave oven 9 yr after purchase if the purchase price was $450.
35. Solve for C in terms of V, r, and t.
36. Solve for r in terms of V, C, and t.

For Exercises 37–40, we will use geometric formulas that were introduced in Chapter 5.

37. The perimeter (P) of a rectangle is given by the formula $P = 2L + 2W$ where L = the length and W = the width of the rectangle. (P, L, and W are all measured in the same units.) A picture frame is rectangular. The perimeter of the outside of the frame is 13 ft and the length is 4 ft. Solve the formula for W and determine the width of the frame.
38. The area (A) of a trapezoid is given by the formula $A = \dfrac{1}{2}h(b + B)$ where h = the height, b = the length of the shorter base, and B = the length of the longer base. (h, b, and B are all measured in the same units; A is measured in the unit squared.) Solve the formula for h and determine the height of a trapezoid if $b = 13$ cm, $B = 18$ cm, and $A = 124$ cm².
39. The circumference (C) of a circle is given by the formula $C = 2\pi r$ where r = the radius. (C and r are measured in the same units.) Solve the formula for r and determine the radius of a circle whose circumference is 28 yd. (Leave your answer in terms of π.)
40. The volume (V) of a rectangular solid is given by the formula $V = LWh$ where L = length, W = width, and h = height. (L, W, and h are all measured in the same units; V is measured in the unit cubed.) Solve the formula for W and determine the width of a rectangular solid if $L = 3$ m, $h = 1.8$ m, and $V = 3.24$ m³.

In Exercises 41–50, a formula is given. Solve for the indicated variable.

41. $s = 16t^2 + vt$; v
42. $p = \dfrac{kT}{V}$; T
43. $A = \dfrac{1}{2}bh$; b
44. $V = 4\pi r^2 h$; h

45. $A = \dfrac{B}{C}; C$ 46. $3pq - 2r = t; q$

47. $L = a + (n - 1)d; d$ 48. $A = P(1 + rt); r$

49. $a = 2b + 3c^2d; b$ 50. $a = 2b + 3c^2d; d$

8.8 Ratio and Proportion

OBJECTIVES

After completing this section, you should be able to:
1. Find the ratio of one quantity to another.
2. Set up and solve proportions.

When ordering two numbers, we are comparing them to determine which number is larger. This comparison can also be done using the operation of subtraction to indicate how much larger one number is than another. For instance, $8 - 5 = 3$ means that 8 is 3 more than 5.

Comparing numbers by ordering or by subtracting doesn't always provide us with enough information about the numbers. For instance, the difference between 8 and 2 is 6, and the difference between 12 and 6 is also 6. We note, however, that 8 is 4 times as large as 2, whereas 12 is only twice as large as 6. Here, we are comparing the numbers by division. The comparison of numbers by division is called a "ratio."

DEFINITION

A **ratio** of the number a to the number b is the quotient of the two numbers and is written as $\dfrac{a}{b}$ or $a:b$ or a to b.

Practice Exercise 1. Write the ratio of 9 to 7.

EXAMPLE 1 Write the ratio of 5 to 3.

SOLUTION: The ratio of 5 to 3 is written $\dfrac{5}{3}$ or $5:3$ or 5 to 3 and means that 5 is $\dfrac{5}{3}$ $\left(\text{or } 1\dfrac{2}{3}\right)$ times larger than 3.

Practice Exercise 2. Write the ratio of 23 to 12.

EXAMPLE 2 Write the ratio of 20 to 30.

SOLUTION: The ratio of 20 to 30 is written $\dfrac{20}{30}$ or $20:30$ or 20 to 30 and means that 20 is $\dfrac{20}{30}$ $\left(\text{or } \dfrac{2}{3}\right)$ as large as 30.

Practice Exercise 3. Write the ratio of the number of passing grades to the number of failing grades, if 17 students in a class of 22 students received a passing grade.

EXAMPLE 3 In a class, there are 11 girls and 14 boys. Write (a) the ratio of boys to girls and (b) the ratio of girls to the total number of students.

SOLUTION: (a) The ratio of boys to girls is $\dfrac{14}{11}$. (b) The total number of students is 11 (girls) + 14 (boys), or 25. Hence, the ratio of girls to the total number of students is $\dfrac{11}{25}$.

Practice Exercise 4. Write as a ratio, the number of hits to times at bat if a player gets 8 hits out of 21 times at bat.

EXAMPLE 4 Bill bought 16 light bulbs and later determined that 4 of them were defective. Write the ratio of defective light bulbs to nondefective light bulbs, in simplified form.

SOLUTION: The ratio of defective light bulbs to nondefective light bulbs is 4 to 12 or 1 to 3. That is, for every 1 defective light bulb, there were 3 nondefective light bulbs.

When using a ratio to compare two quantities, always compare like quantities. For instance, the ratio of 23 cm to 3 m can be determined by the following two methods:

- Changing both units to centimeters, using the conversion that 100 cm = 1 m. Hence, 3 m = 300 cm. The ratio of 23 cm to 3 m is $\frac{23 \text{ m}}{300 \text{ cm}}$ or 23 to 300

- Changing both units to meters, using the conversion that 1 cm = $\frac{1}{100}$ m. Hence, 23 cm = $\frac{23}{100}$ m. The ratio of 23 cm to 3 cm is $\left(\frac{23}{100} \text{ m}\right) \div (3 \text{ m})$ or $\frac{23}{300}$ or 23 to 300.

In either conversion, the resulting ratios are the same.

The ratio $\frac{4}{7}$ is read "4 to 7" and *not* "four-sevenths." It is customary to simplify a ratio. For instance, instead of the ratio $\frac{6}{4}$ (read "6 to 4"), we would write the ratio $\frac{3}{2}$ (read "3 to 2").

Practice Exercise 5. Simplify the ratio 14 to 21.

EXAMPLE 5 Simplify the ratio 6 to 100.

SOLUTION: The ratio 6 to 100, simplified, is 3 to 50.

Practice Exercise 6. Simplify the ratio 4 ft to 4 yd.

EXAMPLE 6 Simplify the ratio 5 centimeters to 3 meters.

SOLUTION: Since the units are not the same, we will change one to the other. Namely, we will change meters to centimeters, using the conversion that 1 m is equal to 100 cm. Hence,

$$\frac{5 \text{ cm}}{3 \text{ m}} = \frac{5 \text{ cm}}{(3)(100) \text{ cm}} = \frac{5}{300} = \frac{1}{60}$$

Hence, the simplified ratio of 5 cm to 3 m is 1 to 60.

ANSWERS TO PRACTICE EXERCISES

1. $\frac{9}{7}$ or 9 : 7. 2. $\frac{23}{12}$. 3. 17 to 5. 4. 8 to 21. 5. 2 to 3. 6. 1 to 3.

Since a ratio is a quotient of two numbers and since a fraction is a quotient, we may think of a ratio as being a fraction. Further, since we encountered equivalent fractions earlier, it is not surprising to learn that we also have equal ratios. For instance, the ratio 1 : 3 is equal to the ratio 5 : 15 since the fractions $\frac{1}{3}$ and $\frac{5}{15}$ are equivalent. When we have two ratios that are equal, we have a "proportion."

59. In Little City, 4 out of 15 multiapartment buildings inspected failed to have smoke detectors installed. At that rate, how many of the 225 multiapartment buildings in Little City do have smoke detectors installed?
60. Eleven of 13 people surveyed prefer toothpaste with fluoride. At that rate, how many people among 3042 people prefer fluoride toothpaste?
61. A 64-oz bottle of apple juice costs $1.50 and a 48-oz bottle of the same brand of apple juice costs $1.10. Which is the better buy?
62. Company A hired 11 minority workers among 68 new employees. Company B hired 9 minority workers among 53 new employees. Which company hired the higher rate of minority workers among its new employees?
63. Bill drove 2100 mi in 6 days while Sue drove 1300 mi in 5 days. Who had the higher rate of miles driven per day?
64. A 22-oz bottle of liquid soap detergent sells for $1.19, while a 12-oz bottle of the same brand detergent is on sale for $0.73. Which is the better buy?
65. Bette walked 26 mi in 3 days, while Darlene walked 33 mi in 4 days. Who had the higher rate of miles walked per day?
66. The Hot Shot basketball team won 8 games of 14 games played. The Hot Rod basketball team won 7 of 13 games played. Which team had the better record?
67. John drove 265 mi on 13 gal of gasoline, while Alexa drove 187 mi on 9.5 gal of gasoline. Who had better gas mileage?
68. The Best Deal in Town rental agency rented 116 cars in 7 days. The Most for Your Dollar rental agency rented 131 cars in 9 days. Which rental agency had the higher rental rate per day?
69. Which is the better buy, a 40-lb bushel of Ida Red apples for $6.85 or a 15-lb basket of the same kind of apples for $3.75?
70. The Savings Bank requires a $5000 downpayment for a $62,000 mortgage, while the Consumer Credit Union requires a $5500 downpayment for a $75,000 mortgage. Who has the higher requirement for the downpayment?
71. Which is the greater savings: a $3 rebate on a toaster selling for $19.95 or a $5 rebate on a toaster oven selling for $26.95?
72. Which is the better buy: a 6.4-oz tube of toothpaste selling for $1.49 or an 8.2-oz size of the same brand of toothpaste selling for $1.97?
73. Working on a production line, Sally can weld 4167 widgets in 8 days, while Bruce can weld 5203 of the same type of widgets in 9 days. Who has the greater rate of welds per day?
74. Seminar A has 87 female participants among its 139 participants. Seminar B has 102 female participants among its 161 participants. Which seminar has the greater rate of male participants?
75. The 168 people in Group A were served 92 lb of cheesecake for dessert while the 219 people in Group B were served 112 lb of the same kind of cheesecake for dessert. Which group had the greater amount of cheesecake per person?

Chapter Summary

In this chapter, you were introduced to simple linear equations and their solutions. Several types of word problems that can be solved by using linear equations were considered. Ratios and proportions were also discussed.

The **key words** introduced in this chapter are listed here in the order in which they appeared in the chapter. You should know what these words mean before proceeding to the next chapter.

equation (p. 334)
variables (p. 334)
linear equation in a single variable (p. 334)
equivalent equations (p. 335)
consecutive integers (p. 356)
formula (p. 369)
ratio (p. 377)
proportion (p. 379)
extremes (p. 380)
means (p. 380)

8.1 Solving Linear Equations Using Addition or Subtraction

Solve each of the following equations for the given variable. Check your results in the given equations.

1. $x + 7 = 2$
2. $y - 5 = -1$
3. $u - (-1.2) = -2.3$
4. $t + (-2.7) = 3.1$
5. $4 = t - 5$
6. $17 = w + 3$

8.2 Solving Linear Equations Using Multiplication or Division

Solve each of the following equations for the given variable. Check your results in the given equations.

7. $3t = 5$
8. $-4p = -3.2$
9. $\dfrac{z}{3} = -5.1$
10. $p \div \dfrac{2}{3} = \dfrac{1}{4}$
11. $\dfrac{x}{-7} = \dfrac{1}{2}$
12. $\dfrac{y}{0.2} = 4.7$

8.3 Solving Linear Equations Using More than One Operation

Solve each of the following equations for the given variable. Check your results in the given equations.

13. $2x + 7 = 4$
14. $3x - 4 = 5$
15. $\dfrac{y-3}{2} = \dfrac{1}{4}$
16. $2(y - 3) = 6$
17. $\dfrac{3(p-1)}{4} = 2$
18. $6 - \dfrac{4}{7}t = 5$

8.4 Simplifying First before Solving Linear Equations

Solve each of the following equations for the given variable. Check your results in the given equations.

19. $3(y + 1) = 2(y - 3)$
20. $\dfrac{x-3}{4} = \dfrac{2-x}{5}$
21. $2w - 3 = 3(w + 2)$
22. $\dfrac{2(y-3)}{5} = y + 3$
23. $\dfrac{3(x-1)}{2} = \dfrac{-2(x+4)}{3}$
24. $\dfrac{2}{3}x + 1 = \dfrac{1}{2}x - 5$

8.5 Applications Involving Linear Equations: Part 1

25. Dan is older than Karen. The sum of their ages is 64 yr. Five years from now, Dan will be twice as old as Karen was 9 yr ago. How old is each now?
26. A parking lot behind the Student Union is in the form of a rectangle. Determine the width of the lot if its length is 48.5 m and its area is 1678.1 m².
27. According to nutrition information provided by McDonald's, a Filet-O-Fish® sandwich has 215 more calories than an order of regular fries. A vanilla milkshake has 132 more calories than an order of regular fries. A meal of an order of regular fries, a Filet-O-Fish® sandwich, and a vanilla milkshake contains 1007 calories. How many calories are in each of the three items?

8.6 Applications Involving Linear Equations: Part 2

28. Brian has a collection of coins consisting of pennies, nickels, dimes, and quarters. He has half as many quarters as dimes; the number of nickels he has is 2 less than 3 times the number of dimes; the number of pennies he has is 3 more than the number of quarters. If his total collection is worth $2.97, how many of each kind of coin does Brian have?
29. A child weighing 70 lb sits 5 ft from the fulcrum on a teeterboard. Another child weighing 85 lb sits on the teeterboard 6 ft from the other side of the fulcrum. Will the teeterboard be balanced? Why (not)?
30. At 3:45 P.M., Jessica left the Riverside Mall on her 10-speed bike and traveled west at 7 mph. Exactly 2 hr later, Jim left the mall on his motorcycle and traveled west at 25 mph. Assuming that Jessica was still riding her bike, at what time would Jim catch up to her?

8.7 Formulas

31. Using the formula $A = \dfrac{a+b+c+d}{4}$, solve for d in terms of A, a, b, and c. Then evaluate d when $A = 82$, $a = 86$, $b = 78$, and $c = 79$.
32. Using the formula $C = \dfrac{400d + 9m}{100}$, solve for d in terms of C and m. Then evaluate d when $C = \$33.20$ and $m = 280$.
33. Using the formula $C = \dfrac{aD}{a+12}$, solve for D in terms of C and a. Then evaluate D when $C = 15$ and $a = 12$.

8.8 Ratio and Proportion

Write each of the statements in Exercises 34–36 as a ratio.

34. The number of inches in a yard to the number of centimeters in a meter.
35. The number of students bused to the number of students not bused in a school district with 11,603 enrolled students and 5123 students *not* bused
36. The number of questions answered correctly on an 80-question test to the number of questions answered incorrectly if 4 questions were left blank and 12 questions were answered incorrectly
37. Simplify the ratio: 9 to 3.
38. Simplify the ratio: 6 to 126.

In Exercises 39–42, solve each of the given proportions for n.

39. $\dfrac{4}{7} = \dfrac{n}{21}$
40. $\dfrac{6}{9} = \dfrac{24}{n}$
41. $\dfrac{3n}{5} = \dfrac{45}{75}$
42. $\dfrac{0.6}{n} = \dfrac{4.2}{84}$

Teasers

1. Solve the equation $\dfrac{1}{4}(s - 3) = \dfrac{7}{6} - \dfrac{1}{3}(s + 4)$ for s.
2. Solve the equation $6.2 + 7y = 2(4.9 - 1.8y)$ for y.
3. Solve the equation $\dfrac{4p}{5} - \dfrac{5p - 1}{6} = \dfrac{3p - 2}{7} - \dfrac{p}{2}$ for p.
4. Paula mailed a package at the priority rate of $2.90. She also mailed a package at the express rate of $9.95. How many $0.29 stamps could she buy after these two mailings if she has a total of $21.13?
5. You decide to rent a car from the ABC Car Rental Agency for a week and are offered two plans. The first plan is for $95 for the week plus $0.15 per mile for each mile after the first 150 miles. The second plan is for $80 per week plus $0.11 for each mile driven. You plan to travel 800 miles during the week. Which plan should you choose? What are the savings?
6. The sum of three integers is 43. The largest integer is 3 times the smallest integer. The third integer is 6 less than the largest integer. Determine the integers.
7. The dimensions of a rectangle are in the ratio of 1 to 3. If the perimeter of the rectangle is 144 ft, determine the area of the rectangular region.
8. The sides of one square are each 9 yd longer than the sides of another square. Determine the dimensions of the two squares if the sum of their perimeters is 600 ft.
9. Jodi invested a total of $14,000. For one investment, she earned 8%, and for the other, she lost 6%. How much was each investment if her earnings were equal to her losses?
10. The corresponding dimensions of two rectangles are in the ratio 2 to 3. Determine the combined areas of the two rectangular regions if the total of the two perimeters is 120 ft.

CHAPTER 8 TEST

This test covers materials from Chapter 8. Read each question carefully and answer all questions. If you miss a question, review the appropriate section of the text.

In Exercises 1–14, solve each question for the given variable. Check your results.

1. $x + 5 = 3$
2. $y + 6 = 0$
3. $u - 7 = 5$
4. $w - \dfrac{2}{3} = \dfrac{1}{2}$
5. $z + 1.6 = 0.7$
6. $t - 1.9 = 1.2$
7. $2x = -5$
8. $3y = \dfrac{1}{4}$
9. $\dfrac{t}{-5} = 1.6$
10. $\dfrac{u}{7} = \dfrac{-2}{3}$
11. $3x - 2 = 1$
12. $5 - \dfrac{1}{2}y = 4$
13. $2(y - 1) = 3(y + 2)$
14. $\dfrac{t - 1}{5} = \dfrac{2 - t}{3}$

15. Simplify the ratio: 4 to 20.
16. Simplify the ratio: 121 to 11.
17. Solve for n: $\dfrac{4}{12} = \dfrac{6}{n}$.
18. Determine three consecutive integers whose sum is -69.
19. The sum of two numbers is 13. The smaller number is 8 less than twice the larger number. What are the numbers?
20. A collection of coins consists of nickels, dimes, and quarters. The number of quarters in the collection is 1 more than the number of nickels. The number of nickels is 2 less than the number of dimes. If the total collection is worth $2.05, how many of each kind of coin are in the collection?

CUMULATIVE REVIEW: CHAPTERS 1–8

Perform the indicated operations in Exercises 1–6.

1. $-4^2 \div (-2)^3$
2. $\dfrac{11}{12} - \dfrac{2}{9}$
3. 1.26% of 0.32
4. $\left|\dfrac{-1}{2}\right|^2 \times \left(\dfrac{1}{2}\right)^2$
5. $(-16 \div 2) - (-27 \div 3)$
6. $2.3 + 1.2 \times (-3.4)$

7. Round $\dfrac{1}{2}$ of 2.37% to the nearest thousandth.
8. Round $|-29| \div 4.09$ to the nearest tenth.
9. Determine the perimeter of a rectangle with length 14.9 cm and width 11.9 cm.
10. Evaluate the expression $-2x^2y$ if $x = -2$ and $y = 0.5$.
11. Simplify the expression $2u - 3[v - 2(u - v)] - 4v$ by combining like terms.
12. Simplify the expression $-2\{4a - 2[a - 3(a + b) + b] - 4(a - b)\}$ by combining like terms.
13. Write an algebraic expression for the cube of the product of twice x and the square of y.
14. Write an algebraic expression for the quotient of p and q added to the product of twice p and 3 times q.
15. If twice r is diminished by 7, then the result is 3 more than r. Determine the value of r.
16. Solve the equation $4(u - 3) = -3(1 - 2u)$ for u.
17. Solve the equation $\dfrac{-2}{3}(p - 2) = \dfrac{-3}{4}(3 - 2p)$ for p.
18. Simplify the ratio: 2.3 to 0.69.
19. Solve for w: $\dfrac{3}{4w} = \dfrac{-2}{7}$.
20. The sum of two numbers is -28. The smaller number is 8 more than twice the larger number. What are the numbers?
21. Convert the fraction $\dfrac{319}{1000}$ to a percent.
22. Estimate the product of 299.2 and 189.8 by rounding each number to the nearest ten.
23. If $3500 is invested at 6.5% per year, determine the amount in the account after 18 months.
24. Determine the base of a triangular region whose height is 11.8 yd and area is 103.84 yd^2.
25. How many seconds are in the month of September?

CHAPTER 9
Exponents and Scientific Notation

In this chapter, we will continue our discussion of exponents that was begun in Chapter 1. We will work further with positive integer exponents and introduce zero and negative integer exponents. The use of exponential notation enables us to write an algebraic expression more concisely.

We will discuss the multiplication of powers of the same base, powers of a power of a base, and division of powers of the same base. Scientific notation will also be introduced as an application of exponents.

CHAPTER 9 PRETEST

This pretest covers material on integer exponents and scientific notation. Read each question carefully and try to answer all questions. Each question is keyed to the chapter section in which the particular topic is discussed. Check your answers with those given in the back of the text.

Questions 1–6 pertain to Section 9.1. Simplify each of the given expressions.

1. $(x^3)(x^2)(x^4)$

2. $(y^3)^2$

3. $(-3)^4 \div (-3)^7$

4. $(p^3)(2p^4) \div (4p^3)$

5. $(9p^4q^2) \div (-3p^2q^5)$

6. $(q^4)^3(q^3)^2 \div (q^3)^5$

Questions 7–12 pertain to Section 9.2. Simplify each of the given expressions.

7. $(x^2y^3)^3$

8. $(2p^3q^2)^4$

9. $(-3s^3)(-2s^2)^3$

10. $(4s^2t)^2(-3s^3t^4)^3$

11. $(x^4y^2)^3 \div (2x^2y)^2$

12. $(-5m^3n^2) \div (-2m^2n)^3$

Questions 13–18 pertain to Section 9.3. Simplify each of the given expressions, and write your results without negative exponents.

13. a^3a^{-7}

14. $p^2p^{-4}p^{-3}$

15. $(-4x^2y^{-3}z)^0$

16. $(-2r^2s^{-1}t^0) \div (-3r^{-4}s^0t^{-3})$

17. $(-3u^{-2}v^4w^0) \div (2u^3v^0w^{-1})^0$

18. $(2a^{-2}b^0c^3)^{-2} \div (-3a^0b^{-3}c^2)^2$

Questions 19–27 pertain to Section 9.4.

In Exercises 19–22, rewrite each of the given numbers in scientific notation.

19. 36.2

20. 0.0097

21. 1.2×2.3

22. 396.2×10^{-3}

In Exercises 23–25, determine the number N for each of the numbers given in scientific notation.

23. 2.9×10^4

24. 9.6×10^{-3}

25. 1.063×10^{-4}

26. Round 39.62 to three significant digits.

27. Round 4.31×10^{-2} to two significant digits.

9.1 More on Positive Integer Exponents

OBJECTIVE

After completing this section, you should be able to use the basic rules for working with positive integer exponents.

In Section 1.7, you were introduced to exponents and learned that an exponent is a number to which a base is raised. Recall in the expression x^3, the exponent is 3 and the base is x. In the expression $4y^5$, the exponent is 5, the base is y (*not* $4y$), and 4 is the numerical coefficient of y^5.

In this section, you will learn more about integer exponents and the basic rules of exponents. In general, if x is any number and n is a positive integer greater than 1, then the expression x^n is the product of n factors all equal to x. If $n = 1$, then $x^n = x^1 = x$ where the exponent 1 is understood.

DEFINITION

If x is any number and n is a positive integer greater than 1, then

$$x^n = \underbrace{x \cdot x \cdot x \cdots x}_{n \text{ times}}$$

If $n = 1$, then $x^n = x^1 = x$.

- $b^3 = b \cdot b \cdot b$
- $5^4 = 5 \cdot 5 \cdot 5 \cdot 5$
- $u^2 = u \cdot u$
- $(-2)^6 = (-2)(-2)(-2)(-2)(-2)(-2)$
- $t^8 = t \cdot t \cdot t \cdot t \cdot t \cdot t \cdot t \cdot t$

What happens if we try to multiply two powers of the same base? For instance, what is $x^3 \cdot x^4$? Since

$$x^3 = x \cdot x \cdot x$$
and
$$x^4 = x \cdot x \cdot x \cdot x$$
then
$$x^3 \cdot x^4 = (x \cdot x \cdot x)(x \cdot x \cdot x \cdot x)$$
$$= x \cdot x \cdot x \cdot x \cdot x \cdot x \cdot x$$
$$= x^7$$

Hence, $x^3 \cdot x^4 = x^7$. (Observe that $7 = 3 + 4$.)

What about $y^4 \cdot y^6$?

Since
$$y^4 = y \cdot y \cdot y \cdot y$$
and
$$y^6 = y \cdot y \cdot y \cdot y \cdot y \cdot y$$
then
$$y^4 \cdot y^6 = (y \cdot y \cdot y \cdot y)(y \cdot y \cdot y \cdot y \cdot y \cdot y)$$
$$= y \cdot y \cdot y \cdot y \cdot y \cdot y \cdot y \cdot y \cdot y \cdot y$$
$$= y^{10}$$

Therefore, $y^4 \cdot y^6 = y^{10}$. (Observe that $10 = 4 + 6$.)

If we multiply two powers of the *same* base, then the product is also a power of the *same* base and its exponent is the sum of the two exponents in the factors.

9.1 More on Positive Integer Exponents

Product Rule for Exponents If x represents any number and m and n are positive integers, then

$$x^m \cdot x^n = x^{m+n}$$

Using the product rule for exponents, simplify each of the following. Leave your answers in exponential form.

Practice Exercise 1. $(x^3)(x^4)$

EXAMPLE 1 Simplify: $a^3 \cdot a^2$.

SOLUTION: $\quad a^3 \cdot a^2 = a^{3+2} = a^5$

Practice Exercise 2. $(y^6)(y^5)$

EXAMPLE 2 Simplify: $(6^4)(6^9)$.

SOLUTION: $\quad (6^4)(6^9) = 6^{4+9} = 6^{13}$

Practice Exercise 3. $(-2)^4(-2)^3$

EXAMPLE 3 Simplify: $m^3 \cdot m$.

SOLUTION: $\quad m^3 \cdot m = m^{3+1} = m^4$

Practice Exercise 4. $(2.7)^4(2.7)^5$

EXAMPLE 4 Simplify: $(-3)^5(-3)^8$.

SOLUTION: $\quad (-3)^5(-3)^8 = (-3)^{5+8} = (-3)^{13}$

Practice Exercise 5. $(u^2)(u^3)(u^4)$

EXAMPLE 5 Simplify: $z^7 \cdot z^{12}$.

SOLUTION: $\quad z^7 \cdot z^{12} = z^{7+12} = z^{19}$

Observe that in Example 2, we multiplied the 4th power of 6 by the 9th power of 6 and got the 13th power of 6. We did *not* multiply the bases. Similarly, in Example 4, we multiplied the 5th power of -3 by the 8th power of -3 and got the 13th power of -3. We did *not* multiply the bases.

CAUTION To use the product rule for exponents, remember that you are *multiplying* powers of the *same* base. Hence,

- $x^2 + x^3$ **cannot** be simplified further since the expression is a *sum* of two powers of the same base.
- $x^2 \cdot y^3$ **cannot** be simplified further since x and y are two *different bases*.

Of course, the product rule for exponents can be extended to include a product of three or more factors, all of which are powers of the same base.

Practice Exercise 6. $z(z^4)(z^2)(z^6)$

EXAMPLE 6 Simplify: $x^2 \cdot x^3 \cdot x^4$.

SOLUTION:
$$\begin{aligned} x^2 \cdot x^3 \cdot x^4 &= (x^2 \cdot x^3) \cdot x^4 \\ &= (x^{2+3}) \cdot x^4 \\ &= x^5 \cdot x^4 \\ &= x^{5+4} \\ &= x^9 \end{aligned}$$

Therefore, $x^2 \cdot x^3 \cdot x^4 = x^9$. (Observe that $9 = 2 + 3 + 4$.)

Practice Exercise 7.
$(p^6)(p^8)(p^2)(p^3)p$

EXAMPLE 7 Simplify: $(7^4)(7^5)(7^2)(7^3)$.

SOLUTION:
$$(7^4)(7^5)(7^2)(7^3) = 7^{4+5+2+3}$$
$$= 7^{14}$$

ANSWERS TO PRACTICE EXERCISES
1. x^7. 2. y^{11}. 3. $(-2)^7$. 4. $(2.7)^9$. 5. u^9. 6. z^{13}. 7. p^{20}.

Now, consider the expression $(x^3)^2$. Note that we do *not* have a product of powers of x. Instead, we have a power of a power of x. But

$$(x^3)^2 = x^3 \cdot x^3$$
$$= x^{3+3}$$
$$= x^6$$

Therefore, $(x^3)^2 = x^6$. (Observe that $6 = 3 \times 2$.)

What about $(y^3)^4$?

$$(y^3)^4 = y^3 \cdot y^3 \cdot y^3 \cdot y^3$$
$$= y^{3+3+3+3}$$
$$= y^{12}$$

Therefore, $(y^3)^4 = y^{12}$. (Observe that $12 = 3 \times 4$.)
Further,

$$(u^3)^5 = u^3 \cdot u^3 \cdot u^3 \cdot u^3 \cdot u^3$$
$$= u^{3+3+3+3+3}$$
$$= u^{15}$$

Therefore, $(u^3)^5 = u^{15}$. (Observe that $15 = 3 \times 5$.)

Power Rule for Exponents

If x represents any number and m and n are positive integers, then

$$(x^m)^n = x^{mn}$$

Using the rules for exponents, simplify each of the following.

Practice Exercise 8. $(x^4)^5$

EXAMPLE 8 Simplify: $(x^2)^3$.

SOLUTION:
$$(x^2)^3 = x^{(2)(3)} = x^6$$

Practice Exercise 9. $(3^2)^5$

EXAMPLE 9 Simplify: $[(-4)^4]^4$.

SOLUTION:
$$[(-4)^4]^4 = (-4)^{(4)(4)} = (-4)^{16}$$

Practice Exercise 10. $[(-9)^6]^3$

EXAMPLE 10 Simplify: $(u^7)^4$.

SOLUTION:
$$(u^7)^4 = u^{(7)(4)} = u^{28}$$

9.1 More on Positive Integer Exponents

Practice Exercise 11. $(y^3)(y^4)(y)$

EXAMPLE 11 Simplify: $(10^6)^5$.

SOLUTION:
$$(10^6)^5 = 10^{(6)(5)} = 10^{30}$$

Practice Exercise 12. $(4^3)^2$

EXAMPLE 12 Simplify: $(t^8)^4$.

SOLUTION:
$$(t^8)^4 = t^{(8)(4)} = t^{32}$$

Practice Exercise 13. $[(-2)^3]^4$

EXAMPLE 13 Simplify: (a) $(x^2)(x^3)$; (b) $(x^2)^3$.

SOLUTION:
(a) $(x^2)(x^3) = x^{2+3} = x^5$ (We use the product rule for this example.)
(b) $(x^2)^3 = x^{(2)(3)} = x^6$ (We use the power rule for this example.)

Practice Exercise 14. $(y^7)^5$

EXAMPLE 14 Simplify: (a) $(y^4)(y^5)$; (b) $(y^4)^5$.

SOLUTION:
(a) $(y^4)(y^5) = y^{4+5} = y^9$ (We use the product rule for this example.)
(b) $(y^4)^5 = y^{(4)(5)} = y^{20}$ (We use the power rule for this example.)

Practice Exercise 15. $[(a+b)^4]^3$

EXAMPLE 15 Simplify: (a) $(3^6)(3^7)$; (b) $(3^6)^7$.

SOLUTION:
(a) $(3^6)(3^7) = 3^{6+7} = 3^{13}$ (We use the product rule for this example.)
(b) $(3^6)^7 = 3^{(6)(7)} = 3^{42}$ (We use the power rule for this example.)

ANSWERS TO PRACTICE EXERCISES

8. x^{20}. **9.** 3^{10}. **10.** $(-9)^{18}$. **11.** y^8. **12.** 4^6. **13.** $(-2)^{12}$. **14.** y^{35}. **15.** $(a+b)^{12}$.

We have considered simplifying $x^m x^n$, which is a product of two powers of x. We also considered simplifying the expression $(x^m)^n$, which is a power of a power of x. We will now consider simplifying the expression $x^m \div x^n$, which is a quotient of powers of x.

What happens if we try to divide two powers of the same base? For instance, what is $4^5 \div 4^3$?

$$\frac{4^5}{4^3} = \frac{(\cancel{4})(\cancel{4})(\cancel{4})(4)(4)}{(\cancel{4})(\cancel{4})(\cancel{4})}$$
$$= (4)(4)$$
$$= 4^2$$

Hence, $4^5 \div 4^3 = 4^2$ (Observe that $2 = 5 - 3$.)

What about $5^2 \div 5^6$?

$$\frac{5^2}{5^6} = \frac{(\cancel{5})(\cancel{5})}{(\cancel{5})(\cancel{5})(5)(5)(5)(5)}$$
$$= \frac{1}{(5)(5)(5)(5)}$$
$$= \frac{1}{5^4}$$

Hence, $5^2 \div 5^6 = 1 \div 5^4$ (Observe that $4 = 6 - 2$.)

Consider, now, the following. What is $x^m \div x^n$, if m and n are positive integers with $m > n$?

$$\frac{x^m}{x^n} = \frac{\overbrace{(x)(x)(x)\cdots(x)}^{m \text{ times}}}{\underbrace{(x)(x)(x)\cdots(x)}_{n \text{ times}}}$$

$$= \frac{\overbrace{(x)(x)\cdots(x)}^{n \text{ times}}\overbrace{(x)(x)\cdots(x)}^{(m-n) \text{ times}}}{\underbrace{(x)(x)\cdots(x)}_{n \text{ times}}} \quad \text{(Since } m > n\text{)}$$

$$= \overbrace{(x)(x)(x)\cdots(x)}^{(m-n) \text{ times}} \quad \text{(Provided } x \neq 0\text{)}$$

$$= x^{m-n}$$

Hence, if $m > n$ and $x \neq 0$, then $x^m \div x^n = x^{m-n}$.

What is $x^m \div x^n$, if m and n are positive integers with $m < n$?

$$\frac{x^m}{x^n} = \frac{\overbrace{(x)(x)(x)\cdots(x)}^{m \text{ times}}}{\underbrace{(x)(x)(x)\cdots(x)}_{n \text{ times}}}$$

$$= \frac{\overbrace{(x)(x)(x)\cdots(x)}^{m \text{ times}}}{\underbrace{(x)(x)(x)\cdots(x)}_{m \text{ times}}\underbrace{(x)(x)(x)\cdots(x)}_{(n-m) \text{ times}}} \quad \text{(Since } m < n\text{)}$$

$$= \frac{1}{\underbrace{(x)(x)(x)\cdots(x)}_{(n-m) \text{ times}}} \quad \text{(Provided } x \neq 0\text{)}$$

$$= \frac{1}{x^{n-m}}$$

Hence, if $m < n$ and $x \neq 0$, then $x^m \div x^n = \frac{1}{x^{n-m}}$.

Of course, if $x \neq 0$ and m and n are both *equal* positive integers, then $x^m \div x^n = 1$. This will be discussed further in Section 9.3.

Quotient Rule for Exponents

If x represents any *nonzero* number and m and n are *positive integers,* then:

1. $x^m \div x^n = x^{m-n}$, if $m > n$
2. $x^m \div x^n = \dfrac{1}{x^{n-m}}$, if $m < n$
3. $x^m \div x^n = 1$, if $m = n$

9.1 More on Positive Integer Exponents

Using the rules for exponents, simplify each of the following. Leave your answers in exponential form.

Practice Exercise 16.
$(-4)^6 \div (-4)^8$

EXAMPLE 16 Simplify: $3^7 \div 3^4$.

SOLUTION:

$$3^7 \div 3^4 = 3^{7-4} = 3^3 \quad \text{(Part 1, quotient rule)}$$

Practice Exercise 17.
$(1.3)^7 \div (1.3)^3$

EXAMPLE 17 Simplify: $y^2 \div y^4$.

SOLUTION:

$$y^2 \div y^4 = \frac{1}{y^{4-2}} = \frac{1}{y^2} \quad \text{if } y \neq 0 \quad \text{(Part 2, quotient rule)}$$

Practice Exercise 18.
$(-11)^5 \div (-11)^5$

EXAMPLE 18 Simplify: $u^4 \div u^4$.

SOLUTION:

$$u^4 \div u^4 = 1 \quad \text{if } u \neq 0 \quad \text{(Part 3, quotient rule)}$$

Practice Exercise 19.
$(2x)^8 \div (2x)^5$

EXAMPLE 19 Simplify: $(-3)^6 \div (-3)^4$.

SOLUTION:

$$(-3)^6 \div (-3)^4 = (-3)^{6-2} = (-3)^2 \quad \text{(Part 1, quotient rule)}$$

Practice Exercise 20. $\dfrac{(u^4)^2}{(u^4)(u^2)^4}$

EXAMPLE 20 Simplify the answers given in (a) Example 16, and (b) Example 19.

SOLUTION:
(a) The answer given in Example 16 is 3^3:

$$3^3 = (3)(3)(3) = 27$$

(b) The answer given in Example 19 is $(-3)^2$:

$$(-3)^2 = (-3)(-3) = 9$$

Practice Exercise 21. $\dfrac{(-2)^3(-2)^4}{[(-2)^3]^4}$

EXAMPLE 21 Simplify $\dfrac{(a^3)^2(a^2)^3}{a^4}$, if $a \neq 0$.

SOLUTION:

$$\frac{(a^3)^2(a^2)^3}{a^4} = \frac{(a^6)(a^6)}{a^4} \quad \text{(Power rule)}$$

$$= \frac{a^{12}}{a^4} \quad \text{(Product rule)}$$

$$= a^8 \quad \text{(Part 1, quotient rule)}$$

Practice Exercise 22. $\dfrac{[(-3)^3(-3)^2]^2}{(-3)^7}$

EXAMPLE 22 Simplify $\dfrac{[(-4)^2(-4)^3]^2}{(-4)^{13}}$.

SOLUTION:

$$\frac{[(-4)^2(-4)^3]^2}{(-4)^{13}} = \frac{[(-4)^5]^2}{(-4)^{13}} \quad \text{(Product rule)}$$

$$= \frac{(-4)^{10}}{(-4)^{13}} \quad \text{(Power rule)}$$

$$= \frac{1}{(-4)^3} \quad \text{(Part 2, quotient rule)}$$

$$= \frac{1}{-64} \quad \text{(Rewrite)}$$

$$= \frac{-1}{64} \quad \text{(Simplifying)}$$

ANSWERS FOR PRACTICE EXERCISES

16. $\frac{1}{16}$. 17. $(1.3)^4$. 18. 1. 19. $8x^3$, if $x \neq 0$. 20. $\frac{1}{u^4}$, if $u \neq 0$. 21. $\frac{1}{(-2)^5}$. 22. $(-3)^3$.

Note
- $4x^2$ means $(4)(x)(x)$.
- $(4x)^2 = (4x)(4x) = 16x^2$.
- -2^2 means $(-1)(2)(2) = -4$.
- $(-2)^2$ means $(-2)(-2) = 4$.
- $(2x)(3x) = (2)(3)(x)(x) = 6x^2$.

EXERCISES 9.1

Use the rules for exponents given in this section to simplify each of the given expressions, if possible. If not possible, give an appropriate reason.

1. $(x^2)(x^4)$
2. $(3^5)(3^6)$
3. $x^3 \div x$
4. $y^6 \div y^2$
5. $8^7 \div 8^4$
6. $x^2 + x^3$
7. $(u^5)(u^6)$
8. $(5^2)^3$
9. $(x^3)^2$
10. $\left(\frac{2}{3}\right)^3 \div \left(\frac{2}{3}\right)^2$
11. $x^2 + y^2$
12. $(1.9)^3(1.9)^4$
13. $-(p^2)^3$
14. $-2(q^3)^4$
15. $t^8 t^5 t^6$
16. $(y^2)(y^3)^2$
17. $(-2)^3(-2)^4(-2)$
18. $(-6)^2 \div (-6)^5$
19. $(-4)^3 \div (-4)^2$
20. $(-2)^3(-2)^2$
21. $(0.6)^2(0.6)^3(0.6)^4$
22. $(0.23)^4(0.23)(0.23)^6$
23. $\left(\frac{1}{3}\right)^3 \left(\frac{1}{3}\right)^4 \left(\frac{1}{3}\right)^2$
24. $p^2 p^4 p^3 p^5$
25. $q^7 q^2 q^6 q^3 q^4$
26. $4(p^2)^3(q^3)^2$

27. $(p^3)^4 - 3(q^3)^5$
28. $(4x^4)(-y^5)$
29. $(2x^3)(-3y^4)$
30. $(3x^2)(-2x^4)(5x^3)$
31. $(4p^3q^2) \div (-2p^2q^7)$
32. $(y^4)^3 \div (y^5)^2$
33. $(x^2)^3 \div (x^3)^2$
34. $(u^4)(u^3)^3 \div (u^2)^4$
35. $(m^2)^3 \div [(m^2)(m^3)]^4$
36. $(n^3)^2(n^4) \div (n^3)^2(n^5)$
37. $(-5)^3(-5)^2(-5)^4(-5)^5$
38. $(10^3)(10^3)(10^3)(10^3)$
39. $(0.2)^4(0.2)^5(0.2)(0.2)^6$
40. $(-9)^{10}(-9)^7(-9)^{11}(-9)^3(-9)^4$
41. $(a + 2b)^3(a + 2b)^5$
42. $[(x - y)^2]^3$
43. $[(2p + 3q)^7]^4$
44. $(v^3)^7 \div [(v^3)^2(v^4)]^2$
45. $(x^2 + y)^4(x^2 + y)^5$
46. $(2^3 + 3^2)(2^3 - 3^2)$
47. $(5^4 - 4^3) \div (4^3 - 3^2)$
48. $(2r^2s)(-3r^3s^2)(-2r^4s^2)$
49. $(x + y)^3(x + y)^5 \div (x + y)^4$
50. $[(2p - 3q)^3]^4 \div [(2p - 3q)^4]^5$

9.2 Powers of Products and Quotients

OBJECTIVE

After completing this section, you should be able to use basic rules for working with powers of products and quotients involving positive integer exponents.

In Section 9.1, you learned some basic rules for working with positive integer exponents. These involved determining the product or quotient of powers of the same base and also finding a power of a power of a base. In this section, you will learn additional basic rules for working with positive integer exponents.

Let's consider how we can simplify the expression $(xy)^3$:

$$(xy)^3 = (xy)(xy)(xy)$$
$$= x \cdot y \cdot x \cdot y \cdot x \cdot y$$
$$= (x \cdot x \cdot x)(y \cdot y \cdot y)$$
$$= x^3y^3$$

Therefore, $(xy)^3 = x^3y^3$. Observe that *each factor* in the *product xy* is raised to the third power.

What about $(2x)^3$?

$$(2x)^3 = (2x)(2x)(2x)$$
$$= 2 \cdot x \cdot 2 \cdot x \cdot 2 \cdot x$$
$$= (2 \cdot 2 \cdot 2)(x \cdot x \cdot x)$$
$$= 2^3x^3$$
$$= 8x^3$$

Therefore, $(2x)^3 = 8x^3$. Again, *each factor* in the *product 2x* is raised to the third power.

What about $(x^2y^3)^2$?

$$(x^2y^3)^2 = (x^2y^3)(x^2y^3)$$
$$= x^2y^3x^2y^3$$
$$= (x^2x^2)(y^3y^3)$$
$$= (x^{2+2})(y^{3+3}) \quad \text{(Product rule)}$$
$$= x^4y^6$$

Therefore, $(x^2y^3)^2 = x^4y^6$. *Each factor* (x^2 and y^3) in the *product* (x^2y^3) is squared.

Power of a Product Rule for Exponents

If x and y represent any numbers and m, n, and p are positive integers, then

$$(xy)^m = x^m y^m$$

and

$$(x^m y^n)^p = x^{mp} y^{np}$$

Using the rules for exponents, simplify each of the following expressions.

Practice Exercise 1. $(3^4)^3$

EXAMPLE 1 Simplify: $(xy)^4$.
SOLUTION:
$$(xy)^4 = x^4 y^4$$

Practice Exercise 2. $(4y^3)^2$

EXAMPLE 2 Simplify: $(3x^2)^3$.
SOLUTION:
$$(3x^2)^3 = (3)^3(x^2)^3 = 27x^6$$

Practice Exercise 3. $(-2x)^2(3y)^2$

EXAMPLE 3 Simplify: $(u^4 v^5)^3$.
SOLUTION:
$$(u^4 v^5)^3 = (u^4)^3(v^5)^3 = u^{12} v^{15}$$

Practice Exercise 4. $(p^2 q^3)^3$

EXAMPLE 4 Simplify: $(-2b^3)^5$.
SOLUTION:
$$(-2b^3)^5 = (-2)^5(b^3)^5 = -32b^{15}$$

Practice Exercise 5. $(-2m^4 n^5)^3$

EXAMPLE 5 Simplify: $(m^4 n^2)^7$.
SOLUTION:
$$(m^4 n^2)^7 = (m^4)^7(n^2)^7 = m^{28} n^{14}$$

Practice Exercise 6. $[(3u^2)^2(v^3)]^3$

EXAMPLE 6 Simplify: $(x^3 x^2)^4 (x^2 y^3)^2$.
SOLUTION:

$$\begin{aligned}
(x^3 x^2)^4 (x^2 y^3)^2 &= (x^{3+2})^4 (x^2 y^3)^2 & \text{(Product rule)} \\
&= (x^5)^4 (x^2 y^3)^2 & \text{(Simplifying)} \\
&= x^{(5)(4)} (x^2 y^3)^2 & \text{(Power rule)} \\
&= x^{20} (x^2 y^3)^2 & \text{(Simplifying)} \\
&= x^{20} (x^4 y^6) & \text{(Power of a product rule)} \\
&= (x^{20} x^4) y^6 & \text{(Regrouping)} \\
&= x^{24} y^6 & \text{(Product rule)}
\end{aligned}$$

Practice Exercise 7. $(-3x^2)^3(2y^4)^2$

EXAMPLE 7 Simplify: $(2x^3)^2 (3y^4)^3$.
SOLUTION:

$$\begin{aligned}
(2x^3)^2 (3y^4)^3 &= (2^2 x^6)(3y^4)^3 & \text{(Power of a product rule)} \\
&= (2^2 x^6)(3^3 y^{12}) & \text{(Power of a product rule)} \\
&= (4x^6)(27 y^{12}) & \text{(Simplifying)} \\
&= (4)(27) x^6 y^{12} & \text{(Regrouping)} \\
&= 108 x^6 y^{12} & \text{(Simplifying)}
\end{aligned}$$

9.2 Powers of Products and Quotients

ANSWERS TO PRACTICE EXERCISES

1. 3^{12}. 2. $16y^6$. 3. $36x^2y^2$. 4. p^6q^9. 5. $-8m^{12}n^{15}$. 6. $729u^{12}v^9$. 7. $-108x^6y^8$.

Using the power of a product rule for exponents, you learned that a power of a product is equal to the product of the powers. What can we say about a power of a quotient? In particular, what is $(x \div y)^n$, if n is a positive integer and $y \neq 0$?

$$\left(\frac{x}{y}\right)^n = \underbrace{\left(\frac{x}{y}\right)\left(\frac{x}{y}\right)\left(\frac{x}{y}\right) \cdots \left(\frac{x}{y}\right)}_{n \text{ times}}$$

$$= \frac{\overbrace{(x)(x)(x) \cdots (x)}^{n \text{ times}}}{\underbrace{(y)(y)(y) \cdots (y)}_{n \text{ times}}} \quad \text{(Multiplication of fractions)}$$

$$= \frac{x^n}{y^n}$$

Hence, if n is a *positive* integer and $y \neq 0$, then $\left(\frac{x}{y}\right)^n = \frac{x^n}{y^n}$.

Power of Quotient Rule for Exponents

If x represents any number, y represents any *nonzero* number, and n is a *positive* integer, then

$$\left(\frac{x}{y}\right)^n = \frac{x^n}{y^n}$$

Using the rules for exponents, simplify each of the following.

Practice Exercise 8. $\left(\frac{1}{2}\right)^3\left(\frac{2}{3}\right)^2$

EXAMPLE 8 Simplify: $\left(\frac{x}{y}\right)^4$, if $y \neq 0$.

SOLUTION: $\left(\frac{x}{y}\right)^4 = \frac{x^4}{y^4}$ if $y \neq 0$

Practice Exercise 9. $\left(\frac{2}{5}\right)^3\left(\frac{1}{4}\right)^2$

EXAMPLE 9 Simplify: $\left(\frac{4}{7}\right)^5$.

SOLUTION: $\left(\frac{4}{7}\right)^5 = \frac{4^5}{7^5}$

Practice Exercise 10. $\left(\frac{2x}{y}\right)^4\left(\frac{3y}{5x}\right)^2$

EXAMPLE 10 Simplify: $\left(\frac{a}{b}\right)^9$, if $b \neq 0$.

SOLUTION: $\left(\frac{a}{b}\right)^9 = \frac{a^9}{b^9}$ if $b \neq 0$

Practice Exercise 11. $\dfrac{(3a^2b)^3}{(2ab^3)^3}$

EXAMPLE 11 Simplify: $\left(\dfrac{-3}{4}\right)^{10}$.

SOLUTION:
$$\left(\dfrac{-3}{4}\right)^{10} = \dfrac{(-3)^{10}}{4^{10}}$$

Practice Exercise 12. $\dfrac{(2xy)^2(x^2y^3)^2}{(3x^2y)^3}$

EXAMPLE 12 Simplify: $\left(\dfrac{x^3}{2y}\right)^3\left(\dfrac{2x}{y^2}\right)^2$, if $y \neq 0$.

SOLUTION:

$$\left(\dfrac{x^3}{2y}\right)^3\left(\dfrac{2x}{y^2}\right)^2 = \left(\dfrac{(x^3)^3}{(2y)^3}\right)\left(\dfrac{(2x)^2}{(y^2)^2}\right) \quad \text{(Power of a quotient rule)}$$

$$= \left(\dfrac{x^9}{(2y)^3}\right)\left(\dfrac{(2x)^2}{y^4}\right) \quad \text{(Power rule)}$$

$$= \left(\dfrac{x^9}{2^3y^3}\right)\left(\dfrac{2^2x^2}{y^4}\right) \quad \text{(Power of a product rule)}$$

$$= \left(\dfrac{x^9}{8y^3}\right)\left(\dfrac{4x^2}{y^4}\right) \quad \text{(Simplifying)}$$

$$= \dfrac{(x^9)(4x^2)}{(8y^3)(y^4)} \quad \text{(Multiplying fractions)}$$

$$= \dfrac{4x^{11}}{8y^7} \quad \text{(Product rule)}$$

$$= \dfrac{x^{11}}{2y^7} \quad \text{(Simplifying)}$$

ANSWERS TO PRACTICE EXERCISES

8. $\dfrac{1}{18}$.   9. $\dfrac{1}{250}$.   10. $\dfrac{144x^2}{25y^2}$, if $x \neq 0$, $y \neq 0$.   11. $\dfrac{27a^3}{8b^6}$, if $a \neq 0$, $b \neq 0$.   12. $\dfrac{4y^5}{27}$, if $x \neq 0$, $y \neq 0$.

EXERCISES 9.2

Use the rules for exponents given in this section to simplify each of the given expressions, if possible. If not possible, give an appropriate reason.

1. $(xy)^2$
2. $(2x^3)^3$
3. $(-4a^3b)^2$
4. $(r^2s^3)^4$
5. $(uv^3)^4$
6. $-2(p^2q)^3$
7. $(xy^2z^3)^2$
8. $(a^2b^3c^4)^3$
9. $(x^2y)^3 \div (x^2y^3)$
10. $(2ab^2)^3 \div (3ab)^2$
11. $(a^2bc^4)^3(a^2b^3)^2$
12. $(x^3y^4)^2 + (x^4y^2)^3$
13. $(-2p^2q^3)^3 \div (-3pq^2)^2$
14. $(2x^3y)^2 \div (-3xy^2)^3$
15. $(2a^3b)(-3ab^2)^3$
16. $(-2x^3y^2)(-3xy^3) \div (2xy)^3$
17. $(2p^2)^2(-q^4)^3(3p^2q^3)^2$
18. $(2pq)^2(-p^2q)^3(3pq^2)^2$
19. $(x^3y^2)^2 \div (-2xy^3)(-x^3)^4$
20. $(u^3v^2)^2(vw^4)^3(u^2w)^2$
21. $(5^2 \times 4)^3$
22. $(6^2 \times 4^3)^2$
23. $(5^2 \div 5^3)^4$
24. $(2^3 + 3^2)^2$

25. $(4^2 - 3^3)^2$
26. $(2a^2b)^2 + [9a^8b^3 \div (-3a^2b)]$
27. $(-3x^2y^3)^3 - (2x^2y^5)(x^2y^2)^2$
28. $(a^2b)^3(-2ab^3) \div [(-ab^2)^3(3a^3b)]$
29. $[(a + b)^3(b - c)^2]^3$
30. $[(p + q)^3(p - q)^2]^4$
31. $[(3u - v)^4(2u - 3v)^5]^3$
32. $(2x - y)^3(2x - y)^4 \div [(2x - y)^2]^2$
33. $(x^2 + y)^5(x - y^2)^4 \div [(x - y^2)^2(x^2 + y)^3]$
34. $[(2p - q)^4(q - 2p)^3]^2 \div [(2p - q)^5(2p - q)^2]$
35. a. Compute the value of: $3^2 \times 2^3$.
    b. Compute the value of: $(3 \times 2)^6$.
36. a. Compute the value of: $2^4 \times 3^2$.
    b. Compute the value of: $(2 \times 3)^8$.
37. a. Compute the value of: $2^3 \times 4^2$.
    b. Compute the value of: $(2 \times 4)^6$.
38. a. Compute the value of: $4^4 \div 2^2$.
    b. Compute the value of: $(4 \div 2)^2$.
39. a. Compute the value of: $9^3 \div 3$.
    b. Compute the value of: $(9 \div 3)^3$.
40. a. Compute the value of: $(-4)^6 \div (-2)^2$.
    b. Compute the value of: $[(-4) \div (-2)]^3$.

9.3 Zero and Negative Integer Exponents

OBJECTIVE

After completing this section, you should be able to use the basic rules for working with zero and negative integer exponents.

What meaning, if any, may be given to the expression a^0? To answer the question, we will consider the product a^0a^n where a is a *nonzero* number and n is a positive integer. We want to maintain the properties in the earlier part of this chapter for positive integer exponents. In particular, we want the product rule to hold for a^0a^n. Then,

$$a^0a^n = a^{0+n} = a^n$$

Now, since $a \neq 0$ and $n > 0$, then $a^n \neq 0$. Therefore, we may divide both sides of the preceding equation by a^n and obtain

$$a^0a^n = a^n$$
$$\frac{a^0a^n}{a^n} = \frac{a^n}{a^n}$$
$$a^0\left(\frac{a^n}{a^n}\right) = \frac{a^n}{a^n}$$
$$(a^0)(1) = 1$$
$$a^0 = 1$$

Zero Power Rule for Exponents

Concerning the **zero exponent**, if a is any *nonzero* number, then $a^0 = 1$.

Simplify each of the following.

Practice Exercise 1. $(1.9)^0$

EXAMPLE 1 Simplify: 2^0.

SOLUTION: $\qquad 2^0 = 1 \qquad$ (since $2 \neq 0$)

Practice Exercise 2. $(-3.2)^0$

EXAMPLE 2 Simplify: $(-3)^0$.

SOLUTION: $\qquad (-3)^0 = 1 \qquad$ (since $-3 \neq 0$)

Practice Exercise 3. u^0

EXAMPLE 3 Simplify: x^0.

SOLUTION: $\qquad x^0 = 1 \qquad$ if $x \neq 0$

Practice Exercise 4. $(-3p)^0$

EXAMPLE 4 Simplify: $(2y)^0$.

SOLUTION: $\qquad (2y)^0 = 1 \qquad$ if $y \neq 0$

Practice Exercise 5. $(rs)^0$

EXAMPLE 5 Simplify: $(ab)^0$.

SOLUTION: $\qquad (ab)^0 = 1 \qquad$ if $a \neq 0$ and $b \neq 0$

Practice Exercise 6. $(-5.7ab)^0$

EXAMPLE 6 Simplify: $(-4xy)^0$.

SOLUTION: $\qquad (-4xy)^0 = 1 \qquad$ if $x \neq 0$ and $y \neq 0$

Practice Exercise 7. $-4y^0$

EXAMPLE 7 Simplify: $2x^0$.

SOLUTION:
$$2x^0 = 2(x^0)$$
$$= 2(1) \qquad \text{if } x \neq 0$$
$$= 2 \qquad \text{if } x \neq 0$$

ANSWERS TO PRACTICE EXERCISES

1. 1. 2. 1. 3. 1, if $u \neq 0$. 4. 1, if $p \neq 0$. 5. 1, if $r \neq 0$ and $s \neq 0$. 6. 1, if $a \neq 0$ and $b \neq 0$.
7. -4, if $y \neq 0$.

So far, we have considered positive integers and zero exponents. What abut negative integer exponents? If n is a positive integer, then $-n$ is a negative integer. What meaning, if any, may be given to the expression a^{-n}? To answer this question, we will consider the product $a^n a^{-n}$ where a is a *nonzero* number and n is a positive integer. Again, we want to maintain the properties for positive integer exponents. In particular, we want the product rule for exponents to hold for $a^n a^{-n}$. Then,

$$a^n a^{-n} = a^{n+(-n)}$$
$$= a^0$$
$$= 1 \qquad \text{(since } a \neq 0\text{)}$$

or

$$a^n a^{-n} = 1$$

Now, since $a^n \neq 0$, we divide both sides of the preceding equation by a^n, obtaining

$$\frac{a^n a^{-n}}{a^n} = \frac{1}{a^n}$$

9.3 Zero and Negative Integer Exponents

or

$$a^{-n} = \frac{1}{a^n}$$

Negative Integer Power Rule for Exponents

Concerning **negative integer exponents,** if *a* is any *nonzero* number and *n* is a positive integer exponent, then

$$a^{-n} = \frac{1}{a^n}$$

Simplify the given expressions and write the results without negative exponents.

Practice Exercise 8. $(4^3)(4^{-2})$

EXAMPLE 8 Simplify: 2^{-3}.

SOLUTION: $$2^{-3} = \frac{1}{2^3} = \frac{1}{8}$$

Practice Exercise 9. $(y^{-3})(y^{-2})$

EXAMPLE 9 Simplify: 3^{-2}.

SOLUTION: $$3^{-2} = \frac{1}{3^2} = \frac{1}{9}$$

Practice Exercise 10. $(5y^0)^{-2}$

EXAMPLE 10 Simplify: 5^{-1}.

SOLUTION: $$5^{-1} = \frac{1}{5}$$

Practice Exercise 11. $(-3x^4)^0$

EXAMPLE 11 Simplify: 2^{-4}.

SOLUTION: $$2^{-4} = \frac{1}{2^4} = \frac{1}{16}$$

In Examples 8–11, we applied the negative integer power rule with a constant base. In the next set of examples, the base is a variable in each case. The rule can be used, provided the base is not equal to zero.

Practice Exercise 12. $2^{-1} + 3^{-1}$

EXAMPLE 12 Simplify: x^{-1}.

SOLUTION: $$x^{-1} = \frac{1}{x} \quad \text{if } x \neq 0$$

Practice Exercise 13. $(2a^0b^{-2})^3$

EXAMPLE 13 Simplify: u^{-5}.

SOLUTION: $$u^{-5} = \frac{1}{u^5} \quad \text{if } u \neq 0$$

Practice Exercise 14. $(2x^2y^{-1})^{-2}$

EXAMPLE 14
Simplify: $(ab)^{-2}$.

SOLUTION: $$(ab)^{-2} = \frac{1}{(ab)^2} \quad \text{if } a \neq 0 \text{ and } b \neq 0$$

> **Practice Exercise 15.**
> $(3p^{-2}q^{-3})^{-4}$

EXAMPLE 15

Simplify: $2x^{-3}$.

SOLUTION: $\qquad 2x^{-3} = 2\left(\dfrac{1}{x^3}\right) = \dfrac{2}{x^3} \quad$ if $x \neq 0$

ANSWERS TO PRACTICE EXERCISES

8. 4. 9. $\dfrac{1}{y^5}$, if $y \neq 0$. 10. $\dfrac{1}{25}$, if $y \neq 0$. 11. 1, if $x \neq 0$. $\dfrac{5}{6}$. 13. $\dfrac{8}{b^6}$, if $a \neq 0$, $b \neq 0$.

14. $\dfrac{y^2}{4x^4}$, if $x \neq 0$, $y \neq 0$. 15. $\dfrac{p^8 q^{12}}{81}$, if $p \neq 0$, $q \neq 0$.

Rewrite each of the following without negative exponents.

> **Practice Exercise 16.** $-3a^3b^{-4}c^0$

EXAMPLE 16 Rewrite $2x^{-3}y^2z^{-5}$ without negative exponents.

SOLUTION:

$$2x^{-3}y^2z^{-5} = (2)(x^{-3})(y^2)(z^{-5})$$

$$= (2)\left(\dfrac{1}{x^3}\right)(y^2)\left(\dfrac{1}{z^5}\right) \quad \text{(If } x \neq 0 \text{ and } z \neq 0)$$

$$= \dfrac{(2)(1)(y^2)(1)}{(x^3)(z^5)}$$

$$= \dfrac{2y^2}{x^3z^5} \quad \text{(If } x \neq 0 \text{ and } z \neq 0)$$

> **Practice Exercise 17.**
> $\dfrac{5p^{-2}q^5r^{-1}}{-2p^{-4}q^0r^{-3}}$

EXAMPLE 17 Rewrite $\dfrac{a^3b^{-3}c^4}{d^{-1}e^3f^{-2}}$ without negative exponents.

SOLUTION:

$$\dfrac{a^3b^{-3}c^4}{d^{-1}e^3f^{-2}} = \dfrac{(a^3)\left(\dfrac{1}{b^3}\right)(c^4)}{\left(\dfrac{1}{d}\right)(e^3)\left(\dfrac{1}{f^2}\right)} \quad \text{(If } b \neq 0, d \neq 0, e \neq 0, f \neq 0)$$

$$= \dfrac{\dfrac{a^3c^4}{b^3}}{\dfrac{e^3}{df^2}}$$

$$= \left(\dfrac{a^3c^4}{b^3}\right)\left(\dfrac{df^2}{e^3}\right)$$

$$= \dfrac{a^3c^4df^2}{b^3e^3} \quad (b \neq 0, d \neq 0, e \neq 0, f \neq 0)$$

ANSWERS TO PRACTICE EXERCISES

16. $\dfrac{-3a^3}{b^4}$, if $b \neq 0$ and $c \neq 0$. 17. $\dfrac{-5p^2q^5r^2}{2}$, if $p \neq 0$, $q \neq 0$, and $r \neq 0$.

EXERCISES 9.3

In Exercises 1–42, simplify the given expressions and write the results without negative exponents.

1. 4^{-2}
2. $(-3)^{-1}$
3. $\left(\dfrac{1}{2}\right)^0$
4. $(5^2)(5^{-3})$
5. $(-2)^3(-2)^{-4}$
6. $(2^{-1})(-3)^0(4^{-2})$
7. $3^{-1} + 5^{-1}$
8. $2^{-2} + (-2)^2$
9. $-(4^{-2}) - 4^2$
10. $[(5^{-2})(-3)^{-2}(-4)^6]^0$
11. x^{-5}
12. y^{-6}
13. $x^4 x^{-2}$
14. $y^{-5} y^{-4}$
15. $x^2 x^{-3} x^5$
16. $y^{-2} y^{-3} y^4$
17. $(x^{-3})^2$
18. $(2y^3)^{-2}$
19. $(-2a^{-1})^3$
20. $(4x^0 y^{-1})^2$
21. $(2x^{-3}y^2)(-3x^4y^{-3})$
22. $(-2a^3b^{-2}c^4)(-3a^{-2}b^5c^{-6})$
23. $(x^2y^{-3})(-4x^{-4}y^5)$
24. $(-p^2q^{-3}r)(3p^{-4}qr^3)$
25. $y^{-3} \div y^{-2}$
26. $(-5m^2n^{-1}) \div (2m^{-4}n^3)$
27. $(a^2b^3c^{-4}) \div (m^2p^{-3})$
28. $(p^{-3}q^{-2}r^{-5}) \div (s^3t^{-1})$
29. $(-6a^2b^0c^{-3}) \div (3a^{-3}b^4c^0)$
30. $(r^2st^{-1})^{-2} \div (5r^2s^0t^3)$
31. $[(x^2y^{-3}) \div (u^{-4})][(x^{-4}y) \div (u^3)]$
32. $[(ab^{-2}c) \div (m^{-2}n^3)][(a^3b^{-3})(mn)]$
33. $(3x)^0 \div 3x^0$
34. $(7p^2q^3)^0(-7p^{-2}q^0)$
35. $(2a^0b^3c^{-4})^2(-6a^2b^0c^4)^0$
36. $(4p^{-3}q^0r^2)^0 \div (-pq^{-2}r^0)^2$
37. $(3^0 3^2 3^{-3}) \div (3^{-1} 3^3 3^{-2})$
38. $(3)^2(-3)^3 \div [(-3)^{-2}(3)^{-3}]$
39. $(2a)^3(3b)^{-2}(5c)^0(-4d)^2$
40. $4^2 + 4^0 + 4^{-2}$
41. $(3xy)^0 - (-4x^2y^{-3})^0$
42. $(3x^0y^2)^{-3}(2x^2y^{-3})^3$

In Exercises 43–50, rewrite each of the given expressions without negative exponents. Do not simplify your results.

43. $a + b^{-1}$
44. $x^{-1} + y^{-1}$
45. $(a + b^{-1}) \div b$
46. $(x^{-1} + y^{-1}) \div x$
47. $(a + b) \div b^{-1}$
48. $(x^{-1} + y^{-1}) \div (xy)$
49. $(p + q)^{-1}(p^{-1} + q^{-1})$
50. $(p + q)^{-1} \div (p^{-1} + q^{-1})$

9.4 Scientific Notation

OBJECTIVE

After completing this section, you should be able to rewrite a number in scientific notation.

We often encounter very large numbers or very small numbers in various applications. For instance, in physics, we learn that a light-year, which is a unit of length used to measure astronomical distances, is equivalent to the distance that light travels in a mean solar year. A light-year is equal to 5,880,000,000,000 mi or 9,461,000,000,000 km. The sun is 93,000,000 mi from Earth. Also, Pluto, the farthest planet from the sun, is between 2,800,000,000 mi and 4,600,000,000 mi from the sun.

In chemistry, one also learns, for example, that the mass of a hydrogen atom is 0.00000000000000000000000016734 g. Numbers such as these can be written with fewer digits, using what is called **scientific notation.** This involves an application of integer exponents.

Scientific Notation

If N represents a *positive* number, then we may write

$$N = p \times 10^k$$

where p is a positive number such that $1 \leq p < 10$ and k is an integer.

To rewrite a positive number N in scientific notation:

1. Determine the value of p by placing a decimal point in the number N after the first *nonzero* digit from the left.
2. Determine the integer k by counting the number of digits that the decimal point must be moved to obtain the original number.
   a. If the decimal point is moved to the *right*, then k will be *positive*.
   b. If the decimal point is moved to the *left*, then k will be *negative*.
   c. If the decimal point is not moved at all, then k will be 0.

Rewrite each of the following in scientific notation.

Practice Exercise 1. 129

EXAMPLE 1 Rewrite 236 in scientific notation.

SOLUTION: In scientific notation,

$$236 = 2.36 \times 100$$
$$= 2.36 \times 10^2$$

(The decimal point in 2.36 has to be moved 2 places to the *right* to get back to 236.)

Practice Exercise 2. 0.34

EXAMPLE 2 Rewrite 19,156 in scientific notation.

SOLUTION: In scientific notation,

$$19,156 = 1.9156 \times 10,000$$
$$= 1.9156 \times 10^4$$

(The decimal point in 1.9156 has to be moved 4 places to the *right* to get back to 19,156.)

Practice Exercise 3. 1037

EXAMPLE 3 Rewrite 0.00046 in scientific notation.

SOLUTION: In scientific notation,

$$0.00046 = 4.6 \times \frac{1}{10,000}$$
$$= 4.6 \times 10^{-4}$$

(The decimal point in 4.6 has to be moved 4 places to the *left* to get back to 0.00046.)

9.4 Scientific Notation

Practice Exercise 4. 0.0097

EXAMPLE 4 Rewrite 0.001763 in scientific notation.

SOLUTION: In scientific notation,

$$0.001763 = 1.763 \times \frac{1}{1{,}000}$$
$$= 1.763 \times 10^{-3}$$

(The decimal point in 1.763 has to be moved 3 places to the *left* to get back to 0.001763.)

Practice Exercise 5. 257×10^7

EXAMPLE 5 Rewrite 231,000,000,000 in scientific notation.

SOLUTION: In scientific notation,

$$231{,}000{,}000{,}000 = 2.31 \times 100{,}000{,}000{,}000$$
$$= 2.31 \times 10^{11}$$

(The decimal point in 2.31 has to be moved 11 places to the *right* to get back to 231,000,000,000.)

Practice Exercise 6. 0.19×10^6

EXAMPLE 6 Rewrite 12.34×10^{-9} in scientific notation.

SOLUTION: In scientific notation,

$$12.34 \times 10^{-9} = (1.234 \times 10) \times 10^{-9}$$
$$= 1.234 \times (10 \times 10^{-9})$$
$$= 1.234 \times 10^{-8}$$

Practice Exercise 7. 896,100,000,000

EXAMPLE 7 Rewrite 0.39×10^{13} in scientific notation.

SOLUTION: In scientific notation,

$$0.39 \times 10^{13} = (3.9 \times 10^{-1}) \times 10^{13}$$
$$= 3.9 \times (10^{-1} \times 10^{13})$$
$$= 3.9 \times 10^{12}$$

ANSWERS TO PRACTICE EXERCISES

1. 1.29×10^2. 2. 3.4×10^{-1}. 3. 1.037×10^3. 4. 9.7×10^{-3}. 5. 2.57×10^9. 6. 1.9×10^5.
7. 8.961×10^{11}.

If a number is given in scientific notation, we can rewrite it without the use of exponents, as indicated in the following examples.

Rewrite each of the following numbers without exponents.

Practice Exercise 8. 3.16×10^{-2}

EXAMPLE 8 Rewrite 2.79×10^2 without exponents.

SOLUTION: $2.79 \times 10^2 = 279$

Practice Exercise 9. 5.03×10^0

EXAMPLE 9 Rewrite 1.06×10^{-3} without exponents.

SOLUTION: $1.06 \times 10^{-3} = 0.00106$

Practice Exercise 10. 8×10^{-4}

EXAMPLE 10 Rewrite 5×10^4 without exponents.

SOLUTION: $5 \times 10^4 = 50{,}000$

> **Practice Exercise 11.**
> 9.81×10^5

EXAMPLE 11 Rewrite 7.99×10^0 without exponents.

SOLUTION: $\quad 7.99 \times 10^0 = 7.99$

> **Practice Exercise 12.**
> 8.534×10^{-6}

EXAMPLE 12 Rewrite 9.01×10^{-7} without exponents.

SOLUTION: $\quad 9.01 \times 10^{-7} = 0.000000901$

ANSWERS TO PRACTICE EXERCISES

8. 0.0316.   9. 5.03.   10. 0.0008.   11. 981,000.   12. 0.000008534.

The digits of N required for writing the factor p in the expression $N = p \times 10^k$ are called the **significant digits.** For instance,

$$23$$
$$0.00023$$
$$2300$$
and $\quad 2{,}300{,}000$

all have 2 significant digits since

$$23 = 2.3 \times 10^1$$
$$0.00023 = 2.3 \times 10^{-4}$$
$$2300 = 2.3 \times 10^3$$
and $\quad 2{,}300{,}000 = 2.3 \times 10^6$

where, in each case, $p = 2.3$ has 2 digits.

If a number has 2 significant digits, we refer to the number as being accurate to 2 places or we say that it has 2-place accuracy. If a number has 3 significant digits, we refer to the number as being accurate to 3 places or as having 3-place accuracy. In general, if a number has n significant digits, we refer to the number as being accurate to n places or that it has n-place accuracy.

> **Determine the number of significant digits in each of the following.**
>
> **Practice Exercise 13.** 463

EXAMPLE 13 Determine the place accuracy for the number 235.

SOLUTION: The number 235 has 3-place accuracy since $235 = 2.35 \times 10^2$ and $p = 2.35$ has 3 digits.

> **Practice Exercise 14.** 0.019

EXAMPLE 14 Determine the place accuracy for the number 0.000019.

SOLUTION: The number 0.000019 has 2-place accuracy since $0.000019 = 1.9 \times 10^{-5}$ and $p = 1.9$ has 2 digits.

> **Practice Exercise 15.**
> 16,100,000,000

EXAMPLE 15 Determine the place accuracy for the number 65,100.

SOLUTION: The number 65,100 has 3-place accuracy since $65{,}100 = 6.51 \times 10^4$ and $p = 6.51$ has 3 digits.

> **Practice Exercise 16.**
> 0.394×10^{-23}

EXAMPLE 16 Determine the place accuracy for the number 19,620,000,000.

SOLUTION: The number 19,620,000,000 has 4-place accuracy since $19{,}620{,}000{,}000 = 1.962 \times 10^{10}$ and $p = 1.962$ has 4 digits.

9.4 Scientific Notation

Practice Exercise 17. 0.0003091×10^{-4}

EXAMPLE 17 Determine the place accuracy for the number 102,800.

SOLUTION: The number 102,800 has 4-place accuracy since $102,800 = 1.028 \times 10^5$. (Notice that the 0 in 1.028 is significant.)

ANSWERS TO PRACTICE EXERCISES

13. 3.   14. 2.   15. 3.   16. 3.   17. 4.

It is not always necessary to maintain the number of significant digits given in a particular number. The number may be *rounded* to fewer digits, using the rules of rounding given in Section 3.3. For example, the number 1.3438 has 5 significant digits. Expressed in terms of only 3 significant digits, we write 1.3438 as 1.34 by dropping the digits 3 and 8. However, the number 1.3468 also has 5 significant digits but, expressed in terms of only 3 significant digits, we write 1.3468 as 1.35 by dropping the digits 6 and 8 but increasing the third significant digit by 1.

Practice Exercise 18. Round 1.6497 to 2 significant digits.

EXAMPLE 18 Round 4.91476 to 3 significant digits.

SOLUTION: 4.91476 rounded to 3 significant digits is 4.91.

Practice Exercise 19. Round 0.0069746 to 3 significant digits.

EXAMPLE 19 Round 63.396 to 3 significant digits.

SOLUTION: 63.396 rounded to 3 significant digits is 63.4.

Practice Exercise 20. Round 465 to 1 significant digit.

EXAMPLE 20 Round 428.53 to 4 significant digits.

SOLUTION: 428.53 rounded to 4 significant digits is 428.5.

Practice Exercise 21. Round 634.72 to 2 significant digits.

EXAMPLE 21 Round 92.750 to 2 significant digits.

SOLUTION: 92.750 rounded to 2 significant digits is 93.

Practice Exercise 22. Round 2.39×10^4 to 3 significant digits.

EXAMPLE 22 Round 0.00686500 to 3 significant digits.

SOLUTION: 0.00686500 rounded to 3 significant digits is 0.00687.

ANSWERS TO PRACTICE EXERCISES

18. 1.6.   19. 0.00697.   20. 500.   21. 630.   22. 23,900.

EXERCISES 9.4

In Exercises 1–24, rewrite each of the indicated numbers in scientific notation.

1. 32.6
2. 269.7
3. 17
4. 2.16
5. 0.00418
6. 0.4318

7. 36,000,000
8. 2006
9. 9^2
10. 11^3
11. 4×10^{-2}
12. 169^0
13. $7^2 \times 3$
14. 1.2×3.4
15. 60.74×10^9
16. 69.2×10^2
17. 123.4×10^6
18. 6.9×2^3
19. 12.1×10
20. 0.6×10^3
21. 0.9×10^{-2}
22. 169×10^{-4}
23. $10^4 \times 10^8$
24. $10^7 \times 10^{-3}$

In Exercises 25–48, determine the number of significant digits in each of the numbers given in Exercises 1–24.

In Exercises 49–63, determine the number N for each of the numbers given in scientific notation.

49. 2.3×10^3
50. 3.19×10^{-4}
51. 1.234×10^4
52. 8.63×10^7
53. 2.54×10^9
54. 2.694×10^{-10}
55. 7.146×10^{12}
56. 4×10^{15}
57. 2.3×10^{17}
58. 8.13×10^{-8}
59. 1×10^{10}
60. 5×10^{12}
61. 6.907×10^{13}
62. 2.7×10^{-11}
63. 6.01×10^0

64. In chemistry, one learns that the number of molecules in a gram molecule is approximately equal to 6.023×10^{23}. This is known as Avogadro's number and, according to Avogadro's law, it must be the same for all substances. Write out Avogadro's number without the use of exponents.
65. A unified atomic mass unit is defined to be approximately 1.66057×10^{-27} kg. Write this number without the use of exponents.
66. One square meter is approximately 3.861022×10^{-7} square miles. Write this number without the use of exponents.
67. One gram is approximately 1.102331×10^{-6} ton. Write this number without the use of exponents.
68. One cubic meter is equal to 6.102374×10^4 in.³. Write this number without the use of exponents.
69. One square mile is equal to 4.014490×10^9 in.². Write this number without the use of exponents.
70. The mass of the electron in a neutral atom is 9.11×10^{-28} g. Write this number without the use of exponents.
71. The mass of the neutron in a neutral atom is 1.672×10^{-24} g. Write this number without the use of exponents.
72. Show that, in a neutral atom, the mass of the neutron is approximately 1836 times that of the mass of the electron. (*Hint:* See Exercises 70 and 71.)
73. To convert from British thermal units, or Btu's, to ergs, we multiply by 1.054×10^{10}. Write this number without the use of exponents.
74. To convert from ergs to foot-pounds, we multiply by 7.376×10^{-8}. Write this number without the use of exponents.

In Exercises 75–84, round each of the numbers given to the number of significant digits indicated.

75. 38.345; 3
76. 38.345; 4
77. 0.001378; 2
78. 7.83249; 5
79. 20.03; 3
80. 69,900; 1
81. 65; 1
82. 75; 1
83. 0.00435; 2
84. 0.01020030005; 8

In Exercises 85–90, perform the indicated operations and write your results in scientific notation.

85. $(5 \times 10^4)(7.2 \times 10^{-3})$
86. $(1.2 \times 10^{-7})(28.1 \times 10^{10})$
87. $(70.62 \times 10^5) \div (0.2354 \times 10^8)$
88. $(487.2 \times 10^{-6}) \div (81.2 \times 10^4)$
89. $\dfrac{15{,}000{,}000}{(0.0003)(0.0000005)}$
90. $\dfrac{(23{,}400)(0.0046)}{(1{,}170{,}000)(0.000069)}$

Chapter Summary

In this chapter, the discussion of exponents was extended to include zero and negative integer exponents. Scientific notation, significant digits, and rounding were also discussed.

The **key words** introduced in this chapter are listed here in the order in which they appeared in the chapter. You should know what they mean before proceeding to the next chapter.

zero exponent (p. 405)
negative integer exponents (p. 407)
scientific notation (p. 410)
significant digits (p. 412)

Review Exercises

9.1 More on Positive Integer Exponents

Simplify each of the given expressions, if possible, using the rules of exponents. If not possible, give an appropriate reason.

1. $x^3 x^4$
2. $(x^2)^3$
3. $y^2 + y^3$
4. $(y^4)^2 \div y^3$
5. $(z^3)(z^2) \div z^7$
6. $-5(q^2)^5(r^3)^4$
7. $(6p^2 q^3) \div (-3p^4 q)$
8. $(u^3)^4 (u^2)^3 \div (u^3)^5$

9.2 Powers of Products and Quotients

Simplify each of the given expressions, using the rules of exponents.

9. $(a^2 b)^3 \div (2ab^2)$
10. $(p^3 q^4)^2$
11. $(-2x^2)(-3x^3)^2$
12. $(-5x^3 y^2)(-2x^2 y)^3$
13. $(x^3 y^2)^2 (yz^4)^3 (x^2 z)^2$
14. $[(u - 3v)^2 (2u - v)^3]^4$

9.3 Zero and Negative Integer Exponents

Simplify each of the given expressions and write the results without negative exponents.

15. $b^3 b^{-5}$
16. $x^3 x^{-4} x^{-2}$
17. $(2x^{-3})^2$
18. $(-2a^{-3} b^2)(-3ab^{-4})$
19. $(x^2 y^{-3} z) \div (-2xy^4 z^{-4})$
20. $(5a^2 b^{-3} c^0 d^4)^0$

9.4 Scientific Notation

In Exercise 21–26, rewrite each of the given numbers in scientific notation.

21. 29.6
22. 0.00134
23. 49,600,000
24. 8^3
25. 3.1×4.2
26. 457.6×10^6

In Exercises 27–30, determine the number N for each of the numbers given in scientific notation.

27. 3.6×10^5
28. 9.1×10^{-2}
29. 5.67×10^6
30. 8.975×10^{-5}
31. Round 43.568 to 3 significant digits.
32. Round 0.000169500 to 3 significant digits.

Teasers

1. Explain the difference between $3x^0$ and $(3x)^0$, if $x \neq 0$.
2. Are the expressions $x^{-1} + y^{-1}$ and $(x + y)^{-1}$ equivalent expressions? Answer the question by substituting particular values for x and y.
3. Show that $(a + b)^2 = a^2 + b^2$ if $a = 0$ or $b = 0$ but, general, $(a + b)^2 \neq a^2 + b^2$.
4. In your own words, explain why the expression $(x^2)(y^3)$ cannot be simplified further.
5. Evaluate $x^{-2}(x^2 + x^0)$
6. Evaluate $(y^{-3} - y^{-1})(y^3 - y^2)$, if $y = 0.1$.
7. Multiply 730,000,000 by 0.000023 by first rewriting each number in scientific notation. Write your answer in scientific notation.
8. Divide 0.0013 by 52,000 by first rewriting each number in scientific notation. Write your answer in scientific notation.
9. Add 8.9×10^4 and 7.3×10^6. (*Hint:* First rewrite the numbers so that they each have the same power of 10.)
10. Perform the indicated operations: $(6.3 \times 10^3) + (4.9 \times 10^5) - (5.2 \times 10^2)$.

CHAPTER 9 TEST

This test covers material from Chapter 9. Read each question carefully and answer all questions. If you miss a question, review the appropriate section in the text.

In Exercises 1–10, use rules of exponents to simplify each of the given expressions. Write all results without negative exponents.

1. $x^3 x^{-2} x^7$
2. $x(x^3)^2$
3. $(u^2 v^{-3})^2$
4. $(y^3)^{-2}(y^2 y^3)$
5. $(2a^2 b)(-3a^0 b^3)$
6. $(x^{-1} y^2)(-2x^2)^0(-x)^3$
7. $(2x^4) \div (6x^2)$
8. $(9y^2) \div (-3y^5)$
9. $(-2a^2 b)^3 \div (3ab^2)^2$
10. $(3u^2 v)^3 \div (-uv^2)^3$

In Exercises 11–14, determine the value of the given expressions.

11. $2^3 \times 3^2$
12. $2^{-3} + 3^2$
13. $3^2 \div (-3)^3$
14. $(2^3)^2 \div (2^4)$

In Exercises 15–18, write the given number in scientific notation.

15. 126.9
16. 0.000329
17. 0.36×10^{-21}
18. 5.6×0.4
19. Round 3.765 to 3 significant digits.
20. Round 0.0031942 to 2 significant digits.

CUMULATIVE REVIEW: CHAPTERS 1–9

Perform the indicated operations in Exercises 1–8.

1. $(-3)^2 + (-3^2)$
2. $\left(\dfrac{-3}{4}\right) - \left(\dfrac{-2}{5}\right)$
3. $\left(\dfrac{-1}{2}\right)^2 \times \left(\dfrac{-1}{3}\right)^3$
4. 2.06% of 1.6
5. $(-2.9 + 0.12) - (0.3)^2$
6. $|-3|^2 \div (3)^{-2} - 6$
7. $4^{-2} + (-2)^4$
8. $(-5)^2(-5)^{-3}(-5)^0(-5)^{-2}$
9. Round $|-2.01| \div |0.39|$ to the nearest thousandth.
10. Round 2.1×3.01 to 2 significant digits.
11. Round 0.3162×10^{-3} to 3 significant digits.
12. Evaluate $-3p^2q$ if $p = 1.5$ and $q = -2$.
13. Evaluate $(-2r^3s)^{-2}$ if $r = -1$ and $s = 2$.
14. Simplify the expression $-3s - 2[t - 3(s - 2t)] - 3t$ by combining like terms.
15. Simplify the following expression and write your result without negative exponents: $(-2x^2y^{-1})^{-2} \div (x^{-3}y)^3$.
16. Simplify the following expression and write your result without negative exponents: $(3p^{-2}q^3)^2(-2p^3q^{-2})^0(-p^3q^{-4})^{-3}$.
17. Determine the perimeter (in feet) of a triangle whose sides have lengths 3.4 ft, 1.2 yd, and 2.9 ft.
18. Determine the volume of a cube each of whose edges measures 1.3 cm.
19. Write an algebraic expression for the square of the sum of y and twice z.
20. Write an algebraic expression for the square of the product of $(x + y)$ and $(x - 2y)$.
21. Solve the equation $-2(y - 2) = 4 - (2 - 3y)$ for y.
22. Solve the equation $0.2(t - 3) - 0.3(1 - 2t) = 1.9$ for t.
23. Rewrite 269.2 in scientific notation.
24. Rewrite 0.00397 in scientific notation.
25. Determine the number N for 6.01×10^{-4}.

CHAPTER 10 Polynomials

In Chapter 7, we discussed algebraic expressions and the evaluation of algebraic expressions. Some basic terminology associated with algebraic expressions was introduced. In this chapter, you will learn about special algebraic expressions called polynomials. You will also learn to add, subtract, multiply, and divide polynomials.

CHAPTER 10 PRETEST

This pretest covers material on polynomials. Read each question carefully and try to answer all questions. Each question is keyed to the chapter section in which the particular topic is discussed. Check your answers with those given in the back of the text.

Questions 1–4 pertain to Section 10.1.

1. What is the degree of the third term in the polynomial $2x^3y - 3x^2y^2 + 6x^3y^4$?

2. What is the degree of the polynomial $2p^3q + (-3p^2q)^2 + 5p^2q^4$?

3. Rewrite the polynomial $y^2 - 4y^3 + 5 - 6y + 3y^4$ in descending powers of y.

4. Rewrite the polynomial $2p^2q - 3pq^2 + p^4q^2 - 4p^3q^2$ in descending powers of p.

Questions 5–7 pertain to Section 10.2.
Add.

5. $(3y^2 - y^3 + 4y + 6) + (2 - 3y^3 + 2y - y^2)$

6. $(2a^3b^2 - 3ab^3 + 4a^2b^2) + (2ab^3 - 6a^2b^2 + 6a^2b^3)$

7. $(2u^3 - u^4 + 9u - 16) + (4u^2 - 3u + u^4 - 6) + (7 - 3u^3 + 7u^2 + 6u)$

Questions 8–10 pertain to Section 10.3.

8. Subtract $3x - 2x^2 + 6 - 5x^3$ from $7x^2 - 4x^3 + 1 - 9x$.

9. From $p^5 - 6p^2 - 4p^3 - 3$, subtract $4p^4 + 7p - 3p^2 - 8p^5$.

10. Subtract $2u^2v - 3uv + uv^2$ from the sum of $uv - 3uv^2$ and $6u^2v + 7uv$.

Questions 11–13 pertain to Section 10.4.
Multiply.

11. $(-3x^2y)(2xy - xy^3)$

12. $(u^2 - 2v)(2u + 3v)$

13. $(3r^4 - r^2 + 6)(2r^3 - 5r - 7)$

Questions 14–16 pertain to Section 10.5.

14. Using the FOIL method, multiply: $(2w - 1)(w + 7)$.

15. Using the FOIL method, multiply: $(2x - 3y)(x + 4y)$.

16. Square the binomial $x + 3y$.

Questions 17–20 pertain to Section 10.6.
Divide.

17. $(8x^4 - 6x^2 + 7x - 8) \div (-4)$

18. $(9x^2y^4 - 3xy^3 + 11x^3y + 3x^4) \div (-3xy)$, if $x \neq 0$ and $y \neq 0$

19. $(2u^3 - 7u^2 + 7u - 2) \div (2u - 1)$, if $u \neq \dfrac{1}{2}$

20. $(2v^4 - 4v^3 + 7) \div (v - 2)$, if $v \neq 2$.

10.1 Basic Terminology Associated with Polynomials

OBJECTIVES

After completing this section, you should be able to:
1. Identify the terms of a polynomial.
2. Identify a polynomial as being a monomial, a binomial, a trinomial, or none of these.
3. Determine the degree of each term of a polynomial.
4. Determine the degree of a polynomial.
5. Arrange a polynomial in descending order of a variable.

In Chapter 7, you were introduced to some of the basic terminology associated with algebraic expressions. Recall that an algebraic expression may be a *sum* or a *product*. (Subtraction and division are treated as algebraic addition and multiplication, respectively.) The parts of a sum are called "terms," and the parts of a product are called "factors."

Practice Exercise 1. Classify the expression $-2p + 3q + r^2$.

EXAMPLE 1 Classify the expression $3x + 4y$.

SOLUTION: The algebraic expression $3x + 4y$ is a *sum* of two *terms*. The terms are $3x$ and $4y$. The term $3x$ is a *product* of the two *factors*, 3 and x. Similarly, the term $4y$ is a *product* of the two *factors*, 4 and y.

Practice Exercise 2. Classify the expression $-3a(b - c)(b + c)$.

EXAMPLE 2 Classify the expression $x(y - 3z)$.

SOLUTION: The algebraic expression $x(y - 3z)$ is a *product* of two *factors*. The factors are x and $(y - 3z)$. The factor $(y - 3z)$ is a *sum* of two *terms*, y and $-3z$. The expression $x(y - 3z)$, however, is still considered to be a *single* term.

ANSWERS TO PRACTICE EXERCISES

1. A sum of three terms which are $-2p$, $3q$, and r^2. 2. A product of four factors which are -3, a, $b - c$, and $b + c$.

We will now consider special algebraic expressions called polynomials.

DEFINITION

A **polynomial** *in the variable x* is an algebraic expression that is a term or a sum of terms of the form ax^n where a is any nonzero number and n is a nonnegative integer.

- $2x^{-3}$ is *not* a polynomial since the exponent, -3, is not a *positive* integer or 0.
- $5x^{1/2}$ is *not* a polynomial since the exponent, $\frac{1}{2}$, is not a positive *integer* or 0.
- $\dfrac{-2}{x^4}$ is *not* a polynomial since $\dfrac{-2}{x^4} = -2x^{-4}$ and the exponent, -4, is not a *positive* integer or 0.

10.1 Basic Terminology Associated with Polynomials

We can classify polynomials according to the number of terms in the polynomial.

> **DEFINITION**
> A polynomial that contains exactly one term is called a **monomial.**

- $5y^2$ is a monomial in the variable y ($a = 5$, $n = 2$).
- $\frac{9}{16}z^4$ is a monomial in the variable z $\left(a = \frac{9}{16}, n = 4\right)$.
- 5 is a monomial in the variable x since $5 = 5x^0 = 5(1) = 5$. It is also a monomial in the variable y since $5 = 5y^0$, or in the variable z since $5 = 5z^0$, etc.

> **DEFINITION**
> A polynomial that contains exactly two terms is called a **binomial.**

- $2x^3 - x$ is a binomial in the variable x. [Note that $2x^3 - x = 2x^3 + (-x)$.]
- $3t^4 + 5t^2$ is a binomial in the variable t.
- $4u - 3$ is a binomial in the variable u. [Note that $4u - 3 = 4u + (-3)$.]

> **DEFINITION**
> A polynomial that contains exactly three terms is called a **trinomial.**

- $3x^2 - 4x + 1$ is a trinomial in the variable x.
- $2y - y^4 - 5y^2$ is a trinomial in the variable y.
- $\frac{1}{2}u^3 - \frac{1}{3}u + \frac{4}{7}$ is a trinomial in the variable u.

We will not give special names to polynomials of more than three terms. However, the following are examples of polynomials containing more than three terms.

- $x^3 - 2x^2 + x + 1$ is a polynomial, in the variable x, containing four terms.
- $y^4 - 2 + 4y - y^3 + 4y^2$ is a polynomial, in the variable y, containing five terms.
- $u^6 - 5u^4 + u^3 - 6u^2 - 10$ is a polynomial, in the variable u, containing five terms.
- $t - 3t^4 + t^{19} - 3 - t^{23} + t^8 - 2t^{12}$ is a polynomial, in the variable t, containing seven terms.

All of the polynomials encountered so far in this chapter are polynomials in a single variable. There are also polynomials in two or more variables.

> **DEFINITION**
> A *polynomial in the variables x and y* is an algebraic expression that is a term or the sum of terms of the form ax^my^n, where a is any number and m and n are *nonnegative* integers.

- $2x^2y$ is a monomial in the variables x and y.
- $3u^2v - 4uv^3$ is a binomial in the variables u and v.

- $x^2 - 2xy + y^2$ is a trinomial in the variables x and y.
- $s^3t^2 - 4s^3 + 2t - 2$ is a polynomial, containing four terms, in the variables s and t.

Polynomials can also be classified according to degrees. The degree of a polynomial is defined in terms of the degrees of the terms in the polynomial.

DEFINITION

The **degree of a term** in a polynomial is the sum of the exponents of the variable factors.

- $-4y^5$ is a monomial, in y, of degree 5.
- 17 is a monomial, in any variable, of degree 0, since $17 = 17x^0$ or $17 = 17y^0$ or $17 = 17u^0$, etc.
- $2x^3y^2$ is a monomial, in x and y, of degree 5 since the sum of the exponents, 3 and 2, is 5.
- $-5uv^7$ is a monomial, in u and v, of degree 8 since the sum of the exponents, 1 and 7, is 8.
- 0 is a monomial in any variable since $0 = (0)(x)$ or $0 = (0)(y^2)$ or $0 = (0)(t^3)$, etc. The monomial 0 has degree 0. The degree of any constant term is 0.

DEFINITION

The **degree of a polynomial** is the degree of the highest term in the polynomial, after all like terms have been combined.

Recall that like terms have the same variable parts, with the same exponents. They can differ only by their coefficients. For example, $3y^2$ and $-5y^2$ are like terms. $2x^3$ and $3x^2$ are not like terms. Also, $4x^2y^3$ and $5x^3y^2$ are not like terms.

Practice Exercise 3.
$2 - t^2 + 4t$

EXAMPLE 3 Determine the degree of the polynomial $2x^4 - 3x^2 + 1$.

SOLUTION: $2x^4 - 3x^2 + 1$ is a polynomial of degree 4 since the terms have degrees 4, 2, and 0, respectively.

Practice Exercise 4.
$7u^3 - 2u + u^4 - 6$

EXAMPLE 4 Determine the degree of the polynomial $2y - 4y^2 + y^6 - y^3$.

SOLUTION: $2y - 4y^2 + y^6 - y^3$ is a polynomial of degree 6 since the terms have degrees 1, 2, 6, and 3, respectively.

Practice Exercise 5.
$st^2 - s^2t + s^2t^3 + 2st^3$

EXAMPLE 5 Determine the degree of the polynomial $2u^3v - 3uv^2 + u^2v^2$.

SOLUTION: $2u^3v - 3uv^2 + u^2v^2$ is a polynomial of degree 4 since the terms have degrees 4, 3, and 4, respectively.

Practice Exercise 6.
$-2x^2y + 3xy^3 - 4x^2y^3$

EXAMPLE 6 Determine the degree of the polynomial $t^2 - 3t^3 + 2 - 5t + 3t^3$.

SOLUTION: $t^2 - 3t^3 + 2 - 5t + 3t^3$ is a polynomial of degree 2 since $t^2 - 3t^3 + 2 - 5t + 3t^3 = t^2 - 5t + 2$, after combining like terms.

ANSWERS TO PRACTICE EXERCISES

3. (a) 0, 2, 1; (b) 2.   **4.** (a) 3, 1, 4, 0; (b) 4.   **5.** (a) 3, 3, 5, 4; (b) 5.   **6.** (a) 3, 4, 5; (b) 5.

Polynomials are generally written in **descending powers** of one of the variables. That is, the term containing the highest power of the variable is written to the left. The remaining terms are written in decreasing powers of the variable.

- The polynomial $3x^3 - x^2 + 2x - 7$ is in descending powers of x.
- The polynomial $2u^3v - 3u^2v^3 + 4uv^2$ is in descending powers of u.

Practice Exercise 7. Arrange the polynomial $2p^3q^2 - 4pq^4 + 5p^3q - 6p^2q^2$ in descending powers of p.

EXAMPLE 7   Arrange the polynomial $2s^2t^2 + 3s^3t^2 - 4st^3 - 1$ in descending powers of s.

SOLUTION: In descending powers of the variable s, the polynomial $2s^2t^2 + 3s^3t^2 - 4st^3 - 1$ is written as $3s^3t^2 + 2s^2t^2 - 4st^3 - 1$.

Practice Exercise 8. Arrange the polynomial $2p^3q^2 - 4pq^4 + 5p^3q - 6p^2q^2$ in descending powers of q.

EXAMPLE 8   Arrange the polynomial $2s^2t^2 + 3s^3t^2 - 4st^3 - 1$ is descending powers of t.

SOLUTION: In descending powers of the variable t, the polynomial $2s^2t^2 + 3s^3t^2 - 4st^3 - 1$ is written as $-4st^3 + 3s^3t^2 + 2s^2t^2 - 1$. Notice that there are two terms with the same power of t, $3s^3t^2$ and $2s^2t^2$. The term $3s^3t^2$ is written first since the term is of degree 5 whereas the term $2s^2t^2$ is of degree 4.

ANSWERS TO PRACTICE EXERCISES

7. $2p^3q^2 + 5p^3q - 6p^2q^2 - 4pq^4$.   **8.** $-4pq^4 + 2p^3q^2 - 6p^2q^2 + 5p^3q$.

As noted in Example 8, there may be two or more terms with the same power of the variable when rearranging a polynomial in descending order of that variable. In general, we rearrange those terms with the same power of the variable according to the degree of the term containing the variable. The highest degree term is written first.

EXERCISES 10.1

In Exercises 1–20, determine whether the given algebraic expression is or is not a polynomial. If it is a polynomial, indicate in what variable(s).

1. $2x + 3$
2. $3y^4 - 1$
3. $2u^3$
4. $w^{1/2} + 2$
5. $2s^2t - 3st^3$
6. $-2x^3y^2z^4$
7. $\dfrac{2}{u-3}$
8. $\dfrac{t^2 + 1}{9}$
9. $3x^2y - xy^4 - 5y$
10. $5u - 7v + u^2v^4 - 1$
11. $2x - 6 - 3x - x^2$
12. $\dfrac{x}{y} - \dfrac{y}{x}$

13. $3x - 9y - 2xy$
14. $8u^3z - 5$
15. $1 - 3u + 4v - 5uv$
16. $\dfrac{2y^2 + 3y + 1}{5y}$
17. $2x^3 - x^2 + 4x - x^{-1}$
16. $6x - x^2 - 3$
19. $8p^3q^2 - 5pq^3 - 7p^4$
20. $3(uv)^2 - 3u^3v + 4(u^2v)^4 - 1$

In Exercises 21–40, write the degree for each term in the given polynomial. Then, determine the degree of the polynomial.

21. $2x^2 - 3$
22. $2y - y^3 + 2y^4$
23. $3u - 4u^4 + 2u^3$
24. $t^3 - 2t + 1 - t^5$
25. $2x^2 - 3xy + y^2$
26. $3uv^4 - 5u^2v + 1 + u^3$
27. $s^3t - s^2t^2 + 3s + 4t$
28. $p^3 - 2p^4q + q^5$
29. $x^2y^3 - 2xy^2 - 3x^3y$
30. $4u^3 - 5v^2 + u^2v - 1$
31. $m^2n^3 - 3m^3n - 4mn^5$
32. $2x^2y - x^3 + 4y^3 + 5xy^2$
33. $2x - 6y - 4y^2 + x^3$
34. $2m^2n - 3n^2 - 5m^2n^3$
35. $p^4 - 3p^3 + (2p^2)^3 - 1$
36. $(3x^2y)^2 - (4x^3y)^0 - 7x$
37. $2(x^2y)^3 - 3x^3y^2$
38. $(2uv^2)^3 - (3u^3v)^4$
39. $(3p^2q)^2 - (4p^3)^2 + p^3q^2$
40. $(m^2n^2)^4 - 2(m^4n)^3 - (2m^2n^5)^2$

In Exercises 41–50, rewrite the given polynomial in descending powers of the indicated variable.

41. $2x - 3x^2 + x^4 - 1;\ x$
42. $2y^3 - 5y + y^4 - 2y^2;\ y$
43. $2 - 3u + 4u^2 - 5u^6;\ u$
44. $p^6 - 7p^4 + (2p)^7 - 5;\ p$
45. $x^3y - xy^4 + 2x^4y^3 - x^2y^2;\ x$
46. $x^3y - xy^4 + 2x^4y^3 - x^2y^2;\ y$
47. $u^3v^2 + u^4v^3 - uv + u^3v^3;\ v$
48. $m^2n^3 - (m^3n)^2 + 2(m^2n^4)^2;\ n$
49. $pq^4 - (2p^2q)^3 - p^4q^3 + p^5q^2;\ p$
50. $-2m^5n^2 + m^2n^6 - (7m^6n^4)^0;\ m$

10.2 Addition of Polynomials

OBJECTIVE

After completing this section, you should be able to add polynomials.

Addition of polynomials can be done horizontally or vertically. The procedure is the same as that for addition of algebraic expressions, which was discussed in Chapter 7.

Practice Exercise 1. Add the given polynomials horizontally: $u^3 - 2u - 3;\ 4u^2 + 5u + 1;\ 2u^3 - 3$.

EXAMPLE 1 Add the polynomials $2x^2 + 3x + 1$ and $x^2 - 5x + 7$.

SOLUTION: Adding *horizontally* we have

$(2x^2 + 3x + 1) + (x^2 - 5x + 7)$
$= 2x^2 + 3x + 1 + x^2 - 5x + 7$ (Removing parentheses)
$= 3x^2 + 3x + 1 - 5x + 7$ (Adding the 2nd-degree terms)
$= 3x^2 - 2x + 1 + 7$ (Adding the 1st-degree terms)
$= 3x^2 - 2x + 8$ (Adding the 0-degree, or constant, terms)

10.2 Addition of Polynomials

Practice Exercise 2. Add the given polynomials vertically:
$2 - 3r^3$; $4r + 5 - r^4$;
$3r^4 - 2r^2 + 6r$;
$5r^3 - r^2 + 9r - 11$.

EXAMPLE 2 Add the polynomials $2y^3 - 3y + 7$, $4y^2 - 3$, and $y^3 - 5y^2 + y - 2$.

SOLUTION: Adding *vertically,* we write the like terms under each other (leaving space for any missing terms) and then add their numerical coefficients. Hence,

$$\begin{array}{r} 2y^3 \phantom{{}-5y^2} - 3y + 7 \\ 4y^2 \phantom{{}+y} - 3 \\ \text{Add:}\quad y^3 - 5y^2 + y - 2 \\ \hline 3y^3 - y^2 - 2y + 2 \end{array}$$

Practice Exercise 3. Add the given polynomials either way:
$2x^2y - 3xy^2 + x^2y^2$;
$4x^2y^2 - 5xy^2 + 6xy$;
$4xy^2 - xy - 5x^2y - 6x^2y^2$.

EXAMPLE 3 Add the polynomials $2p^2q^3 - 3p^3q$, $4p^3q - 5p^2q^2$, $p^2q^3 + p^2q^2$, and $-4p^2q^3 + p^3q - 5p^2q^2$.

SOLUTION: Adding vertically, we write the like terms under each other (leaving space for any missing terms) and then add their numerical coefficients. Hence,

$$\begin{array}{r} 2p^2q^3 - 3p^3q \phantom{{}-5p^2q^2} \\ 4p^3q - 5p^2q^2 \\ p^2q^3 \phantom{{}+4p^3q} + p^2q^2 \\ \text{Add:}\quad -4p^2q^3 + p^3q - 5p^2q^2 \\ \hline -p^2q^3 + 2p^3q - 9p^2q^2 \end{array}$$

ANSWERS TO PRACTICE EXERCISES

1. $3u^3 + 4u^2 + 3u - 5$. 2. $2r^4 + 2r^3 - 3r^2 + 19r - 4$. 3. $-x^2y^2 - 3x^2y - 4xy^2 + 5xy$.

EXERCISES 10.2

In Exercises 1–40, add the given polynomials.

1. $3x^4 - x^2 + 1$; $2x^4 - 4x^2 + 5$
2. $y^4 - y^2 + 2$; $y^4 - 2y^2 - 3$
3. $1.1u^3 - 1.2u^2 + 1.4u - 0.1$; $1.5u^2 - 1.2u^3 + 1.6$
4. $2p^5 + p^4 + 2p$; $-3p^4 + 4p^3 - 5p^2 - 10$
5. $\frac{1}{3}q^4 - \frac{1}{2}q^2 + 9$; $\frac{-4}{7}q^3 + \frac{2}{3}q^4 - 1$
6. $2x^4 - 3x^2 + 1$; $3x^4 - x^3 - x + 2$; $4x^3 - 2x^2 - 5x - 6$
7. $5y^4 - 4y^3 + 2$; $2y^3 - 4y^2 + 3y$; $y^4 - 2y^2 + 5y - 7$
8. $0.3t^3 - 0.2t^2 + 0.4t - 0.4$; $-0.2t^3 + 0.1t^2 - 0.5t + 0.1$; $0.5t^3 - 0.4t^2 + 0.1t + 0.5$
9. $z^6 - 2z^4 + z^2 - 7$; $4z^2 - z^6 - 2z^4 + 1$; $3z^2 - 2z^6 + 4z^4 - 9$
10. $t^9 - 3t^6 + t^2 - 6$; $t^6 - 3t^4 - 2t^2 + 1$; $t^4 + 2 - 4t^2 - 2t^9$
11. $\frac{1}{2}x^2 - \frac{2}{3}xy + y^2$; $\frac{3}{4}x^2 + \frac{4}{5}xy - \frac{1}{3}y^2$
12. $3m^2 + mn - n^2$; $4n^2 - 2mn - m^2$
13. $4.1x^2y - 2.3xy^2 + 5.4x^2y^2$; $1.6xy^2 - 7.2x^2y^2 - 2.5x^2y$
14. $u^3v^2 + 2u^2v^2 - uv^4$; $2uv^3 - 5u^2v^2 + 2uv^4$
15. $x^3y^2 - x^2y^2 + xy^3$; $2x^3y^2 - x^2y^2 - x^2y^3$

16. $p^2q - 2p + p^3q^2$; $2p^2q - 3p + 2q - pq$; $4p - 3q - p^3q^2 - pq$
17. $\frac{5}{6}m^2n^2 - \frac{1}{2}m^2n + \frac{3}{5}mn$; $m^2n - mn - m^2n^2$; $\frac{1}{2}mn - \frac{3}{4}m^2n^2 - \frac{1}{6}mn^2$
18. $2p^2q^3 - 11pq^2 + 8pq$; $-9p^2q^3 + 7pq^2 - 6pq$; $8p^2q^3 + 5pq^2 - 4pq$
19. $2.6s^3t - 4.5s^2t^3 + s^2t^2$; $6.2s^2t^3 - 3.1$; $5 - 2s^3t$; $9.7s^2t^2 - 2.3s^3t + 2.5$
20. $x^3y^2 - x^2y^3 + 4x - 5y$; $xy^3 - x^2y^3 - 4x$; $2x^3y^2 - 3xy^3 + 5y$

21. $\quad 2x^3 - 3x^2 + x + 1$
    $\quad \underline{x^3 + 2x^2 - 3x - 7}$

22. $\quad 3y^4 - y^3 + 2y - 3$
    $\quad \underline{-5y^4 + 2y^3 - 6y + 1}$

23. $\quad 4x^3 - 5x^2 + 4$
    $\quad \underline{ 2x^2 - 5x - 3}$

24. $\quad 3u^5 - 2u^3 - u$
    $\quad \underline{ 4u^4 - 2u^2 - 3u}$

25. $\quad 2x^3y^2 - 3x^2y + xy^3$
    $\quad \underline{-5x^3y^2 + 7x^2y - 2xy^3}$

26. $\quad s^2t^3 - 2st^4 + 3st$
    $\quad \underline{-7s^2t^3 + 5st^4 - 7st}$

27. $\frac{2}{3}x^4 - \frac{4}{5}x^2 + 1$; $\frac{1}{2}x^3 - \frac{1}{3}x^2 - \frac{3}{5}$; $x^4 + \frac{1}{2}x^2 - \frac{1}{5}x - 7$
28. $0.3y^5 - 0.4y^2 + 0.1y$; $0.2y^4 - 0.3y^2 - 0.2y$; $0.5y^5 - 0.3y^3 - 0.1y$
29. $u^4 - 2u^3 + u$; $2u^3 - 4u^2 + 5u$; $-5u^4 - u^2 - 3u$
30. $p^6 - 5p^4 + 2p - 1$; $p^5 - 3p^2 + 4p$; $-4p^6 + p^4 - 3p^2 + 4$
31. $2 - 3s + s^2 - 3s^4$; $s^3 - 3s^2 - s^4$; $2s^4 - 5 + s^3 + 7s^2$
32. $\frac{1}{2}x^3 - x^4 + \frac{5}{7}$; $x^2 - \frac{2}{3}x - \frac{4}{9}$; $2 - \frac{2}{3}x^4 + x^2$; $3x - x^3 - \frac{3}{5}x^4$
33. $y^3 - y - 1$; $2y^4 + y^2 - 3y$; $3y^2 - y^4 + 2$; $2y^3 - 7 - 5y^4$
34. $9.2 - u - 4.3u^2$; $1.3 + u^2 + 8.7u^3$; $u^4 - 5.3u - 6.1$; $u^3 - 2.4u - 2.9$
35. $t^6 - t^3 - t$; $2 - 3t + 4t^3$; $2t^6 - 4t - 1$; $2t^3 - 6$
36. $x^2y - xy^2 + xy$; $3xy - 4x^2y + 2xy^2$
37. $\frac{2}{3}u^2v^3 - \frac{1}{2}u^3v + \frac{3}{4}u$; $\frac{4}{5}v - \frac{1}{4}u^2v^3 + \frac{1}{3}u^3v$
38. $2x^2 - 3xy + y^2$; $3y^2 - x^2 + xy$; $4xy - 3x^2$
39. $p^2q - 3pq^3 - 4p^2q^2$; $3q^3 - 2p^2q + p^2q^2$; $3p^2q - 2p^2q^2$
40. $0.3st - 0.1s^2t + 0.2st^4$; $0.2s^2t - 0.3s^3t - 0.4st$; $0.4s^3t - 0.7st - 0.1st^4$

10.3 Subtraction of Polynomials

OBJECTIVE

After completing this section, you should be able to subtract polynomials.

Subtraction of polynomials can be done horizontally or vertically. The procedure is the same as that for subtraction of algebraic expressions, which was discussed in Chapter 7.

In Practice Exercises 1–2, subtract horizontally.

Practice Exercise 1. Subtract $2u^2 - 3u + 4$ from $u^2 + 2u - 5$.

EXAMPLE 1 Subtract the polynomial $2x^3 - x^2 + 4x - 5$ from the polynomial $3x^3 - 5x + 1$.

SOLUTION: We will subtract horizontally. Note that the *first* polynomial given is being subtracted from the second polynomial.

10.3 Subtraction of Polynomials

$$(3x^3 - 5x + 1) - (2x^3 - x^2 + 4x - 5) \quad \text{(Remember, order is important.)}$$
$$= 3x^3 - 5x + 1 - 2x^3 + x^2 - 4x + 5 \quad \text{(Using rules for removing parentheses)}$$
$$= x^3 - 5x + 1 + x^2 - 4x + 5 \quad \text{(Adding the } x^3 \text{ terms)}$$
$$= x^3 + x^2 - 5x + 1 - 4x + 5 \quad \text{(Adding the } x^2 \text{ terms)}$$
$$= x^3 + x^2 - 9x + 1 + 5 \quad \text{(Adding the } x \text{ terms)}$$
$$= x^3 + x^2 - 9x + 6 \quad \text{(Adding the constant terms)}$$

In Example 1, there were two sets of parentheses to be removed. The first set was preceded by a + sign (understood) and the parentheses were simply removed. The second set was preceded by a − sign. Here, the parentheses were removed *and* the sign of *every* term within the parentheses was changed.

Practice Exercise 2. Subtract $p^3 + 2p + 9$ from $3p^3 - p^2 + 4p$.

EXAMPLE 2 From the polynomial $2y - 3y^2 + 4 - y^4$, subtract the polynomial $4y^3 - 1 - 5y + 3y^4 - 2y^2$.

SOLUTION: Note that the *second* polynomial given is being subtracted from the first polynomial.

$$(2y - 3y^2 + 4 - y^4) - (4y^3 - 1 - 5y + 3y^4 - 2y^2) \quad \text{(Again, order is important.)}$$
$$= 2y - 3y^2 + 4 - y^4 - 4y^3 + 1 + 5y - 3y^4 + 2y^2 \quad \text{(Removing parentheses)}$$
$$= -4y^4 - 4y^3 - y^2 + 7y + 5 \quad \text{(Combining like terms and arranging in descending order)}$$

To subtract polynomials vertically, write the polynomial being subtracted directly under the polynomial from which it is being subtracted, writing the like terms under each other. Then, subtract term by term.

In Practice Exercises 3–4, subtract vertically.

Practice Exercise 3. From $2 - 3y + y^2 - y^3$, subtract $4y^2 - 3y + 5y^3$.

EXAMPLE 3 Subtract the polynomial $u^2 - 3u + u^4 + 1$ from the polynomial $4u^2 + 3u^4 - 7 - 3u^3$.

SOLUTION: To subtract the *first* polynomial given from the second polynomial vertically, proceed as follows, noting that the like terms are lined up under each other.

$$\begin{array}{r} 3u^4 - 3u^3 + 4u^2 - 7 \\ \text{Subtract:} \quad \underline{u^4 + u^2 - 3u + 1} \\ 2u^4 - 3u^3 + 3u^2 + 3u - 8 \end{array}$$

(Rearranging the polynomials in descending powers of u)
(Subtracting term by term)

Practice Exercise 4. From $3r^2 - 4r + r^4$, subtract $-r^2 + 3r^3 - 7$.

EXAMPLE 4 From the polynomial $x^3y^2 - 4x^2y^2 + xy^3 - 2x^2y^3$, subtract the polynomial $2x^2y^2 - xy^3 - 4x^3y^2$.

SOLUTION: To subtract the *second* polynomial given from the first polynomial vertically, proceed as follows:

$$\begin{array}{r} x^3y^2 - 2x^2y^3 - 4x^2y^2 + xy^3 \\ \text{Subtract:} \quad \underline{-4x^3y^2 + 2x^2y^2 - xy^3} \\ 5x^3y^2 - 2x^2y^3 - 6x^2y^2 + 2xy^3 \end{array}$$

(Rearranging the polynomials in descending powers of x)
(Subtracting term by term)

Practice Exercise 5. Subtract either horizontally or vertically: Subtract $2p^2q - pq^2 + p^2q^3 + 5p^2q^2$ from $p^2q^2 - 4p^2q - 5p^2q^3$.

EXAMPLE 5 From the sum of the polynomials $2x^3 - 2x - 3x^2 + 1$ and $2 - x^3 + 2x - 4x^2$, subtract the polynomial $5x - 2x^3 + 7 - x^2$.

SOLUTION: There are two parts to this solution. First, we must find the sum of the polynomials being added, and then we must subtract a third polynomial from this sum.

Add:
$$\begin{aligned} 2x^3 - 3x^2 - 2x + 1 \\ -x^3 - 4x^2 + 2x + 2 \\ \hline x^3 - 7x^2 + 3 \end{aligned}$$ (Rearranging the polynomials in descending powers of x)

(The *sum* of the two polynomials)

Subtract:
$$\begin{aligned} -2x^3 - x^2 + 5x + 7 \\ \hline 3x^3 - 6x^2 - 5x - 4 \end{aligned}$$ (The required solution)

ANSWERS TO PRACTICE EXERCISES

1. $-u^2 + 5u - 9$. **2.** $2p^3 - p^2 + 2p - 9$. **3.** $-6y^3 - 3y^2 + 2$. **4.** $r^4 - 3r^3 + 4r^2 - 4r + 7$.
5. $-6p^2q^3 - 4p^2q^2 - 6p^2q + pq^2$.

EXERCISES 10.3

In Exercises 1–30, subtract the given polynomials as indicated.

1. Subtract $2x^4 - 3x^2 + 2x - 4$ from $2x^3 - x^4 - 3x + 1$.
2. Subtract $3y - y^3 + 4y^2 - 5$ from $4 - 2y + 4y^3 - 2y^4$.
3. Subtract $3.4u^5 - 2.3u^3 + u^2 - 5.7$ from $6.3u^2 - 7.9 - u^4 + 8.5u$.
4. Subtract $\frac{-2}{3}p^4 - \frac{4}{5}p^3 + \frac{1}{2}p + \frac{1}{9}$ from $\frac{3}{4}p^2 - \frac{1}{3}p - \frac{7}{11} + \frac{3}{7}p^4$.
5. Subtract $2q^2 - 7q + 10$ from $q^3 - 2q$.
6. From $m^4 - 3m^2 + 1$, subtract $2m^3 - 2m^2 - 4m$.
7. From $\frac{3}{5}t^7 - \frac{2}{3}t^5 + \frac{1}{4}t^2 - \frac{7}{8}t$, subtract $\frac{1}{2} - \frac{4}{5}t^2 - \frac{3}{5}t^6 + \frac{7}{9}t^7 + \frac{1}{3}t^3$.
8. From $z^3 - z - 1$, subtract $z^2 + 2$.
9. From $7.5x^3 - x^4 + 9.2x - x^2$, subtract $8.3x^4 - 5.2x^2 - 2 + 6.5x$.
10. From $u^3 - u^2 + u - 1$, subtract $1 - u + u^2 - u^3$.
11. Subtract $0.2x^2y^3 - 0.4xy^2 + 0.1x^3y - 0.2x$ from $0.5y - 0.1xy^2 - 0.4x^2y^3$.
12. Subtract $u^2v^3 - 2uv^4 + 3u^3v^2$ from $4u - 3uv^4 - 5u^2v^3$.
13. From $2p^3q^2 - p^2q^2 - 4p + 5q$, subtract $6p - 7q + 2p^3q^2 - 1$.
14. From $m^4n^3 - \frac{5}{6}m^2n + m^3n^3 - mn^4$, subtract $\frac{1}{2}m^2n - mn^4 - \frac{3}{5}m^4n^3$.
15. From $3s^4t^2 + 7s^2t^3 - 11st^4$, subtract $2st^4 - 3s^2t^4 + 5s^4t^2$.
16. Subtract $2x^3 - x^4 - 3x + 1$ from $2x^4 - 3x^2 + 2x - 4$.
17. Subtract $9.4 - 8.2y + 1.4y^3 - 5.2y^4$ from $11.3y - y^3 + 1.4y^2 - 10.5$.
18. Subtract $3u^2 - 9 - u^4 + 5u$ from $4u^5 - 3u^3 + u^2 - 7$.
19. Subtract $\frac{3}{4}p^2 - \frac{1}{3}p - \frac{7}{11} + \frac{4}{7}p^4$ from $\frac{-2}{3}p^4 - \frac{2}{5}p^3 + \frac{1}{2}p + \frac{8}{9}$.
20. Subtract $q^3 - 2q$ from $2q^2 - 7q + 10$.
21. From $2m^3 - 2m^2 - 4m$, subtract $m^4 - 3m^2 + 1$.
22. From $11.2 - 10.5t^2 + 12.3t^3 - 6.5t^6 + 0.9t^7$, subtract $17.5t^7 - 10.3t^3 + 8.4t^2 - 3.8t$.
23. From $z^2 + 2$, subtract $z^3 - z - 1$.
24. From $3x^4 - 2x^2 - 2 + 5x$, subtract $5x^3 - x^4 + 2x - x^2$.
25. From $1 - u + u^2 - u^3$, subtract $u^3 - u^2 + u - 1$.
26. Subtract $\frac{4}{5}y - xy^2 - \frac{3}{4}x^2y^3$ from $\frac{1}{2}x^2y^3 - \frac{1}{4}xy^2 + x^3y - \frac{1}{2}x$.

27. Subtract $4u - 3uv^4 - 5u^2v^3$ from $u^2v^3 - 2uv^4 + 3u^3v^2$.
28. From $6p - 7q + 2p^3q^2 - 1$, subtract $2p^3q^2 - p^2q^2 - 4p + 5q$.
29. From $2m^2n - mn^4 - 5m^4n^3$, subtract $7m^4n^3 - 6m^2n + m^3n^3 - mn^4$.
30. From $0.2st^4 - 0.3s^2t^4 + 0.5s^3t^2$, subtract $0.3s^4t^2 + 0.7s^2t^3 - 0.11st^4$.
31. From the sum of $2x^2 - 3x + 1$ and $x^3 - 4x - 3$, subtract $4x^3 - x^2 + 5x - 10$.
32. Subtract the sum of $2y^3 - y^2 + 4y - 1$ and $y^3 - 5y + 9$ from $2y - 4y^3 - 7$.
33. From the sum of $2u^2 - 3uv + v^2$ and $3v^2 + 2uv - 4u^2$, subtract $5uv - 3u^2 + v^2$.
34. Subtract the sum of $\frac{1}{2}p^2q^3 - \frac{2}{3}p^3q^2 - pq$ and $\frac{1}{2}pq - p^2q^3$ from $\frac{2}{3}pq - \frac{4}{5}p^3q^2 + pq^2$.
35. Subtract $2u^2 - 4u + 1$ from $3u^2 - u - 2$. From the result obtained, subtract $5u^3 - 2u + 7u^2 - 9$.
36. From the sum of $3x^3 - 2x + 1$ and $x^2 + 5x - 7$, subtract the sum of $6x - 2 - x^3$ and $x^2 - 4x + 9$.
37. Subtract the sum of $9.2y - 8.1y^3 + 1.6$ and $10.2y^3 - 13.4$ from the sum of $7.2y^2 - 6.1$ and $17.4y^3 + y$.

10.4 Multiplication of Polynomials

OBJECTIVES

After completing this section, you should be able to:
1. Multiply a monomial by a monomial.
2. Multiply a polynomial by a monomial.
3. Multiply a polynomial by a polynomial.

The product of two monomials, or **multiplication of polynomials,** can be determined by using the commutative and associative properties of multiplication, together with rules for exponents. The procedure is outlined in the box.

Multiplying a Monomial by a Monomial

To multiply a monomial by a monomial:
1. Multiply the numerical coefficients.
2. Multiply the variable factors with the same base.
3. Form the product of the results obtained in Steps 1 and 2.

The procedure for multiplying a monomial by a monomial is illustrated in Examples 1–3.

Multiply.

Practice Exercise 1. $(2x^2)(-5x^4)$

EXAMPLE 1 Multiply: $(2x^3)(-3x^2)$.

SOLUTION:
$$(2x^3)(-3x^2) = (2)(-3)(x^3)(x^2)$$
$$= (-6)(x^5)$$
$$= -6x^5$$

Practice Exercise 2. $(9u^3)(-5u^4)$

EXAMPLE 2 Multiply: $(-4y^5)(-5y^6)$.

SOLUTION:
$$(-4y^5)(-5y^6) = (-4)(-5)(y^5)(y^6)$$
$$= (20)(y^{11})$$
$$= 20y^{11}$$

Practice Exercise 3. $(0.3uv^2)(-4u^3v)$

EXAMPLE 3 Multiply: $(2xy^2)(-7x^3y^2)$.

SOLUTION:
$$(2xy^2)(-7x^3y^2) = (2)(-7)(x)(x^3)(y^2)(y^2)$$
$$= (-14)(x^4)(y^4)$$
$$= -14x^4y^4$$

The basis for multiplication of polynomials is the distributive property for multiplication over addition, which will be repeated in the box.

Distributive Property for Multiplication over Addition

If *a*, *b*, and *c* represent any real numbers, then

$$a(b + c) = ab + ac$$

The left-hand side of the equation given for the distributive property for multiplication over addition is a *product* of a *monomial* and a *binomial*. The right-hand side of the same equation is obtained by multiplying each term in the binomial by the monomial factor. This property can be extended to multiply any polynomial by a monomial.

Multiplying a Polynomial by a Monomial

To multiply a polynomial by a monomial, multiply *each* term of the polynomial by the monomial (that is, apply the distributive property) and then add the results.

This rule is illustrated in the following examples.

Practice Exercise 4. $\dfrac{-3}{7}u\left(u^2 - \dfrac{2}{3}u + \dfrac{4}{9}\right)$

EXAMPLE 4 Multiply: $2x(x + 2)$.

SOLUTION:
$$2x(x + 2) = (2x)(x) + (2x)(2)$$
$$= 2x^2 + 4x$$

Practice Exercise 5. $p^2q(2pq^3 - 4p^2q - 2)$

EXAMPLE 5 Multiply: $\dfrac{1}{3}y\left(2y^2 - \dfrac{1}{4}\right)$

SOLUTION:
$$\dfrac{1}{3}y\left(2y^2 - \dfrac{1}{4}\right) = \left(\dfrac{1}{3}y\right)(2y^2) + \left(\dfrac{1}{3}y\right)\left(-\dfrac{1}{4}\right)$$
$$= \dfrac{2}{3}y^3 + \left(-\dfrac{1}{12}y\right)$$
$$= \dfrac{2}{3}y^3 - \dfrac{1}{12}y$$

10.4 Multiplication of Polynomials

Practice Exercise 6.
$-3m^2n^3(2m^3 - 4mn^2 + m^3n - 7n^4)$

EXAMPLE 6 Multiply: $-4u(u^2 - 2u + 4)$.

SOLUTION:

$$-4u(u^2 - 2u + 4) = (-4u)(u^2) + (-4u)(-2u) + (-4u)(4)$$
$$= -4u^3 + 8u^2 - 16u$$

Practice Exercise 7.
$2rs^2(3r^2s - 4r^3s^2 + 5rs^6)$

EXAMPLE 7 Multiply: $m^2n^3(1 - 2mn^2 + m^3n - 2m^2n^3)$.

SOLUTION:

$$m^2n^3(1 - 2mn^2 + m^3n - 2m^2n^3)$$
$$= (m^2n^3)(1) + (m^2n^3)(-2mn^2) + (m^2n^3)(m^3n) + (m^2n^3)(-2m^2n^3)$$
$$= m^2n^3 - 2m^3n^5 + m^5n^4 - 2m^4n^6$$

ANSWERS TO PRACTICE EXERCISES

1. $-10x^6$. 2. $-45u^7$. 3. $-1.2u^4v^3$. 4. $\dfrac{-3}{7}u^3 + \dfrac{2}{7}u^2 - \dfrac{4}{21}u$. 5. $2p^3q^4 - 4p^4q^2 - 2p^2q$.
6. $-6m^5n^3 + 12m^3n^5 - 3m^5n^4 + 21m^2n^7$. 7. $6r^3s^3 - 8r^4s^4 + 10r^2s^8$.

Let's now consider multiplying two binomials. To do so, we make use of the distributive property twice. This is illustrated in Example 8.

Multiply.

Practice Exercise 8.
$(2y - 1)(y + 2)$

EXAMPLE 8 Multiply: $(u - 2)(u + 3)$.

SOLUTION:

$(u - 2)(u + 3) = (u - 2)(u) + (u - 2)(3)$ (Distributive property)
$\qquad\qquad\qquad = (u)(u - 2) + (3)(u - 2)$ (Commutative property for multiplication)
$\qquad\qquad\qquad = (u)(u) + (u)(-2) + (3)(u) + (3)(-2)$ (Distributive property)
$\qquad\qquad\qquad = u^2 + (-2u) + 3u + (-6)$
$\qquad\qquad\qquad = u^2 + u - 6$

In Example 8, *each* term in the binomial $(u + 3)$ was multiplied by *every* term of the binomial $(u - 2)$. This multiplication could also be done in a vertical manner, following the procedure used for multiplication of whole numbers with two or more digits. This procedure is illustrated in Example 9.

Practice Exercise 9.
$(3p + 4)(2p - 1)$

EXAMPLE 9 Multiply: $(u - 2)(u + 3)$.

SOLUTION: Using a vertical format to multiply $(u - 2)(u + 3)$, we have:

$$\begin{array}{r} u + 3 \\ \text{Multiply:} \quad u - 2 \\ \hline u^2 + 3u \\ -2u - 6 \\ \hline u^2 + u - 6 \end{array}$$

    [Multiplying $(u + 3)$ by u]
    [Multiplying $(u + 3)$ by -2 *and* lining up like terms]
    (Adding like terms)

The procedures used in Examples 8 and 9 can be used to multiply any two polynomials.

Multiplying a Polynomial by a Polynomial

To multiply a polynomial by a polynomial, multiply *each* term in one of the polynomials by *every* term in the other polynomial. Then add the results.

Practice Exercise 10.
$(y^2 - 3)(2y + 5)$

EXAMPLE 10 Multiply $2x^2 - 3x + 1$ by $3x + 4$.

SOLUTION:

$$(3x + 4)(2x^2 - 3x + 1) = 3x(2x^2 - 3x + 1) + 4(2x^2 - 3x + 1)$$
$$= (3x)(2x^2) + (3x)(-3x) + (3x)(1) + (4)(2x^2) + (4)(-3x) + (4)(1)$$
$$= 6x^3 - 9x^2 + 3x + 8x^2 - 12x + 4$$
$$= 6x^3 - x^2 - 9x + 4$$

Practice Exercise 11.
$(3u - 1)(u^2 + 2u - 3)$

EXAMPLE 11 Multiply $y^3 - 3y^2 + 4y - 1$ by $4y - 5$.

SOLUTION: Using a vertical format, we have:

Multiply:
$$\begin{array}{r} y^3 - 3y^2 + 4y - 1 \\ 4y - 5 \\ \hline 4y^4 - 12y^3 + 16y^2 - 4y \\ - 5y^3 + 15y^2 - 20y + 5 \\ \hline 4y^4 - 17y^3 + 31y^2 - 24y + 5 \end{array}$$

$(y^3 - 3y^2 + 4y - 1)(4y)$
$(y^3 - 3y^2 + 4y - 1)(-5)$
(Adding the results)

In the preceding examples, both horizontal and vertical methods were used to multiply polynomials. Either method may be used, but you may prefer the horizontal method over the vertical method. If so, use it unless you are instructed otherwise.

Practice Exercise 12.
$(v^2 - 2v + 3)(2v^3 - v^2 + 7v + 1)$

EXAMPLE 12 Multiply $x^2 - 2x + 3$ and $x^2 + 3x - 5$.

SOLUTION:

$$(x^2 - 2x + 3)(x^2 + 3x - 5)$$
$$= (x^2)(x^2 + 3x - 5) + (-2x)(x^2 + 3x - 5) + (3)(x^2 + 3x - 5)$$
$$= (x^2)(x^2) + (x^2)(3x) + (x^2)(-5) + (-2x)(x^2) + (-2x)(3x) + (-2x)(-5) + (3)(x^2) + (3)(3x) + (3)(-5)$$
$$= x^4 + 3x^3 - 5x^2 - 2x^3 - 6x^2 + 10x + 3x^2 + 9x - 15$$
$$= x^4 + x^3 - 8x^2 + 19x - 15$$

Practice Exercise 13.
$(3s^2 - s^3 + 1)(2 - 3s^2 + s^4)$

EXAMPLE 13 Multiply: $2y^4 - y^2 + 7$ and $2y^3 - 3y - 2$.

SOLUTION: We will do this multiplication vertically.

Multiply:
$$\begin{array}{r} 2y^4 - y^2 + 7 \\ 2y^3 - 3y - 2 \\ \hline 4y^7 - 2y^5 + 14y^3 \\ - 6y^5 + 3y^3 - 21y \\ - 4y^4 + 2y^2 - 14 \\ \hline 4y^7 - 8y^5 - 4y^4 + 17y^3 + 2y^2 - 21y - 14 \end{array}$$

$(2y^4 - y^2 + 7)(2y^3)$
$(2y^4 - y^2 + 7)(-3y)$
$(2y^4 - y^2 + 7)(-2)$

Add:

10.4 Multiplication of Polynomials

ANSWERS TO PRACTICE EXERCISES

8. $2y^2 + 3y - 2$. 9. $6p^2 + 5p - 4$. 10. $2y^3 + 5y^2 - 6y - 15$. 11. $3u^3 + 5u^2 - 11u + 3$.
12. $2v^5 - 5v^4 + 15v^3 - 16v^2 + 19v + 3$. 13. $2 + 3s^2 - 2s^3 - 8s^4 + 3s^5 + 3s^6 - s^7$.

EXERCISES 10.4

Multiply in Exercises 1–66.

1. $(2x)(-5x^2)$
2. $(-3y^2)(4y^3)$
3. $(0.1u^3)(-0.3u^4)$
4. $(-2.1v^3)(-3.2v^7)$
5. $(-2x^2y^3)(6x^3y)$
6. $(-4uv^2)(-5u^3v)$
7. $\left(\dfrac{1}{2}p^2q^2\right)\left(\dfrac{-1}{3}pq^3\right)$
8. $\left(\dfrac{-3}{7}m^4n^2\right)\left(\dfrac{-4}{9}m^2n^5\right)$
9. $(2pq^2r^2)(-3p^3qr^4)$
10. $(-4r^2s^3t^4)(-0.2r^3st^2)$
11. $(2xy^2)(-3x^2y)(-4x^2y^3)$
12. $\left(\dfrac{1}{2}p^2q\right)\left(\dfrac{-1}{3}pq^3\right)\left(\dfrac{4}{5}p^3q\right)$
13. $3(x + 2y)$
14. $-3(6u - 5)$
15. $4u(2u^2 - 1)$
16. $3y(x - 3y)$
17. $5t(t^4 - 3)$
18. $-2y(y^3 - 5y)$
19. $0.2x(1.3x + 0.7)$
20. $xy(x^2y - xy^2)$
21. $p^2q^3(pq^2 - p^2q^4)$
22. $x^2y(2x^3 - 4xy^3)$
23. $-2p^2q(3p - 4pq^3)$
24. $0.2r^5s^2(1.7rs^2 - 2.5r^3s^2)$
25. $6(x^3 - 5x + 7)$
26. $-8(2t^7 - t^4 + 3t^2)$
27. $2.3(0.3w^5 - w^2 - 0.7)$
28. $-3v(2 - v - v^3)$
29. $8s(s^2 + s^4 - 6s^7)$
30. $m^4n^3(2m^2n - 3mn + 4m^2 - 7)$
31. $(x - 3)(x + 2)$
32. $(x + 4)(x + 6)$
33. $(y - 7)(y - 5)$
34. $(y + 7)(y + 9)$
35. $(2u + 1)(3u - 2)$
36. $(3u - 7)(3u - 7)$
37. $\left(\dfrac{2}{3}p - 3\right)\left(2p - \dfrac{1}{5}\right)$
38. $\left(3p - \dfrac{1}{4}\right)\left(2p + \dfrac{1}{7}\right)$
39. $\left(\dfrac{-1}{3}t + 1\right)\left(\dfrac{2}{5}t - 3\right)$
40. $(4r - 5)(3r - 2)$
41. $(7r - 3)(1 - 2r)$
42. $(2x^2 + 3)(x - 4)$
43. $(x^2 - 1)(x^2 + 4)$
44. $(0.2u - 3)(4 - 1.5u)$
45. $(xy + 5)(xy - 3)$
46. $(2 - pq)(pq - 6)$
47. $(p^2q + 3p)(2q - pq^2)$
48. $(u^2v - 3u)(uv^2 + 4v)$
49. $(x^2y + 2xy)(xy^3 - 3xy)$
50. $(2pq - 3p^2q)(pq^2 - 2p^2q^2)$
51. $(2y + 6)(3y^2 - 4y + 1)$
52. $(3y - 2)(4y^2 - 5y - 6)$
53. $(3u - 2)(u^2 + 4u + 8)$
54. $(1 - 2u)(2 - 3u + u^2)$
55. $(2x - 3)(x^3 - 2x^2 + 4x + 5)$
56. $(3x + 1)(x^3 + 7x^2 + 5x - 20)$
57. $(3y^2 - 4)(2y^3 - 5y + 4)$
58. $(5y^2 - 3y)(y^4 - 2y^2 + 6)$
59. $(p^2 - 2p + 1)(2p^2 - 3p + 4)$
60. $(3 - p - p^2)(2 + 4p - 2p^2)$
61. $(x^2 - 2x + 3)(2x^2 + x - 7)$
62. $(2u - 3v + 1)(3v + 2u - 3)$
63. $(y^2 - 2y + 3)(2y^2 - 4y + 5)$
64. $(3t^3 - 4t + 1)(2 - 3t + 4t^2)$
65. $(w^4 - 2w^2 + 3)(2w^4 + w^2 - 8)$
66. $(3r^3 - 4r^2 + 7)(2r^4 + r^2 + 1)$

In Exercises 67–72, perform the indicated operations.

67. $(2x - 5)[(2x^2 - 5x + 1) + (2x^3 - 5x - 3)]$
68. $(1 - 2y)[(3y^2 - y + 1) - (4y^2 + 5y - 3)]$
69. $(2u + 7)[3(2u^4 - u^3 + 1) + 4(u^3 - 5u^2 - 6u)]$
70. $(5 - 2t)[-2(t^5 - 4t^3 + t) - 5(t^5 + t^2 - 3t)]$
71. $(2x^2 - 3x + 1)[(x^2 + x + 7) + (3x^2 - 2x + 1)]$
72. $[(2y^4 - 3y^2 - 4y + 1) + (y^3 - 2y - 4)][(y^4 - 4y^3 + 4) - (y^3 - 2y^2 - y)]$

10.5 Multiplying Binomials

OBJECTIVES

After completing this section, you should be able to:
1. Multiply two linear binomials, using the FOIL method.
2. Square a binomial.

In Section 10.4, you learned that to multiply two polynomials, you multiply each term in one polynomial by every term in the other polynomial. In Example 9 of the previous section, we multiplied the two binomials $u + 3$ and $u - 2$ and obtained the trinomial $u^2 + u - 6$. The first term of the trinomial was obtained by multiplying the *first* terms of the binomials. The last term of the trinomial was obtained by multiplying the *last* terms of the binomials. The middle term of the trinomial was obtained by adding the product of the *outer* terms and the product of the *inner* terms of the binomials.

We will illustrate the inner and outer products for this multiplication as follows:

$$\text{Outer product, } -2u$$
$$(u + 3)(u - 2)$$
$$\text{Inner product, } 3u$$

$$\text{Outer product } + \text{ inner product } = -2u + 3u = u$$

Expressing the product of two binomials as the sum of the products of the First, Outer, Inner, and Last terms is known as the **FOIL method.**

FOIL Method for Determining the Product of Two Linear Binomials

To determine the product of two linear binomials, proceed as follows:

1. To form the *first* term of the product, multiply the first terms of the two binomials.
2. To form the *middle* term of the product, determine the *sum* of the *outer* and *inner* products of the two binomials.
3. To form the *last* term of the product, multiply the last terms of the two binomials.

10.5 Multiplying Binomials

The FOIL method is illustrated in the following examples.

Use the FOIL method to multiply.

Practice Exercise 1.
$(y + 4)(y + 6)$

EXAMPLE 1 Multiply: $(x + 2)(x + 3)$.

SOLUTION: Multiplying $x + 2$ and $x + 3$, we obtain.

$$\underbrace{(x+2)(x+3)}_{x^2 \text{ First term}} + \underbrace{(x+2)(x+3)}_{\substack{2x \\ +3x \\ 5x \\ \text{Middle term}}} + \underbrace{(x+2)(x+3)}_{6 \text{ Last term}}$$

Hence, $(x + 2)(x + 3) = x^2 + 5x + 6$.

Practice Exercise 2.
$(2p - 1)(3p + 8)$

EXAMPLE 2 Multiply: $(2u + 4)(2u - 3)$.

SOLUTION: Multiplying $2u + 4$ and $2u - 3$, we obtain:

$$\underbrace{(2u+4)(2u-3)}_{4u^2 \text{ First term}} + \underbrace{(2u+4)(2u-3)}_{\substack{8u \\ +(-6u) \\ 2u \\ \text{Middle term}}} + \underbrace{(2u+4)(2u-3)}_{(-12) \text{ Last term}}$$

Hence, $(2u + 4)(2u - 3) = 4u^2 + 2u - 12$.

Practice Exercise 3.
$(3 - w)(4 + 2w)$

EXAMPLE 3 Multiply: $(2y - 1)(3y + 4)$.

SOLUTION: Multiplying $2y - 1$ and $3y + 4$, we obtain:

$$\underbrace{(2y-1)(3y+4)}_{6y^2 \text{ First term}} + \underbrace{(2y-1)(3y+4)}_{\substack{-3y \\ +8y \\ 5y \\ \text{Middle term}}} + \underbrace{(2y-1)(3y+4)}_{(-4) \text{ Last term}}$$

Hence, $(2y - 1)(3y + 4) = 6y^2 + 5y - 4$.

ANSWERS TO PRACTICE EXERCISES

1. $y^2 + 10y + 24$. **2.** $6p^2 + 13p - 8$. **3.** $12 + 2w - 2w^2$.

The FOIL method can be shortened, as illustrated in the next two examples.

Multiply.

Practice Exercise 4.
$(t + 3)(t - 9)$

EXAMPLE 4 Multiply: $(p - 7)(2p + 3)$.

SOLUTION: $(p - 7)(2p + 3) = 2p^2 + 3p - 14p - 21$
$ \text{First} \text{Outer} \text{Inner} \text{Last}$

$ = 2p^2 - 11p - 21$

Practice Exercise 5.
$(2s - 3)(3s + 2)$

EXAMPLE 5 Multiply: $(3m - 4)(2 - 5m)$.

SOLUTION: $(3m - 4)(2 - 5m) = \underset{\text{First}}{6m} - \underset{\text{Outer}}{15m^2} - \underset{\text{Inner}}{8} + \underset{\text{Last}}{20m}$

$= -15m^2 + 26m - 8$

Practice Exercise 6.
$(2x + 3)(2x - 3)$

EXAMPLE 6 Multiply: $(x - 4)(x + 4)$.

SOLUTION: $(x - 4)(x + 4) = \underset{\text{First}}{x^2} + \underset{\text{Outer}}{4x} - \underset{\text{Inner}}{4x} - \underset{\text{Last}}{16}$

$= x^2 + (4x - 4x) - 16$
$= x^2 + 0 - 16$
$= x^2 - 16$

Practice Exercise 7.
$(t^2 + 1)(t^2 - 1)$

EXAMPLE 7 Multiply: $(2y - 1)(2y + 1)$.

SOLUTION: $(2y - 1)(2y + 1) = \underset{\text{First}}{4y^2} + \underset{\text{Outer}}{2y} - \underset{\text{Inner}}{2y} - \underset{\text{Last}}{1}$

$= 4y^2 + (2y - 2y) - 1$
$= 4y^2 + 0 - 1$
$= 4y^2 - 1$

In Examples 6 and 7, the factors in the product are the *sum* and the *difference* of the same terms. The product is the difference of the squares of the terms. More will be said about this in Section 11.3.

The FOIL method can always be used to determine the product of two binomials. However, sometimes the two binomials are the same. That is, we have the square of a binomial such as $(x + y)(x + y) = (x + y)^2$. To square a binomial means to multiply the binomial by itself, just as squaring a number means to multiply the number by itself.

Determine the square of each of the following binomials.

Practice Exercise 8. $p + 4$

EXAMPLE 8 Determine the square of $(y + 3)$.

SOLUTION: $(y + 3)^2 = (y + 3)(y + 3) = y^2 + 3y + 3y + 9$
$= y^2 + 6y + 9$

Practice Exercise 9. $w - 5$

EXAMPLE 9 Determine the square of $(2u - 3)$.

SOLUTION: $(2u - 3)^2 = (2u - 3)(2u - 3) = 4u^2 - 6u - 6u + 9$
$= 4u^2 - 12u + 9$

In Examples 8 and 9, the binomials were squared using the FOIL method. However, did you notice that, in each case, the inner and outer products were the same? This leads to the following rule.

Rule for Squaring a Binomial

To square a binomial, proceed as follows to obtain the product:

1. To form the *first* term of the product, square the first term of the binomial.
2. To form the *middle* term of the product, take *twice* the product of the two terms of the binomial, *including* the signs of the coefficients.

10.5 Multiplying Binomials

3. To form the *last* term of the product, square the last term of the binomial.
4. The square of a binomial is *always* a trinomial.

In general, we have the following:

$$(a + b)^2 = a^2 + 2ab + b^2$$
$$(a - b)^2 = a^2 - 2ab + b^2$$

The preceding rule is illustrated in the following examples.

Practice Exercise 10. $2y + 3$

EXAMPLE 10 Square $(x + 4)$.

SOLUTION:
$$(x + 4)^2 = (x)^2 + 2(x)(4) + (4)^2$$
$$= x^2 + 8x + 16$$

Practice Exercise 11. $2x - 3y$

EXAMPLE 11 Square $\left(\frac{1}{2}y - 5\right)$.

SOLUTION:
$$\left(\frac{1}{2}y - 5\right)^2 = \left(\frac{1}{2}y\right)^2 + 2\left(\frac{1}{2}y\right)(-5) + (-5)^2$$
$$= \frac{1}{4}y^2 - 5y + 25$$

Practice Exercise 12. $1.2p - 1.5q$

EXAMPLE 12 Square $(1 - 3y)$.

SOLUTION:
$$(1 - 3y)^2 = (1)^2 + 2(1)(-3y) + (-3y)^2$$
$$= 1 - 6y + 9y^2$$

Practice Exercise 13. $0.3r + 0.7s$

EXAMPLE 13 Square $(1.1x + 2.3y)$.

SOLUTION:
$$(1.1x + 2.3y)^2 = (1.1x)^2 + 2(1.1x)(2.3y) + (2.3y)^2$$
$$= 1.21x^2 + 5.06xy + 5.29y^2$$

ANSWERS TO PRACTICE EXERCISES

4. $t^2 - 6t - 27$. 5. $6s^2 - 5s - 6$. 6. $4x^2 - 9$. 7. $t^4 - 1$. 8. $p^2 + 8p + 16$.
9. $w^2 - 10w + 25$. 10. $4y^2 + 12y + 9$. 11. $4x^2 - 12xy + 9y^2$. 12. $1.44p^2 - 3.6pq + 2.25q^2$.
13. $0.09r^2 + 0.42rs + 0.49s^2$.

EXERCISES 10.5

In Exercises 1–40, multiply using the FOIL method.

1. $(x + 1)(x + 2)$
2. $(x + 2)(x + 4)$
3. $(x + 3)(x - 3)$
4. $(x - 2)(x - 4)$
5. $(y + 3)(y + 7)$
6. $(y - 4)(y + 4)$
7. $(1.1u - 5)(0.3u + 1)$
8. $(3.1v - 1)(1.1v + 5)$
9. $(u - 2)(2u + 1)$
10. $(3u - 1)(2u + 1)$

11. $\left(\dfrac{4}{7}u + 1\right)\left(\dfrac{2}{5}u - 3\right)$
12. $\left(\dfrac{3}{7}u + 2\right)\left(\dfrac{2}{9}u - 5\right)$
13. $(2a - b)(3a + b)$
14. $(2x - 3y)(3x + 2y)$
15. $(4x + 5y)(2x - 3y)$
16. $(2a + 5b)(2a + 5b)$
17. $\left(\dfrac{2}{3}x - \dfrac{2}{5}y\right)\left(\dfrac{1}{3}x + \dfrac{2}{7}y\right)$
18. $\left(\dfrac{3}{4}p - \dfrac{1}{5}q\right)\left(\dfrac{1}{4}p - \dfrac{4}{5}q\right)$
19. $(2r + 3s)(2r + 3s)$
20. $(4s - 3t)(4s - 3t)$
21. $(2.3x - 3.1y)(1.9x + 0.3y)$
22. $(2u - 1.7v)(3.8v - 3u)$
23. $(2 - 3p)(4 + 2p)$
24. $(3x^2 - 3y)(5x + 4y)$
25. $(7s + 4t)(3s + 5t)$
26. $(3q - 9p)(2q - 7p)$
27. $\left(\dfrac{3}{4}m + \dfrac{4}{5}n\right)\left(\dfrac{1}{3}n + \dfrac{3}{7}m\right)$
28. $\left(\dfrac{1}{4}t - \dfrac{2}{3}u\right)\left(\dfrac{3}{4}u + \dfrac{4}{5}t\right)$
29. $(x - 2y)(x + 2y)$
30. $(3s^2 - 5t)(4s + 7t)$
31. $(x - 3y)(7x + y)$
32. $(8x + y)(x + 5y)$
33. $(2p - 3q)(2p + 3q)$
34. $(2p - 3q)(2p - 3q)$
35. $(r - 6s)(5r + s)$
36. $(2s - 3t)(t + 5s)$
37. $(2p + 3q)(p - 4q)$
38. $(11r - 12s)(9r + 7s)$
39. $(-4x - 5y)(-2x - 7y)$
40. $(17a - 15b)(12a + 13b)$

In Exercises 41–60, square the given binomial.

41. $x - y$
42. $2a + b$
43. $3x - y$
44. $2x - 3y$
45. $2m - 3n$
46. $4p - 1$
47. $3p + 2q$
48. $4m - 3n$
49. $1 - 2x$
50. $5 - 2y$
51. $1.2p - 1.3$
52. $4.3p - 5.1$
53. $0.1 - 0.4q$
54. $1.5q + 0.7$
55. $11.2x - 1$
56. $x - 6.2$
57. $4a - 3b$
58. $7.1a + 3.2b$
59. $10x - 25y$
60. $40y - 35x$

In Exercises 61–70, perform the indicated operations.

61. $(2x - 1)^2 - (3x + 2)^2$
62. $(3y - 1)^2 + (2y - 5)^2$
63. $(5p + 3q)^2 + (2p - 5q)^2$
64. $(4r + 5s)^2 - (9r - 8s)^2$
65. $(2x - 3y)(x + 4y) + (4x + 5y)(2x - y)$
66. $(3r - 5s)(2r + 7s) - (r + 9s)(8r - s)$
67. $(0.8m + 1.2n)(3.1m - 0.8n) - (0.6m - 0.5n)^2$
68. $(3u - v)^2 - (u + 4v)(5u + 7v)$
69. $(x + y)^2 + (x - y)^2 + (2x - y)^2$
70. $(2p - q)^2 - (p - 3q)^2 - (3p - 4q)^2$

10.6 Division of Polynomials

OBJECTIVES

After completing this section, you should be able to:
1. Divide a polynomial by a monomial.
2. Divide polynomials by first-degree binomials.

You learned to divide a number by a number, starting first with a single-digit divisor. You will learn **division of polynomials** in a similar manner. The first part of this section will involve dividing a polynomial by a monomial, using the following rule.

Rule for Dividing a Polynomial by a Monomial

To divide a polynomial by a nonzero monomial, divide *each* term of the polynomial by the monomial and then simplify the results.

To divide a polynomial by a monomial, you must recall the rules for exponents and also remember that division by zero is not possible.

Practice Exercise 1. Divide $9y^2 - 3y + 12$ by -3.

EXAMPLE 1 Divide $4x^2 - 12x + 20$ by 4.

SOLUTION:

$$\frac{4x^2 - 12x + 20}{4} \quad \text{(A polynomial divided by a monomial)}$$

$$= \frac{4x^2}{4} - \frac{12x}{4} + \frac{20}{4} \quad \text{(Dividing } \textit{each} \text{ term of the polynomial by the monomial)}$$

$$= x^2 - 3x + 5 \quad \text{(Simplifying)}$$

Practice Exercise 2. If $x \neq 0$, divide $6x^3 - 4x^2 + 8x$ by $4x$.

EXAMPLE 2 Divide $3y^4 - 2y^3 + 6y^2$ by $3y$, if $y \neq 0$.

SOLUTION: If $y \neq 0$, then

$$\frac{3y^4 - 2y^3 + 6y^2}{3y} \quad \text{(A polynomial divided by a monomial)}$$

$$= \frac{3y^4}{3y} - \frac{2y^3}{3y} + \frac{6y^2}{3y} \quad \text{(Dividing } \textit{each} \text{ term of the polynomial by the monomial)}$$

$$= y^3 - \frac{2}{3}y^2 + 2y \quad \text{(Simplifying)}$$

In Example 2, note that $y \neq 0$ since division by 0 is not possible.

Practice Exercise 3. If $y \neq 0$, divide $27y^5 - 9y^4 + 7y^2$ by $-3y^2$.

EXAMPLE 3 Divide $4u^3 - 2u^2 + 5u - 7$ by $3u^2$.

SOLUTION: If $u \neq 0$, then

$$\frac{4u^3 - 2u^2 + 5u - 7}{3u^2}$$

$$= \frac{4u^3}{3u^2} - \frac{2u^2}{3u^2} + \frac{5u}{3u^2} - \frac{7}{3u^2}$$

$$= \frac{4}{3}u - \frac{2}{3} + \frac{5}{3}u^{-1} - \frac{7}{3}u^{-2} \quad \text{(Note that the result is } not \text{ a polynomial.)}$$

In Example 3, a polynomial was divided by a monomial, and the quotient was *not* a polynomial. However, the division is still possible, provided that $u \neq 0$.

ANSWERS TO PRACTICE EXERCISES

1. $-3y^2 + y - 4$. 2. $\frac{3}{2}x^2 - x + 2$. 3. $-9y^3 + 3y^2 - \frac{7}{3}$.

Dividing a polynomial by a polynomial is done in a manner similar to dividing numbers when the divisor has two or more digits. We will *now* examine the division of a polynomial by a linear polynomial. The procedure is outlined in Example 4.

Practice Exercise 4. If $x \neq -2$, divide $x^2 - 3x - 10$ by $x + 2$.

EXAMPLE 4 Divide $x^2 + 5x + 6$ by $x + 3$, if $x \neq -3$.

SOLUTION: Note that we have $x \neq -3$ so that the divisor, $x + 3$, will not be 0. We will now proceed with the division.

STEP 1: Arrange both dividend and divisor in *descending* powers of x:

$$x + 3 \overline{\smash{)}x^2 + 5x + 6}$$

STEP 2: Divide the *first* term of the dividend, x^2, by the *first* term of the divisor, x, obtaining x. Write this result as the *first* term of the quotient above the x-term in the dividend:

$$\begin{array}{r} x \\ x + 3 \overline{\smash{)}x^2 + 5x + 6} \end{array}$$

STEP 3: Multiply the *first* term of the quotient, x, by the entire divisor, $x + 3$. Write the result below the dividend, as illustrated in the following, and subtract like terms:

$$\begin{array}{r} x \\ x + 3 \overline{\smash{)}x^2 + 5x + 6} \\ \underline{x^2 + 3x } \quad \text{(Subtracting)} \\ 2x + 6 \end{array}$$

STEP 4: Divide $2x$, the *first* term of the remainder, by x, the *first* term of the divisor, obtaining 2. Write this result as the *second* term of the quotient above the constant term in the dividend, as illustrated here:

10.6 Division of Polynomials

$$\begin{array}{r} x + 2 \\ x + 3 \overline{\smash{)}x^2 + 5x + 6} \\ \underline{x^2 + 3x } \\ 2x + 6 \end{array}$$

STEP 5: Multiply the *second* term of the quotient, 2, by the entire divisor, $x + 3$. Write the result below $2x + 6$, as indicated here. Subtract like terms.

$$\begin{array}{r} x + 2 \\ x + 3 \overline{\smash{)}x^2 + 5x + 6} \\ \underline{x^2 + 3x } \\ 2x + 6 \\ \underline{2x + 6} \quad \text{(Subtracting)} \\ 0 \quad \text{(Remainder)} \end{array}$$

Since the remainder is 0, we are done. Hence, $(x^2 + 5x + 6) \div (x + 3) = x + 2$, if $x \neq -3$.

Just as in arithmetic, we can check this division by multiplying the quotient $(x + 2)$ by the divisor $(x + 3)$ and adding the remainder (0) to determine if we get the dividend $(x^2 + 5x + 6)$.

CHECK:

$$\begin{array}{r} x + 2 \quad \text{(Quotient)} \\ \underline{x + 3} \quad \text{(Divisor)} \\ x^2 + 2x \\ \underline{3x + 6} \\ x^2 + 5x + 6 \end{array}$$

Add: $\quad \underline{ 0} \quad$ (Remainder)

$\phantom{\text{Add:}} \quad x^2 + 5x + 6 \quad$ (Dividend) ✓

Practice Exercise 5. If $y \neq \dfrac{2}{3}$, divide $5y - 6 + 6y^2$ by $3y - 2$.

EXAMPLE 5 Divide $7 + 3y^2 + 2y^3 - 5y$ by $2y - 3$, if $y \neq \dfrac{3}{2}$.

SOLUTION: Rearrange both the dividend and the divisor in *descending* powers of y. Then proceed, using the steps outlined in Example 4:

$$\begin{array}{r} y^2 + 3y + 2 \\ 2y - 3 \overline{\smash{)}2y^3 + 3y^2 - 5y + 7} \\ \underline{2y^3 - 3y^2 } \\ 6y^2 - 5y + 7 \\ \underline{6y^2 - 9y } \\ 4y + 7 \\ \underline{4y - 6} \\ 13 \end{array}$$

(Multiplying $2y - 3$ by y^2 and subtracting)

(Multiplying $2y - 3$ by $3y$ and subtracting)

(Multiplying $2y - 3$ by 2 and subtracting)

(Remainder)

CHECK:

$$\begin{array}{r} y^2 + 3y + 2 \quad \text{(Quotient)} \\ \underline{2y - 3} \quad \text{(Divisor)} \\ 2y^3 + 6y^2 + 4y \\ \underline{- 3y^2 - 9y - 6} \\ 2y^3 + 3y^2 - 5y - 6 \\ \underline{+ 13} \quad \text{(Remainder)} \\ 2y^3 + 3y^2 - 5y - 7 \quad \text{(Dividend)} ✓ \end{array}$$

When dividing a polynomial by a polynomial, arrange the terms of both the dividend and the divisor in descending power of the variable. Further, if there is a missing term in the dividend (such as the y^2 term in $3y^3 - 4y + 7$), leave a space for it. This is done to line up the terms when actually dividing.

Chapter 10 Polynomials

Practice Exercise 6. If $p \neq \frac{1}{2}$, divide $4p^5 + p^3 + 2p^2 - 7$ by $2p - 1$.

EXAMPLE 6 Divide $y^3 - 3y^2 + 4$ by $y - 2$, if $y \neq 2$.

SOLUTION:

STEP 1: Arrange both the dividend and the divisor in *descending* powers of y. Also, leave a space in the dividend for the missing y term:

$$y - 2 \overline{\smash{\big)}\, y^3 - 3y^2 + 4}$$

STEP 2: Divide y^3, the *first* term of the dividend, by y, the *first* term of the divisor, obtaining y^2. Write this result as the *first* term of the quotient, above the y^2 term in the dividend:

$$\begin{array}{r} y^2 \\ y + 2 \overline{\smash{\big)}\, y^3 - 3y^2 + 4} \end{array}$$

STEP 3: Multiply y^2, the *first* term of the quotient, by the entire divisor. Write the result below the dividend, as illustrated, and subtract like terms:

$$\begin{array}{r} y^2 \\ y - 2 \overline{\smash{\big)}\, y^3 - 3y^2 + 4} \\ \underline{y^3 - 2y^2 } \quad \text{(Subtracting)} \\ -y^2 + 4 \end{array}$$

STEP 4: Divide $-y^2$, the *first* term of $-y^2 + 4$, by y, the *first* term of the divisor, obtaining $-y$. Write this result as the *second* term of the quotient, above the space for the missing y term in the dividend:

$$\begin{array}{r} y^2 - y \\ y - 2 \overline{\smash{\big)}\, y^3 - 3y^2 + 4} \\ \underline{y^3 - 2y^2 } \\ -y^2 + 4 \end{array}$$

STEP 5: Multiply $-y$, the *second* term of the quotient, by the entire divisor. Write the result below the dividend, as illustrated, and subtract like terms:

$$\begin{array}{r} y^2 - y \\ y - 2 \overline{\smash{\big)}\, y^3 - 3y^2 + 4} \\ \underline{y^3 - 2y^2 } \\ -y^2 + 4 \\ \underline{-y^2 + 2y } \\ -2y + 4 \end{array}$$

STEP 6: Divide $-2y$, the *first* term of $-2y + 4$, by y, the *first* term of the divisor, obtaining -2. Write this result as the *third* term of the quotient, above the constant term in the dividend:

$$\begin{array}{r} y^2 - y - 2 \\ y - 2 \overline{\smash{\big)}\, y^3 - 3y^2 + 4} \\ \underline{y^3 - 2y^2 } \\ -y^2 + 4 \\ \underline{-y^2 + 2y } \\ -2y + 4 \end{array}$$

STEP 7: Multiply -2, the *third* term of the quotient, by the entire divisor. Write the result below the dividend, as illustrated, and subtract like terms:

10.6 Division of Polynomials

$$\begin{array}{r} y^2 - y - 2 \\ y-2 \overline{\smash{)}y^3 - 3y^2 + 4} \\ \underline{y^3 - 2y^2 } \\ -y^2 + 4 \\ \underline{-y^2 + 2y } \\ -2y + 4 \\ \underline{-2y + 4} \\ 0 \quad \text{(Remainder)} \end{array}$$

Since the remainder is 0, we are done. Hence, $(y^3 - 3y^2 + 4) \div (y - 2) = y^2 - y - 2$, if $y \neq 2$. The check is left to you.

ANSWERS TO PRACTICE EXERCISES

4. $x - 5$.   **5.** $2y + 3$.   **6.** $\left(2p^4 + p^3 + p^2 + \dfrac{3}{2}p + \dfrac{3}{4}\right) R\left(\dfrac{-25}{4}\right)$.

In Examples 7 and 8, we will illustrate division of polynomials in more than one variable.

Practice Exercise 7. Divide $p^2 - 3pq + 4q^2$ by $p + q$, if $p \neq -q$.

EXAMPLE 7   Divide $x^2 + 3xy + y^2$ by $x - y$, if $x \neq y$.

STEP 1: Arrange both the dividend and the divisor in *descending* powers of one of the variables (we will take x):

$$x - y \overline{\smash{)}x^2 + 3xy + y^2}$$

STEP 2: Divide x^2, the *first* term of the dividend, by x, the *first* term of the divisor, obtaining x. Write this result as the *first* term of the quotient, above the xy term in the dividend:

$$\begin{array}{r} x \\ x - y \overline{\smash{)}x^2 + 3xy + y^2} \end{array}$$

STEP 3: Multiply x, the *first* term of the quotient, by the entire divisor. Write the result below the dividend, as illustrated, and subtract like terms:

$$\begin{array}{r} x \\ x - y \overline{\smash{)}x^2 + 3xy + y^2} \\ \underline{x^2 - xy } \quad \text{(Subtracting)} \\ 4xy + y^2 \end{array}$$

STEP 4: Divide $4xy$, the *first* term of $4xy + y^2$, by x, the *first* term of the divisor, obtaining $4y$. Write this result as the *second* term of the quotient, above the y^2 term in the dividend:

$$\begin{array}{r} x + 4y \\ x - y \overline{\smash{)}x^2 + 3xy + y^2} \\ \underline{x^2 - xy } \quad \text{(Subtracting)} \\ 4xy + y^2 \end{array}$$

STEP 5: Multiply $4y$, the *second* term of the quotient, by the entire divisor. Write this result below the dividend, as illustrated, and subtract like terms:

$$\begin{array}{r} x + 4y \\ x - y \overline{\smash{\big)}\, x^2 + 3xy + y^2} \\ \underline{x^2 - xy} \\ 4xy + y^2 \\ \underline{4xy - 4y^2} \\ 5y^2 \end{array}$$ (Subtracting)

Note that the divisor is first degree in x and the remainder, $5y^2$, is zero degree in x. When the remainder is of lesser degree than the degree of the divisor (for the lead variable), the division stops.

CHECK:
Multiply:

$$\begin{array}{rl} x + 4y & \text{(Quotient)} \\ \underline{x - y} & \text{(Divisor)} \\ x^2 + 4xy & \end{array}$$

Add:

$$\begin{array}{r} -xy - 4y^2 \\ \hline x^2 + 3xy - 4y^2 \end{array}$$

Add:

$$\begin{array}{rl} 5y^2 & \text{(Remainder)} \\ \hline x^2 + 3xy + y^2 \checkmark & \text{(Dividend)} \end{array}$$

> **Practice Exercise 8.** Divide $6s^4 - 5s^2t^2 + 6st^3 - t^4$ by $2s - t$, if $t \neq 2s$.

EXAMPLE 8 Divide $4p^3 + 5pq^2 - 7q^3$ by $2p - q$, if $q \neq 2p$.

SOLUTION:

STEP 1: Arrange both the dividend and divisor in *descending* powers of one of the variables. (We will take p.) Also, leave a space in the dividend for the missing p^2q term:

$$2p - q \overline{\smash{\big)}\, 4p^3 + 5pq^2 - 7q^3}$$

STEP 2: Divide $4p^3$, the *first* term of the dividend, by $2p$, the *first* term of the divisor, obtaining $2p^2$. Write this result as the *first* term of the quotient, above the space for the missing term in the dividend:

$$\begin{array}{r} 2p^2 \\ 2p - q \overline{\smash{\big)}\, 4p^3 + 5pq^2 - 7q^3} \end{array}$$

STEP 3: Multiply $2p^2$, the *first* term of the quotient, by the entire divisor. Write this result below the dividend, as illustrated, and subtract like terms:

$$\begin{array}{r} 2p^2 \\ 2p - q \overline{\smash{\big)}\, 4p^3 + 5pq^2 - 7q^3} \\ \underline{4p^3 - 2p^2q } \\ 2p^2q + 5pq^2 - 7q^3 \end{array}$$ (Subtracting)

STEP 4: Divide $2p^2q$, the *first* term of $2p^2q + 5pq^2 - 7q^3$, by $2p$, the *first* term of the divisor, obtaining pq. Write this result as the *second* term of the quotient, above the pq^2 term in the dividend:

$$\begin{array}{r} 2p^2 + pq \\ 2p - q \overline{\smash{\big)}\, 4p^3 + 5pq^2 - 7q^3} \\ \underline{4p^3 - 2p^2q } \\ 2p^2q + 5pq^2 - 7q^3 \end{array}$$ (Subtracting)

STEP 5: Multiply pq, the *second* term of the quotient, by the entire divisor. Write this result below the dividend, as illustrated, and subtract like terms:

10.6 Division of Polynomials

$$
\begin{array}{r}
2p^2 + pq \\
2p - q {\overline{\smash{\big)}\,4p^3 + 5pq^2 - 7q^3}} \\
\underline{4p^3 - 2p^2q } \\
2p^2q + 5pq^2 - 7q^3 \\
\underline{2p^2q - pq^2 } \quad \text{(Subtracting)} \\
6pq^2 - 7q^3
\end{array}
$$

STEP 6: Divide $6pq^2$, the *first* term of $6pq^2 - 7q^3$, by $2p$, the *first* term of the quotient, obtaining $3q^2$. Write this result as the *third* term of the quotient, above the q^3 term in the dividend:

$$
\begin{array}{r}
2p^2 + pq + 3q^2 \\
2p - q {\overline{\smash{\big)}\,4p^3 + 5pq^2 - 7q^3}} \\
\underline{4p^3 - 2p^2q } \\
2p^2q + 5pq^2 - 7q^3 \\
\underline{2p^2q - pq^2 } \quad \text{(Subtracting)} \\
6pq^2 - 7q^3
\end{array}
$$

STEP 7: Multiply $3q^2$, the *third* term of the quotient, by the entire divisor. Write this result below the dividend, as illustrated, and subtract like terms:

$$
\begin{array}{r}
2p^2 + pq + 3q^2 \\
2p - q {\overline{\smash{\big)}\,4p^3 + 5pq^2 - 7q^3}} \\
\underline{4p^3 - 2p^2q } \\
2p^2q + 5pq^2 - 7q^3 \\
\underline{2p^2q - pq^2 } \\
6pq^2 - 7q^3 \\
\underline{6pq^2 - 3q^3} \quad \text{(Subtracting)} \\
-4q^3
\end{array}
$$

Since the remainder is 0 degree in p and the divisor is 1 degree in p, the division stops.

CHECK:
Multiply:

$$
\begin{array}{r}
2p^2 + pq + 3q^2 \quad \text{(Quotient)} \\
2p - q \quad \text{(Divisor)} \\
\hline
4p^3 + 2p^2q + 6pq^2 \\
\end{array}
$$

Add:
$$
\begin{array}{r}
- 2p^2q - pq^2 - 3q^3 \\
\hline
4p^3 + 5pq^2 - 3q^3 \\
\end{array}
$$

Add:
$$
\begin{array}{r}
- 4q^3 \quad \text{(Remainder)} \\
\hline
4p^3 + 5pq^2 - 7q^3 \checkmark \quad \text{(Dividend)}
\end{array}
$$

ANSWERS TO PRACTICE EXERCISES

7. $(p - 4q)$ R $(8q^2)$. 8. $\left(3s^3 + \dfrac{3}{2}s^2t - \dfrac{7}{4}st^2 + \dfrac{17}{8}t^3\right)$ R $\left(\dfrac{9}{8}t^4\right)$.

EXERCISES 10.6

In Exercises 1–35, divide as indicated.

1. Divide $8x^2 - 4x + 6$ by 2.
2. Divide $6y^3 - 4y^2 + 2y - 5$ by 3.

3. $(9u^4 - 8u^2 + 7) \div 4$
4. $(6y^4 - 2y^2 + y - 7) \div (-3)$
5. Divide $2w^7 - 3w^5 + 4w^3 - 7w$ by -6.
6. Divide $3p^3 - p^5 + 4p^8 - p^{10}$ by -5.
7. Divide $5p^6 - 2p^4 + 3p^2 - 7p$ by $7p$, if $p \neq 0$.
8. $(t^4 - 3t^3 + 2t^2 - 11t + 14) \div (3t)$, if $t \neq 0$
9. $(3m^5 - 4m^3 - 2m - 7) \div (2m^2)$, if $m \neq 0$
10. $(x^3y^2 - x^2y + xy^4) \div x$, if $x \neq 0$
11. $(3y^4 - 2y^3 + 4y^2) \div (3y^2)$, if $y \neq 0$
12. Divide $5u^6 - 6u^4 + u^3 - 7u + 8$ by $-5u^3$, if $u \neq 0$.
13. Divide $7t^8 - 6t^5 + t^3 - 2t + 1$ by $7t^3$, if $t \neq 0$.
14. Divide $x^3y^2 - x^2y + xy^4$ by xy, if $x \neq 0$ and $y \neq 0$.
15. Divide $4p^3q^2 - 5p^2q^2 + 7p - 9q$ by $2p^2q$, if $p \neq 0$ and $q \neq 0$.
16. $(3u^4v^3 - 5u^3v^2 + 2u^2v) \div (u^4v^3)$, if $u \neq 0$ and $v \neq 0$
17. $(2a^3b^2 - 3a^2b + a^5b^4) \div (-7ab^3)$, if $a \neq 0$ and $b \neq 0$
18. $(3r^3s^4 + r^4s^4 - 5r^6s^2 + rs^3) \div (-3^4rs^4)$, if $r \neq 0$ and $s \neq 0$
19. Divide $s^3t^3 - 4s^4t^3 + 5s^3t^5 - t^2s^3$ by $2s^3t^4$, if $s \neq 0$ and $t \neq 0$.
20. Divide $x^2yz^3 - 3xy^2z^4 + 4x^4y^3z^2 - 5x^6y^3z$ by $3x^2y^3z^4$, if $x \neq 0$, $y \neq 0$, and $z \neq 0$.
21. $(x^2 + 5x + 6) \div (x + 3)$, if $x \neq -3$
22. $(y^2 - 5y - 6) \div (y + 1)$, if $y \neq -1$
23. $(u^2 - 2u + 1) \div (u - 2)$, if $u \neq 2$
24. $(4x^3 - 2x^2 + 4x - 5) \div (2x - 3)$, if $x \neq \dfrac{3}{2}$
25. Divide $6y^2 - y^3 + 8y^4 + 7$ by $9 - 4y$, if $y \neq \dfrac{9}{4}$.
26. Divide $w^3 - 2w + 7$ by $w - 4$, if $w \neq 4$.
27. Divide $3p + 4p^3 - 7 - 6p^4$ by $2p - 3$, if $p \neq \dfrac{3}{2}$.
28. Divide $6 - 4r^2 + 8r^4 - r^3$ by $2 - 4r$, if $r \neq \dfrac{1}{2}$.
29. Divide $2t - t^3 + t^5 - 6 + t^4$ by $t - 3$, if $t \neq 3$.
30. Divide $x^2 - y^2$ by $x - y$, if $y \neq x$.
31. Divide $3x^2 - 5xy + 7y^2$ by $x - 2y$, if $x \neq 2y$.
32. Divide $4u^2 + 7uv - 5v^2$ by $2v - u$, if $u \neq 2v$.
33. $(u^3 - v^3) \div (u - v)$, if $u \neq v$
34. $(4r^3 - 2r^2s + rs^2 - 6s^3) \div (2r - s)$, if $s \neq 2r$
35. $(2x^4 - x^3y + xy^3 - y^4) \div (2x - 3y)$, if $x \neq \dfrac{3}{2}y$

In Exercises 36–40, perform the indicated operations.

36. $[(x^2 + 3x - 1) + (2x^2 - 3)] \div (x - 1)$, $(x \neq 1)$
37. $[(y^2 + 5y + 6) \div (y + 2)] + [(y^2 - 5y - 6) \div (y + 1)]$, $(y \neq -2, -1)$
38. $[(6t^2 + t - 2) \div (2t - 1)](1 - 4t)$, $\left(t \neq \dfrac{1}{2}\right)$
39. $[(x^3 + x^2 - 14x - 24) \div (x - 4)] \div (x + 2)$, $(x \neq -2, 4)$
40. $[(2p^2 - p - 15) \div (2p + 5)][(3p^2 + 11p - 4) \div (p + 4)]$, $\left(p \neq -4, \dfrac{-5}{2}\right)$

Chapter Summary

In this chapter, you were introduced to polynomials as special types of algebraic expressions. Basic terminology associated with polynomials was discussed. You learned to add, subtract, multiply, and divide polynomials.

Review Exercises

The **key words** introduced in this chapter are listed here in the order in which they appeared. You should know what they mean before proceeding to the next chapter.

polynomial (p. 422)
monomial (p. 423)
binomial (p. 423)
trinomial (p. 423)
degree of a term (p. 424)
degree of a polynomial (p. 424)

descending powers (p. 425)
addition of polynomials (p. 426)
subtraction of polynomials (p. 428)

multiplication of polynomials (p. 431)
FOIL method (p. 436)
division of polynomials (p. 441)

Review Exercises

10.1 Basic Terminology Associated with Polynomials

In Exercises 1–8, indicate whether the given algebraic expression is or is not a polynomial. If it is a polynomial, indicate in what variable(s).

1. $3v - 4$
2. $2x^{1/2}$
3. $y^3 - 4y$
4. $\dfrac{2}{u - 3}$
5. $u^2v - uv^3 + 1$
6. $p^{-3}q - 4p^2q^{-2}$
7. $\dfrac{w^3 - 1}{9}$
8. $4m^2(5m - 2n)$

In Exercises 7–14, write the degree for each term in the given polynomial. Then, determine the degree of the polynomial.

9. $2x - 3x^4 + 7x^3$
10. $u^3 - 5u + 1 - u^4$
11. $4y^5 - 2y + 11y^2 - 5y^4$
12. $s^3t^2 - 3s^2t^3 + s^5$
13. $2x^3y - 3x^2y^3 - 5x^3y^3$
14. $(3u^2v)^3 - (2uv^3)^2$

In Exercises 15–18, rewrite the given polynomial in descending powers of the indicated variable.

15. $4u - 3u^3 + 7 - u^5 - 6u^4$; u
16. $3s^7 - 2s^3 + 3s^2 - s^4 + 1 - 7s^5$; s
17. $m^2n^3 - 7m^3n + 8mn^4 + 6m^3n^2 - 8m^2n^2$; m
18. $m^2n^3 - 7m^3n + 8mn^4 + 6m^3n^2 - 8m^2n^2$; n

10.2 Addition of Polynomials

In Exercises 19–26, add the given polynomials.

19. $2x^3 - 4x + 5x^2 - 7$; $x^2 - 9x - 5x^3$
20. $y^3 - 2y + 1$; $y^2 - 3y - 5$; $2 - y^2 - 5y^3$
21. $u^4 + 3u^2 - 7$; $u^3 - 2u + 1$; $u^2 - 5u^4 - u$; $3u^2 - 13$
22. $x^2y^3 - 4xy^4 - 5x^3y$; $2x^3y - xy^4 - 3x^2y^3$; $xy^3 - x^3y$
23. $1 - 3q^3 + 7q^4$; $3q^4 - 2q^2 + 11$
24. $5u^3 - 4u^2 + u + 5$; $u^3 - 2u^2 + 4u - 4$; $-2u^3 + u^2 - 5u + 1$
25. $s^4 - 4s^2 + 2 - 2s^7$; $s^6 - 3s^4 - 2s^2 + 1$; $s^7 - 3s^6 + s^2 - 6$
26. $6pq^2 - 2p^2q^2 - 5p^2q$; $p^2q - 3pq^2 + 4p^2q^2$

10.3 Subtraction of Polynomials

In Exercises 27–34, subtract the polynomials as indicated.

27. Subtract $3u^5 - 3u^3 + 2u^2 - 4u$ from $2u^4 - u^5 - 3u^2 + 1$.
28. Subtract $x^3y^4 - 2x^2y^5 + 3x^4y^2$ from $4x^2y - 3x^2y^5 - 5x^3y^4$.
29. From $3s^7 + 4s^2 + s^6 - 7s + s^4$, subtract $2s^5 - s^6 - 5s^3 + 2$.
30. From $p^3q - 2p^2q^2 - 4pq^7$, subtract $2pq - 3p^2q^2 + 5p^3q^4$.

31. Subtract $5v^3 - 2v^2 + 7v - 11$ from $7v^2 - 5v + 7$.
32. Subtract $m^3n^2 - 3m^2n^3 + 7m^2n^2$ from $8m^2n^2 - 5m^3n^2 - m^4n^3$.
33. From $y^7 - y^5 + y^4 - 3y^2$, subtract $y^6 - 5y^4 + y^3 - 7y + 1$.
34. From $u^4v^5 - 2u^2v^3 + 4u - 1$, subtract $3u^2v^3 - 5v + 6 - 4u$.

10.4 Multiplication of Polynomials

In Exercises 35–42, multiply as indicated.

35. $2x(3x + 1)$
36. $2y^2(3y^2 - 4y + 1)$
37. $(2u - 3)(4u + 5)$
38. $(2p - 3q)(2p - 3q)$
39. $(2p - 3q)[2(4p - q) - (2q - 5p)]$
40. $(2x - 5)[(x - 3) - (1 - 2x)]$
41. $(2s - 3t)(2s^2 - 4st + 4t^2)$
42. $(u^2 - 3u + 4)(2u^3 - u^2 + 3u - 1)$

10.5 Multiplying Binomials

In Exercises 43–46, multiply using the FOIL method.

43. $(u - 3)(u + 4)$
44. $(2y - 3)(3y - 5)$
45. $(1 - 3w)(2 - 5w)$
46. $(2.1x + 1)(3.2x - 3)$

In Exercises 47–50, square the given binomial.

47. $y + 7$
48. $2p - 3$
49. $5p - 4q$
50. $3x + 4y$

10.6 Division of Polynomials

51. Divide $6u^4 - 4u^2 + 10$ by 2.
52. Divide $7s^3 - 4s^2 + 5s$ by $10s$, if $s \neq 0$.
53. Divide $2p^2q^2 - 3p^2q^4 + p^3q^2$ by $4pq$, if $p \neq 0$ and $q \neq 0$.
54. Divide $u^3 - 8$ by $u - 2$, if $u \neq 2$.
55. Divide $s^2 - 6s - 27$ by $s - 9$, if $s \neq 9$.
56. Divide $3y^5 - 4y^4 + y^2 - y + 11$ by $3y - 1$, if $y \neq \frac{1}{3}$.

Teasers

1. The dimensions of a rectangular region are $2x - 7$ yd and $5x + 2$ yd. Determine the perimeter of the region.
2. For the rectangular region in Exercise 1, can x be equal to 3? Explain your answer.
3. Determine the area of the region in Exercise 1.
4. The sides of a triangular region measure $y^3 + 1$ m, $2y^2 + 5y$ m, and $9y - 2$ m. Determine the perimeter of the region.
5. Determine the perimeter of the triangular region in Exercise 4, if $y = 1.2$.
6. Perform the indicated operations and simplify your results:

$$(2p^2 + 3p - 1)[(2p - 3)(4 - p) - (p^2 - 3p - 2)]$$

7. Perform the indicated operations and simplify your results:

$$[(2r - 3s)(3r + 2s) - (r + 5s)(4s - 2r)]^2$$

8. Perform the indicated operations and simplify your results:

$$[(2p^2q^2 - 3p^2q^4 + p^3q^2) \div 4pq] - (3pq + 4p^2q^3 - 5p^2q)$$

9. Without actually substituting, determine whether 0 is a solution of the equation $3x^4 - 5x^3 + 2x = 0$. Explain your answer.
10. Without actually substituting, determine whether 0 is a solution of the equation $y^3 - 2y + 5 - y^6 = 0$. Explain your answer.

CHAPTER 10 TEST

This test covers material from Chapter 10. Read each question carefully, and answer all questions. If you miss a question, review the appropriate section of the text.

1. What is the degree of the second term in the polynomial $2x^3y + 4x^2y^3 - 7x^3y^3$?
2. What is the degree of the polynomial $2u^3v - 5u^2v^2 + 6u^3v^4$?
3. Rewrite the polynomial $u^2 - 3 - u^3 + 4u + 5u^4$ in descending powers of u.
4. Rewrite the polynomial $m^2n^3 - 7m^3n + 8mn^4 + 6m^3n^2 - 8m^2n^2$ in descending powers of n.

In Exercises 5–8, perform the indicated operations.

5. $(3x^2 - 5x + x^3 - 9) + (4x - 5x^3 + 9 - 3x^2)$
6. $(2s^2t + 3st^2 - s^2t^2) + (2s^2t - 3s^2t^2 - 5st)$
7. $(2u - u^2 + 7 - 4u^3) - (2u^2 - 5u^3 + 7 - 9u)$
8. $(t^6 - 5t^4 + 6t^2 - 2) - (5t^4 - t^2 + 7t + 5t^3)$
9. Simplify by combining like terms: $(4y^2 - 3y + y^4 - 6) + (7 - 3y^3 + 7y^2 + 6y) - (2y^3 - y^4 + 9y - 16)$.

In Exercises 10–13, multiply the polynomials.

10. $2a^2b(3ab^3 - a^3b^2)$
11. $(2s - 1)(2 - 3s)$
12. $(x^2 + y)(x - 2y)$
13. $(3x^4 - 5x^2 + 6)(2x^3 - 7x - 2)$
14. Square the binomial $2x + 5y$.

In Exercises 15–20, divide as indicated.

15. $(7u^2 - 5u + 11) \div (-6)$
16. $(5y^6 - 2y^5 + y^3 - 2y + 10) \div (3y)$, if $y \neq 0$
17. $(9x^3y^2 - 6x^2y^4 + 27x^5y^3) \div (3x^2y^2)$, if $x \neq 0$ and $y \neq 0$
18. $(6y^2 - 13y - 5) \div (2y - 5)$, if $y \neq \dfrac{5}{2}$
19. $(v^3 + 2v^2 + 9) \div (v + 3)$, if $v \neq -3$
20. $(u^4 - 3u^2 + 7) \div (u - 4)$, if $u \neq 4$

CUMULATIVE REVIEW: CHAPTERS 1–10

In Exercises 1–6, perform the indicated operations.

1. $\dfrac{2}{11} - \dfrac{3}{14}$
2. $12.04 - 3.9 - 15.07$
3. $9.16\% \times (-3)^3$
4. $-2(5 - 9) - (2 - 7)^3 \div (-5^2)$
5. $(-1.6)(2.1) - (1.44) \div (0.3)^2$
6. $|-6.36| - 3.01 \times 6.02\%$

In Exercises 7–8, evaluate each of the given expressions.

7. $a^2b - 3ab^3$, if $a = \dfrac{1}{2}$ and $b = \dfrac{-1}{8}$
8. $3xy - 2x^2 - 4(x - 2y)$, if $x = -2$ and $y = -3$

In Exercises 9–11, round as indicated.

9. $9\dfrac{1}{7}$, to the nearest tenth
10. 50.6074, to the nearest hundredth
11. 329,473, to two significant digits
12. Rewrite 2049 in scientific notation.
13. Rewrite 0.0004369 in scientific notation.

In Exercises 14–16, simplify by combining like terms.

14. $2(3u - 2v) - 3(v + 4u) + 7(3u - 5v)$
15. $2x(x^2 - 3x + 1) + x^3(2x - 3) - (2x - 1)(x^2 + 1)$
16. $3p^3(2p^2 - 3p + 1) - (2p + 3)^2 - p(p + 1)(p^2 - 1)$

In Exercises 17–18, solve the given equation for the indicated variable.

17. $4r - 5 = 7 - 9r$, for r
18. $2(x - 3) + 3x = 3 - 4(1 - 2x)$, for x
19. Determine the perimeter (in centimeters) of a triangle whose sides measure 16.3 cm, 2.1 dm, and 14.9 cm.
20. Write an algebraic expression for the cube of the sum of twice x and 3 times y.
21. Which is larger: $30.9\% \times 14.2$ or $\dfrac{2}{3} \times \dfrac{1}{5}$?
22. Determine the area of a rectangular region that is $16\dfrac{1}{3}$ yd wide and $23\dfrac{1}{5}$ yd long.
23. Determine the number N for 5.09×10^{-5}.
24. Solve the proportion $\dfrac{4}{5} = \dfrac{n}{7}$ for n.
25. Convert the decimal 209.6 to a simplified fraction.

CHAPTER 11: Factoring and Special Products

In this chapter, we will discuss factoring algebraic expressions and working with special products. Prime and composite positive integers and the prime factorization of positive integers will also be reviewed. This review is necessary since prime factorization is involved in factoring algebraic expressions.

The distributive property for multiplication over addition will be used in factoring that involves a common monomial factor. Other types of factoring will include factoring the difference of two squares, factoring some quadratic trinomials, and factoring by grouping.

The chapter will conclude with the solution of word problems, using the method of factoring to solve the associated equations.

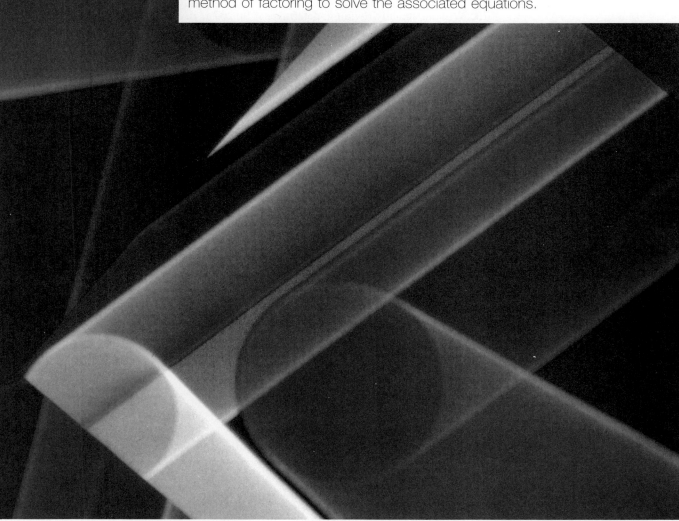

CHAPTER 11 PRETEST

This pretest covers materials on factoring and special products. Read each question carefully, and try to answer all questions. Each question is keyed to the chapter section in which the particular topic is discussed. Check your answers with those given in the back of the text.

Questions 1–3 pertain to Section 11.1.

1. Determine a prime factorization for 266.

2. Without actually dividing, determine if 46,107 is divisible by 9.

3. Without actually dividing, determine if 96,207 is divisible by 4.

Questions 4–7 pertain to Section 11.2. Factor each of the given expressions.

4. $26t - 18$

5. $6p^3 - 18p^2$

6. $27a^2bc^4 - 81a^3b^2c^3$

7. $20s^4t^3u^2 - 15s^2t^4u^5 + 35s^3t^4u^2$

Questions 8–11 pertain to Section 11.3. Factor each of the given expressions.

8. $y^2 - 81$

9. $2u^2 - 50$

10. $v^4 - w^4$

11. $a^5b - ab^5$

Questions 12–13 pertain to Section 11.4. Factor each of the given expressions.

12. $y^2 - y - 42$

13. $w^2 - 6w - 187$

Questions 14–17 pertain to Section 11.5. Factor each of the given expressions.

14. $6x^2 + 5x - 4$

15. $3 - 8y + 4y^2$

16. $7p^2 + 28p - 35$

17. $6t^2 - 15t - 9$

Questions 18–19 pertain to Section 11.6. Factor each of the given expressions by grouping.

18. $xy - 3x - 2y + 6$

19. $m^2 + 2mn - 2n - m$

Questions 20–21 pertain to Section 11.7. Factor each of the given expressions completely.

20. $2x^2y + 10xy + 12y$

21. $2x^3 + 3x^2 - 2x - 3$

Questions 22–24 pertain to Section 11.8. Solve each of the following equations by factoring.

22. $2y^2 + 3y = 2$

23. $(y^2 - 9)(3y - 4) = 0$

24. $12 - 32p = -16p^2$

Question 25 pertains to Section 11.9.

25. The larger of two positive integers is 4 less than 3 times the smaller integer. If their product is 207, what are the numbers?

11.1 Review of Prime Factorization

After completing this section, you should be able to determine the prime factorization for a positive integer, greater than 1, using tests for divisibility by 2, 3, 4, 5, 6, 8, 9, and 10

In this chapter, you will learn to rewrite an algebraic expression as a product of two or more expressions.

DEFINITION

Factoring an algebraic expression means to find two or more expressions whose product is the given expression.

In Section 10.4, you learned to multiply polynomials. In order to multiply a polynomial by a monomial, the distributive property was used. For example,

$$2y^2(3y^2 - 4y + 1) = 6y^4 - 8y^3 + 2y^2$$

The expressions $2y^2$ and $3y^2 - 4y + 1$ are *factors* of the polynomial $6y^4 - 8y^3 + 2y^2$.

To factor the expression $6y^4 - 8y^3 + 2y^2$, we use the distributive property but in reverse. Hence,

$$6y^4 - 8y^3 + 2y^2 = 2y^2(3y^2 - 4y + 1)$$

The monomial $2y^2$ is the greatest common *divisor* of $6y^4$, $-8y^3$, and $2y^2$. The monomial $2y^2$ is called the "greatest common factor" of the polynomial $6y^4 - 8y^3 + 2y^2$. Finding the greatest common factor of an algebraic expression will be discussed in Section 11.2.

When we factor an algebraic expression, we always look for prime factors. We will start, then, with a review of prime and composite numbers and the prime factorization of positive numbers. Recall that if N is a positive integer greater than 1, then:

- N is a *prime* number if the only positive divisors of N are N and 1.
- N is a *composite* number if N is not prime.
- The number 1 is neither prime nor composite.
- Every *composite* positive integer, greater than 1, can be written as a product of prime factors. The factorization, except for the order in which the factors are written, is unique. If the integer is *prime*, it can be written as the product of 1 and itself.

Without actually dividing, determine the following.

Practice Exercise 1. Determine a prime factorization of 264.

EXAMPLE 1 Determine a prime factorization of 42.

SOLUTION: The prime factorization of 42 is

$$42 = 2 \times 3 \times 7$$

Practice Exercise 2. Determine a prime factorization of 485.

EXAMPLE 2 Determine a prime factorization of 484.

11.1 Review of Prime Factorization

SOLUTION: The prime factorization of 484 is

$$484 = 2 \times 2 \times 11 \times 11$$
$$= 2^2 \times 11^2$$

When finding the prime factorization of a positive integer, there are certain tests for divisibility that can be used. Some of these tests were given in Section 2.3. Those tests and some additional ones are listed in the box.

Divisibility Tests

Let N be a positive integer. The following tests apply:

- *Divisibility by 2:* N is divisible by 2 if the ones digit of N is 0, 2, 4, 6, or 8. (A number divisible by 2 is said to be *even*.)
- *Divisibility by 3:* N is divisible by 3 if the sum of the digits of N is divisible by 3.
- *Divisibility by 4:* N is divisible by 4 if the two-digit number formed by using the tens and ones digits of N is divisible by 4.
- *Divisibility by 5:* N is divisible by 5 if the ones digit of N is 0 or 5.
- *Divisibility by 6:* N is divisible by 6 if N is even and is also divisible by 3.
- *Divisibility by 8:* N is divisible by 8 if the 3-digit number formed by using the hundreds, tens, and ones digits of N is divisible by 8.
- *Divisibility by 9:* N is divisible by 9 if the sum of the digits of N is divisible by 9.
- *Divisibility by 10:* N is divisible by 10 if the ones digit of N is 0.

Without actually dividing, determine the following.

Practice Exercise 3. Is 23,569 divisible by 3?

EXAMPLE 3 Determine whether each of the following numbers is divisible by 2: (a) 164; (b) 437.

SOLUTION: (a) 164 is divisible by 2 since the ones digit is 4; (b) 437 is *not* divisible by 2 since the ones digit is *not* 0, 2, 4, 6, or 8.

Practice Exercise 4. Is 60,708 divisible by 4?

EXAMPLE 4 Determine whether each of the following numbers is divisible by 3: (a) 123; (b) 1675.

SOLUTION: (a) 123 is divisible by 3 since $1 + 2 + 3 = 6$ is divisible by 3; (b) 1675 is *not* divisible by 3 since $1 + 6 + 7 + 5 = 19$ is *not* divisible by 3.

Practice Exercise 5. Is 123,498 divisible by 6?

EXAMPLE 5 Determine whether each of the following numbers is divisible by 4: (a) 1328; (b) 4518.

SOLUTION: (a) 1328 is divisible by 4 since 28 is divisible by 4; (b) 4518 is *not* divisible by 4 since 18 is *not* divisible by 4.

Practice Exercise 6. Is 730,972 divisible by 8?

EXAMPLE 6 Determine whether each of the following numbers is divisible by 5: (a) 2055; (b) 41,234.

SOLUTION: (a) 2055 is divisible by 5 since the ones digit is 5; (b) 41,234 is *not* divisible by 5 since the ones digit is *not* 0 or 5.

Practice Exercise 7. Is 42,516,171 divisible by 9?

EXAMPLE 7 Determine whether each of the following numbers is divisible by 6: (a) 492; (b) 587; (c) 3128.

SOLUTION: (a) 492 is divisible by 6 since 492 is even and 492 is divisible by 3 (since $4 + 9 + 2 = 15$ is divisible by 3); (b) 587 is *not* divisible by 6 since 587 is *not* even; (c) 3128 is *not* divisible by 6 since 3128 is *not* divisible by 3 (since $3 + 1 + 2 + 8 = 14$ is *not* divisible by 3).

Practice Exercise 8. Is 999,876 divisible by 10?

EXAMPLE 8 Determine whether each of the following numbers is divisible by 8: (a) 42,128; (b) 53,109.

SOLUTION: (a) 42,128 is divisible by 8 since 128 is divisible by 8; (b) 53,109 is *not* divisible by 8 since 109 is *not* divisible by 8.

Practice Exercise 9. Is 362,012 divisible by 2?

EXAMPLE 9 Determine whether each of the following numbers is divisible by 9: (a) 369; (b) 2845.

SOLUTION: (a) 369 is divisible by 9 since $3 + 6 + 9 = 18$ is divisible by 9; (b) 2845 is *not* divisible by 9 since $2 + 8 + 4 + 5 = 19$ is *not* divisible by 9.

Practice Exercise 10. Is 2,016,987 divisible by 3?

EXAMPLE 10 Determine whether each of the following numbers is divisible by 10: (a) 6220; (b) 8725.

SOLUTION: (a) 6220 is divisible by 10 since the ones digit is 0; (b) 8725 is *not* divisible by 10 since the ones digit is *not* 0.

ANSWERS TO PRACTICE EXERCISES

1. $2^3 \times 3 \times 11$. 2. 5×97. 3. No. 4. Yes. 5. Yes. 6. No. 7. Yes. 8. No. 9. Yes. 10. Yes.

Prime factorization will be used in the next section in determining the greatest common factor of two or more algebraic expressions.

Practice Exercise 11. Determine all of the positive integer factors of 84.

EXAMPLE 11 Determine all of the positive integer factors of 28.

SOLUTION: The positive integer factors of 28 are 1, 2, 4, 7, 14, and 28 since each of these positive integers divides 28.

Practice Exercise 12. Determine all of the positive integer factors of 144.

EXAMPLE 12 Determine all of the positive integer factors of 85.

SOLUTION: The positive integer factors of 85 are 1, 5, 17, and 85 since each of these positive integers divides 85.

ANSWERS TO PRACTICE EXERCISES

11. 1, 2, 3, 4, 6, 7, 12, 14, 21, 28, 42, 84.   12. 1, 2, 3, 4, 6, 8, 9, 12, 16, 18, 24, 36, 48, 72, 144.

EXERCISES 11.1

In Exercises 1–12, determine all of the positive integers that are divisors for each of the given numbers.

1. 27
2. 32
3. 40
4. 50
5. 57
6. 62
7. 75
8. 93
9. 110
10. 320
11. 450
12. 684

In Exercises 13–24, write a prime factorization for each of the given numbers.

13. 78	**14.** 99	**15.** 120	**16.** 135
17. 144	**18.** 210	**19.** 761	**20.** 1462
21. 4090	**22.** 5633	**23.** 6194	**24.** 8649

In Exercises 25–32, test each of the given numbers for divisibility by 2.

25. 38	**26.** 141	**27.** 250	**28.** 469
29. 1234	**30.** 2021	**31.** 6666	**32.** 30,405

In Exercises 33–40, test each of the given numbers for divisibility by 3.

33. 48	**34.** 97	**35.** 465	**36.** 2964
37. 54,954	**38.** 87,903	**39.** 174,668	**40.** 317,946

In Exercises 41–48, test each of the given numbers for divisibility by 4.

41. 56	**42.** 98	**43.** 116	**44.** 232
45. 962	**46.** 2988	**47.** 13,752	**48.** 201,692

In Exercises 49–52, test each of the given numbers for divisibility by 5.

49. 203	**50.** 965	**51.** 13,005	**52.** 27,986

In Exercises 53–60, test each of the given numbers for divisibility by 6.

53. 239	**54.** 468	**55.** 972	**56.** 1031
57. 6123	**58.** 7452	**59.** 12,315	**60.** 23,904

In Exercises 61–68, test each of the given numbers for divisibility by 8.

61. 1896	**62.** 2072	**63.** 5682	**64.** 7912
65. 9130	**66.** 10,560	**67.** 17,844	**68.** 25,648

In Exercises 69–76, test each of the given numbers for divisibility by 9.

69. 369	**70.** 4572	**71.** 5678	**72.** 6318
73. 9876	**74.** 12,798	**75.** 35,973	**76.** 234,567

In Exercises 77–80, test each of the given numbers for divisibility by 10.

77. 18,230	**78.** 20,003	**79.** 45,998	**80.** 54,860

11.2 Common Monomial Factoring

OBJECTIVE

After completing this section, you should be able to factor a polynomial all of whose terms have a common factor.

The distributive property for multiplication over addition involves both a product and a sum. When using this property, we generally start with the factors and form the product. For instance,

$$3(y + 4) = 3y + 12$$

indicates that the product of the factors 3 and $y + 4$ is $3y + 12$, formed by multiplying each term of $y + 4$ by the monomial 3. However,

$$3y + 12 = 3(y + 4)$$

indicates that the sum of $3y + 12$ can be expressed as a product of two factors. Also, note that 3 is a common monomial factor in *each* term of the sum.

Factor each of the following expressions.

Practice Exercise 1. $3x - 6$

EXAMPLE 1 Factor: $4u + 10$.

SOLUTION: To factor the expression $4u + 10$, note that

$$4u + 10 = (2)(2)(u) + (2)(5) \quad \text{(Prime factorization of 4 and 10)}$$
$$\qquad\qquad\qquad\text{(2 is a common factor)}$$
$$= 2(2u + 5) \qquad \text{(Solution)}$$

Therefore, $4u + 10 = 2(2u + 5)$.

CHECK: $2(2u + 5) = (2)(2u) + (2)(5)$
$$= 4u + 10 \checkmark$$

In Example 1, note that 2 is a factor of $4u + 10$. To obtain the other factor of $4u + 10$, we could simply divide $(4u + 10)$ by 2. In general, if a is a factor of n, then $(n \div a)$ is another factor of n.

Practice Exercise 2.
$15u^2 - 20u^3$

EXAMPLE 2 Factor: $9y^2 - 5y$.

SOLUTION: To factor the expression $9y^2 - 5y$, note that

$$9y^2 - 15y = (3)(3)(y)(y) - (3)(5)(y) \quad \text{[Prime factorizations; also } y^2 \text{ means } (y)(y)\text{]}$$
$$\qquad\qquad\qquad\text{(3 is a common factor.)}$$
$$\qquad\qquad\qquad\text{(y is a common factor.)}$$
$$= 3y(3y - 5) \qquad \text{(Solution)}$$

Therefore, $9y^2 - 15y = 3y(3y - 5)$.

CHECK: $3y(3y - 5) = (3y)(3y) + (3y)(-5)$
$$= 9y^2 - 15y \checkmark$$

Again, in Example 2, note that $3y$ is a factor of $9y^2 - 15y$. Also, $(9y^2 - 15y) \div (3y) = 3y - 5$ is another factor of $9y^2 - 15y$.

Practice Exercise 3.
$18u^3v^4 - 36u^2v^3 + 30u^4v^6$

EXAMPLE 3 Factor: $2x^2y^2 - 4xy^2 + 8x^2y^3$.

SOLUTION: To factor the expression $2x^2y^2 - 4xy^2 + 8x^2y^3$, note that

$$2x^2y^2 - 4xy^2 + 8x^2y^3 = (2)(x^2)(y^2) - (2^2)(x)(y^2) + (2^3)(x^2)(y^3)$$

Observe that:

- 2 is the greatest common numerical factor in the three terms.
- The greatest power of x that is common to all three terms is x.
- The greatest power of y that is common to all three terms is y^2.

Therefore, $2xy^2$ is the greatest common factor for all three terms. Hence, $2x^2y^2 - 4xy^2 + 8x^2y^3 = 2xy^2(x - 2 + 4xy)$.

11.2 Common Monomial Factoring

CHECK: $2xy^2(x - 2 + 4xy) = (2xy^2)(x) + (2xy^2)(-2) + (2xy^2)(4xy)$
$= 2x^2y^2 - 4xy^2 + 8x^2y^3$ ✓

> **ANSWERS TO PRACTICE EXERCISES**
> 1. $3(x - 2)$. 2. $5u^2(3 - 4u)$. 3. $6u^2v^3(3uv - 6 + 5u^2v^3)$.

Let's look at the three preceding examples again. In Example 1, when the expression $4u + 10$ was factored, we noted that 2 was the only common factor for the two terms of the given expression. In Example 2, when $9y^2 - 15y$ was factored, we noted that both 3 and y were common factors of the two terms of the expression; therefore, the product, $3y$, is also a common factor. The factor $3y$ is the *greatest common factor* of the two terms in the given expression.

Similarly, in Example 3, when the expression $2x^2y^2 - 4xy^2 + 8x^2y^3$ was factored, we noted that 2, x, and y^2 were all common factors of the three terms of the given expression and that $2xy^2$ was the *greatest* common factor.

Whenever we look for a common factor in the terms of an expression, we should always look for the greatest common factor.

> **DEFINITION**
> The **greatest common factor (GCF)** of two or more terms in an algebraic expression is the greatest factor that is common to *all* of the terms of the expression.

We will write the greatest common factor of a and b as GCF(a,b).

Determine the GCF for each of the following.

Practice Exercise 4. 6, 12, 42

EXAMPLE 4 Determine: GCF(6,10).

SOLUTION: Since $6 = 2 \times 3$ and $10 = 2 \times 5$, then the GCF(6,10) = 2.

Practice Exercise 5. 30, 75, 90

EXAMPLE 5 Determine: GCF(8,12,16).

SOLUTION: Since $8 = 4 \times 2$, $12 = 4 \times 3$, and $16 = 4 \times 4$, then the GCF(8,12,16) = 4.

Practice Exercise 6. $15xy^3$, $90x^2y^4$

EXAMPLE 6 Determine: GCF($2x, 6x^3$).

SOLUTION: Since $2x = (2x)(1)$ and $6x^3 = (2x)(3x^2)$, then the GCF($2x,6x^3$) = $2x$.

Practice Exercise 7. $32a^2b$, $24a^3b^2$, $40b^3$

EXAMPLE 7 Determine: GCF($3x^2y, 4xy^3$).

SOLUTION: Since $3x^2y = (xy)(3x)$ and $4xy^3 = (xy)(4y^2)$, then the GCF($3x^2y, 4xy^3$) = xy.

Rule for Finding the Greatest Common Factor, GCF

To find the greatest common factor of two or more expressions, use the following procedure.

1. Determine the prime factorization of *each* term of the algebraic expression. Write repeated factors using exponents.
2. Form the product of all the *different* prime factors appearing in Step 1.

3. Raise each of the prime factors in the product to the *lowest* power in which it is found in the prime factorization in Step 1.
4. The GCF is the product of all the powers determined in Step 3.

Practice Exercise 8. $18u^3v^2w$, $14u^2v^4w^5$, $24u^4v^3w$

EXAMPLE 8 Determine: GCF(42,63,210).

SOLUTION:

STEP 1: Determine the prime factorization of each number:

$$42 = 2 \times 3 \times 7$$
$$63 = 3^2 \times 7$$
$$210 = 2 \times 3 \times 5 \times 7$$

STEP 2: Form the product of all the *different* prime factors determined in Step 1:

$$2 \times 3 \times 5 \times 7$$

STEP 3: Raise *each* of the prime factors in the product in Step 2 to the *lowest* power in which it is found in the prime factorizations in Step 1:

$$2^0 \times 3^1 \times 5^0 \times 7^1$$

STEP 4: The GCF(42,63,210) = $1 \times 3 \times 1 \times 7 = 21$.

Practice Exercise 9. $7s^2t^3$, $21s^2t^2$, $49s^4t^5$, $14s^6t^4$

EXAMPLE 9 Determine: GCF($6y^4, 12y^3, 42y^2$).

SOLUTION:

STEP 1: Determine the prime factorization of each expression:

$$6y^4 = (2)(3)(y^4)$$
$$12y^3 = (2^2)(3)(y^3)$$
$$42y^2 = (2)(3)(7)(y^2)$$

STEP 2: Form the product of all the *different* prime factors determined in Step 1:

$$(2)(3)(7)(y)$$

STEP 3: Raise *each* of the prime factors in the product in Step 2 to the *lowest* power in which it is found in the prime factorizations in Step 1:

$$(2^1)(3^1)(7^0)(y^2)$$

STEP 4: The GCF($6y^4, 12y^3, 42y^2$) = $(2)(3)(1)(y^2) = 6y^2$.

ANSWERS TO PRACTICE EXERCISES

4. 6.   5. 15.   6. $15xy^3$ 7. $8b$.   8. $2u^2v^2w$.   9. $7s^2t^2$.

We will now consider factoring polynomials all of whose terms have a common factor.

11.2 Common Monomial Factoring

Rule for Factoring a Polynomial

To factor a polynomial *all* of whose terms have a common factor:

1. Determine the greatest common factor, GCF, for all of the terms of the polynomial.
2. Divide the polynomial by the GCF to determine the second factor.

Practice Exercise 10. Factor $5x^3 - 10x^2 + 35x^4$.

EXAMPLE 10 Factor $6y^4 - 12y^3 + 42y^2$.

SOLUTION:

Step 1: First, determine that the $\text{GCF}(6y^4, 12y^3, 42y^2) = 6y^2$ (from Example 9).

Step 2: Divide the given polynomial by the GCF:

$$\frac{6y^4 - 12y^3 + 42y^2}{6y^2} = y^2 - 2y + 7$$

Step 3: Therefore, $6y^4 - 12y^3 + 42y^2 = 6y^2(y^2 - 2y + 7)$.

Practice Exercise 11. Factor $6st^2 - 24s^3t^3 + 12s^3t^4$.

EXAMPLE 11 Factor $4x^4y - 6x^3y^2 + 9x^2y^4$.

SOLUTION:

Step 1: First, determine that the $\text{GCF}(4x^4y, 6x^3y^2, 9x^2y^4) = x^2y$. (Verify this.)

Step 2: Divide the given polynomial by the GCF:

$$\frac{4x^4y - 6x^3y^2 + 9x^2y^4}{x^2y} = 4x^2 - 6xy + 9y^3$$

Step 3: Therefore, $4x^4y - 6x^3y^2 + 9x^2y^4 = x^2y(4x^2 - 6xy + 9y^3)$.

Practice Exercise 12. Factor $12a^3b^2 - 108a^2b^3 + 48a^4b^5 - 36a^2b^7$.

EXAMPLE 12 Factor $24u^3v^2 - 36u^2v^4 + 27u^3v^4 - 42u^4v^3$.

SOLUTION:

Step 1: First, determine that the $\text{GCF}(24u^3v^2, 36u^2v^4, 27u^3v^4, 42u^4v^3) = 3u^2v^2$. (Verify this.)

Step 2: Divide the given polynomial by the GCF:

$$\frac{24u^3v^2 - 36u^2v^4 + 27u^3v^4 - 42u^4v^3}{3u^2v^2} = 8u - 12v^2 + 9uv^2 - 14u^2v$$

Step 3: Therefore, $24u^3v^2 - 36u^2v^4 + 27u^3v^4 - 42u^4v^3 = 3u^2v^2(8u - 12v^2 + 9uv^2 - 14u^2v)$.

ANSWERS TO PRACTICE EXERCISES

10. $5x^2(x - 2 + 7x^2)$. **11.** $6st^2(1 - 4s^2t + 2s^2t^2)$. **12.** $12a^2b^2(a - 9b + 4a^2b^3 - 3b^5)$.

EXERCISES 11.2

Factor each of the following expressions by first determining the greatest monomial factor.

1. $2x - 4$
2. $3y + 9$
3. $5u + 60$
4. $6t - 15$
5. $21a - 49$
6. $36b + 84$
7. $3x^2 - 51x$
8. $8y^2 + 24y$
9. $5u^3 + 20u^2$
10. $6t^4 + 27t^7$
11. $2x^2y - 4xy^2$
12. $3m^2n^3 + 9m^3n^4$
13. $7p^4q^3 - 21p^2q$
14. $10c^2d^4 - 25c^3d^3$
15. $a^2b^3 - 2ab^2$
16. $12m^3n^2 - 18m^4n^3$
17. $27s^4t + 81st^4$
18. $4p^3q^2 + 9p^2q$
19. $15a^2b^5 + 60a^4b^3$
20. $33x^4y - 121xy^2$
21. $a^2b - 3ab^2 + 4a^2b^2$
22. $2c^4d^2 - 4c^3d^2 + 8c^2d^4$
23. $4x^3y - 5xy^2 + 6x^3y^2$
24. $36r^3s^4 - 72r^2s^3 + 60r^4s^6$
25. $33u^4v^6 + 44u^3v^5 - 66u^5v^3$
26. $48s^2t^3 - 12s^3t^3 + 18s^4t^6$
27. $10a^2b - 15a + 25b^3$
28. $6t^2 - 18s^3 + 24s^2t^3$
29. $24m^4n^6 - 32m^5n^4 + 48m^7n^9$
30. $6y^3z^2 - 8y^2z^3 + 12y^4z$
31. $6a^3b^2c^4 - 9a^2b^5c^3$
32. $4u^4v^7w^9 - 9u^3v^4w^6$
33. $16p^3q^2t - 18pq^4t^3$
34. $10c^4d^5e^3 - 25c^2de^7$
35. $32x^2y^3z^2 - 36xy^2z^6$
36. $44m^4n^2p^3 + 24m^2n^3p^4$
37. $2a^2b^3 - 3a^3c^4 + 4a^5b^5$
38. $10x^3y^4 - 25x^2z^3 + 40y^3z^5$
39. $12s^3t^4u^5 + 24s^2t^3u - 36$
40. $6rs^2t^4 - 29r^3st^5 + 27r^2s^4t^2$
41. $4a - 12b + 18c - 20d + 24ac$
42. $2x^2y^3 - 6xy^2 + 8x^3y^2 - 10x^3y^4$
43. $6a^2b^3 - 9ab^2 + 12a^3b^4 - 36a^4b$
44. $10u^3v^2 + 15u^2v^2 - 20u^4v^3 + 50u^3v^5$
45. $6r^3s^2 - 9r^4s^2 + 24r^2s^4 - 42r^5s^4$
46. $16r^3s + 48rt^3 - 32s^4t^3 - 96rs^2t^3$
47. $6a^2b^2c^3 - 12ab^3c^2 - 18a^3bc^4 - 36a^3b^4c^6$
48. $7m^2n^4p^3 + 12m^3n^4p^2 + 49m^4n^2p^3 - 98m^5n^7p^9$
49. $12a^3b^4 - 24a^2b^3 + 48a^4b^5 - 96a^2b^5 + 192a^5b^7$
50. $18y^2z^3 - 12y^4z^2 - 36y^5z^7 - 48y^3z^7 - 90y^9z^6$

11.3 Factoring the Difference of Two Squares

OBJECTIVE

After completing this section, you should be able to factor a binomial that is the difference of two squares.

Before learning to factor a binomial that is the difference of two squares, we will look at some special products.

Multiply.

Practice Exercise 1.
$(x + 7)(x - 7)$

EXAMPLE 1 Multiply: $(x + 2)(x - 2)$.

11.3 Factoring the Difference of Two Squares

SOLUTION: Using the FOIL method, determine that

$$(x + 2)(x - 2) = x^2 - 2x + 2x - 4$$
$$= x^2 - 4$$

Practice Exercise 2.
$(2x + 5)(2x - 5)$

EXAMPLE 2 Multiply: $(y - 3)(y + 3)$.

SOLUTION: Using the FOIL method, determine that

$$(y - 3)(y + 3) = y^2 + 3y - 3y - 9$$
$$= y^2 - 9$$

Practice Exercise 3.
$(4 - y)(4 + y)$

EXAMPLE 3 Multiply: $(2u + 1)(2u - 1)$.

SOLUTION: $(2u + 1)(2u - 1) = 4u^2 - 2u + 2u - 1$
$$= 4u^2 - 1$$

Practice Exercise 4.
$(3p + 2q)(3p - 2q)$

EXAMPLE 4 Multiply: $(3x - 2y)(3x + 2y)$.

SOLUTION: $(3x - 2y)(3x + 2y) = 9x^2 + 6xy - 6xy - 4y^2$
$$= 9x^2 - 4y^2$$

In each of the preceding examples, we formed the product of two binomials. The binomials were the sum and the difference of the identical terms. The product was the difference of the squares of these terms. We have

$(x + 2)(x - 2) = x^2 - 4 = (x)^2 - (2)^2$ (From Example 1)
$(y - 3)(y + 3) = y^2 - 9 = (y)^2 - (3)^2$ (From Example 2)
$(2u + 1)(2u - 1) = 4u^2 - 1 = (2u)^2 - (1)^2$ (From Example 3)
$(3x - 2y)(3x + 2y) = 9x^2 - 4y^2 = (3x)^2 - (2y)^2$ (From Example 4)

Since the product of the sum and the difference of two identical terms is the difference of the squares of the terms, then this suggests a procedure for factoring the **difference of two squares.**

Factoring the Difference of Two Squares

$$a^2 - b^2 = (a)^2 - (b)^2 = (a + b)(a - b)$$

Factor.

Practice Exercise 5. $y^2 - 64$

EXAMPLE 5 Factor: $x^2 - y^2$.

SOLUTION: $x^2 - y^2 = (x)^2 - (y)^2$
$$= (x + y)(x - y)$$

Therefore, $x^2 - y^2 = (x + y)(x - y)$.

Practice Exercise 6. $121 - 4u^2$

EXAMPLE 6 Factor: $9u^2 - 4v^2$.

SOLUTION: $9u^2 - 4v^2 = (3u)^2 - (2v)^2$
$$= (3u + 2v)(3u - 2v)$$

Therefore, $9u^2 - 4v^2 = (3u + 2v)(3u - 2v)$.

Practice Exercise 7. $9u^2 - 64$

EXAMPLE 7 Factor: $16a^2 - 25b^2$.

SOLUTION:
$$16a^2 - 25b^2 = (4a)^2 - (5b)^2$$
$$= (4a + 5b)(4a - 5b)$$

Therefore, $16a^2 - 25b^2 = (4a + 5b)(4a - 5b)$.

Notice that when we factor the difference of two squares, each term in the difference is the product of some expression multiplied by itself. An expression that is multiplied by itself is called a "perfect square." Examples of perfect squares are:

- 9 since $9 = (3)(3)$
- x^2 since $x^2 = (x)(x)$
- $16p^2$ since $16p^2 = (4p)(4p)$
- $25u^2v^4$ since $25u^2v^4 = (5uv^2)(5uv^2)$

Hence, when we factor a difference of squares, we will consider factoring the difference of perfect squares. We therefore will not factor

- $x^2 + y^2$ since $x^2 + y^2$ is a *sum* of perfect squares
- $y^2 - 2$ since 2 is not a *perfect* square

Practice Exercise 8. $32a^2 - 98b^2$

EXAMPLE 8 Factor $8u^2 - 18v^2$.

SOLUTION: Notice that $8u^2$ is not a perfect square, and neither is $18v^2$. However, the expression $8u^2 - 18v^2$ has a common factor of 2. Hence,

$$8u^2 - 18v^2 = 2(4u^2 - 9v^2)$$

The factor $4u^2 - 9v^2$ *is* a difference of two squares and can be factored as

$$4u^2 - 9v^2 = (2u)^2 - (3v)^2$$
$$= (2u + 3v)(2u - 3v)$$

Therefore, $8u^2 - 18v^2 = 2(4u^2 - 9v^2) = 2(2u + 3v)(2u - 3v)$.

Practice Exercise 9. $p^4 - q^4$

EXAMPLE 9 Factor $x^4 - y^4$.

SOLUTION: The given expression *is* the difference of two squares since

$$x^4 - y^4 = (x^2)^2 - (y^2)^2$$

Hence,

$$x^4 - y^4 = (x^2)^2 - (y^2)^2$$
$$= (x^2 + y^2)(x^2 - y^2)$$

Now, observe that the factor $x^2 - y^2$ is also the difference of two squares. We have

$$x^2 - y^2 = (x + y)(x - y)$$

Therefore, $x^4 - y^4$ is factored as

$$x^4 - y^4 = (x^2 + y^2)(x + y)(x - y) \quad (x^2 + y^2 \text{ is } not \text{ factorable.})$$

11.3 Factoring the Difference of Two Squares

ANSWERS TO PRACTICE EXERCISES

1. $x^2 - 49$. 2. $4x^2 - 25$. 3. $16 - y^2$. 4. $9p^2 - 4q^2$. 5. $(y + 8)(y - 8)$.
6. $(11 + 2u)(11 - 2u)$. 7. $(3u + 8)(3u - 8)$. 8. $2(4a + 7b)(4a - 7b)$.
9. $(p^2 + q^2)(p + q)(p - q)$.

Note, in Example 8, that the expression $8u^2 - 18v^2$ contained the common factor 2 in each of its two terms. When we divided $8u^2 - 18v^2$ by 2, we obtained the factor $4u^2 - 9v^2$. We then noted that $4u^2 - 9v^2$ is further factorable as the difference of squares. Therefore, $8u^2 - 18v^2 = 2(4u^2 - 9v^2)$ is not factored completely.

Similarly, in Example 9, $x^4 - y^4 = (x^2 + y^2)(x^2 - y^2)$ is not factored completely since $x^2 - y^2$ is further factorable. Always factor as completely as possible.

When attempting to factor an algebraic expression, always determine first if there is a common monomial factor (other than 1); if so, factor the expression accordingly. Then, determine if the other factor can be factored further.

Practice Exercise 10. Factor $16x^4 - 81y^4$.

EXAMPLE 10   Factor $7p^5 - 112pq^4$.

SOLUTION:

STEP 1: Determine the greatest common monomial factor:

$$7p^5 - 112pq^4 = 7p(p^4 - 16q^4)$$

STEP 2: Determine if the factor $p^4 - 16q^4$ is further factorable. Note that this factor is the difference of two squares. Hence,

$$p^4 - 16q^4 = (p^2)^2 - (4q^2)^2$$
$$= (p^2 + 4q^2)(p^2 - 4q^2)$$

STEP 3: Determine if either factor obtained in Step 2 is further factorable. Note that the factor $p^2 + 4q^2$ is a *sum* of squares and is, therefore, *not* factorable. However, the factor $p^2 - 4q^2$ is the difference of two squares and is factorable. We have

$$p^2 - 4q^2 = (p)^2 - (2q)^2$$
$$= (p + 2q)(p - 2q)$$

STEP 4: Since neither of the factors obtained in Step 3 is further factorable, we now have

$$7p^5 - 112pq^4 = 7p(p^4 - 16q^4)$$
$$= 7p(p^2 + 4q^2)(p^2 - 4q^2)$$
$$= 7p(p^2 + 4q^2)(p + 2q)(p - 2q)$$

Therefore, $7p^5 - 112pq^4 = 7p(p^2 + 4q^2)(p + 2q)(p - 2q)$.

ANSWER TO PRACTICE EXERCISE

10. $(4x^2 + 9y^2)(2x + 3y)(2x - 3y)$

EXERCISES 11.3

Factor each of the following expressions as completely as possible.

1. $p^2 - q^2$
2. $u^2 - 16$
3. $m^2 - n^2$
4. $x^2 - 81$
5. $16 - y^2$
6. $1 - t^2$
7. $4a^2 - b^2$
8. $a^2 - 9b^2$
9. $4a^2 - 9b^2$
10. $8a^2 - 18b^2$
11. $9x^2 - 16$
12. $9y^2 - 729$
13. $2a^2 - 8b^2$
14. $12u^2 - 3v^2$
15. $x^3 - xy^2$
16. $p - pq^2$
17. $64t - 4t^3$
18. $20ab^3 - 45a^3b$
19. $x^2y^2 - u^2v^2$
20. $4 - a^2b^6$
21. $x^5y - x^3y^3$
22. $4m^2n^2 - 9p^2q^2$
23. $x^4 - y^4$
24. $v^4 - u^4$
25. $16a^4 - 81b^4$
26. $p^4 - 256q^4$
27. $3p^3 - 3pq^2$
28. $6 - 6t^4$
29. $a^5b - ab^5$
30. $u^3v^6 - u^7v^2$
31. $4x^4y^4 - 1$
32. $a^2b^4 - a^2$
33. $5a^5b - 20ab^3$
34. $12s^2 - 108t^4$
35. $98u^6 - 72v^4$
36. $2x^5y - 32xy^5$
37. $100p^4 - 81q^6$
38. $3p^5q - 768pq^5$
39. $1 - 4a^2b^2c^4$
40. $81m^5n - 100mn^7$

11.4 Factoring Trinomials of the Form $x^2 + bx + c$

OBJECTIVE

After completing this section, you should be able to factor quadratic trinomials having a leading coefficient of 1.

A **quadratic trinomial** is a 2nd-degree trinomial. The **leading coefficient** of a quadratic trinomial is the coefficient of the 2nd-degree term.

- The leading coefficient of $3x^2 - 2x + 1$ is 3.
- The leading coefficient of $2y^2 + 4y - 5$ is 2.
- The leading coefficient of $u^2 - 3u + 1$ is 1.
- The leading coefficient of $2 - 3x - x^2$ is -1.
- The leading coefficient of $2 - 2u^2 - 3u$ is -2.

If a quadratic trinomial is factorable, then the factors will always be *two* **linear** *(1st-degree)* **binomials.** Quadratic trinomials with leading coefficients of 1 are the easiest to factor.

If the quadratic trinomial $x^2 + (p + q)x + pq$ is factorable, then the two linear factors obtained will be of the form $(x + p)$ and $(x + q)$. That is,

$$x^2 + (p + q) + pq = (x + p)(x + q)$$

where:

1. p and q are factors of pq such that:
   a. If the constant term pq is positive, then the factors p and q will both have the *same* algebraic sign as the algebraic sign of the middle term of the trinomial.
   b. If the constant term pq is negative, then the factor p and q will have *opposite* signs.

11.4 Factoring Trinomials of the Form $x^2 + bx + c$

2. The sum of the inner and outer products of the two linear binomials must equal the middle term of the trinomial. That is, $px + qx = (p + q)x$.

Factor each of the following expressions, if possible.

Practice Exercise 1.
$x^2 + 3x - 4$

EXAMPLE 1 Factor $y^2 + 5y + 6$.

SOLUTION:

STEP 1: The leading coefficient of the trinomial is 1. The constant term of the trinomial is *positive*. The middle term has a *positive* coefficient. Hence, the two linear factors will be of the form

$$(y +)(y +)$$

STEP 2: Determine *positive* factors of 6 such that their sum is 5, the coefficient of the middle term of the trinomial.

Positive Factors of 6	Sum of the Factors
1 and 6	7
2 and 3	5 ✔

Choose 2 and 3.

STEP 3: Therefore, $y^2 + 5y + 6 = (y + 2)(y + 3)$.

CHECK: $(y + 2)(y + 3) = y^2 + 2y + 3y + 6$
$= y^2 + 5y + 6$ ✔

Practice Exercise 2.
$y^2 + 10y + 21$

EXAMPLE 2 Factor $u^2 - 9u + 20$.

SOLUTION:

STEP 1: The leading coefficient of the trinomial is 1. The constant term is *positive*. The middle term has a *negative* coefficient. Hence, the two linear factors will be of the form

$$(u -)(u -)$$

STEP 2: Determine *negative* factors of 20 such that their sum is -9, the coefficient of the middle term of the trinomial.

Negative Factors of 20	Sum of the Factors
-1 and -20	-21
-2 and -10	-12
-4 and -5	-9 ✔

Choose -4 and -5.

STEP 3: Therefore, $u^2 - 9u + 20 = (u - 4)(u - 5)$.

CHECK: $(u - 4)(u - 5) = u^2 - 5u - 4u + 20$
$= u^2 - 9u + 20$ ✔

Practice Exercise 3.
$t^2 + 8t - 33$

EXAMPLE 3 Factor $x^2 + x - 6$.

SOLUTION:

STEP 1: The leading coefficient is 1. The constant term is *negative*. Hence, the factors will be of the form

$$(x +)(x -)$$

STEP 2: Determine the factors of -6, with *opposite* signs, such that their sum is 1, the coefficient of the middle term.

Factors of -6	Sum of Factors
1 and -6	-5
-1 and 6	5
2 and -3	-1
-2 and 3	1 ✔

Choose -2 and 3.

STEP 3: Therefore, $x^2 + x - 6 = (x + 3)(x - 2)$.

CHECK: $(x + 3)(x - 2) = x^2 - 2x + 3x - 6$
$= x^2 + x - 6$ ✔

Practice Exercise 4.
$p^2 + 13p + 42$

EXAMPLE 4 Factor $t^2 - 5t - 36$.

SOLUTION:

STEP 1: The leading coefficient is 1. The constant term is *negative*. Hence, the factors will be of the form

$$(t +)(t -)$$

STEP 2: Determine the factors of -36, with *opposite* signs, whose sum is -5, the coefficient of the middle term.

Factors of -36	Sum of Factors
1 and -36	-35
-1 and 36	35
2 and -18	-16
-2 and 18	16
3 and -12	-9
-3 and 12	9
4 and -9	-5 ✔
-4 and 9	5
6 and -6	0

Choose 4 and -9.

STEP 3: Therefore, $t^2 - 5t - 36 = (t + 4)(t - 9)$.

CHECK: $(t + 4)(t - 9) = t^2 - 9t + 4t - 36$
$= t^2 - 5t - 36$ ✔

Practice Exercise 5.
$r^2 - 3r - 28$

EXAMPLE 5 Factor $s^2 - 4s + 22$.

SOLUTION:

STEP 1: The leading coefficient is 1. The constant term is *positive*. The middle term has a *negative* coefficient. Hence, the two linear factors will be of the form

$$(s -)(s -)$$

STEP 2: Determine *negative* factors of 22 whose sum is -4, the coefficient of the middle term.

11.4 Factoring Trinomials of the Form $x^2 + bx + c$

Negative Factors of 22	Sum of Factors
-1 and -22	-23
-2 and -11	-13

Neither set of factors works!

Step 3: Conclude that $s^2 - 4s + 22$ is *not* factorable.

ANSWERS TO PRACTICE EXERCISES

1. $(x + 4)(x - 1)$. 2. $(y + 3)(y + 7)$. 3. $(t + 11)(t - 3)$. 4. $(p + 6)(p + 7)$.
5. $(r - 7)(r + 4)$.

In summary, if a quadratic (2nd-degree) trinomial is factorable, the factors will always be two linear (1st-degree) binomials. As was illustrated in the preceding examples, factoring such trinomials involves trial-and-error efforts. As you gain more experience in factoring trinomials, the trial-and-error effort will be reduced. Also, as was illustrated in Example 5, some quadratic trinomials are not factorable.

EXERCISES 11.4

Factor each of the following expressions as completely as possible. If the expression is not factorable, indicate why not.

1. $x^2 - x - 2$
2. $x^2 + 4x - 5$
3. $y^2 - 4y + 3$
4. $y^2 + y + 5$
5. $u^2 - 7u + 12$
6. $u^2 + 3u + 2$
7. $r^2 - 5r + 6$
8. $r^2 + 5r - 6$
9. $s^2 + 4s + 5$
10. $s^2 - 5s + 4$
11. $x^2 - 5x - 6$
12. $x^2 + 3x - 28$
13. $y^2 + 7y - 9$
14. $y^2 - 6y - 27$
15. $u^2 + 8u + 15$
16. $u^2 - 13u + 36$
17. $s^2 - s - 6$
18. $s^2 + 13s + 22$
19. $t^2 - 18t + 77$
20. $t^2 + 15t + 54$
21. $r^2 - 11r - 26$
22. $r^2 + 3r - 40$
23. $w^2 - 12w + 34$
24. $w^2 + 11w + 24$
25. $x^2 - 7x - 18$
26. $x^2 + x - 42$
27. $y^2 - 7y - 30$
28. $y^2 + 15y + 56$
29. $u^2 + 3u - 54$
30. $u^2 - 12u + 32$
31. $z^2 + 10z + 25$
32. $z^2 + 8z + 14$
33. $w^2 - 15w + 36$
34. $w^2 + 2w - 99$
35. $r^2 + 14r + 45$
36. $r^2 - 12r + 36$
37. $m^2 + 15m + 50$
38. $m^2 - 18m + 80$
39. $p^2 - 11p - 42$
40. $p^2 - 17p + 52$
41. $x^2 + 10x + 21$
42. $x^2 + 5x - 36$
43. $y^2 + y - 30$
44. $y^2 - 13y + 22$
45. $z^2 - 3z - 54$
46. $z^2 + 9z - 52$
47. $p^2 + 2p - 99$
48. $p^2 - 19p + 84$
49. $q^2 - 21q - 100$
50. $q^2 - 22q + 117$
51. $u^2 - 13u - 68$
52. $u^2 + 23u + 132$
53. $v^2 - 19v + 70$
54. $v^2 - 11v - 80$
55. $w^2 + 21w + 108$
56. $w^2 + 10w - 96$
57. $s^2 + 3s - 154$
58. $s^2 - 25s + 156$
59. $t^2 + 32t + 255$
60. $t^2 - t - 240$

11.5 Factoring Quadratic Trinomials of the Form $ax^2 + bx + c$

OBJECTIVE

After completing this section, you should be able to factor quadratic trinomials having leading coefficients other than 1.

In Section 11.4, we factored quadratic trinomials with leading coefficients of 1. We will now factor quadratic trinomials with leading coefficients other than 1.

In general, a quadratic trinomial in the variable x is of the form

$$ax^2 + bx + c \quad (a \neq 0)$$

with a being the leading coefficient. To factor such trinomials, consider the following cases:

- If $a = 1$, follow the procedure given in the previous section.
- If $a \neq 1$, follow the procedure illustrated in the following examples.

Factor.

Practice Exercise 1.
$2y^2 + 7y - 4$

EXAMPLE 1 Factor $2x^2 + 3x + 1$.

SOLUTION: If the given trinomial is factorable, the factors will be *two linear binomials* in x.

STEP 1: Determine the first terms of the binomial factors. These will be *linear* factors of the first term of the trinomial. The only linear factors of $2x^2$ are $2x$ and x. Write

$$(2x \quad)(x \quad)$$

STEP 2: Determine the last terms of the binomial factors. These will be factors of the last term of the trinomial. Since the last term is 1, the factors could be 1 and 1 or -1 and -1. Write

$$(2x - 1)(x - 1) \quad \text{or} \quad (2x + 1)(x + 1)$$

STEP 3: Check the middle term; we want $3x$.

$(2x - 1)(x - 1)$ $\qquad$ $(2x + 1)(x + 1)$

$\quad -x$ $\qquad\qquad\qquad$ x

$+\dfrac{-2x}{-3x}$ $\qquad\qquad$ $+\dfrac{2x}{3x}$

Wrong middle term $\qquad$ Right middle term

STEP 4: Conclude: $2x^2 + 3x + 1 = (2x + 1)(x + 1)$.

CHECK: $(2x + 1)(x + 1) = 2x^2 + 2x + x + 1$
$\qquad\qquad\qquad\qquad\quad = 2x^2 + 3x + 1$ ✓

Practice Exercise 2.
$6u^2 - 5u - 6$

EXAMPLE 2 Factor $3y^2 + 10y - 8$.

SOLUTION: Again, if the trinomial is factorable, the factors must be *two linear binomials* in y.

11.5 Factoring Quadratic Trinomials of the Form $ax^2 + bx + c$

STEP 1: Determine the first terms of the binomial factors. These will be *linear* factors of the first term of the trinomial. The only linear factors of $3y^2$ are $3y$ and y. Write

$$(3y\quad)(y\quad)$$

STEP 2: Determine the last terms of the binomial factors. These will be factors of the last term of the trinomial. Since the last term of the trinomial is -8, the factors could be 1 and -8 or -1 and 8 or 2 and -4 or -2 and 4. Write

$$(3y + 1)(y - 8);\ (3y - 1)(y + 8);\ (3y + 2)(y - 4);\ (3y - 2)(y + 4)$$

STEP 3: Check the middle term; we want $10y$.

$(3y + 1)(y - 8)$
y
$+\dfrac{-24y}{-23y}$
Wrong middle term

$(3y - 1)(y + 8)$
$-y$
$+\dfrac{24y}{23y}$
Wrong middle term

$(3y + 2)(y - 4)$
$2y$
$+\dfrac{-12y}{-10y}$
Wrong middle term

$(3y - 2)(y + 4)$
$-2y$
$+\dfrac{12y}{10y}$
Right middle term

STEP 4: Conclude: $3y^2 + 10y - 8 = (3y - 2)(y + 4)$.

CHECK: $(3y - 2)(y + 4) = 3y^2 + 12y - 2y - 8$
$= 3y^2 + 10y - 8$ ✓

Practice Exercise 3.
$12p^2 - 17p - 5$

EXAMPLE 3 Factor $6u^2 - u - 2$.

SOLUTION: If the trinomial is factorable, the factors must be *two linear binomials* in u.

STEP 1: Determine the first terms of the binomial factors. These will be *linear* factors of the first term of the trinomial. These factors are $6u$ and u or $2u$ and $3u$. (Note that only the *positive* factors of 6 are considered here.) Write

$$(6u\quad)(u\quad)\quad\text{or}\quad(2u\quad)(3u\quad)$$

STEP 2: Determine the last terms of the binomial factors. These will be factors of the last term of the trinomial. The factors of -2 will be 1 and -2 or -1 and 2. Write

$$(6u + 1)(u - 2);\ (2u + 1)(3u - 2);\ (6u - 1)(u + 2);\ (2u - 1)(3u + 2)$$

STEP 3: Check the middle term; we want $-u$.

$(6u + 1)(u - 2)$
u
$+\dfrac{-12u}{-11u}$
Wrong middle term

$(2u + 1)(3u - 2)$
$3u$
$+\dfrac{-4u}{-u}$
Right middle term

$$(6u - 1)(u + 2) \qquad (2u - 1)(3u + 2)$$
$$-u \qquad\qquad -3u$$
$$+\frac{12u}{11u} \qquad\qquad +\frac{4u}{u}$$
$$\text{Wrong middle term} \qquad \text{Wrong middle term}$$

STEP 4: Conclude: $6u^2 - u - 2 = (2u + 1)(3u - 2)$.

CHECK: $(2u + 1)(3u - 2) = 6u^2 - 4u + 3u - 2$
$= 6u^2 - u - 2$ ✔

Practice Exercise 4.
$18x^2 - 3x - 3$

EXAMPLE 4 Factor $60t^2 - 4t - 24$.

SOLUTION: For this trinomial, 4 is a common monomial factor. Hence,

$$60t^2 - 4t - 24 = 4(15t^2 - t - 6)$$

The factor $15t^2 - t - 6$ is a quadratic trinomial. If it is factorable, the factors will be *two linear binomials* in t.

STEP 1: Determine the first terms of the binomial factors. These will be *linear* factors of $15t^2$ which are $15t$ and t or $3t$ and $5t$. Write

$$(15t \quad)(t \quad) \qquad \text{or} \qquad (3t \quad)(5t \quad)$$

STEP 2: Determine the last terms of the binomial factors. The factors of -6 are 1 and -6 or -1 and 6 or 2 and -3 or -2 and 3.

STEP 3: Check the middle term; we want $-t$. Following the procedure of the previous examples, we select $(3t - 2)(5t + 3)$.

STEP 4: Conclude: $15t^2 - t - 6 = (3t - 2)(5t + 3)$ and therefore

$$60t^2 - 4t - 24 = 4(3t - 2)(5t + 3)$$

CHECK: $4(3t - 2)(5t + 3) = 4(15t^2 + 9t - 10t - 6)$
$= 4(15t^2 - t - 6)$
$= 60t^2 - 4t - 24$ ✔

Note When attempting to factor a quadratic trinomial, always look for a common monomial factor first.

It is not always necessary to consider the product of all of the possible binomial factors for a quadratic trinomial, as was done in the preceding examples. Clearly, once the *right* middle term is determined, you stop. With experience, getting the right middle term becomes easier.

If a quadratic trinomial has a negative leading coefficient, factor out the common factor -1 first and then proceed as illustrated in the preceding examples.

Practice Exercise 5.
$-6p^2 - 5p + 4$

EXAMPLE 5 Factor $-6p^2 + 7p + 3$.

SOLUTION: Since the leading coefficient of the trinomial is negative, first factor out -1 from all terms.

$$-6p^2 + 7p + 3 = -(6p^2 - 7p - 3)$$

11.5 Factoring Quadratic Trinomials of the Form $ax^2 + bx + c$

The expression $6p^2 - 7p - 3$ is a quadratic trinomial. If it is factorable, the factors will be *two linear binomials* in p.

STEP 1: Determine the first terms of the binomial factors. These will be *linear* factors of $6p^2$ which are p and $6p$ or $2p$ and $3p$. Write

$$(p\quad)(6p\quad) \quad \text{or} \quad (2p\quad)(3p\quad)$$

STEP 2: Determine the last terms of the binomial factors. These will be factors of -3 which are 1 and -3 or -1 and 3.

STEP 3: Check the middle term; we want $-7p$. Select $(2p - 3)(3p + 1)$.

STEP 4: Conclude: $6p^2 - 7p - 3 = (2p - 3)(3p + 1)$ and, therefore,

$$-6p^2 + 7p + 3 = -(2p - 3)(3p + 1)$$

CHECK: $\quad -(2p - 3)(3p + 1) = -(6p^2 + 2p - 9p - 3)$
$\qquad\qquad\qquad\qquad\qquad = -(6p^2 - 7p - 3)$
$\qquad\qquad\qquad\qquad\qquad = -6p^2 + 7p + 3$ ✔

Sometimes, the quadratic trinomial is *not* written in descending powers of the variable. In such cases, rewrite the trinomial in descending powers of the variable. This makes it easier to factor the expression, if the trinomial is factorable.

Practice Exercise 6.
$30 - 3t - 6t^2$

EXAMPLE 6 Factor $10y + 3y^2 - 8$.

SOLUTION: The given expression is a quadratic trinomial in the variable y. Rewriting the expression in *descending* powers of y, we have

$$10y + 3y^2 - 8 = 3y^2 + 10y - 8$$

and proceed to factor the right-hand side expression. But, from Example 2, we determine that

$$3y^2 + 10y - 8 = (3y - 2)(y + 4)$$

Therefore, $10y + 3y^2 - 8 = (3y - 2)(y + 4)$.

Practice Exercise 7.
$8 + 8t - 6t^2$

EXAMPLE 7 Factor $3 - 6p^2 + 7p$.

SOLUTION: The given expression is a quadratic trinomial in the variable p. Rewriting the expression in *descending* powers of p, we have

$$3 - 6p^2 + 7p = -6p^2 + 7p + 3$$

and proceed to factor the resulting expression. But, from Example 5, we determine that

$$-6p^2 + 7p + 3 = -(2p - 3)(3p + 1)$$

Therefore, $3 - 6p^2 + 7p = -(2p - 3)(3p + 1)$.

ANSWERS TO PRACTICE EXERCISES

1. $(2y - 1)(y + 4)$. **2.** $(2u - 3)(3u + 2)$. **3.** $(3p - 5)(4p + 1)$. **4.** $3(2x - 1)(3x + 1)$.
5. $-(2p - 1)(3p + 4)$. **6.** $-3(t - 2)(2t + 5)$. **7.** $-2(t - 2)(3t + 2)$.

EXERCISES 11.5

Factor each of the following trinomials. Remember to look for a common monomial factor.

1. $2x^2 - 5x - 3$
2. $2x^2 - x - 3$
3. $3y^2 + 10y - 8$
4. $6y^2 - 28y - 10$
5. $12u^2 - 12u + 3$
6. $4u^2 - 4u - 15$
7. $6p^2 + 5p + 1$
8. $6p^2 + 7p - 20$
9. $12x^2 - 4x - 1$
10. $12x^2 - 5x - 2$
11. $20y^2 - 29y + 5$
12. $8y^2 - 2y - 28$
13. $10m^2 - 34m + 12$
14. $9m^2 + 6m + 1$
15. $21u^2 + u - 2$
16. $18u^2 - 18u - 20$
17. $16s^2 + 16s + 3$
18. $16s^2 - 32s + 12$
19. $49t^2 + 14t + 1$
20. $16t^2 + 66t - 27$
21. $25x^2 - 10x + 1$
22. $24x^2 + 22x - 112$
23. $25y^2 - 25y + 6$
24. $28y^2 + 8y - 36$
25. $24u^2 + 31u - 15$
26. $60u^2 - 57u - 84$
27. $45p^2 + 13p - 2$
28. $10p^2 - 67p - 21$
29. $22q^2 + 93q - 27$
30. $26q^2 - 57q - 20$
31. $6 + 29t + 35t^2$
32. $15 - 22r + 8r^2$
33. $42 - 41t + 5t^2$
34. $12 - 14p + 4p^2$
35. $15 - 17w - 4w^2$
36. $12 + 8s - 15s^2$
37. $-6 + 21y - 18y^2$
38. $72 - 43m + 6m^2$
39. $36 - 3p - 3p^2$
40. $20 - 8p - 12p^2$

11.6 Factoring by Grouping

OBJECTIVE

After completing this section, you should be able to factor some algebraic expressions with four terms, by rearranging and grouping the terms.

In Section 11.2, we considered common *monomial* factoring. For example, we noted that 2 is a common monomial factor in the expression $4u + 10$. If we divide $4u + 10$ by 2, we get $2u + 5$, which is another factor of $4u + 10$.

A common factor does not have to be a monomial. It could be a binomial, a trinomial, or any polynomial. Consider the expression

$$(a + b)(x + y) + (a + b)(z)$$

which is a sum of two terms. Each term in the sum is a product of two factors. In the first term, the two factors are $(a + b)$ and $(x + y)$. In the second term, the two factors are $(a + b)$ and z. Note that $(a + b)$ is a factor common to both terms. Hence, the expression

$$(a + b)(x + y) + (a + b)(z)$$

if factorable. We have

$$(a + b)(x + y) + (a + b)(z) = (a + b)[(x + y) + z]$$
$$= (a + b)(x + y + z)$$

In this section, we consider another method of factoring that may involve a common factor that is not a monomial.

11.6 Factoring by Grouping

The expression $ap + aq + bp + bq$ consists of four terms. There is no common factor. The expression is not a difference of squares nor is it a quadratic trinomial. However, the expression can be factored by rearranging and grouping its terms as follows:

$$ap + aq + bp + bq$$
$$= (ap + aq) + (bp + bq) \quad \text{(Grouping terms)}$$
$$= a(p + q) + (bp + bq) \quad \text{(Common factor, } a\text{, in first group)}$$
$$= a(p + q) + b(p + q) \quad \text{Common factor, } b\text{, in second group)}$$
$$= (a + b)(p + q) \quad \text{(Common factor, } p + q\text{)}$$

Therefore, $ap + aq + bp + bq = (a + b)(p + q)$.

This factoring represents what is known as *factoring by grouping*. Some algebraic expressions can be factored in this manner. The key to success depends upon getting the right grouping of terms within the expression. However, keep in mind that this method does not always work.

Factor each of the following by grouping.

Practice Exercise 1.
$a + b + ab + b^2$

EXAMPLE 1 Factor $2ax - 2ay + bx - by$.

SOLUTION:

$$2ax - 2ay + bx - by$$
$$= (2ax - 2ay) + (bx - by) \quad \text{(Grouping terms)}$$
$$= 2a(x - y) + (bx - by) \quad \text{(Common factor, } 2a\text{, in first group)}$$
$$= 2a(x - y) + b(x - y) \quad \text{(Common factor, } b\text{, in second group)}$$
$$= (2a + b)(x - y) \quad \text{(Common factor, } x - y\text{)}$$

Therefore, $2ax - 2ay + bx - by = (2a + b)(x - y)$.

Practice Exercise 2.
$x^3 + 3x^2 + 4x + 12$

EXAMPLE 2 Factor $ab - 2 + a - 2b$.

SOLUTION:

$$ab - 2 + a - 2b$$
$$= (ab + a) + (-2 - 2b) \quad \text{(Rearranging and grouping terms)}$$
$$= a(b + 1) + (-2 - 2b) \quad \text{(Common factor, } a\text{, in first group)}$$
$$= a(b + 1) - 2(1 + b) \quad \text{(Common factor, } -2\text{, in second group)}$$
$$= a(b + 1) - 2(b + 1) \quad \text{(Commutative property for addition)}$$
$$= (b + 1)(a - 2) \quad \text{(Common factor, } b + 1\text{)}$$

Therefore, $ab - 2 + a - 2b = (b + 1)(a - 2)$.

Practice Exercise 3.
$cu - 1 - c + u$

EXAMPLE 3 Factor $x^2 + 3x - y^2 - 3y$.

SOLUTION: (First attempt)

$$x^2 + 3x - y^2 - 3y$$
$$= (x^2 + 3x) + (-y^2 - 3y) \quad \text{(Grouping terms by same variables)}$$
$$= x(x + 3) + (-y^2 - 3y) \quad \text{(Common factor, } x\text{, in first group)}$$
$$= x(x + 3) - y(y + 3) \quad \text{(Common factor, } -y\text{, in second group)}$$

We *cannot* proceed further in factoring the given expression. (*Caution:* This last result is *not* in factored form since the expression is a *sum* of two terms. An expression is in factored form when it is a *product* of two or more factors.)

SOLUTION: (Second attempt)

$$x^2 + 3x - y^2 - 3y$$
$$= (x^2 - y^2) + (3x - 3y) \quad \text{(Grouping terms by degrees)}$$
$$= (x + y)(x - y) + (3x - 3y) \quad \text{(Factoring in first group)}$$
$$= (x + y)(x - y) + 3(x - y) \quad \text{(Common factor, 3, in second group)}$$
$$= (x - y)(x + y + 3) \quad \text{(Common factor, } x - y\text{)}$$

Therefore, $x^2 + 3x - y^2 - 3y = (x - y)(x + y + 3)$.

Practice Exercise 4.
$2p + 4p^2 - q - q^2$

EXAMPLE 4 Factor $x^3 - x^2 - x + 1$.

SOLUTION:
$$x^3 - x^2 - x + 1$$
$$= x^3 - x^2 + (-x) + 1 \quad \text{(Rewriting)}$$
$$= (x^3 - x^2) + (-x + 1) \quad \text{(Grouping terms)}$$
$$= x^2(x - 1) + (-x + 1) \quad \text{(Common factor, } x^2\text{, in first group)}$$
$$= x^2(x - 1) + (-1)(x - 1) \quad \text{(Common factor, } -1\text{, in second group)}$$
$$= (x - 1)(x^2 - 1) \quad \text{(Common factor, } x - 1\text{)}$$
$$= (x - 1)(x + 1)(x - 1) \quad \text{(Factoring } x^2 - 1\text{)}$$
$$= (x - 1)^2(x + 1) \quad \text{(Rewriting)}$$

Therefore, $x^3 - x^2 - x + 1 = (x - 1)^2(x + 1)$.

ANSWERS TO PRACTICE EXERCISES

1. $(a + b)(1 + b)$. 2. $(x + 3)(x^2 + 4)$. 3. $(c + 1)(u - 1)$. 4. $(2p - q)(2p + q + 1)$.

EXERCISES 11.6

Factor each of the given expressions by grouping.

1. $am + bm + an + bn$
2. $ah + ch - ak - ck$
3. $xy + 3x + 2y + 6$
4. $ac + ad - bc - bd$
5. $2ab + 2a + b + 1$
6. $2mp + mq - 2pn - nq$
7. $x^2 - y^2 + 4x + 4y$
8. $a^2 + b - b^2 - a$
9. $pq + p - q - 1$
10. $px - qx - py + qy$
11. $mx - my - nx + ny$
12. $xy - y + x - 1$
13. $2mn + 8n - 3m - 12$
14. $3pq - 14 + 6p - 7q$
15. $3st - 8 - 4s + 6t$
16. $x^2 - 2y - 2x + xy$
17. $2m^2 - n + mn - 2m$
18. $x^2 + x - y^2 + y$
19. $2ab - 20 + 8a - 5b$
20. $2pq + 4p - 6q - 3q^2$
21. $p^2 + 4p - q^2 + 4q$
22. $10 - 2u + 5u^2 - u^3$
23. $3y^2 - 12 + 3xy - 6x$
24. $r^2 - rs + 5s - 25$
25. $x^2 - 16 + xy + 4y$
26. $w^3 - 2w^2 + 9w - 18$
27. $t^3 + 27 - 9t - 3t^2$
28. $au^2 - bv^2 + av^2 - bu^2$
29. $w^2 - 5v + vw - 25$
30. $q^2 + qt - 9 - 3t$

11.7 Factoring Completely

OBJECTIVE

After completing this section, you should be able to factor an algebraic expression completely so that no factors can be factored further.

The expression $2x^2 - 2y^2$ can be factored as $2(x^2 - y^2)$ by factoring out the common factor 2. However, the remaining factor, $x^2 - y^2$, can be factored further as $(x + y)(x - y)$. Hence, the complete factorization of $2x^2 - 2y^2$ is $2(x + y)(x - y)$.

When no factors can be factored further, we say that the given expression is **factored completely**. Factoring completely was actually used in the previous sections on factoring. We will now summarize a strategy for factoring an algebraic expression completely.

Strategy for Factoring an Algebraic Expression Completely

To factor an algebraic expression completely:
1. Simplify the expression by combining all like terms.
2. Factor out any common factor(s).
3. Look for the difference of two squares; factor accordingly.
4. Look for a factorable quadratic trinomial; factor accordingly.
5. Look for an expression consisting of four terms that can be factored by grouping; factor accordingly.
6. Determine if any factors already obtained can be factored again.

Factor each of the following expressions completely.

Practice Exercise 1. $3x^2 - 48$

EXAMPLE 1 Factor $3x^2 - 3y^2$ completely.

SOLUTION:

$$3x^2 - 3y^2 = 3(x^2 - y^2) \quad \text{(Common factor, 3)}$$
$$= 3(x + y)(x - y) \quad \text{(Difference of squares)}$$

The factors 3, $x + y$, and $x - y$ cannot be factored further. Hence, the given expression is factored completely. Therefore, $3x^2 - 3y^2 = 3(x + y)(x - y)$.

Practice Exercise 2. $12u^2 - 27v^2$

EXAMPLE 2 Factor $5y^2 + 10y - 15$ completely.

SOLUTION:

$$5y^2 + 10y - 15 = 5(y^2 + 2y - 3) \quad \text{(Common factor, 5)}$$
$$= 5(y + 3)(y - 1) \quad \text{(Factorable quadratic trinomial)}$$

The factors 5, $y + 3$, and $y - 1$ cannot be factored further. Hence, the given expression is factored completely. Therefore, $5y^2 + 10y - 15 = 5(y + 3)(y - 1)$.

Practice Exercise 3. $-12p^2 - 14p + 6$

EXAMPLE 3 Factor $6ab - 6a + 9b - 9$ completely.

SOLUTION:

$$6ab - 6a + 9b - 9$$
$$= 3(2ab - 2a + 3b - 3) \quad \text{(Common factor, 3)}$$
$$= 3[(2ab - 2a) + (3b - 3)] \quad \text{(Grouping terms)}$$
$$= 3[2a(b - 1) + (3b - 3)] \quad \text{(Common factor, 2a)}$$
$$= 3[2a(b - 1) + 3(b - 1)] \quad \text{(Common factor, 3)}$$
$$= 3[(b - 1)(2a + 3)] \quad \text{(Common factor, } b - 1\text{)}$$

The given expression is factored completely. Therefore, $6ab - 6a + 9b - 9 = 3(b - 1)(2a + 3)$.

Practice Exercise 4.
$6pq - 6p + 12q - 12$

EXAMPLE 4 Factor $x^2 + 2xy + y^2 - 9$ completely.

SOLUTION:

$$x^2 + 2xy + y^2 - 9$$
$$= (x^2 + 2xy + y^2) - 9 \quad \text{(Grouping terms)}$$
$$= (x + y)(x + y) - 9 \quad \text{(Factorable quadratic trinomial)}$$
$$= (x + y)^2 - 9 \quad \text{(Rewriting)}$$
$$= (x + y + 3)(x + y - 3) \quad \text{(Difference of squares)}$$

The factors $x + y + 3$ and $x + y - 3$ are not further factorable. Therefore, $x^2 + 2xy + y^2 - 9 = (x + y + 3)(x + y - 3)$.

Practice Exercise 5.
$p^2 - 6pq + 9q^2 - 25$

EXAMPLE 5 (Optional) Factor $2x^4 - 26x^2 + 72$ completely.

SOLUTION:

$$2x^4 - 26x^2 + 72$$
$$= 2(x^4 - 13x^2 + 36) \quad \text{(Common factor, 2)}$$
$$= 2[(x^2)^2 - 13(x^2) + 36] \quad \text{(Rewriting as a quadratic trinomial in } x^2\text{)}$$
$$= 2[(x^2 - 9)(x^2 - 4)] \quad \text{(Factoring the trinomial)}$$
$$= 2[(x + 3)(x - 3)(x^2 - 4)] \quad \text{(Difference of squares)}$$
$$= 2[(x + 3)(x - 3)(x + 2)(x - 2)] \quad \text{(Difference of squares)}$$

The factors 2, $x + 3$, $x - 3$, $x + 2$, and $x - 2$ cannot be further factored. Therefore, $2x^4 - 26x^2 + 72 = 2(x + 3)(x - 3)(x + 2)(x - 2)$.

ANSWERS TO PRACTICE EXERCISES

1. $3(x + 4)(x - 4)$. 2. $3(2u + 3v)(2u - 3v)$. 3. $-2(2p + 3)(3p - 1)$.
4. $6(p + 2)(q - 1)$. 5. $(p - 3q + 5)(p - 3q - 5)$.

EXERCISES 11.7

Factor each of the following expressions completely.

1. $3x^2 - 12y^2$
2. $3a^2 - 75b^2$
3. $8x^2 - 18y^2$
4. $27a^2 - 12b^2$
5. $a^4 - b^4$
6. $2x^4 - 32y^6$

7. $3x^2 - 12x - 63$
8. $10m^2 + 30m + 20$
9. $8p^2 + 12p - 80$
10. $30u^2 + 50u - 20$
11. $24t^2 + 10t - 4$
12. $72s^2 + 102s - 30$
13. $3ax + 4bx + 3ay + 4by$
14. $3x^2 + 9x - 3y^2 - 9y$
15. $7pq - 7q - 7 + 7p$
16. $5w^3 - 20w + 40 - 10w^2$
17. $12b + 60 + 20b^2 + 4b^3$
18. $2a^3 - 18a + 54 - 6a^2$
19. $9x^2 - 4x - 9y^2 + 4y$
20. $6m^4 - 3m^2 + 3n^2 - 6n^4$
21. $a^2 + 4ab + 4b^2 - 36$
22. $3p^2 + 6pq + 3q^2 - 48$
23. $4p^2 - 49 - 12pq + 9q^2$
24. $9s^2 + t^2 - 4 - 6st$

Exercises 25–30 are optional. They are similar to Example 5.

25. $y^4 - 5y^2 + 4$
26. $s^4 - 34s^2 + 225$
27. $3t^4 - 30t^2 + 27$
28. $-4p^4 + 104p^2 - 100$
29. $-100 + 45w^2 - 5w^4$
30. $8q^4 - 64q^2 + 128$

11.8 Solving Equations by Factoring

OBJECTIVE

After completing this section, you should be able to solve some equations by factoring.

We will now look at an application of factoring as a means of solving an equation. We will use the fact that if the product of two numbers is 0, then at least one of the numbers must be 0. That is, if

$$ab = 0$$

then

$$a = 0 \quad \text{or} \quad b = 0$$

We will use this property to solve equations. Recall that to solve an equation means to find all the replacement values for the variable(s) that will make a true statement. You are reminded to check your solutions in the *original* equation.

Use the method of factoring to solve the following equations.

Practice Exercise 1.
$(2x + 5)(2x - 5) = 0$

EXAMPLE 1 Solve the equation $(x + 2)(x - 5) = 0$ for x.

SOLUTION: Since $(x + 2)(x - 5) = 0$, then either $x + 2 = 0$ or $x - 5 = 0$.

If $x + 2 = 0$, then $x = -2$ and $(x + 2)(x - 5) = (0)(x - 5) = 0$.
If $x - 5 = 0$, then $x = 5$ and $(x + 2)(x - 5) = (x + 2)(0) = 0$.

Therefore, there are two solutions for the given equation. The solutions are -2 and 5.

CHECK: If $x = -2$, then $(x + 2)(x - 5)$
$= (-2 + 2)(-2 - 5)$
$= (0)(-7)$
$= 0$ ✓

$$\text{If } x = 5, \text{ then } (x + 2)(x - 5)$$
$$= (5 + 2)(5 - 5)$$
$$= (7)(0)$$
$$= 0 \checkmark$$

Practice Exercise 2.
$2y(3y - 1) = 0$

EXAMPLE 2 Solve the equation $(y + 4)(y - 1) = 0$ for y.

SOLUTION: Since $(y + 4)(y - 1) = 0$, then either $y + 4 = 0$ or $y - 1 = 0$.

If $y + 4 = 0$, then $y = -4$ and $(y + 4)(y - 1) = (0)(y - 1) = 0$.
If $y - 1 = 0$, then $y = 1$ and $(y + 4)(y - 1) = (y + 4)(0) = 0$.

Therefore, there are two solutions for the equation. The solutions are -4 and 1.

CHECK:
$$\text{If } y = -4, \text{ then } (y + 4)(y - 1)$$
$$= (-4 + 4)(-4 - 1)$$
$$= (0)(-5)$$
$$= 0 \checkmark$$

$$\text{If } y = 1, \text{ then } (y + 4)(y - 1)$$
$$= (1 - 4)(1 - 1)$$
$$= (5)(0)$$
$$= 0 \checkmark$$

Practice Exercise 3.
$p(2p + 1)(3p - 5) = 0$

EXAMPLE 3 Solve the equation $u(u + 5)(u - 7) = 0$ for u.

SOLUTION: In the given equation, we have a product of three factors that is equal to 0. The product will be 0 if any one of the three factors is 0. In this and the following examples, we will arrange the solutions as follows:

$$u(u + 5)(u - 7) = 0$$
$$u = 0 \quad \text{or} \quad u + 5 = 0 \quad \text{or} \quad u - 7 = 0$$
$$u = -5 \quad\quad\quad u = 7$$

The three solutions to this equation are -5, 0, and 7.

CHECK:
$$\text{If } u = -5, \text{ then } u(u + 5)(u - 7)$$
$$= -5(-5 + 5)(-5 - 7)$$
$$= -5(0)(-12)$$
$$= 0 \checkmark$$

$$\text{If } u = 0, \text{ then } u(u + 5)(u - 7)$$
$$= 0(0 + 5)(0 - 7)$$
$$= 0(5)(-7)$$
$$= 0 \checkmark$$

$$\text{If } u = 7, \text{ then } u(u + 5)(u - 7)$$
$$= 7(7 + 5)(7 - 7)$$
$$= 7(12)(0)$$
$$= 0 \checkmark$$

Practice Exercise 4.
$(t + 1)(4t + 5)(3t - 7) = 0$

EXAMPLE 4 Solve the equation $(2t + 3)(3t - 1)(5t - 2) = 0$ for t.

11.8 Solving Equations by Factoring

SOLUTION: $(2t + 3)(3t - 1)(5t - 2) = 0$

$$2t + 3 = 0 \quad \text{or} \quad 3t - 1 = 0 \quad \text{or} \quad 5t - 2 = 0$$
$$2t = -3 \qquad\qquad 3t = 1 \qquad\qquad 5t = 2$$
$$t = \frac{-3}{2} \qquad\qquad t = \frac{1}{3} \qquad\qquad t = \frac{2}{5}$$

Therefore, the three solutions for this equation are $\frac{-3}{2}, \frac{1}{3},$ and $\frac{2}{5}$.

CHECK:

If $t = \frac{-3}{2}$, then $(2t + 3)(3t - 1)(5t - 2)$

$$= \left[2\left(\frac{-3}{2}\right) + 3\right]\left[3\left(\frac{-3}{2}\right) - 1\right]\left[5\left(\frac{-3}{2}\right) - 2\right]$$

$$= (-3 + 3)\left(\frac{-9}{2} - 1\right)\left(\frac{-15}{2} - 2\right)$$

$$= (0)\left(\frac{-11}{2}\right)\left(\frac{-19}{2}\right)$$

$$= 0 \checkmark$$

If $t = \frac{1}{3}$, then $(2t + 3)(3t - 1)(5t - 2)$

$$= \left[2\left(\frac{1}{3}\right) + 3\right]\left[3\left(\frac{1}{3}\right) - 1\right]\left[5\left(\frac{1}{3}\right) - 2\right]$$

$$= \left(\frac{2}{3} + 3\right)(1 - 1)\left(\frac{5}{3} - 2\right)$$

$$= \left(\frac{11}{3}\right)(0)\left(\frac{-1}{3}\right)$$

$$= 0 \checkmark$$

If $t = \frac{2}{5}$, then $(2t + 3)(3t - 1)(5t - 2)$

$$= \left[2\left(\frac{2}{5}\right) + 3\right]\left[3\left(\frac{2}{5}\right) - 1\right]\left[5\left(\frac{2}{5}\right) - 2\right]$$

$$= \left(\frac{4}{5} + 3\right)\left(\frac{6}{5} - 1\right)(2 - 2)$$

$$= \left(\frac{19}{5}\right)\left(\frac{1}{5}\right)(0)$$

$$= 0 \checkmark$$

For the rest of the examples in this section, we will leave the checking of the solutions to you. Remember, always check your solutions!

If the preceding examples, we had a product of two or more factors set equal to 0. What happens when we have an expression that is not already factored? As will be illustrated in the following examples, we first factor the expression and then rewrite the equation so that we have a product of factors set equal to 0.

Practice Exercise 5.
$2r^2s - 10rs = 12s$

EXAMPLE 5 Solve the equation $x^2 + 5x = 0$ for x.

SOLUTION: To solve this equation, we must first factor the expression on the left-

11.9 Applications Involving Equations

OBJECTIVE

After completing this section, you should be able to solve some word problems using the method of factoring of the associated equation.

We will now examine some word problems that can be solved using the method of factoring.

Practice Exercise 1. The sum of a positive number and its square is 90. What is the number?

EXAMPLE 1 If a number is subtracted from its square, then the answer is 6. What is the number?

SOLUTION: We will let n represent the number. The verbal statement is

The square of a number less the number is equal to 6.

Substituting n for the number and inserting appropriate mathematical symbols, we have the equation

$$n^2 - n = 6$$

Solving this equation, we have

$$\begin{aligned} n^2 - n &= 6 \\ n^2 - n - 6 &= 0 \quad \text{(Getting one side equal to 0)} \\ (n - 3)(n + 2) &= 0 \quad \text{(Factoring)} \\ n - 3 &= 0 \quad \text{or} \\ n &= 3 \end{aligned}$$

Hence, the required number can be either $n = -2$ or $n = 3$. Check each of these solutions in the original statement.

Practice Exercise 2. The product of two positive integers is 120. One integer is 1 less than twice the other integer. What are the two integers?

EXAMPLE 2 Determine the value of two consecutive integers whose product is 552.

SOLUTION: Consecutive integers differ by 1. We could

Let p = one of the integers
Then, $p + 1$ = the other integer

The verbal statement is

The product of the integers is 552.

Substituting p and $p + 1$ for the integers and inserting appropriate mathematical symbols, we have

$$p(p + 1) = 552$$

Solving this equation, we have

$$\begin{aligned} p(p + 1) &= 552 \\ p^2 + p &= 552 \quad \text{(Distributive property)} \\ p^2 + p - 552 &= 0 \quad \text{(Getting one side equal to 0)} \end{aligned}$$

11.9 Applications Involving Equations

$$(p + 24)(p - 23) = 0 \quad \text{(Factoring)}$$
$$p + 24 = 0 \quad \text{or} \quad p - 23 = 0$$
$$p = -24 \qquad\qquad p = 23$$

Therefore, the solutions are the consecutive integers -24 and -23 or the consecutive integers 23 and 24. In either case, the product is 552.

Some geometric problems lead to equations that can be solved by factoring. When solving word problems pertaining to geometric figures, draw a diagram, if possible, and label it with the given information. We will illustrate this technique in the following examples.

> **Practice Exercise 3.** The perimeter of a triangle is 19 yd. The longest side of the triangle is twice the length of the shortest side. The third side is 1 yd less than the longest side. What are the lengths of the three sides of the triangle?

EXAMPLE 3 The length of a rectangle is 3 cm more than twice its width. The area of the rectangular region is 44 cm². What are the dimensions of the rectangle?

SOLUTION: In this problem, there are two unknowns, the length and the width of the rectangle. If we

$$\text{Let } W = \text{width (in cm)}$$
$$\text{Then, } 2W + 3 = \text{length (in cm)}$$

since the length is 3 cm more than twice its width. (See diagram.) Since the area of the rectangular region is given by the formula

$$\text{Area} = \text{length} \times \text{width}$$

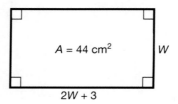

we have, with substitution,

$$W(2W + 3) = 44$$

Solving this equation, we have

$$W(2W + 3) = 44$$
$$2W^2 + 3W = 44$$
$$2W^2 + 3W - 44 = 0$$
$$(2W + 11)(W - 4) = 0$$
$$2W + 11 = 0 \quad \text{or} \quad W - 4 = 0$$
$$2W = -11 \qquad\qquad W = 4 \text{ cm}$$
$$W = \frac{-11}{2} \text{ cm}$$

We seem to have two solutions for W. However, since the width of a rectangle cannot be negative, we reject $W = \dfrac{-11}{2}$ as a solution. Therefore,

and
$$\text{Width} = 4 \text{ cm}$$
$$\text{Length} = 2W + 3 = 11 \text{ cm}$$

Hence, the rectangle has a width of 4 cm and a length of 11 cm. Note that the area of the rectangular region with these dimensions is 44 cm². Therefore, the solution checks.

(*Note:* In the preceding example, we let W = width of the rectangle and determined the length in terms of W. We could have started with L = the length and determined the width in terms of L. However, that would not be as easy. You are encouraged to try it.)

> **DEFINITION**
> In a right triangle, the longest side is called its "hypotenuse." The other two sides are called its "legs."

Practice Exercise 4. One leg of a right triangle is 2 ft less than twice the other leg. If the area of the triangular region is 42 ft², determine the dimensions of the two legs.

EXAMPLE 4 One leg of a right triangle is 1 in. less than twice the other leg, and the area of the triangular region is 14 in.². Find the dimensions of the two legs.

SOLUTION: There are two unknowns in this problem, the lengths of the two legs. If we

Let L = length of the shorter leg (in inches)
Then, $2L - 1$ = length of the other leg (in inches)

since the longer leg is 1 in. less than twice the length of the other leg. (See diagram.)

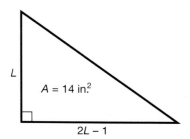

The area of this triangular region is given by the formula

$$\text{Area} = \frac{1}{2} \times \text{base} \times \text{height}$$

We have, with substitution,

$$\frac{1}{2}L(2L - 1) = 14$$

Solving this equation, we have

$$\frac{1}{2}L(2L - 1) = 14$$
$$L(2L - 1) = 28 \quad \text{(Multiplying } both \text{ sides by 2)}$$
$$2L^2 - L = 28$$
$$2L^2 - L - 28 = 0$$
$$(2L + 7)(L - 4) = 0$$
$$2L + 7 = 0 \quad \text{or} \quad L - 4 = 0$$
$$2L = -7 \quad\quad\quad L = 4 \text{ in.}$$
$$L = \frac{-7}{2} \text{ in.}$$

Again, since L represents the length of a side of the triangle, L cannot be negative.

11.9 Applications Involving Equations

Therefore,
$$L = 4 \text{ in.}$$
and
$$2L - 1 = 7 \text{ in.}$$

Hence, the dimensions of the two legs of the right triangle are 4 in. and 7 in. Check this solution in the original problem.

Practice Exercise 5. The base of a triangular region is 3 cm more than twice the height. If the area of the region is 45 cm², determine the dimension of its base.

EXAMPLE 5 Determine the height and the base of a triangular region whose height is 9 cm less than its base and whose area is 126 cm².

SOLUTION: Again, we have two unknowns in this problem, the height and the base of the triangular region. If we

Let b = the length of the base (in cm)
Then, $b - 9$ = the height (in cm)

since the height is 9 cm less than the base. (See diagram.)

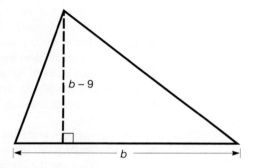

Since the area of the triangular region is given by the formula

$$\text{Area} = \frac{1}{2} \times \text{base} \times \text{height}$$

we have, with substitution,

$$\frac{1}{2}b(b - 9) = 126$$

Solving this equation, we have

$$\frac{1}{2}b(b - 9) = 126$$
$$b(b - 9) = 252$$
$$b^2 - 9b = 252$$
$$b^2 - 9b - 252 = 0$$
$$(b + 12)(b - 21) = 0$$
$$b + 12 = 0 \quad \text{or} \quad b - 21 = 0$$
$$b = -12 \text{ cm} \qquad b = 21 \text{ cm}$$
$$\text{(Reject)}$$

Therefore,
$$b = 21 \text{ cm}$$
and
$$b - 9 = 12 \text{ cm}$$

Hence, the dimensions of the triangular region are base = 21 cm and height = 12 cm.

ANSWERS TO PRACTICE EXERCISES

1. 9.   2. 8 and 15   3. 4 yd, 7 yd, and 8 yd.   4. 7 ft and 12 ft.   5. 15 cm.

EXERCISES 11.9

Solve each of the following word problems by setting up an appropriate equation and using the method of factoring to solve it. Remember to draw a diagram for the problem, if possible, and label it with the information given.

1. One number is 3 more than a second number. Their product is 108. What are the numbers?
2. The sum of a positive number and its square is 42. What is the number?
3. The product of two numbers is 44. One number is 3 more than twice the other number. What are the numbers?
4. The product of three positive integers is 264. If one of the integers is 6, and the second integer is 7 more than the third integer, what are the integers?
5. The product of three positive integers is 105. If one of the integers is 5, and the second integer is 1 more than twice the third integer, what are the three integers?
6. Five less than the square of a positive number is equal to 4 times the number. What is the number?
7. The square of 1 more than twice a negative integer is equal to 1. What is the integer?
8. A rectangular patio has an area of 180 ft^2. The length of the patio is 3 ft more than its width. What are the dimensions of the patio?

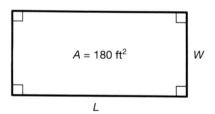

9. A driveway is in the shape of a rectangle, and the length is 5 m greater than twice its width. If its area is 348 m^2, determine the dimensions of the driveway.
10. The width of a rectangular plot is 4 ft less than twice its length. The area of the plot is 646 ft^2. What are the dimensions of the plot?
11. A flower bed is in the form of a right triangle. One leg of the flower bed is

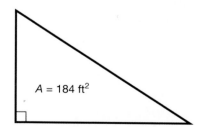

7 ft less than the length of the other leg. If the flower bed has an area of 184 ft^2, determine the dimensions of the two legs.
12. The base of a triangular region is 12 cm more than its height. Determine the dimensions of the region if its area is 54 cm^2.

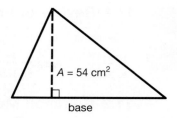

13. If the length of the side of a square is increased by 4 cm, then the area of the new square region formed will be 72 cm² larger than the area of the original square region. Determine the length of the side of the original square.
14. The length of a rectangle is twice its width. The length of the side of a square is equal to the width of the rectangle. If the area of the rectangular region is 25 yd² more than the area of the square region, determine the dimensions of the rectangle.

In Exercises 15–17, use the fact that the sum, S, of the first n consecutive positive integers is given by the formula $S = \frac{n}{2}(n + 1)$. $\left(\text{For example, } 1 + 2 + 3 + 4 + 5 + 6 + 7 = \frac{7}{2}(7 + 1) = \frac{7}{2}(8) = 28.\right)$

15. Determine n if $S = 276$.
16. Determine n if $S = 465$.
17. Determine n if $S = 378$.

In Exercises 18–20, use the fact that the area of a trapezoidal region is given by the formula Area $= \frac{1}{2} \times$ height $\times$ (Base + base). (See diagram.)

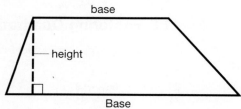

18. Determine the dimensions for the height, base, and Base of a trapezoidal region whose area is 28 yd², if the base is 2 yd more than the height, and the Base is twice the height.
19. Determine the dimensions for the height, base, and Base of a trapezoidal region whose area is 84 cm² if the height is 3 cm less than the base, and the Base is 1 cm more than twice the base.
20. Determine the dimensions for the height, base, and Base of a trapezoidal region whose area is 60 m² if the height is one-half of the Base, and the base is 4 m less than the Base.

Chapter Summary

In this chapter, you were introduced to factoring and special products. We reviewed prime and composite positive integers and the prime factorization of a positive integer. The common monomial factor was discussed, followed by factoring the difference of two squares, factoring some quadratic trinomials, and factoring by grouping. The chapter was concluded by solving equations using the method of factoring. Word problems were included as applications involving such equations.

The **key words** introduced in this chapter are listed here in the order in which they appeared in the text. You should know what they mean before proceeding to Chapter 12.

factoring an algebraic expression (p. 456)
greatest common factor (GCF) (p. 461)
difference of two squares (p. 465)
quadratic trinomial (p. 468)
leading coefficient (p. 468)
linear binomials (p. 468)
factored completely (p. 479)
proper divisor (p. 493)
perfect number (p. 493)

Review Exercises

11.1 Review of Prime Factorization

In Exercises 1–6, determine a prime factorization for each number given.

1. 21
2. 52
3. 99
4. 260
5. 1000
6. 1525

11.2 Common Monomial Factoring

In Exercises 7–12, factor each of the given expressions by first determining the greatest common monomial factor.

7. $10t - 40$
8. $8u^3 - 12u^2$
9. $44x^3y^2 - 121x^2y^4$
10. $16s^2t^3 + 8s^3t^4 - 24st^2$
11. $4a^2b - 8ab^2c + 12b^3c^4$
12. $25m^2n^3p^4 - 10m^4n^2p^3 + 125mn^3p^2$

11.3 Factoring the Difference of Two Squares

In Exercises 13–18, factor each of the given expressions as completely as possible.

13. $4x^2 - 9y^2$
14. $32 - 18p^2$
15. $8p^3q - 18pq^3$
16. $m^4 - n^4$
17. $s^5t - st^5$
18. $2a^3b - 32ab^5$

11.4 Factoring Trinomials of the Form $x^2 + bx + c$

In Exercises 19–26, factor each of the following trinomials.

19. $x^2 + 3x - 18$
20. $y^2 - 14y + 49$
21. $u^2 + 10u + 21$
22. $v^2 - 13v + 36$
23. $s^2 - 10s - 56$
24. $t^2 - 2t - 63$
25. $r^2 - 17r + 52$
26. $w^2 - 24w + 143$

11.5 Factoring Quadratic Trinomials of the Form $ax^2 + bx + c$

In Exercises 27–34, factor each of the given trinomials as completely as possible.

27. $2t^2 - 5t - 3$
28. $4s^2 - 4s - 15$
29. $16x^2 + 66x - 27$
30. $60y^2 - 57y - 84$
31. $7y^2 + 28u - 35$
32. $16p^2 - 32p + 12$
33. $18m^3 + 12m^2 + 2m$
34. $6r^3 + 30r^2 - 36r$

11.6 Factoring by Grouping

In Exercises 35–40, factor each of the given expressions by grouping.

35. $ac - ad + bc - bd$
36. $12m + 2mn - 3n - 18$
37. $2xy - 6x + y - 3$
38. $2ru - sv + 2su - rv$
39. $xy - 3x + y^2 - 3y$
40. $q^2 + 2pq - 2p - q$

11.7 Factoring Completely

In Exercises 41–46, factor each of the given expressions completely.

41. $4x^2 - 4x - 24$
42. $3a^2 - 12b^2$
43. $5a^2 + 5ab - 10b^2$
44. $m^4 - 16n^4$
45. $18p^3 + 3p^2q - 3pq^2$
46. $16x^2 + 76xy - 20y^2$

11.8 Solving Equations by Factoring

In Exercises 47–52, solve each of the given equations for the indicated variable. Use the method of factoring. Check your results.

47. $x^2 + 5x - 6 = 0$
48. $2y^2 - y = 3$
49. $p^3 - p^2 - 2p = 0$
50. $(x^2 - 4)(2x - 3) = 0$
51. $y^3 - 6y^2 = 7y$
52. $18p^3 + 3p^2 - 3p = 0$

11.9 Applications Involving Equations

53. The smaller of two positive integers is 7 less than the larger integer. If their product is 144, what are the two positive integers?
54. The length of a rectangular region is 5 m more than 3 times its width. If the area of the region is 68 m^2, determine the dimensions of the region.
55. The height of a triangular region is 3 ft less than its base. The area of the region is 170 ft^2. Determine the dimensions of the height and the base of the region.

Teasers

1. Let N be a counting number that is divisible by 9. If the digits of N are written in reverse order, is the new number formed also divisible by 9? Explain your answer.

> **DEFINITION**
>
> The number a is a **proper divisor** of the number b if and only if b is exactly divisible by a and $a < b$. (Example: The proper divisors of 6 are 1, 2, and 3.)

2. Determine all of the proper divisors of 28.
3. Determine the sum of all of the proper divisors of 42.

> **DEFINITION**
>
> If a counting number is equal to the sum of all of its proper divisors, then the number is called a **perfect number.**

4. Determine which of the following are perfect numbers:

   6, 18, 28, 42, 125, 496

5. If one factor of the expression $2y^2 + 5y - 3$ is $2y - 1$, indicate how you could determine another factor of the expression.
6. Could the factors of $5u^2 - 7u + 5$ be $2u - 5$ and $3u - 1$? Explain your answer.
7. Is GCF(42,63,210) exactly divisible by GCF(4,10,15)? Explain your answer.
8. Add: GCF($16a^2b, 18a^3b^2, 20b^3$) + GCF($10a^3b^2, 24a^2b^2, 30ab^3$).
9. Factor completely: $uv + 3u^2 - 5v - 15u$.
10. Factor completely: $p^{16} - q^{16}$.

CHAPTER 11 TEST

This test covers materials from Chapter 11. Read each question carefully and answer all questions. If you miss a question, review the appropriate section of the text.

1. Determine a prime factorization for 1225.
2. Without actually dividing, determine if 23,658 is divisible by 3.
3. Without actually dividing, determine if 501,648 is divisible by 8.

In Exercises 4–15, factor each expression completely.

4. $16t - 32$
5. $10y^3 - 5y^2$
6. $4x^2 - 16$
7. $y^2 - 15y + 44$
8. $10a^2b^3c^4 - 125a^4b^2c^3 + 25ab^3c^2$
9. $pq - ps + tq - st$
10. $t^2 - 16s^4$
11. $r^2 + 2rs - 2s - r$
12. $4y^2 - 4y - 15$
13. $8a^3b - 18ab^3$
14. $5s^2 + 5st - 10t^2$
15. $w^2 - 17w + 52$

In Exercises 16—21, solve each of the given equations by factoring.

16. $y^2 - 5y + 6 = 0$
17. $4t^2 - 4t - 15 = 0$
18. $x^3 - 5x^2 - 6x = 0$
19. $(y^2 - 9)(2y - 5) = 0$
20. $16s^2 + 12 = 32s$
21. $7t^2 + 28t = 35$

In Exercises 22–23, solve the word problem.

22. The larger of two positive integers is 1 less than twice the smaller integer. If their product is 231, what are the two numbers?
23. The length of a rectangle is 2 m less than 3 times its width. Determine the dimensions, if the area of the rectangular region is 408 m^2.

CUMULATIVE REVIEW: CHAPTERS 1–11

In Exercises 1–4, perform the indicated operations.

1. $11.04 - 3.1 \times 0.3 \div (-1)^3$
2. $|-3(6 - 10) - (4 - 9)^3 \div (-5^2)|$
3. $0.23\% \times 23.9$
4. $\dfrac{-2}{7} - \dfrac{2}{3} \div \dfrac{-3}{8}$

In Exercises 5–7, evaluate each of the given expressions.

5. $2x^2 - 3xy + 5y^2$, if $x = -2$ and $y = -1$.
6. $4a^2b - 3bc^2$, if $a = -0.1$, $b = -0.2$, and $c = 0.4$.
7. $2p^3 - 3p^2 + 4p$, if $p = \dfrac{-1}{3}$.

In Exercises 8–9, round as indicated.

8. $8\dfrac{4}{9}$, to the nearest hundredth
9. 399,687, to 2 significant digits

In Exercises 10–11, rewrite the given numbers in scientific notation.

10. 23642×10^4
11. 0.0003691

In Exercises 12–13, simplify by combining like terms.

12. $2x^2(3x - 4) - x^3(2 - 3x) + (x^2 - 4)(x^2 - 1)$
13. $2a - 3[b - 3(a - 2c) - (b - 4a) + 5b]$

In Exercises 14–18, solve each equation for the indicated variable.

14. $2(y - 3) + 7 = 3(6 - 5y)$
15. $p^2 - 5p - 6 = 0$
16. $(x^2 - 4)(3x + 5) = 0$
17. $(2r - 1)(3r + 2) = r^2 + 2$
18. $p^3 - 2p^2 - 4p + 8 = 0$
19. Write an algebraic equation for the verbal statement: If 3 times a number is subtracted from the square of the number, then the result is equal to $\dfrac{1}{2}$ the sum of the number and 5. (Let $n =$ the number.)
20. Determine the area of the rectangular region if the length is 3 cm more than its width and the perimeter is 38 cm.
21. Determine the number N represented by 1.09×10^{-6}.
22. Solve the proportion $\dfrac{2}{n} = \dfrac{1.5}{2.4}$ for n.
23. Factor $2x^3 - 8x$.
24. Factor: $6p^2 + pq - 2q^2$.
25. The sides of a triangle have measures of 32.9 cm, 40.6 cm, and 44.3 cm. Estimate the perimeter, P, of the triangle by rounding each measure to the nearest whole number.

CHAPTER 12 Rational Expressions

In this chapter, we will discuss rational expressions and operations on them. This will be an extension of the work we did with rational numbers. When working with rational expressions, you must be able to use the techniques learned in Chapter 11 for factoring polynomials.

In this chapter, you will notice more similarities between arithmetic and algebra. These similarities reinforce the statement made earlier that algebra is a generalization of arithmetic.

CHAPTER 12 PRETEST

This pretest covers materials on rational expressions. Read each question carefully, and try to answer all questions. Each question is keyed to the chapter section in which the particular topic is discussed. Check your answers with those given in the back of the text.

Questions 1–5 pertain to Section 12.1.

In Exercises 1–2, list the excluded values for the given rational expression.

1. $\dfrac{9y}{3y - 5}$

2. $\dfrac{t + 7}{t^2 + 5t + 6}$

In Exercises 3–5, simplify each of the given rational expressions and list all excluded values.

3. $\dfrac{-3x^2 y}{9x^3 y^2}$

4. $\dfrac{p^2 - 16}{2p + 8}$

5. $\dfrac{(2r - 1)(r^2 - 9)}{2r^2 + 5r - 3}$

Questions 6–9 pertain to Section 12.2.

Perform the indicated operations.

6. $\dfrac{2ab^2}{3b} \cdot \dfrac{9a^3 b}{8a}$

7. $\dfrac{6p - 6q}{pq^2} \cdot \dfrac{p^3 q}{p^2 - q^2}$

8. $\dfrac{(s - 1)^2}{t^3} \div \dfrac{3s - 3}{7t^5}$

9. $\dfrac{x^2 + x - 20}{x^2 - 8x + 16} \div \dfrac{x^2 + 11x + 30}{x^2 + 8x + 12}$

Questions 10–11 pertain to Section 12.3.

Perform the indicated operations.

10. $\dfrac{3}{2y - 3} + \dfrac{-7}{2y - 3}$

11. $\dfrac{2a + b}{m + 2} - \dfrac{a - 2b}{m + 2}$

Question 12 pertains to Section 12.4.

12. Determine the least common multiple for $y^2 + 3y + 2$, $y^2 + 4y + 4$, and $y^2 + 5y + 6$.

Questions 13–14 pertain to Section 12.5.

Perform the indicated operations.

13. $\dfrac{x}{2x+1} + \dfrac{3}{4x^2-1}$

14. $\dfrac{3}{s^2-4} - \dfrac{5}{s^2+3s-10}$

Questions 15–16 pertain to Section 12.6.

Simplify each of the given complex rational expressions and list the excluded values for each.

15. $\dfrac{\dfrac{1}{x}}{x + \dfrac{1}{x}}$

16. $\dfrac{\dfrac{2}{ab}}{\dfrac{a}{b} + \dfrac{b}{a}}$

Questions 17–18 pertain to Section 12.7.

Solve each of the following equations for the indicated variable. Check your results in the original equation.

17. $\dfrac{x}{2} - 3 = \dfrac{2x+1}{5}$

18. $\dfrac{4}{p-2} - \dfrac{2}{p+2} = \dfrac{3p}{p^2-4}$

Questions 19–20 pertain to Section 12.8.

19. One number is 5 more than a second number. The reciprocal of the smaller number is twice the reciprocal of the larger number. What are the numbers?

20. The denominator of a fraction is 3 times its numerator. If 1 is added to both the numerator and the denominator of the fraction, the result is $\dfrac{1}{2}$. What is the original fraction?

12.1 Rational Expressions and Their Simplification

OBJECTIVES

After completing this section, you should be able to:
1. Identify a rational expression.
2. Determine the excluded values for rational expressions.
3. Simplify rational expressions that have common *nonzero* factors in their numerator and denominator.

In Chapter 6, you learned that a rational number is a quotient of two integers such that the denominator is not equal to zero. In this section, we will introduce a generalization of rational numbers.

DEFINITION
A **rational expression** is a quotient of two polynomials in the same variable(s).

- $\dfrac{x+2}{2x-3}$ is a rational expression in the variable x.

- $\dfrac{2y^3+4}{y^2+7}$ is a rational expression in the variable y.

Recall that a rational number is a quotient of an integer by a *nonzero* integer. Similarly, the denominator of a rational expression cannot be zero. Therefore, any value of the variable that makes the polynomial in the denominator equal to zero must be excluded. Such a value is called an **excluded value**. To find the excluded values of a rational expression, set its denominator equal to 0 and solve for the variable.

Determine the excluded value(s), if any, for each of the following.

Practice Exercise 1. $\dfrac{x+1}{x-2}$

EXAMPLE 1 Determine all excluded values for the expression $\dfrac{7}{y+1}$.

SOLUTION: The rational expression $\dfrac{7}{y+1}$ has an excluded value of $y=-1$ since, if $y=-1$, then $y+1=0$.

Practice Exercise 2. $\dfrac{y^3+1}{y^2-9}$

EXAMPLE 2 Determine all excluded values for the expression $\dfrac{(7t+1)(t-2)}{(2t-1)(t+5)}$.

SOLUTION: The rational expression $\dfrac{(7t+1)(t-2)}{(2t-1)(t+5)}$ has the excluded values $t=\dfrac{1}{2}$ and $t=-5$ since the denominator will be 0 for either of these values.

Practice Exercise 3.
$\dfrac{2-5u+u^2}{u^2-5u+6}$

EXAMPLE 3 Determine all excluded values for the expression $\dfrac{5x^7-3x^2+1}{x^2+5x+6}$.

12.1 Rational Expressions and Their Simplification

SOLUTION: The rational expression $\dfrac{5x^7 - 3x^2 + 1}{x^2 + 5x + 6}$ has a denominator that is factorable. Since $\dfrac{5x^7 - 3x^2 + 1}{x^2 + 5x + 6} = \dfrac{5x^7 - 3x^2 + 1}{(x + 2)(x + 3)}$, the given rational expression has excluded values $x = -2$ and $x = -3$.

Practice Exercise 4.
$\dfrac{w^2 - 3w + 7}{2w^2 + 9w - 5}$

EXAMPLE 4 Determine all excluded values for the expression $\dfrac{2x - 1}{x^2 + 4}$.

SOLUTION: The rational expression $\dfrac{2x - 1}{x^2 + 4}$ has *no* excluded values since there are *no* values of x for which $x^2 + 4 = 0$.

ANSWERS TO PRACTICE EXERCISES

1. $x = 2$.   2. $y = -3$ and $y = 3$.   3. $u = 2$ and $u = 3$.   4. $w = -5$ and $w = \dfrac{1}{2}$.

In arithmetic, we simplified fractions by using the fundamental principle of fractions. The same principle is used to simplify rational expressions, so we will restate it.

Fundamental Principle of Fractions

If both the numerator and the denominator of a fraction are *divided* by the same *nonzero* factor, an equivalent fraction is obtained. That is, if $b \neq 0$ and $c \neq 0$, then

$$\frac{ac}{bc} = \frac{a}{b}$$

Simplify each of the following rational expressions, if possible.

Practice Exercise 5. $\dfrac{3xy^2}{4xy^3}$

EXAMPLE 5 Simplify: $\dfrac{3x^2y}{7x}$.

SOLUTION: To simplify $\dfrac{3x^2y}{7x}$, divide *both* numerator and denominator by the *common factor* x. Therefore,

$$\frac{3x^2y}{7x} = \left(\frac{3}{7}\right)\left(\frac{x^2}{x}\right)\left(\frac{y}{1}\right) = \left(\frac{3}{7}\right)\left(\frac{x}{1}\right)\left(\frac{y}{1}\right) = \frac{3xy}{7}$$

provided that $x \neq 0$.

Practice Exercise 6. $\dfrac{2y - 8}{3y - 12}$

EXAMPLE 6 Simplify: $\dfrac{y^2 - 9}{y - 3}$.

SOLUTION: To simplify $\dfrac{y^2 - 9}{y - 3}$, first factor the numerator. Hence,

$$\frac{y^2 - 9}{y - 3} = \frac{(y + 3)(y - 3)}{(y - 3)}$$

Divide *both* numerator and denominator by the *common factor* $(y - 3)$. Therefore,

$$\frac{y^2 - 9}{y - 3} = \frac{(y + 3)(y - 3)}{(y - 3)} = \frac{(y + 3)}{1}\left(\frac{\cancel{y - 3}}{\cancel{y - 3}}\right) = \frac{y + 3}{1} = y + 3$$

provided $y \neq 3$.

Practice Exercise 7.
$$\frac{x^2 - 4}{x^2 - 4x + 4}$$

EXAMPLE 7 Simplify: $\dfrac{u + 3}{u - 5}$.

SOLUTION: The rational expression $\dfrac{u + 3}{u - 5}$ *cannot* be simplified. Neither polynomial can be factored.

We will now summarize the procedures for simplifying rational expressions.

Simplifying Rational Expressions

To simplify a rational expression,

1. Factor the numerator and the denominator, if possible.
2. Divide *both* the numerator and the denominator by all *nonzero factors* common to both.

Simplifying rational expressions will be illustrated in the following examples.

Practice Exercise 8.
$$\frac{u^2 - 5u + 6}{u^2 - 9}$$

EXAMPLE 8 Simplify: $\dfrac{x^2 - 16}{x + 4}$.

SOLUTION:
$$\frac{x^2 - 16}{x + 4} = \frac{(x + 4)(x - 4)}{x + 4}$$

$$= \frac{\cancel{(x + 4)}(x - 4)}{\cancel{x + 4}}$$

$$= x - 4$$

provided $x \neq -4$.

Practice Exercise 9.
$$\frac{6t^2 - t - 1}{3t^2 - 11t - 4}$$

EXAMPLE 9 Simplify: $\dfrac{y^2 - 2y}{y^2 + y - 6}$.

SOLUTION:
$$\frac{y^2 - 2y}{y^2 + y - 6} = \frac{y(y - 2)}{(y + 3)(y - 2)}$$

$$= \frac{y\cancel{(y - 2)}}{(y + 3)\cancel{(y - 2)}}$$

$$= \frac{y}{y + 3}$$

provided $y \neq -3, 2$.

12.1 Rational Expressions and Their Simplification

Practice Exercise 10.
$$\frac{s^2 - 3s + 2}{s^2 + s - 6}$$

EXAMPLE 10 Simplify: $\dfrac{u^2 + 3u - 4}{2u^2 + u - 3}$.

SOLUTION:
$$\frac{u^2 + 3u - 4}{2u^2 + u - 3} = \frac{(u - 1)(u + 4)}{(u - 1)(2u + 3)}$$

$$= \frac{\cancel{(u - 1)}(u + 4)}{\cancel{(u - 1)}(2u + 3)}$$

$$= \frac{u + 4}{2u + 3}$$

provided $u \neq \dfrac{-3}{2}, 1$.

Practice Exercise 11. $\dfrac{y - x}{x^2 - y^2}$

EXAMPLE 11 Simplify: $\dfrac{b - a}{a - b}$.

SOLUTION: For this example, we first note that
$$b - a = b + (-a) = -a + b = (-1)(a - b)$$

We now have
$$\frac{b - a}{a - b} = \frac{-(a - b)}{a - b}$$

$$\frac{b - a}{a - b} = \frac{-\cancel{(a - b)}}{\cancel{a - b}}$$

$$= \frac{-1}{1} = -1$$

provided $a \neq b$.

ANSWERS TO PRACTICE EXERCISES

5. $\dfrac{3}{4y}$, if $x \neq 0$, $y \neq 0$.   6. $\dfrac{2}{3}$, if $y \neq 4$.   7. $\dfrac{x + 2}{x - 2}$, if $x \neq 2$.   8. $\dfrac{u - 2}{u + 3}$, if $u \neq -3, 3$.   9. $\dfrac{2t - 1}{t - 4}$, if $t \neq \dfrac{-1}{3}, 4$.   10. $\dfrac{s - 1}{s + 3}$, if $s \neq -3, 2$.   11. $\dfrac{-1}{x + y}$, if $y \neq -x$, x

EXERCISES 12.1

In Exercises 1–10, list the excluded value(s) for each of the given rational expressions. (*Hint:* Factor the denominator, whenever possible.)

1. $\dfrac{5}{x - 9}$

2. $\dfrac{x}{2x - 5}$

3. $\dfrac{y^2 + 7}{y}$

4. $\dfrac{6}{y^2 + 3}$

5. $\dfrac{u}{(u^2 + 1)(u - 2)}$

6. $\dfrac{19u - 7}{2u^2 + 3u - 5}$

7. $\dfrac{5}{t^3 - t}$

8. $\dfrac{7t^{19} + 6t^6}{(t + 3)(t^2 - 16)}$

9. $\dfrac{(s + 5)(s - 3)}{s^3 + 6s^2 - 7s}$

10. $\dfrac{(s^2 - 1)^3}{(s - 3)^2(s^2 + 3)}$

In Exercises 11–50, simplify each of the given rational expressions. Determine the excluded values.

11. $\dfrac{2x^3}{4x}$

12. $\dfrac{5a^2b}{25a^3b^2}$

13. $\dfrac{-3x^3y}{9x^2y^3}$

14. $\dfrac{-12x^2y^3}{16x^3y}$

15. $\dfrac{2u^2v^3w}{-6vw^2}$

16. $\dfrac{18a^3b^2c^4}{6ab^3c^4}$

17. $\dfrac{2x + 6}{x + 3}$

18. $\dfrac{-3y + 12}{y - 4}$

19. $\dfrac{5u + 10}{7u + 14}$

20. $\dfrac{4t^2 - 12t}{2t - 6}$

21. $\dfrac{5 - 2s}{2s - 5}$

22. $\dfrac{a^2x + a^2y}{bx + by}$

23. $\dfrac{u^2 - 4}{u + 2}$

24. $\dfrac{y^2 - 25}{y - 5}$

25. $\dfrac{x^2 - 4}{(x - 2)^2}$

26. $\dfrac{z^2 - 81}{3z - 27}$

27. $\dfrac{s^3}{2s^2 + 3s}$

28. $\dfrac{x^2 - y^2}{4x - 4y}$

29. $\dfrac{x^2 - 5x}{2x^3 - 10x^2}$

30. $\dfrac{2y^4 - 10y^3}{25y - 5y^2}$

31. $\dfrac{a^2 - b^2}{5b - 5a}$

32. $\dfrac{x^2 - x - 2}{x^2 + x - 6}$

33. $\dfrac{2y^2 - y - 1}{3y^2 - y - 2}$

34. $\dfrac{s^2 - s - 6}{s^2 - 9}$

35. $\dfrac{6u^2 + 7u - 3}{3u^2 + 20u - 7}$

36. $\dfrac{2t^2 - 10t - 28}{t^2 - 4}$

37. $\dfrac{3y^2 - 75}{y^2 + 8y + 15}$

38. $\dfrac{(2u - 1)(u^2 - 9)}{2u^2 + 5u - 3}$

39. $\dfrac{(t + 7)(t^2 - 4)}{3t^2 + 15t - 42}$

40. $\dfrac{3t^2 - 15t + 18}{2t^2 - 2t - 12}$

41. $\dfrac{5t^5 - 20t^3}{10t^3 + 40t^2 + 40t}$

42. $\dfrac{(x - 1)(x^2 + 5x + 6)}{(2x + 4)(x^2 - 3x + 2)}$

43. $\dfrac{x^2 + xy - 2y^2}{2x^2 + 3xy - 2y^2}$

44. $\dfrac{2x^2 - 7xy + 3y^2}{2y^2 - 3xy - 2x^2}$

45. $\dfrac{2p^2 - 5pq + 3q^2}{3p^2 - 5pq + 2q^2}$

46. $\dfrac{2s^2 - 11st + 5t^2}{3s^2 - 11st - 20t^2}$

47. $\dfrac{3s^2 + st - 2t^2}{6s^2 - 13st + 6t^2}$

48. $\dfrac{2p^2 - 5pq - 3q^2}{p^2 - 6pq + 9q^2}$

49. $\dfrac{3r^2 + rs - 2s^2}{6r^2 - 7rs + 2s^2}$

50. $\dfrac{u^2 - 2uv + v^2}{2u^2 - uv - v^2}$

12.2 Multiplication and Division of Rational Expressions

OBJECTIVES

After completing this section, you should be able to:
1. Multiply two or more rational expressions.
2. Divide a rational expression by a rational expression.

In arithmetic, you learned to multiply fractions by multiplying their numerators to form the numerator of the new fraction and multiplying their denominators to form the denominator of the new fraction. In algebra, the procedure is basically the same. As in arithmetic, we first try to simplify the rational expressions before multiplying.

Rule for Multiplying Rational Expressions

To multiply rational expressions:

1. Factor the numerator and denominator of each rational expression completely.
2. Multiply all numerators to obtain the numerator of the result.
3. Multiply all denominators to obtain the denominator of the result.
4. Divide the numerator and denominator of the product by all *nonzero* factors common to both.

We will examine multiplication of rational expressions in the following examples.

Multiply, and simplify your results.

Practice Exercise 1. $\dfrac{3x^3y^2}{4y^3} \cdot \dfrac{8x}{9x^2y}$

EXAMPLE 1 Multiply: $\dfrac{3}{x-y} \cdot \dfrac{x^2-y^2}{x+1}$.

SOLUTION:
$$\dfrac{3}{x-y} \cdot \dfrac{x^2-y^2}{x+1} = \dfrac{3}{x-y} \cdot \dfrac{(x+y)(x-y)}{x+1}$$
$$= \dfrac{3(x+y)(x-y)}{(x-y)(x+1)}$$
$$= \dfrac{3(x+y)\cancel{(x-y)}}{\cancel{(x-y)}(x+1)}$$
$$= \dfrac{3(x+y)}{x+1}$$

The excluded values are $x = y$ and $x = -1$.

Practice Exercise 2.
$\dfrac{-2uv^2}{3v^3} \cdot \dfrac{5u^2v}{4u^4} \cdot \dfrac{6u^3v^4}{15uv}$

EXAMPLE 2 Multiply: $\dfrac{p+q}{2p^2} \cdot \dfrac{8p^3}{p^2+2pq+q^2}$

SOLUTION: $\dfrac{p+q}{2p^2} \cdot \dfrac{8p^3}{p^2+2pq+q^2} = \dfrac{p+q}{2p^2} \cdot \dfrac{8p^3}{(p+q)(p+q)}$
$$= \dfrac{8p^3(p+q)}{2p^2(p+q)(p+q)}$$

$$= \frac{\overset{4p}{\cancel{8p^3}}(\cancel{p+q})}{\cancel{2p^2}(\cancel{p+q})(p+q)}$$

$$= \frac{4p}{p+q}$$

The excluded values are $p = 0$ and $p = -q$.

EXAMPLE 3 Multiply: $\dfrac{3x+6}{y-3} \cdot \dfrac{y^2-9}{2y+6} \cdot \dfrac{x-2}{x^2-4}$.

SOLUTION:

$$\frac{3x+6}{y-3} \cdot \frac{y^2-9}{2y+6} \cdot \frac{x-2}{x^2-4} = \frac{3(x+2)}{y-3} \cdot \frac{(y+3)(y-3)}{2(y+3)} \cdot \frac{x-2}{(x+2)(x-2)}$$

$$= \frac{3(x+2)(y+3)(y-3)(x-2)}{2(y-3)(y+3)(x+2)(x-2)}$$

$$= \frac{3\cancel{(x+2)}\cancel{(y+3)}\cancel{(y-3)}\cancel{(x-2)}}{2\cancel{(y-3)}\cancel{(y+3)}\cancel{(x+2)}\cancel{(x-2)}}$$

$$= \frac{3}{2}$$

The excluded values are $x = -2, 2$ and $y = -3, 3$.

Practice Exercise 3.
$\dfrac{8}{y^2-9} \cdot \dfrac{y+3}{2y+4}$

EXAMPLE 4 Multiply: $\dfrac{s^2+5s+6}{s-4} \cdot \dfrac{s^2-4s}{s^2+s-2} \cdot \dfrac{s-5}{s^2-6s+5}$

SOLUTION:

$$\frac{s^2+5s+6}{s-4} \cdot \frac{s^2-4s}{s^2+s-2} \cdot \frac{s-5}{s^2-6s+5}$$

$$= \frac{(s+2)(s+3)}{(s-4)} \cdot \frac{s(s-4)}{(s+2)(s-1)} \cdot \frac{(s-5)}{(s-5)(s-1)}$$

$$= \frac{(s+2)(s+3)(s)(s-4)(s-5)}{(s-4)(s+2)(s-1)(s-5)(s-1)}$$

$$= \frac{\cancel{(s+2)}(s+3)s\cancel{(s-4)}\cancel{(s-5)}}{\cancel{(s-4)}\cancel{(s+2)}(s-1)\cancel{(s-5)}(s-1)}$$

$$= \frac{s(s+3)}{(s-1)^2}$$

The excluded values are $s = -2, 1, 4, 5$.

Practice Exercise 4.
$\dfrac{(x^2-4)(2x^2+9x-5)}{(x^2-25)(2x^2-5x+2)}$

ANSWERS TO PRACTICE EXERCISES

1. $\dfrac{2x^2}{3y^2}$ $(x \neq 0, y \neq 0)$. 2. $\dfrac{-uv^3}{3}$ $(u \neq 0, v \neq 0)$. 3. $\dfrac{4}{y^2-y-6}$ $(y \neq -3, -2, 3)$.
4. $\dfrac{x+2}{x-5}$ $\left(x \neq -5, \dfrac{1}{2}, 2, 5\right)$.

In arithmetic, to divide a fraction by a fraction, we multiply the first fraction by the reciprocal of the second fraction. The same procedure is used in algebra.

12.2 Multiplication and Division of Rational Expressions

Rule for Dividing Rational Expressions

To divide a rational expression by a rational expression:
1. Multiply the first expression by the *reciprocal* of the second expression.
2. Follow the procedure for multiplying rational expressions.

Division of rational expressions is illustrated in the following examples.

Divide and simplify your results.

Practice Exercise 5.
$$\frac{2ab^3}{5bc^2} \div \frac{-8b^2}{10bc}$$

EXAMPLE 5 Divide: $\frac{6}{x^2} \div \frac{3}{x}$.

SOLUTION:
$$\frac{6}{x^2} \div \frac{3}{x} = \frac{6}{x^2} \cdot \frac{x}{3}$$
$$= \frac{6x}{3x^2}$$
$$= \frac{\overset{2}{\cancel{6}}\overset{1}{\cancel{x}}}{\underset{1}{\cancel{3}}\underset{x}{\cancel{x^2}}}$$
$$= \frac{2}{x}$$

The excluded value is $x = 0$.

Practice Exercise 6.
$$\frac{-4p^3q^2}{5q^3r} \div \frac{-12pq}{15pr^3}$$

EXAMPLE 6 Divide: $\frac{5}{x-2} \div \frac{x}{x^2-4}$.

SOLUTION:
$$\frac{5}{x-2} \div \frac{x}{x^2-4} = \frac{5}{x-2} \cdot \frac{x^2-4}{x}$$
$$= \frac{5}{x-2} \cdot \frac{(x+2)(x-2)}{x}$$
$$= \frac{5(x+2)(x-2)}{x(x-2)}$$
$$= \frac{5(x+2)\cancel{(x-2)}^1}{x\cancel{(x-2)}_1}$$
$$= \frac{5(x+2)}{x}$$

The excluded values are $x = -2, 0, 2$.

Practice Exercise 7.
$$\frac{s^2-4}{s^2-5s+6} \div \frac{2s+4}{s^2-9}$$

EXAMPLE 7 Divide: $\frac{y^2+7y+12}{x^2-x-6} \div \frac{y^2-16}{x^2-9}$.

SOLUTION:
$$\frac{y^2+7y+12}{x^2-x-6} \div \frac{y^2-16}{x^2-9} = \frac{y^2+7y+12}{x^2-x-6} \cdot \frac{x^2-9}{y^2-16}$$
$$= \frac{(y+3)(y+4)}{(x-3)(x+2)} \cdot \frac{(x+3)(x-3)}{(y+4)(y-4)}$$

$$= \frac{(y+3)(y+4)(x+3)(x-3)}{(x-3)(x+2)(y+4)(y-4)}$$

$$= \frac{(y+3)(y+4)(x+3)(x-3)}{(x-3)(x+2)(y+4)(y-4)}$$

$$= \frac{(y+3)(x+3)}{(x+2)(y-4)}$$

The excluded values are $x = -3, -2, 3$ and $y = -4, 4$.

Practice Exercise 8.
$$\frac{10p^2 - 11p - 6}{3p^2 + 3p} \div \frac{6p^2 - 13p + 6}{3p^2 - 3p - 6}$$

EXAMPLE 8 Divide: $\dfrac{s^4}{s+5} \div \dfrac{s^2-1}{s^2+6s+5}$.

SOLUTION:
$$\frac{s^4}{s+5} \div \frac{s^2-1}{s^2+6s+5} = \frac{s^4}{s+5} \cdot \frac{s^2+6s+5}{s^2-1}$$

$$= \frac{s^4}{s+5} \cdot \frac{(s+5)(s+1)}{(s+1)(s-1)}$$

$$= \frac{s^4(s+5)(s+1)}{(s+5)(s+1)(s-1)}$$

$$= \frac{s^4(s+5)(s+1)}{(s+5)(s+1)(s-1)}$$

$$= \frac{s^4}{s-1}$$

The excluded values are $s = -5, -1, 1$.

ANSWERS TO PRACTICE EXERCISES

5. $\dfrac{-ab}{2c}$ $(b \neq 0, c \neq 0)$. 6. $\dfrac{p^3 r^2}{q^2}$ $(p \neq 0, q \neq 0, r \neq 0)$. 7. $\dfrac{s+3}{2}$ $(s \neq -3, -2, 2, 3)$.

8. $\dfrac{5p^2 - 8p - 4}{3p^2 - 2p}$ $\left(p \neq -1, 0, \dfrac{2}{3}, \dfrac{3}{2}, 2\right)$.

In Example 9, we consider the division of a rational expression by a polynomial. In Example 10, we consider the division of a polynomial by a rational expression. Both examples are treated as special cases of what we have been doing in Examples 5–8.

Divide.

Practice Exercise 9.
$$\frac{u^2 - 6u - 7}{2u - 5} \div (u^2 - 8u + 7).$$

EXAMPLE 9 Divide: $\dfrac{w^2 - 9}{2w + 3} \div (2w^2 + 7w + 3)$.

SOLUTION:

$$\frac{w^2 - 9}{2w + 3} \div (2w^2 + 7w + 3) = \frac{w^2 - 9}{2w + 3} \div \frac{2w^2 + 7w + 3}{1}$$

$$= \frac{w^2 - 9}{2w + 3} \cdot \frac{1}{2w^2 + 7w + 3}$$

$$= \frac{(w+3)(w-3)}{2w+3} \cdot \frac{1}{(2w+1)(w+3)}$$

$$= \frac{(w+3)(w-3)}{(2w+3)(2w+1)(w+3)}$$

12.2 Multiplication and Division of Rational Expressions

$$= \frac{(\cancel{w+3})(w-3)}{(2w+3)(2w+1)(\cancel{w+3})}$$

$$= \frac{w-3}{(2w+3)(2w+1)}$$

$$= \frac{w-3}{4w^2+8w+3}$$

The excluded values are $w = -3, \frac{-3}{2}, \frac{-1}{2}$.

Practice Exercise 10.
$(2t^2 + 9t - 5) \div \frac{2t-1}{t^2+3}$.

EXAMPLE 10 Divide: $(u^2 + 5u - 14) \div \frac{u^2 - 49}{3u + 12}$.

SOLUTION:

$$(u^2 + 5u - 14) \div \frac{u^2 - 49}{3u + 12} = \frac{u^2 + 5u - 14}{1} \div \frac{u^2 - 49}{3u + 12}$$

$$= \frac{u^2 + 5u - 14}{1} \cdot \frac{3u + 12}{u^2 - 49}$$

$$= \frac{(u+7)(u-2)}{1} \cdot \frac{3(u+4)}{(u+7)(u-7)}$$

$$= \frac{3(u+7)(u-2)(u+4)}{(u-7)(u+7)}$$

$$= \frac{3(\cancel{u+7})(u-2)(u+4)}{(u-7)(\cancel{u+7})}$$

$$= \frac{3(u-2)(u+4)}{u-7}$$

$$= \frac{3(u^2 + 2u - 8)}{u-7}$$

The excluded values are $u = -7, -4,$ and 7.

ANSWERS TO PRACTICE EXERCISES

9. $\frac{u+1}{2u^2 - 7u + 5}$, if $u \ne 1, \frac{5}{2}, 7$. 10. $t^3 + 5t^2 + 3t + 15$, if $t \ne \frac{1}{2}$.

EXERCISES 12.2

In Exercises 1–30, multiply and simplify your results. Include the excluded values.

1. $\dfrac{2x}{3y} \cdot \dfrac{9y}{8x}$

2. $\dfrac{2u^2}{v} \cdot \dfrac{v^3}{3u}$

3. $\dfrac{6s^2}{5t^3} \cdot \dfrac{10st}{6s^3}$

4. $\dfrac{-10a^3 b}{25ab^2} \cdot \dfrac{15a^2 b^3}{40ab}$

5. $\dfrac{2a^3}{3b} \cdot \dfrac{-6ab^2}{5a^3} \cdot \dfrac{10b}{8ab}$

6. $\dfrac{-3xy^2}{4y^3} \cdot \dfrac{5x^2y}{9x^4} \cdot \dfrac{6x^3y^4}{10xy}$

7. $\dfrac{x^2-4}{6} \cdot \dfrac{2x-4}{x+2}$

8. $\dfrac{3y+9}{14y} \cdot \dfrac{y^3}{y^2-9}$

9. $\dfrac{6x-6y}{xy^2} \cdot \dfrac{x^3y}{x^2-y^2}$

10. $\dfrac{2s^2-18}{4t^2} \cdot \dfrac{12t^5}{s-3}$

11. $\dfrac{(x+y)^2}{z^3} \cdot \dfrac{-6z}{4x+4y}$

12. $\dfrac{u(v-w)^2}{5v} \cdot \dfrac{-10w}{u(w-v)}$

13. $\dfrac{(x-4)^2}{-3y} \cdot \dfrac{6y^4}{8-2x}$

14. $\dfrac{t^2-25}{2s-6} \cdot \dfrac{s-3}{5-t}$

15. $\dfrac{t^2-25}{4s^2-9} \cdot \dfrac{2s+3}{5-t}$

16. $\dfrac{w^2-49}{(u+7)^2} \cdot \dfrac{3u+21}{6w+42}$

17. $\dfrac{x^2-5x+6}{3x-6} \cdot \dfrac{-3x^2}{(x-3)^2}$

18. $\dfrac{x+y}{x-y} \cdot \dfrac{x^2-y^2}{x^2-2xy+y^2}$

19. $\dfrac{s+t}{t-s} \cdot \dfrac{s^2-t^2}{(s+t)^2}$

20. $\dfrac{a^2+8a+7}{a^2-6a+5} \cdot \dfrac{a^2-3a-10}{a^2+3a+2}$

21. $\dfrac{2+x-x^2}{8x-4x^2} \cdot \dfrac{6+11x-10x^2}{6x^2-13x+6}$

22. $\dfrac{9b^2+27b}{b^2+4b-5} \cdot \dfrac{b^2+2b-15}{b^2-9}$

23. $\dfrac{y^2-8y+16}{y^2+8y+12} \cdot \dfrac{y^2+11y+30}{y^2+y-20}$

24. $\dfrac{s^2+6st+9t^2}{s^2-st-2t^2} \cdot \dfrac{s^2-4t^2}{3s^2+9st}$

25. $\dfrac{x-y}{2(z-y)} \cdot \dfrac{y-z}{x-z} \cdot \dfrac{z-x}{y-x}$

26. $\dfrac{2x+3y}{6x-9y} \cdot \dfrac{-3x^2y}{4x^2-9y^2} \cdot \dfrac{(2x-3y)^2}{12xy^3}$

27. $\dfrac{x^2-y^2}{y^2-6y-7} \cdot \dfrac{y-7}{2y+2} \cdot \dfrac{(y+1)^2}{y-x}$

28. $\dfrac{x^4-16}{x^2+2x-3} \cdot \dfrac{x^2+x-6}{x^2+x-2} \cdot \dfrac{x-1}{x^2+4}$

29. $\dfrac{a^2-b^2}{2a^2-3ab-2b^2} \cdot \dfrac{(a-2b)^2}{a^2+2ab+b^2} \cdot \dfrac{2a+b}{(a+b)^2}$

30. $\dfrac{2r^2-5rs+2s^2}{9r^2-4s^2} \cdot \dfrac{3r^2+rs-2s^2}{r^2-4s^2} \cdot \dfrac{3r+2s}{2r-2s}$

In Exercises 31–60, divide and simplify your results. Include the excluded values.

31. $\dfrac{6xy^2}{5yz^2} \div \dfrac{12y^3}{10y^3z}$

32. $\dfrac{5x^2y^3}{-6y^2z} \div \dfrac{-15xz^3}{12xy}$

33. $\dfrac{x^2-y^2}{-7} \div \dfrac{x-y}{14}$

34. $\dfrac{x^2-9}{5x^2} \div \dfrac{3-x}{25x^3}$

35. $\dfrac{(y-1)^2}{x^3} \div \dfrac{3y-3}{7x^5}$

36. $\dfrac{2s^3}{t^2-16} \div \dfrac{-6s}{(t-4)^2}$

37. $\dfrac{-3x}{x^2-3x-10} \div \dfrac{9x^3}{3x-15}$

38. $\dfrac{18-3u}{2u+12} \div \dfrac{u^2-36}{6(u+6)}$

39. $\dfrac{a^2-a-2}{a^2+2a+1} \div \dfrac{(a-2)^2}{a^2-5a+6}$

40. $\dfrac{s^2+2s-3}{s^2-4s+3} \div \dfrac{2s^2+8s+6}{6+s-s^2}$

41. $\dfrac{2s^2-5s-3}{3s^2-8s-3} \div \dfrac{4s^2-1}{12s^2+13s+3}$

42. $\dfrac{4x^2-9y^2}{(2x-3y)^2} \div \dfrac{(2x+3y)^2}{2x^2+xy-3y^2}$

43. $\dfrac{d^2-4}{d^2-2d-3} \div \dfrac{d^2-3d+2}{1-d^2}$

44. $\dfrac{6m^2-m-2}{3m^2+m-2} \div \dfrac{m^2-6m+5}{m^2-1}$

45. $\dfrac{x^4-y^4}{(2x+y)^2} \div \dfrac{(x^2+y^2)(x-y)^2}{2x^2-xy-y^2}$

46. $\dfrac{2y^2-162}{5y^2-90y+405} \div \dfrac{8(y+9)^2}{25y+100}$

47. $\dfrac{6t^2-13t+6}{3t^2-3t-6} \div \dfrac{10t^2-11t-6}{3t^2+3t}$

48. $\dfrac{10-12s+2s^2}{s^2+8s+7} \div \dfrac{s^2-3s-10}{4s^2+12s+8}$

49. $\dfrac{w^2 + w - 20}{w^2 - 8w + 16} \div \dfrac{w^2 + 11w + 30}{w^2 + 8w + 12}$ 50. $\dfrac{2m^2 + 3m + 1}{3m^2 - 5m + 2} \div \dfrac{m^2 + 2m + 1}{3m^2 - 5m + 2}$

51. $\dfrac{5x^2 + 14x - 3}{2x^2 + 15x + 7} \div (5x^2 + 19x - 4)$ 52. $\dfrac{x^4 - 1}{x^2 - 2x - 3} \div (x^3 - 3x^2 + x - 3)$

53. $\dfrac{2y^2 + 3y - 9}{2y^2 + 24y + 54} \div (6y^2 - 11y + 3)$ 54. $\dfrac{4y^2 + 27y - 7}{3y^2 + 12y - 63} \div (12y^3 + y^2 - y)$

55. $\dfrac{10u^3 - 7u^2 - 12u}{6u^2 - 11u + 3} \div (5u^4 + 9u^3 + 4u^2)$

56. $(2x^2 + 3x - 2) \div \dfrac{x^2 - 4}{x^2 + x - 6}$

57. $(3y^3 - 20y^2 - 7y) \div \dfrac{3y^2 + 4y + 1}{y^2 - 49}$

58. $(12t^2 - 3) \div \dfrac{4t^2 + 4t - 3}{2t^2 - 7t - 15}$

59. $(5 - 7p + 2p^2) \div \dfrac{2p^2 - p - 1}{6p^2 - p - 2}$

60. $(20q^2 - 19q^3 + 3q^4) \div \dfrac{3q^3 - 14q^2 - 5q}{3q^2 - 8q - 3}$

12.3 Addition and Subtraction of Like Rational Expressions

OBJECTIVE

After completing this section, you should be able to add or subtract like rational expressions.

In arithmetic, you learned that like fractions are fractions with the same denominators. In algebra, we have a similar definition for like rational expressions.

DEFINITION

Like rational expressions are rational expressions with the same denominators.

- $\dfrac{2}{x}$ and $\dfrac{y}{x}$ are like rational expressions with the same denominator, x.

- $\dfrac{4x - 3}{x^2 + 1}$ and $\dfrac{xy}{x^2 + 1}$ are like rational expressions with the same denominator, $x^2 + 1$.

- $\dfrac{2s^2 + 3st + 9}{3s - t}$ and $\dfrac{s^4 - t^4}{3s - t}$ are like rational expressions with the same denominator, $3s - t$.

To add (or subtract) like fractions in arithmetic, we add (or subtract) their numerators and write the result over the common denominator. Once again, our rules for rational expressions are simply generalizations of what we know from arithmetic. Therefore, we have the following rule.

Rule for Adding and Subtracting Like Rational Expressions

To add (or subtract) like rational expressions:

1. Add (or subtract) their numerators.
2. Write the result over the common denominator.
3. Simplify the result by dividing the numerator and the denominator by any *nonzero* factors that are common to both.

We will illustrate this rule in the following examples.

Perform the indicated operations and simplify your results.

Practice Exercise 1. $\dfrac{5}{2u} + \dfrac{7}{2u}$

EXAMPLE 1 Add: $\dfrac{3}{y} + \dfrac{x}{y}$.

SOLUTION:
$$\dfrac{3}{y} + \dfrac{x}{y} = \dfrac{3 + x}{y}$$

The excluded value is $y = 0$.

Practice Exercise 2. $\dfrac{5a}{3x} - \dfrac{4b}{3x}$

EXAMPLE 2 Add: $\dfrac{3}{x-1} + \dfrac{6}{x-1} + \dfrac{8}{x-1}$.

SOLUTION:
$$\dfrac{3}{x-1} + \dfrac{6}{x-1} + \dfrac{8}{x-1} = \dfrac{3+6+8}{x-1}$$
$$= \dfrac{17}{x-1}$$

The excluded value is $x = 1$.

Practice Exercise 3. $\dfrac{2p - 3}{x + 4} + \dfrac{3 - 4p}{x + 4}$

EXAMPLE 3 Add: $\dfrac{-7}{s-7} + \dfrac{s}{s-7}$.

SOLUTION:
$$\dfrac{-7}{s-7} + \dfrac{s}{s-7} = \dfrac{-7 + s}{s-7}$$
$$= \dfrac{s-7}{s-7}$$
$$= \dfrac{\cancel{s-7}^{1}}{\cancel{s-7}_{1}}$$
$$= 1$$

The excluded value is $s = 7$.

Practice Exercise 4. $\dfrac{3m - 1}{2p - 3} - \dfrac{4 - 5m}{2p - 3}$

EXAMPLE 4 Subtract: $\dfrac{6}{a} - \dfrac{7}{a}$.

12.3 Addition and Subtraction of Like Rational Expressions

SOLUTION: $\dfrac{6}{a} - \dfrac{7}{a} = \dfrac{6-7}{a}$

$= \dfrac{-1}{a}$

The excluded value is $a = 0$.

Practice Exercise 5.
$\dfrac{2x}{x+y} + \dfrac{3y}{x+y} - \dfrac{x-2y}{x+y}$

EXAMPLE 5 Subtract: $\dfrac{6x}{x-6} - \dfrac{x^2}{x-6}$.

SOLUTION: $\dfrac{6x}{x-6} - \dfrac{x^2}{x-6} = \dfrac{6x - x^2}{x-6}$

$= \dfrac{x(6-x)}{x-6}$

$= \dfrac{-x(x-6)}{x-6}$

$= \dfrac{-x(\cancel{x-6})}{\cancel{x-6}}$

$= -x$

The excluded value is $x = 6$.

Practice Exercise 6.
$\dfrac{m-n}{2p-q} - \dfrac{n-m}{2p-q} - \dfrac{4m-n}{2p-q}$

EXAMPLE 6 Subtract: $\dfrac{y^2+1}{2y-3} - \dfrac{3y-7}{2y-3}$.

SOLUTION: $\dfrac{y^2+1}{2y-3} - \dfrac{3y-7}{2y-3} = \dfrac{(y^2+1)-(3y-7)}{2y-3}$

$= \dfrac{y^2 - 3y + 8}{2y-3}$

The excluded value is $y = \dfrac{3}{2}$.

ANSWERS TO PRACTICE EXERCISES

1. $\dfrac{6}{u}$ $(u \ne 0)$. 2. $\dfrac{5a-4b}{3x}$ $(x \ne 0)$. 3. $\dfrac{-2p}{x+4}$ $(x \ne -4)$. 4. $\dfrac{8m-5}{2p-3}$ $\left(p \ne \dfrac{3}{2}\right)$. 5. $\dfrac{x+5y}{x+y}$ $(y \ne -x)$.

6. $\dfrac{-2m-n}{2p-q}$ $(2p \ne q)$.

EXERCISES 12.3

Add or subtract as indicated and simplify your results. List the excluded values.

1. $\dfrac{2}{x} + \dfrac{3}{x}$

2. $\dfrac{3}{y} + \dfrac{x}{y}$

3. $\dfrac{2a}{x} + \dfrac{3b}{x}$

4. $\dfrac{4y}{b} - \dfrac{1}{b}$

5. $\dfrac{a}{5x} - \dfrac{b}{5x}$

6. $\dfrac{4m}{7b} + \dfrac{5n}{7b}$

7. $\dfrac{2}{y} - \dfrac{3}{y} + \dfrac{4}{y}$

8. $\dfrac{-5}{2a} + \dfrac{1}{2a} - \dfrac{7}{2a}$

9. $\dfrac{x}{5c} - \dfrac{y}{5c} - \dfrac{2z}{5c}$

10. $\dfrac{2a}{3x} + \dfrac{b}{3x} - \dfrac{4c}{3x}$

11. $\dfrac{p}{6u} - \dfrac{5q}{6u} + \dfrac{7r}{6u}$

12. $\dfrac{2a}{xy} - \dfrac{4b}{xy} - \dfrac{5c}{xy}$

13. $\dfrac{x}{pq} + \dfrac{2y}{pq} - \dfrac{7}{pq}$

14. $\dfrac{2}{x-3} + \dfrac{5}{x-3}$

15. $\dfrac{6}{x+1} - \dfrac{8}{x+1}$

16. $\dfrac{1}{2y-1} + \dfrac{5}{2y-1}$

17. $\dfrac{2a}{w+1} - \dfrac{3b}{w+1}$

18. $\dfrac{4}{2t-3} + \dfrac{s}{2t-3}$

19. $\dfrac{7}{x^2-5} + \dfrac{5}{x^2-5}$

20. $\dfrac{2y}{y^2+7} - \dfrac{3y}{y^2+7}$

21. $\dfrac{5}{3m+2} + \dfrac{1}{3m+2} - \dfrac{2}{3m+2}$

22. $\dfrac{2a}{4p+1} - \dfrac{3b}{4p+1} + \dfrac{c}{4p+1}$

23. $\dfrac{x}{p^2+2q} - \dfrac{2y}{p^2+2q} - \dfrac{3z}{p^2+2q}$

24. $\dfrac{ab}{m^3+3} + \dfrac{cd}{m^3+3} - \dfrac{de}{m^3+3}$

25. $\dfrac{2x+1}{a-1} + \dfrac{3-x}{a-1}$

26. $\dfrac{2a+b}{y+2} - \dfrac{a-2b}{y+2}$

27. $\dfrac{3(a+1)}{x+5} + \dfrac{2(a+3)}{x+5}$

28. $\dfrac{4(b-1)}{y+6} - \dfrac{3(b-1)}{y+6}$

29. $\dfrac{2a-5}{2y+3} + \dfrac{7-4a}{2y+3}$

30. $\dfrac{9y+6}{3m+1} - \dfrac{7y+5}{3m+1}$

31. $\dfrac{2x+3xy}{x+y} + \dfrac{y^2-xy}{x+y}$

32. $\dfrac{a^2-3ab}{a-2b} - \dfrac{ab-b^2}{a-2b}$

33. $\dfrac{x}{x^2-y^2} + \dfrac{3y}{x^2-y^2} - \dfrac{2x-y}{x^2-y^2}$

34. $\dfrac{y}{x-y} - \dfrac{2x-y}{x-y} - \dfrac{x+3y}{x-y}$

35. $\dfrac{3p-q}{2p+1} - \dfrac{p+2q}{2p+1} + \dfrac{4p-5q}{2p+1}$

36. $\dfrac{m^2+2m}{m+4} + \dfrac{3m-m^2}{m+4} - \dfrac{6m}{m+4}$

37. $\dfrac{2p^2-1}{q+7} - \dfrac{2+3p^2}{q+7} - \dfrac{1-p^2}{q+7}$

38. $\dfrac{5a}{2a-3b} - \dfrac{b}{2a-3b} - \dfrac{b-2a}{2a-3b}$

39. $\dfrac{2m^2+3m}{m^2-4} - \dfrac{m^2-2}{m^2-4} - \dfrac{m^2+2m}{m^2-4}$

40. $\dfrac{4y^2+y}{y^2-9} - \dfrac{y^2-2y}{y^2-9} - \dfrac{6y}{y^2-9}$

12.4 Least Common Multiple

OBJECTIVE

After completing this section, you should be able to find the least common multiple (LCM) of two or more polynomials.

When adding two or more unlike fractions in arithmetic, we use a procedure involving the least common denominator (LCD). In Section 12.5, you will learn how to add two or more unlike rational expressions. To do so, we will use the *least common multiple* (LCM) of polynomials. It needs to be emphasized that finding the LCM of polynomials is a generalization of finding the LCD in arithmetic.

Determining the Least Common Multiple

To determine the **least common multiple (LCM)** of two or more polynomials:

1. Factor *each* polynomial completely, writing repeated factors using exponents.
2. Form the product of *all the different* factors found in the polynomial factorizations.
3. Raise each of these factors to the *highest* power in which the factor is found in any of the factorizations.

The resulting product is the LCM.

You should refer back to Chapter 2 to see the similarity between these steps for finding the LCM with those for finding the LCD of ordinary fractions.

Determine the least common multiple for each of the following groups of algebraic expressions.

Practice Exercise 1. $5p^2$; $15p^4$

EXAMPLE 1 Determine the LCM of $2x$ and $5x^2$.

SOLUTION:

STEP 1: Factor each polynomial completely:

$$2x = (2)(x)$$
$$5x^2 = (5)(x)(x) = (5)(x)^2$$

STEP 2: Form the product of all the different factors that appear in Step 1:

$$(2)(5)(x)$$

STEP 3: Raise each factor in Step 2 to the *highest* power in which it appears in Step 1:

$$(2)^1(5)^1(x)^2$$

Therefore, the LCM of $2x$ and $5x^2$ is $(2)(5)(x)^2 = 10x^2$.

Practice Exercise 2. $-2x^2y^3$; $4x^3y^2$; $5x^2y^4$

EXAMPLE 2 Determine the LCM of $3x^2y^3$, $6x^3y^2$, and $18xy^4$.

SOLUTION:

STEP 1: Factor each polynomial completely:

$$3x^2y^3 = (3)(x)^2(y)^3$$
$$6x^3y^2 = (2)(3)(x)^3(y)^2$$
$$18xy^4 = (2)(3)^2(x)(y)^4$$

STEP 2: Form the product of all the different factors that appear in Step 1:

$$(2)(3)(x)(y)$$

STEP 3: Raise each factor in Step 2 to the *highest* power in which it appears in Step 1:

$$(2)^1(3)^2(x)^3(y)^4$$

Therefore, the LCM for the given expressions is $2(9)(x^3)(y^4) = 18x^3y^4$.

Practice Exercise 3. $2u - 4$; $2u^2 - 8$

EXAMPLE 3 Determine the LCM of $x^2 - 25$, $6(x - 5)^2$, and $8(x + 5)$.

SOLUTION:

STEP 1: Factor each polynomial completely:

$$x^2 - 25 = (x + 5)(x - 5)$$
$$6(x - 5)^2 = (2)(3)(x - 5)^2$$
$$8(x + 5) = (2)^3(x + 5)$$

STEP 2: Form the product of all the different factors that appear in Step 1:

$$(2)(3)(x + 5)(x - 5)$$

STEP 3: Raise each factor in Step 2 to the *highest* power in which it appears in Step 1:

$$(2)^3(3)^1(x + 5)^1(x - 5)^2$$

Therefore, the LCM for the given expressions is $(8)(3)(x + 5)(x - 5)^2 = 24(x + 5)(x - 5)^2$.

Practice Exercise 4. $2a^2 - 2ab$; $3ab - 3b^2$; $a^2 - b^2$

EXAMPLE 4 Determine the LCM of $u^2 - 9$, $7u^2 - 21u$, and $u^2 + 6u + 9$.

SOLUTION:

STEP 1: Factor each polynomial completely:

$$u^2 - 9 = (u + 3)(u - 3)$$
$$7u^2 - 21u = 7u(u - 3) = (7)(u)(u - 3)$$
$$u^2 + 6u + 9 = (u + 3)(u + 3) = (u + 3)^2$$

STEP 2: Form the product of all the different factors that appear in Step 1:

$$(7)(u)(u + 3)(u - 3)$$

STEP 3: Raise each factor in Step 2 to the *highest* power in which it appears in Step 1:

$$(7)^1(u)^1(u + 3)^2(u - 3)^1$$

Therefore, the LCM for the given expressions is $7u(u - 3)(u + 3)^2$.

12.4 Least Common Multiple

Practice Exercise 5.
$p^2 + 5p + 6$; $p^2 - 4$; $p^2 - 9$

EXAMPLE 5 Determine the LCM for $2a^2b$, $3a(b - 2)^2$, and $6b^2(b^2 - 4)$.

SOLUTION:

STEP 1: Factor each polynomial completely:

$$2a^2b = (2)(a)^2(b)$$
$$3a(b - 2)^2 = (3)(a)(b - 2)^2$$
$$6b^2(b^2 - 4) = (2)(3)(b)^2(b + 2)(b - 2)$$

STEP 2: Form the product of all the different factors that appear in Step 1:

$$(2)(3)(a)(b)(b + 2)(b - 2)$$

STEP 3: Raise each factor in Step 2 to the *highest* power in which it appears in Step 1:

$$(2)^1(3)^1(a)^2(b)^2(b + 2)^1(b - 2)^2$$

Therefore, the LCM for the given expressions is $6a^2b^2(b + 2)(b - 2)^2$.

ANSWERS TO PRACTICE EXERCISES

1. $15p^4$. 2. $20x^3y^4$. 3. $2(u^2 - 4)$. 4. $6ab(a^2 - b^2)$ 5. $(p^2 - 4)(p^2 - 9)$.

EXERCISES 12.4

Determine the least common multiple for each of the following groups of algebraic expressions.

1. $3x$; $6x^2$
2. $-4y$; $8y^3$
3. $2u$; $-3u$; $5u^3$
4. $2ab$; $3a^2b$; $4ab^3$
5. $3y - 15$; $5y - 25$
6. $u^2 - 9$; $2u + 6$
7. $2t - 4$; $-5(t - 2)$
8. $2s + 1$; $3s - 2$
9. $3s - 6$; $2s - 4$
10. y; $y + 2$; y^3
11. $(3p + 4)^2$; $21p + 28$
12. $10p - 5$; $(1 - 5p)^2$
13. $a^2 - b^2$; $b - a$
14. $s^2 - st$; $st - t^2$
15. $a^2 - b^2$; $3a - 3b$
16. $2p^3 - 3p^2q$; $3q^2 - 2pq$
17. $x^2 - 4$; $3x - 6$
18. $x^2 + 2x - 3$; $x^2 + 5x + 6$
19. $y^2 - 36$; $2y + 12$
20. $p^2 - 9$; $p^2 + 4p - 21$
21. $m^2 - 9$; $m^2 + 7m + 12$
22. $t^2 - 3t - 4$; $t^2 - 9t + 20$
23. $q^2 - 5q + 4$; $q^2 + 6q - 7$
24. $y^2 - 4y - 45$; $y^2 + 7y + 10$
25. $u^2 + 8u + 15$; $u^2 + 2u - 15$
26. $r^2 - 5r - 36$; $r^2 + 6r + 8$
27. $y^2 - 9$; $4y^2 - 12y$; $y^2 + 6y + 9$
28. $x^2 - 4$; $x^2 + 4x + 4$; $(x - 2)^3$
29. $s^2 + 3s + 2$; $s^2 + 4s + 4$; $s^2 + 5s + 6$
30. $2t^3$; $3t^3 - 3t^2 - 6t$; $t^2 + 2t + 1$
31. $2m^2 - m - 1$; $(m - 1)^3$; $4m^2 - 1$
32. $-3(x + 2)^4$; $x^2 + 4x + 4$; $2x^2 - 8$
33. $(2p - q)^2$; $4p^2 - q^2$; $2p + q$
34. $m + 2$; $m^2 + m - 2$; $m^2 - 2m + 1$
35. $2(t - 5)^2$; $50 - 2t^2$; $t^3 - 5t^2$
36. $4s^3$; $s^3 - 6s^2 + 9s$; $s^2 - 9$
37. $3(u + 7)^2$; $u^2 - 49$; $2u^2 + 16u + 2$
38. $p^2 - 3p + 2$; $6 - p - p^2$; $p^2 + 6p + 9$
39. $y + 3$; $y^2 - 3y - 18$; $y^2 - 12y + 36$
40. $x^2 - 2x + 1$; $x^2 - 25$; $x^2 + 4x - 5$

12.5 Addition and Subtraction of Unlike Rational Expressions

OBJECTIVE

After completing this section, you should be able to use the procedure of finding the least common multiple to add or subtract unlike rational expressions.

In arithmetic, you learned that unlike fractions are fractions with different denominators. As a generalization of this, we have unlike rational expressions.

DEFINITION

Unlike rational expressions are rational expressions with different denominators.

- $\dfrac{x}{y}$ and $\dfrac{y}{x}$ are unlike rational expressions; the denominators are different.

- $\dfrac{2}{x-2}$ and $\dfrac{3}{x+2}$ are unlike rational expressions.

- $\dfrac{xy}{x^2-y^2}$ and $\dfrac{x^2-y^2}{xy}$ are unlike rational expressions.

You learned to add and subtract unlike fractions by rewriting the fractions in equivalent form with like denominators. The lowest common denominator was used. To add and subtract unlike rational expressions, we use a similar procedure. Simplify each of the rational expressions first, if necessary.

Rule for Adding and Subtracting Unlike Rational Expressions

To add (or subtract) unlike rational expressions.

1. Determine the LCM for the denominators involved.
2. Rewrite all rational expressions in equivalent form, with the LCM as the common denominator.
3. Add (or subtract) the resulting *like* rational expressions.
4. Simplify your results, if possible.

This rule will be illustrated in the following examples.

Perform the indicated operations and simplify your results. Include all of the excluded values.

Practice Exercise 1.
$\dfrac{2}{x} - \dfrac{3}{x^2} + \dfrac{4}{x^4}$

EXAMPLE 1 Add: $\dfrac{3}{a} + \dfrac{2}{a^4}$.

SOLUTION:

STEP 1: The LCM for a and a^4 is a^4. (Do you agree?)

STEP 2:

$\dfrac{3}{a} = \dfrac{(3)(a^3)}{(a)(a^3)} = \dfrac{3a^3}{a^4}$ (Multiplying both numerator and denominator by a^3 to get the LCM)

12.5 Addition and Subtraction of Unlike Rational Expressions

Step 3:

$$\frac{3}{a} + \frac{2}{a^4} = \frac{3a^3}{a^4} + \frac{2}{a^4}$$
$$= \frac{3a^3 + 2}{a^4}$$

Therefore, $\frac{3}{a} + \frac{2}{a^4} = \frac{3a^3 + 2}{a^4}$. The excluded value is $a = 0$.

Practice Exercise 2.
$$\frac{3}{y+4} + \frac{4}{y-4}$$

EXAMPLE 2 Add: $\frac{2}{a+1} + \frac{5}{a+3}$.

SOLUTION:

Step 1: The LCM for $a + 1$ and $a + 3$ is $(a + 1)(a + 3)$.

Step 2: Use the LCM to rewrite the rational expressions as like rational expressions:

$$\frac{2}{a+1} = \frac{2(a+3)}{(a+1)(a+3)}$$

$$\frac{5}{a+3} = \frac{5(a+1)}{(a+3)(a+1)} = \frac{5(a+1)}{(a+1)(a+3)}$$

Step 3: Add the like rational expressions:

$$\frac{2}{a+1} + \frac{5}{a+3} = \frac{2(a+3)}{(a+1)(a+3)} + \frac{5(a+1)}{(a+1)(a+3)}$$
$$= \frac{2(a+3) + 5(a+1)}{(a+1)(a+3)}$$

Step 4: Simplify:

$$\frac{2(a+3) + 5(a+1)}{(a+1)(a+3)} = \frac{2a + 6 + 5a + 5}{(a+1)(a+3)}$$
$$= \frac{7a + 11}{(a+1)(a+3)}$$

Therefore, $\frac{2}{a+1} + \frac{5}{a+3} = \frac{7a+11}{(a+1)(a+3)}$. The excluded values are $a = -3, -1$.

Practice Exercise 3.
$$\frac{p}{p+7} - \frac{2p}{p-7} + \frac{2}{p^2-49}$$

EXAMPLE 3 Subtract: $\frac{x}{x^2-4} - \frac{3}{2x+4}$.

SOLUTION:

Step 1: The LCM for $x^2 - 4$ and $2x + 4$ is $2(x + 2)(x - 2)$.

Step 2: Use the LCM to rewrite the rational expressions as like rational expressions:

$$\frac{x}{x^2-4} = \frac{x}{(x+2)(x-2)}$$
$$= \frac{x(2)}{(x+2)(x-2)(2)}$$
$$= \frac{2x}{2(x+2)(x-2)}$$

$$\frac{3}{2x+4} = \frac{3}{2(x+2)}$$
$$= \frac{3(x-2)}{2(x+2)(x-2)}$$

Step 3: Subtract the like rational expressions:

$$\frac{x}{x^2-4} - \frac{3}{2x+4} = \frac{2x}{2(x+2)(x-2)} - \frac{3(x-2)}{2(x+2)(x-2)}$$
$$= \frac{2x - 3(x-2)}{2(x+2)(x-2)}$$

Step 4: Simplify:

$$\frac{2x - 3(x-2)}{2(x+2)(x-2)} = \frac{2x - 3x + 6}{2(x+2)(x-2)}$$
$$= \frac{-x+6}{2(x+2)(x-2)}$$
$$= \frac{6-x}{2(x^2-4)}$$

Therefore, $\frac{x}{x^2-4} - \frac{3}{2x+4} = \frac{6-x}{2(x^2-4)}$. The excluded values are $x = -2, 2$.

Practice Exercise 4.
$$\frac{1}{s^2+4s+4} + \frac{3}{s^2+5s+6} - \frac{2}{s^2+6s+9}$$

EXAMPLE 4 Perform the indicated operations and simplify your results:

$$\frac{3}{2y-6} - \frac{y}{y^2-9} + \frac{5}{y+3}$$

SOLUTION:

Step 1: The LCM of the given denominators is $2(y+3)(y-3)$.

Step 2: Use the LCM to rewrite the rational expressions as like rational expressions:

$$\frac{3}{2y-6} = \frac{3}{2(y-3)}$$
$$= \frac{3(y+3)}{2(y-3)(y+3)}$$

$$\frac{y}{y^2-9} = \frac{y}{(y+3)(y-3)}$$
$$= \frac{y(2)}{(y+3)(y-3)(2)}$$
$$= \frac{2y}{2(y-3)(y+3)}$$

$$\frac{5}{y+3} = \frac{5(2)(y-3)}{(y+3)(2)(y-3)}$$
$$= \frac{10(y-3)}{2(y-3)(y+3)}$$

Step 3: Combine the like rational expressions:

$$\frac{3}{2y-6} - \frac{y}{y^2-9} + \frac{5}{y+3}$$

12.5 Addition and Subtraction of Unlike Rational Expressions

$$= \frac{3(y+3)}{2(y-3)(y+3)} - \frac{2y}{2(y-3)(y+3)} + \frac{10(y-3)}{2(y-3)(y+3)}$$

$$= \frac{3(y+3) - 2y + 10(y-3)}{2(y-3)(y+3)}$$

STEP 4: Simplify:

$$\frac{3(y+3) - 2y + 10(y-3)}{2(y-3)(y+3)} = \frac{3y + 9 - 2y + 10y - 30}{2(y-3)(y+3)}$$

$$= \frac{11y - 21}{2(y-3)(y+3)}$$

$$= \frac{11y - 21}{2(y^2 - 9)}$$

Therefore, $\frac{3}{2y-6} - \frac{y}{y^2-9} + \frac{5}{y+3} = \frac{11y - 21}{2(y^2-9)}$. The excluded values are $y = -3, 3$.

ANSWERS TO PRACTICE EXERCISES

1. $\frac{2x^3 - 3x^2 + 4}{x^4}$ $(x \neq 0)$. 2. $\frac{7y + 4}{y^2 - 16}$ $(y \neq -4, 4)$. 3. $\frac{2 - 21p - p^2}{p^2 - 49}$ $(p \neq -7, 7)$.

4. $\frac{2s^2 + 13s + 19}{(s+2)^2(s+3)^2}$ $(s \neq -3, -2)$.

EXERCISES 12.5

Do the indicated additions or subtractions in each of the following exercises. Simplify your results. List the excluded values in each exercise.

1. $\dfrac{2}{3a} + \dfrac{5}{6a^2}$

2. $\dfrac{x}{2b} - \dfrac{y}{b^4}$

3. $\dfrac{6}{5x} + \dfrac{1}{x^2} - \dfrac{2}{x^3}$

4. $\dfrac{a}{y^3} - \dfrac{b}{2y} - \dfrac{2c}{y^2}$

5. $\dfrac{3}{a^2b} + \dfrac{2}{ab} - \dfrac{5}{ab^2}$

6. $\dfrac{7}{xy} - \dfrac{3}{x^3y^2} + \dfrac{1}{x^2y^3}$

7. $\dfrac{1}{x-1} + \dfrac{2}{x-2}$

8. $\dfrac{3}{y+4} - \dfrac{5}{y-2}$

9. $\dfrac{2}{y-3} - \dfrac{3}{y+3}$

10. $\dfrac{-3}{s+2} + \dfrac{4}{s-5}$

11. $\dfrac{7}{a+5} + \dfrac{1}{a-4}$

12. $\dfrac{17}{b-6} - \dfrac{3}{b+1}$

13. $\dfrac{8}{2b-1} + \dfrac{1}{3b+2}$

14. $\dfrac{4}{2a-3b} - \dfrac{7}{3b-2a}$

15. $\dfrac{2x+1}{3x+3} + \dfrac{5x-3}{4x+4}$

16. $\dfrac{3p+q}{p-q} + \dfrac{3q}{2p}$

17. $\dfrac{6t}{5s} - \dfrac{2s+5t}{s-2t}$

18. $\dfrac{3}{x-2} + \dfrac{4}{x^2-4}$

19. $\dfrac{5y}{y^2-9} - \dfrac{2}{3y-9}$

20. $\dfrac{x}{2x+1} - \dfrac{3}{4x^2-1}$

21. $\dfrac{2}{x+1} + \dfrac{4}{x^2-1}$

22. $\dfrac{5}{y-2} - \dfrac{2}{y^2-4}$

23. $\dfrac{x}{x+3} - \dfrac{2x}{x^2-9}$

24. $\dfrac{2x}{(x-y)^2} - \dfrac{1}{x+y}$

25. $\dfrac{u+2}{u^2-16} + \dfrac{u}{2u+8}$

26. $\dfrac{t^2}{t^2-25} - \dfrac{3t}{5-t}$

27. $\dfrac{3}{x^2-1} + \dfrac{4}{x^2+2x+1}$

28. $\dfrac{5}{y^2-5y+6} - \dfrac{2}{y^2-4y+4}$

29. $\dfrac{1}{(u-7)^2} - \dfrac{1}{(u-7)^3}$

30. $\dfrac{4x}{(x-1)(x+2)} + \dfrac{2x-7}{(x+2)(x-3)}$

31. $\dfrac{3}{s^2-4} - \dfrac{5}{s^2+3s-10}$

32. $\dfrac{2x+1}{x^2-xy-2y^2} - \dfrac{3y-2}{x^2+xy-6y^2}$

33. $\dfrac{2a}{a^2-b^2} + \dfrac{1}{a^2+ab-2b^2}$

34. $\dfrac{2y-x}{3x+12y} - \dfrac{x+2y}{x^2+3xy-4y^2}$

35. $\dfrac{5}{2t^2-8} + \dfrac{7}{2t-t^2}$

36. $\dfrac{3}{2p^2-18} + \dfrac{1}{p^2+2p-15}$

37. $\dfrac{2}{4p^2+4p-3} - \dfrac{5}{12p^2-27}$

38. $\dfrac{2r}{r^2+11r+18} - \dfrac{3}{r^2-81}$

39. $\dfrac{7}{t^2-t-30} - \dfrac{5}{t^2-36}$

40. $\dfrac{2x}{x^2-16} + \dfrac{3x}{3x-12}$

12.6 Complex Rational Expressions

OBJECTIVE

After completing this section, you should be able to simplify complex rational expressions.

Just as we have complex fractions in arithmetic, we have complex rational expressions in algebra.

DEFINITION

A **complex rational expression** is an expression having a rational expression in its numerator or in its denominator or in both.

- $\dfrac{\dfrac{3x}{x+1}}{x^2-4}$ is a complex rational expression with a rational expression in the numerator.

- $\dfrac{y+1}{\dfrac{2}{y-3}}$ is a complex rational expression with a rational expression in the denominator.

- $\dfrac{\dfrac{x-2}{x^2-4}}{\dfrac{x+1}{2x-7}}$ is a complex rational expression with a rational expression in the numerator and also the denominator.

12.6 Complex Rational Expressions

- $\dfrac{x - \dfrac{2}{x-3}}{2x+1}$ is a complex rational expression. The numerator is a sum of two terms which, when added, give a rational expression.

To simplify a complex rational expression, we follow a procedure similar to simplifying complex fractions.

Rule for Simplifying Complex Rational Expressions

To simplify a complex rational expression:
1. Write the numerator as a single simplified rational expression.
2. Write the denominator as a single simplified rational expression.
3. Divide the simplified numerator by the simplified denominator.
4. Simplify the result.

Simplify each of the following complex rational expressions.

Practice Exercise 1. $\dfrac{2 + \dfrac{x}{3}}{5}$

EXAMPLE 1 Simplify: $\dfrac{\dfrac{2}{x} - \dfrac{3}{x^2}}{x+1}$.

SOLUTION:

STEP 1: Write the numerator as a single simplified rational expression. The LCM for x and x^2 is x^2. Hence,

$$\frac{2}{x} - \frac{3}{x^2} = \frac{2x}{x^2} - \frac{3}{x^2}$$
$$= \frac{2x - 3}{x^2}$$

STEP 2: The denominator of the given expression is already simplified.

STEP 3: Divide the simplified numerator by the denominator:

$$\frac{\dfrac{2x-3}{x^2}}{x+1} = \frac{\dfrac{2x-3}{x^2}}{\dfrac{x+1}{1}} \quad \left(\text{Rewriting } x+1 = \frac{x+1}{1}\right)$$

$$= \frac{2x-3}{x^2} \cdot \frac{1}{x+1} \quad \text{(Multiplying by the reciprocal of the denominator)}$$

$$= \frac{2x-3}{x^2(x+1)}$$

Therefore, $\dfrac{\dfrac{2}{x} - \dfrac{3}{x^2}}{x+1} = \dfrac{2x-3}{x^2(x+1)}$. The excluded values are $x = -1, 0$.

Practice Exercise 2. $\dfrac{\dfrac{1}{x}}{1 + \dfrac{1}{x}}$

EXAMPLE 2 Simplify: $\dfrac{\dfrac{5}{y+3}}{\dfrac{15}{3y+9}}$.

SOLUTION:

STEP 1: The numerator is already simplified.

STEP 2: Write the denominator as a single simplified rational expression:

$$\frac{15}{3y+9} = \frac{15}{3(y+3)}$$

$$= \frac{\overset{5}{\cancel{15}}}{\underset{1}{\cancel{3}}(y+3)}$$

$$= \frac{5}{y+3}$$

STEP 3: Divide the numerator by the simplified denominator:

$$\frac{\frac{5}{y+3}}{\frac{5}{y+3}} = \frac{5}{y+3} \cdot \frac{y+3}{5}$$

$$= \frac{\overset{1}{\cancel{5}}}{\underset{1}{\cancel{y+3}}} \cdot \frac{\overset{1}{\cancel{y+3}}}{\underset{1}{\cancel{5}}}$$

$$= 1$$

Therefore, $\dfrac{\frac{5}{y+3}}{\frac{15}{3y+9}} = 1$. The excluded value is $y = -3$.

Practice Exercise 3.

$$\frac{\dfrac{2}{p-2} - \dfrac{3}{p+2}}{\dfrac{1}{p^2-4}}$$

EXAMPLE 3 Simplify: $\dfrac{\dfrac{a^2}{b^2} - 1}{\dfrac{a}{b} - 1}$.

SOLUTION:

STEP 1: Write the numerator as a single simplified rational expression:

$$\frac{a^2}{b^2} - 1 = \frac{a^2}{b^2} - \frac{1}{1} \quad \text{(Rewriting)}$$

$$= \frac{a^2}{b^2} - \frac{b^2}{b^2} \quad \text{(LCM of } b^2 \text{ and 1 is } b^2\text{)}$$

$$= \frac{a^2 - b^2}{b^2}$$

STEP 2: Write the denominator as a single simplified rational expression:

$$\frac{a}{b} - 1 = \frac{a}{b} - \frac{1}{1} \quad \text{(Rewriting)}$$

$$= \frac{a}{b} - \frac{b}{b} \quad \text{(LCM of } b \text{ and 1 is } b\text{)}$$

$$= \frac{a-b}{b}$$

12.6 Complex Rational Expressions

STEP 3: Divide the simplified numerator by the simplified denominator:

$$\frac{\frac{a^2-b^2}{b^2}}{\frac{a-b}{b}} = \frac{a^2-b^2}{b^2} \cdot \frac{b}{a-b} \quad \text{(Multiplying by the reciprocal of the divisor)}$$

$$= \frac{(a+b)(a-b)}{b^2} \cdot \frac{b}{a-b} \quad \text{(Factoring)}$$

$$= \frac{(a+b)(a-b)^{1}}{b^{2}_{b}} \cdot \frac{\overset{1}{b}}{a-b^{1}}$$

$$= \frac{a+b}{b}$$

Therefore, $\dfrac{\frac{a^2}{b^2}-1}{\frac{a}{b}-1} = \dfrac{a+b}{b}$. The excluded values are $b = 0$ and $b = a$.

ANSWERS TO PRACTICE EXERCISES

1. $\dfrac{6+x}{15}$.   2. $\dfrac{1}{x+1}$ ($x = -1, 0$).   3. $10 - p$ ($p \neq -2, 2$).

EXERCISES 12.6

Simplify each of the following complex rational expressions. Include the excluded values for each expression.

1. $\dfrac{4-\frac{1}{3}}{5}$

2. $\dfrac{5+\frac{1}{2}}{7}$

3. $\dfrac{6}{5-\frac{1}{7}}$

4. $\dfrac{8}{9-\frac{2}{5}}$

5. $\dfrac{4+\frac{1}{3}}{2-\frac{1}{2}}$

6. $\dfrac{5-\frac{1}{7}}{2+\frac{3}{4}}$

7. $\dfrac{\frac{x^2}{y}}{x^3}$

8. $\dfrac{2m^3}{\frac{4m}{n^5}}$

9. $\dfrac{\frac{2x}{3y^2}}{\frac{4x^3}{9y}}$

10. $\dfrac{a-\frac{1}{a}}{a^2}$

11. $\dfrac{\frac{1}{b}}{b+\frac{1}{b}}$

12. $\dfrac{\frac{2xy}{x+y}}{\frac{4yz}{x+y}}$

13. $\dfrac{\frac{1}{x}+\frac{1}{y}}{\frac{1}{x}-\frac{1}{y}}$

14. $\dfrac{\frac{1}{a^2}-\frac{1}{b^2}}{\frac{3}{ab}}$

15. $\dfrac{\frac{2}{st}}{\frac{s}{t}+\frac{t}{s}}$

16. $\dfrac{\frac{2}{a}-\frac{3}{b}}{c-\frac{4}{d}}$

17. $\dfrac{\dfrac{a}{b}+\dfrac{b}{a}}{\dfrac{a}{b}-\dfrac{b}{a}}$

18. $\dfrac{\dfrac{2}{y}-\dfrac{3}{y^2}}{\dfrac{1}{y^4}-\dfrac{1}{y^3}}$

19. $\dfrac{\dfrac{1}{a}+\dfrac{2}{a^2}}{1-\dfrac{5}{a^2}}$

20. $\dfrac{2+\dfrac{1}{5b}}{1-\dfrac{1}{10b}}$

21. $\dfrac{4-\dfrac{3}{8c}}{5+\dfrac{3}{4c}}$

22. $\dfrac{\dfrac{1}{x}-3}{\dfrac{3}{2x}+4}$

23. $\dfrac{\dfrac{3}{y}+1}{y-\dfrac{1}{y}}$

24. $\dfrac{1+\dfrac{1}{u}}{1-\dfrac{1}{u^2}}$

25. $\dfrac{\dfrac{1}{x}+\dfrac{1}{xy}}{\dfrac{1}{xy}}$

26. $\dfrac{\dfrac{2}{y}-\dfrac{1}{xy}}{\dfrac{3}{x^2y^2}}$

27. $\dfrac{\dfrac{2}{z}}{z-\dfrac{1}{z}}$

28. $\dfrac{\dfrac{1}{xy^2}+\dfrac{1}{x^2y}}{\dfrac{1}{x^3y^3}}$

29. $\dfrac{\dfrac{a}{b}-\dfrac{b}{a}}{\dfrac{1}{ab}}$

30. $\dfrac{\dfrac{s-t}{s}}{\dfrac{s+t}{t}}$

31. $\dfrac{1-\dfrac{1}{x}}{1+\dfrac{1}{x^2}}$

32. $\dfrac{\dfrac{p}{q}-\dfrac{q}{p}}{\dfrac{p+q}{p^2q^3}}$

33. $\dfrac{\dfrac{2}{x-y}}{\dfrac{1}{x^2-y^2}}$

34. $\dfrac{\dfrac{5}{p^2-q^2}}{\dfrac{3}{p+q}}$

35. $\dfrac{\dfrac{x}{x-y}}{\dfrac{xy}{x^2-y^2}}$

36. $\dfrac{\dfrac{m-n}{m}}{\dfrac{m^2-n^2}{mn}}$

37. $\dfrac{\dfrac{3}{(p-q)^2}}{\dfrac{4}{p^2-q^2}}$

38. $\dfrac{\dfrac{a}{a-b}}{\dfrac{a}{a^2-b^2}}$

39. $\dfrac{\dfrac{2}{x}-\dfrac{3}{y}}{\dfrac{5}{x^2}-\dfrac{1}{y^2}}$

40. $\dfrac{\dfrac{1}{a+b}-\dfrac{2}{a-b}}{\dfrac{2}{(a-b)^2}}$

12.7 Equations Involving Rational Expressions

OBJECTIVE

After completing this section, you should be able to solve some equations involving rational expressions.

You have already learned to solve linear and some quadratic polynomial equations. You also learned to solve equations of the form $\dfrac{x-3}{2}=\dfrac{x}{3}$ during the discussion of proportions. We will review this in the following example.

Solve each of the following equations for the indicated variable. Determine all excluded values, and check your results in the original equation.

Practice Exercise 1. $\dfrac{3y-1}{4}=\dfrac{2}{3}$

EXAMPLE 1 Solve the equation $\dfrac{x-3}{2}=\dfrac{x}{3}$ for x.

SOLUTION: $\dfrac{x-3}{2}=\dfrac{x}{3}$ (This is a proportion.)

$3(x-3)=2(x)$ (Product of means = product of extremes)
$3x-9=2x$ (Removing parentheses)

12.7 Equations Involving Rational Expressions

$$x - 9 = 0 \quad \text{(Subtracting 2x from both sides)}$$
$$x = 9 \quad \text{(Adding 9 to both sides)}$$

CHECK:
$$\frac{x-3}{2} = \frac{x}{3}$$
$$\frac{9-3}{2} \stackrel{?}{=} \frac{9}{3}$$
$$\frac{6}{2} \stackrel{?}{=} \frac{9}{3}$$
$$3 = 3 \checkmark$$

Therefore, $x = 3$ is the solution.

How can we solve equations involving rational expressions if the equation is not in the form of a proportion? The answer is as follows.

Consider the equation

$$\frac{2}{u+1} + 1 = \frac{u}{u-3} \quad (u \neq -1, u \neq 3)$$

which we want to solve for u.

The LCM of *all* denominators involved in the equation is $(u + 1)(u - 3)$. Multiplying *both* sides of the equation by this LCM will give us an equivalent equation (since the LCM $\neq 0$). We have

$$(u+1)(u-3)\left(\frac{2}{u+1} + 1\right) = (u+1)(u-3)\left(\frac{u}{u-3}\right) \quad (u \neq -1, u \neq 3)$$

Using the distributive property for multiplication over addition on the left-hand side of the equation, we obtain the equivalent equation

$$(u+1)(u-3)\left(\frac{2}{u+1}\right) + (u+1)(u-3)(1) = (u+1)(u-3)\left(\frac{u}{u-3}\right)$$

or

$$(u+\cancel{1})(u-3)\left(\frac{2}{\cancel{u+1}}\right) + (u+1)(u-3)(1) = (u+1)(u\cancel{-3})\left(\frac{u}{\cancel{u-3}}\right)$$

$$2(u-3) + (u+1)(u-3) = u(u+1)$$

This last equation can now be solved for u:

$$2(u-3) + (u+1)(u-3) = u(u+1)$$
$$2u - 6 + (u^2 - 2u - 3) = u^2 + u$$
$$2u - 6 + u^2 - 2u - 3 = u^2 + u$$
$$u^2 - 9 = u^2 + u$$
$$-9 = u$$

Checking this result in the *original* equation, we determine that -9 is a solution. (You are encouraged to check this solution.)

Note that multiplying *both* sides of the equation by the LCM of *all* denominators involved resulted in:

- *Every* term in the equation being multiplied by the LCM
- Obtaining an equivalent equation *without* fractions

Rule for Solving Equations Involving Rational Expressions

To solve equations involving rational expressions:

1. Multiply *every* term on *both* sides of the equation by the LCM of *all* denominators involved. (The objective here is to remove fractions.)
2. Proceed as you did when solving equations earlier in the text.

Practice Exercise 2.
$$\frac{4u}{u-1} - 4 = \frac{2}{u+1}$$

EXAMPLE 2 Solve the equation $\frac{x-2}{4} + \frac{x}{3} = \frac{1}{12}$ for x.

SOLUTION:

STEP 1: The LCM of 4, 3, and 12 is 12. Multiply *every* term on *both* sides of the equation by 12. We have

$$12\left(\frac{x-2}{4}\right) + 12\left(\frac{x}{3}\right) = 12\left(\frac{1}{12}\right)$$

$$\overset{3}{\cancel{12}}\left(\frac{x-2}{\cancel{4}}\right) + \overset{4}{\cancel{12}}\left(\frac{x}{\cancel{3}}\right) = \overset{1}{\cancel{12}}\left(\frac{1}{\cancel{12}}\right)$$

$$3(x-2) + 4x = 1$$

STEP 2: Solve the resulting equation:

$$3(x-2) + 4x = 1$$
$$3x - 6 + 4x = 1$$
$$7x - 6 = 1$$
$$7x = 7$$
$$x = 1$$

CHECK: Check in the *original* equation:

$$\frac{x-2}{4} + \frac{x}{3} = \frac{1}{12}$$

$$\frac{1-2}{4} + \frac{1}{3} \;?\; \frac{1}{12}$$

$$\frac{-1}{4} + \frac{1}{3} \;?\; \frac{1}{12}$$

$$\frac{-3}{12} + \frac{4}{12} \;?\; \frac{1}{12}$$

$$\frac{1}{12} = \frac{1}{12} \checkmark$$

Therefore, $x = 1$ is the solution.

Practice Exercise 3.
$$\frac{6}{w+5} - \frac{9}{w-5} = \frac{3w}{w^2 - 25}$$

EXAMPLE 3 Solve the equation $\frac{y}{y-3} - 1 = \frac{2}{y+3}$ for y, if $y \neq -3$ or 3.

SOLUTION: The LCM for $y - 3$, 1, and $y + 3$ is $(y - 3)(y + 3)$. Multiplying *every* term on *both* sides of the equation by the LCM, we have

12.7 Equations Involving Rational Expressions

$$(y-3)(y+3)\left(\frac{y}{y-3}\right) - (y-3)(y+3)(1) = (y-3)(y+3)\left(\frac{2}{y+3}\right)$$

$$(\cancel{y-3})(y+3)\left(\frac{y}{\cancel{y-3}}\right) - (y-3)(y+3)(1) = (y-3)(\cancel{y+3})\left(\frac{2}{\cancel{y+3}}\right)$$

$$(y+3)(y) - (y-3)(y+3) = (y-3)(2)$$
$$y^2 + 3y - (y^2 - 9) = 2y - 6$$
$$y^2 + 3y - y^2 + 9 = 2y - 6$$
$$3y + 9 = 2y - 6$$
$$y + 9 = -6$$
$$y = -15$$

CHECK: Checking this result in the *original* equation, we have

$$\frac{y}{y-3} - 1 = \frac{2}{y+3}$$

$$\frac{-15}{-15-3} - 1 \; ? \; \frac{2}{-15+3}$$

$$\frac{15}{18} - 1 \; ? \; \frac{2}{-12}$$

$$\frac{-3}{18} \; ? \; \frac{2}{-12}$$

$$\frac{-1}{6} = \frac{-1}{6} \; \checkmark$$

Therefore, $y = -15$ is the solution.

Practice Exercise 4.
$$\frac{2x}{9x^2-1} + \frac{1}{3x-1} = \frac{4}{3x+1}$$

EXAMPLE 4 Solve the equation $\dfrac{5}{x-1} - \dfrac{2}{x-3} = \dfrac{3x}{x^2-1}$ for x, if $x \neq -1, 1,$ or 3.

SOLUTION: The LCM of $x-1$, $x-3$, and x^2-1 is $(x^2-1)(x-3)$. Multiplying *every* term on *both* sides of the equation by the LCM, we have

$$(x^2-1)(x-3)\left(\frac{5}{x-1}\right) - (x^2-1)(x-3)\left(\frac{2}{x-3}\right) = (x^2-1)(x-3)\left(\frac{3x}{x^2-1}\right)$$

$$(\cancel{x^2-1}^{x+1})(x-3)\left(\frac{5}{\cancel{x-1}}\right) - (x^2-1)(\cancel{x-3})\left(\frac{2}{\cancel{x-3}}\right) = (\cancel{x^2-1})(x-3)\left(\frac{3x}{\cancel{x^2-1}}\right)$$

$$5(x+1)(x-3) - 2(x^2-1) = 3x(x-3)$$
$$5(x^2 - 2x - 3) - 2x^2 + 2 = 3x^2 - 9x$$
$$5x^2 - 10x - 15 - 2x^2 + 2 = 3x^2 - 9x$$
$$3x^2 - 10x - 13 = 3x^2 - 9x$$
$$-10x - 13 = -9x$$
$$-13 = x$$

Therefore, $x = -13$ is the solution. Checking this solution in the *original* equation is left to you.

ANSWERS TO PRACTICE EXERCISES

1. $y = \dfrac{11}{9}$. 2. $u = -3$. 3. $w = \dfrac{-25}{2}$. 4. $x = \dfrac{5}{7}$.

Sometimes an equation involving rational expressions has no solution. This may result from the excluded values for one or more of the rational expressions, as will be illustrated in the next example. Therefore, it is always necessary to check your results in the *original* equation.

EXAMPLE 5 Solve the equation $\dfrac{y}{y-2} = 3 + \dfrac{2}{y-2}$ for y, if $y \neq 2$.

SOLUTION: The LCM of $y - 2$ and 1 is $y - 2$. Multiplying *every* term of *both* sides of the equation by the LCM, we have

$$(y-2)\left(\dfrac{y}{y-2}\right) = (y-2)(3) + (y-2)\left(\dfrac{2}{y-2}\right)$$

$$y = 3(y-2) + 2$$
$$y = 3y - 6 + 2$$
$$y = 3y - 4$$
$$-2y = -4$$
$$y = 2$$

It would appear that $y = 2$ is our solution. However, it is *not*. Observe that $y = 2$ is an excluded value for each of the two rational expressions involved in the equation. Hence, there is *no* solution for the given equation. If the result of solving an equation is an excluded value for even one of the rational expressions in the equation, then it is *not* a solution. The result does *not* need to be an excluded value for *all* of the rational expressions in the equation.

Practice Exercise 5.
$\dfrac{-1}{t+1} = \dfrac{t}{t+1} + 5$

ANSWER TO PRACTICE EXERCISE

5. No solution.

EXERCISES 12.7

Solve each of the following equations for the indicated variable. Determine all excluded values, and check your results in the original equations.

1. $\dfrac{x-3}{2} = 4$

2. $\dfrac{y+6}{3} = -1$

3. $\dfrac{2x+1}{3} = \dfrac{2}{5}$

4. $\dfrac{3y-2}{3} = \dfrac{-1}{2}$

5. $\dfrac{x-3}{4} = \dfrac{x}{2}$

6. $\dfrac{y+1}{5} = \dfrac{2y}{3}$

7. $\dfrac{2p-1}{3} = \dfrac{-3p}{4}$

8. $\dfrac{9q+7}{2} = \dfrac{4q}{-3}$

9. $\dfrac{2x-1}{4} = \dfrac{x+2}{3}$

10. $\dfrac{y-3}{5} = \dfrac{2y+5}{2}$

11. $\dfrac{s}{2} + 3 = \dfrac{s}{3} - 7$

12. $\dfrac{2t}{3} + 1 = \dfrac{3t}{4} - 5$

13. $\dfrac{r}{5} - 5 = \dfrac{7r}{4} + 2$

14. $\dfrac{2t}{3} - \dfrac{1}{2} = \dfrac{4t}{5} + \dfrac{1}{3}$

15. $\dfrac{x}{2} - 3 = \dfrac{2x + 1}{5}$

16. $\dfrac{2y - 3}{7} - \dfrac{1}{2} = \dfrac{4y}{3}$

17. $\dfrac{4t - 5}{3} - \dfrac{2t + 7}{2} = 5$

18. $\dfrac{t + 1}{2} + \dfrac{2t + 3}{3} = 8$

19. $\dfrac{6s - 5}{2} - \dfrac{s - 3}{3} = 4$

20. $6 - \dfrac{2r + 5}{5} = \dfrac{r - 6}{4}$

21. $\dfrac{x}{2} - \dfrac{2x - 1}{3} = \dfrac{4x + 5}{4}$

22. $\dfrac{2y - 1}{3} - \dfrac{y}{6} = \dfrac{4y + 3}{5}$

23. $\dfrac{2s}{3} = \dfrac{s + 4}{2} - \dfrac{2s - 1}{3}$

24. $\dfrac{s}{4} - \dfrac{2s}{3} = \dfrac{4s}{3} - \dfrac{s}{5}$

25. $\dfrac{x - 1}{2} - \dfrac{2x + 1}{3} = \dfrac{3x - 2}{4}$

26. $\dfrac{2y - 1}{9} + \dfrac{y - 4}{5} = \dfrac{4y - 5}{3}$

27. $\dfrac{3r - 1}{4} = \dfrac{2r + 3}{6} - \dfrac{r + 7}{12}$

28. $\dfrac{6t - 1}{5} = \dfrac{2t + 3}{7} + \dfrac{t + 10}{3}$

29. $\dfrac{3x}{x + 2} - 3 = \dfrac{1}{x - 2}$

30. $\dfrac{3y}{y + 3} - \dfrac{4}{y - 3} = 3$

31. $\dfrac{5y}{y - 4} + \dfrac{6}{y + 4} = 5$

32. $\dfrac{-y}{y + 5} + 1 = \dfrac{6}{y - 5}$

33. $\dfrac{4}{s - 2} - \dfrac{2}{s + 2} = \dfrac{3s}{s^2 - 4}$

34. $\dfrac{5}{t + 3} + \dfrac{1}{t - 3} = \dfrac{7t}{t^2 - 9}$

35. $\dfrac{p}{4p^2 - 1} + \dfrac{1}{2p - 1} = \dfrac{3}{2p + 1}$

36. $\dfrac{7s}{9s^2 - 1} - \dfrac{2}{3s + 1} = \dfrac{5}{3s - 1}$

37. $\dfrac{8t}{2t^2 - 8} + \dfrac{3}{t + 2} = \dfrac{5}{3t - 6}$

38. $\dfrac{3}{2s - 10} - \dfrac{4}{3s + 15} = \dfrac{5s}{2s^2 - 50}$

39. $\dfrac{3}{x^2 - 5x + 6} - \dfrac{4}{x^2 - 2x - 3} = \dfrac{2}{x^2 - x - 2}$

40. $\dfrac{3s}{s^2 - s - 12} + \dfrac{2}{s^2 - 2s - 15} = \dfrac{3s + 1}{s^2 - 9s + 20}$

12.8 Applications Involving Rational Expressions

OBJECTIVE

After completing this section, you should be able to solve some word problems whose associated equations involve rational expressions.

You have learned to solve word problems by translating the verbal statement into an algebraic equation. The equation was then solved to determine the solution to the verbal problem. In this section, we will examine additional word problems. The associated equations for some of them will involve rational expressions. Typical word problems will be discussed in the following examples.

Practice Exercise 1. One pipe can fill a water tank in 35 min. A second pipe can fill the same tank in 50 min. If the tank is empty, how long should it take both pipes, operating together, to fill the tank?

EXAMPLE 1 Brian can paint a certain house in 4 days. Gina can paint the same house in 5 days. How long should it take them, working together, to paint the house?

SOLUTION: We could

Let t = the number of days that it would take both of them working together to paint the house

Then,

$$\frac{1}{t} = \text{the portion of the house that would be painted in one day}$$

(Note that $t > 0$. Why?)

Since Brian can paint the house in 4 days working alone, he would be able to paint $\frac{1}{4}$ of the house in one day. Similarly, Gina would be able to paint $\frac{1}{5}$ of the house in 1 day, working alone. (Keep in mind that these are average rates.) Then, working together, Gina and Brian would be able to paint

$$\frac{1}{4} + \frac{1}{5}$$

of the house in one day. Thus, we have the equation

$$\frac{1}{4} + \frac{1}{5} = \frac{1}{t}$$

for this problem. The LCM for 4, 5, and t is $20t$. Multiplying *every* term of *both* sides of the equation by the LCM, we have

$$20t\left(\frac{1}{4}\right) + 20t\left(\frac{1}{5}\right) = 20t\left(\frac{1}{t}\right)$$

$$5t + 4t = 20$$
$$9t = 20$$
$$t = \frac{20}{9}$$
$$= 2\frac{2}{9} \text{ days}$$

Therefore, it should take Brian and Gina $2\frac{2}{9}$ days, working together, to paint the house.

Practice Exercise 2. The length of a rectangle is 12 yd more than its width. The sum of $\frac{1}{2}$ the width and $\frac{1}{4}$ the length is 8 yd. What are the dimensions of the rectangle?

EXAMPLE 2 An oil tank can be filled by three different pipes. The first pipe alone can fill the tank in 12 hr. The second pipe alone can fill the tank in 15 hr. The third pipe alone can fill the tank in 18 hr. How long should it take to fill the tank with all three pipes operating?

SOLUTION: We could

Let t = the number of hours it would take all three pipes, operating together, to fill the tank

12.8 Applications Involving Rational Expressions

Then,

$$\frac{1}{t} = \text{the amount of the tank filled in one hour} \quad (t > 0)$$

$\frac{1}{12}$ of the tank would be filled by the first pipe, operating alone, in 1 hr; $\frac{1}{15}$ by the second pipe alone in 1 hr; and $\frac{1}{18}$ by the third pipe alone in 1 hr. Therefore,

$$\frac{1}{12} + \frac{1}{15} + \frac{1}{18}$$

of the tank would be filled in 1 hr by all three pipes operating together. Thus, the equation representing this problem is

$$\frac{1}{12} + \frac{1}{15} + \frac{1}{18} = \frac{1}{t}$$

The LCM of 12, 15, 18, and t is $180t$. Multiplying *every* term of *both* sides of this equation by the LCM, we have

$$180t\left(\frac{1}{12}\right) + 180t\left(\frac{1}{15}\right) + 180t\left(\frac{1}{18}\right) = 180t\left(\frac{1}{t}\right)$$

$$15t + 12t + 10t = 180$$
$$37t = 180$$
$$t = \frac{180}{37}$$
$$= 4\frac{32}{37} \text{ hr}$$

Hence, it should take $4\frac{32}{37}$ hr for all three pipes, working together, to fill the tank.

Practice Exercise 3. If 21 is added to two consecutive positive integers, the average of the three numbers is 32. What are the numbers?

EXAMPLE 3 One number is 2 more than a second number. The reciprocal of the smaller number is 3 times the reciprocal of the larger number. What are the numbers?

SOLUTION: Let

$$n = \text{the smaller number}$$

Then,

$$n + 2 = \text{the larger number}$$

The reciprocals are $\frac{1}{n}$ and $\frac{1}{n+2}$, respectively. Since the reciprocal of the smaller number is 3 times the reciprocal of the larger number, we have the equation

$$\frac{1}{n} = 3\left(\frac{1}{n+2}\right)$$

or
$$\frac{1}{n} = \frac{3}{1}\left(\frac{1}{n+2}\right)$$
$$= \frac{3}{n+2}$$

The LCM of n and $n+2$ is $n(n+2)$. Multiplying *every* term on *both* sides of the equation, we have

$$n(n+2)\left(\frac{1}{n}\right) = n(n+2)\left(\frac{3}{n+2}\right)$$

$$\overset{1}{\cancel{n}}(n+2)\left(\frac{1}{\underset{1}{\cancel{n}}}\right) = n(\overset{1}{\cancel{n+2}})\left(\frac{3}{\underset{1}{\cancel{n+2}}}\right)$$

$$n + 2 = 3n$$
$$2 = 2n$$
$$1 = n$$

Therefore,

$n = 1$ (The smaller number)

and $n + 2 = 3$ (The larger number)

Check that these numbers satisfy the word problem.

ANSWERS TO PRACTICE EXERCISES

1. $20\frac{10}{17}$ min. 2. 10 yd by 12 yd. 3. 37 and 38.

EXERCISES 12.8

Solve each of the following word problems by using appropriate equations. Check your results.

1. One number is 3 more than a second number. The reciprocal of the smaller number is 4 times the reciprocal of the larger number. What are the numbers?
2. One number is 5 less than a second number. The reciprocal of the smaller number is 6 times the reciprocal of the larger number. What are the numbers?
3. One number is 3 more than twice a second number. The reciprocal of the smaller number is 3 times the reciprocal of the larger number. What are the numbers?
4. The denominator of a fraction is 7 more than its numerator. If 1 is subtracted from both the numerator and the denominator, the result is $\frac{1}{8}$. What is the original number?
5. The sum of two numbers is 75. One number is two-thirds of the other number. What are the numbers?
6. The sum of two numbers is 48. If the larger number is added to one-half of the smaller number, the sum is 40. What are the numbers?
7. The denominator of a fraction is 1 less than twice its numerator. If 1 is added to both the numerator and the denominator of the fraction, the result is $\frac{3}{5}$. What is the original fraction?

8. One number is 1 less than twice another number. Their average is 10. What are the numbers? (*Hint:* To find the average of two numbers, divide their sum by 2.)
9. The average of three consecutive odd numbers is 15. What are the numbers?
10. If 29 is added to two consecutive even numbers, the average of the three numbers is 21. What are the two numbers?
11. Sue can mow a lawn in 1 hr. Don can mow the same lawn in 50 min. If they work together, how long should it take them to mow the lawn?
12. One pipe can fill a water tank in 20 min. A second pipe can fill the same tank in 35 min. If the tank is empty, how long should it take both pipes, operating together, to fill the tank?
13. One pipe can fill an oil tank in 40 min. A second pipe can fill the same tank in 90 min. A third pipe can fill the same tank in 60 min. If the tank is empty and the three pipes are operating together, how long should it take to fill the tank?
14. The width of a rectangle is one-half its length. If the perimeter of the rectangle is 216 m, what are its dimensions?
15. The length of a rectangle is 5 more than its width. The sum of one-half the width and one-third the length is 10 yd. What are the dimensions of the rectangle?
16. An old assembly machine takes 4 times as long to do a job as a new machine. With both machines working together, the job can be done in 8 hr. How many hours should it take the new machine, working alone, to do the job?
17. A plumber and her assistant, working together, can do a job in 3 hr. Working alone, the plumber can do the job in 4 hr. How long should it take her assistant, working alone, to do the same job?
18. Barbara can type a report in 2 hr. Dick can type the same report in 6 hr. Barbara typed alone for 1 hr and stopped typing to relieve the switchboard operator. How long should it take Dick to finish the job?
19. One pipe can fill a petroleum tank in 3 hr. A second pipe can empty the filled tank in 6 hr. If the tank is empty and both pipes are operating, how many hours should it take to fill the tank?
20. A large storage tank can be filled by one pipe in 20 hr. The same tank can be filled by a second pipe in 15 hr. It can also be drained by a third pipe in 18 hr. If the tank is empty and all three pipes are operating, how long should it take to fill the tank?

Chapter Summary

In this chapter, we discussed rational expressions and operations on them. The least common multiple of polynomials was introduced and used in adding and subtracting unlike rational expressions. Complex rational expressions were considered, as were equations involving rational expressions. Applications were also examined.

The **key words** introduced in the chapter are listed here in the order in which they appeared in the text. You should know what they mean before proceeding to the next chapter.

rational expressions (p. 500)
excluded value (p. 500)
like rational expressions (p. 511)
least common multiple (LCM) (p. 515)
unlike rational expressions (p. 518)
complex rational expression (p. 522)

Review Exercises

12.1 Rational Expressions and Their Simplification

In Exercises 1–6, list the excluded values for the given rational expressions.

1. $\dfrac{2}{y+3}$
2. $\dfrac{y^2}{2y-6}$
3. $\dfrac{x+3}{3x-7}$
4. $\dfrac{16}{4x^2-9}$
5. $\dfrac{s+3}{s^2+2s+1}$
6. $\dfrac{t-7}{t^2+5t+6}$

In Exercises 7–14, simplify each of the given rational expressions. List all excluded values.

7. $\dfrac{-6x^2y}{18x^3y^4}$

8. $\dfrac{3s^3t^4u^2}{9st^6u^4}$

9. $\dfrac{3y+9}{y+3}$

10. $\dfrac{b^2s+b^2t}{cs+ct}$

11. $\dfrac{rt^2-2st^2}{2r^2s-4rs^2}$

12. $\dfrac{2r^2-2s^2}{4r+4s}$

13. $\dfrac{2p^2+3pq-2q^2}{p^2+pq-2q^2}$

14. $\dfrac{3x^2-11xy-20y^2}{2x^2-11xy+5y^2}$

12.2 Multiplication and Division of Rational Expressions

In Exercises 15–22, perform the indicated operations and simplify your results. List all excluded values.

15. $\dfrac{2xy^3}{3x^4} \cdot \dfrac{6x^3y}{10xy^2}$

16. $\dfrac{y^2-9}{14} \cdot \dfrac{7y-21}{y+3}$

17. $\dfrac{st^2}{6s-6t} \cdot \dfrac{s^2-t^2}{s^3t}$

18. $\dfrac{a^2-ab-2b^2}{a^2+6ab+9b^2} \cdot \dfrac{3a^2+9ab}{a^2-4b^2}$

19. $\dfrac{a^3}{(b-1)^2} \div \dfrac{7a^5}{3b-3}$

20. $\dfrac{6(s+6)}{s^2-36} \div \dfrac{18-3s}{2s+12}$

21. $\dfrac{x^2-1}{x^2-6x+5} \div \dfrac{6x^2-x-2}{3x^2+x-2}$

22. $\dfrac{y^2+8y+12}{y^2+11y+30} \div \dfrac{y^2-8y+16}{y^2+y-20}$

12.3 Addition and Subtraction of Like Rational Expressions

In Exercises 23–28, add or subtract as indicated and simplify your results. List all excluded values.

23. $\dfrac{3}{y+1}+\dfrac{2}{y+1}$

24. $\dfrac{6}{4t-3}+\dfrac{5}{4t-3}+\dfrac{8}{4t-3}$

25. $\dfrac{2a+b}{3x-1}+\dfrac{2b-1}{3x-1}$

26. $\dfrac{9p+2q}{4s+5}-\dfrac{7q-2p}{4s+5}$

27. $\dfrac{4a}{2x-3y}+\dfrac{b}{2x-3y}-\dfrac{2b-3a}{2x-3y}$

28. $\dfrac{6s+1}{s^2-4}-\dfrac{1-s^2}{s^2-4}-\dfrac{2s^2-7}{s^2-4}$

12.4 Lease Common Multiple

In Exercises 29–36, determine the least common multiple for each of the following groups of algebraic expressions.

29. $3y;\ 18y^4$

30. $6x^2y;\ -3xy^3;\ 9x^3y^2$

31. $2s-4;\ -5(s-2)$

32. $x^2-y^2;\ y-x;\ (x+y)^2$

33. $y^2-9;\ y^2+6y+9;\ (y-3)^3$

34. $-5(s+2)^4;\ s^2+4s+4;\ 3s^2-12$

35. $3s^5;\ s^4-6s^3+9s^2;\ s^2-9$

36. $x^2+6x+9;\ 6-x-x^2;\ x^2-4x+4$

12.5 Addition and Subtraction of Unlike Rational Expressions

In Exercises 37–42, add or subtract as indicated and simplify your results. List all excluded values.

37. $\dfrac{5}{4y^2}+\dfrac{1}{y^3}-\dfrac{4}{y}$

38. $\dfrac{6}{x-3}-\dfrac{5}{x+1}$

39. $\dfrac{6t}{t^2-16}-\dfrac{5}{3t-12}$

40. $\dfrac{1}{y-3}-\dfrac{2}{y+3}-\dfrac{4}{y^2-9}$

41. $\dfrac{6s}{s^2+s-2}+\dfrac{2s-7}{s^2-s-6}$

42. $\dfrac{2r}{r^2-s^2}-\dfrac{3s}{(r+s)^2}+\dfrac{1}{r^2+rs-2s^2}$

12.6 Complex Rational Expressions

In Exercises 43–48, simplify each of the indicated complex rational expressions. List all excluded values.

43. $\dfrac{\tfrac{-3a}{4b^2}}{\tfrac{6a^3}{16b}}$

44. $\dfrac{x+\tfrac{2}{x}}{x^3}$

45. $\dfrac{\tfrac{1}{t}}{t-\tfrac{1}{t}}$

46. $\dfrac{\dfrac{3}{x} - \dfrac{2}{y}}{\dfrac{1}{x^2 y^3}}$

47. $\dfrac{\dfrac{s}{t} + \dfrac{t}{s}}{\dfrac{s}{t} - \dfrac{t}{s}}$

48. $\dfrac{\dfrac{-3}{a^2 - b^2}}{\dfrac{1}{a+b} - \dfrac{3}{a-b}}$

12.7 Equations Involving Rational Expressions

In Exercises 49–54, solve the given equations for the indicated variable. Determine all excluded values and check your results in the original equation.

49. $\dfrac{y+2}{5} = \dfrac{2y}{3}$

50. $\dfrac{3y-1}{2} = \dfrac{2-3y}{4}$

51. $\dfrac{x}{3} + 2 = \dfrac{x}{2} - 5$

52. $\dfrac{s}{3} - \dfrac{2s+3}{2} = \dfrac{4s+1}{5}$

53. $\dfrac{t}{4t^2 - 1} - \dfrac{3}{2t+1} = \dfrac{1}{1-2t}$

54. $\dfrac{5}{3y-6} - \dfrac{8y}{2y^2 - 8} = \dfrac{3}{y+2}$

12.8 Applications Involving Rational Expressions

55. One number is 8 more than a second number. The reciprocal of the smaller number is 5 times the reciprocal of the larger number. What are the numbers?
56. The sum of two numbers is 40. The sum of one-fifth of the smaller number and one-half of the larger number is 10. What are the numbers?
57. A farmer can plow a field, working alone, in 6 hr. His son, working alone, can plow the same field in 8 hr. Working together, how long should it take both of them to plow the same field?
58. An old copying machine takes 3 times as long to do a job as a new machine. With both machines working together, the job can be done in 50 min. How many minutes should it take the old machine, working alone, to do the job?
59. The length of a rectangle is 1 cm more than twice its width. The sum of the width and one-fifth of the length is 10 cm. What are the dimensions of the rectangle?

Teasers

1. Write a rational expression, in the variable x, such that the only excluded values of the expression are -2 and 1.
2. Write a rational expression, in the variable y, such that the numerator is equal to 0 when $y = 3$ and the denominator is equal to 0 when $y = -4$.
3. Write a rational expression, in the variable u, that has no excluded values.
4. Determine LCM(a, b) where $a = $ LCM$(y^2 - 36, 2y + 12)$ and $b = $ LCM$(y^2 + 4y - 12, 3y + 18)$.
5. Show that $\left(\dfrac{-1}{3}\right)\left(\dfrac{2}{5} + \dfrac{-2}{7}\right) = \left(\dfrac{-1}{3}\right)\left(\dfrac{2}{5}\right) + \left(\dfrac{-1}{3}\right)\left(\dfrac{-2}{7}\right)$.
6. Verify that $\left(\dfrac{a}{b}\right)\left(\dfrac{c}{d} - \dfrac{e}{f}\right) = \left(\dfrac{a}{b}\right)\left(\dfrac{c}{d}\right) - \left(\dfrac{a}{b}\right)\left(\dfrac{e}{f}\right)$, if $\dfrac{a}{b} = \dfrac{2}{3}, \dfrac{c}{d} = \dfrac{-1}{7}$, and $\dfrac{e}{f} = \dfrac{3}{4}$.
7. a. Solve the equation $\dfrac{2}{3-x} = 4$ on the set of integers. If no solutions exist, indicate that fact.
   b. Solve the equation in part a on the set of rational numbers. If no solutions exist, indicate that fact.
8. a. Is the average of two integers always an integer? Explain your answer.
   b. Is the average of two rational numbers always a rational number? Explain your answer.
9. Show that if $\dfrac{a}{b} < \dfrac{c}{d}$, then $\left(\dfrac{a}{b}\right)\left(\dfrac{p}{q}\right) < \left(\dfrac{c}{d}\right)\left(\dfrac{p}{q}\right)$ is not always true. Give a particular example to substantiate your answer.
10. a. How many whole numbers are there between the integers -2 and 3?
    b. How many integers are there between the integers -2 and 3?
    c. How many rational numbers are there between the integers -2 and 3?

CHAPTER 12 TEST

This test covers materials from Chapter 12. Read each question carefully and answer all questions. If you miss a question, review the appropriate section of the text.

In Exercises 1–4, determine the excluded values for the given rational expressions.

1. $\dfrac{3}{x+5}$

2. $\dfrac{s^4-6}{s^2+5s+6}$

3. $\dfrac{p^2+1}{p^2-1}$

4. $\dfrac{3t^5+8}{(t^2-1)(2t+3)}$

In Exercises 5–8, simplify each of the given rational expressions and determine the excluded values.

5. $\dfrac{4x^3y^2z^4}{16x^2y^3z}$

6. $\dfrac{5t+15}{t+3}$

7. $\dfrac{a^2x+a^2y}{bx+by}$

8. $\dfrac{uv^2-2wv^2}{2u^2w-4uw^2}$

In Exercises 9–14, perform the indicated operations and simplify your results. Include all excluded values in each case.

9. $\dfrac{x^2-4}{9}\cdot\dfrac{3x+6}{x-2}$

10. $\dfrac{3c+d}{2x-7}+\dfrac{2d-c}{2x-7}$

11. $\dfrac{5}{y+3}-\dfrac{3}{y+1}$

12. $\dfrac{p^2-8p+16}{p^2+p-20}\div\dfrac{p^2+8p+12}{p^2+11p+30}$

13. $\dfrac{-3}{x^2-y^2}\div\left(\dfrac{1}{x+y}-\dfrac{3}{x-y}\right)$

14. $\dfrac{t}{4t^2-1}-\dfrac{3}{2t+1}+\dfrac{1}{2t-1}$

In Exercises 15–18, solve the given equation for the indicated variable. Check your results in the original equation.

15. $\dfrac{2x-1}{3}=\dfrac{4-3x}{2}$

16. $\dfrac{2y}{5}-3=\dfrac{y}{2}+1$

17. $\dfrac{5}{3t-6}-\dfrac{3}{t+2}=\dfrac{8t}{2t^2-8}$

18. $\dfrac{s-1}{4}-\dfrac{2s-1}{5}=\dfrac{5s+2}{3}$

19. One number is 3 less than a second number. The reciprocal of the smaller number is twice the reciprocal of the larger number. What are the numbers?

20. Kathi can wax a car, working alone, in 6 hr. Her brother, working alone, can wax the same car in 8 hr. Working together, how long should it take both of them to wax the same car?

CUMULATIVE REVIEW: CHAPTERS 1–12

In Exercises 1–3, perform the indicated operations.

1. $1.21 \div (-1.1)^2 + (-3^2)$
2. $\dfrac{-3}{8} + \dfrac{2}{7} \div \dfrac{-1}{3}$
3. $0.57\% \times |23.7 - 37.9|$

In Exercises 4–5, evaluate each of the given expressions.

4. $2x^2 - 3x + 4$, if $x = \dfrac{-1}{4}$
5. $3s^2 - 2st - 4t^2$, if $s = -2.1$ and $t = -1.6$
6. Round $7\dfrac{3}{8}$ to the nearest tenth.
7. Round 60,987,031 to two significant digits.
8. Rewrite 0.000839 in scientific notation.
9. Rewrite 6987.9×10^5 in scientific notation.

In Exercises 10–11, simplify by combining like terms.

10. $2p - 3\{q - 3[2p - q] - 5p\} - 4q$
11. $y^2(3y - 4) - y^2(1 - 2y) + (y^2 - 1)(2y + 3)$

In Exercises 12–15, solve each equation for the indicated variable.

12. $3(x - 2) + 5 = 2(7 - 3x)$
13. $p^3 - 5p^2 - 6p = 0$
14. $\dfrac{2t - 3}{3} = \dfrac{3t + 1}{4}$
15. $\dfrac{x^2 + 2}{2x - 1} = 3x + 2$

In Exercises 16–17, perform the indicated operations and simplify your results. Determine all excluded values in each exercise.

16. $\dfrac{-2}{q - 3} + \dfrac{3}{q + 3} - \dfrac{5}{q^2 - 9}$
17. $\dfrac{x^2 - 9}{2y + 1} \div \dfrac{2x + 6}{4y^2 - 1}$

18. Determine the perimeter of a pentagon having two sides each of measure 9.8 cm, two sides each of measure 11.7 cm, and a fifth side of measure 1.8 dm.
19. Determine the volume of a rectangular solid whose dimensions are 41.3 cm long, 20.1 cm wide, and 8.6 cm high.
20. The sum of two numbers is 23. The sum of one-half of the smaller number and one-third of the larger number is 9. What are the numbers?
21. Which is larger: $49.6\% \times (-23.1)$ or $\dfrac{-2}{3} \times \dfrac{3}{7}$?
22. Determine the number given by 3.09×10^{-4}.
23. Solve the proportion $\dfrac{-3}{4.2} = \dfrac{n}{84}$ for n.
24. Determine whether the rational numbers $\dfrac{-7}{11}$ and $\dfrac{-12}{17}$ are equivalent.
25. Factor completely: $2t^3 + 9t^2 - 5t$.

CHAPTER 13 Graphing Linear Equations and Linear Inequalities

In Chapter 8, you learned to solve linear equations in a single variable. In this chapter, we will consider linear equations in two variables, together with their graphs and solutions. We will also discuss linear inequalities in one and two variables, together with their graphs.

As you continue your study of mathematics, you may encounter applications of linear inequalities in two variables. One of the more interesting and useful applications involves linear programming. Linear programming plays a significant role in economic theory.

There are other applications of linear equations and inequalities in two variables. Some of these will be discussed in this chapter.

CHAPTER 13 PRETEST

This pretest covers materials on graphing linear equations and linear inequalities. Read each question carefully, and try to answer all questions. Each question is keyed to the chapter section in which the particular topic is discussed. Check your answers with those given in the back of the text.

Questions 1–2 pertain to Section 13.1.

1. In which quadrant does the point $(2, -3)$ lie?

2. In which quadrant does the point $(-3, -5)$ lie?

Questions 3–6 pertain to Section 13.2.

3. Solve the equation $3x - 4y = 5$ for y.

4. Solve the equation $2(y - 3) - 4(x + 1) = 7 - 5(y + 1)$ for y.

5. Given the equation $3y - 5x + 7 = 0$, determine a value for x for which $y = 0$.

6. Given the equation $2(x - 1) - 5 = 3(y + 4)$, determine a value for y for which $x = -1$.

Questions 7–11 pertain to Section 13.3

7. Determine, by inspection, whether the line associated with the equation $7x + 5 = 0$ is horizontal, vertical, or neither.

8. Determine, by inspection, whether the line associated with the equation $5x - 9 = 6y$ is horizontal, vertical, or neither.

9. Graph the equation $y + 5 = 0$.

10. Graph the equation $2x + 5y = 10$.

11. Determine the y-intercept associated with the graph of the equation $3y - 2x = 10$.

Questions 12–14 pertain to Section 13.4.

12. Determine the slope of the line passing through the points $(-3, 0)$ and $(4, -7)$.

13. Determine by inspection whether the slope of the line whose equation is $3x + 4y = 9$ is positive, negative, zero, or undefined.

14. Determine by inspection whether the slope of the line whose equation is $2x - 7 = 0$ is positive, negative, zero, or undefined.

Question 15 pertains to Section 13.5.

15. Graph the inequality $2x - 3 \geq 6$ on the number line.

Questions 16–18 pertain to Section 13.6.

16. Solve the inequality $2p - 3q > q - 7$ for q.

17. Solve the inequality $2(s - 3) - 3t < 5s + 2(1 - 3t)$ for t.

18. Given the inequality $5x - 3 < 6y$, determine three different values for y, if $x = -2$.

Questions 19–20 pertain to Section 13.7.

19. Graph the inequality $x - y > 3$.

20. Graph the inequality $y \leq 1 - 2x$.

13.1 Rectangular Coordinate System

OBJECTIVES

After completing this section, you should be able to:
1. Determine where a particular point lies in a quadrant or on a coordinate axis.
2. Determine the coordinates of a plotted point.
3. Set up a rectangular coordinate system and plot points whose coordinates are given.

In previous chapters, we used the number line to represent numbers. To do so, we located the number 0 on the line at the point 0 (called the **origin**). (See Figure 13.1.) Next, we located the number 1 on the line and to the right of 0 at the point A. A **scale** is now determined. Using this scale, we can now locate any number on the line. For instance, in Figure 13.2, the number 2 is located at the point B such that 2 is as far to the right of 1 as 1 is to the right of 0. Similarly, the number -1 is located at the point C such that -1 is as far to the *left* of 0 as 1 is to the *right* of 0. The remaining integers can now also be located by extending the number line in either direction.

FIGURE 13.1

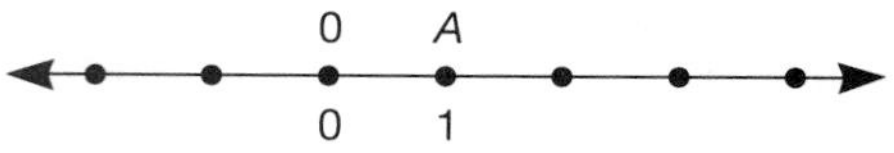

FIGURE 13.2

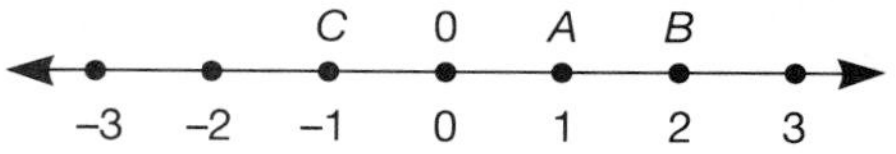

Rational numbers can also be located on the line. For instance, $\frac{1}{2}$ would be located halfway between 0 and 1; $\frac{8}{5}$ would be located three-fifths of the way from 1 to 2; $\frac{-5}{3}$ would be located one-third of the way from -2 to -1.

Suppose we take two number lines, one horizontal and one vertical, and join them at their origins so that the lines are perpendicular to each other (that is, they meet to form right angles). Then these number lines determine a **plane.** (See Figure 13.3.) The horizontal number line is called the **horizontal axis,** the vertical line is called the **vertical axis,** and the point of intersection of the axes (plural of axis) is called the *origin* of the plane. It is customary (but not necessary) to label the horizontal axis the x-axis and the vertical axis the y-axis; the plane would be called the xy-plane. The plane is identified using the horizontal axis name first, followed by the vertical axis name. If the horizontal axis is labeled r and the vertical axis s, then the plane is called the rs-plane. Therefore, in the uv-plane, the horizontal axis is the u-axis and the vertical axis is the v-axis.

Although each point in the plane has a name (or address), it would be impossible to label every point in the diagram. We generally label only certain integer values on both axes. Then these values are used to label all points in the plane.

Figure 13.4 shows the xy-plane with the points A, B, C, D, and E. We could get from the origin to point A by moving one unit to the *right* and then two units *up*. To get from the origin to point B, we could move four units to the *right* and stop (since point B is on the x-axis). For point C, we could move three units to the *left* and two units *down*. To get to point D, move two units to the *left* and three units *up*. Finally, for point E, move two units *down* and stop (since point E is on the y-axis).

Now recall that on the horizontal number line, left of 0 is *negative* and right of 0 is *positive*. Similarly, on a vertical line, up is *positive* and down is *negative*. Using this conven-

13.1 Rectangular Coordinate System

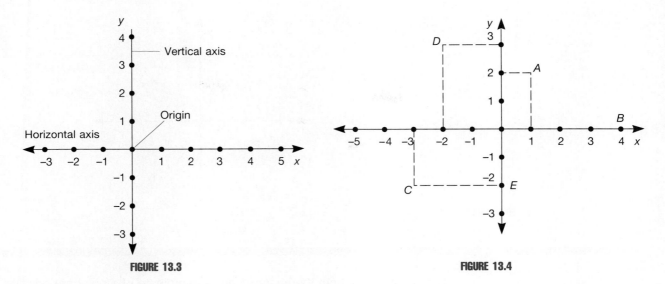

FIGURE 13.3

FIGURE 13.4

tion, we could now represent a point in the plane by a pair of numbers called an **ordered pair.** For instance, point A in Figure 13.4 is represented by using the notation (1, 2), which means that point A is located 1 unit to the *right* of the y-axis and 2 units *above* the x-axis. Point C is represented by (−3, −2), which menas 3 units to the *left* of the y-axis and 2 units *below* the x-axis.

The order in which we write the numbers in a pair is important. That is, (1, 2) is different from (2, 1), (−5, 6) is different from (6, −5), and so forth. Each number in the pair is called a **coordinate.** The first coordinate, called the **abscissa,** indicates the horizontal position of the point right or left of the vertical axis. The second coordinate, called the **ordinate,** indicates the vertical position of the point above or below the horizontal axis.

Determine (a) the abscissa and (b) the ordinate.

Practice Exercise 1. (−3, 4)

EXAMPLE 1 Determine the abscissa and the ordinate for the point (−2, 1).

SOLUTION:

- The abscissa is −2.
- The ordinate is 1.

Practice Exercise 2. (1.9, −2.3)

EXAMPLE 2 Determine the abscissa and the ordinate for the point (4, −3).

SOLUTION:

- The abscissa is 4.
- The ordinate is −3.

ANSWERS TO PRACTICE EXERCISES

1. (a) −3; (b) 4. **2.** (a) 1.9; (b) −2.3.

We have now described what is called a **rectangular coordinate system.** In a rectangular coordinate system, the two axes divide the plane into four regions called **quadrants.** The quadrants are named I, II, III, and IV, as indicated in Figure 13.5. The points on the axes are not considered as being in a quadrant.

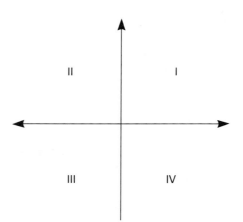

FIGURE 13.5

1. In Quadrant I, both the abscissa and the ordinate of a point are positive.
2. In Quadrant II, the abscissa of a point is negative and its ordinate is positive.
3. In Quadrant III, both the abscissa and the ordinate of a point are negative.
4. In Quadrant IV, the abscissa of a point is positive and its ordinate is negative.
5. On the horizontal axis, the ordinate of a point is 0.
6. On the vertical axis, the abscissa of a point is 0.

Determine the quadrant in which each of the indicated points lies. If the point is on an axis, indicate which one.

Practice Exercise 3. $(7, -9)$

EXAMPLE 3 Determine the quadrant in which the point $(2, 4)$ lies.

SOLUTION: In the graph at left below, the point $(2, 4)$ lies in Quadrant I since both the abscissa, 2, and the ordinate, 4, are positive.

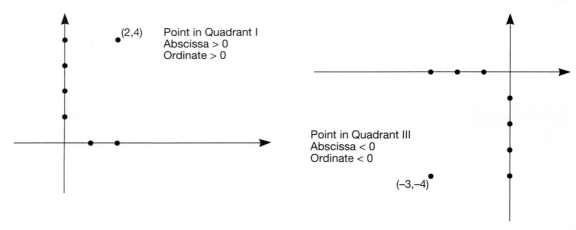

Practice Exercise 4. $(-2, -7)$

EXAMPLE 4 Determine the quadrant in which the point $(-3, -4)$ lies.

SOLUTION: In the graph at right above, the point $(-3, -4)$ lies in Quadrant III since both the abscissa, -3, and the ordinate, -4, are negative.

Practice Exercise 5. $(0, -2)$

EXAMPLE 5 Determine the quadrant in which the point $(4, -3)$ lies.

SOLUTION: The point $(4, -3)$ lies in Quadrant IV since the abscissa, 4, is positive and the ordinate, -3, is negative.

13.1 Rectangular Coordinate System

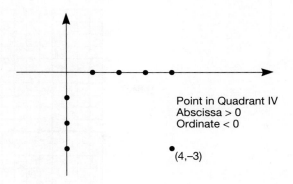

Practice Exercise 6. $(-5, 7)$

EXAMPLE 6 Determine the quadrant in which the point $(-5, 2)$ lies.

SOLUTION: The point $(-5, 2)$ lies in Quadrant II since the abscissa, -5, is negative and the ordinate, 2, is positive.

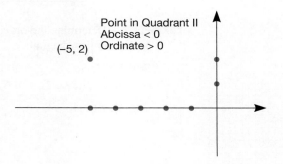

ANSWERS TO PRACTICE EXERCISES
3. IV. 4. III. 5. Vertical axis. 6. II.

To plot (or graph) a point means to locate it in the plane. To plot the point $(4, 3)$, we move 4 units to the *right* of the origin and then *up* 3 units. To plot the point $(-5, -2)$, we move 5 units to the *left* of the origin and then *down* 2 units. (See Figure 13.6.)

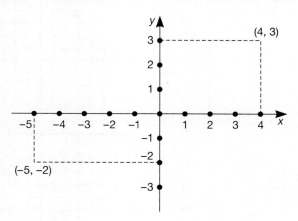

FIGURE 13.6

Practice Exercise 7. Plot the points $A(-3, 2)$, $B(0, 4)$, $C(-5, -4)$, $D(3, -5)$, $E(-3, 0)$, and $F(3, 4)$.

EXAMPLE 7 Plot the points $A(1, 2)$, $B(-3, 4)$, $C(3, 0)$, $D(-4, -2)$, $E(0, -5)$, and $F(5, -3)$.

SOLUTION: To plot points, it is helpful to use coordinate (or graph) paper as illustrated.

Chapter 13 Graphing Linear Equations and Linear Inequalities

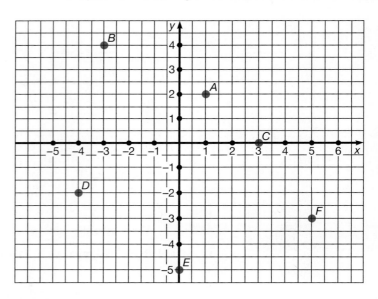

Practice Exercise 8. Determine the ordered pairs represented by the points G, H, I, J, and K in the accompanying diagram.

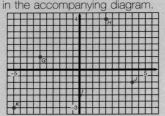

EXAMPLE 8 Determine the ordered pairs represented by the points G, H, I, J, and K in the accompanying diagram.

SOLUTION:
$G(-4, 4)$, $H(7, -2)$, $I(5, 6)$, $J(0, 4)$, $K(-4, -4)$

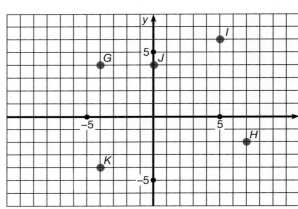

ANSWERS TO PRACTICE EXERCISES

7.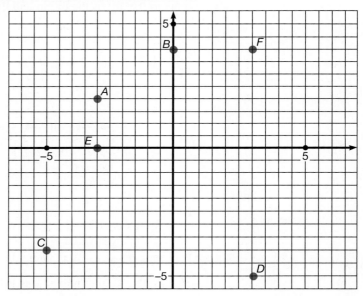

8. $G(-3, 1)$, $H(2, 4)$, $I(0, -2)$, $J(4, -1)$, $K(-5, -3)$.

13.1 Rectangular Coordinate System

EXERCISES 13.1

In Exercises 1–20, determine the quadrant in which each of the indicated points lies. If the point lies on one of the coordinate axes, indicate which one.

1. $(2, 3)$
2. $(4, 1)$
3. $(5, 3)$
4. $(-3, 2)$
5. $(0, 6)$
6. $(-4, -1)$
7. $(6, -3)$
8. $(-6, 5)$
9. $(7, 0)$
10. $\left(\dfrac{1}{2}, \dfrac{-1}{3}\right)$
11. $(0, -3)$
12. $\left(\dfrac{17}{3}, 0\right)$
13. $\left(-5, \dfrac{1}{2}\right)$
14. $\left(2, \dfrac{-4}{5}\right)$
15. $(17, -19)$
16. $(-19, -23)$
17. $\left(\dfrac{-1}{2}, 5\right)$
18. $\left(\dfrac{-1}{4}, \dfrac{-2}{3}\right)$
19. $\left(0, \dfrac{-2}{5}\right)$
20. $\left(\dfrac{-1}{9}, \dfrac{-1}{7}\right)$

In Exercises 21–40, determine the coordinates of each of the indicated points.

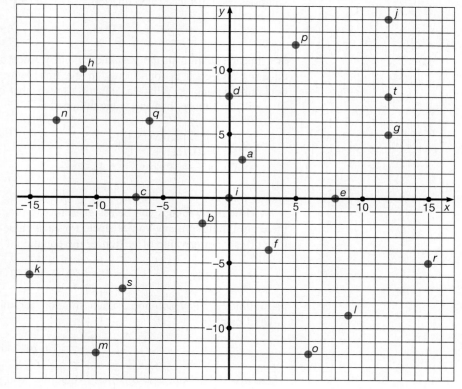

21. a
22. b
23. c
24. d
25. e
26. f
27. g
28. h
29. i
30. j
31. k
32. l
33. m
34. n
35. o
36. p
37. q
38. r
39. s
40. t

In Exercises 41–60, set up a rectangular coordinate system and plot each of the following points. Label the points on your graph according to the letters given.

41. $A(2, 3)$
42. $B(7, 1)$
43. $C(0, 3)$
44. $D(-3, 4)$
45. $E(1, 5)$
46. $F(-2, 3)$
47. $G(3, 0)$
48. $H(0, -4)$
49. $I(5, 7)$
50. $J(-1, -3)$
51. $K(-4, 0)$
52. $L(0, 0)$
53. $M(5, 5)$
54. $N(-4, -4)$
55. $O(6, -2)$
56. $P(4, -6)$
57. $Q(9, -8)$
58. $R(-5, 8)$
59. $S(-7, -6)$
60. $T(8, 7)$

13.2 Linear Equations in Two Variables

OBJECTIVES

After completing this section, you should be able to work with a linear equation in two variables and:
1. Solve the equation for one variable in terms of the other.
2. Determine whether an ordered pair of numbers is a solution for the equation.
3. Evaluate one variable, given the value for the other variable.

In Chapter 8, you learned to solve simple linear equations in one variable such as the equation $2x + 3 = 0$. In general, if we have the equation $ax + b = 0$ with $a \neq 0$, then you learned that there is exactly one solution to the equation, which is $x = \dfrac{-b}{a}$.

In this section, you will be introduced to **linear equations in two variables** and their solutions. For instance, the equation

$$y = x + 3$$

is a linear equation in the variables x and y since only those variables are present in the equation and each is found only in the first degree. Further, there are no denominators containing variables. What this equation says is that if we assign a value to x, then the value of y will be 3 more than the value assigned to x. For example,

- If $x = 3$, then $y = 6$.
- If $x = 0$, then $y = 3$.
- If $x = -4$, then $y = -1$.

and so forth.

How many values can be assigned to x to produce corresponding values of y? Surely there are infinitely many values since, for each value assigned to x, you simply add 3 to get the corresponding value of y. Hence, the equation

$$y = x + 3$$

has infinitely many solutions. Each solution for this equation can be represented by an ordered pair

$$(a, b)$$

where a is the x-value and b is the y-value that will satisfy the equation. The ordered pair

$$(2, 5)$$

is a solution of the equation since if

$$x = 2$$

then
$$\begin{aligned} y &= x + 3 \\ &= 2 + 3 \\ &= 5 \checkmark \end{aligned}$$

The ordered pair (2, 7) is *not* a solution for the equation $y = x + 3$ since 7 (the y-value) is not 3 more than 2 (the x-value).

13.2 Linear Equations in Two Variables

Practice Exercise 1. If $y = 2x - 1$, determine the value for y if $x = -2$.

EXAMPLE 1 Determine five different solutions for the equation $y = x - 4$.

SOLUTION: Since the equation states that y is 4 less than x, we could assign any five different values to x and determine the corresponding values of y. The ones we selected are listed in the following table:

x	y
-3	-7
-6	-10
1	-3
$\frac{1}{3}$	$\frac{-11}{3}$
-0.5	-4.5

Hence, $(-3, -7)$, $(-6, -10)$, $(1, -3)$, $\left(\frac{1}{3}, \frac{-11}{3}\right)$ and $(-0.5, -4.5)$ are all solutions for the given equation. These are only five possible solutions, not the only five, and there are many other solutions.

Practice Exercise 2. If $2y = 4x$, determine the value for y if $x = \frac{1}{2}$.

EXAMPLE 2 Solve the equation $2x - 3y = 7$ for y.

SOLUTION: To solve the equation $2x - 3y = 7$ for y, we must isolate y on one side of the equation. This can be done using the procedure discussed in Chapter 8.

$$2x - 3y = 7$$
$$-3y = 7 - 2x \quad \text{(Subtract } 2x \text{ from both sides.)}$$
$$y = \frac{7 - 2x}{-3} \quad \text{(Divide both sides by } -3.)$$
$$y = \frac{2x - 7}{3} \quad \text{(Multiply both numerator and denominator by } -1.)$$

Hence, $y = \dfrac{2x - 7}{3}$ is the required solution.

In Example 2, to find a value for y, we could assign any value to x, multiply that value by 2, subtract 7 from the product, $2x$, and then divide the difference, $2x - 7$, by 3. Therefore, if

$$x = 2$$

then

$$y = \frac{2(2) - 7}{3}$$
$$= -1$$

Hence, $(2, -1)$ is a solution for the equation $2x - 3y = 7$.

CHECK: We have that if

$$x = 2 \quad \text{and} \quad y = -1$$

then

$$2x - 3y = 7$$
$$2(2) - 3(-1) \stackrel{?}{=} 7$$
$$4 - (-3) \stackrel{?}{=} 7$$
$$4 + 3 \stackrel{?}{=} 7$$
$$7 = 7 \checkmark$$

Practice Exercise 3. If $3x - y = 5$, determine the value for x if $y = 4$.

EXAMPLE 3 Is $(1, 2)$ a solution for the equation $5x - 2y = 6$?

SOLUTION: If $(1, 2)$ is a solution for the given equation, then when 1 is substituted for x and 2 for y, the left-hand side of the equation should equal 6.

CHECK:
$$5x - 2y = 6$$
$$5(1) - 2(2) \ ? \ 6$$
$$5 - 4 \ ? \ 6$$
$$1 \neq 6$$

Therefore, $(1, 2)$ is *not* a solution for the given equation. ✔

ANSWERS TO PRACTICE EXERCISES

1. $y = -5$. **2.** $y = 1$. **3.** $x = 3$.

EXERCISES 13.2

In Exercises 1–10, solve the given equation for y.

1. $x + y = 3$
2. $x - y = 4$
3. $x + 3y = 0$
4. $2x - y = 1$
5. $x - 2y = 5$
6. $2y = 1 - 3x$
7. $4x = 3y - 1$
8. $x + 6y = 2$
9. $2(x - 1) + 3(y + 2) = 1$
10. $3(x + 2) = 2(y - 7) - 4$

In Exercises 11–15, use the equation $2x - 3y = 1$ and determine the value of y for the indicated value of x.

11. $x = -2$
12. $x = 0$
13. $x = \dfrac{1}{3}$
14. $x = 3$
15. $x = 1.5$

In Exercises 16–20, use the equation $y = 3x + 7$ and determine the value of y for the indicated value of x.

16. $x = \dfrac{-1}{2}$
17. $x = -1$
18. $x = 0$
19. $x = -3.2$
20. $x = -9$

In Exercises 21–25, use the equation $3x - y = 6$ and determine the value of x for the indicated value of y.

21. $y = -2$
22. $y = \dfrac{-1}{3}$
23. $y = 0$
24. $y = -1.7$
25. $y = 7$
26. Given the equation $2x + y = 3$:
    a. For what value of x is $y = 0$?
    b. For what value of y is $x = 0$?

27. Given the equation $3x - y = 7$:
    a. For what value of x is $y = 1$?
    b. For what value of y is $x = -2$?
28. Given the equation $2(x - 1) = 3(y + 2)$:
    a. For what value of x is $y = 0$?
    b. For what value of y is $x = -1$?

13.3 Graphing Straight Lines

OBJECTIVES

After completing this section, you should be able to:
1. Graph the line associated with a linear equation in two variables.
2. Determine the intercepts for the graph of such equations.

An equation of the form $ax + by = c$ (a and b cannot both be 0) has infinitely many solutions. Each one of these solutions can be written as an ordered pair of the form (x, y).

The graph of such a linear equation is a straight line. Every line has infinitely many points. It should not be surprising, then, that each solution (x, y) of the equation corresponds to the coordinates for a point on the line.

When we say "graph the equation," we actually mean "determine the graph associated with the equation." For instance, the graph of the equation $2x + 3y = 4$ gives us the graph of all the ordered pairs (x, y) that are solutions of the given equation.

Rule for Graphing a Linear Equation in Two Variables

To graph a linear equation in two variables:
1. Determine at least two solutions (ordered pairs) for the equation.
2. Plot the points that correspond to these solutions.
3. Draw and label a straight line through these points.

When graphing a linear equation, it is only necessary to determine two points through which the line passes. However, a third point should also be determined as a check.

Graph each equation in the *xy*-plane.

Practice Exercise 1. $y = x + 6$

EXAMPLE 1 Graph the equation $y = x + 4$ in the *xy*-plane.

SOLUTION:

STEP 1: Determine three solutions of the equation. Select values for x or y as indicated in the table on the left. Then determine the corresponding values for the other variable as indicated in the table on the right.

x	y
0	
-2	
	0

x	y
0	4
-2	2
-4	0

STEP 2: Plot the points $(0, 4)$, $(-2, 2)$, and $(-4, 0)$ (below left).

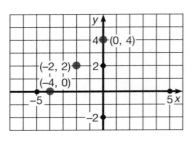
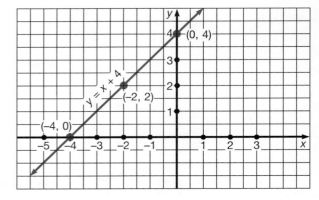

STEP 3: Draw and label a straight line through these points (above right).

The three-step procedure for graphing a linear equation in two variables will be shortened, as indicated in the following examples.

Practice Exercise 2.
$2x + 3y = 6$

EXAMPLE 2 Graph the equation $x - 3y = 6$ in the xy-plane.

SOLUTION: We determine three solutions of the equation, as listed in the accompanying table. The corresponding points are plotted and the line drawn as indicated.

x	y
-3	-3
0	-2
6	0

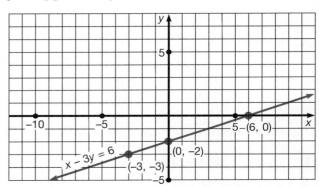

ANSWERS TO PRACTICE EXERCISES

1.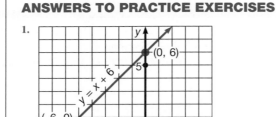

2.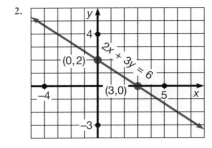

Note that, in Examples 1 and 2, we used the points where the line crosses the two coordinate axes. These were not only convenient points to use but involve what are called the intercepts of the graph and are usually used when graphing equations.

13.3 Graphing Straight Lines

> **DEFINITIONS**
>
> Given the linear equation $ax + by = c$ with $a \neq 0$ and $b \neq 0$.
> 1. The *x-intercept* is the *abscissa* of the point where the graph of the equation crosses the *x*-axis (that is, where $y = 0$).
> 2. The *y-intercept* is the *ordinate* of the point where the graph of the equation crosses the *y*-axis (that is, where $x = 0$).

In Example 1, the *x*-intercept $= -4$ [the line crosses the *x*-axis at the point $(-4, 0)$], and the *y*-intercept $= 4$ [the line crosses the *y*-axis at the point $(0, 4)$]. In Example 2, the *x*-intercept $= 6$ [the line crosses the *x*-axis at the point $(6, 0)$], and the *y*-intercept $= -2$ [the line crosses the *y*-axis at the point $(0, -2)$].

In Practice Exercises 3–4, determine the x-intercept and the y-intercept for the graph of each of the given equations.

Practice Exercise 3. $3x + y = 6$

EXAMPLE 3 Determine both the *x*-intercept and the *y*-intercept for the graph of $2x - 3y = 6$.

SOLUTION: Since the *x*-intercept is determined when $y = 0$, we solve the equation $2x - 3y = 6$ for x if $y = 0$:

$$2x - 3y = 6$$
$$2x - 3(0) = 6$$
$$2x - 0 = 6$$
$$2x = 6$$
$$x = 3 \quad (x\text{-intercept})$$

Since the *y*-intercept is determined when $x = 0$, we solve the equation $2x - 3y = 6$ for y if $x = 0$.

$$2x - 3y = 6$$
$$2(0) - 3y = 6$$
$$0 - 3y = 6$$
$$-3y = 6$$
$$y = -2 \quad (y\text{-intercept})$$

Practice Exercise 4.
$5x - 3y = 10$

EXAMPLE 4 Determine the points on the graph of $2x - 3y = 6$ associated with the *x*-intercept and the *y*-intercept.

SOLUTION: From Example 3, we determined that the *x*-intercept $= 3$ and the *y*-intercept $= -2$. Since the *x*-intercept is the *abscissa* of the point where $y = 0$, we have the point $(3, 0)$. Since the *y*-intercept is the *ordinate* of the point where $x = 0$, we have the point $(0, -2)$. Therefore, the required points are $(3, 0)$ and $(0, -2)$.

Using the intercepts is the easiest way to graph a linear equation in two variables. In general, if the *x*-intercept $= a$ and the *y*-intercept $= b$, then the graph passes through the points $(a, 0)$ and $(0, b)$. (See Figure 13.7.)

Intercept Method for Graphing Linear Equations in Two Variables

To use the intercept method to graph a linear equation in two variables *x* and *y*, follow this procedure:

1. To determine the *x*-intercept, set $y = 0$ and solve the resulting equation.
2. To determine the *y*-intercept, set $x = 0$ and solve the resulting equation.
3. The graph of a linear equation in two variables is a line.

a. If the x-intercept = a, then the line passes through the point with coordinates (a, 0).
b. If the y-intercept = b, then the line passes through the point with coordinates (0, b).
c. Plot the two points (a, 0) and (0, b) and draw a straight line through them.

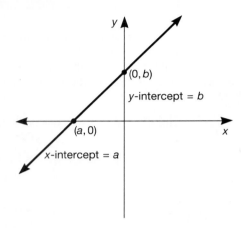

FIGURE 13.7

In Practice Exercises 5–6, graph the equations using the intercept method.

Practice Exercise 5.
$4x = 1 - 2y$

EXAMPLE 5 Graph the equation $2x - 5y = 10$ in the xy-plane.

SOLUTION: Using the *intercept method*, we have the following:

- *x-intercept:* Setting $y = 0$, we have $2x - 5(0) = 10$, or $x = 5$. Therefore, we have the point $(5, 0)$.
- *y-intercept:* Setting $x = 0$, we have $2(0) - 5y = 10$, or $y = -2$. Therefore, we have the point $(0, -2)$.

Plot these points and draw a straight line through them:

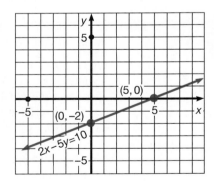

Practice Exercise 6.
$4x - 5y = 20$

EXAMPLE 6 Graph the equation $2x + 3y = 9$ in the xy-plane.

SOLUTION: Using the *intercept method*, we have the following:

- *x-intercept:* Setting $y = 0$, we have $2x + 3(0) = 9$, or $x = 4.5$. Therefore, we have the point $(4.5, 0)$.
- *y-intercept:* Setting $x = 0$, we have $2(0) + 3y = 9$, or $y = 3$. Therefore, we have the point $(0, 3)$.

13.3 Graphing Straight Lines

Plot these points and draw a straight line through them:

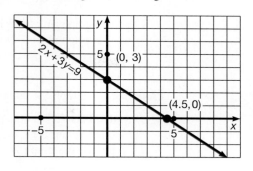

ANSWERS TO PRACTICE EXERCISES

3. x-intercept $= 2$, y-intercept $= 6$. 4. x-intercept $= 2$, y-intercept $= \dfrac{-10}{3}$.

5. 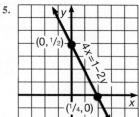   6.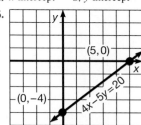

The intercept method should be used whenever possible to graph an equation. However, the two points determined using this method are sometimes very close together. When that happens, or when both intercepts are 0, determine at least one other point through which the line passes.

For some lines, there is only one intercept, as will be illustrated in the following examples.

Describe the graph, in the *xy*-plane.

Practice Exercise 7. $y = -4$

EXAMPLE 7 Graph the equation $y = 4$ in the *xy*-plane.

SOLUTION: The equation $y = 4$ means that for *all* values of x, $y = 4$. Selecting three values for x, we get the points as listed in the accompanying table. Plotting these points, the graph is a *horizontal* line, as indicated.

x	y
-2	4
0	4
3	4

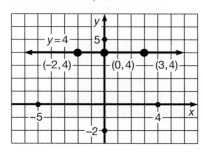

In general, the graph in the *xy*-plane of an equation of the form $y = k$ is a horizontal line and is

- k units *above* the x-axis if $k > 0$
- $|k|$ units *below* the x-axis if $k < 0$
- the x-axis if $k = 0$

In the *xy*-plane:

- The graph of $y = 3$ is a horizontal line 3 units *above* the *x*-axis.
- The graph of $y = 0$ is the *x*-axis.
- The graph of $y = -2$ is a horizontal line 2 units *below* the *x*-axis.

Practice Exercise 8. $x = -3.1$

EXAMPLE 8 Graph the equation $x = -3$ in the *xy*-plane.

SOLUTION: The equation $x = -3$ means that for *all* values of y, $x = -3$. Selecting three values for y, we get the points listed in the accompanying table. Plotting these points, the graph is a *vertical* line, as illustrated.

x	y
-3	-4
-3	0
-3	3

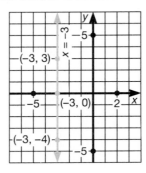

In general, the graph, in the *xy*-plane, of an equation of the form $x = h$ is a vertical line and is

- h units to the *right* of the *y*-axis if $h > 0$
- $|h|$ units to the *left* of the *y*-axis if $h < 0$
- the *y*-axis if $h = 0$

In the *xy*-plane:

- The graph of $x = 4$ is a vertical line 4 units to the *right* of the *y*-axis.
- The graph of $x = 0$ is the *y*-axis.
- The graph of $x = -5$ is a vertical line 5 units to the *left* of the *y*-axis.

ANSWERS TO PRACTICE EXERCISES

7. Horizontal line 4 units below the *x*-axis. 8. Vertical line 3.1 units to the left of the *y*-axis.

EXERCISES 13.3

In Exercises 1–20, graph each of the given equations by determining at least three points through which the graph passes.

1. $x + y = 2$
2. $2x - 3 = y$
3. $x = 4$
4. $y = 3x$
5. $y = x + 5$
6. $y + 3 = 0$
7. $y = 2x - 3$
8. $x - 4y = 7$
9. $2y = x$
10. $x + y = 0$
11. $x + 3 = 0$
12. $y = x$
13. $y = 4x$
14. $2x + 7y = 3$
15. $y = -5x$

16. $2y + 5 = 0$ **17.** $x = 2y + 7$ **18.** $x = 3y$
19. $3y - 2x = 7$ **20.** $x + 9 = 0$

In Exercises 21–30, determine the *x*-intercept and the *y*-intercept for the graph for each of the given equations.

21. $x + y = 3$ **22.** $x - 2y = 5$ **23.** $y = 4x$
24. $2x = 9 + y$ **25.** $y = 2x - 5$ **26.** $2x + y = 4$
27. $x - 3y = 6$ **28.** $x - 14 = 7y$ **29.** $4y = 8 - 2x$
30. $2x = 3 - y$

In Exercises 31–50, use the intercept method to graph each of the equations.

31. $x + y = 2$ **32.** $x + y = 6$ **33.** $x - y = 2$
34. $x - y = 6$ **35.** $x + 2y = 4$ **36.** $x - 3y = 6$
37. $2x + 3y = 6$ **38.** $3x - 12y = 12$ **39.** $4x - 2y = 8$
40. $4x + 2y = 6$ **41.** $4x + 5y = 10$ **42.** $3x - 5y = 15$
43. $x + 7y = 14$ **44.** $5x - y = 10$ **45.** $x - 2y = 5$
46. $2x = 3 - y$ **47.** $y = x + 5$ **48.** $x = y - 7$
49. $\frac{1}{2}x + \frac{1}{3}y = 2$ **50.** $\frac{2}{3}x - \frac{1}{4}y = 1$

13.4 Slope of a Line

OBJECTIVES

After completing this section, you should be able to:
1. Determine the slope of a line passing through two distinct points.
2. Determine the slope of a line, given its equation.
3. Determine whether the slope of a line is positive, negative, zero, or undefined, by inspection of its equation.

When graphing lines in the previous section, you may have noticed that some lines that are neither horizontal nor vertical are "steeper" than others. Further, in some cases, as the *x*-values increase, the *y*-values also increase. In other cases, as the *x*-values increase, the *y*-values decrease. We can characterize these lines by their "steepness" or what is called their slope.

DEFINITION
Consider the line that passes through the two points *A* and *B*. Then, the *slope* of the line is given by

$$\text{Slope} = \frac{\text{change in the ordinates from } A \text{ to } B}{\text{change in the abscissas from } A \text{ to } B}$$

To determine the slope of a line, we need to know the coordinates of two points on the line. In Figure 13.8, we illustrate a line passing through the two points $P_1(x_1, y_1)$ and

$P_2(x_2, y_2)$ in the xy-plane. The change in the ordinates from P_1 to P_2 is given by $y_2 - y_1$. The change in the abscissas from P_1 to P_2 is given by $x_2 - x_1$. The slope of the line is given by

$$\text{Slope} = \frac{\text{change in ordinates from } P_1 \text{ to } P_2}{\text{change in abscissas from } P_1 \text{ to } P_2} = \frac{y_2 - y_1}{x_2 - x_1}$$

Slope is usually denoted by the letter m.

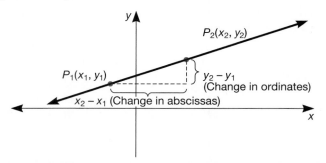

FIGURE 13.8

DEFINITION
The slope, m, of a line, in the xy-plane, that passes through the points with coordinates (x_1, y_1) and (x_2, y_2) is given by

$$m = \frac{y_2 - y_1}{x_2 - x_1} \quad \text{or} \quad m = \frac{y_1 - y_2}{x_1 - x_2} \quad (x_1 \neq x_2)$$

In the above definition for slope, you may start with the y-value of either point and subtract the y-value of the other point to form the numerator. However, you must start with the x-value of the same point in the denominator as was used in the numerator. In other words,

$$\frac{y_2 - y_1}{x_1 - x_2} \quad \text{and} \quad \frac{y_1 - y_2}{x_2 - x_1}$$

cannot be used to determine the slope.

In the preceding discussion, points $P_1(x_1, y_1)$ and $P_2(x_2, y_2)$ were used. The notation P_1 and P_2 refers to different points. That is, P_1 and P_2 are used instead of A and B, for instance. Similarly, (x_1, y_1) and (x_2, y_2) represent coordinates of different points. This will be illustrated in the following examples.

Determine the slope of the line that passes through the two given points.

Practice Exercise 1. $(0, 3)$ and $(4, -2)$

EXAMPLE 1 Determine the slope of the line that passes through the points with coordinates $(-3, -4)$ and $(2, 5)$.

SOLUTION: Let $(x_1, y_1) = (-3, -4)$ and $(x_2, y_2) = (2, 5)$. Then,

$$m = \frac{y_2 - y_1}{x_2 - x_1}$$

$$= \frac{5 - (-4)}{2 - (-3)}$$

$$= \frac{9}{5}$$

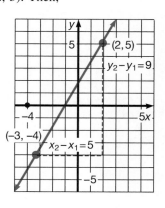

13.4 Slope of a Line

Therefore, the slope of the line is $\frac{9}{5}$. This slope is positive, and the line "rises" from left to right.

In this example, we let $(x_1, y_1) = (-3, -4)$ and $(x_2, y_2) = (2, 5)$. We could have taken $(x_1, y_1) = (2, 5)$ and $(x_2, y_2) = (-3, -4)$. The result would be the same.

Practice Exercise 2. $(-3, -2)$ and $(5, 4)$

EXAMPLE 2 Determine the slope of the line that passes through the points with coordinates $(-1, 2)$ and $(6, -3)$.

SOLUTION: Let $(x_1, y_1) = (-1, 2)$ and $(x_2, y_2) = (6, -3)$. Then,

$$m = \frac{y_2 - y_1}{x_2 - x_1}$$
$$= \frac{-3 - 2}{6 - (-1)}$$
$$= \frac{-5}{7}$$

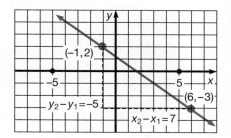

Therefore, the slope of the line is $\frac{-5}{7}$. This slope is negative, and the line "falls" from left to right.

ANSWERS TO PRACTICE EXERCISES

1. $m = \frac{-5}{4}$. 2. $m = \frac{3}{4}$.

If the equation for the line is given, we can determine its slope by first determining the coordinates for two points through which the line passes. Using intercepts to determine these points is convenient.

Determine the slope of the line with the given equation.

Practice Exercise 3. $x + 4y = 5$

EXAMPLE 3 Determine the slope of the line with equation $2x - 3y = 6$.

SOLUTION: Using intercepts, we can determine two points on the line. When $x = 0$, then $y = -2$. Let $(x_1, y_1) = (0, -2)$. When $y = 0$, then $x = 3$. Let $(x_2, y_2) = (3, 0)$. We now have the following:

$$m = \frac{y_2 - y_1}{x_2 - x_1}$$
$$= \frac{0 - (-2)}{3 - 0}$$
$$= \frac{2}{3}$$

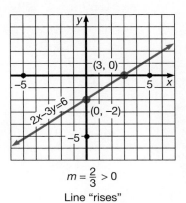

$m = \frac{2}{3} > 0$
Line "rises"

Therefore, the slope of the line is $\frac{2}{3}$.

Practice Exercise 4.
$3x - y = 2$

EXAMPLE 4 Determine the slope of the line with equation $3x + y = 6$.

SOLUTION: Using intercepts, we can determine two points on the line. When $x = 0$, then $y = 6$. Let $(x_1, y_1) = (0, 6)$. When $y = 0$, then $x = 2$. Let $(x_2, y_2) = (2, 0)$. We now have the following:

$$m = \frac{y_2 - y_1}{x_2 - x_1}$$
$$= \frac{0 - 6}{2 - 0}$$
$$= \frac{-6}{2}$$
$$= -3$$

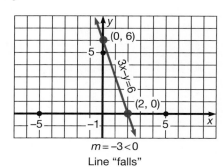
$m = -3 < 0$
Line "falls"

Therefore, the slope of the line is -3.

In the next two examples, we will examine the slope of a horizontal line and a vertical line.

Practice Exercise 5. $x = -9$

EXAMPLE 5 Determine the slope of the line with equation $y = 3$.

SOLUTION: The given line is horizontal. For *all* values of x, $y = 3$. For example, if we let $(x_1, y_1) = (-2, 3)$ and $(x_2, y_2) = (4, 3)$, then we have the following:

$$m = \frac{y_2 - y_1}{x_2 - x_1}$$
$$= \frac{3 - 3}{4 - (-2)}$$
$$= \frac{0}{6}$$
$$= 0$$

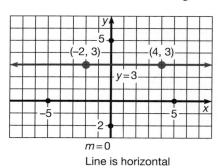
$m = 0$
Line is horizontal

Therefore, the slope of the line is 0.

For a horizontal line, the change in the ordinates of any two points on the line is 0. Therefore, the slope of any horizontal line is 0.

Practice Exercise 6.
$y = 2x + 7$

EXAMPLE 6 Determine the slope of the line with equation $x = 4$.

SOLUTION: The given line is vertical. For *all* values of y, $x = 4$. Let $(x_1, y_1) = (4, 5)$ and $(x_2, y_2) = (4, -3)$. We now have the following:

$$m = \frac{y_2 - y_1}{x_2 - x_1}$$
$$= \frac{-3 - 5}{4 - 4}$$
$$= \frac{-8}{0}$$

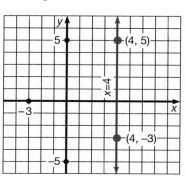
m is undefined
Line is vertical

which is undefined. Therefore, the slope of the given line is undefined.

ANSWERS TO PRACTICE EXERCISES

3. $m = \dfrac{-1}{4}$.   4. $m = 3$.   5. Undefined.   6. $m = 2$.

For a vertical line, the change in the abscissas of any two points on the line is 0. Therefore, the denominator in the expression for the slope is 0, and the slope is undefined. Hence, the slope of any vertical line is undefined.

We will now summarize some of the observations made about the slope of a line in the previous examples.

Slope of a Line

Consider the equation $ax + by + c = 0$. Let m be the slope of the line which is the graph of the given equation.

1. If $a = 0$ and $b \neq 0$, the graph is a *horizontal* line; $m = 0$.
2. If $a \neq 0$ and $b = 0$, the graph is a *vertical* line; m is undefined.
3. If $a \neq 0$ and $b \neq 0$, the graph is an *oblique* line (that is, neither horizontal nor vertical); m is either positive or negative.
   a. If a and b both have the *same* algebraic signs, then $m < 0$.
   b. If a and b have *different* algebraic signs, then $m > 0$.

- The slope of the line with equation $y = 7$ is 0 since the line is horizontal.
- The slope of the line with equation $x = 9$ is undefined since the line is vertical.
- The slope of the line with equation $2x + y = 9$ is negative since $a = 2$ and $b = 1$ both have the same algebraic signs (both positive).
- The slope of the line with equation $x - y = 5$ is positive since $a = 1$ and $b = -1$ have different algebraic signs.

We will conclude this section by determining the graph of the line that passes through a given point and has a known slope.

Graph the line that passes through the given point and has the indicated slope.

Practice Exercise 7. $(-2, -1)$; $m = 1$

EXAMPLE 7 Graph the line that passes through the point (2, 3) and has a slope of 3.

SOLUTION: First, plot the point (2, 3) in the xy-plane. Since the slope is 3, we have the following:

$$m = 3$$
$$= \frac{3}{1}$$
$$= \frac{\text{Change in ordinates}}{\text{Change in abscissas}}$$

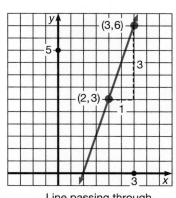

Line passing through the point (2, 3) with $m = 3$

We can, therefore, determine a second point on the line as follows:

1. Start at the point (2, 3).
2. Move 1 unit horizontally to the *right* and then 3 units vertically *up*.

We now have the second point (3, 6). Finally, draw the line between the points (2, 3) and (3, 6). (Check that the slope of the line through these two points is 3.)

Practice Exercise 8. $(3, -2)$; $m = \dfrac{-2}{3}$

EXAMPLE 8 Graph the line that passes through the point $(-3, 4)$ and has a slope of $\dfrac{-1}{2}$.

SOLUTION: First, plot the point $(-3, 4)$ in the *xy*-plane. Since the slope is $\dfrac{-1}{2}$, we have the following:

$$m = \dfrac{-1}{2}$$
$$= \dfrac{\text{Change in ordinates}}{\text{Change in abscissas}}$$

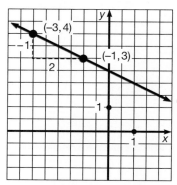

Line passing through the point (–3, 4) with $m = \dfrac{-1}{2}$

We can, therefore, determine a second point on the line as follows:

1. Start at the point $(-3, 4)$.
2. Move 1 unit vertically *down* and then 2 units horizontally to the *right*.

We now have the second point $(-1, 3)$. Finally, draw the line between the points $(-3, 4)$ and $(-1, 3)$. (Check that the slope of the line through these two points is $\dfrac{-1}{2}$.)

ANSWERS TO PRACTICE EXERCISES

7.

8.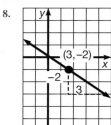

EXERCISES 13.4

In Exercises 1–20, determine the slope of the line passing through the given pair of points.

1. $(2, 3)$ and $(0, 1)$
2. $(-1, 2)$ and $(7, 3)$
3. $(-4, 5)$ and $(1, 1)$
4. $(-3, 4)$ and $(0, 2)$
5. $(2, 0)$ and $(4, -3)$
6. $(3, 7)$ and $(3, -2)$
7. $(-4, 5)$ and $(7, 6)$
8. $(0, -7)$ and $(0, 8)$
9. $(2, 3)$ and $(-2, 0)$
10. $(4, 5)$ and $(5, 4)$
11. $(0, 0)$ and $(8, -3)$
12. $(1, 1)$ and $(4, 4)$
13. $(-5, -1)$ and $(2, -1)$
14. $(-3, -3)$ and $(-2, -1)$
15. $(-2, 3)$ and $(0, 0)$
16. $(4, 5)$ and $(4, 0)$
17. $(2, 3)$ and $(-7, 3)$
18. $(-1, 2)$ and $(-1, 3)$
19. $(1, 0)$ and $(0, 1)$
20. $(-1, 0)$ and $(0, -1)$

In Exercises 21–40, determine the slope of the line with the given equation.

21. $x + y = 7$
22. $3x - y = 7$
23. $y + 3 = 0$
24. $y = 3x - 8$
25. $4x - 5y = 20$
26. $x - 3 = 0$
27. $x = 6y - 1$
28. $3x + 7y = 0$
29. $y = 4x$
30. $3x - 8y = 12$
31. $x = -1$
32. $y = -4x + 7$
33. $3y - x = 5$
34. $y = 0$
35. $2x - y = 9$
36. $7x + 3y = 21$
37. $y = -6$
38. $x = 5y - 3$
39. $4x - 5 = 0$
40. $3y = 2x - 9$

In Exercises 41–60, determine whether the slope of the line is positive, zero, negative, or undefined, by inspection of the given equation.

41. $y = 3x - 4$
42. $y = 1 - 5x$
43. $2y = 3x + 2$
44. $y = 3$
45. $3x = y + 7$
46. $x - 2y = 3$
47. $x = 7$
48. $y = 2x - 3$
49. $4x + 3y - 2 = 0$
50. $-5x = 2y - 3$
51. $3y - 4x = -1$
52. $2x - 5y = 0$
53. $y = -4x$
54. $3x + 7 = 0$
55. $3y - 5 = 0$
56. $2x = 3y$
57. $x + 3y = 8$
58. $3x - 4y - 5 = 0$
59. $7y - 8x = -9$
60. $-3x = 7y$

13.5 Linear Inequalities in One Variable and Their Graphs

OBJECTIVE

After completing this section, you should be able to solve and graph linear inequalities in a single variable.

You have already learned about ordering of numbers. For instance, if a and b are any two numbers such that a is greater than b, we write $a > b$. Similarly, the notation $x < y$ means that the number x is less than the number y. Each of the statements $a > b$ and $x < y$ is called an *inequality* and the symbols $>$ and $<$ are called *inequality signs*.

In Chapter 8, you learned to solve linear equations in one variable. Recall that such an equation has exactly one solution. In this section, you will learn to solve **linear inequalities in one variable**. The solutions for such inequalities will be infinitely many.

Graph each of the following inequalities.

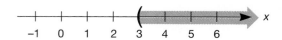

EXAMPLE 1 Determine the solutions for $x > 3$.

SOLUTION: The expression $x > 3$ is a linear inequality in the variable x. The solutions for this inequality can be displayed by shading that portion of the number line to the right of, but not including, 3. Hence, *all* values of x that are greater than 3 are solutions of the given inequality. We use the parenthesis to indicate that 3 is *not* included.

EXAMPLE 2 Determine the solutions for $x < 4$.

SOLUTION: The solution for the linear inequality $x < 4$ is the set of *all* numbers that are less than 4. These can be displayed by shading that portion of the number line to the left of, but not including, 4. We use the parenthesis to indicate that 4 is *not* included

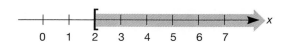

EXAMPLE 3 Determine the solutions for $x \geq 2$.

SOLUTION: The solution for the linear inequality $x \geq 2$ is the set of *all* numbers that are greater than or equal to 2. These can be displayed by shading that portion of the number line to the right of *and* including 2. We use the square bracket, [, to indicate that 2 *is* included.

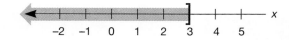

EXAMPLE 4 Determine the solutions for $x \leq 3$.

SOLUTION: The solution for the linear inequality $x \leq 3$ is the set of *all* numbers that are less than or equal to 3. These can be displayed by shading that portion of the number line to the left of *and* including 3. We use the square bracket,], to indicate that 3 *is* included.

- The inequality associated with the graph is $x > -3$:

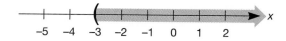

- The inequality associated with the graph is $u \leq 5$:

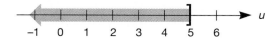

13.5 Linear Inequalities in One Variable and Their Graphs

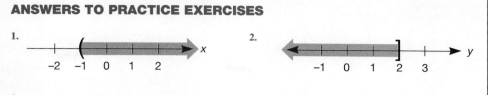

ANSWERS TO PRACTICE EXERCISES

1.
2.
3.
4.

To solve the linear equation $x - 3 = 4$, we add 3 to *both* sides of the equation to isolate x. We get the equation $x = 7$.

What happens to the inequality sign when we add a positive number to both sides of it? Let's consider what happens when 2 is added to *both* sides of the inequality $4 < 7$.

$$4 < 7$$
$$4 + 2 \;?\; 7 + 2 \quad \text{(2 is positive)}$$
$$6 < 9$$

Note that the inequality sign remains unchanged.

What happens if we add a negative number to *both* sides of the inequality $4 < 7$?

$$4 < 7$$
$$4 + (-3) \;?\; 7 + (-3) \quad (-3 \text{ is negative})$$
$$1 < 4$$

Again, the inequality sign remains unchanged.

Addition Property of Inequalities

If the same number is added to *both* sides of an inequality, then the inequality sign remains unchanged. That is, if a, b, and c are *any* numbers, we have the following:

1. If $a < b$, then $a + c < b + c$.
2. If $a > b$, then $a + c > b + c$.

These statements are true whether c is positive, negative, or zero.

Solve each of the inequalities for the indicated variable. Graph your solutions.

Practice Exercise 5. $x - 7 < 1$

EXAMPLE 5 Solve the inequality $x - 1 > 7$ for x.

SOLUTION:

$$x - 1 > 7$$
$$(x - 1) + 1 > 7 + 1 \quad \text{(Add 1 to \textit{both} sides.)}$$
$$x > 8$$

The solution for the given inequality is graphed below:

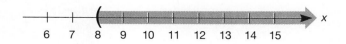

Practice Exercise 6. $2y + 1 \geq 0$

EXAMPLE 6 Solve the inequality $3y - 1 \leq 2(y + 3)$ for y.

SOLUTION:

$$3y - 1 \leq 2(y + 3)$$
$$3y - 1 \leq 2y + 6 \qquad \text{(Remove parentheses.)}$$
$$3y \leq 2y + 7 \qquad \text{(Add 1 to both sides.)}$$
$$y \leq 7 \qquad \text{(Add } -2y \text{ to both sides.)}$$

The solution for the given inequality is graphed below.

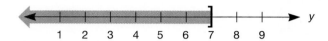

What happens when both sides of the inequality $4 < 7$ are multiplied by a positive number?

$$4 < 7$$
$$4(5) \; ? \; 7(5) \qquad \text{(5 is positive.)}$$
$$20 < 35$$

Again, the inequality sign remains unchanged.

Now, let's examine what happens when both sides of the inequality $4 < 7$ are multiplied by a *negative* number:

$$4 < 7$$
$$4(-5) \; ? \; 7(-5) \qquad (-5 \text{ is negative.})$$
$$-20 > -35$$

Notice that the inequality sign is *reversed*.

Multiplication Property of Inequalities

1. If *both* sides of an inequality are multiplied by the same *positive* number, then the inequality sign remains unchanged. That is, if a and b are any numbers and $c > 0$, we have the following:
   a. If $a < b$, then $ac < bc$.
   b. If $a > b$, then $ac > bc$.
2. If *both* sides of an inequality are multiplied by the same *negative* number, then the inequality sign is *reversed*. That is, if a and b are any numbers and $c < 0$, we have the following:
   a. If $a < b$, then $ac > bc$.
   b. If $a > b$, then $ac < bc$.

Practice Exercise 7. $-2(x + 1) \leq 6$

EXAMPLE 7 Solve the inequality $0.2w \geq 0.5$ for w.

SOLUTION:

$$0.2w \geq 0.5$$
$$(10)(0.2w) \geq (10)(0.5) \qquad (10 > 0; \text{ sign remains unchanged})$$
$$2w \geq 5$$
$$w \geq \frac{5}{2}$$

13.5 Linear Inequalities in One Variable and Their Graphs

The graphical solution for the given inequality is given below:

Practice Exercise 8.
$2(p - 3) \leq -3(p + 4)$

EXAMPLE 8 Solve the inequality $\dfrac{-1}{9}u > \dfrac{2}{3}$ for u.

SOLUTION:

$$\dfrac{-1}{9}u > \dfrac{2}{3}$$

$$(9)\left(\dfrac{-1}{9}u\right) > (9)\left(\dfrac{2}{3}\right) \quad \text{[LCM (3, 9) = 9. 9 > 0; sign remains unchanged]}$$

$$-u > 6$$

$$(-1)(-u) < (-1)(6) \quad (-1 < 0; \text{ sign reverses})$$

$$u < -6$$

The graphical solution of the given inequality is given below:

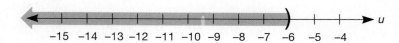

When solving linear equations in Chapter 8, it was sometimes necessary to use more than one operation. Similarly, to solve linear inequalities, it is sometimes necessary to use both the addition and the multiplication properties of inequalities.

Practice Exercise 9.
$0.4 - 0.3x > 1.3$

EXAMPLE 9 Solve the inequality $2x - 1 > 7$ for x.

SOLUTION:

$$2x - 1 > 7$$

$$(2x - 1) + 1 > 7 + 1 \quad \text{(Add 1 to both sides; sign remains unchanged.)}$$

$$2x > 8$$

$$\left(\dfrac{1}{2}\right)(2x) > \left(\dfrac{1}{2}\right)(8) \quad \left(\text{Multiply both sides by } \dfrac{1}{2} > 0; \text{ sign remains unchanged.}\right)$$

$$x > 4$$

Practice Exercise 10. $0.7u + 0.9 - 0.5u \geq 0.3u + 0.5$

EXAMPLE 10 Solve the inequality $4(t - 1) + 2 > 6t + 9$ for t.

SOLUTION:

$$4(t - 1) + 2 > 6t + 9$$

$$4t - 4 + 2 > 6t + 9 \quad \text{(Remove parentheses.)}$$

$$4t - 2 > 6t + 9 \quad \text{(Combine like terms.)}$$

$$4t > 6t + 11 \quad \text{(Add 2 to both sides; sign remains unchanged.)}$$
$$-2t > 11 \quad \text{(Add } -6t \text{ to both sides; sign remains unchanged.)}$$
$$t < \frac{-11}{2} \quad \left(\text{Multiply both sides by } \frac{-1}{2} < 0\text{; sign reverses.}\right)$$

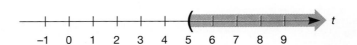

Practice Exercise 11.
$\dfrac{y-3}{5} \geq \dfrac{y+1}{4}$

EXAMPLE 11 Solve the inequality $\dfrac{t-2}{3} + 1 > \dfrac{t+3}{4}$ for t.

SOLUTION:

$$\frac{t-2}{3} + 1 > \frac{t+3}{4}$$

$$(12)\left(\frac{t-2}{3} + 1\right) > (12)\left(\frac{t+3}{4}\right) \quad [\text{LCM }(3,4) = 12.\ 12 > 0;\ \text{sign remains unchanged}]$$

$$\begin{aligned}
4(t-2) + 12 &> 3(t+3) \\
4t - 8 + 12 &> 3t + 9 &&\text{(Removing parentheses)} \\
4t + 4 &> 3t + 9 &&\text{(Combining like terms)} \\
t + 4 &> 9 &&\text{(Subtracting } 3t \text{ from both sides)} \\
t &> 5 &&\text{(Subtracting 4 from both sides)}
\end{aligned}$$

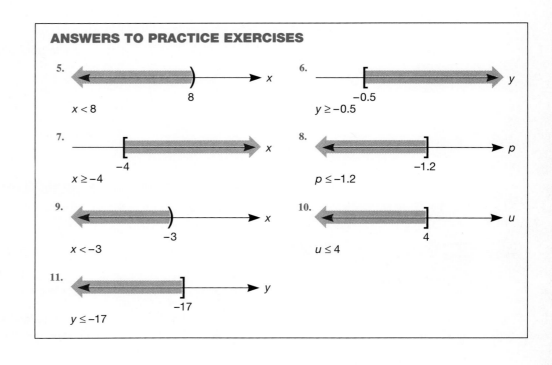

ANSWERS TO PRACTICE EXERCISES

5. $x < 8$

6. $y \geq -0.5$

7. $x \geq -4$

8. $p \leq -1.2$

9. $x < -3$

10. $u \leq 4$

11. $y \leq -17$

EXERCISES 13.5

In Exercises 1–10, graph each of the given inequalities on the number line.

1. $x > 2$
2. $x > -3$
3. $x < 0$
4. $x < \dfrac{2}{3}$
5. $y > -2$
6. $y > -3.5$
7. $u \leq -2.5$
8. $u > \dfrac{-13}{5}$
9. $t < 0$
10. $t \geq 0$

In Exercises 11–40, solve each of the inequalities for the indicated variable.

11. $y - 2 > 6$
12. $x + 3 \leq 0$
13. $2u + 1 < 7$
14. $0.5t - 0.6 \geq -0.1$
15. $4 - 3x > 9$
16. $11 \leq 6 - 5m$
17. $2p + 1 < 3p - 2$
18. $4t - 5 \geq 6 - 2t$
19. $7x + 9 - 5x > 3x + 5$
20. $0.2u - 1.5 \leq 1.3u + 2.7 - 0.5u$
21. $\dfrac{2}{9}(u + 3) \leq \dfrac{-1}{3}(u - 4)$
22. $5(p + 1) - 2 \leq 6(1 - 2p)$
23. $\dfrac{x - 3}{4} > \dfrac{2 - x}{5}$
24. $\dfrac{4 - 2y}{3} \leq 4(1 - y)$
25. $\dfrac{t + 7}{-11} \geq \dfrac{3 - 2t}{2}$
26. $\dfrac{3n - 1}{-5} \leq \dfrac{2 + 4n}{-4}$
27. $2(x + 3) - 4 \leq 5 - 3(2x - 1)$
28. $\dfrac{2}{15}(y - 1) - \dfrac{1}{5}(y + 2) \geq \dfrac{1}{3}(1 - 3y) - \dfrac{7}{15}$
29. $4 - 3(t + 2) > 5(t - 2) + 1$
30. $2.9 - 1.6(4u - 3) \leq 0.2(3 - u) - 3.4(5 - 3u)$
31. $\dfrac{5y + 28}{0.4} > \dfrac{2y + 7}{0.3}$
32. $-0.22u - 0.08 \leq (-0.1)u + 0.08$
33. $\dfrac{4(6p - 11)}{7} + \dfrac{7p}{5} < 10$
34. $1.5(3 - 2r) \geq 1.8(3r - 4) - 1.4(7r + 1)$
35. $\dfrac{3q - 7}{2} - \dfrac{7}{3} \leq \dfrac{2q - 5}{6}$
36. $\dfrac{16w + 1}{0.7} > \dfrac{4w + 25}{0.3}$
37. $\dfrac{2.1(u + 1)}{5} - \dfrac{2}{3} < \dfrac{3.2(2u - 1)}{4}$
38. $2.3(2 - 5v) - 6.1 > 3.1(2v + 3) + 5.9$
39. $\dfrac{3x - 4}{2} - \dfrac{2x + 1}{3} > \dfrac{x - 9}{4} - \dfrac{5 - 2x}{6}$
40. $\dfrac{2p + 1}{7} + \dfrac{3 - 4p}{3} \leq \dfrac{1 - 3p}{5} + \dfrac{6 - p}{2}$

13.6 Linear Inequalities in Two Variables

OBJECTIVE

After completing this section, you should be able to solve a linear equality in two variables, for one variable in terms of the other.

In Section 13.2, you learned to solve linear equations in two variables. Each such equation has infinitely many solutions. For instance, the equation

$$y = x + 3$$

has infinitely many solutions since there are infinitely many values that can be assigned to x. For each value of x, there is a corresponding value of y.

In this section, you will be introduced to linear inequalities in two variables and their solutions. A **linear inequality in two variables** resembles a linear equation in two variables with the $=$ symbol replaced by one of the symbols $<$, $\leq$, $>$, or $\geq$. The addition and multiplication properties of inequalities used to solve linear inequalities in a single variable are also used to solve linear inequalities in two variables.

The inequality

$$y < x + 3$$

is a linear inequality in the two variables x and y. There are infinitely many values that can be assigned to x which will give corresponding values for y. Moreover, for *each* value assigned to x, there are infinitely many values for y. For instance, if

$$x = 3$$
then $\quad x + 3 = 6$
and $\quad y < x + 3$
becomes $\quad y < 6$

This means that when $x = 3$, y will take *all* values that are less than 6. The ordered pairs $(3, 5)$, $(3, 0)$, $(3, -0.5)$, and $(3, 5.9)$ are a few of these solutions.

For the inequality

$$y \leq x + 4$$
if $\quad x = 2$
then $\quad x + 4 = 6$
and $\quad y \leq x + 4$
becomes $\quad y \leq 6$

This means that when $x = 2$, y will take *all* values that are less than or equal to 6. The ordered pairs $(2, 6)$, $(2, 4.5)$, $(2, 0)$, $(2, -13)$, and $(2, -17)$ are a few of these solutions.

Observe that the inequality

$$y \geq x + 4$$

means that

$$y > x + 4 \quad \text{or} \quad y = x + 4$$

whereas the inequality

$$y > x + 4$$

does not allow for y to be equal to $x + 4$.

13.6 Linear Inequalities in Two Variables

In a similar manner, the inequality

$$y \leq 2x - 1$$

means that

$$y < 2x - 1 \quad \text{or} \quad y = 2x - 1$$

Solve each inequality for y.

Practice Exercise 1. $2x + y \geq 3$

EXAMPLE 1 Solve the inequality $2x - 3y < 1$ for y.

SOLUTION:

$$2x - 3y < 1$$
$$-3y < 1 - 2x \quad \text{(Add } -2x \text{ to both sides; sign remains unchanged.)}$$
$$y > \frac{2x - 1}{3} \quad \left(\text{Multiply both sides by } \frac{-1}{3} < 0; \text{ sign reverses.}\right)$$

For the inequality in Example 1, if $x = 2$, then

$$y > \frac{2(2) - 1}{3}$$
$$> \frac{4 - 1}{3}$$
$$> 1$$

Hence, if $x = 2$, then *any* value of y that is greater than 1 will satisfy the inequality $2x - 3y < 1$.

Practice Exercise 2.
$2 - 3x \geq 1 - 2y$

EXAMPLE 2 Solve the inequality $3y + 2 < 5x - 4y$ for y.

SOLUTION:

$$3y + 2 < 5x - 4y$$
$$7y + 2 < 5x \quad \text{(Add } 4y \text{ to both sides; sign remains unchanged.)}$$
$$7y < 5x - 2 \quad \text{(Add } -2 \text{ to both sides; sign remains unchanged.)}$$
$$y < \frac{5x - 2}{7} \quad \left(\text{Multiply both sides by } \frac{1}{7} > 0; \text{ sign remains unchanged.}\right)$$

Practice Exercise 3.
$2(1 - x) + 3(y - 2) > -1$

EXAMPLE 3 Solve the inequality $2x + 3y - 4 \leq 5x - 2y$ for y.

SOLUTION:

$$2x + 3y - 4 \leq 5x - 2y$$
$$2x + 5y - 4 \leq 5x \quad \text{(Add } 2y \text{ to both sides; sign remains unchanged.)}$$
$$5y - 4 \leq 3x \quad \text{(Add } -2x \text{ to both sides; sign remains unchanged.)}$$
$$5y \leq 3x + 4 \quad \text{(Add 4 to both sides; sign remains unchanged.)}$$
$$y \leq \frac{3x + 4}{5} \quad \left(\text{Multiply both sides by } \frac{1}{5} > 0; \text{ sign remains unchanged.}\right)$$

> **ANSWERS TO PRACTICE EXERCISES**
>
> 1. $y \geq 3 - 2x$. 2. $y \geq \dfrac{3x - 1}{2}$. 3. $y > \dfrac{2x + 3}{3}$.

EXERCISES 13.6

In Exercises 1–20, solve each of the given inequalities for y.

1. $x + y < 3$
2. $x - y > 4$
3. $x + 3y \leq 0$
4. $2x - y \geq 1$
5. $x - 2y < 5$
6. $2y > 1 - 3x$
7. $4x \geq 3y - 1$
8. $x + 6y \leq 2$
9. $6x + 5y \leq 7$
10. $4x - 7y + 2 > 0$
11. $x > 4 - 2y$
12. $4 - 5x \geq 2 - 8y$
13. $2x + 3 - 3y < x - 2y$
14. $4 - 3y \leq 5 + 6y - x$
15. $\dfrac{x + 4}{2} > \dfrac{y - 3}{5}$
16. $\dfrac{2x + 1}{4} \leq \dfrac{3y + 4}{-3}$
17. $3x - y \leq \dfrac{2y + 1}{3}$
18. $2(x - 1) + 3(y + 2) < 1$
19. $3(x + 2) \geq 2(y - 7) - 4$
20. $-3(1 - 2x) \geq \dfrac{3y - 8}{2}$

21. Given the inequality $2x - 3y < 1$, determine five different values for y, if $x = 2$.
22. Given the inequality $y \geq 3x + 7$, determine the smallest value of y, if $x = -3$.
23. Given the inequality $3y \leq 1 - 2x$, determine the largest value of y, if $x = 4$.
24. Does the ordered pair $(1, -3)$ satisfy the inequality $x - 2y > 3$? Why (not)?
25. Does the ordered pair $(-2, 4)$ satisfy the inequality $2y - 3x < 0$? Why (not)?

13.7 Graphs of Linear Inequalities in Two Variables

OBJECTIVE

After completing this section, you should be able to graph linear inequalities in two variables.

In Section 13.3, you learned that the graph of a linear equation in two variables is a straight line. The line was determined by plotting at least two points through which it passes.

If we have the equation

$$2x + 3y = 6$$

its graph can be readily determined, using intercepts, as follows:

1. If $x = 0$, then $2x + 3y = 6$ becomes $3y = 6$ or $y = 2$. Hence, the point $(0, 2)$ lies on the line.

13.7 Graphs of Linear Inequalities in Two Variables

2. If $y = 0$, then $2x + 3y = 6$ becomes $2x = 6$ or $x = 3$. Hence, the point $(3, 0)$ lies on the line.
3. Plot the points $(0, 2)$ and $(3, 0)$ and draw a line through them.

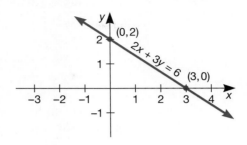

Observe that the line separates the xy-plane into two **half-planes,** one above the line and the other below the line. The line itself is the **boundary of each of these half-planes** but is not included in either of them. Each of these half-planes can be described by an inequality. In Figure 13.9, the shaded half-plane is described by the inequality $2x + 3y < 6$; the boundary is dashed to indicate that it is not included as part of the graph.

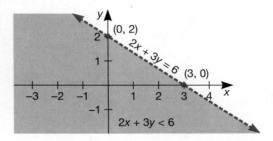

FIGURE 13.9

The unshaded half-plane in Figure 13.9 is described by the inequality $2x + 3y > 6$.

To determine the half-plane that is the graph of an inequality, we must first determine the line that separates the plane into two half-planes. That is, we must look at the graph of the corresponding equation.

Graph each of the following inequalities in the *xy*-plane.

Practice Exercise 1. $y > 1 - 3x$

EXAMPLE 1 Graph the inequality $y < 2x + 5$ in the xy-plane.

SOLUTION:

STEP 1: Determine the line that separates the plane. That is, consider the graph of the corresponding equation, $y = 2x + 5$. If $x = 0$, then $y = 5$. If $y = 0$, then $x = -2.5$. Hence, the line passes through the two points $(0, 5)$ and $(-2.5, 0)$.

STEP 2: Graph the equation $y = 2x + 5$ as the *dashed* line in the xy-plane as indicated in the graph. The dashed line is the boundary of the half-plane but is *not* part of the graph.

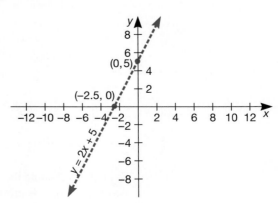

Step 3: Determine the correct half-plane. Since the required solution is one of the two half-planes determined, we can select any point and test it in the inequality to see if its coordinates satisfy the inequality. For instance, using (0, 0), we note that (0, 0) does satisfy the inequality since

$$y < 2x + 5$$
$$0 \; ? \; 2(0) + 5$$
$$0 \; ? \; 0 + 5$$
$$0 < 5$$

Therefore, the half-plane containing the point (0, 0) is the required solution and is given as the shaded portion in the accompanying diagram.

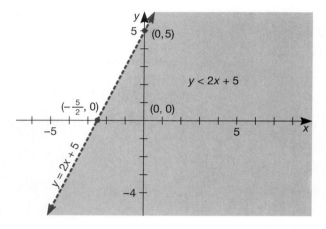

Practice Exercise 2.
$3x - 5y < -15$

EXAMPLE 2 Graph the inequality $7x + 4y \geq -28$.

SOLUTION:

Step 1: Determine the line that separates the plane. That is, consider the graph of the equation $7x + 4y = -28$. If $x = 0$, then $y = -7$. If $y = 0$, then $x = -4$. Hence, the line passes through the two points $(0, -7)$ and $(-4, 0)$. (Of course, other points could have been determined by taking *any* value for x or y and solving for the other variable. However, using intercepts is more convenient.)

Step 2: Graph the equation $7x + 4y = -28$ as the *solid* line in the xy-plane, as indicated in the diagram. The line is the boundary of the half-plane and *is* part of the required solution.

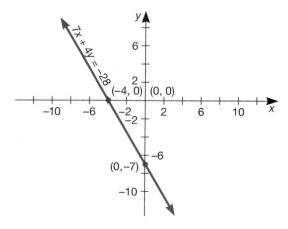

Step 3: Determine the correct half-plane. Since the required solution is one of the two half-planes determined, we can select any point and test it in the inequality.

13.7 Graphs of Linear Inequalities in Two Variables

Again, for convenience, we use the point (0, 0) and note that (0, 0) does satisfy the inequality since

$$7x + 4y \geq -28$$
$$7(0) + 4(0) \; ? \; -28$$
$$0 + 0 \; ? \; -28$$
$$0 \geq -28$$

Therefore, the half-plane containing the point (0, 0) is the required solution and is given as the shaded portion in the diagram.

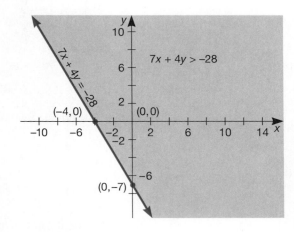

We will now summarize the procedure for graphing a linear inequality in two variables.

Graphing a Linear Inequality in Two Variables

To graph a linear inequality in two variables, proceed as follows:

1. First, determine the graph of the corresponding equation. This is the line that separates the plane into two half-planes, one of which will be the required solution.
   a. If the inequality contains either symbol < or >, graph the equation as a *dashed* line to indicate that the boundary is not part of the solution.
   b. If the inequality contains either symbol ≤ or ≥, graph the equation as a *solid* line to indicate that the boundary is part of the solution.
2. To determine the required half-plane that is the graphical solution of the inequality, pick *any* point on one side of the boundary. [The point (0, 0) is the easiest to use, if the point is not on nor very close to the boundary.]
   a. If the coordinates of the point selected satisfy the given inequality, the required solution is the half-plane containing the point.
   b. If the coordinates of the point selected do not satisfy the given inequality, the required solution is the half-plane that does *not* contain the point.
3. Shade the required solution.

Practice Exercise 3.
$4x - 5y \leq 0$

EXAMPLE 3 Graph the inequality $2x - 3y > 4$.

SOLUTION:

STEP 1: Determine the line that separates the *xy*-plane. That is, consider the graph of $2x - 3y = 4$. Verify that the line passes through the points (2, 0) and (−1, −2).

STEP 2: Graph the equation $2x - 3y = 4$ as the *dashed* line in the xy-plane as indicated in the diagram. This line becomes the boundary of the required solution but is *not* part of the solution (since the inequality contains the symbol $>$).

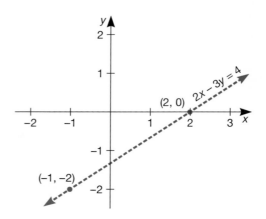

STEP 3: Determine the correct half-plane. Using $(0, 0)$ as the test point, we have

$$2x - 3y > 4$$
$$2(0) - 3(0) \; ? \; 4$$
$$0 - 0 \; ? \; 4$$
$$0 < 4$$

Since the coordinates of the point $(0, 0)$ do *not* satisfy the given inequality, the required solution is the half-plane that does *not* contain the point $(0, 0)$.

STEP 4: The required solution is given as the shaded portion of the xy-plane in the graph.

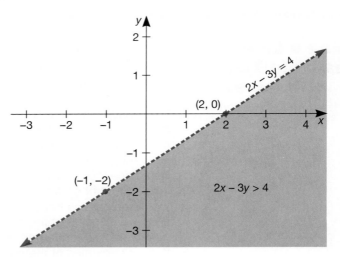

Practice Exercise 4. $y \leq 3x + 1$

EXAMPLE 4 Graph the inequality $y \leq 2x$.

SOLUTION:

STEP 1: Determine the line that separates the xy-plane. That is, consider the graph of $y = 2x$. The line passes through the two points $(0, 0)$ and $(2, 4)$. (Do you agree?)

STEP 2: Graph the equation $y = 2x$ as the *solid* line in the xy-plane as indicated in the diagram. This line becomes the boundary of the required solution and *is* part of the solution (since the inequality contains the symbol $\leq$).

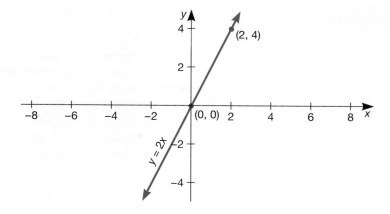

STEP 3: Determine the correct half-plane. We *cannot* use the point (0, 0) for a test point since the line contains that point. We can, however, use any point in the plane that is not on the line $y = 2x$. For convenience, select a point on one of the coordinate axes, other than (0, 0). We will use the point (2, 0) as our test point. We have

$$y \leq 2x$$
$$0 \;?\; 2(2)$$
$$0 \leq 4$$

Since the coordinates of the point (2, 0) *do* satisfy the given inequality, the required solution is the half-plane that *does* contain the point (2, 0).

STEP 4: The required solution is given as the shaded portion of the *xy*-plane in the graph.

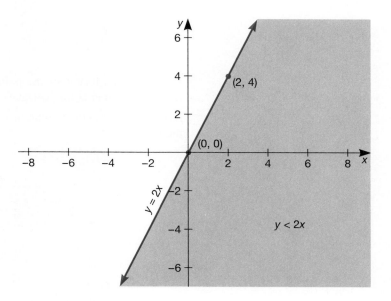

When working with linear inequalities in the variables *x* and *y*, it is possible that the coefficient of *x* or the coefficient of *y* (but not both coefficients) will be 0. If the coefficient of *x* is 0, then the graph of the inequality will have a boundary that is a horizontal line. Similarly, if the coefficient of *y* is 0, then the graph of the inequality will have a boundary that is a vertical line. In Example 5, we illustrate one of these cases.

Practice Exercise 5. $y \leq 3$

EXAMPLE 5 Graph the inequality $3x - 2 > 1$ in the xy-plane.

SOLUTION:

STEP 1: Determine the line that separates the xy-plane. That is, consider the graph of $3x - 2 = 1$ or, equivalently, $3x = 3$. The equation $3x = 3$ or, equivalently, $x = 1$, means that for *all* values of y, $x = 1$. Hence, the line passes through *all* points in the xy-plane with an abscissa 1. We will use the points $(1, 0)$ and $(1, 3)$.

STEP 2: Graph the equation $3x - 2 = 1$ as the *dashed* line in the xy-plane as indicated in the diagram. This line becomes the boundary of the required solution but is *not* part of the solution (since the inequality contains the symbol $>$).

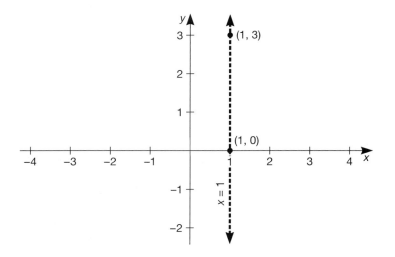

STEP 3: Using the point $(0, 0)$ as a test point, we have

$$3x - 2 > 1$$
$$3(0) - 2 \;?\; 1$$
$$0 - 2 \;?\; 1$$
$$-2 < 1$$

Since the coordinates of the point $(0, 0)$ do *not* satisfy the given inequality, the required solution is the half-plane that does *not* contain the point $(0, 0)$.

STEP 4: The required solution is given as the shaded portion of the xy-plane in the graph.

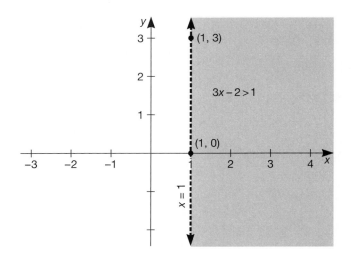

13.7 Graphs of Linear Inequalities in Two Variables

ANSWERS TO PRACTICE EXERCISES

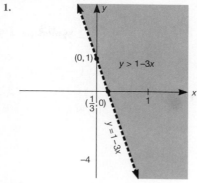

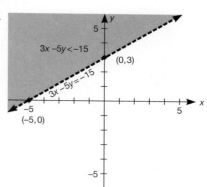

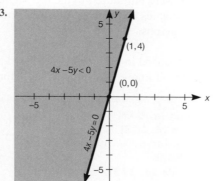

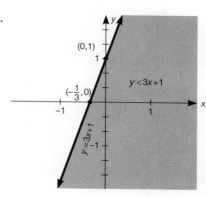

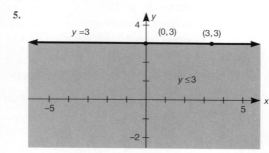

EXERCISES 13.7

Graph each of the following inequalities in the xy-plane.

1. $y > 2x + 1$
2. $y < 2x - 3$
3. $3x + 2y > 0$
4. $2x < 3y$
5. $3x - 5y > -2$
6. $3x + 2y \geq 1$
7. $4x - 3y \leq -2$
8. $4 \geq y$
9. $3 - 4y \geq x$
10. $3x + 1 \leq 3 - 2y$
11. $3x + 5y \leq 15$
12. $x \leq 2y + 6$
13. $5x - 5y \geq 10$
14. $3x > 4y$
15. $4x + 7y > 28$
16. $2x + 5y \leq 18$
17. $x - 5y < 10$
18. $y \leq 3x$
19. $2y > x - 7$
20. $4x \geq y - 2$
21. $3x - y \geq 2$

22. $x + 3y \leq 3$
23. $2x + 5y > 10$
24. $5x - 2y \leq 10$
25. $x + 2 \geq 0$
26. $y - 3 \leq 0$
27. $2x + 3y > 7$
28. $3x - 2y < 8$
29. $4x \geq 5y - 2$
30. $3 - 4y \leq 3x$
31. $0.4x - 1.3y < 5.2$
32. $2.3x - 4.6y > 9.2$
33. $y \leq \frac{3}{2}x - 4$
34. $4 + \frac{1}{3}x \geq y$
35. $x - 7 < 0$
36. $y + 5 \geq 0$
37. $2(1 - 3x) < 3(2y - 1)$
38. $4(y - 1) \geq 3(x + 1)$
39. $\frac{-1}{5}(x + 2) < \frac{1}{6}(1 - y)$
40. $\frac{2}{3}(y - 2) \geq \frac{-1}{2}(x + 1)$

Chapter Summary

In this chapter, you were introduced to the rectangular coordinate system in the plane, linear equations in two variables and their graphs, and linear inequalities in one and two variables, together with their graphs.

The **key words** introduced in this chapter are listed here in the order in which they appeared in the text. You should know what they mean before proceeding to the next chapter.

origin (p. 544)
scale (p. 544)
plane (p. 544)
horizontal axis (p. 544)
vertical axis (p. 544)
ordered pair (p. 545)
coordinate (p. 545)
abscissa (p. 545)
ordinate (p. 545)
rectangular coordinate system (p. 545)
quadrants (p. 545)
linear equations in two variables (p. 550)
linear inequalities in one variable (p. 565)
linear inequality in two variables (p. 572)
half-planes (p. 575)
boundary of each half-plane (p. 575)

Review Exercises

13.1 Rectangular Coordinate System

In Exercises 1–6, determine in which quadrant each of the indicated points lies. If the point lies on one of the coordinate axes, indicate which one.

1. $(3, -4)$
2. $(-4, -5)$
3. $(0, 6)$
4. $(-4, 2)$
5. $(-7, 0)$
6. $(6, 3)$

In Exercises 7–12, set up a rectangular coordinate system and plot each of the following points. Label the points according to the letters given.

7. $A(2, 5)$
8. $B(6, 0)$
9. $C(-7, -4)$
10. $D(-4, 2)$
11. $E(5, -3)$
12. $F(0, -5)$

13.2 Linear Equations in Two Variables

In Exercises 13–15, solve the given equations for y.

13. $2x + 3y = 4$
14. $7x - 2(y + 1) = 6$
15. $4(y - 5) - 2(x + 5) = 6 - 2(y + 1)$
16. Given the equation $3x + 2y = 12$, determine a value of x for which $y = 0$.
17. Given the equation $5x - 4y = 13$, determine a value of y for which $x = 0$.

13.3 Graphing Straight Lines

In Exercises 18–21, determine by inspection if the line associated with each equation is horizontal, vertical, or neither.

18. $2x - 3y = 7$
19. $4x = 5$
20. $5 - 3y = 0$
21. $6x = 2 - 3y$

In Exercises 22–25, graph each of the given equations by determining at least three points through which the graph passes.

22. $x + 2y = 4$
23. $y = 5x$
24. $x - 3 = 0$
25. $2x = 3y - 1$

In Exercises 26–31, use the intercept method to determine the graph of each of the given equations.

26. $x + 2y = 4$
27. $x - y = 5$
28. $2x - 3y = 6$
29. $3x + 2y = 9$
30. $3x + 5y + 15 = 0$
31. $x - 3y + 9 = 0$

13.4 Slope of a Line

In Exercises 32–37, determine the slope of the line passing through the given pair of points.

32. $(-2, 3)$ and $(1, -2)$
33. $(0, 0)$ and $(3, -4)$
34. $(3, 4)$ and $(-7, 4)$

35. $(3, -4)$ and $(-5, 2)$
36. $(5, 0)$ and $(5, -5)$
37. $\left(\frac{1}{2}, 0\right)$ and $\left(-1, \frac{1}{3}\right)$

In Exercises 38–43, determine, by inspection of the equation, whether the slope of the line is positive, negative, zero, or undefined.

38. $2x - y = 5$
39. $y = 3x + 5$
40. $y = 6$
41. $1 - 2x = 3y$
42. $x + 2y = 7$
43. $x = -6$

13.5 Linear Inequalities in One Variable and Their Graphs

In Exercises 44–47, graph each of the given inequalities on the number line.

44. $x < 3$
45. $x \geq -2$
46. $y \leq 4$
47. $y > -5$

In Exercises 48–51, solve each of the inequalities for the indicated variable.

48. $y + 3 \geq 1$
49. $x - 2 < 5.1$
50. $3t + 1 < 2 - 4t$
51. $4(2p - 1) + 3 \geq 5 - 3(6 - 5p)$

13.6 Linear Inequalities in Two Variables

In Exercises 52–55, solve the given inequalities for y.

52. $2x - 3y > 1$
53. $x + 7y \geq 3$
54. $3(x + 1) + 2(1 - y) \leq 1$
55. $4(x + 1) < 5 - 4(y - 2)$
56. Given the inequality $3x + 1 < 2y$, determine four different values for y, if $x = 3$.
57. Does the ordered pair $(-2, 3)$ satisfy the inequality $2x + 3y < 2$? Why (not)?

13.7 Graphs of Linear Inequalities in Two Variables

Determine the graph of each of the given inequalities.

58. $y < 3x + 1$
59. $2x - 3y \leq 6$
60. $x > 2y - 5$

Teasers

1. If two lines in the same plane are parallel, then their slopes are equal. If two lines in the same plane have the same slopes, are they necessarily parallel? Explain your answer.
2. Determine the equation of the line, in the xy-plane, that has a slope of 1.07 and passes through the point of intersection of the lines with equations $2x - y = 3$ and $y = 4$.
3. If three or more points in the same plane lie on the same line, the points are said to be "collinear." Given three points in the same plane, explain how you could determine if they are collinear. (*Hint:* Consider their slopes.)
4. A line in the xy-plane has an x-intercept of $2\sqrt{3}$ and a y-intercept of $-3\sqrt{2}$. Determine the equation of the line.
5. Determine the y-intercept of the line, in the xy-plane, that passes through the points $A(-2, 3)$ and $B(1, -4)$.

In Exercises 6–8, we have two inequalities connected by the word "and." A solution of such an inequality statement must be a solution of both inequalities.

6. Solve the following for x: $2x - 3 \leq 4$ *and* $3x - 7 < 5$.
7. Solve the following for y: $3(2 - y) > 0$ *and* $2y \leq y + 5$.
8. Solve the following for p: $3(5p + 4) > -3$ *and* $3(5p + 4) \leq 6$.

In Exercises 9–10, we have two inequalities connected by the word "or." A solution of such an inequality statement must be a solution of *at least one* of the inequalities.

9. Solve the following for u: $3(u - 1) \leq 5$ *or* $2(u + 1) > u - 5$.
10. Solving the following for q: $(2q + 3) - 4q < 0$ *or* $2(3q + 1) > 8q$.

CHAPTER 13 TEST

This test covers materials from Chapter 13. Read each question carefully, and answer all questions. If you miss a question, review the appropriate section of the text.

1. In which quadrant does the point $(-3, 4)$ lie?
2. In which quadrant does the point $(3, -4)$ lie?
3. Solve the equation $3x - 2y = 4$ for y.
4. Solve the equation $2(x - 3) - 3y = 4 - 2(y + x)$ for y.
5. Given the equation $4x - 3y = 12$, determine a value of x for which $y = 2$.
6. Given the equation $5x - 2(y + 3) = 4$, determine a value of y for which $x = 0$.
7. Determine by inspection if the line associated with the equation $4 - 3y = 0$ is horizontal, vertical, or neither.
8. Determine by inspection if the line associated with the equation $3x = 2 - y$ is horizontal, vertical, or neither.
9. Graph the equation $x + 6 = 0$.
10. Graph the equation $3x = 2y - 1$.
11. Determine the x-intercept associated with the graph of the equation $3x - 2y = 6$.
12. Determine the y-intercept associated with the graph of the equation $2x - 5y - 10 = 0$.
13. Determine the slope of the line passing through the two points $(-3, 2)$ and $(4, -1)$.
14. Determine the slope of the line passing through the two points $(-5, 6)$ and $(7, 6)$.
15. Graph the inequality $x \geq -3$ on the number line.
16. Graph the inequality $x < 4$ on the number line.
17. Solve the inequality $2(x - 1) - 3(y + 2) \leq 1$ for y.
18. Does the ordered pair $(-1, -2)$ satisfy the inequality $4x - 5y < 3$? Why (not)?
19. Does the ordered pair $(2, -3)$ satisfy the inequality $7x - y \geq 8$? Why (not)?
20. Graph the inequality $3x - 5y \geq 6$.

CUMULATIVE REVIEW: CHAPTERS 1-13

1. Evaluate: $-5^2 + (-3)^3 \div (-9) - [(-2) + (-4) \times (-12)]$.
2. Which is larger, 1.23% of 63 or $|123\% \text{ of } -0.63|$?
3. Evaluate $(2a - 3b)^2$, if $a = 1.1$ and $b = -2.2$.
4. Evaluate $4p^2 - 3q^2$, if $p = -2^2$ and $q = (-2)^2$.
5. Round 23.0579 to the nearest hundredth.
6. Round 3097.98 to two significant digits.
7. Round $9\frac{2}{7}$ to the nearest thousandth.
8. Rewrite 0.00376×10^{-4} in scientific notation.
9. Simplify by combining like terms: $4x - 5\{y - 3[y - 2x] - y\} - 6x$.
10. Simplify by combining like terms: $x^2(2x^2 - x + 4) - x(5x^2 - x) + 5x^2$.
11. Factor completely: $6p^3 + 15p^2 - 9p$.

In Exercises 12–14, solve each equation for the indicated variable.

12. $-2(p - 3) + 6 = 4(2 - 5p)$
13. $2u^2 - 10u + 12 = 0$
14. $3r^3 = 48r$
15. Determine the excluded value(s) for the rational expression
$$\frac{5t}{t^2 - 9} + \frac{t^2 - 4}{t^2 + 8} - \frac{5t^3}{2t + 7}.$$
16. Divide: $\dfrac{2p + 6}{4q^2 - 1} \div \dfrac{p^2 - 9}{2q + 1}$.

17. Determine the perimeter of a triangular region whose sides measure $2\frac{1}{3}$ yd, $4\frac{1}{2}$ ft, and $5\frac{1}{4}$ ft.

18. The sum of two numbers is 4. The sum of their reciprocals is $\dfrac{-1}{8}$. What are the numbers?

19. Does the ordered pair $(-2, 4)$ satisfy the inequality $4y - 3x < 7$? Why (not)?
20. Solve the inequality $2(s - 3t) + 4t \geq 5 - 4s$ for t.
21. Solve the proportion $\dfrac{-3}{p} = \dfrac{10}{17}$ for p.

22. The area of a rectangular region is $225\frac{5}{24}$ ft². Determine the width of the region if its length is $19\frac{1}{6}$ ft.

23. Graph the inequality $x \leq 5$ on the number line.
24. Determine 6.9% of -12.1×3.4 and round the result to the nearest hundredth.
25. Determine the commission if Karen receives 11.5% on net sales if her gross sales were $2873.60 and returns were $307.10.

CHAPTER 14: Systems of Linear Equations

In this chapter, we will introduce systems of two linear equations in two variables and find their solutions. A solution for such a system of equations consists of values for each of the two variables that are solutions for *both* of the given equations. You will learn that such systems have either no solution, exactly one solution, or infinitely many solutions.

Methods—both graphical and algebraic—for solving such systems of equations will be discussed. Applications that can be solved using the methods of this chapter will be examined.

CHAPTER 14 PRETEST

This pretest covers material on systems of linear equations. Read each question carefully, and try to answer all questions. Each question is keyed to the chapter section in which the particular topic is discussed.

Questions 1–4 pertain to Section 14.1.

In Exercises 1–2, solve the given systems of equations graphically.

1. $\begin{cases} x - y = 4 \\ x + y = 4 \end{cases}$

2. $\begin{cases} y = 2 - x \\ y = 3x \end{cases}$

In Exercises 3–4, rewrite the equations in each system of equations in the slope-intercept form. Then, determine if the graph for the pair of equations consists of intersecting, parallel, or coinciding lines.

3. $\begin{cases} 2x + 3y = 5 \\ x - 2y = 1 \end{cases}$

4. $\begin{cases} x - 3y = 2 \\ 2x = 4 + 6y \end{cases}$

Questions 5–6 pertain to Section 14.2.

Solve each of the following systems of equations using the method of elimination by addition or subtraction.

5. $\begin{cases} 2x - 3y = 1 \\ 3x + 4y = 8 \end{cases}$

6. $\begin{cases} y = 3x - 4 \\ 2x + y = 1 \end{cases}$

Questions 7–8 pertain to Section 14.3.

Solve each of the following systems of equations using the method of substitution.

7. $\begin{cases} 3x - y = 4 \\ 2x + y = 1 \end{cases}$

8. $\begin{cases} 2x + 3y = 1 \\ 3x - 4y = 4 \end{cases}$

Questions 9–10 pertain to Section 14.4.

Solve each of the following using a system of two equations in two variables.

9. The sum of two numbers is 45. The sum of one-half the smaller and twice the larger is 63. What are the numbers?

10. The sum of the digits of a 2-digit number is 7. If the digits are reversed, the new number formed is 45 more than the original number. What is the original number?

14.1 Graphical Method for Solving a System of Linear Equations

OBJECTIVE

After completing this section, you should be able to solve a system of two linear equations in two variables using a graphical method.

You have learned to solve a linear equation in one variable. Examples of such equations are

$$2x = 1$$
$$3y + 1 = 4$$
$$2(u - 3) = 2 - 3(1 - 2u)$$

Each such equation has exactly one solution.

Also, given a linear equation in two variables, you learned to solve for one of the variables in terms of the other. Examples of such equations are

$$x + y = 6$$
$$2u - 3v = 7$$
$$2(s - 1) + 3t = 5 - 4(s + t)$$

Each such equation has infinitely many solutions. You further learned that the graph of such an equation is a straight line and that every point on the line corresponds to a solution of the equation.

In this chapter, you will learn to solve a system of two linear equations in the same two variables, such as the system

$$\begin{cases} x + y = 1 \\ 2x - y = 4 \end{cases}$$

There are three methods that we will use to solve such a system of equations. In this section, you will learn about a graphical method. In subsequent sections of this chapter, you will study two algebraic methods.

The equation $y = 2x + 3$ is a linear equation in the two variables x and y. Its graph is the line in Figure 14.1 passing through the two points $(0, 3)$ and $(-3, -3)$. The ordered pairs $(0, 3)$ and $(-3, -3)$ represent two of the infinitely many solutions for the equation $y = 2x + 3$.

The equation $y = 8 - 3x$ is also a linear equation in the two variables x and y. Its graph is the line in Figure 14.2 passing through the points $(0, 8)$ and $(4, -4)$. The ordered pairs $(0, 8)$ and $(4, -4)$ represent two of the infinitely many solutions for the equation $y = 8 - 3x$.

The two equations taken together

$$\begin{cases} y = 2x + 3 \\ y = 8 - 3x \end{cases}$$

represent what is called a **system of two linear equations in two variables.** To solve such a system of equations, we must determine all of the ordered pairs (x, y) that satisfy *both* equations. The graph of each equation is a line (Figures 14.1 and 14.2). The graph of the system of the two equations is given in Figure 14.3 as the two intersecting lines. Notice that there is only one point that lies on *both* of the lines: the point of intersection. This point seems to have coordinates $(1, 5)$. This is readily checked by substituting $x = 1$ and $y = 5$ in *each* of the two *given* equations.

14.1 Graphical Method for Solving a System of Linear Equations

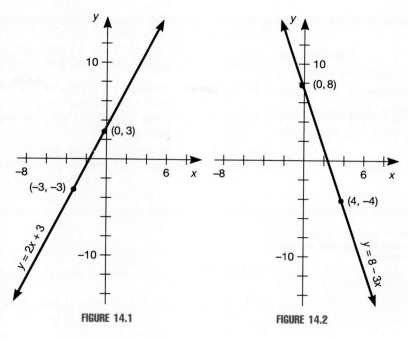

FIGURE 14.1

FIGURE 14.2

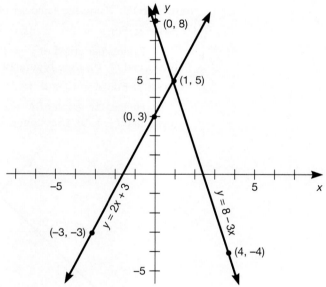

FIGURE 14.3

For
we have
$$y = 2x + 3$$
$$5 \;?\; 2(1) + 3$$
$$5 \;?\; 2 + 3$$
$$5 = 5 \;✓$$

For
we have
$$y = 8 - 3x$$
$$5 \;?\; 8 - 3(1)$$
$$5 \;?\; 8 - 3$$
$$5 = 5 \;✓$$

Hence, the ordered pair (1, 5) is a solution of the given system of equations. It is the only solution since there is only one point of intersection of the two lines which are the graph of the given equations.

The **graphical method** for solving a system of equations sometimes gives only an *approximate* solution. This depends upon how accurately you graph each line and how accurately you read the coordinates of the point of intersection.

Possible Solutions for a System of Two Linear Equations in Two Variables

Given a system of two linear equations in two variables and knowing that the graph of each equation is a line, the system will have:

1. *Exactly one solution* if the graphs are *intersecting lines*. (The solution will be the ordered pair representing the coordinates of the point of intersection.)
2. *No solution* if the graphs are *parallel lines*. (In this case, there are no points common to *both* lines.)
3. *Infinitely many solutions* if the graphs are the *same line*. (The solutions will be all of the ordered pairs representing the coordinates of the points on the common lines.) The lines are called "coinciding" lines.

Solve each of the following systems of equations graphically. Check your solutions in the given equations.

Practice Exercise 1.
$$\begin{cases} y = 3x - 1 \\ 3x - y = 5 \end{cases}$$

EXAMPLE 1 Solve the following system of equations graphically:

$$\begin{cases} y = 1 - x \\ y = x + 1 \end{cases}$$

SOLUTION: Using the same set of coordinate axes, graph each equation as follows:

STEP 1: For $y = 1 - x$: If $x = 0$, then $y = 1$, and if $y = 0$, then $x = 1$.

Hence, the graph of $y = 1 - x$ is the line that passes through the two points $(0, 1)$ and $(1, 0)$. (See Figure 14.4.)

STEP 2: For $y = x + 1$: If $x = 0$, then $y = 1$, and if $y = 0$, then $x = -1$.

Hence, the graph of $y = x + 1$ is the line that passes through the two points $(0, 1)$ and $(-1, 0)$. (See Figure 14.4.)

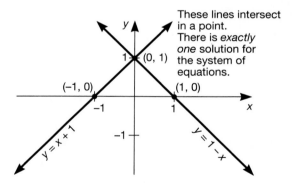

FIGURE 14.4

STEP 3: Observe that the *only* point that lies on *both* lines is the point $(0, 1)$.

STEP 4: Therefore, the only solution for the given system of equations is $(0, 1)$.

CHECK: Checking in the given equations, we have

$$\begin{array}{ccc} y = 1 - x & \text{and} & y = x + 1 \\ 1 \; ? \; 1 - 0 & & 1 \; ? \; 0 + 1 \\ 1 = 1 \checkmark & & 1 = 1 \checkmark \end{array}$$

Practice Exercise 2.
$$\begin{cases} x + y = 5 \\ x - y = 5 \end{cases}$$

EXAMPLE 2 Solve the following system of equations graphically:

$$\begin{cases} y = 2x - 3 \\ y = 2x + 4 \end{cases}$$

14.1 Graphical Method for Solving a System of Linear Equations

SOLUTION: Using the same set of coordinate axes, graph each equation as follows:

STEP 1: For $y = 2x - 3$: If $x = 0$, then $y = -3$, and if $x = 4$, then $y = 5$.

Hence, the graph of $y = 2x - 3$ is the line that passes through the two points $(0, -3)$ and $(4, 5)$. (See Figure 14.5.)

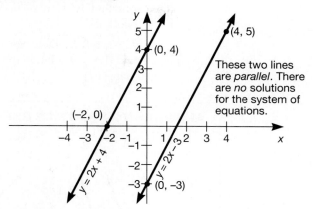

These two lines are *parallel*. There are *no* solutions for the system of equations.

FIGURE 14.5

STEP 2: For $y = 2x + 4$: If $x = 0$, then $y = 4$, and if $y = 0$, then $x = -2$.

Hence, the graph of $y = 2x + 4$ is the line that passes through the two points $(0, 4)$ and $(-2, 0)$. (See Figure 14.5.)

STEP 3: Observe that there are *no* points that lie on both lines; these lines do not intersect and are called "parallel" lines.

STEP 4: Therefore, there are *no* solutions for the given system of equations because the two lines have no points in common.

Practice Exercise 3.
$\begin{cases} 2x - 4y = 2 \\ x = 1 + 2y \end{cases}$

EXAMPLE 3 Solve the following system of equations graphically:

$$\begin{cases} x + y = 4 \\ 2x + 2y = 8 \end{cases}$$

SOLUTION: Using the same set of coordinate axes, graph each equation as follows:

STEP 1: For $x + y = 4$: If $x = 0$, then $y = 4$, and if $y = 0$, then $x = 4$.

Hence, the graph of $x + y = 4$ is the line that passes through the two points $(0, 4)$ and $(4, 0)$. (See Figure 14.6.)

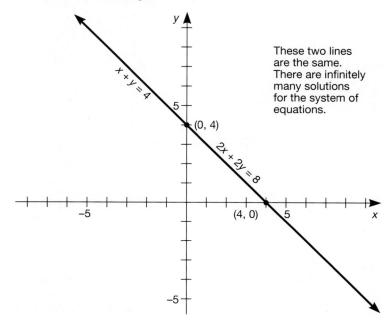

These two lines are the same. There are infinitely many solutions for the system of equations.

FIGURE 14.6

Step 2: For $2x + 2y = 8$: If $x = 0$, then $y = 4$, and if $y = 0$, then $x = 4$.

Hence, the graph of $2x + 2y = 8$ is the line that passes through the two points $(0, 4)$ and $(4, 0)$. (See Figure 14.6.)

Step 3: Observe that both equations have the same line for their graphs. Since there are infinitely many points on a line, there are infinitely many points common to the two lines.

Step 4: Therefore, there are infinitely many solutions for the given system of equations. (Try any solution from one equation and substitute it into the other. What should happen?)

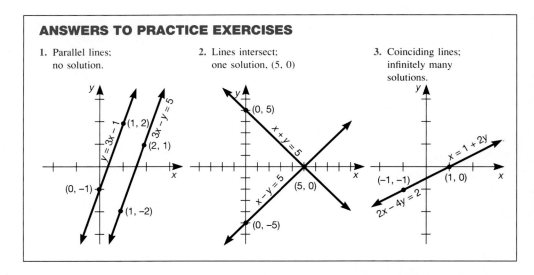

ANSWERS TO PRACTICE EXERCISES

1. Parallel lines; no solution.
2. Lines intersect; one solution, $(5, 0)$.
3. Coinciding lines; infinitely many solutions.

We will now examine the relationships between the solution for a system of two linear equations and the slopes of the lines that are the graphs of the equations in the system.

In Section 13.4, you learned that the slope of a line is the number that indicates the "steepness" of the line. In this section, we are looking at two lines graphed on the same coordinate system. In Example 2, we noted that the graph of the given system of equations was a pair of parallel lines. What can we say about the slopes of these lines?

The graph of $y = 2x + 4$ is the line passing through the points $(-2, 0)$ and $(0, 4)$. The slope of the line is

$$m = \frac{4 - 0}{0 - (-2)}$$
$$= \frac{4}{2}$$
$$= 2$$

The graph of $y = 2x - 3$ is the line passing through the points $(0, -3)$ and $(4, 5)$. The slope of the line is

$$m = \frac{5 - (-3)}{4 - 0}$$
$$= \frac{8}{4}$$
$$= 2$$

These two lines have the *same* slope. Also, note that, in each equation with y solved for in terms of x, the slope of the line, 2, is the coefficient of x. Further, examining the graphs in Figure 14.5, we note the following:

14.1 Graphical Method for Solving a System of Linear Equations

- For the equation $y = 2x + 4$, the coefficient of x, 2, is the slope of its graph, and the constant term, 4, is the y-intercept of the line.
- For the equation $y = 2x - 3$, the coefficient of x, 2, is the slope of its graph, and the constant terms, -3, is the y-intercept of the line.
- Both of the equations, $y = 2x + 4$ and $y = 2x - 3$, then are of the form

$$y = (\text{slope})x + (y\text{-intercept})$$

If we let m = slope and b = y-intercept of the line, then the equation

$$y = mx + b$$

is called the "slope-intercept" form of the line.

Referring back to the examples of Section 13.4, note the following:

1. In Example 3, the slope of the line with equation $2x - 3y = 6$ is $\frac{2}{3}$. Rewriting the equation in the slope-intercept form we have

$$2x - 3y = 6$$
$$-3y = -2x + 6$$
$$y = \underset{m}{\tfrac{2}{3}}x \underset{b}{- 2}$$

Check that we found the slope of the line to be $\frac{2}{3}$ and the y-intercept to be -2.

2. In Example 4, the equation $3x + y = 6$ can be rewritten in slope-intercept form as

$$3x + y = 6$$
$$y = \underset{m}{-3x} \underset{b}{+ 6}$$

Check that we found the slope of the line to be -3 and the y-intercept to be 6.

3. In Example 5, the equation $y = 3$ is already in the slope-intercept form since $y = 3$ is the same as

$$y = \underset{m}{(0)x} \underset{b}{+ 3}$$

Check that we found the slope of the line to be 0 and the y-intercept to be 3.

4. In Example 6, the equation $x = 4$ *cannot* be put in the slope-intercept form. The line is vertical and its slope is undefined.

The slope-intercept form of the equation of a line is easy to use. Rewriting both equations in a system of linear equations enables us to determine if the lines are intersecting, parallel, or coinciding.

Use the slope-intercept form for the equation of a line to determine whether the graph of each system consists of intersecting, parallel, or coinciding lines.

Practice Exercise 4.
$$\begin{cases} y = x + 3 \\ x + y = 6 \end{cases}$$

EXAMPLE 4 Consider the system of equations given:

$$\begin{cases} x + y = 7 \\ 2x - 3y = 5 \end{cases}$$

SOLUTION: Rewriting each of the equations in slope-intercept form, we have

$$x + y = 7$$
$$y = -x + 7 \quad (m = -1, b = 7)$$

and

$$2x - 3y = 5$$
$$-3y = -2x + 5$$
$$y = \frac{2}{3}x - \frac{5}{3} \quad \left(m = \frac{2}{3}, b = \frac{-5}{3}\right)$$

Since the slopes of the two lines are *different*, the lines are *intersecting*.

Practice Exercise 5.
$$\begin{cases} 2x + 3y = 0 \\ 6y = 5 - 4x \end{cases}$$

EXAMPLE 5 Consider the system of equations given:

$$\begin{cases} 2y - 6x = 4 \\ y = 3x + 1 \end{cases}$$

SOLUTION: Rewriting the first equation in slope-intercept form, we have

$$2y - 6x = 4$$
$$2y = 6x + 4$$
$$y = 3x + 2 \quad (m = 3, b = 2)$$

The second equation is already in slope-intercept form:

$$y = 3x + 1 \quad (m = 3, b = 1)$$

Since the slopes of the two lines are *equal*, the lines are either parallel or coinciding. However, their *y*-intercepts are different. Hence, the lines are *parallel*.

Practice Exercise 6.
$$\begin{cases} y = 2x \\ 6x - 3y = 0 \end{cases}$$

EXAMPLE 6 Consider the system of equations given:

$$\begin{cases} 3x - y = 4 \\ 2y - 6x = -8 \end{cases}$$

SOLUTION: Rewriting the first equation in slope-intercept form, we have

$$3x - y = 4$$
$$-y = -3x + 4$$
$$y = 3x - 4 \quad (m = 3, b = -4)$$

Rewriting the second equation in slope-intercept form, we have

$$2y - 6x = -8$$
$$2y = 6x - 8$$
$$y = 3x - 4 \quad (m = 3, b = -4)$$

14.1 Graphical Method for Solving a System of Linear Equations

Since the slopes of the two lines are *equal*, the lines are either parallel or coinciding. However, the *y*-intercepts are also *equal*. Hence, the lines are *coinciding*.

> **ANSWERS TO PRACTICE EXERCISES**
> 4. Intersecting. 5. Parallel. 6. Coinciding.

We will now summarize these observations.

Graphs of Systems of Two Linear Equations in Two Unknowns

To determine whether the graph of a system of two linear equations in two variables consists of *intersecting, parallel,* or *coinciding lines:*

1. (Re)write each of the equations in the slope-intercept form.
   a. If the slopes are different, the lines are **intersecting.**
   b. If the slopes are equal and the *y*-intercepts are different, the lines are **parallel.**
   c. If the slopes are equal and the *y*-intercepts are also equal, the lines are **coinciding.**
2. Different horizontal lines are parallel.
3. Different vertical lines are parallel.

EXERCISES 14.1

In Exercises 1–20, solve each of the given systems of equations graphically.

1. $\begin{cases} y = x + 1 \\ y = 2x - 1 \end{cases}$

2. $\begin{cases} y = \dfrac{1}{2}x \\ y = \dfrac{1}{3}x \end{cases}$

3. $\begin{cases} y = 4x - 1 \\ y = 4x + 2 \end{cases}$

4. $\begin{cases} y = 2x + 1 \\ y = 1 - 2x \end{cases}$

5. $\begin{cases} x + y = 2 \\ 0.2x = 0.4 - 0.2y \end{cases}$

6. $\begin{cases} y = 3 - x \\ y = 2x + 1 \end{cases}$

7. $\begin{cases} y = x + 5 \\ y = x - 5 \end{cases}$

8. $\begin{cases} y = 6 + \dfrac{2}{3}x \\ y = 6 - \dfrac{2}{3}x \end{cases}$

9. $\begin{cases} y = 6 \\ y = x \end{cases}$

10. $\begin{cases} y = 3x \\ x = -1 \end{cases}$

11. $\begin{cases} 1.2x = 7.2 \\ 1.3y = 5.2 \end{cases}$

12. $\begin{cases} x = 3 \\ x = 5 \end{cases}$

13. $\begin{cases} y = x \\ y = 6x + 1 \end{cases}$

14. $\begin{cases} y = \dfrac{1}{3}(7 + x) \\ x = \dfrac{1}{4}(7 + y) \end{cases}$

15. $\begin{cases} 4.6x + 4.6y = 0 \\ 3x + 3y = 0 \end{cases}$

16. $\begin{cases} x - y = 1 \\ 2x + y = 0 \end{cases}$

17. $\begin{cases} y = 0 \\ x - y = 1 \end{cases}$

18. $\begin{cases} y = 4x \\ x + y = 0 \end{cases}$

19. $\begin{cases} y = 3x + 1 \\ y = 2 - x \end{cases}$
20. $\begin{cases} y = \dfrac{2}{5}x - 5 \\ y = \dfrac{1}{2}x - 4 \end{cases}$

In Exercises 21–30, rewrite each of the given equations in slope-intercept form.

21. $1.1x + 1.1y = 9.9$
22. $\dfrac{3}{4}x - \dfrac{2}{3}y = 5$
23. $3y - 1 = 0$
24. $x = 2y + 5$
25. $4y + x = 0$
26. $6x - 2y = 7$
27. $5y = 10 - x$
28. $\dfrac{3}{5}y = x + \dfrac{1}{7}$
29. $1 + 6x = y$
30. $7x = 5 - 2y$

In Exercises 31–40, use slopes to determine whether the graph of each of the following systems of equations consists of intersecting, parallel, or coinciding lines.

31. $\begin{cases} x = \dfrac{1}{3}y + 1 \\ y = 2x - 1 \end{cases}$
32. $\begin{cases} y - x = 0 \\ y = 3x \end{cases}$
33. $\begin{cases} 2x - y + 1 = 0 \\ y = 1 - 2x \end{cases}$
34. $\begin{cases} \dfrac{5}{7}y - \dfrac{7}{9} = 0 \\ \dfrac{1}{3}y = 8 \end{cases}$
35. $\begin{cases} x + 1.9 = 0 \\ x = 3.7 \end{cases}$
36. $\begin{cases} y = 2 - x \\ 2x = 4 - 2y \end{cases}$
37. $\begin{cases} x = 9 \\ y = -6 \end{cases}$
38. $\begin{cases} y = -3.8 \\ y = -1.3x \end{cases}$
39. $\begin{cases} y = 2x + 3 \\ x = 2y + 3 \end{cases}$
40. $\begin{cases} 3(x - y) = 4 \\ -3(y - x) = 4 \end{cases}$

14.2 Elimination Method for Solving a System of Linear Equations

OBJECTIVE

After completing this section, you should be able to solve a system of two linear equations in two variables, using the method of elimination.

We will discuss two algebraic methods for solving a system of two linear equations in two variables. The first of these methods, known as the **elimination method,** will be discussed in this section.

Consider the system of equations

$$\begin{cases} x + y = 5 \\ x - y = 1 \end{cases}$$

14.2 Elimination Method for Solving a System of Linear Equations

If we add the left-hand sides of these equations, the sum will be equal to the sum of their right-hand sides. (This follows from our earlier work. If equals are added to equals, the sums are equal.) This is known as adding the equations, term by term. Hence, we have

Add
$$x + y = 5$$
$$x - y = 1$$
$$\overline{2x + 0 = 6}$$

Adding the two equations, the variable y was "eliminated," and the result is an equation in one variable only. The equation

$$2x = 6$$

is readily solved for x, giving

$$x = 3$$

Substituting $x = 3$ in *either* of the two *original* equations (say, the first), we obtain

$$3 + y = 5$$
or
$$y = 2$$

It appears that $x = 3$ and $y = 2$, or the ordered pair $(3, 2)$, is the required solution.

CHECK: Checking in *both* of the *original* equations, we have

$$x + y = 5 \quad \text{and} \quad x - y = 1$$
$$3 + 2 \stackrel{?}{=} 5 \qquad\qquad 3 - 2 \stackrel{?}{=} 1$$
$$5 = 5 \checkmark \qquad\qquad 1 = 1 \checkmark$$

This system was easy to solve by adding both equations to eliminate the variable y. Observe that the coefficient of y in the first equation is 1 and the coefficient of y in the second equation is -1; their sum is 0. That is not always the case as we shall illustrate in the following examples.

Solve using the method of elimination.

Practice Exercise 1.
$$\begin{cases} x + y = 2 \\ x - y = 0 \end{cases}$$

EXAMPLE 1 Solve the system of equations

$$\begin{cases} 3x + y = 3 \\ x + y = 1 \end{cases}$$

SOLUTION: Adding both equations, we obtain

$$4x + 2y = 4$$

and neither variable is eliminated. Note, however, that the y terms in both equations have the same coefficients. The variable y can be eliminated if we subtract the second equation from the first equation:

Subtract
$$3x + y = 3$$
$$x + y = 1$$
$$\overline{2x + 0 = 2}$$

The result is a single equation in only one variable. Solving the equation

$$2x = 2$$

we readily obtain

$$x = 1$$

Substituting $x = 1$ in *either* of the *original* equations (say, the second), we obtain

$$1 + y = 1$$

from which we determine that

$$y = 0$$

It appears that $(1, 0)$ is the required solution.

CHECK: Checking in *both* of the *original* equations, we have

$$\begin{array}{ccc} 3x + y = 3 & \text{and} & x + y = 1 \\ 3(1) + 0 \; ? \; 3 & & 1 + 0 \; ? \; 1 \\ 3 + 0 \; ? \; 3 & & 1 = 1 \; ✓ \\ 3 = 3 \; ✓ & & \end{array}$$

Practice Exercise 2.
$$\begin{cases} 2x - y = 3 \\ 2x + 3y = 1 \end{cases}$$

EXAMPLE 2 Solve the system of equations

$$\begin{cases} x - 2y = 6 \\ 2x + y = 2 \end{cases}$$

SOLUTION: Neither adding the two equations nor subtracting one from the other will cause a variable to be eliminated. (Do you agree?) In this case, we could elect to eliminate the variable x. To eliminate the variable x, we need to have the coefficients of x in the two equations the same (in which case, we subtract the second equation from the first equation) or the opposites of each other (in which case we add the two equations). We will multiply the first equation by 2 which we know, from Chapter 8, will give us an equivalent system of equations:

$$\begin{cases} 2x - 4y = 12 \\ 2x + y = 2 \end{cases}$$

Now, subtracting the second equation from the first, we have

Subtract
$$\begin{array}{r} 2x - 4y = 12 \\ 2x + y = 2 \\ \hline -5y = 10 \end{array}$$

or
$$y = -2$$

Substituting $y = -2$ in *either* of the *original* equations (say, the first), we have

$$\begin{array}{r} x - 2(-2) = 6 \\ x + 4 = 6 \\ x = 2 \end{array}$$

It appears that $(2, -2)$ is the required solution.

CHECK: Checking in *both* of the *original* equations, we have

$$\begin{array}{ccc} x - 2y = 6 & \text{and} & 2x + y = 2 \\ 2 - 2(-2) \; ? \; 6 & & 2(2) + (-2) \; ? \; 2 \\ 2 + 4 \; ? \; 6 & & 4 - 2 \; ? \; 2 \\ 6 = 6 \; ✓ & & 2 = 2 \; ✓ \end{array}$$

14.2 Elimination Method for Solving a System of Linear Equations

Practice Exercise 3.
$$\begin{cases} 4p + 3q = -5 \\ 5p + 3q = 7 \end{cases}$$

EXAMPLE 3 Solve the system of equations

$$\begin{cases} 2x - 3y = 1 \\ 3x + 4y = 2 \end{cases}$$

SOLUTION: Neither adding the two equations nor subtracting one from the other will cause a variable to be eliminated. Suppose that we try to eliminate the variable x. Multiplying the first equation by 3 (the coefficient of x in the second equation) and the second equation by 2 (the coefficient of x in the first equation), we obtain the equivalent system

$$\begin{cases} 6x - 9y = 3 \\ 6x + 8y = 4 \end{cases}$$

Now, subtracting the second equation from the first, we have

$$-17y = -1$$

or

$$y = \frac{1}{17}$$

Substituting $y = \frac{1}{17}$ in *either* of the *original* equations (say, the first), we have

$$2x - 3\left(\frac{1}{17}\right) = 1$$

$$2x - \frac{3}{17} = 1$$

$$2x = \frac{20}{17}$$

$$x = \frac{10}{17}$$

It appears that $\left(\frac{10}{17}, \frac{1}{17}\right)$ is the required solution.

CHECK: Checking in *both* of the *original* equations, we have

$$2x - 3y = 1 \quad \text{and} \quad 3x + 4y = 2$$

$$2\left(\frac{10}{17}\right) - 3\left(\frac{1}{17}\right) \overset{?}{=} 1 \qquad 3\left(\frac{10}{17}\right) + 4\left(\frac{1}{17}\right) \overset{?}{=} 2$$

$$\frac{20}{17} - \frac{3}{17} \overset{?}{=} 1 \qquad \frac{30}{17} + \frac{4}{17} \overset{?}{=} 2$$

$$\frac{17}{17} \overset{?}{=} 1 \qquad \frac{34}{17} \overset{?}{=} 2$$

$$1 = 1 \checkmark \qquad 2 = 2 \checkmark$$

Practice Exercise 4.
$$\begin{cases} 2s - 3t = 0 \\ s + 2t = 4 \end{cases}$$

EXAMPLE 4 Solve the system of equations

$$\begin{cases} 4x - 3y = 1 \\ 5x + 2y = 0 \end{cases}$$

SOLUTION: Suppose we try to eliminate the y variable. This can be done by multiplying the first equation by 2 (the coefficient of y in the second equation) and the second equation by 3 (the coefficient of $-y$ in the first equation). We obtain the equivalent system

$$\begin{cases} 8x - 6y = 2 \\ 15x + 6y = 0 \end{cases}$$

Now, adding the two equations, we obtain

$$23x = 2$$

or

$$x = \frac{2}{23}$$

Substituting $x = \frac{2}{23}$ in *either* of the *original* equations (say, the second), we have

$$5\left(\frac{2}{23}\right) + 2y = 0$$

$$\frac{10}{23} + 2y = 0$$

$$2y = \frac{-10}{23}$$

$$y = \frac{-5}{23}$$

It appears that $\left(\frac{2}{23}, \frac{-5}{23}\right)$ is the required solution.

CHECK: Checking in *both* of the *original* equations, we have

$$4x - 3y = 1 \qquad \text{and} \qquad 5x + 2y = 0$$

$$4\left(\frac{2}{23}\right) - 3\left(\frac{-5}{23}\right) \stackrel{?}{=} 1 \qquad\qquad 5\left(\frac{2}{23}\right) + 2\left(\frac{-5}{23}\right) \stackrel{?}{=} 0$$

$$\frac{8}{23} + \frac{15}{23} \stackrel{?}{=} 1 \qquad\qquad \frac{10}{23} - \frac{10}{23} \stackrel{?}{=} 0$$

$$\frac{23}{23} \stackrel{?}{=} 1 \qquad\qquad 0 = 0 ✓$$

$$1 = 1 ✓$$

ANSWERS TO PRACTICE EXERCISES

1. $(1, 1)$. 2. $\left(\frac{5}{4}, \frac{-1}{2}\right)$. 3. $\left(12, \frac{-53}{3}\right)$. 4. $\left(\frac{12}{7}, \frac{8}{7}\right)$.

Summary of the Elimination Method for Solving a System of Equations

To use the elimination method to solve a system of two linear equations in two variables:

1. Multiply *both* sides of one or both of the equations by a constant so that the coefficients of one of the variables in the resulting system have the same absolute value.

14.2 Elimination Method for Solving a System of Linear Equations

  a. If the coefficients are the same, *subtract* one equation from the other to eliminate the variable.
  b. If the coefficients are opposites of each other, *add* the two equations to eliminate the variable.
2. Solve the resulting equation for the remaining variable.
3. Substitute the value obtained for this variable into one of the *original* equations and solve for the other variable.
4. Check the solution by substituting the values obtained for each of the two variables into *both* of the *original* equations.

Solve using the method of elimination.

Practice Exercise 5.
$$\begin{cases} 2x + y = 3 \\ x - y = 3 \end{cases}$$

EXAMPLE 5 Solve the system of equations

$$\begin{cases} 3u + v = 10 \\ u - v = 2 \end{cases}$$

SOLUTION:

STEP 1: We will eliminate the variable u by multiplying *both* sides of the second equation by 3 and leaving the first equation unchanged. We have

$$\begin{cases} 3u + v = 10 \\ 3u - 3v = 6 \end{cases}$$

STEP 2: Since the coefficients of u are the same in both equations, *subtract* the second equation from the first, obtaining

$$4v = 4$$

STEP 3: Solve the resulting equation for v, obtaining

$$v = 1$$

STEP 4: Substitute $v = 1$ into one of the *original* equations and solve for u. We will substitute into the second equation, obtaining

$$u - 1 = 2$$
$$u = 3$$

CHECK: Check the solution by substituting $u = 3$ and $v = 1$ into *both* of the *original* equations.

$$\begin{array}{ccc} 3u + v = 10 & \text{and} & u - v = 2 \\ 3(3) + 1 \; ? \; 10 & & 3 - 1 \; ? \; 2 \\ 9 + 1 \; ? \; 0 & & 2 = 2 \; \checkmark \\ 10 = 10 \; \checkmark & & \end{array}$$

Therefore, the solution for the given system is $u = 3$ and $v = 1$.

Practice Exercise 6.
$$\begin{cases} x + y = 5 \\ 3x - 2y = 0 \end{cases}$$

EXAMPLE 6 Solve the system of equations

$$\begin{cases} 3p - 5q = 2 \\ 2p + 3q = -5 \end{cases}$$

SOLUTION:

STEP 1: We will eliminate the variable q. This can be done by multiplying *both* sides of the first equation by 3 and *both* sides of the second equation by 5, obtaining

$$\begin{cases} 9p - 15q = 6 \\ 10p + 15q = -25 \end{cases}$$

STEP 2: Since the coefficients of q in the two equations are opposites of each other, *add* the two equations to eliminate the variable. We have

$$19p = -19$$

STEP 3: Solve the resulting equation for p, obtaining

$$p = -1$$

STEP 4: Substitute $p = -1$ into one of the *original* equations and solve for q. We will substitute into the first equation, obtaining

$$3(-1) - 5q = 2$$
$$-3 - 5q = 2$$
$$-5q = 5$$
$$q = -1$$

CHECK: Check the solution by substituting $p = -1$ and $q = -1$ into *both* of the *original* equations.

$$\begin{array}{ccc} 3p - 5q = 2 & \text{and} & 2p + 3q = -5 \\ 3(-1) - 5(-1) \; ? \; 2 & & 2(-1) + 3(-1) \; ? \; -5 \\ -3 + 5 \; ? \; 2 & & -2 + (-3) \; ? \; -5 \\ 2 = 2 \; ✔ & & -5 = -5 \; ✔ \end{array}$$

Therefore, the solution for the given system is $p = -1$ and $q = -1$.

All of the examples in this section so far illustrate systems of equations with exactly one solution. What about systems with no solutions or infinitely many solutions? The next two examples will address this question.

Practice Exercise 7.
$$\begin{cases} 2x - y = 5 \\ 4x - 2y = 1 \end{cases}$$

EXAMPLE 7 Solve the system of equations

$$\begin{cases} x + y = 3 \\ 2x + 2y = 5 \end{cases}$$

SOLUTION: Multiply the first equation by 2 and leave the second equation unchanged; we have the equivalent system

$$\begin{cases} 2x + 2y = 6 \\ 2x + 2y = 5 \end{cases}$$

Now, subtracting the second equation from the first, we obtain

$$0 = 1$$

Since $0 \neq 1$, there is *no solution* for the given system. It should be noted that, if the two original equations are rewritten in the slope-intercept form, both graphs will have a slope of -1 and different y-intercepts; the lines are parallel. This confirms our conclusion of no solution for the given system.

14.2 Elimination Method for Solving a System of Linear Equations

Practice Exercise 8.
$$\begin{cases} 2x - 3y = 6 \\ 3x + 2y = 0 \end{cases}$$

EXAMPLE 8 Solve the system of equations

$$\begin{cases} x + y = 2 \\ 3x + 3y = 6 \end{cases}$$

SOLUTION: Multiplying the first equation by 3 and leaving the second equation unchanged, we have the equivalent system

$$\begin{cases} 3x + 3y = 6 \\ 3x + 3y = 6 \end{cases}$$

Now, subtracting the second equation from the first, we have

$$0 = 0$$

Since $0 = 0$ is a true statement, this means that *any* solution for the first equation is also a solution for the second equation. Therefore, there are *infinitely many solutions* for the given system. Again, note that if the two given equations are rewritten in slope-intercept form, their graphs will have the same slopes and the same y-intercepts; the lines are coinciding. This confirms our conclusion that there are infinitely many solutions for the given system.

ANSWERS TO PRACTICE EXERCISES

5. $(2, -1)$.   6. $(2, 3)$.   7. No solution.   8. $\left(\dfrac{12}{13}, \dfrac{-18}{13}\right)$.

EXERCISES 14.2

Use the method of elimination to solve the following systems of equations. Be sure to check your solutions in both of the original equations.

1. $\begin{cases} x - y = 1 \\ 2x - y = -1 \end{cases}$
2. $\begin{cases} x - y = 0 \\ 3x + y = 0 \end{cases}$
3. $\begin{cases} 4x - y = 1 \\ 4x - y = 2 \end{cases}$

4. $\begin{cases} 2x - \dfrac{1}{3}y = -1 \\ 2x + \dfrac{1}{3}y = 1 \end{cases}$
5. $\begin{cases} x + y = 2 \\ 0.2x + 0.2y = 1 \end{cases}$
6. $\begin{cases} 2x - y = 3 \\ 3x + 2y = 1 \end{cases}$

7. $\begin{cases} x + 3 = 0 \\ y - 4 = 0 \end{cases}$
8. $\begin{cases} 2.2x - 3.3y = 1.1 \\ 3.6x + 4.8y = 0 \end{cases}$
9. $\begin{cases} x - y = 0 \\ 3x + 5y = 1 \end{cases}$

10. $\begin{cases} x - 7y = 1 \\ 2x + 3y = 4 \end{cases}$
11. $\begin{cases} 4x - 3y = 1 \\ 3x + 5y = 2 \end{cases}$
12. $\begin{cases} 1.3x = 3.9 \\ 1.4x - 0.7y = 0.7 \end{cases}$

13. $\begin{cases} 4.5x - 1.8y = 2.7 \\ 1.6x + 2.4y = -3.2 \end{cases}$
14. $\begin{cases} 7x - 6y = 5 \\ 2x + 5y = 1 \end{cases}$
15. $\begin{cases} y = 2x - 3 \\ x + y = 1 \end{cases}$

16. $\begin{cases} 2x - 3y = 5 \\ y = 1 - 2x \end{cases}$
17. $\begin{cases} p = 3q - 1 \\ 2p - q = 0 \end{cases}$
18. $\begin{cases} p - \dfrac{2}{3}q = \dfrac{1}{3} \\ \dfrac{1}{2}p + \dfrac{3}{4}q = \dfrac{1}{2} \end{cases}$

19. $\begin{cases} 2u - v = 5 \\ u + 3v = 4 \end{cases}$
20. $\begin{cases} 3u + 4v = -1 \\ u = 3 - 2v \end{cases}$
21. $\begin{cases} \dfrac{1}{4}s + \dfrac{1}{5}t = 0 \\ \dfrac{1}{3}s - \dfrac{1}{2}t = 4 \end{cases}$

22. $\begin{cases} 6s - 5t = -2 \\ 4s + 2t = 1 \end{cases}$
23. $\begin{cases} 0.5r + 0.3s = 0.6 \\ 0.3r + 0.5s = 0.4 \end{cases}$
24. $\begin{cases} 0.2r + 0.3s = -0.5 \\ 0.3r + 0.5s = 0.2 \end{cases}$

25. $\begin{cases} \dfrac{1}{4}x + \dfrac{1}{3}y = \dfrac{1}{12} \\ \dfrac{1}{3}x + \dfrac{1}{4}y = \dfrac{2}{3} \end{cases}$
26. $\begin{cases} \dfrac{1}{2}x + \dfrac{7}{10}y = \dfrac{3}{10} \\ \dfrac{1}{5}x - \dfrac{3}{10}y = \dfrac{7}{10} \end{cases}$
27. $\begin{cases} 2(x - y) = 3 \\ 3(x + y) = 1 \end{cases}$

28. $\begin{cases} 3(2x - 3y) = 1 \\ 2(3x + y) = 1 \end{cases}$
29. $\begin{cases} 3(x - 1) = y \\ 4x = 3(1 - y) \end{cases}$
30. $\begin{cases} 2x - 3 = 4y + 1 \\ 5y + 2 = 6 - x \end{cases}$

14.3 Substitution Method for Solving a System of Linear Equations

OBJECTIVE

After completing this section, you should be able to solve a system of two linear equations in two variables, using the method of substitution.

Another algebraic method for solving a system of two linear equations in two variables is the **method of substitution.** Consider the system of equations

$$\begin{cases} 2x - 3y = 1 \\ y = x + 1 \end{cases}$$

The second equation already has y solved for in terms of x. If we substitute $x + 1$ for y in the *first* equation, we have

or
$$2x - 3(x + 1) = 1$$
$$2x - 3x - 3 = 1$$

which simplifies to

$$x = -4$$

Next, we substitute $x = -4$ in the *second* equation and solve for y, obtaining

$$y = x + 1$$
$$= -4 + 1$$
$$= -3$$

It now appears that the ordered pair $(-4, -3)$ is the required solution.

14.3 Substitution Method for Solving a System of Linear Equations

CHECK: Checking in *both* of the *original* equations, we have

$$2x - 3y = 1 \quad \text{and} \quad y = x + 1$$
$$2(-4) - 3(-3) \overset{?}{=} 1 \qquad\qquad -3 \overset{?}{=} -4 + 1$$
$$-8 + 9 \overset{?}{=} 1 \qquad\qquad -3 = -3 \checkmark$$
$$1 = 1 \checkmark$$

Solve using the method of substitution. Check your results in the original equations.

Practice Exercise 1.
$$\begin{cases} y = 3x - 1 \\ x + y = 3 \end{cases}$$

EXAMPLE 1 Solve the following system of equations:

$$\begin{cases} 2x + y = 3 \\ 3x - 2y = 1 \end{cases}$$

SOLUTION: In the given system of equations, we note that neither equation is solved for one variable in terms of the other. However, in the *first* equation, the coefficient of y is 1. We could easily solve for y in terms of x, obtaining

$$y = 3 - 2x$$

Now, substituting $3 - 2x$ for y in the *second* equation, we have

$$3x - 2(3 - 2x) = 1$$
or
$$3x - 6 + 4x = 1$$

which simplifies to

$$x = 1$$

Next, we substitute $x = 1$ in the equation

$$y = 3 - 2x$$

and obtain

$$y = 1$$

It now appears that the ordered pair $(1, 1)$ is the required solution. The check is left to you.

Practice Exercise 2.
$$\begin{cases} x - 4y = 0 \\ 2x + 3y = 0 \end{cases}$$

EXAMPLE 2 Solve the following system of equations:

$$\begin{cases} 2x - 3y = 1 \\ 3x + 4y = 3 \end{cases}$$

SOLUTION: In the given system, neither equation has one variable already solved for in terms of the other. Also observe that neither variable in either equation has a coefficient of 1 or -1. In this case, it is easiest to select a variable in either equation whose coefficient has a small value. Therefore, in the *first* equation, we can solve for x since 2 is the least of all the absolute values of the numerical coefficients. We have

$$2x - 3y = 1$$
$$2x = 1 + 3y$$
or
$$x = \frac{1 + 3y}{2}$$

Substituting $\dfrac{1 + 3y}{2}$ for x in the *second* equation, we have

$$3\left(\dfrac{1 + 3y}{2}\right) + 4y = 3$$
$$3(1 + 3y) + 8y = 6 \quad \text{(Multiplying both sides by 2)}$$
$$3 + 9y + 8y = 6$$
$$17y = 3$$
$$y = \dfrac{3}{17}$$

Next, we substitute $y = \dfrac{3}{17}$ in the equation

$$x = \dfrac{1 + 3y}{2}$$

and obtain

$$x = \dfrac{13}{17}$$

It now appears that the ordered pair $\left(\dfrac{13}{17}, \dfrac{3}{17}\right)$ is the required solution.

CHECK: Checking in *both* of the *original* equations, we have

$$2x - 3y = 1 \quad \text{and} \quad 3x + 4y = 3$$
$$2\left(\dfrac{13}{17}\right) - 3\left(\dfrac{3}{17}\right) \overset{?}{=} 1 \qquad 3\left(\dfrac{13}{17}\right) + 4\left(\dfrac{3}{17}\right) \overset{?}{=} 3$$
$$\dfrac{26}{17} - \dfrac{9}{17} \overset{?}{=} 1 \qquad \dfrac{39}{17} + \dfrac{12}{17} \overset{?}{=} 3$$
$$\dfrac{17}{17} \overset{?}{=} 1 \qquad \dfrac{51}{17} \overset{?}{=} 3$$
$$1 = 1 \checkmark \qquad 3 = 3 \checkmark$$

To solve the system of equations given in Example 2, you may find it easier to use the method of elimination rather than the method of substitution. Either method will work, and you should gain experience with both methods.

Practice Exercise 3.
$$\begin{cases} 2x - y = 3 \\ 2y - 4x = 1 \end{cases}$$

EXAMPLE 3 Solve the following system of equations:

$$\begin{cases} x + 2y = 3 \\ 2x + 4y = 5 \end{cases}$$

SOLUTION: Since the coefficient of x in the *first* equation is 1, we can solve for x in terms of y and obtain

$$x = 3 - 2y$$

Substituting $3 - 2y$ for x in the *second* equation, we obtain

$$2(3 - 2y) + 4y = 5$$
$$6 - 4y + 4y = 5$$
$$6 = 5$$

14.3 Substitution Method for Solving a System of Linear Equations

There are *no* values of x and y such that 6 is equal to 5. Therefore, there are *no* solutions to this system of equations. Is the graph for this system of equations a pair of intersecting, parallel, or coinciding lines?

Practice Exercise 4.
$$\begin{cases} 3x - 2y = 1 \\ 4x + 3y = 2 \end{cases}$$

EXAMPLE 4 Solve the following system of equations:

$$\begin{cases} 2x - 3y = 1 \\ 6x - 9y = 3 \end{cases}$$

SOLUTION: In the *first* equation, we can solve for x in terms of y and obtain

$$x = \frac{1 + 3y}{2}$$

Substituting $\frac{1 + 3y}{2}$ for x in the *second* equation, we obtain

$$6\left(\frac{1 + 3y}{2}\right) - 9y = 3$$

$$3(1 + 3y) - 9y = 3$$

which simplifies to

$$3 = 3$$

Since $3 = 3$ for all values of x and y, there are *infinitely many solutions* for this system of equations.

ANSWERS TO PRACTICE EXERCISES

1. (1, 2). **2.** (0, 0). **3.** No solution. **4.** $\left(\frac{7}{17}, \frac{2}{17}\right)$.

It should be noted that the solution for a system of two linear equations in two variables is the same regardless of the method used to solve the system. Further, always check your solutions in *both* of the *original* equations.

EXERCISES 14.3

Use the method of substitution to solve each of the following systems of equations. Check your results in the original equations.

1. $\begin{cases} y = 2x + 1 \\ x + y = 2 \end{cases}$
2. $\begin{cases} 2x - y = 1 \\ x = 1 - y \end{cases}$
3. $\begin{cases} 3x - 4y = 5 \\ y = 1 - 2x \end{cases}$

4. $\begin{cases} x = 3y - 2 \\ 3x + y = 4 \end{cases}$
5. $\begin{cases} 2x + y = 1 \\ 3x - 2y = 0 \end{cases}$
6. $\begin{cases} 1.1x - 3.3y = 7.7 \\ 1.2x + 2.4y = 3 \end{cases}$

7. $\begin{cases} x + 4y = 2 \\ 4x - y = 5 \end{cases}$
8. $\begin{cases} x + y = 2 \\ x + y = 4 \end{cases}$
9. $\begin{cases} 0.2x + 0.3y = 0.5 \\ 0.3x - 0.2y = 0.4 \end{cases}$

10. $\begin{cases} 2.1x - 2.8y = 0 \\ 4.5x + 1.8y = 2.7 \end{cases}$
11. $\begin{cases} 2x - y = 2 \\ 4x - 2y = 4 \end{cases}$
12. $\begin{cases} 3x - 5y = 7 \\ 3x + 4y = 2 \end{cases}$

13. $\begin{cases} 1.8x + 3.6y = 0.6 \\ 1.6x - 0.8y = 2 \end{cases}$
14. $\begin{cases} y = 2(x - 3) \\ 3x + 2y = 1 \end{cases}$
15. $\begin{cases} 2x - 3y = 4 \\ x = 3(1 - 2y) \end{cases}$

16. $\begin{cases} 3x - 2y = 6 \\ 2x - 3y = 3 \end{cases}$
17. $\begin{cases} \dfrac{1}{3}x + y = 3 \\ x - \dfrac{1}{2}y = \dfrac{-9}{2} \end{cases}$
18. $\begin{cases} \dfrac{5}{2}x + 2y = \dfrac{13}{2} \\ \dfrac{2}{3}x - y = \dfrac{-13}{3} \end{cases}$

19. $\begin{cases} 5x - 6y = 2 \\ 10x - 12y = 4 \end{cases}$
20. $\begin{cases} 6x - 4y = 14 \\ 6x - 4y = 9 \end{cases}$
21. $\begin{cases} 3p - 4q = -1 \\ \dfrac{1}{2}p = q - \dfrac{3}{2} \end{cases}$

22. $\begin{cases} 5p - 2q = -11 \\ 9p - q = 1 \end{cases}$
23. $\begin{cases} p = 3q - 13 \\ p = 3 - 5q \end{cases}$
24. $\begin{cases} q = 3p - 4 \\ 15q = 5p + 20 \end{cases}$

25. $\begin{cases} \dfrac{1}{9}s + \dfrac{1}{3}t = \dfrac{5}{3} \\ 2s - 4t = -10 \end{cases}$
26. $\begin{cases} s - 8t = -11 \\ 4s - 7t = 6 \end{cases}$
27. $\begin{cases} 1.1u + 2.2v = 3.3 \\ 2.2u = 11 - 4.4v \end{cases}$

28. $\begin{cases} 0.4u - 0.3v = 1.7 \\ 0.3u + 0.3v = 1.1 \end{cases}$
29. $\begin{cases} \dfrac{1}{2}x - \dfrac{1}{3}y = \dfrac{5}{6} \\ \dfrac{1}{3}x + \dfrac{1}{2}y = \dfrac{1}{6} \end{cases}$
30. $\begin{cases} \dfrac{2}{9}x + \dfrac{1}{7}y = \dfrac{11}{63} \\ \dfrac{5}{9}x - \dfrac{3}{7}y = \dfrac{23}{63} \end{cases}$

14.4 Applications Involving Systems of Equations

OBJECTIVE

After completing this section, you should be able to solve certain types of word problems by using a system of two linear equations in two variables.

In Chapter 8, we solved verbal problems using one equation in a single variable. In this section, we will solve similar problems using a system of equations.

Using a system of two equations in two variables, solve each of the following.

Practice Exercise 1. The sum of two numbers is 46. If the larger number is 7 more than twice the smaller number, what are the numbers?

EXAMPLE 1 If one number is 7 more than another number and their sum is 31, what are the numbers?

SOLUTION: We will present two solutions to this problem.

1. Solution using only *one* variable:
   Let
   $$n = \text{one number}$$
   Then
   $$n + 7 = \text{other number}$$

   Since the sum of the two numbers is 31, we now form the equation

   $$n + (n + 7) = 31$$

14.4 Applications Involving Systems of Equations

Solving this equation for n, we have

$$n + n + 7 = 31$$
$$2n + 7 = 31$$
$$2n = 24$$
$$n = 12$$

and $$n + 7 = 19$$

Hence, the two numbers are 12 and 19.

2. Solution using *two* variables: Since there are two numbers, we let

$$n = \text{one number}$$
and $$p = \text{other number}$$

Since one number is 7 more than the other, we have the equation

$$n = p + 7$$

Since the sum of the two numbers is 31, we also have the equation

$$n + p = 31$$

We now have the system of equations

$$\begin{cases} n = p + 7 \\ n + p = 31 \end{cases}$$

which can be solved using the method of substitution. Substititing $p + 7$ for n in the *second* equation, we have

$(p + 7) + p = 31$ (Note the similarity with the equation
$p + 7 + p = 31$ $n + (n + 7) = 31$ in the first solution.)
$2p + 7 = 31$
$2p = 24$
$p = 12$ (One number)

Substituting 12 for p in the equation

$$n = p + 7$$

we obtain

$$n = 19 \quad \text{(The second number)}$$

Again, the numbers are 12 and 19.

> **Practice Exercise 2.** A plane rectangular region is such that its width is 4 m less than one-half of its length. If the perimeter of the region is 118 m, what are the dimensions of the region?

EXAMPLE 2 A football game was attended by 4730 people, some of whom paid $1.50 each for reserved seats and the rest of whom paid $.90 for general admission tickets. If the total receipts for the game were $5172, how many tickets of each kind were sold?

SOLUTION: Using a system of equations, we let

$$r = \text{the number of reserved seat tickets sold}$$

and

$$g = \text{the number of general admission tickets sold}$$

Since there were 4730 tickets sold, we have the equation

$$r + g = 4730$$

Each r ticket sold for $1.50, and each g ticket sold for $.90. The total receipts were $5172, giving the equation

$$(\$1.50)r + (\$.90)g = \$5172$$

or

$$150r + 90g = 517{,}200$$

We now have the system of equations

$$r + g = 4730$$
$$150r + 90g = 517{,}200$$

Solving for g in the *first* equation, we have

$$g = 4730 - r$$

Substituting $4730 - r$ for g in the *second* equation, we have

$$150r + 90(4730 - r) = 517{,}200$$
$$150r + 425{,}700 - 90r = 517{,}200$$
$$60r = 91{,}500$$
$$r = 1525 \qquad \text{(The number of reserved seat tickets sold)}$$

Substituting 1525 for r in the equation

$$g = 4730 - r$$

we obtain

$$g = 4730 - 1525$$
$$= 3205 \qquad \text{(The number of general admission seat tickets sold)}$$

Hence, there were 1525 reserved seat tickets and 3205 general admission seat tickets sold. (Do these values satisfy all of the conditions of the *original* problem?)

Practice Exercise 3. A piece of wire 37 cm long is cut into two pieces. The longer piece is 8 cm less than twice the shorter piece. What is the length of each piece?

EXAMPLE 3 A rectangular parking lot is to be constructed so that its length is 3 times as long as its width. What are the dimensions of the lot if the perimeter is to be 640 ft?

SOLUTION: Since we are to determine the dimensions of the lot, we let

$$W = \text{width (in feet)}$$

and

$$L = \text{length (in feet)}$$

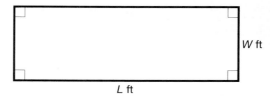

14.4 Applications Involving Systems of Equations

Since the length is 3 times the width, we have

$$L = 3W$$

The perimeter of the rectangular lot is given by the formula

$$P = 2L + 2W$$

Since the perimeter is to be 640 ft, we have the equation

$$2L + 2W = 640 \text{ (in feet)}$$
or
$$L + W = 320$$

We now have the system of equations

$$L = 3W$$
$$L + W = 320$$

Since L is already solved for in terms of W in the first equation, we can use the method of substitution. Substituting $3W$ for L in the second equation, we have

$$3W + W = 320$$
$$4W = 320$$
$$W = 80 \text{ ft}$$

Substituting 80 ft for W in the equation

we have
$$L = 3W$$
$$L = 3(80 \text{ ft})$$
$$= 240 \text{ ft}$$

Hence, the dimensions of the rectangular lot are 80 ft by 240 ft. Check that the perimeter will be 640 ft.

Practice Exercise 4. Joshua has a collection of 59 coins consisting of dimes and quarters. If the collection is worth $11.30, how many of each kind of coin does he have?

EXAMPLE 4 Karole has a collection of 63 coins consisting of dimes and quarters. If her total collection of coins is worth $10.20, how many of each kind of coin does Karole have?

SOLUTION: Since there are dimes and quarters in the collection, we let

$$d = \text{number of dimes}$$

and

$$q = \text{number of quarters}$$

Since there are 63 coins in the collection, we have the equation

$$d + q = 63$$

The value of the dimes plus the value of the quarters equals the value of the collection. Hence, we have the equation

$$\$.10d + \$.25q = \$10.20$$
or
$$10d + 25q = 1020$$

We now have the system of equations

$$d + q = 63$$
$$10d + 25q = 1020$$

Using the method of elimination, we can multiply the first equation by 10 and leave the second equation unchanged. We have the equivalent system

$$10d + 10q = 630$$
$$10d + 25q = 1020$$

Subtracting the first equation from the second, we obtain

$$15q = 390$$

or

$$q = 26$$

Substituting 26 for q in the first of the original equations, we have

$$d + q = 63$$
$$d + 26 = 63$$
$$d = 37$$

Hence, Karole has 37 dimes and 26 quarters in her collection. Check that the total value is $10.20.

We will conclude this section with an example pertaining to 2-digit numbers. If N is a 2-digit number, then N has a units digit (which we will denote by u) and a tens digit (which we will denote by t). Because of place value, we can write

$$N = 10t + u$$

to represent this two-digit number.

Practice Exercise 5. The sum of the digits in a 2-digit number is 12. If the digits are reversed, the new number formed will be 54 less than the original number. What is the original number?

EXAMPLE 5 The sum of the digits of a 2-digit number is 6. If the digits are reversed, then the new number formed will be 36 more than the original number. What is the original number?

SOLUTION: We let N = the original number. Since N is a 2-digit number, we have

$$N = 10t + u$$

Reversing the digits of N, the new number, M, formed is

$$M = 10u + t$$

Since M is 36 more than N, we have

$$M = N + 36$$

or

$$(10u + t) = (10t + u) + 36$$
$$10u + t = 10t + u + 36$$
$$9u - 9t = 36$$
$$u - t = 4$$

for one of the equations. Since the sum of the digits of N is 6, we have the equation

$$u + t = 6$$

We now have the system of equations

$$u - t = 4$$
$$u + t = 6$$

Using the method of elimination, we can add the two equations and obtain

$$2u = 10$$
$$u = 5$$

Substituting 5 for u in the equation

$$u + t = 6$$

we obtain

$$5 + t = 6$$

or

$$t = 1$$

Therefore,

$$N = 10t + u$$
$$= 10(1) + 5$$
$$= 15$$

as the original number. Check that, by reversing the digits of N, we get 51, which is 36 more than 15.

ANSWERS TO PRACTICE EXERCISES

1. 13 and 33. **2.** 17 m by 42 m. **3.** 15 cm and 22 cm. **4.** 23 dimes and 36 quarters. **5.** 93.

EXERCISES 14.4

Solve each of the following exercises, using a system of equations with two variables.

1. The sum of two numbers is 27. If one-fourth of the larger number is 7 less than the smaller number, what are the numbers?
2. The sum of two numbers is 165. If one number is twice as large as the other, what are the numbers?
3. Determine two consecutive whole numbers whose sum is 39.
4. Determine two consecutive odd numbers whose sum is 20.
5. Ann invested $17,000, part at 7% simple interest and the balance at 8.5% simple interest per year. If the total amount of interest earned at the end of the year is $1310, how much was invested at each rate?
6. A butcher has two grades of ground meat, one that sells for $1.49 a pound and another that sells for $1.79 a pound. How much of each kind should the butcher mix in order to have a 100-lb mixture that he can sell during a weekend special at $1.61 a pound?
7. Karen has a collection of 85 coins consisting of nickels and dimes. If her total collection of coins is worth $5.55, how many of each kind of coin does Karen have?
8. Cindy has a collection of 42 coins consisting of quarters and half-dollars. If her collection is worth $15.25, how many of each kind of coin does Cindy have?
9. The sum of two numbers is 146, and their difference is 28. What are the numbers?
10. One-fourth the sum of two numbers is 29, and one-half of their difference is 19. What are the numbers?
11. A plane rectangular region is such that its length is 7 m more than its width. If the perimeter of the region is 182 m, determine its dimensions.

12. A piece of string 24 in. long is cut into two pieces such that one piece is 4 in. less than 3 times the length of the other piece. Determine the length of each piece of string.
13. Two consecutive even integers have a sum of 26. The sum of twice one of the integers and 3 times the other is 66. What are the two consecutive integers?
14. The sum of two numbers is 31. If twice the smaller number is subtracted from 4 times the larger number, the difference is 79. What are the numbers?
15. The difference of two positive numbers is -7. If 4 times the larger number is subtracted from 3 times the smaller number, the difference is -41. What are the two numbers?
16. One size package of dog food costs $1.20 more than another size package. The sum of the costs of the two items is $3.50. How much does each size package cost?
17. A fuel tank was filled twice during one month. The total number of gallons of fuel delivered was 812. If one delivery was 144 gal less than 3 times the other, how many gallons of fuel oil were delivered each time?
18. If 3 times the smaller of two numbers is subtracted from twice the larger number, the difference is 19. If 5 times the smaller number is added to 4 times the larger number, the sum is 16. What are the two numbers?
19. One number is 9 less than a second number. The smaller number is twice 3 less than the larger number. What are the numbers?
20. The number of kW hours of electricity used by a family during a 2-month period was 134. The second month, the amount of electricity used was 25 kW hours less than twice the number of kilowatt hours used the first month. How many kilowatt hours of electricity were used each of the 2 months?
21. The sum of the digits of a 2-digit number is 10. If the digits are reversed, the new number formed will be 18 larger than the original number. What is the original number?
22. The sum of the digits of a 2-digit number is 7. If the digits are reversed, the new number formed will be 27 less than the original number. What is the original number?
23. The units digit of a 2-digit number is 4 more than its tens digit. If the units digit is decreased by 4 and the tens digit is increased by 3, the new number formed will be twice the original number. What is the original number?
24. The units digit of a 2-digit number is 7 less than its tens digit. If the units digit is increased by 6 and the tens digit is decreased by 6, the new number formed will be one-third of the original number. What is the original number?

In this chapter, you were introduced to systems of two linear equations in two variables and their solutions. Such systems have either no solution, exactly one solution, or infinitely many solutions. The methods we discussed for solving such a system are the graphical method, the method of elimination using addition or subtraction, and the method of substitution. Verbal problems that can be solved by using such systems of equations were also discussed.

The **key words** introduced in this chapter are listed in the order in which they appeared in the text. You should know what they mean before proceeding to the next chapter.

system of two linear equations in two variables (p. 590)	intersecting lines (p. 597)	elimination method (p. 598)
graphical method (p. 591)	parallel lines (p. 597)	method of substitution (p. 606)
	coinciding lines (p. 597)	

Review Exercises

14.1 Graphical Method for Solving a System of Linear Equations

Solve each of the following systems of equations graphically.

1. $\begin{cases} y = x + 2 \\ y = 3 - x \end{cases}$
2. $\begin{cases} y = 1 + x \\ x = 2 - y \end{cases}$
3. $\begin{cases} y = 2x + 1 \\ y = 2x + 4 \end{cases}$
4. $\begin{cases} y = 3x \\ y = 1 - 4x \end{cases}$

14.2 Elimination Method for Solving a System of Linear Equations

Solve each of the following systems of equations using the method of elimination by addition or subtraction.

5. $\begin{cases} 2x + y = 1 \\ 3x - y = 4 \end{cases}$
6. $\begin{cases} x + 3y = 2 \\ x - 7y = 1 \end{cases}$
7. $\begin{cases} 2x - 3y = 1 \\ 3x + 4y = 8 \end{cases}$
8. $\begin{cases} 2x - 5y = 3 \\ 4x - 10y + 2 \end{cases}$

14.3 Substitution Method for Solving a System of Linear Equations

Solve each of the following systems of equations using the method of substitution.

9. $\begin{cases} y = 1 - 3x \\ 2x + 3y = 1 \end{cases}$
10. $\begin{cases} 3x + y = 4 \\ 2x - 5y = 6 \end{cases}$
11. $\begin{cases} 2x + 3y = 1 \\ 3x - 4y = 4 \end{cases}$
12. $\begin{cases} 3x - 5y = 6 \\ 6x - 10y = 3 \end{cases}$

14.4 Applications Involving Systems of Equations

Solve each of the following exercises using a system of equations.

13. The sum of two numbers is 153, and their difference is 39. What are the numbers?
14. Daryl has a collection of 59 coins. There are 3 dimes and the others are nickels and quarters. If his total collection is worth $6.50, how many nickels and how many quarters does Daryl have?
15. Brian bought two batteries and paid $2.20 for both. One battery cost $1.00 more than the other. How much did Brian pay for each of the batteries?
16. The Hazardous Production Company lost 56 days of work due to either illness or accidents during one month. The number of days lost to accidents was 11 more than twice the number of days lost due to illness. How many days were lost in each case?
17. The sum of the digits of a 2-digit number is 8. If the digits are reversed, then the new number formed will be 36 more than the original number. What is the original number?

Teasers

1. Determine if $(1, -1)$ is a solution for the following system of equations:

$$\begin{cases} x + 3y = -2 \\ 2x - y = 4 \end{cases}$$

2. Determine if $\left(\dfrac{17}{7}, \dfrac{-9}{7}\right)$ is a solution for the following system of equations:

$$\begin{cases} x - 2y = 5 \\ 2x + 3y = 1 \end{cases}$$

In Exercises 3–6, solve each of the given systems of equations.

3. $\begin{cases} 1.4x - 1.2y = 1 \\ 0.2x + 0.5y = 0.1 \end{cases}$

4. $\begin{cases} 1.5p + 2.1q = 0.9 \\ 0.4p - 0.6q = 1.4 \end{cases}$

5. $\begin{cases} \dfrac{1}{6}y = \dfrac{1}{3}(x - 3) \\ \dfrac{3}{8}x + \dfrac{1}{4}y = \dfrac{1}{8} \end{cases}$

6. $\begin{cases} \dfrac{1}{3}r - \dfrac{7}{12}s = \dfrac{1}{2} \\ \dfrac{-1}{2}s + \dfrac{1}{16}r = \dfrac{-11}{6} \end{cases}$

7. Consider the equation $p = aq + b$. If $(p = -2, q = -10)$ and $(p = 3, q = 5)$ are solutions of the equation, determine the values of a and b.

8. Determine the values of a and b if $\left(x = \dfrac{-1}{13}, y = \dfrac{-5}{13}\right)$ is a solution of the system of equations

$$\begin{cases} ax - by = 1 \\ bx + ay = -1 \end{cases}$$

9. Solve the system of equations given in Exercise 8 for a and b in terms of x and y.
10. The length of a rectangle is 2 less than 3 times the width. If the perimeter of the rectangle is 52 ft, determine the dimensions of the rectangle. (Use a system of two equations in two unknowns to solve the problem.)

CHAPTER 14 TEST

This test covers materials from Chapter 14. Read each question carefully, and answer all questions. If you miss a question, review the appropriate section of the text.

1. Determine if $(x = 1, y = 4)$ is a solution for the system of equations

$$\begin{cases} y = x + 3 \\ y = 2 - x \end{cases}$$

2. Determine the values of a and b such that $(x = 2, y = 1)$ is a solution for the system of equations

$$\begin{cases} y = a + x \\ x = b - y \end{cases}$$

In Exercises 3–4, solve the given systems of equations graphically.

3. $\begin{cases} 3x - y = 4 \\ 2x + y = 1 \end{cases}$
4. $\begin{cases} 3x - 4y = 8 \\ 2x - 3y = 1 \end{cases}$

In Exercises 5–8, solve the given systems of equations algebraically.

5. $\begin{cases} y = x + 2 \\ y = 3 - x \end{cases}$
6. $\begin{cases} y = 2x \\ y = 1 - 4x \end{cases}$
7. $\begin{cases} 3x - 5y = 6 \\ x + y = 1 \end{cases}$
8. $\begin{cases} 2x - 3y = 1 \\ 3x + 2y = 4 \end{cases}$

9. The sum of two numbers is 60, and their difference is 14. What are the two numbers?
10. Vonnie has a collection of 36 coins. There are 4 nickels, and the others are dimes and quarters. If the coin collection is worth $5.20, how many dimes and quarters does Vonnie have?

CUMULATIVE REVIEW: CHAPTERS 1–14

In Exercises 1–5, perform the indicated operations, and simplify your results.

1. $\dfrac{2}{5} - \dfrac{1}{4} + \dfrac{3}{7}$
2. $\dfrac{6}{7} - \dfrac{2}{3} \div \dfrac{5}{6}$
3. $1.204 \div (-0.004)$
4. 3.6% of 26.18
5. $-4^2 - 2[(-3) \times (-2^3) \div (2 - 8)]$

In Exercises 6–7, round as indicated.

6. 30.0672, to the nearest tenth
7. $208{,}463$, to 2 significant digits
8. The perimeter of a triangular region is 41.45 m. Determine the measure of a third side of the region if the other two sides measure $11\dfrac{1}{2}$ m and $13\dfrac{1}{5}$ m.

In Exercises 9–13, perform the indicated operations, and simplify your results.

9. $\dfrac{-12xy^3z^2}{18x^4yz^6}$
10. $(-2p^{-3}q^2)^{-2}$
11. $(2p + 3)(p^2 - 5)$
12. $(r - 3s)(r^2 + rs - 2s^2)$
13. $\dfrac{2q}{q^3 - 4q} \div \dfrac{-5q}{q + 2}$

14. Determine the area of a rectangular region that is 13.6 yd wide and $20\dfrac{2}{5}$ yd long.

In Exercises 15–20, solve each of the given equations or inequalities for the given variable.

15. $1.2y - 4.3 = 4.1y - 1.9$
16. $2(x - 3) + 4 = 5 - 3(1 - 2x)$
17. $3(p - 2q) + 3q \geq 3 - 2p$; p
18. $\dfrac{2}{r - 3} = \dfrac{-5}{7 - r}$
19. $(u^2 - 9)(2u + 5) = 0$
20. $3^2 + 11s - 4 = 0$

In Exercises 21–22, solve the given systems of equations algebraically.

21. $\begin{cases} 3x + 2y = 4 \\ y = 1 - x \end{cases}$
22. $\begin{cases} 3u - 5v = 6 \\ 2u - 3v = 1 \end{cases}$

23. Determine the number N given by 3.96×10^8.
24. Factor completely: $4p^3 + 18p^2 - 10p$.
25. Determine whether $(2, -3)$ is a solution of the inequality $2x - 3y \leq 10$.

CHAPTER 15 Square Roots

In Chapter 1, you learned to square a number by multiplying the number by itself. In this chapter, you will learn the opposite procedure. Given a number, is it the square of another number? The procedure used to answer this question is known as finding the square root of a number.

Simplifying square roots and performing operations on them will be discussed. The solution of simple equations involving square roots will also be examined.

CHAPTER 15 PRETEST

This pretest covers materials on square roots. Read each question carefully, and try to answer all questions. Each question is keyed to the chapter section in which the particular topic is discussed. Check your answers with those given in the back of the book. *For this test, all variables represent positive numbers.*

Questions 1–11 pertain to Section 15.1.

In Exercises 1–6, determine the principal square root for the indicated expression.

1. 49
2. x^6
3. $16a^4$

4. $81x^4y^6$
5. $\dfrac{(2x+y)^2}{16}$
6. $(3ab^2c)^8$

In Exercises 7–11, simplify each of the given expressions.

7. $\sqrt{100}$
8. $-\sqrt{225}$
9. $\sqrt{a^6}$

10. $-\sqrt{x^{10}}$
11. $\sqrt{16x^2y^4z^6}$

Questions 12–14 pertain to Section 15.2.

Simplify each of the given expressions.

12. $\sqrt{300}$
13. $\sqrt{x^3y^4}$
14. $\sqrt{50a^3b^2c^5}$

Questions 15–17 pertain to Section 15.3.

Simplify each of the given expressions. Rationalize all denominators.

15. $\sqrt{\dfrac{32}{162}}$

16. $\sqrt{\dfrac{7}{3}}$

17. $\dfrac{\sqrt{3a}}{\sqrt{2b}}$

Questions 18–20 pertain to Section 15.4.

Combine by adding or subtracting.

18. $2\sqrt{7} - 3\sqrt{7} + 4\sqrt{7}$

19. $3\sqrt{3} + 2\sqrt{18} - 5\sqrt{12}$

20. $\sqrt{x^4 y} - 2x\sqrt{y^5} - 4\sqrt{x^2 y^3}$

Questions 21–23 pertain to Section 15.5.

Perform the indicated operations and simplify.

21. $2\sqrt{3}(\sqrt{2} - 3\sqrt{3})$

22. $(2\sqrt{5} + \sqrt{3})(3\sqrt{2} - \sqrt{5})$

23. $(x\sqrt{27} - y\sqrt{12}) \div (xy\sqrt{3})$

Questions 24–25 pertain to Section 15.6.

Solve each equation for the indicated variable. Check your results.

24. $\sqrt{3t + 1} = 5$

25. $\sqrt{1 - 2p} = \sqrt{p + 5}$

15.1 Square Roots of Nonnegative Numbers

OBJECTIVE

After completing this section, you should be able to evaluate square roots of nonnegative numbers.

In Chapter 1, you learned that to "square a number" means to multiply the number by itself. For instance, the square of 5 is $(5)(5) = 5^2 = 25$.

- $4^2 = (4)(4) = 16$
- $\left(\frac{1}{2}\right)^2 = \left(\frac{1}{2}\right)\left(\frac{1}{2}\right) = \frac{1}{4}$
- $0^2 = (0)(0) = 0$
- $(-3)^2 = (-3)(-3) = 9$
- $\left(\frac{-1}{5}\right)^2 = \left(\frac{-1}{5}\right)\left(\frac{-1}{5}\right) = \frac{1}{25}$

Note that the exponent 2 on the number being squared indicates that the number is a *factor* twice.

We now list some perfect squares:

1 is the square of 1	x^2 is the square of x
4 is the square of 2	y^4 is the square of y^2
9 is the square of 3	w^6 is the square of w^3
16 is the square of 4	t^8 is the square of t^4
25 is the square of 5	u^{10} is the square of u^5
36 is the square of 6	
49 is the square of 7	
64 is the square of 8	
81 is the square of 9	
100 is the square of 10	
121 is the square of 11	
144 is the square of 12	
169 is the square of 13	
196 is the square of 14	
225 is the square of 15	
625 is the square of 25	
1024 is the square of 32	

In this section, we will discuss the opposite (or inverse) of squaring a number. This is called "taking the square root" of a number. We will use the symbol $\sqrt{}$, called a **radical sign**, to denote the square root of a number. The expression under the radical sign is called the **radicand**.

DEFINITION

Let $a \geq 0$. Then the **square root** of a, denoted by $\sqrt{a}$, means one of two *equal* factors of a.

15.1 Square Roots of Nonnegative Numbers

- $\sqrt{16}$ is read "the square root of 16." $\sqrt{16} = 4$ since $4^2 = 16$. Also note that $(-4)^2 = 16$.
- $\sqrt{49}$ is read "the square root of 49." $\sqrt{49} = 7$ since $7^2 = 49$. Also note that $(-7)^2 = 49$.
- $\sqrt{0}$ is read "the square root of 0." $\sqrt{0} = 0$ since $0^2 = 0$.

Looking carefully at the definition for the square root of a nonnegative number, note that we could get two results for the square root. For instance, we note that $\sqrt{16} = 4$ and $\sqrt{16} = -4$. Also, $\sqrt{49} = 7$ and $\sqrt{49} = -7$. Every *positive* number has *two* square roots—one positive and the other negative.

- The two square roots of 36 are $\sqrt{36} = \pm 6$. (± 6 is read "positive and negative 6.") The positive square root is 6, and the negative square root is -6. The symbol $\pm$ is used to denote that both square roots are taken.
- The two square roots of 121 are $\sqrt{121} = \pm 11$. The positive square root is 11, and the negative square root is -11.

DEFINITION

If $a > 0$, then the **principal square root** of a is the positive square root of a.

Note Unless otherwise indicated, the square root of a positive number is *always* taken to be the *principal* square root.

Determine the principal square root of each of the following. All algebraic expressions represent nonnegative numbers.

Practice Exercise 1. 64

EXAMPLE 1 Determine the principal square root of 25.

SOLUTION: $\sqrt{25} = 5$

Practice Exercise 2. 256

EXAMPLE 2 Evaluate: $-\sqrt{81}$.

SOLUTION: $-\sqrt{81} = -9$ since $-\sqrt{81}$ means $-(\sqrt{81})$ and $-(\sqrt{81}) = -(9) = -9$.

For the numbers that we have worked with so far in this text, their squares are either positive or zero. That is, none of these numbers has a square that is negative. Therefore, we cannot take the square root of a negative number. To do so will require a new kind of number called "complex numbers"; these will not be considered in this text. However, as you continue your study of mathematics beyond this course, you will probably encounter the study of complex numbers.

Just as we can take the square root of nonnegative numbers, we can also consider taking the square root of algebraic expressions. However, the algebraic expressions must represent nonnegative numbers.

Practice Exercise 3. 900

EXAMPLE 3 Determine the principal square root of x^2.

SOLUTION: If $x \geq 0$, then $\sqrt{x^2} = x$.

Practice Exercise 4. $9a^2$

EXAMPLE 4 Determine the principal square root of $49y^2$.

SOLUTION: If $y \geq 0$, then $\sqrt{49y^2} = 7y$ since $(7y)^2 = 49y^2$.

Practice Exercise 5.
$25(2x - 3y)^2$

EXAMPLE 5 Determine the principal square root of $(a + b)^4$.

SOLUTION: If $(a + b) \geq 0$, then $\sqrt{(a + b)^4} = (a + b)^2$ since $[(a + b)^2]^2 = (a + b)^4$.

Practice Exercise 6.
$36a^4b^2(c + d)^6$

EXAMPLE 6 Determine: $-\sqrt{16z^4}$.

SOLUTION: If $z \geq 0$, then

$$-\sqrt{16z^4} = -(\sqrt{16z^4})$$
$$= -[\sqrt{(4z^2)^2}]$$
$$= -(4z^2)$$
$$= -4z^2$$

ANSWERS TO PRACTICE EXERCISES

1. 8.   2. 16.   3. 30.   4. $3a$.   5. $5(2x - 3y)$.   6. $6a^2b(c + d)^3$.

EXERCISES 15.1

In Exercises 1–15, square each of the indicated expressions.

1. 5	**2.** 7	**3.** 10	**4.** 14
5. 21	**6.** x	**7.** $3y$	**8.** $x + 2$
9. $y - 3$	**10.** ab	**11.** $2(x - 4)$	**12.** y^2
13. x^2y	**14.** $2xy^2$	**15.** u^3	

In Exercises 16–30, determine the principal square root of the indicated expressions. All algebraic expressions represent nonnegative numbers.

16. 36	**17.** 81	**18.** 100	**19.** 169
20. 200	**21.** 225	**22.** 400	**23.** a^2
24. $4b^2$	**25.** $81c^4$	**26.** $9(y + 2)^2$	**27.** $16(2a + 1)^2$
28. $25(3y - 1)^4$	**29.** $16a^2(b + 1)^2$	**30.** $49c^4(2d + 1)^6$	

In Exercises 31–50, determine the value of each of the given expressions, if possible. If the expression is meaningless, indicate why. All variables represent nonnegative numbers.

31. $\sqrt{36}$	**32.** $-\sqrt{25}$	**33.** $\sqrt{-25}$	**34.** $-\sqrt{49}$
35. $\sqrt{9} - \sqrt{4}$	**36.** $\sqrt{9 - 4}$	**37.** $\sqrt{4} - \sqrt{9}$	**38.** $\sqrt{4 - 9}$
39. $\sqrt{x^2}$	**40.** $-\sqrt{x^2}$	**41.** $\sqrt{-x^2}$	**42.** $\sqrt{x^2} - \sqrt{y^2}$
43. $\sqrt{-(5 - 1)}$	**44.** $\sqrt{-(1 - 5)}$	**45.** $-\sqrt{-81}$	**46.** $\sqrt{36x^2}$
47. $\sqrt{49y^4}$	**48.** $-\sqrt{z^6}$	**49.** $-\sqrt{-x^2y^2}$	**50.** $\sqrt{100t^8}$

15.2 Simplifying Square Roots of Expressions without Fractions

OBJECTIVE

After completing this section, you should be able to simplify square roots of nonnegative expressions that do not involve fractions.

In later sections of this chapter, we will perform arithmetic operations with square roots. However, it sometimes becomes necessary to simplify the square roots, if possible, before performing the indicated operations.

For instance, $\sqrt{256}$ can be simplified by rewriting it as 16, as we will now illustrate:

$$\sqrt{256} = \sqrt{(16)(16)}$$
$$= \sqrt{(16)^2}$$
$$= 16 \quad \text{(By definition)}$$

Simplify each of the following.

Practice Exercise 1. $\sqrt{81}$

EXAMPLE 1 Simplify: $\sqrt{49}$.

SOLUTION:
$$\sqrt{49} = \sqrt{(7)(7)}$$
$$= \sqrt{(7)^2}$$
$$= 7$$

Practice Exercise 2. $\sqrt{400}$

EXAMPLE 2 Simplify: $\sqrt{121}$.

SOLUTION:
$$\sqrt{121} = \sqrt{(11)(11)}$$
$$= \sqrt{(11)^2}$$
$$= 11$$

Practice Exercise 3. $\sqrt{x^{10}}$

EXAMPLE 3 Simplify: $\sqrt{x^4}$.

SOLUTION:
$$\sqrt{x^4} = \sqrt{(x^2)(x^2)}$$
$$= \sqrt{(x^2)^2}$$
$$= x^2$$

Practice Exercise 4. $\sqrt{16x^2y^4}$

EXAMPLE 4 Simplify: $\sqrt{y^6}$.

SOLUTION:
$$\sqrt{y^6} = \sqrt{(y^3)(y^3)}$$
$$= \sqrt{(y^3)^2}$$
$$= y^3$$

Practice Exercise 5. $\sqrt{36a^6b^4c^2}$

EXAMPLE 5 Simplify: $\sqrt{9a^4}$.

SOLUTION:
$$\sqrt{9a^4} = \sqrt{(3a^2)(3a^2)}$$
$$= \sqrt{(3a^2)^2}$$
$$= 3a^2$$

ANSWERS TO PRACTICE EXERCISES

1. 9.   2. 20.   3. x^5.   4. $4xy^2$.   5. $6a^3b^2c$.

As we continue to simplify square roots, the following rule can be used.

Rule for Finding the Square Root of a Product

If $a \geq 0$ and $b \geq 0$, then

$$\sqrt{ab} = \sqrt{a}\sqrt{b}$$

When using this rule, the strategy is to factor the given expression, whenever possible, so that one factor is a square of another expression.

Simplify each of the following.

Practice Exercise 6. $\sqrt{72}$

EXAMPLE 6 Simplify: $\sqrt{18}$.

SOLUTION:
$$\sqrt{18} = \sqrt{(9)(2)}$$
$$= \sqrt{(3^2)(2)}$$
$$= \sqrt{3^2}\sqrt{2}$$
$$= 3\sqrt{2}$$

Practice Exercise 7. $-\sqrt{700}$

EXAMPLE 7 Simplify: $-\sqrt{300}$.

SOLUTION:
$$-\sqrt{300} = -\sqrt{(100)(3)}$$
$$= -\sqrt{(10)^2(3)}$$
$$= -\sqrt{(10)^2}\sqrt{3}$$
$$= -10\sqrt{3}$$

Practice Exercise 8. $\sqrt{128}$

EXAMPLE 8 Simplify: $\sqrt{32}$.

SOLUTION:
$$\sqrt{32} = \sqrt{2^5}$$
$$= \sqrt{(2^4)(2)}$$
$$= \sqrt{(2^2)^2(2)}$$
$$= \sqrt{(2^2)^2}\sqrt{(2)}$$
$$= (2)^2\sqrt{2}$$
$$= 4\sqrt{2}$$

Practice Exercise 9. $\sqrt{y^9}$

EXAMPLE 9 Simplify: $\sqrt{x^7}$.

SOLUTION:
$$\sqrt{x^7} = \sqrt{(x^6)(x)}$$
$$= \sqrt{(x^3)^2(x)}$$
$$= \sqrt{(x^3)^2}\sqrt{x}$$
$$= x^3\sqrt{x}$$

Practice Exercise 10. $-\sqrt{u^{13}}$

EXAMPLE 10 Simplify: $-\sqrt{y^{11}}$.

SOLUTION:
$$-\sqrt{y^{11}} = -\sqrt{(y^{10})(y)}$$
$$= -\sqrt{(y^5)^2(y)}$$
$$= -\sqrt{(y^5)^2}\sqrt{y}$$
$$= -y^5\sqrt{y}$$

15.2 Simplifying Square Roots of Expressions without Fractions

Practice Exercise 11. $\sqrt{9t^7}$

EXAMPLE 11 Simplify: $\sqrt{4y^5}$.

SOLUTION:
$$\sqrt{4y^5} = \sqrt{(4y^4)(y)}$$
$$= \sqrt{(2y^2)^2(y)}$$
$$= 2y^2\sqrt{y}$$

Practice Exercise 12. $-\sqrt{162s^{11}}$

EXAMPLE 12 Simplify: $-\sqrt{32u^7}$.

SOLUTION:
$$-\sqrt{32u^7} = -\sqrt{(16u^6)(2u)}$$
$$= -\sqrt{(4u^3)^2(2u)}$$
$$= -4u^3\sqrt{2u}$$

Remember, all variables used throughout this chapter represent nonnegative numbers unless otherwise indicated.

The preceding rule can also be used for a product of three or more factors.

Practice Exercise 13. $\sqrt{16u^6v^3}$

EXAMPLE 13 Simplify: $\sqrt{4x^4y}$.

SOLUTION:
$$\sqrt{4x^4y} = \sqrt{(2^2)(x^2)^2(y)}$$
$$= 2x^2\sqrt{y}$$

Practice Exercise 14. $\sqrt{25xy^2z^5}$

EXAMPLE 14 Simplify: $\sqrt{9a^3bc^4}$.

SOLUTION:
$$\sqrt{9a^3bc^4} = \sqrt{(3^2)(a^2)(a)(b)(c^2)^2}$$
$$= \sqrt{(3^2)(a^2)(ab)(c^2)^2}$$
$$= (3a)\sqrt{ab}(c^2)$$
$$= 3ac^2\sqrt{ab}$$

In Example 14, notice that we wrote the square root factor in the answer to the right of all the other factors. This is done to avoid confusion as to what really belongs under the radical sign.

Practice Exercise 15. $-\sqrt{200a^3b^6c^7}$

EXAMPLE 15 Simplify: $-\sqrt{50x^6y^4z}$.

SOLUTION:
$$-\sqrt{50x^6y^4z} = -\sqrt{(25)(2)x^6y^4z}$$
$$= -\sqrt{(5^2)(2)(x^3)^2(y^2)^2(x)}$$
$$= -\sqrt{5^2}\sqrt{2}\sqrt{(x^3)^2}\sqrt{(y^2)^2}\sqrt{z}$$
$$= -5\sqrt{2}(x^3)(y^2)\sqrt{z}$$
$$= -5x^3y^2\sqrt{2x}$$

Practice Exercise 16. $-\sqrt{18p^3qr^9s^{11}}$

EXAMPLE 16 Simplify: $-\sqrt{60a^2b^4c^5}$.

SOLUTION:
$$-\sqrt{60a^2b^4c^5} = -\sqrt{(4a^2b^4c^4)(15c)}$$
$$= -\sqrt{(2ab^2c^2)^2(15c)}$$
$$= -\sqrt{(2ab^2c^2)^2}\sqrt{15c}$$
$$= -2ab^2c^2\sqrt{15c}$$

ANSWERS TO PRACTICE EXERCISES

6. $6\sqrt{2}$. 7. $-10\sqrt{7}$. 8. $8\sqrt{2}$. 9. $y^4\sqrt{y}$. 10. $-u^6\sqrt{u}$. 11. $3t^3\sqrt{t}$. 12. $-9s^5\sqrt{2s}$. 13. $4u^3v\sqrt{v}$. 14. $5yz^2\sqrt{xz}$. 15. $-10ab^3c^3\sqrt{2ac}$. 16. $-3pr^4s^5\sqrt{2pqrs}$.

We will now summarize what we have been doing in Exercises 6–16.

When simplifying the square root of a product of two or more factors:

1. Always look for a factor that is an even power of another expression.
2. Write the answer with any factor containing the square root sign at the extreme right in the product.

As an additional aid in simplifying square roots, the prime factorization of the expression may be helpful.

Simplify. All variables represent nonnegative numbers.

Practice Exercise 17. $\sqrt{98}$

EXAMPLE 17 Simplify: $\sqrt{96}$.

SOLUTION:
$\sqrt{96} = \sqrt{(2)(2)(2)(2)(2)(3)}$ (Prime factorization)
$= \sqrt{(2^4)(2)(3)}$ (Rewriting, using even powers)
$= \sqrt{(2^4)(6)}$ (Rewriting)
$= \sqrt{2^4}\sqrt{6}$ (Product rule)
$= 2^2\sqrt{6}$
$= 4\sqrt{6}$

Practice Exercise 18. $\sqrt{147}$

EXAMPLE 18 Simplify: $\sqrt{1620}$.

SOLUTION:
$\sqrt{1620} = \sqrt{(2)(2)(3)(3)(3)(3)(5)}$
$= \sqrt{(2^2)(3^4)(5)}$
$= \sqrt{2^2}\sqrt{3^4}\sqrt{5}$
$= (2)(3^2)\sqrt{5}$
$= 18\sqrt{5}$

Practice Exercise 19. $\sqrt{8a^2b^3}$

EXAMPLE 19 Simplify: $\sqrt{18x^3y^2}$.

SOLUTION:
$\sqrt{18x^3y^2} = \sqrt{(2)(3)(3)(x)(x)(x)(y)(y)}$
$= \sqrt{(2)(3^2)(x^2)(x)(y^2)}$
$= \sqrt{(3^2)(x^2)(y^2)(2x)}$
$= \sqrt{3^2}\sqrt{x^2}\sqrt{y^2}\sqrt{2x}$
$= 3xy\sqrt{2x}$

Practice Exercise 20. $\sqrt{27p^4q^5}$

EXAMPLE 20 Simplify: $\sqrt{50p^3q^4r^3}$.

SOLUTION:
$\sqrt{50p^3q^4r^3} = \sqrt{(2)(5)(5)(p)(p)(p)(q)(q)(q)(q)(r)(r)(r)}$
$= \sqrt{(2)(5^2)(p^2)(p)(q^4)(r^2)(r)}$
$= \sqrt{(5^2)(p^2)(q^4)(r^2)(2pr)}$
$= \sqrt{5^2}\sqrt{p^2}\sqrt{q^4}\sqrt{r^2}\sqrt{2pr}$
$= 5pq^2r\sqrt{2pr}$

> **ANSWERS TO PRACTICE EXERCISES**
> 17. $7\sqrt{2}$. 18. $7\sqrt{3}$. 19. $2ab\sqrt{2b}$. 20. $3p^2q^2\sqrt{3q}$.

EXERCISES 15.2

Simplify each of the following. All variables represent nonnegative numbers.

1. $\sqrt{8}$ 2. $\sqrt{12}$ 3. $\sqrt{18}$ 4. $\sqrt{20}$
5. $\sqrt{24}$ 6. $\sqrt{27}$ 7. $\sqrt{28}$ 8. $\sqrt{32}$
9. $\sqrt{40}$ 10. $\sqrt{44}$ 11. $\sqrt{45}$ 12. $\sqrt{48}$
13. $\sqrt{50}$ 14. $\sqrt{52}$ 15. $\sqrt{54}$ 16. $\sqrt{56}$
17. $\sqrt{60}$ 18. $\sqrt{63}$ 19. $\sqrt{68}$ 20. $\sqrt{72}$
21. $\sqrt{75}$ 22. $\sqrt{76}$ 23. $\sqrt{80}$ 24. $\sqrt{84}$
25. $\sqrt{88}$ 26. $\sqrt{90}$ 27. $\sqrt{92}$ 28. $\sqrt{96}$
29. $\sqrt{99}$ 30. $\sqrt{100}$ 31. $\sqrt{104}$ 32. $\sqrt{108}$
33. $\sqrt{200}$ 34. $\sqrt{250}$ 35. $\sqrt{275}$ 36. $\sqrt{400}$
37. $\sqrt{800}$ 38. $\sqrt{900}$ 39. $\sqrt{960}$ 40. $\sqrt{999}$
41. $\sqrt{x^3}$ 42. $\sqrt{y^4}$ 43. $\sqrt{u^5}$ 44. $\sqrt{s^7}$
45. $\sqrt{t^8}$ 46. $\sqrt{r^{10}}$ 47. $\sqrt{p^{11}}$ 48. $\sqrt{q^{13}}$
49. $\sqrt{x^{14}}$ 50. $\sqrt{y^{15}}$ 51. $\sqrt{z^{17}}$ 52. $\sqrt{w^{19}}$
53. $\sqrt{s^{20}}$ 54. $\sqrt{t^{21}}$ 55. $\sqrt{w^{23}}$ 56. $\sqrt{s^{31}}$
57. $\sqrt{4x^2}$ 58. $\sqrt{9y^4}$ 59. $\sqrt{16z^6}$ 60. $\sqrt{25s^8}$
61. $\sqrt{9y^3}$ 62. $\sqrt{16x^5}$ 63. $\sqrt{49t^7}$ 64. $\sqrt{121u^9}$
65. $\sqrt{4x^4y^3}$ 66. $\sqrt{2x^3y^6}$ 67. $\sqrt{3x^4y^8}$ 68. $\sqrt{5a^5b^6}$
69. $\sqrt{48mn^3}$ 70. $\sqrt{7a^4bc^6}$ 71. $\sqrt{8xy^3}$ 72. $\sqrt{12a^2b^4c^6}$
73. $\sqrt{121x^5y^6}$ 74. $\sqrt{100p^3q^7}$ 75. $\sqrt{250t^7}$ 76. $\sqrt{400m^3n^2}$
77. $\sqrt{2ax^2y^6}$ 78. $\sqrt{3ab^2x^3}$ 79. $\sqrt{90p^3q^4s}$ 80. $\sqrt{a^{11}b^{13}c^{17}d^4}$

15.3 Simplifying Square Roots of Expressions with Fractions

OBJECTIVE

After completing this section, you should be able to simplify square roots of nonnegative expressions that involve fractions.

In Section 15.2, you learned to simplify square roots of nonnegative expressions that do not involve fractions. To simplify square roots of nonnegative expressions that do involve fractions, we can use the following rule.

Rule for Finding the Square Root of a Quotient

If $a \geq 0$ and $b > 0$, then

$$\sqrt{\frac{a}{b}} = \frac{\sqrt{a}}{\sqrt{b}}$$

All variables represent positive numbers. In Practice Exercises 1–3, simplify each of the given expressions.

Practice Exercise 1. $\sqrt{\dfrac{64}{25}}$

EXAMPLE 1 Simplify: $\sqrt{\dfrac{4}{9}}$.

SOLUTION:
$$\sqrt{\frac{4}{9}} = \frac{\sqrt{4}}{\sqrt{9}}$$
$$= \frac{2}{3}$$

Practice Exercise 2. $\sqrt{\dfrac{121}{225}}$

EXAMPLE 2 Simplify: $\sqrt{\dfrac{16}{49}}$.

SOLUTION:
$$\sqrt{\frac{16}{49}} = \frac{\sqrt{16}}{\sqrt{49}}$$
$$= \frac{4}{7}$$

Practice Exercise 3. $\sqrt{\dfrac{y^6}{49}}$

EXAMPLE 3 Simplify: $\sqrt{\dfrac{9}{11}}$.

SOLUTION:
$$\sqrt{\frac{9}{11}} = \frac{\sqrt{9}}{\sqrt{11}}$$
$$= \frac{3}{\sqrt{11}}$$

Practice Exercise 4. $\sqrt{\dfrac{9a^4b^2}{25c^2d^6}}$

EXAMPLE 4 Simplify: $\sqrt{\dfrac{81}{23}}$.

SOLUTION:
$$\sqrt{\frac{81}{23}} = \frac{\sqrt{81}}{\sqrt{23}}$$
$$= \frac{9}{\sqrt{23}}$$

An expression such as $\sqrt{\dfrac{3}{5}}$ appears to be in simplified form. Using the quotient rule, we have

$$\sqrt{\frac{3}{5}} = \frac{\sqrt{3}}{\sqrt{5}}$$

which cannot be simplified further, according to the rules for square roots discussed so far. However, consider the following.

$$\sqrt{\frac{3}{5}} = \sqrt{\frac{3}{5} \times 1} \quad \text{(Multiplicative identity element)}$$

15.3 Simplifying Square Roots of Expressions with Fractions

$$= \sqrt{\frac{3}{5} \times \frac{5}{5}} \quad \left(\text{Rewriting } 1 = \frac{5}{5} \text{ so that the denominator will be perfect square after multiplying under the radical sign}\right)$$

$$= \sqrt{\frac{15}{25}} \quad \text{(Multiplying)}$$

$$= \frac{\sqrt{15}}{\sqrt{25}} \quad \text{(Quotient rule)}$$

$$= \frac{\sqrt{15}}{5} \quad \text{(Simplifying)}$$

We now have that $\sqrt{\frac{3}{5}}$ can be written in equivalent form as $\frac{\sqrt{15}}{5}$ without a radical sign in the denominator.

In Practice Exercises 5–8, rationalize the denominator and express your results in simplest form.

Practice Exercise 5. $\frac{8}{\sqrt{2}}$

EXAMPLE 5 Rewrite the expression $\frac{1}{\sqrt{3}}$ without a radical sign in the denominator.

SOLUTION: $\frac{1}{\sqrt{3}} = \frac{1}{\sqrt{3}} \times 1 \quad \text{(Multiplicative identity element)}$

$$= \frac{1}{\sqrt{3}} \times \frac{\sqrt{3}}{\sqrt{3}} \quad \left(\text{Rewriting } 1 = \frac{\sqrt{3}}{\sqrt{3}}\right)$$

$$= \frac{\sqrt{3}}{\sqrt{3}\sqrt{3}} \quad \text{(Multiplying)}$$

$$= \frac{\sqrt{3}}{3} \quad \text{(Simplifying)}$$

Therefore, $\frac{1}{\sqrt{3}} = \frac{\sqrt{3}}{3}$.

The procedure used in Example 5 is called **rationalizing the denominator** of a fraction. In the example, the procedure could have been shortened by simply multiplying *both* the numerator and denominator of $\frac{1}{\sqrt{3}}$ by $\sqrt{3}$. (Do you agree?)

Practice Exercise 6. $\frac{3}{\sqrt{y}}$

EXAMPLE 6 Rationalize the denominator of $\frac{4}{\sqrt{7}}$.

SOLUTION: $\frac{4}{\sqrt{7}} = \frac{4}{\sqrt{7}} \times \frac{\sqrt{7}}{\sqrt{7}} \quad \text{(Multiplying } both \text{ numerator and denominator by } \sqrt{7}\text{)}$

$$= \frac{4\sqrt{7}}{\sqrt{7}\sqrt{7}}$$

$$= \frac{4\sqrt{7}}{7}$$

Practice Exercise 7. $\frac{x}{\sqrt{5x^3}}$

EXAMPLE 7 Simplify: $\sqrt{\frac{5}{7}}$.

SOLUTION: $\sqrt{\frac{5}{7}} = \frac{\sqrt{5}}{\sqrt{7}} \quad \text{(Quotient rule)}$

$$= \frac{\sqrt{5}}{\sqrt{7}} \times \frac{\sqrt{7}}{\sqrt{7}} \quad \text{(Multiplying } both \text{ numerator and denominator by } \sqrt{7}\text{)}$$

$$= \frac{\sqrt{5}\sqrt{7}}{\sqrt{7}\sqrt{7}}$$

$$= \frac{\sqrt{35}}{7}$$

Therefore, $\sqrt{\frac{5}{7}} = \frac{\sqrt{35}}{7}$.

Practice Exercise 8. $\sqrt{\frac{32a^2b}{2a^3}}$

EXAMPLE 8 Simplify $\sqrt{\frac{4x^3y}{xy^2}}$, if $x > 0$ and $y > 0$.

SOLUTION:
$$\sqrt{\frac{4x^3y}{xy^2}} = \sqrt{\frac{4x^2}{y}} \quad \text{(Simplifying under radical sign)}$$

$$= \frac{\sqrt{4x^2}}{\sqrt{y}} \quad \text{(Quotient rule)}$$

$$= \frac{\sqrt{4}\sqrt{x^2}}{\sqrt{y}} \quad \text{(Product rule)}$$

$$= \frac{2x}{\sqrt{y}}$$

$$= \frac{2x\sqrt{y}}{\sqrt{y}\sqrt{y}} \quad \text{(Rationalizing the denominator)}$$

$$= \frac{2x\sqrt{y}}{y}$$

which is simplified.

ANSWERS TO PRACTICE EXERCISES

1. $\frac{8}{5}$. 2. $\frac{11}{15}$. 3. $\frac{y^3}{7}$. 4. $\frac{3a^2b}{5cd^3}$. 5. $4\sqrt{2}$. 6. $\frac{3\sqrt{y}}{y}$. 7. $\frac{\sqrt{5x}}{5x}$. 8. $\frac{4\sqrt{ab}}{a}$.

EXERCISES 15.3

In Exercises 1–40, simplify each of the given expressions. All variables represent positive numbers.

1. $\sqrt{\frac{4}{25}}$
2. $\sqrt{\frac{9}{16}}$
3. $\sqrt{\frac{36}{25}}$
4. $\sqrt{\frac{81}{49}}$
5. $\sqrt{\frac{25}{64}}$
6. $\sqrt{\frac{64}{100}}$
7. $\sqrt{\frac{81}{16}}$
8. $\sqrt{\frac{25}{121}}$
9. $\sqrt{\frac{121}{169}}$
10. $\sqrt{\frac{100}{144}}$
11. $\sqrt{\frac{121}{361}}$
12. $\sqrt{\frac{100}{225}}$
13. $\sqrt{\frac{x^2}{4}}$
14. $\sqrt{\frac{y^6}{100}}$
15. $\sqrt{\frac{u^6}{49}}$
16. $\sqrt{\frac{25u^6}{81}}$
17. $\sqrt{\frac{4z^4}{9}}$
18. $\sqrt{\frac{25}{x^2}}$
19. $\sqrt{\frac{4}{y^4}}$
20. $\sqrt{\frac{81}{z^6}}$

21. $\sqrt{\dfrac{x^2}{y^4}}$ 22. $\sqrt{\dfrac{y^6}{z^4}}$ 23. $\sqrt{\dfrac{4x^2}{y^4}}$ 24. $\sqrt{\dfrac{9x^4y^2}{16}}$

25. $\sqrt{\dfrac{x^4y^6}{z^2}}$ 26. $\sqrt{\dfrac{9u^6}{x^4y^4}}$ 27. $\sqrt{\dfrac{4a^2b^4}{c^4d^2}}$ 28. $\sqrt{\dfrac{p^4q^6}{81t^8}}$

29. $\sqrt{\dfrac{121u^6}{v^4w^8}}$ 30. $\sqrt{\dfrac{2xy^2}{z^4}}$ 31. $\sqrt{\dfrac{16m^2}{25n^4p^6}}$ 32. $\sqrt{\dfrac{3x^3y^6}{xy^2}}$

33. $\sqrt{\dfrac{7p^3}{s^2t^4}}$ 34. $\sqrt{\dfrac{10t}{5s^4}}$ 35. $\sqrt{\dfrac{4a^3y}{ay^3}}$ 36. $\sqrt{\dfrac{x^7y}{16xy^3}}$

37. $\sqrt{\dfrac{s^2t^5}{16t^3}}$ 38. $\sqrt{\dfrac{a^3b^2c^4}{a^5b}}$ 39. $\sqrt{\dfrac{7xy^3z^4}{x^3yz^3}}$ 40. $\sqrt{\dfrac{s^3t^5u^7}{81stu}}$

In Exercises 41–80, rationalize the denominator and express the results in simplest form. All variables represent positive numbers.

41. $\dfrac{3}{\sqrt{2}}$ 42. $\dfrac{7}{\sqrt{3}}$ 43. $\dfrac{8}{\sqrt{5}}$ 44. $\dfrac{1}{\sqrt{6}}$

45. $\dfrac{9}{\sqrt{7}}$ 46. $\dfrac{x}{\sqrt{8}}$ 47. $\dfrac{y}{\sqrt{10}}$ 48. $\dfrac{5}{\sqrt{x}}$

49. $\dfrac{6}{\sqrt{y}}$ 50. $\dfrac{5}{\sqrt{2x}}$ 51. $\dfrac{1}{\sqrt{3y}}$ 52. $\dfrac{4}{\sqrt{5t}}$

53. $\dfrac{\sqrt{3}}{\sqrt{2}}$ 54. $\dfrac{\sqrt{2}}{\sqrt{3}}$ 55. $\dfrac{\sqrt{5}}{\sqrt{7}}$ 56. $\dfrac{\sqrt{8}}{\sqrt{2}}$

57. $\dfrac{\sqrt{2x}}{\sqrt{3}}$ 58. $\dfrac{\sqrt{3y}}{\sqrt{2}}$ 59. $\dfrac{\sqrt{2x}}{\sqrt{x}}$ 60. $\dfrac{\sqrt{3y}}{\sqrt{y}}$

61. $\dfrac{1}{\sqrt{x^3}}$ 62. $\dfrac{2}{\sqrt{y^5}}$ 63. $\dfrac{3}{\sqrt{z^7}}$ 64. $\dfrac{x}{\sqrt{2x^3}}$

65. $\dfrac{\sqrt{y}}{\sqrt{x}}$ 66. $\dfrac{\sqrt{2x}}{\sqrt{3y}}$ 67. $\dfrac{a}{\sqrt{ab}}$ 68. $\dfrac{x}{2\sqrt{5}}$

69. $\dfrac{y}{3\sqrt{y}}$ 70. $\dfrac{u}{5\sqrt{u}}$ 71. $\dfrac{7}{\sqrt{5y^7}}$ 72. $\dfrac{2}{\sqrt{3x^3}}$

73. $\dfrac{\sqrt{x^3}}{2\sqrt{y^3}}$ 74. $\dfrac{\sqrt{3}}{2\sqrt{y^5}}$ 75. $\dfrac{\sqrt{5}}{\sqrt{2y^3}}$ 76. $\dfrac{\sqrt{11}}{\sqrt{ab^3}}$

77. $\dfrac{\sqrt{32x^3y}}{\sqrt{2xy}}$ 78. $\dfrac{\sqrt{18ab^4}}{2\sqrt{ab}}$ 79. $\dfrac{\sqrt{5c^4d^3}}{3\sqrt{cd}}$ 80. $\dfrac{\sqrt{7p^6q^7}}{7\sqrt{p^3q^2}}$

15.4 Addition and Subtraction of Square Roots

OBJECTIVE

After completing this section, you should be able to add and subtract like square roots.

Just as you learned to combine like algebraic expressions, you will learn how to combine *like square roots* using addition and subtraction.

DEFINITION
Two square roots are said to be **like square roots** if they have the same radicands after the square roots have been simplified.

It should be noted that like square roots can have different coefficients. The following are examples of like square roots:

- $\sqrt{5}$ and $2\sqrt{5}$
- $\sqrt{3}$ and $-4\sqrt{3}$
- $\frac{1}{2}\sqrt{7}$ and $\frac{-2}{3}\sqrt{7}$
- $a\sqrt{x}$ and $b\sqrt{x}$ ($x \geq 0$)
- $x\sqrt{a-b}$ and $y\sqrt{a-b}$ ($a \geq b$)

To combine like square roots means to add or subtract them. To do this, we use the distributive property for multiplication over addition (or subtraction) and add (or subtract) the coefficients of the square roots. Hence, to combine

$$3\sqrt{2} + 5\sqrt{2}$$

we note that $3\sqrt{2}$ means 3 times $\sqrt{2}$ and $5\sqrt{2}$ means 5 times $\sqrt{2}$. Therefore,

$$3\sqrt{2} + 5\sqrt{2} = (3 + 5)\sqrt{2} \quad \text{(Distributive property)}$$
$$= 8\sqrt{2} \quad \text{(Simplifying)}$$

Note $3x + 5x = 8x$ for *any* x. Therefore, $3\sqrt{2} + 5\sqrt{2} = 8\sqrt{2}$ must be true with $x = \sqrt{2}$.

Simplify by combining like terms. All variables represent positive numbers.

Practice Exercise 1.
$5\sqrt{2} + 2\sqrt{2} - \sqrt{2}$

EXAMPLE 1 Rewrite $2\sqrt{5} + 5\sqrt{5} - 4\sqrt{5}$ as a single term.
  SOLUTION: $2\sqrt{5} + 5\sqrt{5} - 4\sqrt{5} = (2 + 5 - 4)\sqrt{5}$
  $= 3\sqrt{5}$

Practice Exercise 2.
$\frac{1}{2}\sqrt{7} - \frac{2}{3}\sqrt{7} - \frac{1}{4}\sqrt{7}$

EXAMPLE 2 Rewrite $3\sqrt{7} - 11\sqrt{7} - 6\sqrt{7} + \sqrt{7}$ as a single term.
  SOLUTION: $3\sqrt{7} - 11\sqrt{7} - 6\sqrt{7} + \sqrt{7} = (3 - 11 - 6 + 1)\sqrt{7}$
  $= -13\sqrt{7}$

Practice Exercise 3.
$7\sqrt{a} + \frac{1}{2}\sqrt{a} + b\sqrt{a}$

EXAMPLE 3 Rewrite $2a\sqrt{x} + 3b\sqrt{x} - c\sqrt{x}$ as a single term.
  SOLUTION: $2a\sqrt{x} + 3b\sqrt{x} - c\sqrt{x} = (2a + 3b - c)\sqrt{x}$

ANSWERS TO PRACTICE EXERCISES

1. $6\sqrt{2}$. 2. $\frac{-5}{12}\sqrt{7}$. 3. $\left(\frac{15}{2} + b\right)\sqrt{a}$.

15.4 Addition and Subtraction of Square Roots

When adding unlike square roots, rewrite all of them in simplified form. Then determine if any of the results are like square roots. If so, combine them.

EXAMPLE 4 Combine: $\sqrt{8} + \sqrt{32}$.

SOLUTION: The two square roots given in this sum are not like square roots. However, upon simplifying, we have

$$\sqrt{8} = \sqrt{4 \times 2} = 2\sqrt{2}$$
and $$\sqrt{32} = \sqrt{16 \times 2} = 4\sqrt{2}$$

Hence,

$$\sqrt{8} + \sqrt{32} = 2\sqrt{2} + 4\sqrt{2}$$
$$= (2 + 4)\sqrt{2}$$
$$= 6\sqrt{2}$$

Simplify by combining like terms. All variables represent positive numbers.

Practice Exercise 4.
$3\sqrt{2} + 2\sqrt{8}$

EXAMPLE 5 Simplify: $2\sqrt{3} - \sqrt{\dfrac{3}{4}} + \sqrt{12}$.

SOLUTION: Since

$$\sqrt{\dfrac{3}{4}} = \dfrac{\sqrt{3}}{\sqrt{4}} = \dfrac{\sqrt{3}}{2} = \dfrac{1}{2}\sqrt{3}$$
and $$\sqrt{12} = \sqrt{4 \times 3} = 2\sqrt{3}$$

we have

$$2\sqrt{3} - \sqrt{\dfrac{3}{4}} + \sqrt{12} = 2\sqrt{3} - \dfrac{1}{2}\sqrt{3} + 2\sqrt{3}$$
$$= \left(2 - \dfrac{1}{2} + 2\right)\sqrt{3}$$
$$= \dfrac{7}{2}\sqrt{3}$$

Practice Exercise 5.
$a\sqrt{3} - b\sqrt{27}$

EXAMPLE 6 Simplify: $\sqrt{\dfrac{1}{2}} - \sqrt{8} + \sqrt{\dfrac{1}{8}}$.

SOLUTION: $\sqrt{\dfrac{1}{2}} - \sqrt{8} + \sqrt{\dfrac{1}{8}} = \sqrt{\dfrac{1}{2} \times \dfrac{2}{2}} - \sqrt{4 \times 2} + \sqrt{\dfrac{1}{8} \times \dfrac{2}{2}}$

$$= \dfrac{\sqrt{2}}{2} - 2\sqrt{2} + \dfrac{\sqrt{2}}{4}$$
$$= \left(\dfrac{1}{2} - 2 + \dfrac{1}{4}\right)\sqrt{2}$$
$$= \dfrac{-5}{4}\sqrt{2}$$

Practice Exercise 6.
$7\sqrt{2} - 5\sqrt{18} + 9\sqrt{32}$

EXAMPLE 7 Simplify: $3\sqrt{x^2 y} - a\sqrt{x^2 y^3} + 2b\sqrt{y}$ ($x > 0$, $y > 0$).

Practice Exercise 7.
$\sqrt{p^2 q} - 3\sqrt{p^2 q} + \sqrt{4p^2 q}$

SOLUTION:
$$3\sqrt{x^2y} - a\sqrt{x^2y^3} + 2b\sqrt{y}$$
$$= 3\sqrt{(x^2)y} - a\sqrt{(x^2)(y^2)y} + 2b\sqrt{y}$$
$$= 3\sqrt{x^2}\sqrt{y} - a\sqrt{x^2}\sqrt{y^2}\sqrt{y} + 2b\sqrt{y}$$
$$= 3x\sqrt{y} - axy\sqrt{y} + 2b\sqrt{y}$$
$$= (3x - axy + 2b)\sqrt{y}$$

Practice Exercise 8.
$\sqrt{a^3b} + 2\sqrt{ab} - 3\sqrt{ab^3}$

EXAMPLE 8 Simplify: $\sqrt{\dfrac{x}{2y}} - \sqrt{\dfrac{2x}{y}} + xy\sqrt{18xy}$ $(x > 0, y > 0)$.

SOLUTION:
$$\sqrt{\frac{x}{2y}} - \sqrt{\frac{2x}{y}} + xy\sqrt{18xy}$$
$$= \sqrt{\left(\frac{x}{2y}\right)\left(\frac{2y}{2y}\right)} - \sqrt{\left(\frac{2x}{y}\right)\left(\frac{y}{y}\right)} + xy\sqrt{(2)(3^2)xy}$$
$$= \sqrt{\frac{2xy}{4y^2}} - \sqrt{\frac{2xy}{y^2}} + xy\sqrt{2(3^2)xy}$$
$$= \frac{\sqrt{2xy}}{\sqrt{4y^2}} - \frac{\sqrt{2xy}}{\sqrt{y^2}} + 3xy\sqrt{2xy}$$
$$= \frac{\sqrt{2xy}}{2y} - \frac{\sqrt{2xy}}{y} + 3xy\sqrt{2xy}$$
$$= \left(\frac{1}{2y} - \frac{1}{y} + 3xy\right)\sqrt{2xy}$$
$$= \left(3xy - \frac{1}{2y}\right)\sqrt{2xy}$$

ANSWERS TO PRACTICE EXERCISES
4. $7\sqrt{2}$. 5. $(a - 3b)\sqrt{3}$. 6. $28\sqrt{2}$. 7. 0. 8. $(a + 2 - 3b)\sqrt{ab}$.

EXERCISES 15.4

Simplify each of the following expressions by combining like terms. All variables represent positive numbers.

1. $2\sqrt{3} + 5\sqrt{3}$
2. $6\sqrt{2} - 4\sqrt{2}$
3. $4\sqrt{5} - 5\sqrt{5}$
4. $9\sqrt{6} + 7\sqrt{6}$
5. $3\sqrt{7} + 5\sqrt{7} - 6\sqrt{7}$
6. $\sqrt{10} - 2\sqrt{10} + 5\sqrt{10}$
7. $2\sqrt{x} + 3\sqrt{x}$
8. $9\sqrt{y} - 2\sqrt{y}$
9. $6\sqrt{2u} - 7\sqrt{2u} + 4\sqrt{2u}$
10. $a\sqrt{x} - b\sqrt{x} + \sqrt{x}$
11. $a\sqrt{p - q} + b\sqrt{p - q}$
12. $x\sqrt{s + t} - y\sqrt{s + t}$
13. $3\sqrt{12} + 2\sqrt{3}$
14. $4\sqrt{3} - 5\sqrt{27}$
15. $3\sqrt{5} - 9\sqrt{45}$
16. $6\sqrt{45} - 7\sqrt{125}$
17. $6\sqrt{2} - 5\sqrt{18}$
18. $3\sqrt{18} + 4\sqrt{98}$

19. $6\sqrt{12} + 5\sqrt{3} - 4\sqrt{27}$
20. $2\sqrt{5} - 6\sqrt{45} - \sqrt{125}$
21. $5\sqrt{2} - 6\sqrt{18} + 9\sqrt{98}$
22. $2\sqrt{5} - 4\sqrt{20} - 5\sqrt{45}$
23. $5\sqrt{6} - \sqrt{\dfrac{2}{3}}$
24. $3\sqrt{\dfrac{1}{2}} - 4\sqrt{18}$
25. $\sqrt{50x} + \sqrt{32x}$
26. $\sqrt{8y} - \sqrt{18y}$
27. $3\sqrt{2b} - 2\sqrt{50b}$
28. $\sqrt{x^2y} - \sqrt{x^2y} + \sqrt{4x^2y}$
29. $2\sqrt{2b} - 3\sqrt{18b} + 5\sqrt{32b}$
30. $\sqrt{xy} + \sqrt{x^3y} - 2\sqrt{xy^3}$
31. $a\sqrt{9b} - 3\sqrt{4a^2b^3} - \sqrt{a^2b^3}$
32. $\sqrt{x^3y^2} - y\sqrt{x} + \sqrt{xy^6}$
33. $\sqrt{\dfrac{a}{b}} - \sqrt{\dfrac{b}{a}} - 2\sqrt{ab}$
34. $\sqrt{\dfrac{x}{2y}} - \sqrt{\dfrac{8y}{x}} + \sqrt{2x^3y^3}$
35. $\sqrt{\dfrac{a}{2}} + \sqrt{18a} - \sqrt{2a^3}$
36. $\sqrt{48a^3b} - \sqrt{27ab^3} - \sqrt{12a^5b^7}$
37. $\sqrt{2xy} + \dfrac{1}{a}\sqrt{32xy^3} + b\sqrt{50x^3y}$
38. $3y\sqrt{y} + \dfrac{1}{2}\sqrt{y^3} + y^2\sqrt{\dfrac{9}{y}}$
39. $\sqrt{a^3bc} - x\sqrt{ab^3c^3} + y\sqrt{abc^3}$
40. $\sqrt{\dfrac{a}{3b}} - \sqrt{\dfrac{3b}{a}} + ab\sqrt{12ab}$
41. $2a\sqrt{18b^3} - 3\sqrt{8b^5}$
42. $4c\sqrt{p^3q} + 2d\sqrt{pq^5}$
43. $4x^2y\sqrt{ab^2c^3} - 2xy^2\sqrt{a^3c}$
44. $7p\sqrt{3p^4q} - 3q\sqrt{12p^2q^3}$
45. $5xy^2\sqrt{8y^3} - 7x^2\sqrt{32x^2y}$
46. $-6rs\sqrt{98r^3s^5} + r^2\sqrt{18rs^7}$
47. $2\sqrt{0.09xy^3z^4} - 3\sqrt{0.16x^3yz^2}$
48. $-2.1\sqrt{18a^2b^4c} - 3.9\sqrt{8b^2c^5}$
49. $3x\sqrt{32p^2q^3} + \dfrac{1}{2}y\sqrt{\dfrac{2}{9}p^2q} - \dfrac{1}{3}z\sqrt{\dfrac{8}{25}q^3}$
50. $-0.7a\sqrt{\dfrac{8}{49}xy^2} + 0.3b\sqrt{\dfrac{2}{9}x^3} - 0.5c\sqrt{\dfrac{18}{25}x^5y^4}$

15.5 Multiplication and Division of Square Roots

OBJECTIVES

After completing this section, you should be able to:
1. Multiply two or more square roots.
2. Divide one square root by another square root.

To multiply two square roots, simply use the product rule for square roots and simplify.

If $a \geq 0$ and $b \geq 0$, then $\sqrt{a}\sqrt{b} = \sqrt{ab}$.

Multiply and simplify your results. All variables represent positive numbers.

Practice Exercise 1. $\sqrt{3}\sqrt{7}$

EXAMPLE 1 Multiply: $\sqrt{2}\sqrt{5}$.

SOLUTION:
$$\sqrt{2}\sqrt{5} = \sqrt{(2)(5)}$$
$$= \sqrt{10}$$

Practice Exercise 2. $\sqrt{12}\sqrt{18}$

EXAMPLE 2 Multiply: $\sqrt{3}\sqrt{6}$.

SOLUTION:
$$\sqrt{3}\sqrt{6} = \sqrt{(3)(6)}$$
$$= \sqrt{18}$$
$$= \sqrt{(9)(2)}$$
$$= \sqrt{9}\sqrt{2}$$
$$= 3\sqrt{2}$$

Practice Exercise 3. $\sqrt{5}\sqrt{10}$

EXAMPLE 3 Multiply: $\sqrt{3}\sqrt{12}$.

SOLUTION:
$$\sqrt{3}\sqrt{12} = \sqrt{(3)(12)}$$
$$= \sqrt{36}$$
$$= 6$$

Practice Exercise 4. $\sqrt{2x}\sqrt{8y}$

EXAMPLE 4 Multiply: $\sqrt{8}\sqrt{12}$.

SOLUTION:
$$\sqrt{8}\sqrt{12} = \sqrt{(8)(12)}$$
$$= \sqrt{96}$$
$$= \sqrt{(16)(6)}$$
$$= \sqrt{16}\sqrt{6}$$
$$= 4\sqrt{6}$$

Practice Exercise 5. $\sqrt{ab}\sqrt{bc}$

EXAMPLE 5 Multiply: $\sqrt{x}\sqrt{y}$.

SOLUTION:
$$\sqrt{x}\sqrt{y} = \sqrt{xy} \quad \text{if } x \geq 0 \text{ and } y \geq 0$$

Practice Exercise 6. $(3\sqrt{2})(2\sqrt{3})$

EXAMPLE 6 Multiply: $\sqrt{ab}\sqrt{ac}$.

SOLUTION:
$$\sqrt{ab}\sqrt{ac} = \sqrt{(ab)(ac)}$$
$$= \sqrt{a^2bc}$$
$$= \sqrt{a^2}\sqrt{bc}$$
$$= a\sqrt{bc} \quad \text{if } a \geq 0, b \geq 0, \text{ and } c \geq 0$$

Practice Exercise 7. $2\sqrt{3}(3\sqrt{3} - 4\sqrt{5})$

EXAMPLE 7 Determine the product of $\sqrt{5}$ and $(\sqrt{2} - \sqrt{8})$.

SOLUTION: *(First approach)* Using the distributive property for multiplication over subtraction, we have

$$\sqrt{5}(\sqrt{2} - \sqrt{8}) = \sqrt{5}\sqrt{2} - \sqrt{5}\sqrt{8} \quad \text{(Multiplying each term in the second factor by } \sqrt{5}\text{)}$$
$$= \sqrt{(5)(2)} - \sqrt{(5)(8)} \quad \text{(Product rule)}$$
$$= \sqrt{10} - \sqrt{40}$$
$$= \sqrt{10} - \sqrt{(4)(10)}$$
$$= \sqrt{10} - \sqrt{4}\sqrt{10} \quad \text{(Product rule)}$$

15.5 Multiplication and Division of Square Roots

$$= \sqrt{10} - 2\sqrt{10}$$
$$= (1 - 2)\sqrt{10} \qquad \text{(Adding like square roots)}$$
$$= -\sqrt{10}$$

(Second approach) First, simplify the second factor.

$$\sqrt{5}(\sqrt{2} - \sqrt{8}) = \sqrt{5}[\sqrt{2} - \sqrt{(4)(2)}]$$
$$= \sqrt{5}(\sqrt{2} - \sqrt{4}\sqrt{2}) \qquad \text{(Product rule)}$$
$$= \sqrt{5}(\sqrt{2} - 2\sqrt{2})$$
$$= \sqrt{5}[(1 - 2)\sqrt{2}] \qquad \text{(Adding like square roots)}$$
$$= \sqrt{5}(-\sqrt{2})$$
$$= -\sqrt{(5)(2)} \qquad \text{(Product rule)}$$
$$= -\sqrt{10}$$

Practice Exercise 8.
$(\sqrt{5a})(\sqrt{10a})$

EXAMPLE 8 Multiply: $(\sqrt{2} - \sqrt{3})$ and $(\sqrt{3} + 3\sqrt{2})$.

SOLUTION: We can multiply the two expressions just as we multiply any two binomials:

$$(\sqrt{2} - \sqrt{3})(\sqrt{3} + 3\sqrt{2}) = (\sqrt{2} - \sqrt{3})\sqrt{3} + (\sqrt{2} - \sqrt{3})(3\sqrt{2})$$
$$= \sqrt{2}\sqrt{3} - \sqrt{3}\sqrt{3} + \sqrt{2}(3\sqrt{2}) - \sqrt{3}(3\sqrt{2})$$
$$= \sqrt{(2)(3)} - \sqrt{(3)(3)} + 3\sqrt{(2)(2)} - 3\sqrt{(3)(2)}$$
$$= \sqrt{6} - \sqrt{9} + 3\sqrt{4} - 3\sqrt{6}$$
$$= \sqrt{6} - 3 + 3(2) - 3\sqrt{6}$$
$$= \sqrt{6} - 3 + 6 - 3\sqrt{6}$$
$$= 3 - 2\sqrt{6}$$

Practice Exercise 9.
$(\sqrt{p} + \sqrt{q})(\sqrt{p} - \sqrt{q})$

EXAMPLE 9 Multiply: $\sqrt{2ab^2}\sqrt{3a^2b}\sqrt{a^2b^3}$ (where $a > 0$, $b > 0$).

SOLUTION:
$$\sqrt{2ab^2}\sqrt{3a^2b}\sqrt{a^2b^3}$$
$$= \sqrt{(2ab^2)(3a^2b)(a^2b^3)} \qquad \text{(Multiplying square roots with positive radicands)}$$
$$= \sqrt{6a^5b^6} \qquad \text{(Simplifying radicand)}$$
$$= \sqrt{6(a^4)(a)(b^6)} \qquad \text{(Rewriting radicand)}$$
$$= \sqrt{6}\sqrt{a^4}\sqrt{a}\sqrt{b^6}$$
$$= \sqrt{6}(a^2)\sqrt{a}(b^3)$$
$$= a^2b^3(\sqrt{6}\sqrt{a})$$
$$= a^2b^3\sqrt{6a}$$

Practice Exercise 10.
$\sqrt{2a^2b}\sqrt{3ab^2}\sqrt{4a^2b^3}$

EXAMPLE 10 Multiply: $\sqrt{\dfrac{p}{q}}\sqrt{\dfrac{q}{p}}\sqrt{\dfrac{p^2}{q^4}}$ (where $p > 0$, $q > 0$).

SOLUTION:
$$\sqrt{\dfrac{p}{q}}\sqrt{\dfrac{q}{p}}\sqrt{\dfrac{p^2}{q^4}} = \sqrt{\left(\dfrac{p}{q}\right)\left(\dfrac{q}{p}\right)\left(\dfrac{p^2}{q^4}\right)}$$
$$= \sqrt{\dfrac{p^3q}{pq^5}}$$
$$= \sqrt{\dfrac{p^2}{q^4}}$$

$$= \sqrt{\left(\frac{p}{q^2}\right)^2}$$
$$= \frac{p}{q^2}$$

ANSWERS TO PRACTICE EXERCISES

1. $\sqrt{21}$.   2. $6\sqrt{6}$.   3. $5\sqrt{2}$.   4. $4\sqrt{xy}$.   5. $b\sqrt{ac}$.   6. $6\sqrt{6}$.   7. $18 - 8\sqrt{15}$.   8. $5a\sqrt{2}$.
9. $p - q$.   10. $2a^2b^3\sqrt{6a}$.

To divide two square roots, we use the quotient rule for square roots and the procedure for rationalizing the denominator.

If $a \geq 0$ and $b > 0$, then $\dfrac{\sqrt{a}}{\sqrt{b}} = \sqrt{\dfrac{a}{b}}$.

Divide and simplify your results. Rationalize all denominators. All variables represent positive numbers.

Practice Exercise 11.
$5\sqrt{12} \div (2\sqrt{3})$

EXAMPLE 11   Divide: $\dfrac{\sqrt{6}}{\sqrt{2}}$.

SOLUTION:
$$\frac{\sqrt{6}}{\sqrt{2}} = \sqrt{\frac{6}{2}}$$
$$= \sqrt{3}$$

Practice Exercise 12.
$2\sqrt{27} \div (3\sqrt{54})$

EXAMPLE 12   Divide: $\dfrac{\sqrt{8}}{\sqrt{3}}$.

SOLUTION:
$$\frac{\sqrt{8}}{\sqrt{3}} = \sqrt{\frac{8}{3}}$$
$$= \sqrt{\frac{8}{3} \times \frac{3}{3}}$$
$$= \sqrt{\frac{24}{9}}$$
$$= \frac{\sqrt{24}}{\sqrt{9}}$$
$$= \frac{\sqrt{(4)(6)}}{3}$$
$$= \frac{2\sqrt{6}}{3}$$

Just as we would divide any binomial by a monomial, we do the same with square roots.

Practice Exercise 13.
$(6\sqrt{8} + 15\sqrt{18}) \div (3\sqrt{2})$

EXAMPLE 13   Divide $(\sqrt{3} - \sqrt{5})$ by $\sqrt{2}$.

15.5 Multiplication and Division of Square Roots

SOLUTION: *(First approach)*

$$\frac{\sqrt{3}-\sqrt{5}}{\sqrt{2}} = \frac{\sqrt{3}}{\sqrt{2}} - \frac{\sqrt{5}}{\sqrt{2}}$$

$$= \sqrt{\frac{3}{2}} - \sqrt{\frac{5}{2}}$$

$$= \sqrt{\frac{3}{2} \times \frac{2}{2}} - \sqrt{\frac{5}{2} \times \frac{2}{2}}$$

$$= \sqrt{\frac{6}{4}} - \sqrt{\frac{10}{4}}$$

$$= \frac{\sqrt{6}}{\sqrt{4}} - \frac{\sqrt{10}}{\sqrt{4}}$$

$$= \frac{\sqrt{6}}{2} - \frac{\sqrt{10}}{2}$$

$$= \frac{1}{2}(\sqrt{6} - \sqrt{10})$$

(Second approach)

$$\frac{\sqrt{3}-\sqrt{5}}{\sqrt{2}} = \frac{\sqrt{3}-\sqrt{5}}{\sqrt{2}} \times \frac{\sqrt{2}}{\sqrt{2}}$$

$$= \frac{(\sqrt{3}-\sqrt{5})\sqrt{2}}{\sqrt{2}\sqrt{2}}$$

$$= \frac{\sqrt{3}\sqrt{2} - \sqrt{5}\sqrt{2}}{2}$$

$$= \frac{\sqrt{6} - \sqrt{10}}{2}$$

> **Practice Exercise 14.**
> $\sqrt{x^4 y^5} \div (\sqrt{x^5 y})$

EXAMPLE 14 Divide: $x\sqrt{xy} \div (y\sqrt{x^3 y})$ (where $x > 0$, $y > 0$).

SOLUTION:
$$x\sqrt{xy} \div (y\sqrt{x^3 y}) = \frac{x\sqrt{xy}}{y\sqrt{x^3 y}}$$

$$= \left(\frac{x}{y}\right)\left(\frac{\sqrt{xy}}{\sqrt{x^3 y}}\right)$$

$$= \left(\frac{x}{y}\right)\left(\sqrt{\frac{xy}{x^3 y}}\right)$$

$$= \left(\frac{x}{y}\right)\sqrt{\frac{1}{x^2}}$$

$$= \left(\frac{x}{y}\right)\sqrt{\left(\frac{1}{x}\right)^2}$$

$$= \left(\frac{x}{y}\right)\left(\frac{1}{x}\right)$$

$$= \left(\frac{\overset{1}{\cancel{x}}}{y}\right)\left(\frac{1}{\underset{1}{\cancel{x}}}\right)$$

$$= \frac{1}{y}$$

Practice Exercise 15.
$(a\sqrt{2} - b\sqrt{3}) \div \sqrt{7}$

EXAMPLE 15 Divide: $\sqrt{\dfrac{x^3}{y}} \div \sqrt{\dfrac{y^5}{x}}$ (where $x > 0$, $y > 0$).

SOLUTION:

$$\sqrt{\dfrac{x^3}{y}} \div \sqrt{\dfrac{y^5}{x}} = \sqrt{\dfrac{x^3}{y}} \sqrt{\dfrac{x}{y^5}} \quad \text{(Multiplying by the reciprocal of the divisor)}$$

$$= \sqrt{\left(\dfrac{x^3}{y}\right)\left(\dfrac{x}{y^5}\right)}$$

$$= \sqrt{\dfrac{x^4}{y^6}}$$

$$= \sqrt{\left(\dfrac{x^2}{y^3}\right)^2}$$

$$= \dfrac{x^2}{y^3}$$

ANSWERS TO PRACTICE EXERCISES

11. 5. **12.** $\dfrac{1}{3}\sqrt{2}$. **13.** 19. **14.** $\dfrac{y^2\sqrt{x}}{x}$. **15.** $\dfrac{a\sqrt{14} - b\sqrt{21}}{7}$.

EXERCISES 15.5

In Exercises 1–40, multiply and simplify your results. All variables represent positive numbers.

1. $\sqrt{2}\sqrt{3}$
2. $\sqrt{3}\sqrt{3}$
3. $\sqrt{3}\sqrt{8}$
4. $\sqrt{5}\sqrt{10}$
5. $\sqrt{7}\sqrt{7}$
6. $\sqrt{8}\sqrt{8}$
7. $(2\sqrt{3})(5\sqrt{3})$
8. $(6\sqrt{2})(4\sqrt{2})$
9. $(4\sqrt{5})(5\sqrt{5})$
10. $(9\sqrt{6})(-7\sqrt{6})$
11. $(3\sqrt{2})(-2\sqrt{3})$
12. $(3\sqrt{5})(-3\sqrt{2})$
13. $(4\sqrt{3})(-2\sqrt{3})$
14. $(2\sqrt{3})(\sqrt{12})$
15. $(2\sqrt{3})(-3\sqrt{2})(\sqrt{6})$
16. $(\sqrt{2})(-\sqrt{3})(2\sqrt{4})$
17. $\sqrt{2}(\sqrt{3} - \sqrt{5})$
18. $2\sqrt{3}(\sqrt{7} - 3\sqrt{5})$
19. $3\sqrt{7}(4\sqrt{7} - \sqrt{2})$
20. $\sqrt{5}(2\sqrt{10} - 3\sqrt{2})$
21. $(\sqrt{2} + \sqrt{3})(2\sqrt{2} - \sqrt{3})$
22. $(\sqrt{5} - \sqrt{7})(\sqrt{5} + \sqrt{7})$
23. $(\sqrt{3} - \sqrt{6})(2 - \sqrt{2})$
24. $(3 - \sqrt{3})(2 + 4\sqrt{3})$
25. $\sqrt{3x}\sqrt{2x}$
26. $\sqrt{x^2y}\sqrt{xy^4}$
27. $\sqrt{2x}\sqrt{4y}$
28. $\sqrt{\dfrac{x}{y}}\sqrt{\dfrac{4y}{x}}$
29. $\sqrt{a}(2\sqrt{a} - 3\sqrt{b})$
30. $\sqrt{a}(\sqrt{ab} - a\sqrt{b})$
31. $(\sqrt{a} + \sqrt{b})(\sqrt{a} - \sqrt{b})$
32. $\sqrt{20ab^2c}\sqrt{5a^3bc}$

15.5 Multiplication and Division of Square Roots

33. $\sqrt{2x^2y}\sqrt{3xy^2}\sqrt{4x^2y^3}$

34. $\sqrt{\dfrac{a}{b}}\sqrt{\dfrac{b}{a}}\sqrt{\dfrac{a^2}{b^4}}$

35. $s\sqrt{2}\sqrt{3s^3}\sqrt{2s^2}$

36. $s\sqrt{3}(\sqrt{3s} - \sqrt{2s^3})$

37. $\sqrt{\dfrac{a^2b}{c}}\sqrt{\dfrac{a}{bc^2}}\sqrt{\dfrac{bc}{a^5}}$

38. $(\sqrt{st} + \sqrt{t})(\sqrt{s} - \sqrt{st})$

39. $(a\sqrt{b} - b\sqrt{a})(a^2\sqrt{b} + b^2\sqrt{a})$

40. $\left(\dfrac{\sqrt{x}}{y} - \dfrac{\sqrt{y}}{x}\right)\left(\dfrac{x}{\sqrt{y}} + \dfrac{y}{\sqrt{x}}\right)$

In Exercises 41–70, divide and simplify your results. Rationalize all denominators. All variables represent positive numbers.

41. $5\sqrt{3} \div (2\sqrt{3})$

42. $5\sqrt{7} \div (3\sqrt{7})$

43. $6\sqrt{8} \div (5\sqrt{8})$

44. $11\sqrt{10} \div (12\sqrt{10})$

45. $5\sqrt{27} \div (2\sqrt{3})$

46. $6\sqrt{12} \div (3\sqrt{3})$

47. $16\sqrt{98} \div (8\sqrt{2})$

48. $3\sqrt{54} \div (2\sqrt{27})$

49. $-6\sqrt{90} \div (3\sqrt{18})$

50. $9\sqrt{20} \div (4\sqrt{10})$

51. $6\sqrt{7} \div (3\sqrt{28})$

52. $5\sqrt{3} \div (8\sqrt{12})$

53. $a\sqrt{ab} \div (b\sqrt{a^3b})$

54. $\sqrt{x^2y} \div (2\sqrt{xy^4})$

55. $y\sqrt{xy} \div (\sqrt{2xy})$

56. $5\sqrt{16x^5y^2} \div (3\sqrt{2xy^4})$

57. $\sqrt{p^6q^7} \div (\sqrt{p^7q})$

58. $\sqrt{\dfrac{x}{y}} \div \sqrt{\dfrac{y}{x}}$

59. $\dfrac{3\sqrt{8} - 9\sqrt{18}}{3\sqrt{2}}$

60. $\dfrac{8\sqrt{27} - 6\sqrt{12}}{2\sqrt{3}}$

61. $\dfrac{\sqrt{16} + 2\sqrt{54}}{\sqrt{2}}$

62. $\dfrac{5\sqrt{9} - 2\sqrt{243}}{3\sqrt{3}}$

63. $\dfrac{7\sqrt{15} - 9\sqrt{10}}{2\sqrt{5}}$

64. $\dfrac{2\sqrt{24} + 5\sqrt{32}}{3\sqrt{8}}$

65. $\dfrac{a\sqrt{8} + b\sqrt{12}}{c\sqrt{2}}$

66. $\dfrac{c\sqrt{32} - d\sqrt{56}}{b\sqrt{8}}$

67. $\dfrac{x\sqrt{2} - y\sqrt{3}}{\sqrt{5}}$

68. $\dfrac{s^2\sqrt{5} + st\sqrt{6}}{t\sqrt{3}}$

69. $\dfrac{\sqrt{128} - 2\sqrt{32} - 3\sqrt{80}}{5\sqrt{32}}$

70. $\dfrac{3\sqrt{15} + \sqrt{20} - 5\sqrt{60}}{2\sqrt{150}}$

15.6 Equations Involving Square Roots

OBJECTIVE: After completing this section, you should be able to solve certain equations containing square roots.

We will conclude this chapter by considering some simple equations that involve square roots. Recall that $\sqrt{x}$ is meaningful only if $x \geq 0$.

The procedure used in solving equations involving square roots depends upon the fact that if two numbers are equal, then their squares are also equal. That is,

$$\text{if } a = b, \text{ then } a^2 = b^2$$

However, if the squares of numbers are equal, the numbers themselves may not be equal. For instance, we know that $25 = 25$. But 25 is the square of both -5 and 5. Hence,

$$\text{if } a^2 = b^2, \text{ then } a = b \text{ or } a = -b$$

Therefore, when solving equations involving square roots, it is very important and necessary to check your results in the *original* equation.

Solve for the indicated variables. Check your results in the original equations.

Practice Exercise 1. $2\sqrt{y} = 5$

EXAMPLE 1 Solve the equation $\sqrt{x} = 2$ for x.

SOLUTION: Solving the equation $\sqrt{x} = 2$ for x, we have

$$\sqrt{x} = 2$$
$$(\sqrt{x})^2 = 2^2 \quad \text{(Squaring both sides)}$$
$$x = 4 \quad \text{(The solution?)}$$

CHECK: Checking in the *original* equation, we have

$$\sqrt{x} = 2$$
$$\sqrt{4} \; ? \; 2$$
$$2 = 2 \checkmark$$

Therefore, 4 is the required solution.

Practice Exercise 2. $\sqrt{4 - 2t} = 7$

EXAMPLE 2 Solve the equation $\sqrt{y - 1} = 3$ for y.

SOLUTION: Solving the equation $\sqrt{y - 1} = 3$ for y, we have

$$\sqrt{y - 1} = 3$$
$$(\sqrt{y - 1})^2 = 3^2 \quad \text{(Squaring both sides)}$$
$$y - 1 = 9$$
$$y = 10 \quad \text{(The solution?)}$$

CHECK: Checking in the *original* equation, we have

$$\sqrt{y - 1} = 3$$
$$\sqrt{10 - 1} \; ? \; 3$$

15.6 Equations Involving Square Roots

$$\sqrt{9} \stackrel{?}{=} 3$$
$$3 = 3 \checkmark$$

Therefore, 10 is the required solution.

Practice Exercise 3.
$\sqrt{2-u} + 3 = 0$

EXAMPLE 3 Solve the equation $\sqrt{2u - 3} + 1 = 0$ for u.

SOLUTION: Solving the equation $\sqrt{2u - 3} + 1 = 0$ for u, we have

$$\sqrt{2u - 3} + 1 = 0$$

Before squaring both sides of the equation, we will first isolate the square root term by subtracting 1 from both sides of the equation:

$$\sqrt{2u - 3} = -1$$
$$(\sqrt{2u - 3})^2 = (-1)^2$$
$$2u - 3 = 1$$
$$2u = 4$$
$$u = 2 \quad \text{(The solution?)}$$

CHECK: Checking in the *original* equation, we have

$$\sqrt{2u - 3} + 1 = 0$$
$$\sqrt{2(2) - 3} + 1 \stackrel{?}{=} 0$$
$$\sqrt{4 - 3} + 1 \stackrel{?}{=} 0$$
$$\sqrt{1} + 1 \stackrel{?}{=} 0$$
$$1 + 1 \stackrel{?}{=} 0$$
$$2 \neq 0$$

which does *not* check. Therefore, 2 is *not* the solution for the given equation. There are no solutions for the equation. Notice that the equation $\sqrt{2u - 3} + 1 = 0$ is equivalent to the equation $\sqrt{2u - 3} = -1$, but the square root of a number cannot be negative.

CAUTION Always check your results in the *original* equation.

Practice Exercise 4.
$4 - \sqrt{2s + 1} = 0$

EXAMPLE 4 Solve the equation $7 - \sqrt{1 + t} = 4$ for t.

SOLUTION: Solving the equation $7 - \sqrt{1 + t} = 4$ for t, we first isolate the square root term. We have

$$7 - \sqrt{1 + t} = 4$$
$$-\sqrt{1 + t} = -3$$
$$\sqrt{1 + t} = 3$$
$$(\sqrt{1 + t})^2 = (3)^2$$
$$1 + t = 9$$
$$t = 8$$

CHECK: Checking in the *original* equation, we have

$$7 - \sqrt{1 + t} = 4$$
$$7 - \sqrt{1 + 8} \stackrel{?}{=} 4$$

$$7 - \sqrt{9} \; ? \; 4$$
$$7 - 3 \; ? \; 4$$
$$4 = 4 \checkmark$$

Therefore, 8 is the required solution.

Practice Exercise 5. $\sqrt{1 - 3x} - \sqrt{2x - 1} = 0$

EXAMPLE 5 Solve the equation $\sqrt{2x - 1} = \sqrt{x + 2}$ for x.

SOLUTION: Solving the equation $\sqrt{2x - 1} = \sqrt{x + 2}$ for x, we have

$$\sqrt{2x - 1} = \sqrt{x + 2}$$
$$(\sqrt{2x - 1})^2 = (\sqrt{x + 2})^2$$
$$2x - 1 = x + 2$$
$$x = 3 \quad \text{(The solution?)}$$

CHECK: Checking in the *original* equation, we have

$$\sqrt{2x - 1} = \sqrt{x + 2}$$
$$\sqrt{2(3) - 1} \; ? \; \sqrt{3 + 2}$$
$$\sqrt{6 - 1} \; ? \; \sqrt{5}$$
$$\sqrt{5} = \sqrt{5} \checkmark$$

Therefore, 3 is the required solution.

ANSWERS TO PRACTICE EXERCISES

1. $\frac{25}{4}$. 2. $\frac{-45}{2}$. 3. No solution. 4. $\frac{15}{2}$. 5. No solution.

EXERCISES 15.6

Solve each of the following equations for the indicated variables. Check your results in the original equations.

1. $\sqrt{x} = 3$
2. $\sqrt{y} = 6$
3. $3\sqrt{u} = 12$
4. $-2\sqrt{w} = -6$
5. $\sqrt{x - 1} = 5$
6. $\sqrt{2 - y} = 7$
7. $\sqrt{1 - 2u} = 0$
8. $\sqrt{3p - 1} = -1$
9. $2 - \sqrt{3x - 1} = 0$
10. $3 - \sqrt{1 - 4y} = 1$
11. $3\sqrt{x} = 2\sqrt{x}$
12. $5\sqrt{y} = 3\sqrt{y} + 3$
13. $2\sqrt{3u} = 4\sqrt{3u} - 1$
14. $\sqrt{2y} + 1 = 2\sqrt{2y}$
15. $\sqrt{1 - 3u} = \sqrt{4u + 3}$
16. $\sqrt{3p - 2} = 2\sqrt{p + 1}$
17. $\sqrt{2q - 1} - \sqrt{1 - 3q} = 0$
18. $\sqrt{3t - 1} + \sqrt{2t + 3} = 0$
19. $\frac{1}{2}\sqrt{u - 1} = \frac{2}{3}\sqrt{u + 3}$
20. $\frac{1}{7}\sqrt{1 - 3p} = \frac{2}{5}\sqrt{2p - 1}$

21. $2\sqrt{a-2} - 7 = 13$
22. $5\sqrt{b+1} + 2 = 17$
23. $0.6 - \sqrt{2y+7} = 0.1$
24. $0.4 + \sqrt{6-5u} = 0.9$
25. $\frac{1}{5}\sqrt{5x-1} = \frac{1}{3}\sqrt{1-3x}$
26. $\frac{2}{3}\sqrt{1-2y} - \frac{1}{4}\sqrt{2y+3} = 0$
27. $-3\sqrt{2p+1} = -4\sqrt{4p-3}$
28. $\frac{1}{9}\sqrt{1+4y} = \frac{1}{7}\sqrt{2q+1}$
29. $\dfrac{\sqrt{1-x}}{5} = \dfrac{\sqrt{2x-1}}{3}$
30. $\dfrac{2\sqrt{2-y}}{3} = \dfrac{3\sqrt{y+1}}{7}$

Chapter Summary

In this chapter, we discussed square roots and operations with them. This included combining square roots by addition and subtraction, and also multiplying and dividing square roots. Rationalizing the denominator was also discussed. The chapter concluded with a brief examination of the solution of simple equations involving square roots.

The **key words** introduced in this chapter are listed in the order in which they appeared in the text. You should know what they mean before proceeding to the next chapter.

radical sign (p. 624)
radicand (p. 624)
square root (p. 624)

principal square root (p. 625)
rationalizing the denominator (p. 633)

like square roots (p. 636)

Review Exercises

All variables used in these exercises represent positive numbers.

15.1 Square Roots of Nonnegative Numbers

In Exercises 1–8, determine the principal square root of the indicated expression.

1. 49
2. 256
3. x^6
4. $36y^4$
5. $(y+3)^2$
6. $25a^4b^6$
7. $(abc)^8$
8. $\dfrac{(x+y)^4}{4}$

In Exercises 9–14, determine the value of the given expression, if possible. If the expression is meaningless, indicate why.

9. $\sqrt{81}$
10. $-\sqrt{49}$
11. $\sqrt{-25}$
12. $\sqrt{x^6}$
13. $-\sqrt{y^4}$
14. $\sqrt{-xy}$

15.2 Simplifying Square Roots of Expressions without Fractions

Simplify each of the given expressions.

15. $\sqrt{18}$
16. $\sqrt{72}$
17. $\sqrt{280}$
18. $\sqrt{x^5}$
19. $\sqrt{9y^6}$
20. $\sqrt{4x^2y^3}$
21. $\sqrt{100a^4b^5}$
22. $\sqrt{3p^3q^4t^5}$

15.3 Simplifying Square Roots of Expressions with Fractions

Simplify each of the given expressions. Rationalize all denominators.

23. $\sqrt{\dfrac{25}{81}}$
24. $\sqrt{\dfrac{3}{2}}$
25. $\dfrac{\sqrt{5}}{\sqrt{7}}$
26. $\sqrt{\dfrac{16}{y^6}}$
27. $\sqrt{\dfrac{9xy^2}{25}}$
28. $\sqrt{\dfrac{3y}{2x}}$
29. $\dfrac{\sqrt{2a}}{\sqrt{3b}}$
30. $\dfrac{p}{\sqrt{pq}}$

15.4 Addition and Subtraction of Square Roots

Simplify each of the given expressions by combining like terms.

31. $3\sqrt{5} - 4\sqrt{5} + 7\sqrt{5}$
32. $2\sqrt{2} + \frac{1}{2}\sqrt{2} - \frac{2}{3}\sqrt{2}$
33. $2a\sqrt{2} - 3b\sqrt{2} + 4c\sqrt{2}$
34. $3\sqrt{50x} - a\sqrt{32x} + b\sqrt{2x}$
35. $\sqrt{a^4 b} - 2a\sqrt{b^5} - 5\sqrt{a^2 b^3}$
36. $\sqrt{\frac{x}{y}} + \sqrt{\frac{y}{x}} - \sqrt{xy}$

15.5 Multiplication and Division of Square Roots

Perform the indicated operations and simplify your results. Rationalize all denominators.

37. $(3\sqrt{2})(-4\sqrt{2})(2\sqrt{3})$
38. $(-2\sqrt{3})(3\sqrt{2})(-4\sqrt{2})$
39. $3\sqrt{5}(\sqrt{6} + 2\sqrt{5})$
40. $(\sqrt{2} - \sqrt{3})(\sqrt{3} + \sqrt{5})$
41. $25\sqrt{18} \div (5\sqrt{2})$
42. $\sqrt{x^3 y^2} \div (2\sqrt{xy})$
43. $(9\sqrt{8} - 6\sqrt{18}) \div (3\sqrt{2})$
44. $(a\sqrt{27} - b\sqrt{12}) \div (ab\sqrt{3})$

15.6 Equations Involving Square Roots

Solve each of the following equations for the indicated variable. Check your results in the original equation.

45. $\sqrt{3x} = 5$
46. $\sqrt{2y - 1} = 3$
47. $\sqrt{1 - 3p} + 4 = 0$
48. $3\sqrt{t} = 7\sqrt{t} - 2$
49. $\sqrt{2u - 4} = \sqrt{5 - 3u}$
50. $\frac{1}{4}\sqrt{3s + 1} = \frac{1}{5}\sqrt{3 - 4s}$

1. In general, $\sqrt{a + b} \neq \sqrt{a} + \sqrt{b}$. Under what conditions, if any, does the equal sign hold?
2. If $p < 0$, under what conditions, if any, does $\sqrt{pq}$ represent a real number?
3. a. Evaluate: $\sqrt{(-4)^2}$.
   b. If $u < 0$, evaluate: $\sqrt{u^2}$.
4. Evaluate: $\sqrt{11 + \sqrt{25}}$.
5. Evaluate: $\sqrt{a^2 \sqrt{a^2 b^2}}$.

In Exercises 6–10, solve the given equations for the indicated variables.

6. $0.2\sqrt{1 - x} = 0.1$
7. $\sqrt{0.1 - 0.3u} = \sqrt{0.4u + 0.3}$
8. $\frac{5}{4}\sqrt{3 + 2r} - \frac{10}{3}\sqrt{1 - 2r} = 0$
9. $\sqrt{2p - 4} = \sqrt{5 - 3p} - 2\sqrt{2p - 4}$
10. $0.75\sqrt{3t + 1} = 0.2\sqrt{3 - 4t} + 0.5\sqrt{3t + 1}$

CHAPTER 15 TEST

This test covers materials from Chapter 15. Read each question carefully and answer all questions. If you miss a question, review the appropriate section of the text.

For this test, all variables used represent positive numbers. In Exercises 1–6, determine the principal square root of the indicated expression.

1. 64
2. y^6
3. $49p^4$
4. $36a^4b^8$
5. $\dfrac{(x-y)^6}{9}$
6. $(a^2bc)^{10}$

In Exercises 7–15, simplify each of the given expressions.

7. $\sqrt{x^8}$
8. $-\sqrt{y^4}$
9. $\sqrt{98}$
10. $\sqrt{16x^3y^2}$
11. $\sqrt{\dfrac{36}{121}}$
12. $\sqrt{\dfrac{5}{2}}$
13. $\sqrt{\dfrac{81}{z^6}}$
14. $\sqrt{\dfrac{8xy}{49}}$
15. $\sqrt{\dfrac{2x^2}{3y}}$

In Exercises 16–23, simplify each of the given expressions.

16. $2\sqrt{5} - 5\sqrt{5} + 8\sqrt{5}$
17. $2\sqrt{50y} - \sqrt{2y} + a\sqrt{32y}$
18. $\sqrt{x^4y} - 2a\sqrt{y^5} - 5\sqrt{x^2y^3}$
19. $\sqrt{\dfrac{y}{x}} - \sqrt{\dfrac{x}{y}} + \sqrt{xy}$
20. $(2\sqrt{3})(-3\sqrt{2})(4\sqrt{3})$
21. $36\sqrt{18} \div (9\sqrt{2})$
22. $\sqrt{a^3b^2} \div (2\sqrt{ab})$
23. $(\sqrt{5} - \sqrt{2})(\sqrt{3} + \sqrt{2})$

In Exercises 24–25, solve the given equations for the indicated variable. Check your results.

24. $\sqrt{3t-1} = 2$
25. $2\sqrt{3x-1} = \sqrt{2-5x}$

CUMULATIVE REVIEW: CHAPTERS 1–15

1. State which property of arithmetic operations is illustrated by the following: $2(3 + 6) = 2(3) + 2(6)$
2. What, if any, is the multiplicative identity element for the integers?
3. The operation of multiplication can be distributed over which two other operations?
4. If a and b represent different numbers, which is greater, the GCF or the LCM of the two numbers?
5. Determine the complete prime factorization of 492.
6. Translate the following into a mathematical expression: the cube of the difference of x and y.
7. How many minutes are there in the month of June?
8. The base of an isosceles triangle is 4.7 m long. The perimeter of the triangle is 23.9 m. Determine the measure of each of the other two sides.
9. Rewrite 11.06 as a simplified fraction.
10. Rewrite 1.04% as a simplified fraction.
11. Simplify: $\dfrac{19}{25} \div \left(\dfrac{2}{3} - \dfrac{11}{15}\right)$.
12. Simplify and write your results without negative exponents: $(2a^0 b^{-3} c^2)^{-1} \div (-3a^2 bc^{-4})^2$.
13. Factor completely: $2u^2 - u - 21$.
14. Factor completely: $16p^4 - 81q^4$.
15. Factor completely: $2t^3 + 3t^2 - 4t - 6$.
16. Simplify: $3\sqrt{50} - 4\sqrt{18}$.
17. Simplify: $4a\sqrt{a^3 b^5} \div (3b\sqrt{ab})$.

In Exercises 18–20, solve the given equations for the indicated variables.

18. $2(x - 1) + 3 = 5 - 3(2 - 5x)$
19. $2s^3 + s^2 - 6s = 0$
20. $\sqrt{2y + 5} = \sqrt{3 - y}$
21. Round 28.962 to the nearest tenth.
22. Rewrite 262,090,700 in scientific notation.
23. Solve the proportion $\dfrac{-11}{23} = \dfrac{s}{69}$ for s.
24. The area of a region in the shape of a parallelogram is $252\dfrac{12}{35}$ in.². If the base of the region is $19\dfrac{1}{5}$ in., determine its height.
25. Solve the inequality $2x - 3y > 4$ for y.

CHAPTER 16: Quadratic Equations and the Pythagorean Theorem

In previous chapters, you learned to solve linear equations in one or two variables. In Section 11.8, you learned to solve equations using the method of factoring. In this chapter, you will learn to solve quadratic equations in a single variable. The methods used will include factoring, using square roots, completing the square, and using the quadratic formula.

We will also continue our discussion of solving word problems. However, in this chapter, the equations used with these applications will be quadratic equations.

The chapter concludes with the Pythagorean theorem, which covers measurement, square roots, and quadratic equations.

CHAPTER 16 PRETEST

This pretest covers material on quadratic equations and the Pythagorean theorem. Read each question carefully, and try to answer all questions. Each question is keyed to the chapter section in which the particular topic is discussed. Check your answers with those given in the back of the text.

Questions 1–2 pertain to Section 16.1.

1. Rewrite the equation $3y^2 - 7 = 6y$ in its standard form.

2. Rewrite the equation $(3u - 1)^2 = 6u + 5$ in its standard form.

Questions 3–6 pertain to Section 16.2.

Solve each of the following equations by the method of factoring.

3. $x^2 + 4x = 21$

4. $2y^2 = 6 - 11y$

5. $7(u^2 - 3) = 20u - 18$

6. $6t^2 + 8t = 28 + 21t$

Questions 7–10 pertain to Section 16.3.

Solve each of the following equations using square roots.

7. $x^2 - 81 = 0$

8. $5y^2 - 125 = 0$

9. $(t - 4)^2 = 16$

10. $(2p - 3)^2 = 81$

Questions 11–14 pertain to Section 16.4

11. What must be added to complete the square for $t^2 + 8t$?

12. What must be added to complete the square for $u^2 + 7u$?

In Exercises 13–14, solve the given equations using the method of completing the square.

13. $x^2 + 5x - 2 = 0$

14. $2u - 5u^2 + 7 = 0$

Questions 15–17 pertain to Section 16.5

15. Compute the discriminant for the equation $4r^2 - 5r = 7$.

In Exercises 16–17, solve the equations using the quadratic formula.

16. $3y^2 + 4y - 1 = 0$

17. $2q - 3q^2 + 7 = 0$

Question 18 pertains to Section 16.6.

18. The length of a rectangular region is 3 cm less than twice its width. The area of the region is 170 cm². Using a quadratic equation, determine the dimensions of the region.

Questions 19–20 pertain to Section 16.7.

19. Let $a = 7$ cm and $b = 11$ cm be the lengths of the legs of a right triangle. Determine the length, c, of the hypotenuse of the triangle.

20. A triangle has sides with measures of 8 ft, 10 ft, and 14 ft. Using the Pythagorean theorem, determine if the triangle is a right triangle.

16.1 The Standard Form of a Quadratic Equation

OBJECTIVE

After completing this section, you should be able to rewrite a quadratic equation in its standard form.

In Chapter 8, you learned to solve simple linear equations in one variable. A linear equation is a linear polynomial set equal to zero. In this chapter, you will learn to solve quadratic equations.

DEFINITION

A **quadratic equation in a single variable** is an equation that can be written as a 2nd-degree polynomial in that variable, set equal to 0.

- $2x^2 - 3x + 1 = 0$ is a quadratic equation in the variable x.
- $3y^2 = 5 + 2y$ is a quadratic equation in the variable y.
- $\frac{2}{3}u^2 - \frac{1}{4}u = \frac{1}{7}$ is a quadratic equation in the variable u.
- $5 - 2r + 3r^2 = 0$ is a quadratic equation in the variable r.

Most of the time, we will write quadratic equations in the standard form.

DEFINITION

The **standard form for a quadratic equation** in the variable x is

$$ax^2 + bx + c = 0$$

where a, b, and c are constants and $a \neq 0$.

In the equation $ax^2 + bx + c = 0$, the requirement that $a \neq 0$ ensures that the polynomial will be 2nd degree; a is called the **leading coefficient**. The other coefficients are b (the coefficient of the linear term) and c (the coefficient of the 0-degree or constant term). It is customary to write the standard form of a quadratic equation such that (a) the leading coefficient is positive, and (b) all coefficients are integers.

Rewrite each equation in its standard form.

Practice Exercise 1.
$2y^2 + 5 = 3y$

EXAMPLE 1 Rewrite the equation $2x - 3x^2 + 7 = 0$ in standard form.

SOLUTION: The equation $2x - 3x^2 + 7 = 0$ can be rewritten in standard form as $3x^2 - 2x - 7 = 0$ with $a = 3$, $b = -2$, and $c = -7$.

Practice Exercise 2.
$3 - 4w^2 = 7w$

EXAMPLE 2 Rewrite the equation $2y^2 = 3 - 4y$ in standard form.

SOLUTION: The equation $2y^2 = 3 - 4y$ can be rewritten in standard form as $2y^2 + 4y - 3 = 0$ with $a = 2$, $b = 4$, and $c = -3$.

16.1 The Standard Form of a Quadratic Equation

Practice Exercise 3.
$2(p + 1) = p(p - 3)$

EXAMPLE 3 Rewrite the equation $\frac{1}{2} - \frac{2}{3}u^2 = \frac{1}{5}u$ in standard form.

SOLUTION: The equation $\frac{1}{2} - \frac{2}{3}u^2 = \frac{1}{5}u$ can be rewritten in standard form as $\frac{2}{3}u^2 + \frac{1}{5}u - \frac{1}{2} = 0$. Multiplying both sides of the equation by 30 (the LCM of 2, 3, and 5), we have

$$30\left(\frac{2}{3}u^2 + \frac{1}{5}u - \frac{1}{2}\right) = 30(0)$$

$$30\left(\frac{2}{3}u^2\right) + 30\left(\frac{1}{5}u\right) - 30\left(\frac{1}{2}\right) = 0$$

$$20u^2 + 6u - 15 = 0$$

This is the standard form of the equation with *integer* coefficients; $a = 20$, $b = 6$, and $c = -15$.

Practice Exercise 4.
$\frac{x^2 - 1}{3} = \frac{2 - 3x}{4}$

EXAMPLE 4 Rewrite the equation $y(y - 4) = 6$ in standard form.

SOLUTION: Rewriting the equation $y(y - 4) = 6$ in standard form, we have

$$y(y - 4) = 6$$
$$y^2 - 4y = 6$$
$$y^2 - 4y - 6 = 0$$

which is in standard form with $a = 1$, $b = -4$, and $c = -6$.

ANSWERS TO PRACTICE EXERCISES

1. $2y^2 - 3y + 5 = 0$. **2.** $4w^2 + 7w - 3 = 0$. **3.** $p^2 - 5p - 2 = 0$. **4.** $4x^2 + 9x - 10 = 0$.

EXERCISES 16.1

Rewrite each of the following equations in its standard form.

1. $3x^2 - 2x = 4$
2. $2x^2 = 5 - x$
3. $y - 4 = y^2$
4. $5 - 2y^2 = 3y$
5. $u(u + 2) = 3$
6. $(u + 2)(u - 3) = 0$
7. $(2r - 1)(3r + 2) = 0$
8. $(1 - 3r)(2 + r) = 0$
9. $2t(t - 1) = t(t + 2)$
10. $3t(t + 1) - 2(t + 4) = t^2$
11. $p^2 - 2p(3p - 3) = 6$
12. $3p^2 - 4 = 2p(1 - 3p)$
13. $x^2 - \frac{2}{3}x + \frac{1}{2} = 0$
14. $\frac{1}{4}x^2 - 2 = \frac{3}{4}x$
15. $\frac{1}{5}y(y + 2) = \frac{2}{3}$
16. $\frac{1}{9}(1 - y) = \frac{1}{3}(y^2 - 1)$
17. $\frac{u - 3}{4} = \frac{2 - 3u^2}{5}$
18. $\frac{u^2 - 2u + 1}{3} = \frac{2u + 1}{4}$
19. $\frac{1}{7}(s^2 - 1) + \frac{2}{3}(4 - s) = 2s^2$
20. $\frac{1}{5}(2s - 3) - \frac{4}{3}(s - s^2) = 2 + 4s^2$

16.2 Solving Quadratic Equations by Factoring

OBJECTIVE

After completing this section, you should be able to solve some quadratic equations by the method of factoring.

In Section 11.8, you learned to solve equations using the method of factoring. In this section, we will review the procedure and consider some special cases for quadratic equations.

Solving Quadratic Equations by Factoring

To **solve a quadratic equation by the method of factoring:**
1. (Re)write the equation in its standard form.
2. Factor the polynomial part, if possible.
3. Set each factor equal to 0 and solve the resulting equations.

Solve each equation by the method of factoring.

Practice Exercise 1.
$2w^2 - w = 6$

EXAMPLE 1 Solve the equation $2x^2 = 3 - 5x$ for x.

SOLUTION:

STEP 1: Rewrite the equation in its standard form:

$$2x^2 + 5x - 3 = 0$$

STEP 2: Factor the polynomial:

$$2x^2 + 5x - 3 = (2x - 1)(x + 3)$$

STEP 3: Set each factor equal to 0 and solve the resulting equations:

$$2x^2 + 5 - 3 = 0$$
$$(2x - 1)(x + 3) = 0$$
$$2x - 1 = 0 \quad \text{or} \quad x + 3 = 0$$
$$2x = 1 \qquad\qquad x = -3$$
$$x = \frac{1}{2}$$

Hence, the required solutions are $x = \frac{1}{2}$ and $x = -3$. You are encouraged to check these results in the original equation. Both values must be checked individually.

Practice Exercise 2.
$\dfrac{2}{y+1} = 1 + \dfrac{y-1}{3}$

EXAMPLE 2 Solve the equation $\dfrac{-6}{y-3} + \dfrac{y+3}{3} = 1$ for y.

SOLUTION:

STEP 1: Rewrite the equation in its standard form:

$$\frac{-6}{y-3} + \frac{y+3}{3} = 1 \qquad \text{(Excluded value is } y = 3.\text{)}$$

16.2 Solving Quadratic Equations by Factoring

$$3(y-3)\left(\frac{-6}{y-3} + \frac{y+3}{3}\right) = 3(y-3)(1) \quad [\text{LCM} = 3(y-3)]$$
$$3(-6) + (y+3)(y-3) = 3(y-3)$$
$$-18 + y^2 - 9 = 3y - 9$$
$$y^2 - 27 = 3y - 9$$
$$y^2 - 3y - 18 = 0$$

STEP 2: Factor the polynomial:

$$y^2 - 3y - 18 = (y-6)(y+3)$$

STEP 3: Set each factor equal to 0 and solve the resulting equations:

$$y^2 - 3y - 18 = 0$$
$$(y-6)(y+3) = 0$$
$$y - 6 = 0 \quad \text{or} \quad y + 3 = 0$$
$$y = 6 \qquad\qquad y = -3$$

Hence, the required solutions are $y = -3$ and $y = 6$. Check each solution in the *original* equation.

ANSWERS TO PRACTICE EXERCISES

1. $w = \frac{-3}{2}, 2$.   2. $y = -4, 1$.

There are special cases of a quadratic equation. For instance, in the equation $ax^2 + bx + c = 0$, a cannot be 0, but b or c or both can be 0. If $c = 0$, then the equation $ax^2 + bx + c = 0$ becomes

$$ax^2 + bx = 0$$

which, upon factoring, is equivalent to

$$x(ax + b) = 0$$

and the solutions are $x = 0$ and $x = \frac{-b}{a}$ obtained as follows:

$$x(ax+b) = 0$$
$$x = 0 \quad \text{or} \quad ax + b = 0$$
$$ax - b$$
$$x = \frac{-b}{a} \quad (\text{Since } a \neq 0)$$

Solve each equation by factoring.

Practice Exercise 3.
$2p - 5p^2 = 0$

EXAMPLE 3 Solve the equation $3y^2 - 6y = 0$ for y.

SOLUTION: Solving the equation $3y^2 - 6y = 0$ for y, we have

$$3y^2 - 6y = 0$$
$$3y(y-2) = 0$$
$$3y = 0 \quad \text{or} \quad y - 2 = 0$$
$$y = 0 \qquad\qquad y = 2$$

Hence, the required solutions for the given equation are $y = 0$ and $y = 2$. You should check these solutions in the original equation. Also note that the solution $y = 2$ is of the form $y = \dfrac{-b}{a}$ with $a = 3$ and $b = -6$.

For the following examples, the check of solutions is left to you.

Practice Exercise 4. $6w^2 = 3w$

EXAMPLE 4 Solve the equation $4u^2 = 5u$ for u.

SOLUTION: Solving the equation $4u^2 = 5u$ for u, we have

$$4u^2 = 5u$$
$$4u^2 - 5u = 0$$
$$u(4u - 5) = 0$$
$$u = 0 \quad \text{or} \quad 4u - 5 = 0$$
$$4u = 5$$
$$u = \frac{5}{4}$$

Hence, the required solutions for the given equation are $u = 0$ and $u = \dfrac{5}{4}$. Again, note that the solution $u = \dfrac{5}{4}$ is of the form $u = \dfrac{-b}{a}$ with $a = 4$ and $b = -5$.

Another special type of quadratic equation is the equation of the type $ay^2 + by + c = 0$ with $b = 0$. Hence, the equation $ay^2 + by + c = 0$ becomes $ay^2 + c = 0$. If either a or c is negative (but not both a and c are negative), then this equation can be solved by factoring.

Practice Exercise 5
$3t(t - 4) = 2t(t - 1)$

EXAMPLE 5 Solve the equation $u^2 - 25 = 0$ for u.

SOLUTION: Solving the equation $u^2 - 25 = 0$ for u, we have

$$u^2 - 25 = 0 \quad (a > 0, b = 0, c < 0)$$
$$(u + 5)(u - 5) = 0 \quad \text{(Factoring)}$$
$$u + 5 = 0 \quad \text{or} \quad u - 5 = 0$$
$$u = -5 \qquad\qquad u = 5$$

Hence, the required solutions for the given equation are $u = -5$ and $u = 5$.

Practice Exercise 6.
$2x^2 - 50 = 0$

EXAMPLE 6 Solve the equation $4t^2 = 9$ for t.

SOLUTION: Solving the equation $4t^2 = 9$ for t, we have

$$4t^2 = 9$$
$$4t^2 - 9 = 0 \quad (a > 0, b = 0, c < 0)$$
$$(2t + 3)(2t - 3) = 0$$
$$2t + 3 = 0 \quad \text{or} \quad 2t - 3 = 0$$
$$2t = -3 \qquad\qquad 2t = 3$$
$$t = \frac{-3}{2} \qquad\qquad t = \frac{3}{2}$$

Hence, the required solutions are $t = \dfrac{-3}{2}$ and $t = \dfrac{3}{2}$.

ANSWERS TO PRACTICE EXERCISES

3. $p = 0, \frac{2}{5}$. 4. $w = 0, \frac{1}{2}$. 5. $t = 0, 10$. 6. $x = -5, 5$.

EXERCISES 16.2

Solve each of the following equations using the method of factoring.

1. $2x^2 + x - 3 = 0$
2. $1 = 7x + 8x^2$
3. $x^2 - 81 = 0$
4. $x^2 - 121 = 0$
5. $3y^2 + 5y = 0$
6. $6y^2 = 7y$
7. $\frac{3}{4}p^2 = 5p$
8. $\frac{1}{3}p^2 = \frac{7}{4}p$
9. $y^2 + 12 = 7y$
10. $6y^2 + 9y = 8y + 12$
11. $\frac{1}{3}t^2 + 3 = \frac{5}{2}t$
12. $\frac{2}{3}t^2 = \frac{t(t+1)}{4}$
13. $2r(r - 1) = 3r(r - 4)$
14. $r(2r - 3) = 3r(1 - 4r)$
15. $2x^2 - 9x = 35$
16. $63x^2 = 10 - 61x$
17. $p^2(p - 2) = p^2(p - 3) + 4p$
18. $2p^2(1 - p) = 7p^2 - p^2(2p + 3)$
19. $v(2v^2 - 3v + 1) = v^2(2v - 5)$
20. $3v^2 - 2v - 6 = 2v^2 - 7v$
21. $2y^2 - 8 = 0$
22. $3y^2 - 27 = 0$
23. $5u^2 - 45 = 0$
24. $y^2 = 9$
25. $3t^3 = 27t^2$
26. $4t^2 = t$
27. $8 - 2t^2 = 0$
28. $121 - t^2 = 0$
29. $\frac{2}{3}p^2 = \frac{4}{7}p$
30. $\frac{5}{4}p - \frac{1}{3}p^2 = 0$

16.3 Solving Quadratic Equations Using Square Roots

OBJECTIVE

After completing this section, you should be able to solve quadratic equations using the method of square roots.

In Section 16.2, you learned to solve the equation $x^2 = 4$ by rewriting it as $x^2 - 4 = 0$ and then using the method of factoring. The solutions are $x = -2$ and $x = 2$.

From Chapter 15, recall that both -2 and 2 are square roots of 4. Hence, we could solve the equation

$$x^2 = 4$$

simply by taking the square root of *both* sides of the equation, obtaining

$$x = \pm 2$$

Solve each equation for the indicated variables, using square roots.

Practice Exercise 1. $r^2 = 49$

EXAMPLE 1 Solve the equation $x^2 = 8$ for x.

SOLUTION: Solving the equation $x^2 = 8$, we have

$$x^2 = 8$$
$$x = \pm\sqrt{8}$$
$$x = \pm 2\sqrt{2}$$

Practice Exercise 2. $5u^2 = 45$

EXAMPLE 2 Solve the equation $3x^2 = 7$ for x.

SOLUTION: Solving the equation $3x^2 = 7$, we have

$$3x^2 = 7$$
$$x^2 = \frac{7}{3}$$
$$x = \pm\sqrt{\frac{7}{3}}$$
$$= \pm\sqrt{\frac{7}{3} \times \frac{3}{3}}$$
$$= \pm\frac{\sqrt{21}}{3}$$

In general, if $c > 0$, then the equation $x^2 = c$ has *two* solutions which are $x = \sqrt{c}$ and $x = -\sqrt{c}$ or, simply, $x = \pm\sqrt{c}$.

This technique can also be applied to a quadratic equation when one side of the equation is the square of a binomial and the other side is a positive constant.

Practice Exercise 3.
$(y - 3)^2 = 4$

EXAMPLE 3 Solve the equation $(x - 2)^2 = 4$ for x.

SOLUTION: Solving the equation $(x - 2)^2 = 4$, we have

$$(x - 2)^2 = 4$$
$$x - 2 = \pm 2 \quad \text{(Taking square root of both sides)}$$
$$x = 2 \pm 2 \quad \text{(Adding 2 to both sides)}$$

We now have

$$x = 2 + 2 \quad \text{and} \quad x = 2 - 2$$
$$= 4 \quad\quad\quad\quad\quad = 0$$

Therefore, the required solutions are $x = 0, 4$.

CHECK: Checking *each* solution in the *original* equation, we have

$$(x - 2)^2 = 4 \quad \text{and} \quad (x - 2)^2 = 4$$
$$(0 - 2)^2 \stackrel{?}{=} 4 \quad\quad\quad (4 - 2)^2 \stackrel{?}{=} 4$$
$$(-2)^2 \stackrel{?}{=} 4 \quad\quad\quad\quad (2)^2 \stackrel{?}{=} 4$$
$$4 = 4 \checkmark \quad\quad\quad\quad\quad 4 = 4 \checkmark$$

16.3 Solving Quadratic Equations Using Square Roots

Practice Exercise 4.
$(2p + 3)^2 = 9$

EXAMPLE 4 Solve the equation $(2u - 3)^2 = 25$ for u.

SOLUTION:

$$(2u - 3)^2 = 25$$
$$2u - 3 = \pm 5 \quad \text{(Taking square root of both sides)}$$
$$2u = 3 \pm 5 \quad \text{(Adding 3 to both sides)}$$
$$u = \frac{3 \pm 5}{2} \quad \text{(Dividing both sides by 2)}$$

We now have

$$u = \frac{3 + 5}{2} \quad \text{and} \quad u = \frac{3 - 5}{2}$$
$$= \frac{8}{2} \qquad\qquad\qquad = \frac{-2}{2}$$
$$= 4 \qquad\qquad\qquad\quad = -1$$

Therefore, the required solutions are $u = -1, 4$.

CHECK: Checking *each* solution in the *original* equation, we have

$$(2u - 3)^2 = 25 \quad \text{and} \quad (2u - 3)^2 = 25$$
$$[2(-1) - 3]^2 \; ? \; 25 \qquad\qquad [2(4) - 3]^2 \; ? \; 25$$
$$(-2 - 3)^2 \; ? \; 25 \qquad\qquad\quad (8 - 3)^2 \; ? \; 25$$
$$(-5)^2 \; ? \; 25 \qquad\qquad\qquad\quad 5^2 \; ? \; 25$$
$$25 = 25 \checkmark \qquad\qquad\qquad\quad 25 = 25 \checkmark$$

ANSWERS TO PRACTICE EXERCISES

1. $r = \pm 7$. 2. $u = \pm 3$. 3. $y = 1, 5$. 4. $p = -3, 0$.

In Section 16.2, we reviewed the technique used in solving quadratic equations by the method of factoring. Sometimes the polynomial involved in the equation is a quadratic that is a **perfect square.** That is, the polynomial is factorable, and its factors are two identical binomials. For instance, we have

$$x^2 + 2x + 1 = (x + 1)(x + 1)$$
$$y^2 + 6y + 9 = (y + 3)(y + 3)$$
$$u^2 - 10u + 25 = (u - 5)(u - 5)$$
$$4r^2 - 12r + 9 = (2r - 3)(2r - 3)$$

Each of these quadratic trinomials is a perfect square.

If we have a perfect square quadratic trinomial set equal to a positive constant, then we can use the technique given in this section to solve the equation.

Solve each equation, using the method of square roots.

Practice Exercise 5.
$x^2 - 8x + 16 = 1$

EXAMPLE 5 Solve the equation $y^2 - 4y + 4 = 16$ for y.

SOLUTION: The left-hand side of the equation is a perfect square since

$$y^2 - 4y + 4 = (y - 2)(y - 2)$$

Hence, we have

$$y^2 - 4y + 4 = 16$$
$$(y - 2)(y - 2) = 16 \quad \text{(Factoring)}$$
$$(y - 2)^2 = 16 \quad \text{(Rewriting)}$$
$$y - 2 = \pm 4 \quad \text{(Taking square root of both sides)}$$
$$y = 2 \pm 4 \quad \text{(Adding 2 to both sides)}$$

We now have

$$y = 2 + 4 \quad \text{and} \quad y = 2 - 4$$
$$= 6 \qquad\qquad\qquad = -2$$

Therefore, the required solutions are $y = -2, 6$. The check of these solutions (as well as those for the following examples) is left to you. Remember to check each solution in the original equation.

> **Practice Exercise 6.**
> $y^2 + 2y + 1 = 9$

EXAMPLE 6 Solve the equation $u^2 - 6u + 9 = 8$ for u.

SOLUTION: Again, observe that the left-hand side of the equation is a perfect square trinomial. We have

$$u^2 - 6u + 9 = 8$$
$$(u - 3)(u - 3) = 8$$
$$(u - 3)^2 = 8$$
$$u - 3 = \pm\sqrt{8}$$
$$u - 3 = \pm 2\sqrt{2}$$
$$u = 3 \pm 2\sqrt{2}$$

Therefore, the required solutions are $u = 3 + 2\sqrt{2}$ and $u = 3 - 2\sqrt{2}$.

> **Practice Exercise 7.**
> $w^2 + 14w + 49 = 3$

EXAMPLE 7 Solve the equation $9u^2 - 6u + 1 = 5$ for u.

SOLUTION: Since the left-hand of the given equation is a perfect square, we have

$$9u^2 - 6u + 1 = 5$$
$$(3u - 1)(3u - 1) = 5$$
$$(3u - 1)^2 = 5$$
$$3u - 1 = \pm\sqrt{5}$$
$$3u = 1 \pm \sqrt{5}$$
$$u = \frac{1 \pm \sqrt{5}}{3}$$

Therefore, the required solutions are $u = \dfrac{1 + \sqrt{5}}{3}$ and $u = \dfrac{1 - \sqrt{5}}{3}$.

ANSWERS TO PRACTICE EXERCISES

5. $x = 3, 5$. 6. $y = -4, 2$. 7. $w = -7 \pm \sqrt{3}$.

EXERCISES 16.3

Solve each of the following equations using the square root method. Check your results in the original equations.

1. $x^2 - 81 = 0$
2. $y^2 - 121 = 0$
3. $2u^2 = 50$
4. $3w^2 = 12$
5. $u^2 - 8 = 0$
6. $4p^2 - 18 = 0$
7. $(x - 3)^2 = 9$
8. $(y + 1)^2 = 36$
9. $(v - 5)^2 = 81$
10. $(w + 9)^2 = 100$
11. $(2p - 1)^2 = 1$
12. $(3q + 4)^2 = 4$
13. $(4t + 5)^2 = 9$
14. $(5r - 3)^2 = 81$
15. $2(x - 3)^2 = 6$
16. $3(y + 7)^2 = 15$
17. $3 - (u - 1)^2 = 0$
18. $8 - (2v + 5)^2 = 0$
19. $4(y + 6)^2 - 20 = 0$
20. $4(q - 9)^2 - 2 = 0$
21. $x^2 + 6x + 9 = 1$
22. $y^2 - 12y + 36 = 4$
23. $u^2 - 18u + 81 = 9$
24. $w^2 + 10w + 25 = 16$
25. $p^2 - 4p + 4 = 11$
26. $q^2 + 14q + 49 = 12$
27. $4t^2 - 4t + 1 = 25$
28. $9s^2 + 12s + 4 = 100$
29. $25r^2 - 20r + 4 = 3$
30. $49u^2 + 28u + 4 = 5$

16.4 Solving Quadratic Equations by Completing the Square

OBJECTIVE

After completing this section, you should be able to solve quadratic equations using the method of completing the square.

What happens when we have a quadratic equation that cannot be solved by the method of factoring and such that the polynomial is not a perfect square? How do we solve such an equation?

In this section, we will discuss a method that can always be used to solve a quadratic equation. The method is called **completing the square.**

Consider

$$(x + a)(x + a) = x^2 + 2ax + a^2$$

The trinomial on the right-hand side of the equation is a perfect square with a leading coefficient of 1. Suppose, now, that we start with

$$x^2 + 4x$$

and wish to add a constant to the expression to obtain a perfect square trinomial. What must we add? We want

$$x^2 + 4x + \underline{\quad ? \quad} = (x + a)(x + a)$$

or

$$x^2 + 4x + \underline{\quad ? \quad} = x^2 + 2ax + a^2$$

from which we have

$$x^2 + 4x = x^2 + 2ax$$
$$4x = 2ax$$
$$4 = 2a$$
$$2 = a$$

Since $a = 2$, then $a^2 = 2^2 = 4$, and $\underline{\quad ? \quad} = 4$. Adding 4 to $x^2 + 4x$, we have

$$x^2 + 4x + 4 = (x + 2)^2$$

which is a perfect square.

Determine what must be added to each expression to complete the square.

Practice Exercise 1. $x^2 + 2x$

EXAMPLE 1 Complete the square for $y^2 + 6y$.

SOLUTION: Completing the square for $y^2 + 6y$, we have

$$y^2 + 6y + \underline{\quad ? \quad} = (y + a)(y + a)$$
$$y^2 + 6y + \underline{\quad ? \quad} = y^2 + 2ay + a^2$$
$$y^2 + 6y = y^2 + 2ay$$
$$6y = 2ay$$
$$6 = 2a$$
$$3 = a$$

Since $a = 3$, then $a^2 = 3^2 = 9$, and $\underline{\quad ? \quad} = 9$. Adding 9 to $y^2 + 6y$, we have

$$y^2 + 6y + 9 = (y + 3)^2$$

which is a perfect square.

Practice Exercise 2. $y^2 - 10y$

EXAMPLE 2 Complete the square for $u^2 - 8u$.

SOLUTION: Completing the square for $u^2 - 8u$, we have

$$u^2 - 8u + \underline{\quad ? \quad} = (u + a)(u + a)$$
$$u^2 - 8u + \underline{\quad ? \quad} = u^2 + 2au + a^2$$
$$u^2 - 8u = u^2 + 2au$$
$$-8u = 2au$$
$$-8 = 2a$$
$$-4 = a$$

Since $a = -4$, then $a^2 = (-4)^2 = 16$, and $\underline{\quad ? \quad} = 16$. Adding 16 to $u^2 - 8u$, we have

$$u^2 - 8u + 16 = (u - 4)^2$$

which is a perfect square.

In each of the two preceding examples, observe that the constant added to each binomial to get a perfect square trinomial was the *square* of *one-half* the coefficient of the *linear* term. Also, note that the leading coefficient of the quadratic in each case is 1.

16.4 Solving Quadratic Equations by Completing the Square

To complete the square for the expression $x^2 + px$:

1. Take one-half the coefficient of x, or $\frac{1}{2}p$.
2. Square $\frac{1}{2}p$ to get $\left(\frac{1}{2}p\right)^2$.
3. Add it to the given expression to obtain

$$x^2 + px + \left(\frac{1}{2}p\right)^2 = x^2 + px + \frac{1}{4}p^2$$

The resulting trinomial is a perfect square with factors $\left(x + \frac{p}{2}\right)\left(x + \frac{p}{2}\right)$, or

$$x^2 + px + \frac{p^2}{4} = \left(x + \frac{p}{2}\right)^2$$

Practice Exercise 3. $u^2 + 6u$

EXAMPLE 3 Complete the square for $x^2 + 10x$.

SOLUTION: Completing the square for $x^2 + 10x$, we have

$$x^2 + 10x + \left[\frac{1}{2}(10)\right]^2 = x^2 + 10x + (5)^2$$
$$= x^2 + 10x + 25$$
$$x^2 + 10x + 25 = (x + 5)^2$$

Practice Exercise 4. $p^2 - 5p$

EXAMPLE 4 Complete the square for $y^2 + 7y$.

SOLUTION: Completing the square for $y^2 + 7y$, we have

$$y^2 + 7y + \left[\frac{1}{2}(7)\right]^2 = y^2 + 7y + \left(\frac{7}{2}\right)^2$$
$$= y^2 + 7y + \frac{49}{4}$$
$$y^2 + 7y + \frac{49}{4} = \left(y + \frac{7}{2}\right)^2$$

ANSWERS TO PRACTICE EXERCISES

1. 1.   2. 25.   3. 9.   4. $\frac{25}{4}$.

We will now solve quadratic equations using the method of completing the square.

Solve each equation by the method of completing the square.

Practice Exercise 5.
$u^2 + 2u - 5 = 0$

EXAMPLE 5 Solve the equation $x^2 + 4x - 5 = 0$ by completing the square.

SOLUTION: Since the left-hand side of the equation is *not* a perfect square, we can add 5 to both sides and then complete the square for $x^2 + 4x$. We have

$$x^2 + 4x - 5 = 0$$
$$x^2 + 4x = 5 \quad \text{(Adding 5 to both sides)}$$
$$x^2 + 4x + \left[\frac{1}{2}(4)\right]^2 = 5 + \left[\frac{1}{2}(4)\right]^2 \quad \text{(Completing the square on the left-hand side } and \text{ adding the same quantity to the right-hand side, since we have an equation)}$$
$$x^2 + 4x + 4 = 5 + 4 \quad \text{(Simplifying)}$$
$$(x + 2)^2 = 9 \quad \text{(Rewriting)}$$
$$x + 2 = \pm 3 \quad \text{(Taking square root of both sides)}$$
$$x = -2 \pm 3 \quad \text{(Adding } -2 \text{ to both sides)}$$

We now have

$$x = -2 + 3 \quad \text{or} \quad x = -2 - 3$$
$$= 1 \qquad\qquad\qquad = -5$$

Therefore, the required solutions are $x = -5, 1$. Check these solutions in the original equation.

The method of completing the square can also be used to solve a quadratic equation with leading coefficient other than 1. This will be illustrated in the following examples.

Practice Exercise 6.
$t^2 - 14t + 3 = 0$

EXAMPLE 6 Solve the equation $2u^2 - 6u + 2 = 0$ by completing the square.

SOLUTION: Since the leading coefficient is *not* 1, we will divide both sides of the equation by 2, obtaining

$$u^2 - 3u + 1 = 0$$

Since the trinomial is *not* a perfect square, we add -1 to both sides, obtaining

$$u^2 - 3u = -1$$

We now complete the square on the left-hand side, add the same quantity to the right-hand side, and proceed. We have

$$u^2 - 3u + \left[\frac{1}{2}(-3)\right]^2 = -1 + \left[\frac{1}{2}(-3)\right]^2$$
$$u^2 - 3u + \left(\frac{-3}{2}\right)^2 = -1 + \left(\frac{-3}{2}\right)^2$$
$$u^2 - 3u + \frac{9}{4} = -1 + \frac{9}{4}$$
$$\left(u - \frac{3}{2}\right)^2 = \frac{5}{4}$$
$$u - \frac{3}{2} = \pm\sqrt{\frac{5}{4}}$$
$$u - \frac{3}{2} = \pm\frac{\sqrt{5}}{2}$$
$$u = \frac{3}{2} \pm \frac{\sqrt{5}}{2}$$
$$= \frac{3 \pm \sqrt{5}}{2}$$

16.4 Solving Quadratic Equations by Completing the Square

Therefore, the required solutions are $u = \dfrac{3 + \sqrt{5}}{2}$ and $u = \dfrac{3 - \sqrt{5}}{2}$.

EXAMPLE 7 Solve the equation $2r^2 + 3r - 2 = 0$ by completing the square.

SOLUTION: Rewrite the given equation as

$$2r^2 + 3r = 2$$

Since the leading coefficient is *not* 1, divide *both* sides of the equation by 2, obtaining

$$r^2 + \frac{3}{2}r = 1$$

Add the *square* of $\dfrac{1}{2}\left(\dfrac{3}{2}\right)$ to *both* sides of the equation and proceed. We have

$$r^2 + \frac{3}{2}r + \left[\frac{1}{2}\left(\frac{3}{2}\right)\right]^2 = 1 + \left[\frac{1}{2}\left(\frac{3}{2}\right)\right]^2$$

$$r^2 + \frac{3}{2}r + \left(\frac{3}{4}\right)^2 = 1 + \left(\frac{3}{4}\right)^2$$

$$\left(r + \frac{3}{4}\right)^2 = 1 + \frac{9}{16}$$

$$\left(r + \frac{3}{4}\right)^2 = \frac{25}{16}$$

$$r + \frac{3}{4} = \pm\sqrt{\frac{25}{16}}$$

$$r + \frac{3}{4} = \pm\frac{5}{4}$$

$$r = \frac{-3}{4} \pm \frac{5}{4}$$

$$= \frac{-3 \pm 5}{4}$$

We now have

$$r = \frac{-3 + 5}{4} \quad \text{or} \quad r = \frac{-3 - 5}{4}$$

$$= \frac{2}{4} \qquad\qquad\qquad = \frac{-8}{4}$$

$$= \frac{1}{2} \qquad\qquad\qquad = -2$$

Therefore, the required solutions are $r = -2, \dfrac{1}{2}$.

Practice Exercise 7.
$3q^2 - 4q - 2 = 0$

ANSWERS TO PRACTICE EXERCISES

5. $u = -1 \pm \sqrt{6}$. 6. $t = 7 \pm \sqrt{46}$. 7. $q = \dfrac{2 \pm \sqrt{10}}{3}$.

We will now summarize the method of completing the square which can be used to solve a quadratic equation.

Solving a Quadratic Equation by Completing the Square Method

To solve the quadratic equation $ax^2 + bx + c = 0$ using the method of completing the square:

1. Rewrite the equation in the form $ax^2 + bx = -c$ with the variable terms on one side of the equation and the constant term on the other side.
2. If a, the coefficient of x^2, is *not* 1, divide *both* sides of the equation by a, obtaining the resulting equation $x^2 + \frac{b}{a}x = \frac{-c}{a}$.
3. Add the square of one-half $\frac{b}{a}$, the resulting coefficient of x, to *both* sides of the equation, obtaining

$$x^2 + \frac{b}{a}x + \left(\frac{1}{2}\frac{b}{a}\right)^2 = \frac{-c}{a} + \left(\frac{1}{2}\frac{b}{a}\right)^2$$

This completes the square on the left-hand side of the equation.

4. Rewrite the left-hand side of the equation as the square of a binomial, and simplify the right-hand side of the equation:

$$\left(x + \frac{b}{2a}\right)^2 = \frac{b^2 - 4ac}{4a^2}$$

5. Take the square root of *both* sides of the equation and solve for x:

$$x = \frac{-b \pm \sqrt{b^2 - 4ac}}{2a}$$

Note: If $b^2 - 4ac < 0$, the solutions of the given equation will be complex numbers.

EXERCISES 16.4

Solve each of the given equations using the method of completing the square. Check your results in the original equations.

1. $x^2 - 6x + 5 = 0$
2. $y^2 - 16y - 48 = 0$
3. $w^2 - 2 = 4 + w$
4. $m^2 + 7m = 18$
5. $x^2 + 2x - 5 = 0$
6. $y^2 + 21 = 10y$
7. $u^2 + 24 = -14u$
8. $x^2 - 3x + 1 = 0$
9. $y^2 + 21 = 10y + 5$
10. $x^2 - 5x - 12 = 0$
11. $8u^2 + 5u = 3$
12. $4p^2 - 15 = -4p$
13. $6t^2 - 5t - 6 = 0$
14. $11s = 15 - 12s^2$
15. $6r^2 + 5r - 6 = 0$
16. $8w^2 + 15w = 2$
17. $2s^2 + 3 = 7s$
18. $2y^2 + 7y = 0$
19. $4t^2 + 20t + 25 = 0$
20. $(2u - 3)(3u + 2) = 7u$

21. $x^2 - 6x + 4 = 0$
22. $4y^2 - 4y = 6$
23. $u^2 + 10u + 19 = 0$
24. $4p^2 - 28p = -24$
25. $9t^2 + 30t + 14 = 0$
26. $49w^2 = 28w + 9$
27. $16q^2 + 14 = 40q$
28. $4r^2 = 12r + 8$
29. $s^2 - 22s + 110 = 0$
30. $x^2 + 24x + 132 = 0$

16.5 The Quadratic Formula

OBJECTIVES

After completing this section, you should be able to:
1. Solve quadratic equations using the quadratic formula.
2. By evaluating the discriminant, determine if the solutions of a quadratic equation exist or are complex numbers.

The method of factoring discussed in Section 16.2 can be used to solve only some equations—those such that the polynomial, which is set equal to 0, is factorable. There are two methods that can be used to solve all quadratic equations. However, the solutions are sometimes not real numbers. The first of these methods, completing the square, was discussed in Section 16.4. The other method involves what is called the **quadratic formula.**

The standard form of a quadratic equation in the variable x is $ax^2 + bx + c = 0$ with $a \neq 0$. The quadratic formula involves only the coefficients a, b, and c.

The Quadratic Formula

Consider the standard form of the quadratic equation in the variable x, which is

$$ax^2 + bx + c = 0 \quad (a \neq 0)$$

Then, the solutions for this equation are given by the formula

$$x = \frac{-b \pm \sqrt{b^2 - 4ac}}{2a}$$

This formula is called the *quadratic formula.*

Using the method of completing the square, we will now derive the quadratic formula using the standard form of the quadratic equation in the variable x.

Step 1: Start with the standard form of the quadratic equation in the variable x:

$$ax^2 + bx + c = 0 \quad (a \neq 0)$$

Step 2: Divide both sides of the equation by a (since the coefficient of the x^2 term must be 1 to use the method of completing the square). We have

$$x^2 + \frac{b}{a}x + \frac{c}{a} = 0$$

STEP 3: Add $\dfrac{-c}{a}$ to *both* sides of the last equation. We have

$$x^2 + \frac{b}{a}x = \frac{-c}{a}$$

STEP 4: Take one-half of $\dfrac{b}{a}$ (the coefficient of the linear term), square it, and add it to *both* sides of the equation. We have

$$x^2 + \frac{b}{a}x + \left(\frac{b}{2a}\right)^2 = \frac{-c}{a} + \left(\frac{b}{2a}\right)^2$$

STEP 5: Write the left-hand side of the last equation as a perfect square. Also, combine the terms on the right-hand side of the equation. We have

$$\left(x + \frac{b}{2a}\right)^2 = \frac{b^2 - 4ac}{4a^2}$$

STEP 6: Solve for $x + \dfrac{b}{2b}$. We have

$$x + \frac{b}{2a} = \frac{\pm\sqrt{b^2 - 4ac}}{2a}$$

STEP 7: Solve for x. We have

$$x = \frac{-b}{2a} \pm \frac{\sqrt{b^2 - 4ac}}{2a}$$

$$x = \frac{-b \pm \sqrt{b^2 - 4ac}}{2a}$$

which is the required solution for the given equation.

Note the symbol $\pm$ in the quadratic formula. The symbol $\pm$ is read "plus and minus" and, in the formula, will generally produce two solutions for the given equation.

Solve each equation by using the quadratic formula.

Practice Exercise 1.
$2y^2 - 3y + 1 = 0$

EXAMPLE 1 Using the quadratic formula, solve the equation $x^2 - 5x + 6 = 0$ for x.

SOLUTION: The equation $x^2 - 5x + 6 = 0$ is already in the standard form with $a = 1$, $b = -5$, and $c = 6$. Hence,

$$x = \frac{-b \pm \sqrt{b^2 - 4ac}}{2a}$$

becomes

$$x = \frac{-(-5) \pm \sqrt{(-5)^2 - 4(1)(6)}}{2(1)}$$

$$= \frac{5 \pm \sqrt{25 - 24}}{2}$$

$$= \frac{5 \pm \sqrt{1}}{2}$$

$$= \frac{5 \pm 1}{2}$$

16.5 The Quadratic Formula

We now have

$$x = \frac{5+1}{2} \quad \text{and} \quad x = \frac{5-1}{2}$$
$$= \frac{6}{2} \qquad\qquad\qquad = \frac{4}{2}$$
$$= 3 \qquad\qquad\qquad = 2$$

Therefore, the required solutions are $x = 2, 3$. You are encouraged to check these results in the original equation. (*Note:* The given equation could also have been solved using the method of factoring. It is always worth trying the method of factoring first to solve a quadratic equation in a single variable. If the method works, there is usually less work involved.)

Practice Exercise 2.
$2t^2 + 5t = 3$

EXAMPLE 2 Using the quadratic formula, solve the equation $2u^2 - 3u + 1 = 0$ for u.

SOLUTION: The given equation is a quadratic equation in the variable u. Hence, the appropriate form of the quadratic formula that we will use is

$$u = \frac{-b \pm \sqrt{b^2 - 4ac}}{2a}$$

with $a = 2$, $b = -3$, and $c = 1$. Substituting in the formula, we have

$$u = \frac{-(-3) \pm \sqrt{(-3)^2 - 4(2)(1)}}{2(2)}$$
$$= \frac{3 \pm \sqrt{9 - 8}}{4}$$
$$= \frac{3 \pm \sqrt{1}}{4}$$
$$= \frac{3 \pm 1}{4}$$

We now have

$$u = \frac{3+1}{4} \quad \text{and} \quad u = \frac{3-1}{4}$$
$$= \frac{4}{4} \qquad\qquad\qquad = \frac{2}{4}$$
$$= 1 \qquad\qquad\qquad = \frac{1}{2}$$

Therefore, the required solutions are $u = \frac{1}{2}, 1$.

Practice Exercise 3.
$1 - 4p + 3p^2 = 0$

EXAMPLE 3 Using the quadratic formula, solve the equation $-5 = 3u + 4u^2$ for u.

SOLUTION: Rewriting the given equation in its standard form, we have

$$4u^2 + 3u + 5 = 0$$

with $a = 4$, $b = 3$, and $c = 5$. Since the equation involves the variable u, we must use the quadratic formula for u. Hence,

$$u = \frac{-b \pm \sqrt{b^2 - 4ac}}{2a}$$

becomes

$$u = \frac{-3 \pm \sqrt{3^2 - 4(4)(5)}}{2(4)}$$
$$= \frac{-3 \pm \sqrt{9 - 80}}{8}$$
$$= \frac{-3 \pm \sqrt{-71}}{8}$$

Note that $\sqrt{-71}$ is not a real number. Hence, we *cannot* complete the solution for this equation.

ANSWERS TO PRACTICE EXERCISES

1. $y = \frac{1}{2}, 1$. 2. $t = -3, \frac{1}{2}$. 3. $p = \frac{1}{3}, 1$.

In Example 3, we were not able to complete the solution for the given equation since the number under the radical sign is negative. Recall that the square root of a negative number is not a real number.

In the quadratic formula, the quantity under the radical sign, $b^2 - 4ac$, is called the **discriminant**. In this course, the solution of a quadratic equation can be determined only if the discriminant is greater than or equal to 0.

Compute the discriminant, and then determine whether the solutions for the given equations exist or are not real numbers.

Practice Exercise 4.
$y^2 - 4y + 1 = 0$

EXAMPLE 4 Determine if the equation $2y^2 - 3y = 7$ can be solved, by first evaluating its discriminant.

SOLUTION: The equation $2y^2 - 3y = 7$ can be rewritten in its standard form as $2y^2 - 3y - 7 = 0$ with $a = 2$, $b = -3$, and $c = -7$. Hence, the discriminant is

$$b^2 - 4ac = (-3)^2 - 4(2)(-7)$$
$$= 9 - (-56)$$
$$= 9 + 56$$
$$= 65$$

which is positive. Therefore, the solutions for the given equation can be determined.

Practice Exercise 5.
$3 - 2p + 5p^2 = 0$

EXAMPLE 5 Determine if the equation $2s - 5s^2 = 7$ can be solved by evaluating its discriminant.

SOLUTION: Rewriting the given equation in its standard form, we have $5s^2 - 2s + 7 = 0$ with $a = 5$, $b = -2$, and $c = 7$. Hence, the discriminant is

$$b^2 - 4ac = (-2)^2 - 4(5)(7)$$
$$= 4 - 140$$
$$= -136$$

which is negative. Therefore, the solutions for the given equation are not real numbers.

> **ANSWERS TO PRACTICE EXERCISES**
>
> 4. 12; exist.   5. −56; not real.

EXERCISES 16.5

In Exercises 1–20, compute the value of the discriminant, and then determine whether the solutions for the given equations exist or are not real numbers.

1. $x^2 - 3x + 2 = 0$
2. $2x^2 - 4x - 5 = 0$
3. $2x - 3x^2 = 7$
4. $9 - 4x^2 = 5x$
5. $1 - 2x - x^2 = 0$
6. $4 - 5x - 3x^2 = 0$
7. $2y^2 - 5y - 7 = 0$
8. $3y^2 + 4y = 6$
9. $5y - 2y^2 = 9$
10. $6 - y^2 = 7y$
11. $5 - 11y^2 = 10y$
12. $6y^2 = 7y + 3$
13. $4u^2 + 5u + 7 = 0$
14. $6u = 3u^2 - 8$
15. $2(p - 3)(4 - 3p) = 0$
16. $p^2 - 3(p + 1)(p - 1) = 4$
17. $(2q - 1)(q + 3) = (q - 5)(3 - q)$
18. $4q^2 - 5(q - 2) = 3 - 2q^2$
19. $(t - 1)^2 = 5t + 4$
20. $(2t - 3)^2 - 4 = 7(t^2 - 1) + 5$

In Exercises 21–40, solve the given equations using the quadratic formula.

21. $x^2 - 5x + 6 = 0$
22. $2x^2 - 3x - 1 = 0$
23. $3 - 5x - 7x^2 = 0$
24. $x^2 - 1 = x$
25. $2y^2 + 6y + 3 = 0$
26. $7 = 4y + y^2$
27. $2(y - 3)(y + 1) = 0$
28. $(2y + 5)(3y - 4) = 0$
29. $3u^2 = u + 2$
30. $3u = 2 - 2u^2$
31. $u^2 + 1 = 4u$
32. $4u^2 - 12u - 5 = 0$
33. $2s^2 + 5s = 3$
34. $s^2 + 4 = 5s$
35. $(3s - 4)(4s + 3) = 0$
36. $4(3s - 1)^2 = 0$
37. $3(t - 2)^2 = (t - 1)(t + 3)$
38. $2(t - 3)^2 - (t + 4)^2 = 0$
39. $3t^2 + 2 = (t + 2)^2$
40. $4t^2 - 3t - (t + 5)^2 = 1$

16.6 Applications Involving Quadratic Equations

OBJECTIVE
After completing this section, you should be able to solve some word problems by using quadratic equations.

There are some word problems that can be solved by setting up and solving quadratic equations. The procedure is the same as we have been using in earlier sections of this text. However, the resulting equations will be quadratic. We will now consider some of these applications.

Practice Exercise 1. Determine the value of two consecutive negative even integers whose product is 1088.

EXAMPLE 1 Determine the value of two consecutive whole numbers whose product is 552.

SOLUTION: For the two consecutive numbers,

let n = one number
and $n + 1$ = the other number

Since their product is 552, we have

$$n(n + 1) = 552$$

Solving this equation, we have

$$n(n + 1) = 552$$
$$n^2 + n = 552$$
$$n^2 + n - 552 = 0$$
$$(n + 24)(n - 23) = 0$$
$$n + 24 = 0 \quad \text{or} \quad n - 23 = 0$$
$$n = -24 \quad \quad n = 23$$

Since this problem involves whole numbers, we reject the solution $n = -24$. Then, using

$$n = 23$$

we have

$$n + 1 = 24$$

Therefore, the required numbers are 23 and 24. Check that their product is 552.

Practice Exercise 2. The width of a rectangular region is 2 yd more than one-half of its length. The area of the region is 240 yd². What are the dimensions of the region?

EXAMPLE 2 A flower bed is in the shape of a rectangle and the length is 5 m more than twice its width. If its area is 348 m², determine the dimensions of the flower bed.

SOLUTION: Since the length is given in terms of the width,

let W = the width (in meters)
Then, $2W + 5$ = the length (in meters)

[Diagram: rectangle with Area = 348 m², width W, length $2W + 5$]

(See the accompanying diagram.) Since the area of the rectangular flower bed is the product of its length and width, we now have

$$W(2W + 5) = 348$$

Solving this equation, we have

$$W(2W + 5) = 348$$
$$2W^2 + 5W - 348 = 0$$
$$(2W + 29)(W - 12) = 0$$
$$2W + 29 = 0 \quad \text{or} \quad W - 12 = 0$$
$$2W = -29 \quad \quad W = 12$$
$$W = \frac{-29}{2}$$

16.6 Applications Involving Quadratic Equations

Since W = width of the flower bed, we reject the solution $W = \dfrac{-29}{2}$; width cannot be negative. Hence, we have

$$W = 12 \text{ m}$$
and
$$2W + 5 = 29 \text{ m}$$

Therefore, the dimensions of the flower bed are 12 m by 29 m.

Practice Exercise 3. The sum of a negative number and 6 is equal to -8 times the reciprocal of the number. What is the number?

EXAMPLE 3 The base of a triangular region is 12 ft more than its height. Determine the dimensions of the region if its area is 54 ft².

SOLUTION: Since the base of the region is given in terms of its height, we

let
$$h = \text{height (in feet)}$$
Then,
$$h + 12 = \text{base (in feet)}$$

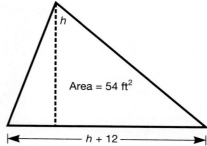

as in the accompanying diagram. Since the area of a triangular region is one-half the product of its height and base, we have

$$\frac{1}{2}h(h + 12) = 54$$

Solving this equation, we have

$$h(h + 12) = 108$$
$$h^2 + 12h = 108$$
$$h^2 + 12h - 108 = 0$$
$$(h - 6)(h + 18) = 0$$
$$h - 6 = 0 \quad \text{or} \quad h + 18 = 0$$
$$h = 6 \qquad\qquad h = -18$$

Since h represents height, we reject the solution $h = -18$. Hence, we have

$$h = 6 \text{ ft}$$
and
$$h + 12 = 18 \text{ ft}$$

Therefore, the triangular region would have a height of 6 ft and a base of 18 ft.

Practice Exercise 4. The base of a triangular region is 8 m more than its height. The area of the region is 120 m². What is the length of the base of the region?

EXAMPLE 4 Jill plans to make a box by cutting equal squares from each corner of a 16 cm by 20 cm piece of cardboard and then folding up the sides. The bottom of the box is to have an area of 96 cm². What is the length of the side of each square that Jill should cut out?

SOLUTION: Referring to the accompanying figure, we let

s = length of side of each square (in centimeters)

Then, $16 - 2s$ = width of bottom of box (in centimeters)

and $20 - 2s$ = length of bottom of box (in centimeters)

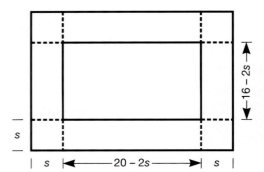

Since the bottom of the box is rectangular and the area is to be 96 cm², we have

$$A = LW$$
$$96 \text{ cm}^2 = (20 - 2s)(16 - 2s) \text{ cm}^2$$
$$96 = (20 - 2s)(16 - 2s)$$
$$96 = 320 - 72s + 4s^2$$
$$4s^2 - 72s + 224 = 0$$
$$s^2 - 18s + 56 = 0$$
$$(s - 14)(s - 4) = 0$$

We now have that either

$$s - 14 = 0 \quad \text{or} \quad s - 4 = 0$$
$$s = 14 \quad\quad\quad\quad s = 4$$

It appears that we have two solutions. However, it is not possible to cut squares, each having a side of length 14 cm, from a piece of cardboard that measures 16 cm by 20 cm. Therefore, we reject the solution $s = 14$. If the side of each square cut is 4 cm (the other solution), then the bottom of the box will have dimensions

Width = $(16 - 2s)$ cm = $(16 - 8)$ cm = 8 cm

and Length = $(20 - 2s)$ cm = $(20 - 8)$ cm = 12 cm

The area of the bottom of the box will be

$$(8 \text{ cm})(12 \text{ cm}) = 96 \text{ cm}^2$$

which checks with what is required. Therefore, the length of the side of each square cut from the cardboard is 4 cm.

Practice Exercise 5. A carpenter and his assistant working together can complete a job in 8 hr. The assistant, working alone, could complete the job in 12 hr more time than the carpenter working alone. How long would it take the carpenter to do the job working alone?

EXAMPLE 5 Working together, Linda and her friend John can paint her living room in 4 hr. Working alone, John can paint the living room in 6 hr more time than it would take Linda working alone. How long would it take each to do the job alone?

SOLUTION: Since the rates of painting the living room referred to are different, we let

x = the number of hours required for Linda to paint the living room working alone

and $x + 6$ = the number of hours required for John to paint the living room working alone

Then, $\dfrac{1}{x}$ and $\dfrac{1}{x+6}$ are the portions of the living room that could be painted by Linda and John, respectively, in 1 hr. In 4 hr (the required time for both working together to paint the room), Linda and John would paint $\dfrac{4}{x}$ and $\dfrac{4}{x+6}$ portions of the room, respectively.

Working *together*, then, we have

$$\text{(Work done by Linda)} + \text{(work done by John)} + \text{(whole job)}$$

or $$\dfrac{4}{x} + \dfrac{4}{x+6} = 1$$

Multiplying both sides of this equation by LCM$(x, x+6) = x(x+6)$, we have

$$x(x+6)\left(\dfrac{4}{x} + \dfrac{4}{x+6}\right) = x(x+6)(1)$$

$$x(x+6)\left(\dfrac{4}{x}\right) + x(x+6)\left(\dfrac{4}{x+6}\right) = x(x+6)(1)$$

$$\overset{1}{\cancel{x}}(x+6)\left(\dfrac{4}{\cancel{x}}\right) + x\cancel{(x+6)}\left(\dfrac{4}{\cancel{x+6}}\right) = x(x+6)(1)$$

$$4(x+6) + 4x = x(x+6)$$
$$4x + 24 + 4x = x^2 + 6x$$
$$8x + 24 = x^2 + 6x$$
$$x^2 - 2x - 24 = 0$$
$$(x+4)(x-6) = 0$$

Therefore, $\quad x + 4 = 0 \quad$ or $\quad x - 6 = 0$
$$x = -4 \qquad\qquad x = 6$$

We reject the solution $x = -4$ since that would give a negative number of hours of work. Hence,

$x = 6$ is the number of hours in which Linda can paint the living room working alone

and $\quad x + 6 = 12$ is the number of hours in which John can paint the living room working alone.

ANSWERS TO PRACTICE EXERCISES

1. -32 and -34. 2. 12 yd by 20 yd. 3. -4. 4. 20 m. 5. 12 hr.

EXERCISES 16.6

Solve each of the following by setting up and solving an appropriate quadratic equation.

1. Determine the value of two consecutive odd integers whose product is 323.
2. Determine the value of two consecutive negative even integers whose product is 528.
3. Determine two consecutive positive odd integers whose product is 783.

4. One positive number is 3 less than twice another positive number. Their product is 527. What are the numbers?
5. If a number is diminished by 9 and the resulting number is squared, the new number is 121. What is the original number?
6. The length of a rectangular region is 4 m more than 3 times its width. The area of the region is 407 m². What are the dimensions of the region?
7. The length of a rectangular region is 7 yd less than twice its width. The area of the region is 204 yd². What are the dimensions of the region?
8. The height of a triangular region is 6 cm less than its base. The area of the region is 56 cm². What are the lengths of the base and the height of the region?
9. The height of a triangular region is one-half its base. The area of the region is 169 in.². What are the dimensions of the height and base of the region?
10. The number of centimeters in the perimeter of a square region is equal to two-thirds the number of square centimeters in the area of the region. Determine the length of a side of the region.
11. The sum of a positive number and 3 is equal to 18 times the reciprocal of the number. What is the number?
12. The sum of a negative number and 6 is equal to 27 times the reciprocal of the number. What is the number?
13. If 1 is subtracted from a positive number, the difference will be equal to 156 times the reciprocal of the number. What is the number?
14. If an object is dropped from the top of a vertical cliff, the distance, s (measured in feet), that the object falls in t seconds is given by the formula $s = 16t^2$. If the cliff is 196 ft high, how long will it take the object to reach bottom?
15. Using the formula in Exercise 14, determine how long it would take an object dropped from a vertical cliff 225 ft high to reach bottom.
16. If 4 times a positive integer is added to the square of the integer, the result will be 45. What is the integer?
17. If a positive integer is subtracted from twice the square of the integer, the result will be 66. What is the integer?
18. The sum of a number and its reciprocal is 2. What is the number?
19. If a positive number is subtracted from its reciprocal, then the result will be 3 times the number. What is the number?
20. If a negative number is added to its reciprocal, then the result will be 10 times the number. What is the number?
21. A rectangular playing field measures 60 ft by 80 ft. Joel is hired to mow a strip of grass around the field so that one-half of the field is left unmowed. If the strip is to have uniform width, how wide a strip around the field should Joel mow?
22. A plumber and her assistant working together can complete a job in 6 hr. The assistant, working alone, could complete the job in 16 hr more time than the plumber working alone. How long would it take each person to do the job working alone?
23. A picture is to be pasted on a rectangular poster board that measures 16 in. by 20 in. The picture covers three-fifths of the poster board. If there is a border around the picture with uniform width, how wide is the border?
24. A group of college students chartered a bus for a trip to Daytona Beach during their spring break. The cost of the bus was $720. If six more students had joined the group, each person's cost would have been decreased by $4. How many students were in the group?
25. Sally sells wedding floral arrangements and receives a commission for each sale. For a certain number of sales, she received a commission of $384. If she had received $16 more commission for each sale, she would have received the same amount of money with four fewer sales. How many sales did Sally make?
26. Dick wants his vegetable garden to be rectangular in shape, with an area of 140 ft². He wants to put up a wire mesh fence to keep the animals out. If he uses 48 ft of wire mesh fencing, what will be the length and width of the garden?
27. Dick's neighbor, Don, also wants his vegetable garden to be rectangular. (See Exercise 26.) He has a plot of ground 14 yd by 20 yd for his garden. However, he wants a uniform

strip of grass around it for a border. If the area of the garden is to be 160 yd², how wide a strip of grass should Don have?

28. Working together, Krystle and Sonya can paint the inside of their garage in 2 hr. Krystle, working alone, can do the job in 3 hr less time than Sonya working alone. How long would it take each to do the job by herself?
29. A group of recreational leadership majors rented a camper for $320 for a one-week camping trip. Daryl was unable to join them but if he had, each person's share of the rental would have been reduced by $16. How many people were in the original group?
30. Dave and Brian install carpets. Working together, they can finish a carpet job in 8 hr. Working alone, Brian could finish the job in 12 hr less time than Dave working alone. How long would it take each person to do the job working alone?

16.7 The Pythagorean Theorem

OBJECTIVES

After completing this section, you should be able to:
1. Determine the length of one side of a right triangle, given the lengths of the other two sides.
2. Determine if a triangle is a right triangle by using the Pythagorean theorem.

We will conclude this chapter with a topic that involves measurement, square roots, and quadratic equations: the Pythagorean theorem. It pertains to right triangles.

In Section 5.2, we defined a rectangle to be a parallelogram with four right angles (that is, square corners). In Figure 16.1, we have the rectangle $ABCD$. The line segment from A to C is called a "diagonal" of the rectangle. This diagonal divides the rectangle into two right triangles. A **right triangle** is a triangle that has a right angle (that is, a square corner).

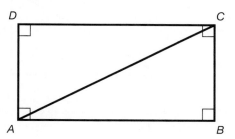

FIGURE 16.1

The longest side of a right triangle is called the **hypotenuse.** The other two sides are called the **legs of the right triangle.** The hypotenuse of a right triangle is the side that is opposite the right angle. (See Figure 16.2.)

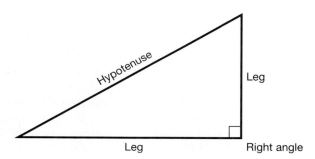

FIGURE 16.2

The Pythagorean Theorem

We will now state the **Pythagorean theorem.** You should note that the Pythagorean theorem pertains only to right triangles.

The Pythagorean Theorem Given a right triangle, the square of the length of the hypotenuse is equal to the sum of the squares of the lengths of the two legs.

Symbolically, if a right triangle has legs of lengths a and b and hypotenuse of length c, then

$$c^2 = a^2 + b^2$$

(See Figure 16.3.)

The Pythagorean theorem can be verified by examining Figure 16.4. The large square has sides of measure $(a + b)$. Therefore, from Section 5.4, the area of the large square region is $(a + b)^2$. The inside square had sides of measure c. Therefore, the area of the inside square region is c^2. In Figure 16.4, there are four triangular regions. Each of these has a base of a units and a height of b units. Therefore, the area of each triangular region is $\frac{1}{2}ab$.

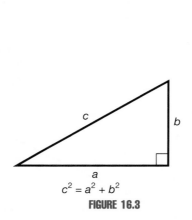

$c^2 = a^2 + b^2$

FIGURE 16.3

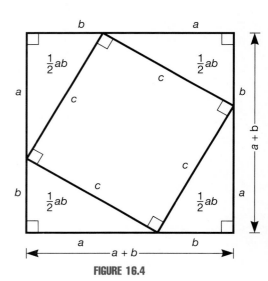

FIGURE 16.4

Now, observe that the area of the large square region, $(a + b)^2$, is equal to the area of the inside square region, c^2, plus the area of the four triangular regions, $4\left(\frac{1}{2}ab\right)$. We have

$$(a + b)^2 = c^2 + 4\left(\frac{1}{2}ab\right)$$
$$a^2 + 2ab + b^2 = c^2 + 2ab$$
$$a^2 + b^2 = c^2 \qquad \text{(Subtracting } 2ab \text{ from both sides)}$$

Notice that a and b are the measures of the legs of each of the four triangular regions and c is the measure of the hypotenuse.

Using the Pythagorean theorem, we can now determine the length of any side of a right triangle if we know the lengths of the other two sides.

16.7 The Pythagorean Theorem

For each of the following, let a and b represent the lengths of the legs of a right triangle, and let c represent the length of its hypotenuse.

Practice Exercise 1. Determine c if $a = 12$ cm and $b = 5$ cm.

EXAMPLE 1 A right triangle has legs of lengths 6 in. and 8 in. Determine the length of its hypotenuse.

SOLUTION: Using the Pythagorean theorem, we have

$$c^2 = a^2 + b^2$$
$$x^2 = 8^2 + 6^2 \quad \text{(Substituting)}$$
$$= 64 + 36$$
$$= 100$$

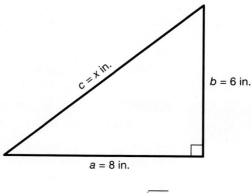

$$x = \pm\sqrt{100}$$
$$= \pm 10 \text{ in.}$$

However, since x represents the length of the hypotenuse, x cannot be negative. Therefore, we take the *principal* square root of 100 and conclude that

$$x = 10 \text{ in.}$$

Practice Exercise 2. Determine b if $a = 7$ ft and $c = 10$ ft.

EXAMPLE 2 A right triangle has a leg of length 5 cm and a hypotenuse with length of 13 cm. Determine the length of the other leg.

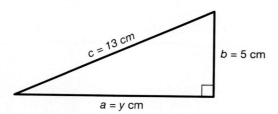

SOLUTION: Using the Pythagorean theorem, we have

$$c^2 = a^2 + b^2$$
$$13^2 = y^2 + 5^2$$
$$169 = y^2 + 25$$
$$144 = y^2$$
$$y = \sqrt{144} \quad \text{(Principal square root)}$$
$$= 12 \text{ cm}$$

Practice Exercise 3. Determine a if $b = 13$ dm and $c = 17$ dm.

EXAMPLE 3 A right triangle has legs of lengths 2 ft and 4 ft. Determine the length of its hypotenuse.

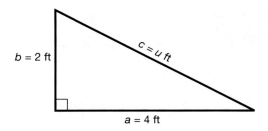

SOLUTION: Using the Pythagorean theorem, we have

$$c^2 = a^2 + b^2$$
$$u^2 = 4^2 + 2^2$$
$$u^2 = 16 + 4$$
$$u^2 = 20$$
$$u = \sqrt{20}$$
$$= 2\sqrt{5} \text{ ft}$$

Practice Exercise 4. Determine c if $a = 10$ yd and $b = 10$ yd.

EXAMPLE 4 A right triangle has a hypotenuse of length 17 yd and a leg of length 11 yd. Determine the length of the other leg.

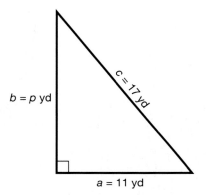

SOLUTION: Using the Pythagorean theorem, we have

$$c^2 = a^2 + b^2$$
$$17^2 = 11^2 + p^2$$
$$289 = 121 + p^2$$
$$p^2 = 168$$
$$p = \sqrt{168}$$
$$= 2\sqrt{42} \text{ yd}$$

Practice Exercise 5. Determine a if $b = 14$ in. and $c = 17$ in.

EXAMPLE 5 A 36-ft ladder is placed against a building, with the foot of the ladder 9 ft from the base of the building. How far above the ground does the top of the ladder lean against the building? (See accompanying figure.)

SOLUTION: To solve this problem, we use the Pythagorean theorem. Referring to the accompanying figure, we let

$d =$ the height of the top of the ladder above the ground (in feet)

Hence, d becomes the length of one leg of the right triangle, 9 ft is the length of the other leg, and 36 ft is the length of the hypotenuse. We now have

16.7 The Pythagorean Theorem

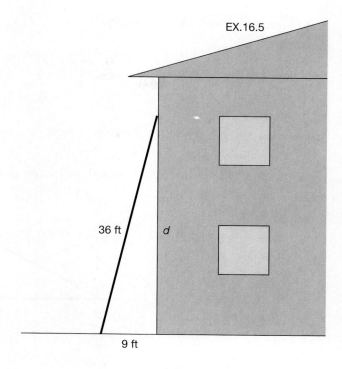

EX.16.5

$(36 \text{ ft})^2 = (d)^2 + (9 \text{ ft})^2$
$1296 \text{ ft}^2 = d^2 + 81 \text{ ft}^2$
$d^2 = 1215 \text{ ft}^2$
$d = \sqrt{1215 \text{ ft}^2}$
$\approx 34.8 \text{ ft}$

Therefore, the top of the ladder leans against the building approximately 34.8 ft above the ground.

ANSWERS TO PRACTICE EXERCISES

1. $c = 13$ cm.   **2.** $b = \sqrt{51}$ ft.   **3.** $a = 2\sqrt{30}$ dm.   **4.** $c = 10\sqrt{2}$ yd.   **5.** $\sqrt{93}$ in.

We can also use the Pythagorean theorem to determine if a given triangle is or is not a right triangle.

For each of the following, let *a* and *b* represent the lengths of the legs of a right triangle and let *c* represent the length of its hypotenuse. Determine if each is or is not a right triangle.

Practice Exercise 6. $a = 4$ ft, $b = 5$ ft, and $c = 6$ ft

EXAMPLE 6 Using the Pythagorean theorem, determine if the triangle in the accompanying figure is a right triangle.

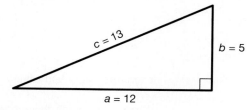

SOLUTION: If the given triangle is a right triangle, we must have $c^2 = a^2 + b^2$.

CHECK: We have

$$c^2 = a^2 + b^2$$
$$13^2 \; ? \; 12^2 + 5^2$$
$$169 \; ? \; 144 + 25$$
$$169 = 169 \checkmark$$

Therefore, the given triangle *is* a right triangle.

Practice Exercise 7. $a = 7$ cm, $b = 8$ cm, and $c = \sqrt{113}$ cm

EXAMPLE 7 Determine if the triangle given in the accompanying figure is a right triangle.

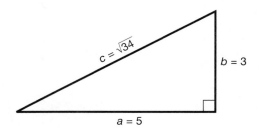

SOLUTION: Does $c^2 = a^2 + b^2$?

CHECK: We have

$$c^2 = a^2 + b^2$$
$$(\sqrt{34})^2 \; ? \; 5^2 + 3^2$$
$$34 \; ? \; 25 + 9$$
$$34 = 34 \checkmark$$

Therefore, the given triangle *is* a right triangle.

Practice Exercise 8. $a = 3$ dm, $b = 11$ dm, and $c = \sqrt{130}$ dm

EXAMPLE 8 Determine if the triangle given in the accompanying figure is a right triangle.

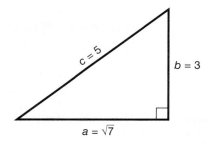

SOLUTION: Does $c^2 = a^2 + b^2$?

CHECK: We have

$$c^2 = a^2 + b^2$$
$$5^2 \; ? \; (\sqrt{7})^2 + 3^2$$
$$25 \; ? \; 7 + 9$$
$$25 \neq 16$$

Therefore, the given triangle is *not* a right triangle.

When using the Pythagorean theorem, the measures for all three sides of the triangle must be in the *same* units. If they are not, then convert one or more of them to a common unit.

16.7 The Pythagorean Theorem

EXAMPLE 9 Determine if the triangle given in the accompanying figure is a right triangle.

Practice Exercise 9. $a = 5$ ft, $b = 4$ yd, and $c = 13$ ft

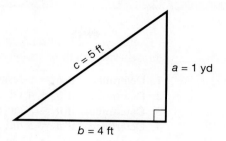

SOLUTION: We first observe that different units of measure are used in this example. Converting 1 yd = 3 ft, we then proceed to determine if $c^2 = a^2 + b^2$.

CHECK: We have

$$c^2 = a^2 + b^2$$
$$5^2 \; ? \; 3^2 + 4^2$$
$$25 \; ? \; 9 + 16$$
$$25 = 25 \; ✔$$

Therefore, the given triangle *is* a right triangle.

EXAMPLE 10 To "square" the wall of a house being built by a carpenter, measurements are taken, as shown in the accompanying figure. Is the wall "square?"

Practice Exercise 10. $a = 8$ yd, $b = 13$ yd, and $c = 16$ yd

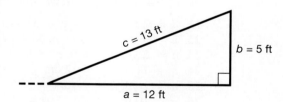

SOLUTION: If the wall is "square," then the triangle indicated in the accompanying figure will be a right triangle and c^2 must be equal to $a^2 + b^2$.

CHECK: We have

$$c^2 = a^2 + b^2$$
$$(13 \text{ ft})^2 \; ? \; (12 \text{ ft})^2 + (5 \text{ ft})^2$$
$$169 \text{ ft}^2 \; ? \; 144 \text{ ft}^2 + 25 \text{ ft}^2$$
$$169 \text{ ft}^2 = 169 \text{ ft}^2 \; ✔$$

Since $c^2 = a^2 + b^2$, then the wall is "square."

ANSWERS TO PRACTICE EXERCISES

6. Not a right triangle. **7.** Is a right triangle. **8.** Is a right triangle. **9.** Is a right triangle. **10.** Not a right triangle.

EXERCISES 16.7

In Exercises 1–20, let a and b represent the lengths of the legs of a right triangle and let c represent the length of its hypotenuse.

1. Determine c if $a = 2$ ft and $b = 3$ ft.
2. Determine c if $a = 4$ yd and $b = 5$ ft.
3. Determine c if $a = 3$ cm and $b = 8$ cm.
4. Determine c if $a = 5$ dm and $b = 12$ dm.
5. Determine c if $a = 11$ in. and $b = 15$ in.
6. Determine c if $a = 6.2$ ft and $b = 7.3$ ft.
7. Determine c if $a = 4.9$ yd and $b = 2.6$ yd.
8. Determine c if $a = 12.4$ cm and $b = 14.1$ cm.
9. Determine a if $b = 7$ in and $c = 10$ in.
10. Determine a if $b = 8$ dm and $c = 17$ dm.
11. Determine a if $b = 5$ ft and $c = 7.8$ ft.
12. Determine a if $b = 6.3$ yd and $c = 9.1$ yd.
13. Determine b if $a = 9$ cm and $c = 13$ cm.
14. Determine b if $a = 5.6$ in. and $c = 11.1$ in.
15. Determine b if $a = 2.3$ ft and $c = 13.9$ ft.
16. Determine b if $a = 13$ dm and $c = 17.5$ dm.
17. Determine c if $a = 3$ ft and $b = 4$ yd.
18. Determine a if $b = 2.6$ cm and $c = 0.84$ dm.
19. Determine b if $a = 13$ in. and $c = 0.8$ yd.
20. Determine c if $a = \sqrt{3}$ ft and $b = \sqrt{17}$ ft.

In Exercises 21–30, let, a, b, and c represent the lengths of the sides of a triangle. Using the Pythagorean theorem, determine if the triangle is or is not a right triangle.

21. $a = 3$ ft, $b = 3$ ft, $c = 3\sqrt{2}$ ft
22. $a = 9$ cm, $b = 10$ cm, $c = 14$ cm
23. $a = 2$ yd, $b = 8$ ft, $c = 10$ ft
24. $a = 11$ dm, $b = 12$ dm, $c = 15$ dm
25. $a = 2$ yd, $b = 2$ yd, $c = 6$ ft
26. $a = 10$ in., $b = 24$ in., $c = 26$ in.
27. $a = 8$ cm, $b = 15$ cm, $c = 17$ cm
28. $a = \sqrt{17}$ ft, $b = \sqrt{17}$ ft, $c = \sqrt{34}$ ft
29. $a = \sqrt{7}$ m, $b = \sqrt{13}$ m, $c = \sqrt{62}$ m
30. $a = 9$ yd, $b = \sqrt{17}$ yd, $c = 7\sqrt{2}$ yd

31. A 24-ft ladder is leaning against the wall of a building. The top of the ladder is 17 ft above the ground. How far away is the foot of the ladder from the base of the building?
32. A ladder is leaning against the wall of a building. The top of the ladder is 15 ft above the ground. The foot of the ladder is 8 ft away from the base of the building. How long is the ladder?
33. An airplane is flying at an altitude of 5 mi on a flight path that is directly over a radar tracking station. If the horizontal distance between the plane and the tracking station is 12 mi, what is the distance between the plane and the station?

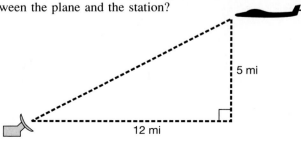

34. A television camera at ground level is filming the lift-off of a space shuttle that is rising vertically. The camera is 1500 ft from the launch pad. How far apart are the camera and the shuttle when the shuttle is 1 mi directly above the launch pad? (*Hint:* 1 mi = 5280 ft.)

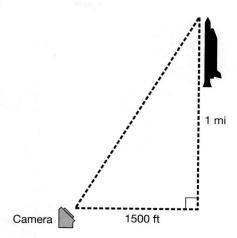

35. An air traffic controller is tracking two planes at the same altitude converging on a point in space as they fly at right angles to each other. One plane is 75 mi from the point when the other plane is 100 mi from the point. How far apart are the two planes?

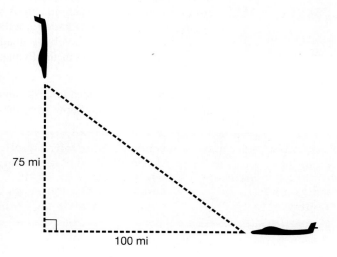

36. Referring to Exercise 35, the plane that is 75 mi from the point is flying at the rate of 450 mph. The plane that is 100 mi from the point is flying at the rate of 600 mph. How much time does the traffic controller have to get one of the planes on a different path to avoid a collision?
37. A baseball diamond is in the shape of a square. The vertices of the square are the three bases and the home plate. If the distance from first base to second base is 90 ft, what is the distance between home plate and the second base? (*Hint:* Consider a sketch of the baseball diamond.)
38. A 36-ft utility pole is sunk 6 ft below ground level. A guy wire supports the pole and is attached to a stake in the ground 18 ft from the base of the pole. If the other end of the wire is attached to a hook on the pole and is 24 ft above the ground, what is the distance between the hook on the pole and the stake in the ground?
39. A jogger started from point A and traveled 4 km south to point B and then 5 km east to point C. After resting for 1 hr, she then traveled 7 km back to point A. (See the accompanying figure.) Did the total path traveled by the jogger form a right triangle?

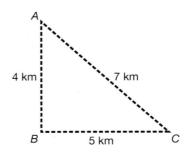

40. A roof rafter is needed to span a length of 15 ft for a roof rise of 8 ft. A carpenter cuts a rafter of length 18 ft. Will the rafter span the distance?

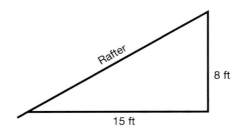

Chapter Summary

In this chapter, you were introduced to quadratic equations and their solutions. The methods discussed for solving such equations were factoring, using square roots, completing the square, and using the quadratic formula. The discriminant of a quadratic equation was also discussed. Applications involving quadratic equations were considered. We concluded the chapter with a discussion of the Pythagorean theorem.

The **key words** introduced in this chapter are listed here in the order in which they appeared in the text. You should know what they mean.

quadratic equation in a single variable (p. 656)	solving a quadratic equation by the method of factoring (p. 658)	discriminant (p. 674)
standard form for a quadratic equation (p. 656)	perfect square (p. 663)	right triangle (p. 681)
leading coefficient (p. 656)	completing the square (p. 665)	hypotenuse (p. 681)
	quadratic formula (p. 671)	leg of the right triangle (p. 681)
		Pythagorean theorem (p. 682)

Review Exercises

16.1 The Standard Form of a Quadratic Equation

Rewrite each of the given equations in its standard form.

1. $2x^2 - 4 = 5x$
2. $2(x - 3)^2 - 5x = 2$
3. $3y^2 - (y + 1)(y - 3) = 0$
4. $y^2 - 4 = (3y - 5)^2$
5. $u - 5u^2 = (u + 1)^2$
6. $u(u + 1)(u - 2) = u(u - 4)^2$

16.2 Solving Quadratic Equations by Factoring

Solve each of the following equations by factoring.

7. $2x^2 + 5x = 12$
8. $3 = 10x + 8x^2$
9. $6y^2 - 21y = 28 - 8y$
10. $15(y^2 - 1) + 25y = 9y$
11. $6t^2 - 44 = 25t$
12. $21t^2 - 13t = 20$

16.3 Solving Quadratic Equations Using Square Roots

Solve each of the following equations using the square root method.

13. $y^2 - 64 = 0$
14. $3p^2 - 243 = 0$
15. $(x - 4)^2 = 9$
16. $(2y - 1)^2 = 16$
17. $w^2 + 6w + 9 = 25$
18. $16p^2 - 8p + 1 = 5$

16.4 Solving Quadratic Equations by Completing the Square

Solve each of the following equations by the method of completing the square.

19. $x^2 + 5x - 2 = 0$
20. $y^2 - 3y = 4$
21. $u^2 + u = 16$
22. $2s^2 - 3s - 4 = 0$
23. $3t^2 = 6 - t$
24. $2p - 5p^2 + 7 = 0$

16.5 The Quadratic Formula

In Exercises 25–28, compute the discriminant, and then determine whether the solutions for the given equations exist or are complex numbers.

25. $2x^2 - 7x = 13$
26. $6y - 4y^2 = 11$
27. $9 - 7t^2 = 5t$
28. $7w - 3w^2 = 0$

In Exercises 29–34, solve the given equations by using the quadratic formula.

29. $(2x - 1)(3x + 2) = 0$
30. $3x^2 - 4 = 0$
31. $y^2 - 5y - 9 = 0$
32. $4y - 5y^2 + 8 = 0$
33. $4u + 3 = 7u^2$
34. $9 - 6u^2 = 7u$

16.6 Applications Involving Quadratic Equations

Solve each of the following word problems by setting up and solving an appropriate quadratic equation.

35. One number is 3 less than another number. Their product is 304. What are the numbers?
36. The width of a rectangular region is 5 cm less than its length. The area of the region is 204 cm^2. What are the dimensions of the rectangular region?
37. The base of a triangular region is 2 yd less than twice its height. The area of the region is 110 yd^2. Determine the height and the base of the region.
38. The sum of a number and 4 is equal to 21 times the reciprocal of the number. What is the number?

16.7 The Pythagorean Theorem

In Exercises 39–42, let a and b represent the lengths of the legs of a right triangle and let c represent the length of its hypotenuse.

39. Determine c if $a = 6$ in. and $b = 10$ in.
40. Determine a if $b = 12$ cm and $c = 20$ cm.
41. Determine b if $a = 3$ yd and $c = 15$ ft.
42. Determine c if $a = 7$ dm and $b = 7$ dm.

In Exercises 43–46, let a, b, and c represent the lengths of the sides of a triangle. Using the Pythagorean theorem, determine if the triangle is a right triangle.

43. $a = 4$ yd, $b = 4$ yd, $c = 4\sqrt{2}$ yd
44. $a = 9$ cm, $b = 11$ cm, $c = 13$ cm
45. $a = \sqrt{11}$ yd, $b = \sqrt{13}$ yd, $c = 2\sqrt{6}$ yd
46. $a = \sqrt{17}$ in., $b = \sqrt{19}$ in., $c = \sqrt{37}$ in.

Teasers

1. The sum of the squares of two consecutive integers is 313. What are the integers?
2. If 5 times an integer is added to the square of the integer, the sum is 126. What is the integer?
3. If one solution of the equation $x^2 - 5x + k = 0$ is 7, determine the other solution and the value of k.
4. Write a quadratic equation, in the variable s, with integer coefficients, if the solutions of the equation are $\frac{-3}{2}$ and 5.
5. The length of a rectangular region is 9 cm less than twice its width. The area of the region is 180 cm². Determine the dimensions of the region.
6. A 26-ft ladder leans against a vertical wall. The foot of the ladder is 10 ft away from the base of the wall. When the top of the ladder is 17 ft above the ground level, how far away is the foot of the ladder from the base of the wall? (Leave your answer in simplified radical form.)
7. The sum of twice a number and its reciprocal is $14\frac{1}{7}$. Determine the number.
8. If the hypotenuse of a right triangle is 26 yd and one side of the triangle measures 10 yd, determine the following:
   a. The length of the other side
   b. The perimeter of the triangle
9. If the two legs of a right triangle measure 9 m and 12 m, respectively, determine the following:
   a. The measure of the hypotenuse
   b. The area of the triangular region
10. In the shortest side of a right triangle measures 7 ft and the area of the triangular region is 84 ft², determine the following:
    a. The measure of the hypotenuse
    b. The perimeter of the triangle

# CHAPTER 16 TEST

This test covers material from Chapter 16. Read each question carefully, and answer all questions. If you miss a question, review the appropriate section of the text.

1. Rewrite the equation $3x^2 - 4 = 6x$ in its standard form.
2. Rewrite the equation $(2y - 1)^2 = 6y^2 - y$ in its standard form.
3. What must be added to complete the square for $x^2 + 20x$?
4. What must be added to complete the square for $y^2 + 9y$?
5. Solve the equation $2y^2 + 7y - 4 = 0$ by the method of factoring.
6. Solve the equation $2x + 7 = 5x^2$ by the method of completing the square.
7. Solve the equation $2u^2 + 5u = 12$ by using the quadratic formula.
8. Compute the discriminant for the equation $5t^2 - 2t + 7 = 0$, and determine if the solutions of the equation exist or are complex numbers.

In Exercises 9–14, solve the given equation using any method.

9. $2x^2 - 5x - 1 = 0$
10. $2y^2 - 11y - 21 = 0$
11. $6t^2 - 20t = 28 - 7t$
12. $y^2 + y - 1 = 0$
13. $2u - 1 = (u - 1)^2$
14. $2s^2 - s + 5 = 1 - 10s$

In Exercises 15–16, use a quadratic equation to solve each of the problems.

15. One number is 3 more than another number. Their product is 70. What are the two numbers?
16. The length of a rectangular region is 4 cm more than its width. If the width of the region is increased by 1 cm and the length is decreased by 1 cm, the area of the new region formed will be 48 cm². What are the dimensions of the original region?

In Exercises 17–18, let a and b represent the lengths of the legs of a right triangle and let c represent the length of its hypotenuse.

17. Determine c if $a = 10$ cm and $b = 15$ cm.
18. Determine a if $b = 11$ ft and $c = 13$ ft.

In Exercises 19–20, let a, b, and c represent the lengths of the sides of a triangle. Using the Pythagorean theorem, determine if the triangle is a right triangle.

19. $a = 3.3$ cm, $b = 4.4$ cm, $c = 5.5$ cm
20. $a = 7$ ft, $b = 11$ ft, $c = 5\sqrt{7}$ ft

CUMULATIVE REVIEW: CHAPTERS 1–16

1. Name two arithmetic operations that are not commutative.
2. Name two arithmetic operations that are associative.
3. Which is larger, 2^3 or 3^2?
4. Which is larger, $(-2)^3$ or 3^{-2}?
5. Round to the nearest hundredth, the product of 2.36 and 1.07.
6. The length of a rectangular region is 3 less than twice its width. If the perimeter of the region is 65.4 ft, determine the dimensions of the region.
7. Solve the equation $2(x - 1) = 4x$ for x.
8. Solve the inequality $1 - 3(3 - 3y) \leq 8$ for y.
9. Solve the equation $2u^2 - 3u - 5 = 0$ for u.
10. Compute the discriminant for the equation $(2p - 3)(3p + 1) = 6$.
11. What must be added to complete the square for $t^2 + 5t$?
12. Given the equation $2x - 3y = 7$, determine the value of y when $x = -3$.
13. Determine the slope of the line which is the graph of the equation $5x + 3y - 8 = 0$.
14. Determine the area of the rectangular region given in Exercise 6 above.
15. Rewrite the equation $4x + 5y - 7 = 0$ in the slope-intercept form.
16. Simplify by combining like terms: $2x(x^2 - 3) + x^2(1 - x) - 5(x - 2x^2)$.
17. Simplify by combining like terms: $2a - 3[b - 2(a - 2b) + 4a] - 6b$.
18. Simplify and write your result without negative exponents:

$$\left(\frac{2x^{-3}y^4z}{3x^4y^{-2}z^0}\right)^{-2}$$

19. Determine all excluded values for $\dfrac{x}{(2x - 3)^2} + \dfrac{3x - 1}{x^2 + 5x + 6}$.
20. Using slopes, determine whether the graph of the following system of equations is a pair of intersecting, parallel, or coinciding lines:

$$\begin{cases} x + 2y = 3 \\ 3x - 2y = 5 \end{cases}$$

21. Determine if the ordered pair $(-2, 3)$ is a solution of the equation $3y - 2x = 0$.
22. If the ordered pair $(3, b)$ is a solution of the equation $4x - 3y = 9$, determine the value of b.
23. Simplify by combining like terms: $\sqrt{18} - 2\sqrt{12} + 3\sqrt{75} - 6\sqrt{8}$.
24. Determine the reciprocal of $\dfrac{-2}{3} + \dfrac{4}{7}$.
25. Using the Pythagorean theorem, determine whether the triangle whose sides have lengths 8 cm, 13 cm, and 17 cm is a right triangle.

CHAPTER 17: Basic Concepts of Geometry

Throughout this text, we have been almost exclusively concerned with the arithmetic and algebra of numbers and their properties.

However, in Chapter 5, we introduced measurement and considered the geometric measures of length, area, and volume. Triangles, squares, rectangles, parallelograms, and trapezoids were considered as special polygons. The formulas for the perimeter and area of each of these figures were used to solve problems related to these geometric measures. The circumference and area of circles were introduced. Further, the surface area and volume of rectangular solids and the volume of other solids were also discussed.

In this chapter, additional basic geometric concepts are discussed. The inclusion of this material will introduce you to another branch of mathematics—geometry—and should help to increase your skills in mathematical reasoning.

CHAPTER 17 PRETEST

This pretest covers material on the basic concepts of geometry. Read each question carefully, and try to answer all questions. Each question is keyed to the chapter section in which the particular topic is discussed. Check your answers with those given in the back of the text.

Questions 1–3 pertain to Section 17.1.

Name each of the symbols given.

1. $\overline{CD}$
2. $\overrightarrow{RS}$
3. $\overleftrightarrow{EF}$

Questions 4–6 pertain to Section 17.2.

4. What is the measure of a right angle?

5. What is the name of an angle whose measure is 39°?

6. If $\overrightarrow{OC}$ bisects $\angle AOB$ and $m(\angle AOC) = 42°$, what is $m(\angle AOB)$?

Questions 7–8 pertain to Section 17.3.

7. What is the measure of the complement of an angle whose measure is 44°?

8. If angles A and B are vertical angles with $m(\angle A) = y°$ and $m(\angle B) = (4y - 9)°$, determine the value of y.

Questions 9–10 pertain to Section 17.4.

9. If two lines are cut by a transversal, are the alternate interior angles always congruent?

10. If the lines p, q, and r are such that $p \parallel q$ and $q \parallel r$, what is the relationship between p and r?

Questions 11–12 pertain to Section 17.5.

11. Given $\triangle JKL$ with $m(\angle J) = 38°$ and $m(\angle K) = 49°$, determine $m(\angle L)$.

12. If the measure of a base angle of an isosceles triangle is 41°, determine the measure of the vertex angle of the triangle.

Questions 13–14 pertain to Section 17.6.

In Exercises 13–14, the measures for parts of $\triangle ABC$ and $\triangle DEF$ are given. Determine if the triangles are similar.

13. $m(\angle A) = 61°$, $m(\angle B) = 43°$; $m(\angle D) = 73°$

14. $AB = 6$ cm, $AC = 10$ cm, $BC = 8$ cm; $DE = 3$ cm, $DF = 4$ cm, $EF = 5$ cm

Questions 15–16 pertain to Section 17.7.

15. Given $\triangle RST$ with $m(\angle R) = 30°$, $m(\angle S) = 90°$, $RT = 20$ yd, determine RS.

16. Given $\triangle NOP$ with $m(\angle N) = m(\angle P) = 45°$, $OP = 8$ cm, determine NP.

Questions 17–18 pertain to Section 17.8.

17. Given $\triangle ABC \cong \triangle DEF$ with $\angle A \cong \angle D$ and $\angle C \cong \angle F$. List the four remaining congruent corresponding parts.

18. Does the diagonal of a parallelogram separate it into two congruent triangles? Give an appropriate reason for your answer.

17.1 Terminology and Notation

OBJECTIVE

After completing this section, you should be familiar with the terminology and notation used to describe points, lines, line segments, rays, and planes.

In the earlier chapters of this text, we made use of the number line and referred to points on the line. We indicated points as dots and named them by capital letters such as *P* and *Q* as in Figure 17.1.

It is to be noted that the dots in Figure 17.1 have some measure but the points they name have no measure. A *point* is the most basic figure in geometry. All other geometric figures consist of sets of points.

> **DEFINITION**
>
> A **line** is a set of points that extends indefinitely in opposite directions.

Any two points on the line can be used to name the line. In Figure 17.2, the line illustrated can be referred to by using the symbol $\overleftrightarrow{RT}$ (which is read "the line *RT*" or "the line passing through the points *R* and *T*"). The symbols $\overleftrightarrow{RS}$, $\overleftrightarrow{ST}$, $\overleftrightarrow{SR}$, $\overleftrightarrow{TR}$, or $\overleftrightarrow{TS}$ could also be used to name the line. A line can also be named by a lower-case letter. In Figure 17.2, the line can be named simply as *l*.

Parts of lines are also named.

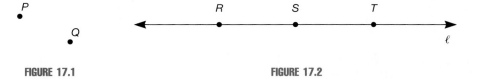

FIGURE 17.1   **FIGURE 17.2**

> **DEFINITION**
>
> A **line segment** is the part of a line consisting of two points on the line and all the points between them. The two points are called its **endpoints**.

FIGURE 17.3

A line segment is named by its endpoints. In Figure 17.3, the line segment with endpoints *C* and *D* is named $\overline{CD}$ (which is read "line segment *CD*") or $\overline{DC}$.

A line segment has *length*, which is the distance between its endpoints. The length of the line segment $\overline{CD}$ is indicated by either the symbol *CD* or $m(\overline{CD})$ where $m(\overline{CD})$ is read "the measure (length) of $\overline{CD}$."

If two line segments have the same length, then they are **congruent**. If the line segments $\overline{EF}$ and $\overline{GH}$ have the same length, we indicate this fact by writing any of the following:

1. $EF = GH$
2. $m(\overline{EF}) = m(\overline{GH})$
3. $\overline{EF} \cong \overline{GH}$ (which is read "$\overline{EF}$ is congruent to $\overline{GH}$")

17.1 Terminology and Notation

For these Practice Exercises, refer to the figure.

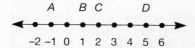

Practice Exercise 1. Write two different symbols for the line segment with endpoints B and D.

EXAMPLE 1 Determine the lengths of $\overline{AB}$ and $\overline{CD}$

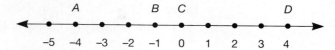

SOLUTION: To determine $m(\overline{AB})$, we could count the number of units from A to B, which is 3. Therefore, $m(\overline{AB}) = 3$. Similarly, $m(\overline{CD}) = 4$.

ALTERNATIVE SOLUTION: To determine $m(\overline{AB})$, we could subtract the coordinate of A from the coordinate of B. Hence, we have

$$m(\overline{AB}) = (-1) - (-4)$$
$$= (-1) + (+4)$$
$$= 3$$

Similarly,
$$m(\overline{CD}) = 4 - 0$$
$$= 4$$

Practice Exercise 2. Determine $m(\overline{AC})$.

EXAMPLE 2 Determine if $\overline{EF}$ and $\overline{GH}$ are congruent.

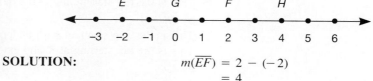

SOLUTION:
$$m(\overline{EF}) = 2 - (-2)$$
$$= 4$$
$$m(\overline{GH}) = 4 - 0$$
$$= 4$$

Since $m(\overline{EF}) = m(\overline{GH})$, then $\overline{EF} \cong \overline{GH}$.

Every line segment contains a point that divides it into two congruent line segments. Such a point is called its **midpoint**. In Figure 17.4, we note that U is the midpoint of $\overline{XY}$. We have that $m(\overline{XU}) = (-1) - (-3) = 2$ and $m(\overline{UY}) = 1 - (-1) = 2$. Therefore, $\overline{XU} \cong \overline{UY}$ and U is the midpoint of $\overline{XY}$.

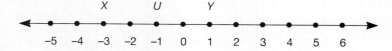

FIGURE 17.4

In Figure 17.5, on the number line, the point A is located at -2 and the point B is located at 4. We determine that $m(\overline{AB}) = 4 - (-2) = 6$. The midpoint of $\overline{AB}$, therefore, would be the point that is located on the line halfway between A and B, or 3 units to the right of A. The midpoint of $\overline{AB}$ would be located at $(-2) + 3 = 1$ on the number line.

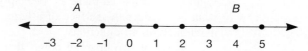

FIGURE 17.5

Now, suppose that we start at a point on a line and extend indefinitely in one of the directions along the line. What is that part of the line?

> **DEFINITION**
>
> A **ray** is the part of a line that starts at a point on the line and extends indefinitely in one direction. The point is called the "endpoint of the ray."

In Figure 17.6a, the ray from A and in the direction of B is denoted by $\overrightarrow{AB}$ (which is read "the ray AB"). Similarly, the ray in Figure 17.6b is denoted by $\overrightarrow{BA}$.

FIGURE 17.6

> **Practice Exercise 3.** What is the midpoint of $\overline{AD}$?

EXAMPLE 3 Name each of the following geometric figures.

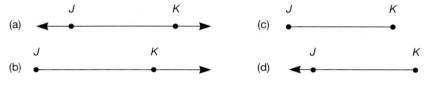

SOLUTION:

a. This is a line determined by the two points, J and K; it is denoted by $\overleftrightarrow{JK}$.
b. This is the ray staring at J and extending in the direction of K; it is denoted by $\overrightarrow{JK}$.
c. This is the line segment with endpoints J and K; it is denoted by $\overline{JK}$.
d. This is the ray starting at K and extending in the direction of J; it is denoted by $\overrightarrow{KJ}$.

> **ANSWERS TO PRACTICE EXERCISES**
>
> 1. $\overline{BD}, \overline{DB}$.   2. 3.   3. C.

In addition to the geometric figures discussed so far, we also have planes. We have already dealt with planes in Chapter 13 when learning about the rectangular coordinate system.

> **DEFINITION**
>
> A **plane** is a set of points lying on a flat surface and extending indefinitely in all directions.

It is impossible to draw a figure that extends indefinitely in all directions. Hence, we will represent a plane by a figure such as that given in Figure 17.7. The plane denoted in Figure 17.7 is named plane P. To help you get a better mental picture of a plane, you may want to consider the surface of a large, flat parking lot and imagine that surface extended indefinitely in all directions.

A set of two or more points that lie on the same plane are called **coplanar points.** Similarly, a set of two or more points that lie on the same line are called **collinear points.**

We will conclude this section with some basic properties of points, lines, and planes.

FIGURE 17.7

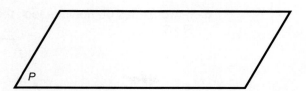

Properties of Points, Lines, and Planes

1. Two different points determine one and only one line.

2. Three *noncollinear* points determine one and only one plane.

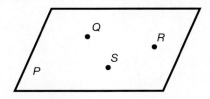

3. If two distinct lines intersect, then their intersection is a point.

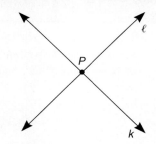

4. If two lines lie in the same plane and do not intersect, then the lines are parallel.

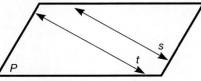

We use the symbol ∥ to denote "is parallel to." Hence, in the preceding figure, $s \parallel t$.

5. If two planes intersect, then their intersection is a line.

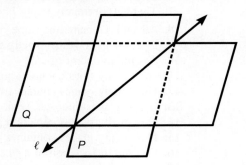

6. If two planes do not intersect, then they are parallel planes. In the figure here, $R \parallel S$.

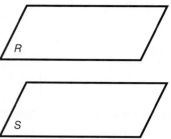

To help you visualize intersecting and parallel planes, consider the following. A wall of a room "intersects" with the floor of the room. Also, two adjacent walls intersect. However, two shelves on a bookshelf do not intersect; they are parallel.

EXERCISES 17.1

For Exercises 1–10, refer to the following figure and classify each statement as being true or false. *Consider all points and lines in plane P and not just those illustrated in the figure.*

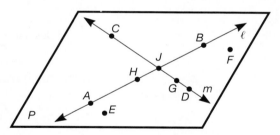

1. $\overleftrightarrow{AB}$ and $\overleftrightarrow{HD}$ name the same line.
2. The points C, D, and J are collinear.
3. $l \parallel m$.
4. The points C and F determine exactly one line.
5. The points A, H, and B are coplanar.
6. There are only two lines in P that contain the point J.
7. There are infinitely many lines in P that contain the point G.
8. $\overrightarrow{AB}$ and $\overrightarrow{AJ}$ name the same ray.
9. The line m contains exactly three points.
10. J is the midpoint of $\overline{HB}$.
11. How many lines contain a given point?
12. How many planes contain two given points?
13. If two planes intersect, what is their intersection?
14. How many planes contain a given line?
15. How many planes contain a given line and a given point not on the line?
16. Can the intersection of a line and a plane be a point?
17. Can the intersection of a line and a plane be a line?

In Exercises 18–29, refer to the following figure.

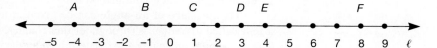

18. Determine AC.
19. Determine BF.
20. Determine $m(\overline{CF})$.
21. Determine $m(\overline{AE})$.
22. Name a line segment that is congruent to $\overline{BD}$. (Answers may vary.)
23. Name two line segments that are congruent to $\overline{AC}$. (Answers may vary.)
24. How many points are there on l whose distance from B is 4?
25. How many points are there on $\overrightarrow{BF}$ whose distance from B is 3?
26. How many points are there on $\overline{BD}$ whose distance from C is 4?
27. How many points are common to both $\overline{AD}$ and $\overline{BD}$?
28. What is the coordinate of the midpoint of $\overline{BE}$?
29. What is the coordinate of the midpoint of $\overline{CF}$?
30. If point R lies in plane T, how many planes are there parallel to T that also contain R?

17.2 Angles

OBJECTIVE

After completing this section, you should be able to work with angles (including special angles) and determine their measures.

In Section 17.1, you learned about rays. Sometimes we have two rays with a common endpoint. When this happens, they form another basic geometric figure: an angle. In this section, we will discuss angles and their measures.

DEFINITION

An **angle** is a geometric figure formed by two rays with a common endpoint. Each ray is called a **side of the angle**. The common endpoint is called the **vertex of the angle**.

The symbol $\angle$ is used to denote an angle. Consider an angle that has its vertex at the point O. When there is only one angle with the given point O as its vertex, then the angle can be referred to as $\angle O$, identifying it by its vertex. If A is a point on one side of the angle and B is a point on the other side of the angle, then the angle is referred to as $\angle AOB$ or as $\angle BOA$. (See Figure 17.8.) Note, however, that when three letters are used to name an angle,

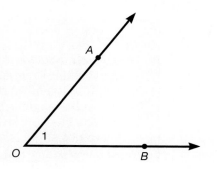

FIGURE 17.8

the letter that names the vertex is always placed between the other two letters. The angle given in Figure 17.8 could also be named as ∠1.

The **measure of an angle** can be given in terms of either a unit called a "degree" or a unit called a "radian." Throughout this chapter, we will use degree measure exclusively. The measure of an angle can be determined by the use of an instrument called a "protractor." A protractor is illustrated in Figure 17.9.

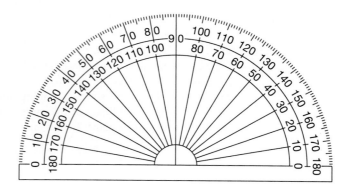

FIGURE 17.9

Notice that a protractor has two scales, an inner scale and an outer scale. Which scale do you use when measuring an angle? If a point on one side of the angle lies to the *right* of O on the horizontal line, then use the inner scale. If a point on one side of the angle lies to the *left* of O on the horizontal line, then use the outer scale. In Figure 17.10, we illustrate that the measure of ∠DOC is 35 degrees (using the outer scale) while the measure of ∠AOB is 50 degrees (using the inner scale). We could also write $m(\angle DOC) = 35°$ (read "35 degrees") and $m(\angle AOB) = 50°$ (read "50 degrees").

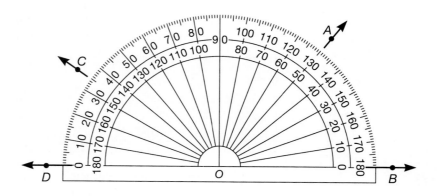

FIGURE 17.10

Angles can be classified according to their measures.

> **DEFINITIONS**
> 1. An **acute angle** is an angle whose measure is greater than 0 degrees but less than 90 degrees.
> 2. A **right angle** is an angle whose measure is 90 degrees.
> 3. An **obtuse angle** is an angle whose measure is greater than 90 degrees but less than 180 degrees.
> 4. A **straight angle** is an angle whose measure is 180 degrees.

17.2 Angles

For Practice Exercises 1 and 2 refer to the following figure.

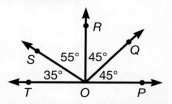

Practice Exercise 1. Determine $m(\angle POR)$.

EXAMPLE 1 Classify the angles $\angle AOB$, $\angle BOC$, $\angle AOC$, $\angle COD$, $\angle AOD$, and $\angle AOE$.

SOLUTION:

- $\angle AOB$ is an acute angle.
- $\angle BOC$ is an acute angle.
- $\angle AOC$ is a right angle.
- $\angle COD$ is an acute angle.
- $\angle AOD$ is an obtuse angle.
- $\angle AOE$ is a straight angle.

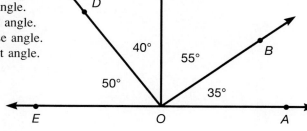

Practice Exercise 2. Determine $m(\angle SOP)$.

EXAMPLE 2 Determine the measures of angles $\angle ROS$, $\angle SOT$, $\angle TOU$, $\angle ROT$, and $\angle SOU$.

SOLUTION:

- $m(\angle ROS) = 55°$
- $m(\angle SOT) = 80°$
- $m(\angle TOU) = 45°$
- $m(\angle ROT) = m(\angle ROS) + m(\angle SOT) = 55° + 80° = 135°$
- $m(\angle SOU) = m(\angle SOT) + m(\angle TOU) = 80° + 45° = 125°$

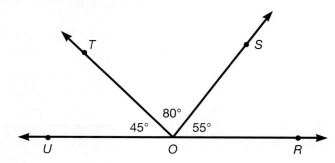

ANSWERS TO PRACTICE EXERCISES

1. 90°. 2. 145°.

In Section 17.1, we discussed congruent line segments. We also have congruent angles.

DEFINITIONS

1. Two angles are **congruent angles** if they have the same measures.
2. If a ray divides an angle into two congruent angles, then the ray is called the **bisector of the angle**.

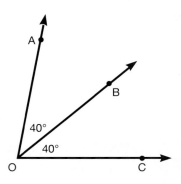

FIGURE 17.11

In Figure 17.11, we illustrate that $\overrightarrow{OB}$ is the bisector of $\angle AOC$. Note that $m(\angle AOB) = m(\angle BOC)$; hence, $\overrightarrow{OB}$ is the bisector of $\angle AOC$.

If two angles, $\angle ABC$ and $\angle DEF$, are congruent, we indicate this fact by writing either

1. $m(\angle ABC) = m(\angle DEF)$
2. $\angle ABC \cong \angle DEF$

EXERCISES 17.2

For Exercises 1–4, refer to the accompanying figure.

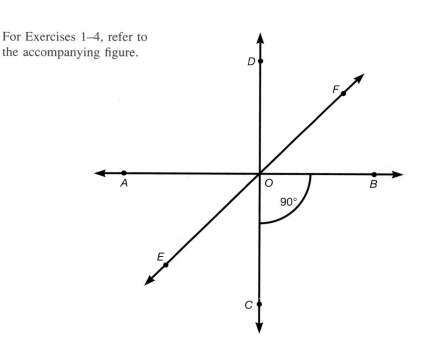

1. Name all acute angles in the figure.
2. Name all obtuse angles in the figure.
3. Name all right angles in the figure.
4. Name all straight angles in the figure.

17.2 Angles

For Exercises 5–10, refer to the accompanying figure.

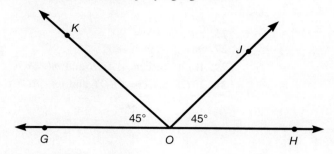

5. Name all acute angles in the figure.
6. Name all obtuse angles in the figure.
7. Name all right angles in the figure.
8. Name all straight angles in the figure.
9. Name three angles that have $\overrightarrow{OG}$ as a side.
10. Name two angles that are congruent.

For Exercises 11–18, refer to the accompanying figure. Determine the measure of the given angle.

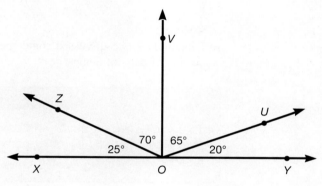

11. ∠YOU
12. ∠VOZ
13. ∠YOV
14. ∠UOZ
15. ∠XOZ
16. ∠XOY
17. ∠UOV
18. ∠ZOY

For Exercises 19–22, use a protractor to measure each of the given angles.

19.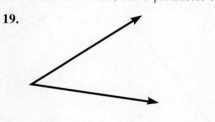

20.

21.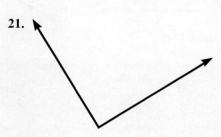

22.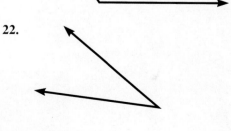

For Exercises 23–28, use a protractor to draw the angles with the given measures.

23. $m(\angle AOB) = 55°$
24. $m(\angle COD) = 68°$
25. $m(\angle EOF) = 90°$
26. $m(\angle GOH) = 125°$
27. $m(\angle JOK) = 180°$
28. $m(\angle LOM) = 115°$
29. If $\overrightarrow{OC}$ bisects $\angle AOB$ and $m(\angle AOC) = 42°$, what is $m(\angle BOC)$?
30. If $\overrightarrow{OS}$ bisects $\angle ROT$ and $m(\angle ROT) = 156°$, what is $m(\angle ROS)$?

17.3 Some Special Pairs of Angles

OBJECTIVE

After completing this section, you should be able to work with adjacent, vertical, complementary, or supplementary angles.

In Section 17.2, we discussed some special individual angles. In this section, we discuss some special pairs of angles.

DEFINITION

Two angles are **adjacent angles** if they have a common vertex and a common side but share no other common points in the plane containing the two angles.

In Figure 17.12, we note that $\angle AOB$ and $\angle BOC$ are adjacent angles. However, $\angle AOB$ and $\angle AOC$ are *not* adjacent angles. (Do you agree?)

When two lines intersect, they form four angles as illustrated in Figure 17.13. Notice that $\angle 1$ and $\angle 2$ are adjacent angles but $\angle 1$ and $\angle 3$ are not adjacent angles. (Why not?)

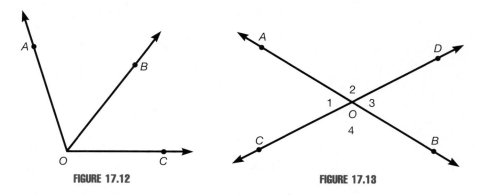

FIGURE 17.12   FIGURE 17.13

DEFINITION

Two nonadjacent angles formed by two intersecting lines are called **vertical angles**.

17.3 Some Special Pairs of Angles

In Figure 17.13, ∠1 and ∠3 are vertical angles. Also, ∠2 and ∠4 are vertical angles. Vertical angles are congruent since they have the same measures. This is illustrated as follows for ∠1 and ∠3:

1. Since ∠AOB is a straight angle, $m(\angle 2) + m(\angle 3) = 180°$.
2. Since ∠COD is a straight angle, $m(\angle 1) + m(\angle 2) = 180°$.
3. $m(\angle 1) + m(\angle 2) = m(\angle 2) + m(\angle 3)$, by substitution.
4. $m(\angle 1) = m(\angle 3)$, by subtracting $m(\angle 2)$ from both sides of the equation.
5. Hence, ∠1 ≅ ∠3.

Practice Exercise 1. If ∠ROS and ∠SOT are adjacent angles, what is their common side?

EXAMPLE 1 Given the accompanying figure,

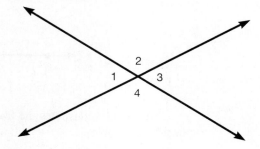

a. Name all pairs of adjacent angles.
b. Name all pairs of vertical angles.

SOLUTION:

a. The adjacent angles are ∠1 and ∠2; ∠2 and ∠3; ∠3 and ∠4; and ∠1 and ∠4.
b. The vertical angles are ∠1 and ∠3; and ∠2 and ∠4.

Some pairs of angles are special because of their measures.

DEFINITIONS

1. If the sum of the measures of two angles is 90°, then the angles are called **complementary angles**. Each angle is called the "complement" of the other.
2. If the sum of the measures of two angles is 180°, then the angles are called **supplementary angles**. Each angle is called the "supplement" of the other.

Practice Exercise 2. If ∠1 and ∠2 are vertical angles with $m(\angle 1) = 57°$, determine $m(\angle 2)$.

EXAMPLE 2 Classify the angles ∠AOB and ∠BOC.

SOLUTION: In the accompanying figure, ∠AOB and ∠BOC are adjacent angles. They are also complementary angles since $m(\angle AOB) = 50°$, $m(\angle BOC) = 40°$, and $50° + 40° = 90°$.

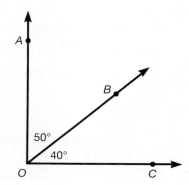

Practice Exercise 3. If ∠3 and ∠4 are complementary angles with $m(\angle 3) = 64°$, determine $m(\angle 4)$.

EXAMPLE 3 Classify the angles ∠COD and ∠EOF.

SOLUTION: In the accompanying figure, ∠COD and ∠EOF are *not* adjacent angles. However, they are complementary angles since $m(\angle COD) + m(\angle EOF) = 60° + 30° = 90°$.

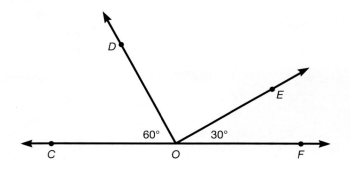

Example 3 illustrates the fact that two angles that are complementary do not have to be adjacent angles.

Practice Exercise 4. If ∠5 and ∠6 are supplementary angles, $m(\angle 5) = (y + 7)°$ and $m(\angle 6) = (2y - 1)°$, determine the numerical value for the measure of each angle.

EXAMPLE 4 If ∠A and ∠B are supplementary angles with $m(\angle A) = 137°$, determine $m(\angle B)$.

SOLUTION: Since ∠A and ∠B are supplementary angles, then

$$m(\angle A) + m(\angle B) = 180°$$
$$137° + m(\angle B) = 180° \quad \text{(Substitution)}$$
$$m(\angle B) = 43° \quad \text{(Subtraction)}$$

Therefore, $m(\angle B) = 43°$.

Practice Exercise 5. If $m(\angle A)$ is equal to 5 times the measure of its complement, what is $m(\angle A)$?

EXAMPLE 5 If ∠C and ∠D are supplementary angles, $m(\angle C) = (x - 4)°$, and $m(\angle D) = (2x + 1)°$, determine the numerical value for the measure of each angle.

SOLUTION: Since ∠C and ∠D are supplementary angles, then

$$m(\angle C) + m(\angle D) = 180°$$
$$(x - 4)° + (2x + 1)° = 180° \quad \text{(Substitution)}$$
$$x° - 4° + (2x)° + 1° = 180°$$
$$(3x)° - 3° = 180°$$
$$(3x)° = 183°$$
$$x = 61$$

Therefore, $m(\angle C) = (x - 4)° = (61 - 4)° = 57°$
and $m(\angle D) = (2x + 1)° = (122 + 1)° = 123°$

Note that $57° + 123° = 180°$.

ANSWERS TO PRACTICE EXERCISES

1. $\overrightarrow{OS}$.   2. 57°.   3. 26°.   4. $m(\angle 5) = 65°, m(\angle 6) = 115°$.   5. 15°.

17.3 Some Special Pairs of Angles

EXERCISES 17.3

For Exercises 1–5, refer to the accompanying figure.

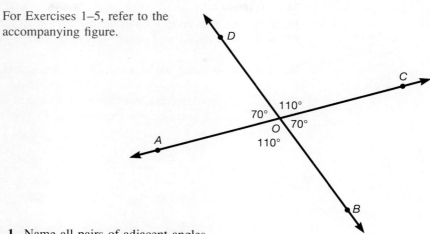

1. Name all pairs of adjacent angles.
2. Name all pairs of vertical angles.
3. Name all pairs of supplementary angles.
4. What is the measure of the complement of ∠AOD?
5. What is the measure of the supplement of ∠COB?

For Exercises 6–15, determine the measure of the complement of an angle with the given measure.

6. 63°
7. 19°
8. 71°
9. 8°
10. $x°$
11. $(2y)°$
12. $(u - 3)°$
13. $(3u + 1)°$
14. $(1 - 2p)°$
15. $(p - 20)°$

For Exercises 16–25, determine the measure of the supplement of an angle with the given measure.

16. 90°
17. 37°
18. 102°
19. 53°
20. $y°$
21. $(x - 20)°$
22. $(2u + 7)°$
23. $\left(\frac{2}{5}p + 3\right)°$
24. $\left(4 - \frac{1}{3}w\right)°$
25. $(5q - 213)°$

For Exercises 26–35, refer to the accompanying figure. If $m(\angle UOV) = 43°$, determine the measure for each of the given angles.

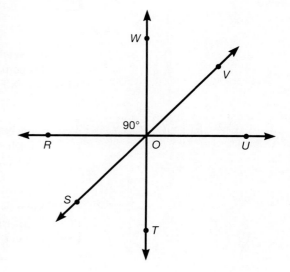

26. ∠WOV
27. ∠ROS
28. ∠SOT
29. ∠TOU
30. ∠ROV
31. ∠SOW
32. ∠SOU
33. ∠WOT
34. ∠TOV
35. ∠ROT
36. If $m(\angle A)$ is equal to 4 times the measure of its complement, what is $m(\angle A)$?
37. If $m(\angle B)$ is equal to 1 degree less than 6 times the measure of its complement, what is $m(\angle B)$?
38. If $m(\angle C)$ is twice the measure of its complement, determine the measure of its complement.
39. If $m(\angle D)$ is 40 degrees less than the measure of its supplement, determine $m(\angle D)$.
40. If $\angle E$ and $\angle F$ are vertical angles with $m(\angle E) = u°$ and $m(\angle F) = (2u - 57)°$, determine the value of u.

17.4 Parallel and Perpendicular Lines

OBJECTIVE

After completing this section, you should be able to work with parallel and perpendicular lines, transversals, and angles formed by lines and transversals.

In Section 17.2, you learned that if two different lines intersect, then they intersect in one and only one point. In Section 17.3, you learned that two intersecting lines form two pairs of vertical angles. There is a special case of intersecting lines.

DEFINITION

Two intersecting lines are **perpendicular lines** if they form a right angle.

The symbol ⊥ is used for "is perpendicular to." In Figure 17.14, $\overleftrightarrow{AB} \perp \overleftrightarrow{CD}$ (read "the line AB is perpendicular to the line CD") since the two lines intersect and form a right angle. Also note, in Figure 17.14, that

- Each angle in a pair of adjacent angles is a right angle.
- Each angle in a pair of vertical angles is a right angle.
- All four angles formed by the intersecting lines are right angles.

When two intersecting lines are perpendicular, we will use a small box to denote that they form a right angle. This is illustrated in Figure 17.15. Sometimes a pair of lines is crossed or cut by another line.

17.4 Parallel and Perpendicular Lines

> **DEFINITION**
> A line that intersects each of two or more coplanar lines (that is, lines that lie on the same plane) is called a **transversal**.

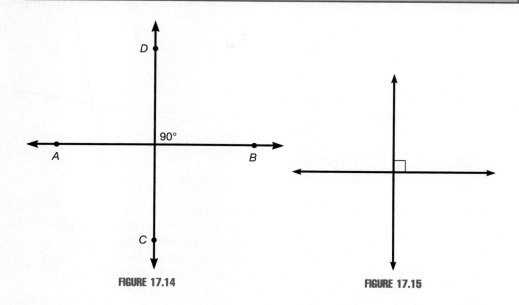

FIGURE 17.14   **FIGURE 17.15**

In Figure 17.16, the line $\overleftrightarrow{PQ}$ is a transversal that intersects the line m at the point P and the line n at the point Q. Notice that $\overleftrightarrow{PQ}$ forms eight different angles with the lines m and n. These angles have names as follows:

1. The angles 3, 4, 5, and 6 are called **interior angles.**
2. The angles 1, 2, 7, and 8 are called **exterior angles.**
3. Two interior angles that are not adjacent angles and that lie on opposite sides of the transversal are called **alternate interior angles.** Hence, $\angle 3$ and $\angle 5$ are alternate interior angles. Also, $\angle 4$ and $\angle 6$ are alternate interior angles.
4. Two exterior angles that are not adjacent angles and that lie on opposite sides of the transversal are called **alternate exterior angles.** Hence, $\angle 1$ and $\angle 7$ are alternate exterior angles. Also, $\angle 2$ and $\angle 8$ are alternate exterior angles.
5. Two angles that lie on the same side of the transversal and such that one is an interior angle and the other is an exterior angle are called **corresponding angles.** Hence, the following pairs of angles are corresponding angles: $\angle 1$ and $\angle 5$; $\angle 2$ and $\angle 6$; $\angle 3$ and $\angle 7$; $\angle 4$ and $\angle 8$.

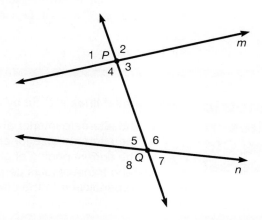

FIGURE 17.16

In Figure 17.17,

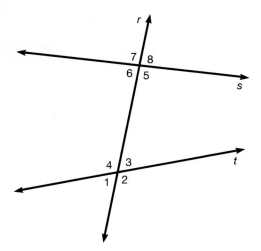

FIGURE 17.17

- ∠2 and ∠8 are exterior angles.
- ∠3 and ∠6 are alternate interior angles.
- ∠1 and ∠6 are corresponding angles.
- ∠5 and ∠7 are vertical angles.
- ∠5 and ∠6 are supplementary angles.

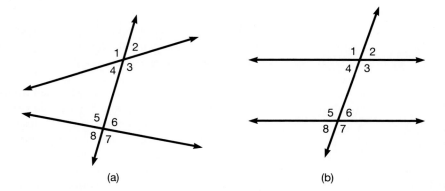

FIGURE 17.18 (a) (b)

In Figure 17.18, we illustrate pairs of lines cut by transversals. Notice that in Figure 17.18a, the alternate interior angles are not congruent, but in Figure 17.18b, they do appear to be congruent. If you measure these angles with your protractor, you will determine that this statement is true. What do you think that we can say about the corresponding angles?

We have the following geometric properties for *any* pair of parallel lines that are cut by a transversal. These will be stated without proof.

Geometric Properties of Parallel Lines Cut by a Transversal

If two **parallel lines** are cut by a transversal, then

- The alternate interior angles are congruent.
- The alternate exterior angles are congruent.
- The corresponding angles are congruent.
- If the transversal is perpendicular to one of the parallel lines, then it is also perpendicular to the other line.

17.4 Parallel and Perpendicular Lines

For Practice Exercises 1–3, refer to the figure. In the figure, $s \parallel t$ and $p \parallel q$.

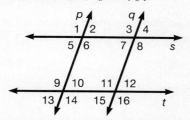

Practice Exercise 1. List all of the angles that are congruent to $\angle 3$.

EXAMPLE 1 In the accompanying figure, $p \parallel q$. If $m(\angle 2) = 52°$, determine the measure of the other seven numbered angles.

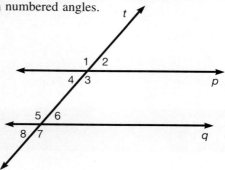

SOLUTION:

1. $\angle 1$ and $\angle 2$ are supplementary angles. Therefore, $m(\angle 1) = 180° - m(\angle 2) = 180° - 52° = 128°$.
2. $\angle 1$ and $\angle 3$ are vertical angles. Therefore, $m(\angle 3) = 128°$.
3. $\angle 4$ and $\angle 2$ are vertical angles. Therefore, $m(\angle 4) = 52°$.
4. $\angle 2$ and $\angle 6$ are corresponding angles. Since $p \parallel q$, they are congruent angles. Therefore, $m(\angle 6) = 52°$.
5. $\angle 7$ and $\angle 1$ are alternate exterior angles. Since $p \parallel q$, they are congruent angles. Therefore, $m(\angle 7) = 128°$.
6. $\angle 5$ and $\angle 3$ are alternate interior angles. Since $p \parallel q$, they are congruent angles. Therefore, $m(\angle 5) = 128°$.
7. $\angle 8$ and $\angle 6$ are vertical angles. Therefore, $m(\angle 8) = 52°$.

Practice Exercise 2. List all of the angles that are supplementary to $\angle 11$.

EXAMPLE 2 In the accompanying figure, $s \parallel t$. Determine the value of y.

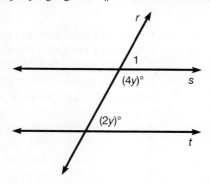

SOLUTION: We determine that $m(\angle 1) = 180° - (4y)° = (180 - 4y)°$. We now have that $(180 - 4y)° = (2y)°$ since these are the measures of corresponding angles and $s \parallel t$. Hence,

$$(180 - 4y)° = (2y)°$$
$$180° - (4y)° = (2y)°$$
$$180° = (6y)°$$
$$y = 30$$

Practice Exercise 3. If $m(\angle 2) + m(\angle 5) = 104°$, determine $m(\angle 14)$.

EXAMPLE 3 Given lines p, q, and r such that $p \parallel q$ and $r \perp p$, what is the relationship between r and q?

SOLUTION: The line r is a transversal for the two parallel lines p and q. Since $r \perp p$ and $p \parallel q$, then $r \perp q$. (You are encouraged to sketch the three lines p, q, and r.)

ANSWERS TO PRACTICE EXERCISES

1. Angles 1, 6, 8, 9, 11, 14, and 16.   2. Angles 2, 4, 5, 7, 10, 12, 13, and 15.   3. 128°.

EXERCISES 17.4

For Exercises 1–18, refer to the accompanying figure. You are given that $s \parallel t$ and $p \parallel q$.

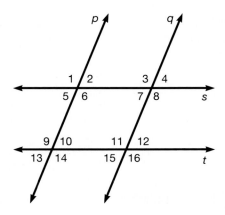

1. List all angles that are acute angles.
2. List all angles that are obtuse angles.
3. List all angles that are adjacent angles to $\angle 1$.
4. List all angles that are adjacent angles to $\angle 15$.
5. List all angles that have the same measure as $\angle 5$.
6. List all angles that have the same measure as $\angle 11$.
7. With p as the transversal for lines s and t, list all of the interior angles.
8. With p as the transversal for lines s and t, list all pairs of alternate interior angles.
9. With q as the transversal for lines s and t, list all of the exterior angles.
10. With q as the transversal for lines s and t, list all pairs of alternate exterior angles.
11. With p as the transversal for lines s and t, list all pairs of corresponding angles.
12. With q as the transversal for lines s and t, list all pairs of corresponding angles.
13. With s as the transversal for lines p and q, list all pairs of alternate interior angles.
14. With t as the transversal for lines p and q, list all pairs of corresponding angles.
15. List all pairs of vertical angles.
16. List all angles that are supplementary to $\angle 3$.
17. List all angles that are supplementary to $\angle 10$.
18. Determine: $m(\angle 9) + m(\angle 4)$

17.4 Parallel and Perpendicular Lines

For Exercises 19–21, refer to the accompanying figure. You are given that $m \parallel n$.

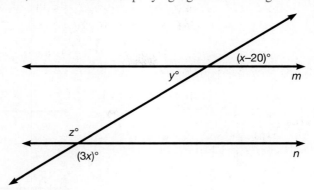

19. Determine the value of x.
20. Determine the value of y.
21. Determine the value of z.

For Exercises 22–24, refer to the accompanying figure. You are given that $s \parallel t$.

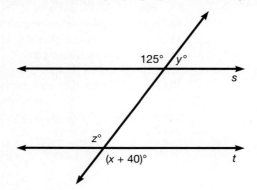

22. Determine the value of x.
23. Determine the value of y.
24. Determine the value of z.

For Exercises 25–30, refer to the accompanying figure. You are given that $p \parallel q$ and $r \parallel s$.

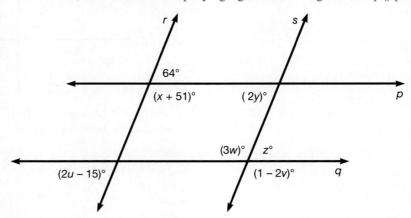

25. Determine the value of x.
26. Determine the value of y.
27. Determine the value of z.
28. Determine the value of u.
29. Determine the value of w.
30. Determine the value of v.

For Exercises 31–38, refer to the accompanying figure. You are given that $p \parallel q$ and $t \perp p$.

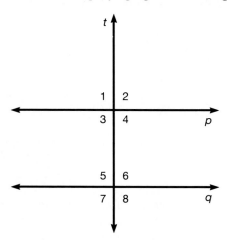

31. What is the relationship between lines t and q?
32. What is the measure of $\angle 2$?
33. What is the measure of $\angle 7$?
34. Name all pairs of corresponding angles.
35. Name all pairs of alternate interior angles.
36. Name all exterior angles.
37. List all angles that are congruent to $\angle 5$.
38. List all angles that are supplementary to $\angle 4$.

For Exercises 39–40, refer to the accompanying figure. You are given that $p \parallel q$, $q \parallel r$, and $t \perp p$.

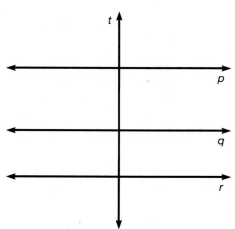

39. What is the relationship between p and r?
40. What is the relationship between t and r?

17.5 Triangles

OBJECTIVES

After completing this section, you should be able to:
1. Classify triangles according to their sides or angles.
2. Work with the properties of special triangles.

In Section 5.2, we introduced polygons as closed geometric figures that consist of line segments. The line segments are called the sides of the polygon. **Triangles** are polygons with three sides and are classified according to the measure (length) of their sides. We will restate the definitions for these special triangles.

DEFINITIONS

A triangle can be classified according to the measures (lengths) of its sides.

1. If all three sides of a triangle have the same measures, the triangle is an *equilateral triangle*.
2. If exactly two sides of a triangle have the same measures, the triangle is an *isosceles triangle*.
3. If no two sides of a triangle have the same measures, the triangle is a *scalene triangle*.

A triangle can be denoted by the use of the symbol △. In Figure 17.19, we illustrate △ABC (read "triangle ABC"). The letters A, B, and C denote the vertices of the three angles of the triangle. Just as triangles can be classified according to the measures of their sides, triangles can also be classified according to the measures of their angles.

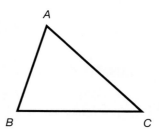

FIGURE 17.19

DEFINITIONS

A triangle can be classified according to the measures of its angles.

- If all three angles of a triangle are congruent (that is, they have the same measures), the triangle is an **equiangular triangle**.
- If all three angles of a triangle are acute angles, the triangle is an **acute triangle**.
- If a triangle contains an obtuse angle, the triangle is an **obtuse triangle**.
- If a triangle contains a right angle, the triangle is a **right triangle**.

720   Chapter 17   Basic Concepts of Geometry

In Practice Exercises 1–4, classify the triangle given according to its angles.

Practice Exercise 1. △ABC with $m(\angle A) = 57°$, $m(\angle B) = 69°$, and $m(\angle C) = 54°$

EXAMPLE 1   Classify the triangle △DEF.

SOLUTION: △DEF is an acute triangle since all of its angles are acute angles.

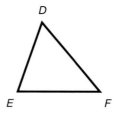

Practice Exercise 2. △DEF with $m(\angle D) = 60°$, $m(\angle E) = 60°$, and $m(\angle F) = 60°$

EXAMPLE 2   Classify the triangle △GHJ.

SOLUTION: △GHJ is an obtuse triangle since ∠H is an obtuse angle.

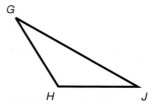

Practice Exercise 3. △GHJ with $m(\angle G) = 22°$, $m(\angle H) = 112°$, and $m(\angle J) = 46°$

EXAMPLE 3   Classify △PQR.

SOLUTION: △PQR is a right triangle since ∠Q is a right angle.

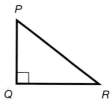

Practice Exercise 4. △KLM with $m(\angle K) = 33°$, $m(\angle L) = 57°$, and $m(\angle M) = 90°$

EXAMPLE 4   Classify △XYZ.

SOLUTION: △XYZ is a equiangular triangle since all three angles have the same measure, $x°$.

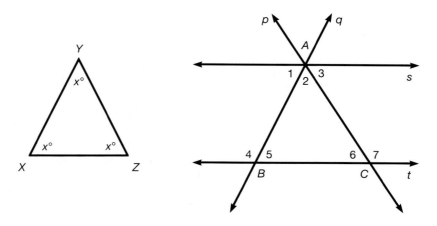

FIGURE 17.20

17.5 Triangles

Consider Figure 17.20, which consists of two parallel lines s and t and two transversals p and q. Notice that the transversals intersect the parallel lines to form $\triangle ABC$. We have

$$m(\angle 1) + m(\angle 2) + m(\angle 3) = 180°$$

since the three angles form a straight angle. Since $s \parallel t$ and q is a transversal, then

$$m(\angle 1) = m(\angle 5)$$

since the two angles are alternate interior angles. Similarly, since $s \parallel t$ and p is a transversal, then

$$m(\angle 3) = m(\angle 6)$$

We now have

$$m(\angle 1) + m(\angle 2) + m(\angle 3) = 180°$$
$$m(\angle 5) + m(\angle 2) + m(\angle 3) = 180° \quad [\text{Since } m(\angle 1) = m(\angle 5)]$$
$$m(\angle 5) + m(\angle 2) + m(\angle 6) = 180° \quad [\text{Since } m(\angle 3) = m(\angle 6)]$$

But $\angle 2$, $\angle 5$, and $\angle 6$ are the three angles of $\triangle ABC$. Hence, we have established the following statement for the sum of the measures of the angles of a triangle.

The sum of the measures of the three angles of a triangle is equal to 180°.

From the preceding statement, we can conclude the following:

1. A triangle can contain exactly one obtuse angle. (If it has two obtuse angles, the sum of the measures of the angles of the triangle will be greater than 180°.)
2. A triangle can contain exactly one right angle. (If it has two right angles, the sum of the measures of the two angles will be 180°. Since the third angle has a positive measure, the sum of the measures of all three angles will be greater than 180°.)

Practice Exercise 5. Given $\triangle NOP$ with $m(\angle N) = 29°$ and $m(\angle O) = 89°$, determine $m(\angle P)$.

EXAMPLE 5 Given $\triangle DEF$ with $m(\angle D) = 62°$ and $m(\angle E) = 51°$, determine $m(\angle F)$.

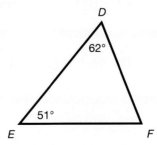

SOLUTION: Since the sum of the measures of the angles of a triangle is equal to 180°, we have

$$m(\angle D) + m(\angle E) + m(\angle F) = 180°$$
$$62° + 51° + m(\angle F) = 180° \quad \text{(Substitution)}$$
$$113° + m(\angle F) = 180°$$
$$m(\angle F) = 67°$$

Practice Exercise 6. Given △QRS with $m(\angle Q) = 46°$ and $m(\angle R) = m(\angle S)$, determine $m(\angle S)$.

EXAMPLE 6 Given △RST with $m(\angle S) = 48°$ and $\angle R \cong \angle T$, determine $m(\angle R)$.

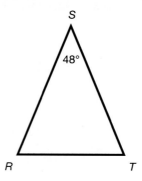

SOLUTION: Since $\angle R \cong \angle T$, $m(\angle R) = m(\angle T)$. Let $m(\angle R) = x°$. We have

$$m(\angle R) + m(\angle S) + m(\angle T) = 180°$$
$$x° + 48° + x° = 180°$$
$$(2x)° + 48° = 180°$$
$$(2x)° = 132°$$
$$x = 66$$

Therefore, $m(\angle R) = 66°$.

ANSWERS TO PRACTICE EXERCISES
1. Acute triangle. 2. Equiangular triangle. 3. Obtuse triangle. 4. Right triangle. 5. 62°. 6. 67°.

We will conclude this section with a discussion of isosceles triangles. A triangle is isosceles if exactly two of its sides are congruent. The third side of an isosceles triangle is called the "base." The angle between the two congruent sides is called the "vertex angle." The other two angles are called the "base angles." In Figure 17.21, we have the following:

1. △RST is an isosceles triangle with $\overline{RS} \cong \overline{RT}$. (This is denoted by the slash marks.)
2. $\angle R$ is the vertex angle.
3. $\angle S$ and $\angle T$ are the base angles.
4. $\overline{ST}$ is the base.

In Figure 17.21, the base angles appear to be congruent, and, indeed, they are. We will state this fact without proof.

The base angles of an isosceles triangle are congruent.

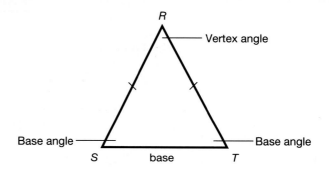

FIGURE 17.21

17.5 Triangles

Practice Exercise 7. Given △ABC with $\overline{AB} \cong \overline{BC}$ and $m(\angle A) = 56°$, determine $m(\angle C)$.

EXAMPLE 7 Given the isosceles △ABC with $\overline{AB} \cong \overline{AC}$ and $m(\angle A) = 36°$, determine $m(\angle B)$.

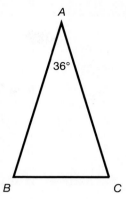

SOLUTION: We know that the sum of the measures of the angles of a triangle is equal to 180°. Therefore,

$$m(\angle A) + m(\angle B) + m(\angle C) = 180°$$
$$36° + m(\angle B) + m(\angle C) = 180°$$
$$m(\angle B) + m(\angle C) = 144°$$

Since the base angles of an isosceles are congruent, we have $m(\angle B) = m(\angle C)$. Hence, we have

$$m(\angle B) + m(\angle B) = 144° \quad \text{(Substitution)}$$
$$2[m(\angle B)] = 144°$$
$$m(\angle B) = 72°$$

Therefore, $m(\angle B) = 72°$.

Practice Exercise 8. Given △DEF with $\overline{EF} \cong \overline{DE}$ and $m(\angle E) = 28°$, determine $m(\angle D)$.

EXAMPLE 8 Given the isosceles △DEF with $\overline{DE} \cong \overline{DF}$ and $m(\angle E) = 40°$, determine $m(\angle D)$.

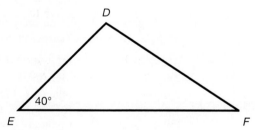

SOLUTION: Since $\overline{DE} \cong \overline{DF}$, then $\angle E \cong \angle F$ and $m(\angle E) = m(\angle F)$. Therefore, $m(\angle F) = 40°$. Hence,

$$m(\angle D) + m(\angle E) + m(\angle F) = 180°$$
$$m(\angle D) + 40° + 40° = 180°$$
$$m(\angle D) + 80° = 180°$$
$$m(\angle D) = 100°$$

Therefore, $m(\angle D) = 100°$.

ANSWERS TO PRACTICE EXERCISES

7. 56°.   8. 76°.

EXERCISES 17.5

For Exercises 1–5, classify each of the given triangles as being isosceles, equilateral, or scalene.

1. $\triangle ABC$ with $m(\overline{AB}) = 2$ cm, $m(\overline{BC}) = 3$ cm, and $m(\overline{AC}) = 4$ cm
2. $\triangle DEF$ with $m(\overline{DE}) = 4.5$ ft, $m(\overline{EF}) = 2.5$ ft, and $m(\overline{DF}) = 2.5$ ft
3. $\triangle GHJ$ with $m(\overline{GH}) = 5$ ft, $m(\overline{HJ}) = 6$ ft, and $m(\overline{GJ}) = 2$ yd
4. $\triangle KLM$ with $m(\overline{KL}) = 4.6$ m, $m(\overline{LM}) = 4.6$ m, and $m(\overline{KM}) = 4.6$ m
5. $\triangle NOP$ with $m(\overline{NO}) = 23$ cm, $m(\overline{OP}) = 230$ mm, and $m(\overline{NP}) = 2.3$ dm

For Exercises 6–10, classify each of the given triangles as being acute, right, or obtuse.

6. $\triangle ABC$ with $m(\angle A) = 62°$, $m(\angle B) = 38°$, and $m(\angle C) = 80°$
7. $\triangle DEF$ with $m(\angle D) = 120°$, $m(\angle E) = 35°$, and $m(\angle F) = 25°$
8. $\triangle GHJ$ with $m(\angle G) = 37°$, $m(\angle H) = 90°$, and $m(\angle J) = 53°$
9. $\triangle KLM$ with $m(\angle K) = 40°$, $m(\angle L) = 40°$, and $m(\angle M) = 100°$
10. $\triangle NOP$ with $m(\angle N) = 45°$, $m(\angle O) = 45°$, and $m(\angle P) = 90°$

For Exercises 11–20, the measures of two angles of a triangle are given. Determine the measure of the third angle of the triangle.

11. 51° and 63°
12. 86° and 47°
13. 71° and 79°
14. 105° and 26°
15. 48° and 48°
16. 19° and 107°
17. $x°$ and $(2x)°$
18. $x°$ and $(x + 6)°$
19. $(y + 4)°$ and $(2y - 3)°$
20. $(3y - 1)°$ and $(y + 5)°$

For Exercises 21–30, you are given the measure of the vertex angle of an isosceles triangle. Determine the measure for each of the base angles of the triangle.

21. 68°
22. 50°
23. 104°
24. 90°
25. 72°
26. 12°
27. $x°$
28. $(3y)°$
29. $(4u - 2)°$
30. $(t + 18)°$

For Exercises 31–40, you are given the measure for each of the base angles of an isosceles triangle. Determine the measure of the vertex angle of the triangle.

31. 41°
32. 55°
33. 29°
34. 73°
35. 87°
36. 45°
37. $y°$
38. $(2x - 3)°$
39. $(4u - 1)°$
40. $(t + 17)°$

For Exercises 41–46, you are given isosceles $\triangle ABC$ with $\overline{AB} \cong \overline{BC}$.

41. If $m(\angle B) = 54°$, determine $m(\angle C)$.
42. If $m(\angle B) = x°$, determine $m(\angle A)$.
43. If $m(\angle A) = 47°$, determine $m(\angle C)$.
44. If $m(\angle A) = (y - 2)°$, determine $m(\angle B)$.
45. If $m(\angle C) = 81°$, determine $m(\angle A)$.
46. If $m(\angle C) = (2u + 1)°$, determine, $m(\angle B)$.

For Exercises 47–54, classify each of the statements as being either true or false.

47. An acute triangle can be an isosceles triangle.
48. An acute triangle can be a right triangle.
49. All equiangular triangles are acute triangles.
50. A right triangle can be an obtuse triangle.
51. A right triangle can be an equiangular triangle.

52. A right triangle can be an isosceles triangle.
53. A triangle can contain both a right angle and an obtuse angle.
54. An equilateral triangle is also an isosceles triangle.
55. The measure of the vertex angle of an isosceles triangle is 18° more than the measure of a base angle of the triangle. Determine the measures of all three angles of the triangle.
56. The measure of the vertex angle of an isosceles triangle is 16° less than twice the measure of a base angle of the triangle. Determine the measures of all three angles of the triangle.
57. The measure of a base angle of an isosceles triangle is 21° more than the measure of the vertex angle of the triangle. Determine the measures of all three angles of the triangle.
58. The measure of a base angle of an isosceles triangle is 8° more than one-half the measure of the vertex angle of the triangle. Determine the measures of all three angles of the triangle.
59. The measure of one angle of a triangle is 6° more than the measure of a second angle of the triangle. If the third angle of the triangle has a measure of 64°, determine the measures of the other two angles of the triangle.
60. In $\triangle ABC$, the measure of $\angle A$ is 1° less than 3 times the measure of $\angle B$. The measure of $\angle C$ is 4° less than the measure of $\angle B$. Determine the measures of all three angles of the triangle.

17.6 Similar Triangles

OBJECTIVES

After completing this section, you should be able to:
1. Determine whether triangles are similar.
2. Determine the measures of corresponding parts of similar triangles.

Some triangles appear to have the same shape. Other triangles appear to have the same shape and the same size. We will now consider triangles that have the same shape but not necessarily the same size.

DEFINITION

Similar triangles have the following properties:
1. Their corresponding angles are congruent.
2. The measures of their corresponding sides are in proportion.

In Figure 17.22, we note that $\triangle ABC$ is similar to $\triangle DEF$. The corresponding angles are A and D, B and E, and C and F, as indicated in the figure. The corresponding sides are also indicated as being $\overline{AB}$ and $\overline{DE}$, $\overline{AC}$ and $\overline{DF}$, and $\overline{BC}$ and $\overline{EF}$. The symbol $\sim$ is used for "is similar to." Hence, in Figure 17.22, we have $\triangle ABC \sim \triangle DEF$.

The measures of the corresponding sides of similar triangles are in proportion. That is, the ratios of the measures of the corresponding sides are equal. Therefore, for $\triangle ABC$ and $\triangle DEF$ in Figure 17.22, we have

$$\frac{m(\overline{AB})}{m(\overline{DE})} = \frac{m(\overline{AC})}{m(\overline{DF})} = \frac{m(\overline{BC})}{m(\overline{EF})}$$

since the triangles are similar.

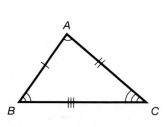
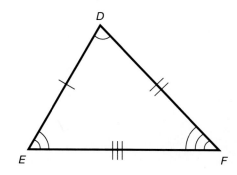

FIGURE 17.22

In Figure 17.23, we have

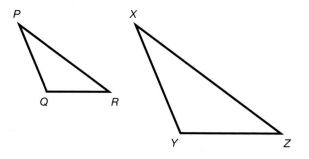

FIGURE 17.23

$\triangle PQR \sim \triangle XYZ$ with $m(\angle P) = m(\angle X)$
$m(\angle Q) = m(\angle Y)$
$m(\angle R) = m(\angle Z)$

and

$$\frac{PQ}{XY} = \frac{PR}{XZ} = \frac{QR}{YZ} \quad \text{[Remember, } PQ = m(\overline{PQ})\text{]}$$

Practice Exercise 1. Given $\triangle ABC \sim \triangle DEF$ with angles A, B, C corresponding, respectively, to angles D, E, F. If $m(\angle A) = 63°$ and $m(\angle B) = 58°$, determine $m(\angle F)$.

EXAMPLE 1 Given $\triangle ABC \sim \triangle DEF$ as indicated in the accompanying figures. If $m(\angle A) = 80°$, $m(\angle B) = 60°$, and $m(\angle C) = 40°$, determine $m(\angle D)$, $m(\angle E)$, and $m(\angle F)$.

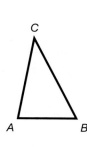
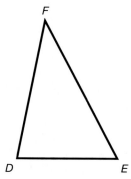

17.6 Similar Triangles

SOLUTION: We are given that $\triangle ABC \sim \triangle DEF$. Since corresponding angles of similar triangles are congruent, we have

$$m(\angle D) = m(\angle A) = 80°$$
$$m(\angle E) = m(\angle B) = 60°$$
$$m(\angle F) = m(\angle C) = 40°$$

Practice Exercise 2. Given $\triangle PQR \sim \triangle STU$ with angles P, Q, R corresponding, respectively, to angles S, T, U. If $PQ = 4$, $QR = 7$, and $TU = 21$, determine ST.

EXAMPLE 2 Given $\triangle GHJ \sim \triangle KLM$ in the accompanying figures, with the measures of the sides of the triangles as indicated. Determine the values of x and y. (All measures are in inches.)

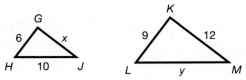

SOLUTION: Since $\triangle GHJ \sim \triangle KLM$, the measures of the corresponding sides are in proportion. Hence, we have

$$\frac{GH}{KL} = \frac{HJ}{LM}$$
$$\frac{6}{9} = \frac{10}{y}$$
$$6y = 90$$
$$y = 15$$

Also, we have

$$\frac{GJ}{HJ} = \frac{KM}{LM}$$
$$\frac{x}{10} = \frac{12}{y}$$
$$\frac{x}{10} = \frac{12}{15} \quad \text{(Since } y = 15\text{)}$$
$$15x = 120$$
$$x = 8$$

Therefore, $x = 8$ in. and $y = 15$ in.

From the definition for similar triangles, we note that two triangles are similar if the three angles of one triangle are congruent to the three corresponding angles of the other triangle and if the measures of the corresponding sides of the triangles are in proportion. However, we really do not need all of those conditions to have similar triangles.

Two triangles are similar if two angles of one triangle are congruent to the two corresponding angles of the other triangle.

Practice Exercise 3. In the figures below, $\triangle GHJ \sim \triangle KLM$ and the measures of the sides of the triangles are as indicated. Determine the values of x and y.

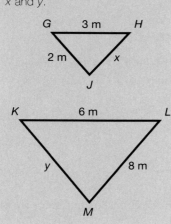

EXAMPLE 3 In the accompanying figure, $\overline{CD} \parallel \overline{AB}$. Show that $\triangle CDE \sim \triangle BAE$.

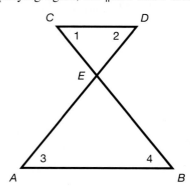

SOLUTION: Since $\overline{CD} \parallel \overline{AB}$, we have

$$\angle 1 \cong \angle 4$$
and
$$\angle 2 \cong \angle 3$$

since alternate interior angles are congruent. Since two angles of $\triangle CDE$ are congruent to the two corresponding angles of $\triangle BAE$, then $\triangle CDE \sim \triangle BAE$.

Practice Exercise 4. In the figure below, $\triangle LOP \sim \triangle ROS$. Determine the value of x.

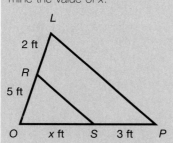

EXAMPLE 4 Given $\triangle RST \sim \triangle RXY$, $RX = 5$, $XS = 2$, $YT = 3$, determine RY.

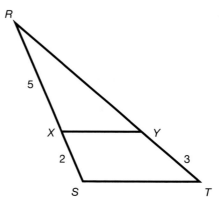

SOLUTION: Since $\triangle RST \sim \triangle RXY$, the measures of the corresponding sides are in proportion. Hence,

$$\frac{RX}{RS} = \frac{RY}{RT}$$

But, $RS = RX + XS$ and $RT = RY + YT$. Therefore, we have

$$\frac{RX}{RX + XS} = \frac{RY}{RY + YT}$$

$$\frac{5}{5 + 2} = \frac{RY}{RY + 3}$$

$$\frac{5}{7} = \frac{RY}{RY + 3}$$

$$5(RY + 3) = 7RY$$
$$5RY + 15 = 7RY$$
$$15 = 2RY$$
$$RY = 7.5$$

ANSWERS TO PRACTICE EXERCISES

1. 59°.   2. 12.   3. $x = 4$ m, $y = 4$ m.   4. 7.5.

EXERCISES 17.6

For Exercises 1–10, classify each of the given statements as being either true or false.

1. All isosceles triangles are similar.
2. All equiangular triangles are similar.
3. All right triangles are similar.
4. If two angles of a triangle are congruent to the two corresponding angles of a second triangle, then the two triangles are similar.
5. If an acute angle of a right triangle is congruent to an acute angle of a second right triangle, then the two triangles are similar.
6. If two triangles are similar, then the measures of their corresponding sides are equal.
7. If two triangles are similar, then the perimeters of the triangles are in the same ratio as the ratio of the measures of their corresponding sides.
8. If $\triangle ABC \sim \triangle DEF$ and the measures of two angles of $\triangle ABC$ are 37° and 102°, then $\triangle DEF$ has an angle with measure 46°.
9. If $\triangle GHJ \sim \triangle KLM$ with $\angle G \cong \angle J$, then $\triangle KLM$ has two angles whose measures are equal.
10. It is possible for an acute triangle to be similar to an obtuse triangle.

For Exercises 11–15, the two triangles illustrated for each exercise are similar, and the measures of their sides are indicated. Determine the values of x and y in each case.

11.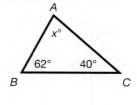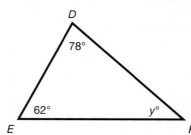

12.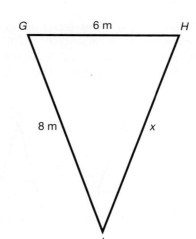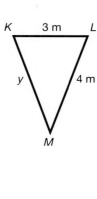

13.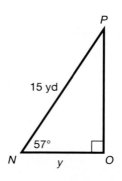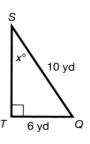

14.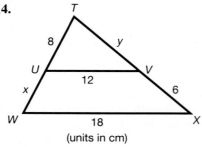
(units in cm)

15.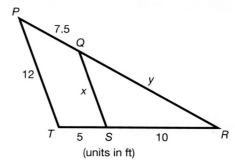
(units in ft)

For Exercises 16–20, the measures for parts of △ABC and △DEF are given. Determine if the triangles are similar.

16. $m(\angle A) = 71°$, $m(\angle B) = 46°$; $m(\angle E) = 46°$, $m(\angle F) = 63°$
17. $m(\angle B) = x°$, $m(\angle C) = x°$; $m(\angle D) = (x + 40)°$ ($x = 45$)
18. $AB = 3$, $BC = 4$, $AC = 5$; $DE = 4$, $EF = 5$, $DF = 6$
19. $AB = 3n$, $BC = 6n$, $AC = 4n$; $DE = 4n + 1$, $EF = 9n$, $DF = 6n$ ($n = 2$)
20. $AB = 10$, $BC = 17$, $AC = 15$; $DE = 15$, $EF = 25.5$, $DF = 22.5$
21. Given the square ABCD in the figure at left below, show that △DAB ~ △BCD. (*Hint:* The opposite sides of a square are parallel.)

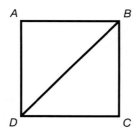 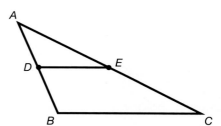

22. Given △ABC in the figure at right above, if D is the midpoint of $\overline{AB}$ and E is the midpoint of $\overline{AC}$, show that △ADE ~ △ABC.
23. Given the isosceles triangles in the figure at left below, with $\overline{AB} \cong \overline{BC}$, $\overline{DE} \cong \overline{EF}$, and $\angle B \cong \angle E$. Show that △ABC ~ △DEF.

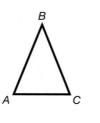  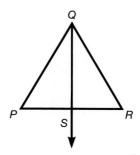

24. Given the isosceles triangle in the figure at right on the previous page, with $\overline{PQ} = \overline{QR}$. If $\overrightarrow{QS}$ is the bisector of $\angle Q$, show that $\triangle PSQ \sim \triangle RSQ$.
25. Given the accompanying figure, with $p \parallel q$, and $q \parallel r$, show that $\triangle ABC \sim \triangle ADE$.

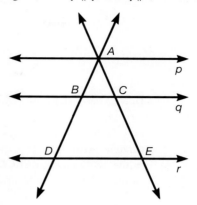

17.7 Special Right Triangles

OBJECTIVES

After completing this section, you should be able to work with:
1. 30°–60°–90° triangles.
2. 45°–45°–90° triangles.

In Section 16.7, we introduced right triangles and the Pythagorean theorem. These topics will be discussed further in this section as we consider some special triangles.

A right triangle that has an acute angle with measure 30° is of special interest. Clearly, the other acute angle, then, has measure 60°. Such a right triangle is referred to as a 30°–60°–90° triangle. In Figure 17.24, we illustrate right $\triangle ABC$ with $m(\angle A) = 30°$ and $m(\angle B) = 90°$, and $m(\angle C) = 60°$

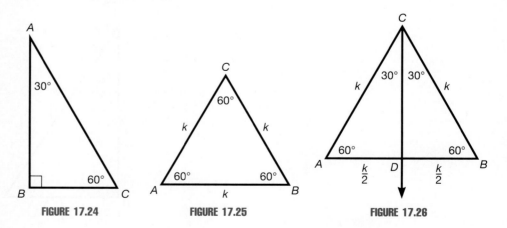

FIGURE 17.24   **FIGURE 17.25**   **FIGURE 17.26**

Recall that the longest side of a right triangle is called the hypotenuse. This is the side opposite the right angle. The other two sides are called the legs of the triangle. The ratios of the measures of the sides of a 30°–60°–90° triangle can easily be determined, as we will now do.

Consider the equilateral triangle whose sides have measure k. (See Figure 17.25.) An equilateral triangle is also equiangular. If each angle of the triangle has measure $x°$, then we have

$$x° + x° + x° = 180°$$
$$(3x)° = 180°$$
$$3x = 180$$
$$x = 60$$

Hence, each angle of $\triangle ABC$, illustrated in Figure 17.25, has measure $60°$. Next, consider $\overrightarrow{CD}$ which is the bisector of $\angle C$. (See Figure 17.26.) Clearly, $\triangle CDB$ is a right triangle since $m(\angle CDB) = 180° - (30° + 60°) = 180° - 90° = 90°$. Similarly, $\triangle ADC$ is a right triangle. Also, $\triangle ADC \sim \triangle BDC$ since the acute angle, ($\angle ACD$), of right $\triangle ADC$ is congruent to the acute angle, ($\angle BCD$), of right $\triangle BDC$. Since the measure of corresponding sides of similar triangles are in proportion, we have

$$\frac{AD}{DB} = \frac{AC}{CB}$$
$$= \frac{k}{k}$$
$$= 1$$

or $\quad m(\overline{AD}) = m(\overline{DB})$

Hence,

$$m(\overline{AD}) + m(\overline{BD}) = k$$
$$m(\overline{BD}) + m(\overline{BD}) = k \quad \text{(Substitution)}$$
$$2[m(\overline{BD})] = k$$
$$m(\overline{BD}) = \frac{k}{2}$$

We have now established that the measure of the hypotenuse of a $30°$–$60°$–$90°$ triangle is twice the measure of the leg opposite the $30°$ angle.

Now, consider the right $\triangle CDB$ in Figure 17.27 with the measures of its sides as indicated.

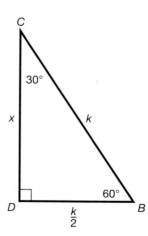

FIGURE 17.27

To determine the value of x, we use the Pythagorean theorem as follows:

$$(x)^2 + \left(\frac{1}{2}k\right)^2 = (k)^2$$

17.7 Special Right Triangles

$$x^2 + \frac{1}{4}k^2 = k^2$$

$$x^2 = \frac{3}{4}k^2$$

$$x = \frac{\pm k\sqrt{3}}{2}$$

But, since x is the measure of a side of the triangle, $x > 0$. Therefore,

$$x = \frac{k\sqrt{3}}{2}$$

We have now established the following statement for all 30°–60°–90° triangles.

In a 30°–60°–90° triangle:

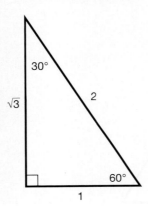

1. The measure of the hypotenuse is twice the measure of the leg opposite the 30° angle (which is the shorter leg).
2. The measure of the leg opposite the 60° angle (which is the longer leg) is one-half the measure of the hypotenuse multiplied by $\sqrt{3}$.

Practice Exercise 1. Given $\triangle ABC$ with $m(\angle B) = 90°$, $m(\angle A) = 60°$, and $AC = 8$ yd, determine AB and BC.

EXAMPLE 1 In the accompanying figure, we have right $\triangle ABC$ with $AC = 10$ ft. Determine AB and BC.

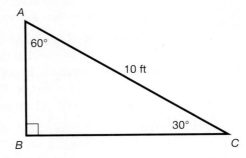

SOLUTION: We have a 30°–60°–90° triangle. AC is the measure of the hypotenuse. AB is the measure of the leg opposite the 30° angle. Therefore,

$$AB = \frac{1}{2}(AC)$$

$$= \frac{1}{2}(10 \text{ ft})$$

$$= 5 \text{ ft}$$

Since BC is the measure of the side opposite the 60° angle, we have

$$BC = \left[\frac{1}{2}(AC)\right]\sqrt{3}$$
$$= \left[\frac{1}{2}(10\text{ ft})\right]\sqrt{3}$$
$$= 5\sqrt{3}\text{ ft}$$

Practice Exercise 2. Given $\triangle DEF$ with $m(\angle E) = 90°$, $m(\angle D) = 30°$, and $DE = 8\sqrt{3}$ cm, determine EF and DF.

EXAMPLE 2 In the figure at left below, we have right $\triangle DEF$ with $m(\angle E) = 90°$, $EF = 3$ ft, and $DF = 6$ ft. Determine $m(\angle D)$.

SOLUTION: $\triangle DEF$ is a right triangle. Since DF, the measure of the hypotenuse, is twice EF, the measure of the shorter leg, then the angle opposite that shorter leg has measure 30°. Therefore, $m(\angle D) = 30°$.

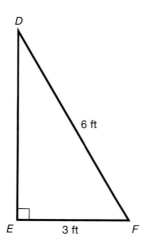

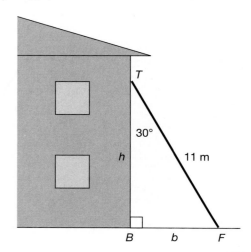

Practice Exercise 3. Given $\triangle GHJ$ with $m(\angle H) = 90°$, $m(\angle J) = 45°$, and $HJ = 10$ in., determine GH and GJ.

EXAMPLE 3 An 11-m ladder leans against a vertical wall of a building. The top of the ladder makes an angle of 30° with the wall. (See figure at right above.) (a) How far from the base of the building is the foot of the ladder? (b) How far above the ground does the ladder lean against the building?

SOLUTION: Since $\triangle TBF$ is a right triangle with $m(\angle BTF) = 30°$, then $m(\angle BFT) = 90° - 30° = 60°$. For this 30°–60°–90° triangle, the measure of the side opposite the 30° angle (b) is one-half the measure of the hypotenuse (11 m). Also, the measure of the side opposite the 60° angle (h) is equal to one-half the measure of the hypotenuse (11 m) multiplied by $\sqrt{3}$. We now have

$$b = \frac{1}{2}(11\text{ m})$$
$$= 5.5\text{ m}$$
$$h = \frac{1}{2}(11\text{ m})\sqrt{3}$$
$$= 5.5\sqrt{3}\text{ m}$$

Therefore, (a) the foot of the ladder is 5.5 m away from the base of the building, and (b) the top of the ladder is $5.5\sqrt{3}$ m (approximately 9.5 m) above the ground.

17.7 Special Right Triangles

Another special triangle is an isosceles right triangle. Since the acute angles of an isosceles right triangle are congruent (do you agree?), then the measure of each acute angle is 45°. An isosceles right triangle is referred to as a 45°–45°–90° triangle. (See Figure 17.28)

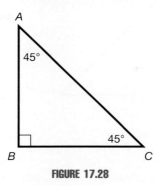

FIGURE 17.28

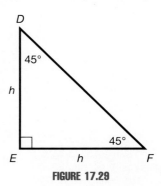

FIGURE 17.29

To determine the ratios of the measures of a 45°–45°–90° triangle, consider the triangle illustrated in Figure 17.29 such that each leg of the triangle has measure h. Then, by the Pythagorean theorem, we have

$$(DE)^2 + (EF)^2 = (DF)^2$$
$$h^2 + h^2 = (DF)^2$$
$$2h^2 = (DF)^2$$
$$DF = \pm h\sqrt{2}$$

Again, since DF is the measure of a side of the triangle, then $DF \geq 0$. Hence,

$$DF = h\sqrt{2}$$

We have established the following statement which is true for all 45°–45°–90° triangles.

In a 45°–45°–90° triangle, the measure of the hypotenuse is equal to the measure of one of its legs multiplied by $\sqrt{2}$.

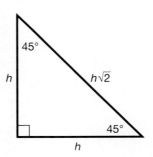

Practice Exercise 4. Given $\triangle KLM$ with $m(\angle L) = 90°$, $m(\angle K) = 45°$, and $KM = 6\sqrt{2}$ dm, determine LK and LM.

EXAMPLE 4 In the figure at left on the next page, we have right $\triangle ABC$ with the measures of some of its parts indicated. Determine the value of x.

SOLUTION: Since $\triangle ABC$ is a 45°–45°–90° triangle, the measure of $\overline{AC}$ is equal to $m(\overline{AB})$ multiplied by $\sqrt{2}$. Therefore,

$$x = (3 \text{ cm})\sqrt{2}$$
$$= 3\sqrt{2} \text{ cm}$$

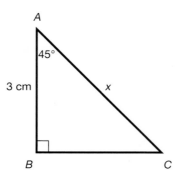
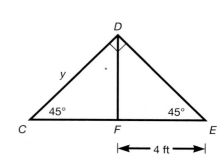

Practice Exercise 5. Given △RST with m(∠S) = 90°, m(∠R) = 60°, and RS = 4 ft, determine ST and RT.

EXAMPLE 5 In the figure at right above, we have right △CDE with the measures of some of its parts indicated. If $\overline{DF} \perp \overline{CE}$, determine the value of y.

SOLUTION: Since $\overline{DF} \perp \overline{CE}$, △DFE is a right triangle. Since m(∠E) = 45°, then m(∠EDF) = 45°. Hence, △FDE is an isosceles right triangle with $\overline{DF} \cong \overline{FE}$ and $m(\overline{DF})$ = 4 ft. Similarly, △CFD is an isosceles right triangle with hypotenuse $\overline{CD}$. We now have

$$m(\overline{CD}) = [m(\overline{DF})]\sqrt{2}$$

or

$$y = (4 \text{ ft})\sqrt{2}$$
$$= 4\sqrt{2} \text{ ft}$$

Practice Exercise 6. Given △PQR with m(∠Q) = 90° and m(∠R) = 45°, determine PQ if PR = 10 m.

EXAMPLE 6 A picture hangs on a vertical wall. A 6-ft person stands 6 ft from the base of the wall. The angle between the line of sight to the lower edge of the picture and a ray, extending horizontally from the eye of the person, has a measure of 45°. The angle between the line of sight of the upper edge of the picture and the horizontal ray has a measure of 60°. What is the height of the picture?

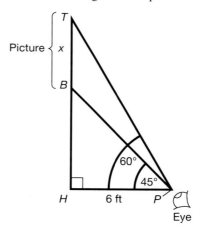

SOLUTION: The height of the picture is denoted by x. But

$$x = m(\overline{HT}) - m(\overline{BH})$$

△BHP is a 45°–45°–90° triangle with $\overline{BH} \cong \overline{HP}$. Since

$$m(\overline{HP}) = 6 \text{ ft}$$

then

$$m(\overline{BH}) = 6 \text{ ft}$$

△THP is a 30°–60°–90° triangle with $m(\overline{HP}) = 6$ ft. $\overline{HT}$ is the side opposite the 60° angle. Therefore,

$$m(\overline{HT}) = [m(\overline{HP})]\sqrt{3} = (6 \text{ ft})\sqrt{3} = 6\sqrt{3} \text{ ft}$$

We now have

$$\begin{aligned} x &= m(\overline{HT}) - m(\overline{BH}) \\ &= (6\sqrt{3} \text{ ft}) - (6 \text{ ft}) \\ &= (6\sqrt{3} - 6) \text{ ft} \\ &= 6(\sqrt{3} - 1) \text{ ft} \\ &\approx 6(1.73 - 1) \text{ ft} \\ &= 6(0.73) \text{ ft} \\ &= 4.4 \text{ ft} \end{aligned}$$

Therefore, the height of the picture is approximately 4.4 ft.

ANSWERS TO PRACTICE EXERCISES

1. $AB = 4$ yd, $BC = 4\sqrt{3}$ yd. 2. $EF = 8$ cm, $DF = 16$ cm. 3. $GH = 10$ in., $GJ = 10\sqrt{2}$ in.
4. $LK = 6$ dm, $LM = 6$ dm. 5. $ST = 4\sqrt{3}$ ft, $RT = 8$ ft. 6. $5\sqrt{2}$ m.

EXERCISES 17.7

1. Given △ABC with $m(\angle A) = 90°$, $m(\angle B) = 60°$, and $BC = 8$ ft, determine $m(\angle C)$, AB, and AC.
2. Given △DEF with $m(\angle E) = 90°$, $EF = 3$ cm, and $DF = 6$ cm, determine $m(\angle D)$, $m(\angle F)$, and DE.
3. Given △GHJ with $m(\angle H) = 90°$, $m(\angle G) = 45°$, and $HJ = 7$ in., determine $m(\angle J)$, GH, and GJ.
4. Given △KLM with $m(\angle K) = m(\angle M) = 45°$ and $KM = 13\sqrt{2}$ m, determine $m(\angle L)$, KL, and LM.
5. Given △NOP with $m(\angle N) = 30°$, $m(\angle O) = 90°$, and $OP = 6$ yd, determine the perimeter of the triangle.
6. Given △RST with $m(\angle R) = 45°$, $m(\angle S) = 90°$, and $RS = 19$ dm, determine the perimeter of the triangle.

For Exercises 7–16, the measure of one side of the right triangle is given. Determine the values for x and y.

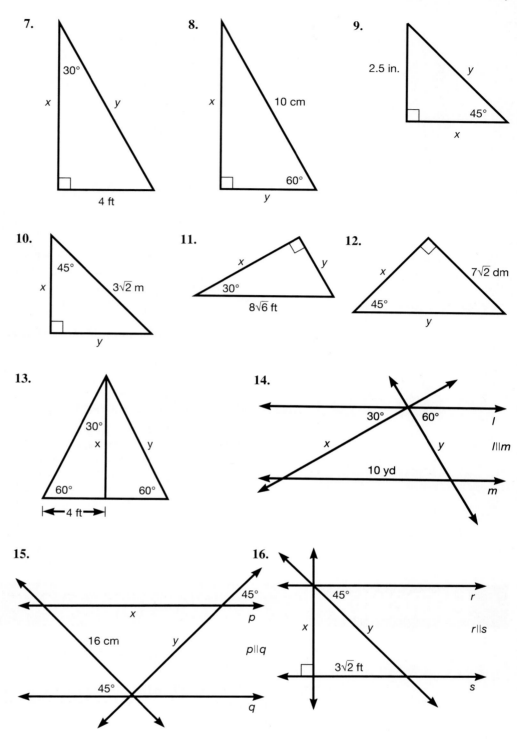

17. Determine the length of a diagonal of a square each of whose sides has measure 12 yd.
18. Determine the length of a diagonal of a square if its perimeter is 64 m.
19. Determine the perimeter of a square if the length of a diagonal of the square is $7\sqrt{2}$ in.
20. What is the ratio of the length of a diagonal of a square to its perimeter?

For Exercises 21–26, refer to the figure at left on the next page.

21. Determine DC.
22. Determine $m(\angle ADB)$.
23. Determine $m(\angle ABD)$.

17.7 Special Right Triangles

24. Determine AB.
25. Determine the perimeter of $\triangle BCD$.
26. Determine the perimeter of $\triangle ACD$.

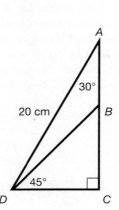
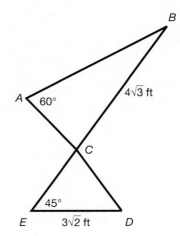

For Exercises 27–30, refer to the figure at right above.

27. Determine AB.
28. Determine EC.
29. Determine the perimeter of $\triangle ABC$.
30. Determine the perimeter of the total figure.
31. A 26-ft ladder leans against a vertical wall. The top of the ladder makes a 30° angle with the wall. How far away from the base of the wall is the foot of the ladder?
32. A plank is flush against a vertical wall. The top of the plank then starts to slide down along the wall. When the foot of the plank is 11 m away from the base of the wall, the angle between the top of the plank and the wall is 60°. How long is the plank?
33. A 55-gal barrel of crude oil is being rolled up a ramp of a loading dock. The ramp makes an angle of 45° with the level ground. If the height of the dock is 6 ft, how long is the ramp? (See figure.)

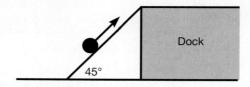

34. In Exercise 33, if the ramp is 9 m long, what is the height of the dock? (Give your answer to the nearest tenth of a meter.)
35. A baseball diamond is in the form of a square, 90 ft on a side. What is the distance from homeplate to second base? (See figure.)

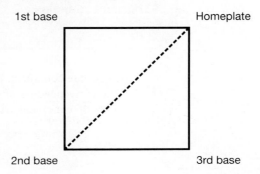

36. Sue wants her flower garden to be in the form of an isosceles right triangle. If the perimeter of the garden is to be 30 yd, how long will the hypotenuse of the garden be? (Give your answer to the nearest tenth of a yard.)

37. A rigid metal rod is to be placed lying flat in a rectangular box. If the base of the box measures 4 ft by $4\sqrt{3}$ ft, what is the length of the longest rod that can be placed in the box? (Disregard the thickness of the rod.)
38. Two joggers start from the same point and jog at right angles to each other at the rate of 3 mph. If they maintain that rate for 75 min, how far apart will they be at that time? (Give your answer to the nearest tenth of a mile.)
39. A kite, attached to a long, heavy string, is being flown by Sarah. The angle between the end of the string held by Sarah and a horizontal ray is 30°. The length of the string between Sarah's hand and the kite is 37 m. A point P is on the horizontal ray and directly below the kite. How high above the point P is the kite?

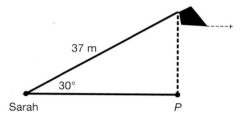

40. The town council decreed that a playing field is to be built in the form of a right triangle and that the area of the playing field is to be no less than 325 m² but no more than 360 m². The field was constructed with the shortest side 20 m long and the hypotenuse 40 m long. Does the area of the field satisfy the council's decree? (Why or why not?)

17.8 Congruent Triangles

OBJECTIVE

After completing this section, you should be able to determine if two triangles are or are not congruent.

In Section 17.6, we discussed similar triangles. You learned that similar triangles have the same shape. Some similar triangles also have the same size.

DEFINITION

Congruent triangles are triangles that have the same shape and the same size.

If triangles are congruent, then they are also similar (since they have the same shape). Therefore, the corresponding angles of congruent triangles are congruent. Also, the corresponding sides of congruent triangles are congruent. That is, if $\triangle ABC$ is congruent to $\triangle DEF$, then

$$\angle A \cong \angle D \qquad \angle B \cong \angle E \qquad \angle C \cong \angle F$$

and

$$\overline{AB} \cong \overline{DE} \qquad \overline{BC} \cong \overline{EF} \qquad \overline{AC} \cong \overline{DF}$$

17.8 Congruent Triangles

We denote the congruence of △ABC and △DEF by

$$\triangle ABC \cong \triangle DEF$$

which is read "△ABC is congruent to △DEF." (See Figure 17.30.)

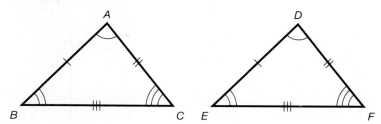

FIGURE 17.30

Sometimes, two triangles appear to be congruent but are not. To show that two triangles are congruent, we must show that their corresponding parts (angles *and* sides) are congruent. That is, we must show that the six pairs of corresponding parts are congruent. And that's a big job! However, if you examine congruent triangles very carefully, you may be able to conclude the following:

1. The lengths of the three sides of one triangle are equal, respectively, to the lengths of the three sides of the other triangle.
2. The lengths of two sides of one triangle and the measure of the angle between those two sides are equal, respectively, to the lengths of two sides of the other triangle and the measure of the angle between those two sides.
3. The measures of two angles of one triangle and the length of the side between those two angles are equal, respectively, to the measures of two angles of the other triangle and the length of the side between those two angles.

It would appear, then, that we don't have to establish the congruence of all six pairs of corresponding parts of the triangle to determine if the two triangles are congruent. And, indeed, that is true.

We will now state, without proof, three basic conditions under which two triangles are congruent. (The proofs depend upon basic geometric constructions, which will not be discussed in this text.)

Side, Side, Side (SSS)

If the three sides of one triangle are congruent, respectively, to the three sides of a second triangle, then the two triangles are congruent.

Determine if the following pairs of triangles are congruent. If they are, give the appropriate reason.

Practice Exercise 1. △ABC and △DEF such that $\overline{AB} \cong \overline{DE}$, $m(\overline{BC}) = m(\overline{EF})$, and $m(\overline{AC}) = m(\overline{DF})$.

EXAMPLE 1 Determine whether △ABC ≅ △DEF.

SOLUTION: Given △ABC and △DEF such that $\overline{AB} \cong \overline{DE}$, $\overline{BC} \cong \overline{EF}$, and $\overline{AC} \cong \overline{DF}$, then △ABC ≅ △DEF by SSS.

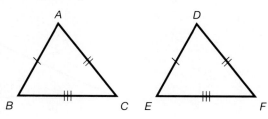

Practice Exercise 2. △GHJ and △KLM with $\overline{GH} \cong \overline{LM}$, $\overline{HJ} \cong \overline{MK}$, and $\overline{GJ} \cong \overline{LK}$.

EXAMPLE 2 Determine if △RST ≅ △XYZ.

SOLUTION: Given △RST with $m(\overline{RS}) = 12$, $m(\overline{ST}) = 5$, and $m(\overline{RT}) = 13$. Given △XYZ with $m(\overline{XY}) = 12$, $m(\overline{YZ}) = 5$, and $m(\overline{XZ}) = 13$. Since $m(\overline{RS}) = m(\overline{XY})$, then $\overline{RS} \cong \overline{XY}$. Similarly, $\overline{ST} \cong \overline{YZ}$ and $\overline{RT} \cong \overline{XZ}$. Therefore, △RST ≅ △XYZ by SSS. (Note that both triangles are right triangles. Why?)

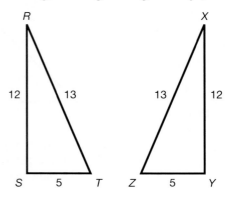

Side, Angle, Side (SAS)

If two sides of one triangle and the angle between those sides are congruent, respectively, to two sides of a second triangle and the angle between those sides, then the two triangles are congruent.

Practice Exercise 3. △NOP and △QRS with $\overline{OP} \cong \overline{RS}$, $\overline{NP} \cong \overline{QS}$, and $m(\angle P) = m(\angle S)$.

EXAMPLE 3 Determine whether △PQR ≅ △STU.

SOLUTION: Given △PQR and △STU such that $\overline{PQ} \cong \overline{ST}$, $\overline{QR} \cong \overline{TU}$, and $\angle Q \cong \angle T$, then △PQR ≅ △STU by SAS.

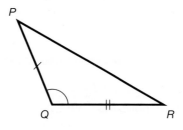 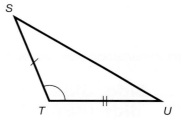

Practice Exercise 4. △TUV and △XYZ with $m(\overline{TU}) = 4.9$, $m(\angle U) = 61°$, $m(\overline{XY}) = 4.9$, $m(\angle Y) = 61°$, and $\overline{UV} \cong \overline{YZ}$.

EXAMPLE 4 Determine whether △GHJ ≅ △KLM.

SOLUTION: Given △GHJ with $m(\overline{GH}) = 10$, $m(\overline{HJ}) = 8$, and $m(\angle H) = 125°$. Given △KLM with $m(\overline{LK}) = 10$, $m(\overline{LM}) = 8$, and $m(\angle L) = 125°$. Since $m(\overline{GH}) = m(\overline{LK})$, then $\overline{GH} \cong \overline{LK}$. Similarly, $\overline{HJ} \cong \overline{LM}$ and $\angle H \cong \angle L$. Therefore, △GHJ ≅ △KLM by SAS.

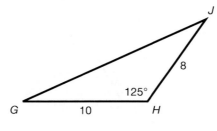 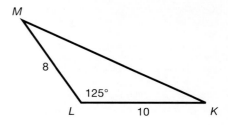

17.8 Congruent Triangles

Angle, Side, Angle (ASA)

If two angles of a triangle and the side between those angles are congruent, respectively, to two angles of a second triangle and the side between those angles, then the two triangles are congruent.

Practice Exercise 5. △CDE and △GHJ with $\overline{CD} \cong \overline{GH}$, ∠D ≅ ∠H, m(∠C) = 62°, and m(∠G) = 68°.

EXAMPLE 5 Determine whether △CDE ≅ △HJK.

SOLUTION: Given △CDE and △HJK such that ∠C ≅ ∠H, ∠D ≅ ∠J, and $\overline{CD} \cong \overline{HJ}$, then △CDE ≅ △HJK by ASA.

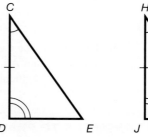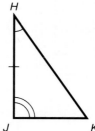

Practice Exercise 6. △LMP and △KJE with $\overline{MP} \cong \overline{JE}$, m(∠M) = m(∠J), and m(∠L) = m(∠K).

EXAMPLE 6 Determine whether △NOP ≅ △MRS.

SOLUTION: Given △NOP with m(∠N) = 32°, m(∠P) = 78°, and m($\overline{NP}$) = 9. Given △MRS with m(∠M) = 32°, m(∠S) = 78°, and m($\overline{MS}$) = 9. Then △NOP ≅ △MRS by ASA.

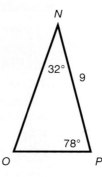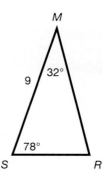

ANSWERS TO PRACTICE EXERCISES

1. △ABC ≅ △DEF by SSS. 2. △GHJ ≅ △KLM by SSS. 3. △NOP ≅ △QRS by SAS. 4. △TUV ≅ △XYZ by SAS. 5. △CDE ≇ △GHJ. 6. △LMP ≅ △KJE by ASA.

Practice Exercise 7. Consider the given figure. If l ∥ m and △AOB ≅ COD, determine all of the pairs of congruent corresponding parts.

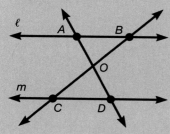

EXAMPLE 7 Given the accompanying figure, if $\overline{AB} \parallel \overline{DE}$ and △ABC ≅ △CDE, determine all of the pairs of congruent corresponding parts.

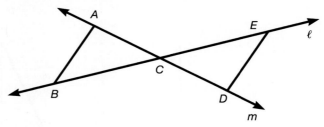

SOLUTION: Since l and m are intersecting lines, then ∠ACB and ∠DCE are vertical angles. Therefore, ∠ACB ≅ ∠DCE. Since $\overline{AB} \parallel \overline{DE}$, then ∠BAC and ∠CDE are alternate interior angles. Therefore, ∠BAC ≅ ∠CDE. Similarly, ∠ABC ≅ ∠CED.

Matching the vertices of the angles of the two congruent triangles, we now determine that $\overline{AB} \cong \overline{DE}$, $\overline{AC} \cong \overline{CD}$, and $\overline{BC} \cong \overline{CE}$.

Practice Exercise 8. Consider the rectangle *ABCD* with diagonals $\overline{AC}$ and $\overline{BD}$. Determine all pairs of congruent triangles.

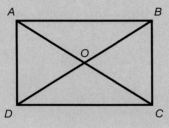

EXAMPLE 8 From the accompanying figure, determine if the two triangles indicated are congruent. If so, given an appropriate reason.

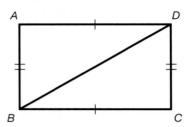

SOLUTION: The two triangles indicated are $\triangle ABD$ and $\triangle BCD$. We are given that $\overline{AD} \cong \overline{BC}$ and $\overline{AB} \cong \overline{DC}$. But $\overline{BD}$ is a side of *both* triangles and $\overline{BD} \cong \overline{BD}$ (since every line segment is congruent to itself). Therefore $\triangle ABD \cong \triangle BCD$ by SSS.

ANSWERS TO PRACTICE EXERCISES

7. $\angle A \cong \angle D$, $\angle AOB \cong \angle COD$, $\angle B \cong \angle C$, $\overline{AB} \cong \overline{CD}$, $\overline{BO} \cong \overline{OC}$, and $\overline{AO} \cong \overline{OD}$. 8. $\triangle AOB \cong \triangle DOC$, $\triangle AOD \cong \triangle BOC$, $\triangle ABC \cong ADC$, and $\triangle BAD \cong BCD$.

EXERCISES 17.8

In Exercises 1–6, determine if the pairs of triangles given are congruent. If they are, give an appropriate reason. If not enough information is given to determine if the triangles are congruent, state that fact.

1.

2.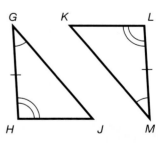

3.

4.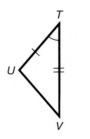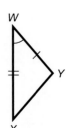

17.8 Congruent Triangles

5.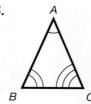
6.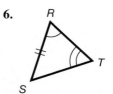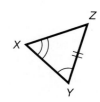

7. In the figure at left below, determine if △ADC is congruent to △ABC. If so, give an appropriate reason.

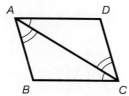 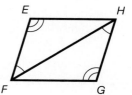

8. In the figure at right above, determine if △EFH is congruent to △FGH. If so, give an appropriate reason.
9. In the figure at left below, determine if △JKL is congruent to △MNL. If so, give an appropriate reason.

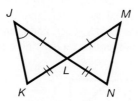 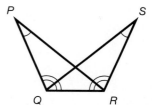

10. In the figure at right above, determine if △PQR is congruent to △QRS. If so, give an appropriate reason.
11. In the figure at left below, $m(\angle XUZ) = m(\angle YUZ)$. Determine if △XUZ is congruent to △YUZ. If so, give an appropriate reason.

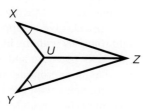 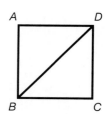

12. Given the square ABCD with diagonal $\overline{BD}$, determine if △ABD is congruent to △BCD. If so, give an appropriate reason. (See figure at right above.)
13. Given the parallelogram EFGH with diagonal $\overline{EG}$. Determine if △EFG is congruent to △EGH. If so, give an appropriate reason. (See figure at left below.)

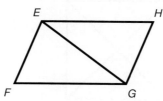 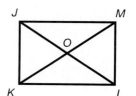

14. Given the rectangle JKLM in the figure at right above, $\overline{JL}$ and $\overline{KM}$ are diagonals of the rectangle. List all pairs of congruent corresponding line segments.
15. Given the trapezoid PQRS with diagonals $\overline{QS}$ and $\overline{PR}$. If △PQR ≅ △SRQ, list six pairs of congruent corresponding angles in the figure at left on the next page.

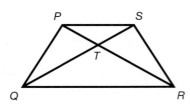 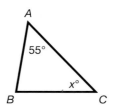 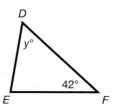

16. In the figures at right above, $\triangle ABC \cong \triangle DEF$ with $m(\angle A) = m(\angle D)$, $m(\angle C) = m(\angle F)$, and $\overline{BC} \cong \overline{EF}$. Determine $m(\angle B)$.

17. Given $\triangle ABC$ with $\overline{AB} \cong \overline{AC}$. $\overrightarrow{AD}$ is the bisector of $\angle BAC$. If $\triangle BDA$ is congruent to $\triangle CDA$, give an appropriate reason. (See figure at left below.)

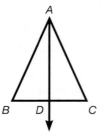 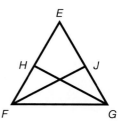

18. Given equilateral $\triangle EFG$, H is the midpoint of $\overline{EF}$, and J is the midpoint of $\overline{EG}$. If $\triangle HFG$ is congruent to $\triangle FJG$, give an appropriate reason. (See figure at right above.)

19. Given the square $KLMN$, P, Q, and R are the midpoints of $\overline{KL}$, $\overline{MN}$, and $\overline{KN}$, respectively. If $\triangle PKR$ is congruent to $\triangle QNR$, give an appropriate reason. (See figure at left below.)

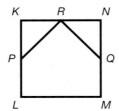 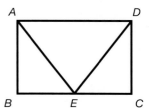

20. Given the rectangle $ABCD$ in the figure at right above, E is the midpoint of $\overline{BC}$. If $\triangle ABE$ is congruent to $\triangle ECD$, give an appropriate reason.

21. Given two right triangles with the hypotenuse and acute angle of one triangle congruent, respectively, to the hypotenuse and acute angle of the other triangle. If the two triangles are congruent, give an appropriate reason.

22. Given two right triangles with the two legs of one triangle congruent, respectively, to the two legs of the other triangle. If the two triangles are congruent, give an appropriate reason.

23. Given the square $ABCD$. Using congruent triangles, prove that $\overline{AC} \cong \overline{BD}$. (See figure at left below.)

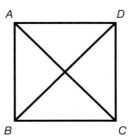 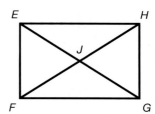

24. Given the rectangle $EFGH$ in the figure at right above and using congruent triangles, prove that J is the midpoint of $\overline{EG}$. (Hint: If J is the midpoint of $\overline{EG}$, then $\overline{EJ} \cong \overline{JG}$.)

17.8 Congruent Triangles

25. A surveyor is hired to determine the distance from X to Y, but cannot do it directly because there is a large pond between the two points. The surveyor proceeds by finding a point W (as indicated in the figure at left below). He then determines a line segment $\overline{XV}$ such that W is the midpoint of $\overline{XV}$. He also determines a line segment $\overline{UY}$ such that W is the midpoint of $\overline{UY}$. Is $m(\overline{XY}) = m(\overline{UV})$? If so, give an appropriate reason.

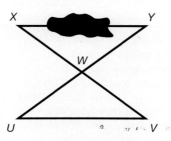
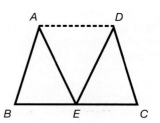

26. A man wishes to build twin sand box areas for his twin daughters. He wants the areas to be next to each other, with a play area in between. He decides that the boundary of the two sand box areas, together with the play area, should be an isosceles trapezoid (trapezoid $ABCD$ in the figure at right above). If E is the midpoint of $\overline{BC}$ and the sand box areas are denoted by $\triangle ABE$ and $\triangle DCE$, would the boundaries of the sand box areas be congruent? If so, give an appropriate reason.

27. An A-frame cabin is built so that the front of the cabin is in the form of an isosceles triangle ($\triangle RST$ in the figure above, with $\overline{RS} \cong \overline{RT}$). The areas denoted by $\triangle RSP$ and $\triangle RTQ$ are glass enclosed and $\overline{SP} \cong \overline{QT}$. Using congruent triangles, show that $\triangle RPQ$ is an isosceles triangle.

28. A vegetable garden is in the shape of a quadrilateral (a four-sided polygon). If a diagonal of the quadrilateral separates the garden into two congruent areas, is the boundary of the garden a parallelogram? Give an appropriate reason for your answer.

29. Four community garden plots are adjacent to each other, and the overall boundary is an equilateral triangle ($\triangle JKL$ in the figure at left below). If M, N, and O are the midpoints, respectively, of $\overline{JK}$, $\overline{JL}$, and $\overline{KL}$, are the triangular plots, $\triangle JMN$ and $\triangle MON$, congruent? If so, give an appropriate reason.

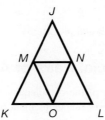
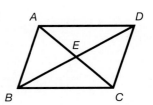

30. Using congruent triangles, prove that the diagonals $\overline{AC}$ and $\overline{BD}$ of the parallelogram $ABCD$ bisect each other in the figure at right above.

Chapter Summary

In this chapter, you learned about some basic concepts of geometry. Lines and parts of lines were discussed as sets of points. Angles, their measures, and classifications were introduced. Special triangles were examined. Similar and congruent triangles were also discussed.

The **key words** introduced in this chapter are listed in the order in which they appeared in the chapter. You should know what they mean.

line (p. 698)
line segment (p. 698)
endpoints (p. 698)
congruent (p. 698)
midpoint (p. 699)
ray (p. 700)
plane (p. 700)
coplanar points (p. 700)
collinear points (p. 700)
angle (p. 703)
side of the angle (p. 703)
vertex of the angle (p. 703)
measure of an angle (p. 704)
acute angle (p. 704)
right angle (p. 704)
obtuse angle (p. 704)
straight angle (p. 704)
congruent angles (p. 705)
bisector of the angle (p. 705)
adjacent angles (p. 708)
vertical angles (p. 708)
complementary angles (p. 709)
supplementary angles (p. 709)
perpendicular lines (p. 712)
transversal (p. 713)
interior angles (p. 713)
exterior angles (p. 713)
alternate interior angles (p. 713)
alternate exterior angles (p. 713)
corresponding angles (p. 713)
parallel lines (p. 714)
triangles (p. 719)
equiangular triangle (p. 719)
acute triangle (p. 719)
obtuse triangle (p. 719)
right triangle (p. 719)
similar triangles (p. 725)
congruent triangles (p. 740)

Review Exercises

17.1 Terminology and Notation

For Exercises 1–5, refer to the figure.

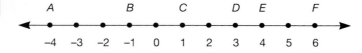

1. Name a line segment that is congruent to $\overline{AB}$.
2. What is $m(\overline{BE})$?
3. What is the midpoint of $\overline{AE}$?
4. Give another name for $\overrightarrow{BD}$.
5. How many points are on the given line that are exactly 5 units away from C?

17.2 Angles

For Exercises 6–9, classify each angle with given measure as being acute, right, obtuse, or straight.

6. 105°   7. 90°   8. 72°   9. 180°

10. If $\overrightarrow{OC}$ bisects $\angle AOB$ and $m(\angle AOB) = 114°$, what is $m(\angle AOC)$?

17.3 Some Special Pairs of Angles

11. What is the measure of the complement of $\angle A$ if $m(\angle A) = 57°$?
12. What is the measure of the supplement of $\angle B$ if $m(\angle B) = (x + 12)°$?
13. If angles C and D are vertical angles and are also complementary angles, what is $m(\angle C)$?
14. If angles E and F are vertical angles with $m(\angle E) = (2u)°$ and $m(\angle F) = (3u - 5)°$, determine the value of u.
15. If $m(\angle G)$ is equal to 6 times the measure of its supplement, what is $m(\angle G)$?

17.4 Parallel and Perpendicular Lines

For Exercises 16–20, refer to the figure, in which $p \parallel q$.

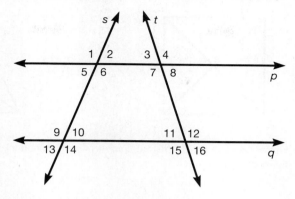

16. List all angles that are adjacent to $\angle 7$.
17. List all angles that are congruent to $\angle 10$.
18. List all angles that are supplementary to $\angle 11$.
19. With s as the transversal for lines p and q, list all alternate exterior angles.
20. With t as the transversal for lines p and q, list all corresponding angles.
21. If lines m and n are parallel and line j is perpendicular to m, what is the relationship between n and j?

17.5 Triangles

22. Given $\triangle ABC$ with $m(\angle A) = 63°$ and $m(\angle C) = 49°$, determine $m(\angle B)$.
23. If $\triangle DEF$ is equiangular, what is $m(\angle F)$?
24. If the measure of a base angle of an isosceles triangle is 37°, what is the measure of the vertex angle of the triangle?
25. If the measure of the vertex angle of an isosceles triangle is 102°, what is the measure for each base angle of the triangle?
26. In $\triangle RST$, the measure of $\angle S$ is 2° more than the measure of $\angle T$. The measure of $\angle R$ is twice the measure of S. Determine the measures of all three angles of the triangle.

17.6 Similar Triangles

27. If $\triangle ABC \sim \triangle DEF$ and the measures of two angles of $\triangle DEF$ are 46° and 57°, can $\triangle ABC$ have an angle with measure of 77°?

For Exercises 28–31, the measures for parts of $\triangle RST$ and $\triangle JKL$ are given. Determine if the triangles are similar.

28. $RS = 6$, $RT = 7$, $ST = 9$; $JK = 8$, $JL = 9$, $KL = 11$
29. $m(\angle J) = 43°$, $m(\angle K) = 60°$; $m(\angle S) = 86°$
30. $RS = 3$, $RT = 4$, $ST = 5$; $JK = 7.5$, $JL = 12.5$, $KL = 10$
31. $JK = JL$; $m(\angle R) = 32°$, $m(\angle S) = 59°$

17.7 Spcial Right Triangles

32. Given $\triangle ABC$ with $m(\angle A) = 30°$, $m(\angle B) = 90°$, and $AC = 12$ cm, determine $m(\angle C)$, AB, and BC.
33. Given $\triangle DEF$ with $m(\angle D) = m(\angle F) = 45°$ and $DF = 8\sqrt{2}$ cm, determine $m(\angle E)$, DE, and EF.
34. Determine the length of a diagonal of a square each of whose sides has measure $3\sqrt{2}$ yd.
35. Determine the perimeter of a square if the length of its diagonal is 10 cm.

17.8 Congruent Triangles

36. Any two equilateral triangles are such that the three angles of one of the triangles are congruent, respectively, to the three angles of the other triangle. Are the two triangles congruent? Give an appropriate reason for your answer.

37. Given the square *ABCD* in the figure at left below, if $\overline{AE} \cong \overline{EC}$, is △*AED* congruent to △*BEC*? If so, give an appropriate reason.

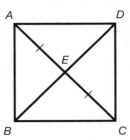 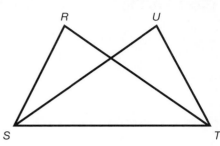

38. Given △*RST* ≅ △*SUT*, list all congruent corresponding parts (angles and line segments) in the figure at right above.

Teasers

1. If the measure of the vertex angle of an isosceles triangle is $(2x - 1)°$, what is the measure of each of the base angles of the triangle?
2. In a 30°–60°–90° triangle whose hypotenuse has measure $(3y + 5)$ m, what is the measure of the longer side of the triangle?
3. In an isosceles right triangle whose legs each have a measure of x cm, what is the perimeter of the triangle?
4. Two triangles are similar and their corresponding sides are in the ratio of 5:9. If the sides of the smaller triangle have measures of 6 cm, 7 cm, and 11 cm, determine the perimeter of the larger triangle.
5. A fence is to be built around a 30°–60°–90° triangular plot of land. The shortest side of the plot measures 8 yd. If the fence cost $7 per foot, how much would it cost to fence the plot of land? (Give your answer to the nearest dollar.)
6. If *P* and *Q* are two distinct points in a plane, how many points in the plane are collinear with the two given points?
7. A point on a line separates the line into three distinct sets. What are those sets?
8. A line in a plane separates the plane into three distinct sets. What are those sets?
9. If two angles in the same plane have a common vertex, are they necessarily adjacent angles? Explain your answer.
10. Consider a rectangular region. How many triangular regions are formed by the two diagonals of the rectangle?

CHAPTER 17 TEST

This test covers material from Chapter 17. Read each question carefully and answer all questions. If you miss a question, review the appropriate section of the text.

1. Write the symbol for the ray from the point P and in the direction of the point R.
2. Write the symbol for the line segment with endpoints G and H.
3. What is the measure of a straight angle?
4. If $\overrightarrow{OD}$ is the bisector of $\angle AOC$ and $m(\angle AOC) = 88°$, what is $m(\angle AOD)$?
5. What is the measure of the supplement of an angle whose measure is 63°?
6. What is the measure of the complement of an angle whose measure is $(2u - 15)°$?
7. If angles A and B are vertical angles and $m(\angle A) = 57°$, what is $m(\angle B)$?
8. If two parallel lines are cut by a transversal, name three pairs of angles that are congruent.
9. If line p is parallel to line q, and line q is perpendicular to line r, what is the relationship between p and r?
10. Given $\triangle HJK$ with $m(\angle H) = 39°$ and $m(\angle J) = 72°$, determine $m(\angle K)$.
11. If the measure of the vertex angle of an isosceles triangle is 54°, what is the measure of each of the base angles of the triangle?
12. What is the measure of each angle in an equiangular triangle?
13. In a 30°–60°–90° triangle whose hypotenuse has measure 20 in., what is the measure of the side opposite the 60° angle?
14. In an isosceles right triangle whose legs each have a measure of 7 ft, what is the measure of the hypotenuse?
15. If $\triangle ABC \sim \triangle DEF$, $AB = 8$ yd, $BC = 16$ yd, $AC = 12$ yd, $DE = 12$ yd, and $EF = 6$ yd, determine DF.
16. If two triangles are congruent, are they also similar? Give an appropriate reason.
17. Given two right triangles with the hypotenuse and acute angle of one congruent, respectively, to the hypotenuse and acute angle of the other, are the two triangles congruent? Give an appropriate reason for your answer.
18. A square has sides each of measure 7 yd. What is the measure of the diagonal of the square?
19. In $\triangle ABC$, $m(\angle A) = 30°$, $m(\angle B) = (2x)°$, and $m(\angle C) = (x + 30)°$. Determine the value of x.
20. In a 45°–45°–90° triangle, each leg has a measure of 13 cm. Determine the perimeter of the triangle. (Give your answer to the nearest tenth of a centimeter.)

CUMULATIVE REVIEW: CHAPTERS 1–17

1. Which is larger, -3^2 or 2^{-3}?
2. Round, to the nearest tenth, the quotient when 3.56 is divided by 1.9.
3. How many significant digits are there in 0.0096×10^2?
4. Solve the equation $3(t + 3) = 4 - 7t$ for t.
5. Solve the equation $p^2 - 5p + 6 = 0$ for p.
6. What must be added to complete the square for $y^2 - 7y$?
7. Rewrite the equation $5 - 3y = 6x$ in the slope-intercept form.
8. Determine the slope of the line which is the graph of the equation $4x - 7y + 1 = 0$.
9. Simplify by combining like terms: $3x - 2[y - 4(x - 2y) + 5x] - 7y$.
10. Simplify by combining like terms: $3a(a^2 - 3) + a^2(2 - 3a) - 6(a^3 - 2a^2)$.
11. Simplify and write your result without negative exponents:
$$\left(\frac{-4a^{-2}b^0c^3}{2ab^{-3}c^0}\right)^{-2}$$
12. Determine if the ordered pair $(4, -3)$ is a solution for the equation $2x - 3y + 17 = 0$.
13. If the ordered pair $(a, -4)$ is a solution for the equation $5x - 2y = 9$, determine the value of a.
14. Solve the system of equations algebraically:
$$\begin{cases} 2x - 3y = 6 \\ x + 4y = 8 \end{cases}$$
15. Simplify by combining like terms: $\sqrt{50} - 3\sqrt{12} + \sqrt{32} - 4\sqrt{27}$.
16. If $\triangle ABC$ is a right triangle with $m(\angle A) = 30°$, $m(\angle B) = 90°$, and $AC = 10$ ft, determine the perimeter of the triangle.
17. If the diagonal of a square has measure $9\sqrt{2}$ yd, what is the area of the square?
18. What is the volume of a triangular prism whose base has area 23.1 yd² and whose height is 4.9 yd?
19. What is the volume of a sphere with radius 4 ft?
20. If $\triangle RST$ is an obtuse triangle and $\triangle XYZ$ is an acute triangle, can $\triangle RST$ be similar to $\triangle XYZ$?
21. Given $\triangle ABC$ with $m(\angle A) = m(\angle C) = 45°$ and $m(\overline{AC}) = 9\sqrt{2}$ yd, determine $m(\angle B)$ and $m(\overline{BC})$.
22. If two triangles are isosceles such that the base angles of one triangle are congruent to the base angles of the other triangle, are the triangles congruent?
23. What is the measure of each of the angles between the diagonals of a square?
24. a. If two triangles are congruent, are they also similar?
    b. If two triangles are similar, are they also congruent?
25. Name three conditions under which two triangles are congruent.

Answers to Odd-Numbered Exercises

EXERCISES 1.1, PAGE 6

1. Hundreds **3.** Ones **5.** Hundred millions **7.** Ten thousands **9.** Ten millions
11. 1450 **13.** 3006 **15.** 15,000,785 **17.** 312,405,006,089 **19.** 6,007,005,004,903
21. Nine hundred sixty-eight
23. Nineteen thousand, two hundred one
25. Eight hundred ninety-six thousand, five
27. Thirteen million, seven hundred sixty-nine thousand, one hundred twenty-three
29. One billion, six hundred nineteen million, seven hundred twenty-five thousand, ten
31. 635 **33.** 5000 **35.** 0 **37.** 5 **39.** 1000

EXERCISES 1.2, PAGE 11

1. < **3.** < **5.** > **7.** > **9.** <
11. 418 **13.** 889 **15.** 44,798 **17.** 906,921 **19.** 13,775,388
21. 17,726 **23.** 1,060,519 **25.** 6133 **27.** 421 **29.** $8350
31. $1253 **33.** $3961 **35.** 896 lb 1 oz **37.** 7 gal 3 qt **39.** 345,999
41. 2 yd 1 ft 3 in. **43.** 24,215,309 **45.** 1468 mm **47.** 61 yd **49.** 126 ft
51. 72 ft **53.** 122 cm **55.** 254 in.

EXERCISES 1.3, PAGE 20

1. 21 **3.** 15 **5.** 524 **7.** 504 **9.** 1413
11. 2612 **13.** 1891 **15.** 3757 **17.** 31 **19.** 2433
21. 563 **23.** 4013 **25.** 2081
27. a. 25 min 40 sec **b.** 49 min 6 sec **29.** 48 sec
31. a. A; 4 work-days **b.** B; 4 work-days **c.** 39 **d.** 39 **e.** Both lost the same number.
33. a. 15 min 35 sec **b.** 8 min 56 sec **c.** 6 min 39 sec **d.** 47 min 10 sec
35. $336 **37.** $547.35

EXERCISES 1.4, PAGE 29

1. 7636 **3.** 8970 **5.** 11,711 **7.** 28,782 **9.** 145,962
11. 452,097 **13.** 5,334,264 **15.** 2,688,213 **17.** 2,608,016 **19.** 3,064,842
21. 17,119,674 **23.** 16,851,456 **25.** 42,543,720 **27.** 11,049 **29.** 6,531,354
31. 8856 **33.** $23,580 **35.** $521,680 **37. a.** $3744 **b.** $744
39. a. 2768 **b.** 7008 **c.** 3670 **41.** 224 in.^{2} **43.** 925 cm^2 **45.** 729 ft^2

EXERCISES 1.5, PAGE 36

1. 15 **3.** 12 **5.** 12 R7 **7.** 636 R13
9. 1173 R31 **11.** 24 R172 **13.** 435 R24 **15.** 157 R335
17. 8043 R551 **19.** 1214 R565 **21.** 245 R22 **23.** 101 R12
25. 344 R717 **27.** 1121 R369 **29.** 237 **31.** 2
33. 24
35. Low gas mileage, approximately 15 mi/gal
37. 829 **39.** 16 **41.** $97 **43.** $2464

753

EXERCISES 1.6, PAGE 39

1. 230	**3.** 690	**5.** 800	**7.** 6000	**9.** 12,000
11. 125,000	**13.** 1,235,000	**15.** 1,000,000	**17.** 6,000,000	**19.** 19,000,000
21. 270	**23.** 2200	**25.** 18,000	**27.** 300	**29.** 6000
31. 5600	**33.** 120,000	**35.** 960,000	**37.** 40	**39.** 3

EXERCISES 1.7, PAGE 46

1. 16	**3.** 121	**5.** 625	**7.** 6561	**9.** 48
11. 2058	**13.** 88	**15.** 380	**17.** 9	**19.** 3
21. 23	**23.** 9	**25.** 94	**27.** 12	**29.** 83
31. 11	**33.** 5	**35.** $288	**37.** $78	**39.** 6

EXERCISES 1.8, PAGE 49

1. 5	**3.** 6	**5.** 0	**7.** 0	**9.** 9
11. Not unique	**13.** Impossible	**15.** 23	**17.** 0	**19.** 0
21. Not unique	**23.** 1	**25.** 0	**27.** 2	**29.** Impossible
31. 0				

REVIEW EXERCISES, PAGE 50

1. Tens **3.** Ones **5.** Hundred millions **7.** 356 **9.** 4,602,019
11. 14,780,933,745,871
13. One hundred thirty-one thousand, sixty-nine
15. Four billion, six hundred ninety-five million, one hundred sixty-nine thousand, seven

17. <	**19.** >	**21.** 585	**23.** 699	**25.** 139
27. 9333	**29.** 6530	**31.** 14 hr 43 min	**33.** 34	**35.** 21
37. 3112	**39.** 1447	**41.** 1177	**43.** 36 min	**45.** 322
47. 53,157,962	**49.** 5963	**51.** 154,088	**53.** 1,224,293	**55.** 399 lb 10 oz
57. 24	**59.** 50	**61.** 45 R16	**63.** 3092 R33	**65.** 2987 R22
67. 127	**69.** 700	**71.** 920,000	**73.** 3200	**75.** 78,300
77. 17	**79.** 1	**81.** 2	**83.** 2	**85.** Not unique
87. 9				

TEASERS, PAGE 52

1. 0 and 13; 1 and 12; 2 and 11; 3 and 10; 4 and 9   **3.** 9   **5.** 30

EXERCISES 2.1, PAGE 63

1. Two-ninths; proper fraction **3.** Eleven-twentieths; proper fraction
5. Nineteen fifty-ones; proper fraction **7.** Twenty-three and four-sevenths; mixed number
9. One hundred sixteen twenty-thirds; improper fraction

11. $\frac{7}{8}$	**13.** $\frac{109}{16}$	**15.** $7\frac{12}{20}$	**17.** $2\frac{1}{11}$	**19.** $5\frac{6}{11}$
21. $6\frac{9}{13}$	**23.** $7\frac{1}{2}$	**25.** $12\frac{3}{19}$	**27.** $8\frac{23}{27}$	**29.** $56\frac{1}{6}$
31. $117\frac{11}{17}$	**33.** $\frac{11}{3}$	**35.** $\frac{45}{7}$	**37.** $\frac{76}{9}$	**39.** $\frac{111}{10}$
41. $\frac{155}{9}$	**43.** $\frac{1223}{12}$	**45.** $\frac{184}{5}$	**47.** $\frac{563}{7}$	**49.** 1
51. 0	**53.** 0	**55.** 102	**57.** 0	**59.** 199
61. $\frac{3}{17}$	**63.** $\frac{3}{8}$	**65.** $\frac{11}{23}$	**65.** $\frac{0}{8}$	**69.** $\frac{23}{120}$
71. $\frac{8}{60}$	**73.** $\frac{7}{18}$	**75.** $\frac{11}{39}$		

Answers to Odd-Numbered Exercises

EXERCISES 2.2, PAGE 70

1. Not equivalent 3. Equivalent 5. Equivalent 7. Equivalent 9. Equivalent
11. Equivalent 13. Equivalent 15. Equivalent 17. Equivalent 19. Equivalent
21. 4 23. 15 25. 8 27. 10 29. 24
31. 4 33. 72 35. 5 37. 0 39. 15

41. $\frac{2}{4}, \frac{3}{6}, \frac{4}{8}, \frac{5}{10}, \frac{6}{12}$ are possible answers.

43. $\frac{6}{8}, \frac{9}{12}, \frac{12}{16}, \frac{15}{20}, \frac{18}{24}$ are possible answers.

45. $\frac{4}{26}, \frac{6}{39}, \frac{8}{52}, \frac{10}{65}, \frac{12}{78}$ are possible answers.

47. $\frac{26}{14}, \frac{39}{21}, \frac{52}{28}, \frac{65}{35}, \frac{78}{42}$ are possible answers.

49. $\frac{16}{10}, \frac{24}{15}, \frac{32}{20}, \frac{40}{25}, \frac{48}{30}$ are possible answers.

51. $\frac{4}{8}$ and $\frac{6}{8}$ 53. $\frac{5}{35}$ and $\frac{21}{35}$ 55. $\frac{18}{66}$ and $\frac{33}{66}$ 57. $\frac{8}{72}$ and $\frac{54}{72}$ 59. $\frac{77}{14}$ and $\frac{8}{14}$

61. $\frac{3}{4}$ 63. $\frac{6}{11}$ 65. $\frac{10}{17}$ 67. $\frac{2}{3}$ 69. $\frac{23}{14}$

71. $\frac{11}{9}$ 73. $\frac{11}{3}$ 75. $\frac{5}{18}$ 77. $\frac{11}{8}$ 79. $\frac{17}{35}$

81. $\frac{28}{17}$ 83. $\frac{13}{17}$ 85. $\frac{2}{3}$ 87. 3 89. $\frac{11}{26}$

91. $\frac{401}{181}$ 93. $\frac{5}{7}$ 95. $\frac{25}{33}$ 97. Yes 99. No

101. No 103. Yes 105. a. $\frac{6}{7}$ b. $\frac{1}{7}$

EXERCISES 2.3, PAGE 77

1. Prime 3. Prime 5. Prime 7. Prime 9. Composite
11. Composite 13. Prime 15. Composite 17. Composite 19. Composite
21. $2^2 \times 3$ 23. $2^2 \times 7$ 25. $2^2 \times 11$ 27. $2^3 \times 7$ 29. $2 \times 3 \times 11$
31. $2^3 \times 11$ 33. $2 \times 3 \times 17$ 35. 5^3 37. 2×71 39. $5^2 \times 7$
41. 2×11^2 43. 3×5^3 45. $2 \times 5^2 \times 11$ 47. $2 \times 5 \times 79$ 49. 5×193
51. a. No b. Yes c. No 53. a. Yes b. Yes c. No
55. a. Yes b. No c. Yes 57. a. Yes b. Yes c. No
59. a. Yes b. No c. Yes 61. a. No b. No c. Yes
63. a. Yes b. No c. Yes 65. a. No b. Yes c. No
67. a. Yes b. Yes c. Yes 69. a. Yes b. No c. No
71. 18 73. 40 75. 140 77. 352 79. 186
81. 48 83. 210 85. 2860 87. 11,550 89. 2520
91. a 93. 0

EXERCISES 2.4, PAGE 84

1. $\frac{3}{5}$ 3. $\frac{2}{3}$ 5. $\frac{11}{13}$ 7. $\frac{7}{9}$ 9. $1\frac{19}{23}$

11. $\frac{9}{10}$ 13. $1\frac{1}{8}$ 15. $\frac{101}{117}$ 17. $\frac{49}{57}$ 19. $\frac{31}{45}$

21. $1\frac{7}{36}$ 23. $\frac{155}{156}$ 25. $\frac{47}{80}$ 27. $\frac{23}{66}$ 29. $1\frac{67}{247}$

31. $1\frac{7}{209}$ 33. $\frac{304}{315}$ 35. $\frac{296}{315}$ 37. $2\frac{1}{210}$ 39. $1\frac{538}{693}$

41. $1\frac{59}{120}$ 43. $\frac{29}{60}$ 45. $2\frac{19}{60}$ 47. $\frac{27}{28}$ 49. $1\frac{79}{504}$

51. $1\frac{17}{162}$ 53. $3\frac{1}{36}$ 55. $1\frac{173}{264}$ 57. $1\frac{1199}{2520}$ 59. $\frac{1207}{2520}$

61. $2\frac{239}{315}$ 63. $\frac{9}{10}$ 65. $\frac{19}{28}$ 67. $\frac{9}{20}$ 69. $\frac{47}{60}$

71. $2\frac{1}{8}$ points

EXERCISES 2.5, PAGE 90

1. $<$ 3. $<$ 5. $<$ 7. $<$ 9. $<$
11. $>$ 13. $>$ 15. $>$ 17. $>$ 19. $<$
21. $\frac{1}{4}$ 23. $\frac{9}{37}$ 25. $\frac{5}{12}$ 27. $\frac{2}{9}$ 29. 0
31. $\frac{1}{33}$ 33. $\frac{73}{238}$ 35. $\frac{47}{117}$ 37. $\frac{84}{575}$ 39. $4\frac{35}{117}$
41. $\frac{7}{12}$ 43. $\frac{1}{4}$ 45. $\frac{17}{36}$ 47. $\frac{34}{45}$ 49. $\frac{181}{1260}$
51. $\frac{7}{17}$ 53. $\frac{1}{10}$ 55. $\frac{2}{15}$

EXERCISES 2.6, PAGE 96

1. $5\frac{1}{2}$ 3. $15\frac{1}{3}$ 5. $12\frac{7}{30}$ 7. $13\frac{25}{63}$ 9. $14\frac{20}{39}$
11. $24\frac{31}{117}$ 13. $141\frac{29}{68}$ 15. $39\frac{25}{42}$ 17. $57\frac{103}{221}$ 19. $20\frac{1}{120}$
21. $10\frac{1}{12}$ 23. $17\frac{137}{180}$ 25. $27\frac{3}{10}$ 27. $27\frac{34}{55}$ 29. $16\frac{1}{4}$
31. $1\frac{3}{8}$ 33. $4\frac{16}{21}$ 35. $6\frac{1}{15}$ 37. $9\frac{11}{30}$ 39. $18\frac{5}{24}$
41. $4\frac{32}{105}$ 43. $6\frac{13}{30}$ 45. $23\frac{32}{39}$ 47. $9\frac{1}{72}$ 49. $16\frac{13}{80}$
51. $6\frac{11}{13}$ 53. $38\frac{19}{42}$ 55. $5\frac{5}{7}$ 57. $19\frac{3}{8}$ 59. $14\frac{49}{72}$
61. $3\frac{23}{24}$ 63. $8\frac{8}{21}$ 65. $6\frac{37}{360}$ 67. $2\frac{38}{105}$ 69. $3\frac{46}{105}$
71. $24\frac{19}{70}$ 73. $14\frac{1}{24}$ 75. $11\frac{23}{120}$ 77. $78\frac{11}{15}$ m   79. $57\frac{3}{8}$

EXERCISES 2.7, PAGE 106

1. $1\frac{1}{2}$ 3. $3\frac{6}{7}$ 5. $\frac{1}{12}$ 7. $\frac{8}{21}$ 9. $\frac{20}{117}$
11. $\frac{4}{15}$ 13. $\frac{1}{18}$ 15. $\frac{13}{15}$ 17. $\frac{93}{140}$ 19. $\frac{11}{189}$
21. $16\frac{1}{3}$ 23. $20\frac{1}{15}$ 25. $59\frac{5}{24}$ 27. $84\frac{7}{32}$ 29. $265\frac{7}{20}$
31. $21\frac{13}{24}$ 33. $34\frac{19}{252}$ 35. $6\frac{11}{18}$ 37. $21\frac{11}{24}$ 39. $7\frac{39}{40}$
41. $\frac{3}{2}$ 43. $\frac{9}{4}$ 45. $\frac{7}{9}$ 47. $\frac{11}{23}$ 49. $\frac{17}{101}$

51. $\frac{5}{3}$ **53.** 7 **55.** Does not exist **57.** 4 **59.** 1
61. 4 **63.** 14 **65.** 90 **67.** $5,046,680.64

EXERCISES 2.8, PAGE 112

1. 12   **3.** $24\frac{3}{4}$ **5.** $2\frac{2}{5}$ **7.** $\frac{24}{35}$ **9.** 0
11. $\frac{4}{5}$ **13.** $2\frac{311}{529}$ **15.** 28   **17.** $\frac{11}{23}$ **19.** $\frac{12}{47}$
21. $\frac{1}{18}$ **23.** $1\frac{2}{5}$ **25.** $\frac{3}{5}$ **27.** $1\frac{9}{25}$ **29.** $1\frac{5}{6}$
31. $6\frac{3}{4}$ **33.** $\frac{28}{51}$ **35.** $\frac{10}{17}$ **37.** $4\frac{1}{2}$ **39.** $1\frac{17}{129}$
41. $2\frac{358}{871}$ **43.** 0   **45.** $\frac{14}{87}$ **47.** $2\frac{19}{117}$ **49.** 0
51. $\frac{55}{122}$ **53.** $2\frac{83}{588}$ **55.** $\frac{207}{703}$ **57.** $\frac{121}{3752}$ **59.** $2\frac{1816}{2277}$
61. 64   **63.** $49\frac{7}{26}$ **65.** 11 oz at 79 cents   **67.** $816\frac{2}{3}$

REVIEW EXERCISES, PAGE 113

1. Two-ninths   **3.** One-seventeenth   **5.** Thirty-seven forty-ninths
7. $\frac{10}{6}$ **9.** $11\frac{2}{9}$ **11.** $100\frac{3}{7}$ **13.** $7\frac{6}{11}$ **15.** $10\frac{13}{17}$
17. $\frac{37}{7}$ **19.** $\frac{35}{3}$ **21.** $\frac{258}{11}$ **23.** $\frac{77}{94}$ **25.** Equivalent
27. Not equivalent   **29.** Not equivalent   **31.** 9   **33.** 0   **35.** 12
37. $\frac{3}{5}$ **39.** $\frac{9}{4}$ **41.** $\frac{9}{13}$ **43.** No   **45.** Composite
47. Composite   **49.** Composite   **51.** 3×23 **53.** $2^6 \times 5$ **55.** 28
57. 6156   **59.** 2790   **61.** $\frac{5}{11}$ **63.** $8\frac{1}{9}$ **65.** $3\frac{19}{36}$
67. $\frac{43}{96}$ **69.** $1\frac{227}{693}$ **71.** $\frac{12}{35}$ **73.** $1\frac{251}{504}$ cm   **75.** <
77. >   **79.** <   **81.** $\frac{49}{247}$ **83.** $\frac{16}{253}$ **85.** $\frac{7}{60}$
87. $\frac{10}{33}$ **89.** $5\frac{2}{3}$ **91.** $14\frac{11}{20}$ **93.** $6\frac{59}{60}$ **95.** $3\frac{8}{15}$
97. $7\frac{32}{63}$ **99.** $53\frac{33}{40}$ ft   **101.** $3\frac{1}{16}$ points   **103.** $\frac{2}{7}$ **105.** $\frac{43}{70}$
107. $18\frac{17}{18}$ **109.** $25\frac{1}{96}$ **111.** $\frac{11}{17}$ **113.** $1\frac{7}{8}$ **115.** $6\frac{9}{26}$
117. $1\frac{1}{3}$ **119.** $31\frac{1}{2}$ **121.** 0   **123.** $97\frac{1}{2}$ **125.** 300

TEASERS, PAGE 117

1. a. $\frac{29}{32}$ **b.** $\frac{5}{48}$ **c.** $\frac{1}{10}$ **3.** $\left(\frac{1}{121}, \frac{1}{11}\right), \left(\frac{1}{10}, \frac{1}{100}\right)$ **5.** No; $n^2 = (n)(n)$

CUMULATIVE REVIEW EXERCISES, CHAPTERS 1–2, PAGE 120

1. 1051
3. $11\frac{41}{42}$
5. 215
7. $2\frac{5}{12}$
9. 7452
11. $34\frac{4}{63}$
13. 1586 R13
15. $\frac{27}{55}$
17. $4\frac{1}{2}$
19. 9
21. $1\frac{5}{12}$
23. Sixteen and eleven-fifteenths
25. $85\frac{1}{2}$ yd^2

EXERCISES 3.1, PAGE 127

1. Hundredths
3. Tenths
5. Tens
7. Ten thousandths
9. Hundred thousandths
11. Ones
13. Tens
15. Hundreds
17. Ones
19. Ones
21. Thirty-nine hundredths
23. Sixteen thousandths
25. Fifteen and seventy-one hundredths
27. One and two hundred seventy-six ten thousandths
29. One thousand, nine hundred eighty-seven and twenty-three hundredths
31. 1.16
33. 50.55
35. 91.0091
37. 17,018.00022
39. 67,750.89
41. <
43. =
45. <
47. <
49. <
51. >
53. >
55. =
57. >
59. >
61. 0.9; 0.813; 0.71; 0.59
63. 0.417; 0.218, 0.2; 0.05
65. 23.203; 23.1; 22.109; 22.09
67. 20.54; 2.4532; 2.453; 0.2567
69. 10.31; 10.103; 10.031; 10.0013

EXERCISES 3.2, PAGE 130

1. 0.88
3. 2.17
5. 241.096
7. 86.79
9. 54.5062
11. 31.109
13. 24.862
15. 68.761
17. 345.598
19. 345.095
21. 26.133
23. 193.132
25. 703.797
27. 2206.7523
29. 26,376.451
31. 32.23
33. 12.59
35. 36.924
37. 278.15
39. 68.4018
41. 22.608
43. 33.46
45. 37.128
47. 0.8544
49. 505.211
51. 36.01
53. 182.291
55. 0.0779
57. 122.91
59. 1952.1044
61. 24.386
63. 28.437
65. 63.1831
67. 103.129
69. 194.321

EXERCISES 3.3, PAGE 134

1. 24.108
3. 718.38
5. 203.2248
7. 1353.3261
9. 5.621434
11. 0.003876
13. 0.000024
15. 1083.62608
17. 43.996086
19. 6190.3
21. 167.3938
23. 79.49868
25. 333.67284
27. 1175.00832
29. 47,057.238
31. 231
33. 17.89
35. 5920
37. 7347.9
39. 14,165.8
41. 789
43. 571,000
45. 87,954.7
47. 2,147,980
49. 23,000
51. 14.952
53. 66.2093
55. 606.2273

EXERCISES 3.4, PAGE 137

1. 2.3
3. 6.9
5. 0.78
7. 6.05
9. 1.206
11. 124.96
13. 123.457
15. 0.123457
17. 5.6
19. 0.02
21. 0.1
23. 1.0
25. 0.60
27. 1.01
29. 0.124
31. 40.123
33. 0.1235
35. 9.9091
37. 1.234568
39. 10.001000
41. 58.508
43. 359.5
45. 45.0
47. 483.57
49. 90.8 in.

EXERCISES 3.5, PAGE 144

1. 2.32
3. 42.3
5. 1440
7. 23.192
9. 1.23
11. 2.469
13. 0.25
15. 2.4175
17. 0.36
19. 112.5
21. 4.775
23. 62.3
25. 23.19
27. 7.32
29. 3.9
31. 2.31
33. 1.789
35. 0.592
37. 73.479
39. 1.41658
41. 7.8912
43. 0.5719
45. 0.00879547
47. 0.214798
49. 0.02345
51. 30.8
53. 0.11
55. 136.85
57. 51.250
59. 82.03
61. 2.01
63. 23.2
65. 1.0512

EXERCISES 3.6, PAGE 149

1. $\frac{3}{10}$
3. $\frac{4}{25}$
5. $\frac{1}{8}$
7. $\frac{83}{5}$
9. $\frac{117}{5000}$
11. $\frac{197}{20}$
13. $\frac{39}{160}$
15. $\frac{951}{50}$
17. $\frac{5051}{250}$
19. $\frac{22,003}{20,000}$
21. $\frac{21}{4000}$
23. $\frac{4321}{250}$
25. $\frac{200,113}{2000}$
27. 0.625
29. 0.4375
31. 2.625
33. 1.6
35. 0.505
37. 2.95
39. 0.5475
41. 1.17
43. 0.56
45. 1.77
47. 4.85
49. 0.28
51. 0.467
53. 1.467
55. 0.463
57. 1.917
59. 2.412

EXERCISES 3.7, PAGE 153

1. $5.34
3. $38.17
5. $3125.24
7. a. $11,520,000   b. $15,840,000   c. $6,480,000   d. $27,360,000   e. $10,800,000
9. $4964
11. $7.12
13. $328.68
15. $22.45
17. $267.81
19. 39.681 g
21. Approx. 1.28 kg
23. $64.25
25. 248.5 kg
27. 66.568 ft
29. 166.7968 in.
31. 16.1 ft
33. 78.0 cm
35. 159.4 in.

REVIEW EXERCISES, PAGE 154

1. Hundredths
3. Thousandths
5. Tenths
7. Three and nine-tenths
9. One hundred one and one hundredth
11. <
13. =
15. >
17. 0.99; 0.909; 0.099; 0.09
19. 15.762
21. 211.0769
23. 8.6427
25. 23.117
27. 190.862
29. 24.1326
31. 96.6952
33. 70.0
35. 0.030
37. 0.2091
39. 0.51
41. 5.446
43. 75.8
45. $\frac{13}{50}$
47. $\frac{1523}{50}$
49. 1.75
51. 0.048
53. 1.8
55. 1.387
57. 122.88
59. 12.2
61. $286.10
63. $338
65. $755.94

TEASERS, PAGE 156

1. More than 3100
3. Yes; consider 2.31 + 2.69 = 5
5. 46.47

CUMULATIVE REVIEW EXERCISES, CHAPTERS 1–3, PAGE 158

1. 238
3. $\frac{21}{4}$
5. $3\frac{1}{15}$
7. 24.65
9. 1072
11. 21.93
13. 3610.5
15. $2\frac{1}{3}$
17. $\frac{93}{112}$
19. 2300
21. 56.108
23. 845.37 cm²
25. 5.67

EXERCISES 4.1, PAGE 166

1. 20%
3. 75%
5. 160%
7. 150%
9. 2%
11. $14\frac{2}{7}\%$
13. 175%
15. 400%
17. 1100%
19. 6300%
21. 250%
23. 720%
25. $1516\frac{2}{3}\%$
27. $\frac{2}{9}$
29. $\frac{17}{20}$
31. $\frac{11}{20}$
33. $\frac{49}{50}$
35. $\frac{23}{25}$
37. 1
39. $\frac{5}{4}$
41. $\frac{37}{20}$
43. $\frac{9}{2}$
45. $\frac{61}{10}$
47. $\frac{161}{20}$
49. $\frac{234}{25}$
51. $\frac{1}{5}$
53. $\frac{19}{200}$
55. $\frac{463}{500}$
57. 37.5%
59. 17.5%

EXERCISES 4.2, PAGE 171

1. $\frac{3}{25}$ 3. $\frac{9}{25}$ 5. $\frac{1}{2}$ 7. $\frac{31}{50}$ 9. $\frac{3}{4}$

11. $\frac{11}{10}$ 13. $\frac{3}{2}$ 15. 2   17. $\frac{27}{10}$ 19. $\frac{81}{20}$

21. $\frac{41}{500}$ 23. $\frac{148}{625}$ 25. $\frac{629}{1000}$ 27. $\frac{1633}{2000}$ 29. $\frac{1197}{500}$

31. $\frac{14,003}{5000}$ 33. $\frac{1811}{500}$ 35. $\frac{7801}{2000}$ 37. $\frac{626}{125}$ 39. $\frac{2801}{400}$

41. 0.36   43. 0.81   45. 1.27   47. 2.56   49. 4.96
51. 7.83   53. 0.069   55. 0.4269   57. 0.0006   59. 0.00087
61. 17%   63. 12.3%   65. 43.59%   67. 82.15%   69. 238%
71. 6375%   73. 12,302%   75. 28,696%   77. 1296.2%   79. 1.07%
81. 20%   83. 140%   85. 12.5%   87. 187.5%   89. 10%
91. 120%   93. 18.75%   95. 7.8125%   97. 16.25%   99. 19.21875%

101. a. 0.12   b. $\frac{3}{25}$ 103. a. 0.816   b. $\frac{102}{125}$

105. a. 36%   b. $\frac{9}{25}$ 107. a. 234%   b. $\frac{117}{50}$

109. a. 0.075   b. 7.5%

EXERCISES 4.3, PAGE 176

1. 4.14   3. 4.52   5. 22.512   7. 124.8   9. 0.3
11. 190.344   13. 4.486   15. 5.523   17. 103.8   19. 20.444

21. 25%   23. 50%   25. 1000%   27. 2000%   29. $133\frac{1}{3}\%$

31. 200%   33. 100%   35. 50%   37. 48.8%   39. 65.93%

41. 50%   43. 200%   45. $41\frac{2}{3}\%$ 47. 25%   49. 60%

51. 72   53. 10   55. 30   57. 2.3   59. 0.05
61. 900   63. 3680   65. 1.5   67. 11.3   69. 2000

71. $16\frac{2}{3}\%$ 73. $55\frac{5}{9}\%, 44\frac{4}{9}\%$ 75. $15   77. 62.5   79. $7020

EXERCISES 4.4, PAGE 181

1. $9844   3. Approx. 10.4%   5. $422.28   7. $8\frac{1}{3}\%$ 9. 9.2%

11. 17%   13. 19.4%   15. 9.0%   17. 10.5%   19. 896
21. 20%   23. 234   25. 88%

EXERCISES 4.5, PAGE 185

1. $.72   3. $1.60   5. $11.93   7. $50.40   9. $32.09
11. $17.68   13. $89.71   15. $355.56   17. $1007.76   19. $1170.40
21. 5%   23. 5%   25. 4%   27. 4%   29. 5%
31. $201.20   33. $633.92   35. $975.12   37. $569.25   39. $1337.85
41. 21.8%   43. 18.6%   45. 4.2%   47. 4.1%   49. 4.5%

EXERCISES 4.6, PAGE 191

1. $135.60   3. $58.32   5. $112.06   7. $116.96   9. $2816.08
11. $10,719   13. $27,756   15. $8000   17. $850   19. $3840
21. $1340   23. $71,900   25. $79,359

Answers to Odd-Numbered Exercises

EXERCISES 4.7, PAGE 194

1. $156.50
3. $405.25
5. $20.50
7. $392
9. $46.20
11. 40%
13. $203.40
15. No, the discount, $69, is less than 20% of the marked price.
17. $1113.90
19. $98.56

EXERCISES 4.8, PAGE 200

1. $49.50
3. $72
5. $8
7. $5.10
9. $45.60
11. 6%
13. 7%
15. 9%
17. 12%
19. 8%
21. $1000
23. $1800
25. $500
27. $1400
29. $5000
31. $630
33. $98.50
35. $54.50
37. $627
39. $3690

REVIEW EXERCISES, PAGE 201

1. 40%
3. 3100%
5. $\frac{3}{50}$
7. $\frac{43}{20}$
9. $\frac{7}{20}$
11. $\frac{473}{20}$
13. 0.089
15. 239%
17. 40%
19. 18.896
21. 25%
23. $66\frac{2}{3}\%$
25. 36
27. 0.005
29. $9331
31. 14.3%
33. $1254
35. $1141
37. $126.50
39. $136.22
41. 25%
43. $44.80
45. $300

TEASERS, PAGE 203

1. B
3. $32,603.90
5. $540.50
7. $527.13
9. Take the 2.9% per year for 48 months.

CUMULATIVE REVIEW EXERCISES, CHAPTERS 1–4, PAGE 204

1. $\frac{1}{14}$
3. 29.106
5. 85
7. 407.40
9. 0.8
11. $\frac{13}{25}$
13. 2.3%
15. $\frac{20}{45}$
17. 0.06
19. 25.608
21. 175
23. $142.50
25. 442 m^2

EXERCISES 5.1, PAGE 216

1. 12 ft
3. 10,560 ft
5. 45 ft
7. 4 lb
9. 4000 lb
11. 30 cm
13. 3000 cm
15. 5000 cm
17. 0.256 kg
19. 1.7 kg
21. 17.7 kg
23. 10 pt
25. 24 pt
27. 3 pt
29. 180 min
31. 30 min
33. 83 min
35. 644,000 cm
37. 221,760 ft
39. 950.4 ft
41. 5.902 kg
43. 70.4 oz
45. 4.73 L
47. 50.88 oz
49. Approx. 1.48 km/min
51. Approx. 1.01 L
53. 23.876 dm

EXERCISES 5.2, PAGE 223

1. Isosceles
3. Equilateral
5. Isosceles
7. 14 in.
9. 15.2 ft
11. 10.2 cm
13. 9.1 dm *or* 91 cm
15. 69.3 cm
17. 155
19. $16.50

EXERCISES 5.3, PAGE 228

1. 75.4 cm
3. 64.06 in.
5. 84.53 yd
7. 33.60 m
9. 61.8 cm
11. 304 dm
13. 18.7 m
15. 153 mm
17. 52.0 cm
19. 48.7 dm

EXERCISES 5.4, PAGE 241

1. $P = 28$ ft; $A = 48$ ft^2
3. $P = 48$ ft or 16 yd; $A = 135$ ft^2 or 15 yd^2
5. $P = 78.4$ in.; $A = 384.16$ in.^{2}
7. $P = 64$ m
9. $A = 291.45$ cm^2
11. $h = 14.5$ in.

13. $A = 171$ cm² or 1.71 dm²   15. $b = 29$ cm
17. 7.5 cm²   19. 22.5 mm²   21. 30 yd²   23. 8.41 cm²
25. 43.5 cm²   27. 114 in²   29. 79.8 yd²   31. 129.6 cm²
33. 36 ft²   35. 7 ft   37. $P = 24$ ft; $A = 24$ ft²
39. $P = 40$ ft; $A = 60$ ft²   41. $P = 12$ cm; $A = 9$ cm²
43. $P = 36$ cm; $A = 72$ cm²   45. 4 times   47. $(28 + 2n)$ cm
49. 11.3 ft²   51. 2306.0 mm²   53. 62.5 dm²   55. 315.3 cm²
57. Approx. 18.14 yd² 59. 455.3 ft²   61. a. 121.5 ft²   b. 54 ft
63. $51\frac{4}{7}$ ft²

EXERCISES 5.5, PAGE 250
1. 840 ft³   3. 2.5 in.   5. 4 cm   7. 7.6 cm   9. 1728 in.³
11. 1584 ft³   13. 221.2 cm³   15. 528.77 ft³   17. 12 cm   19. 39 in.
21. 16π ft³   23. 1453.4π cm³ 25. 2427.6π in.³ 27. 4 ft   29. 6 cm
31. $\frac{2048\pi}{3}$ cm³   33. $\frac{907.924\pi}{3}$ yd³   35. $\frac{4900.172\pi}{3}$ ft³

REVIEW EXERCISES, PAGE 251
1. $5\frac{2}{3}$ yd   3. 216,480 ft   5. 5.796 km   7. 0.125 km   9. 5.3 qt
11. 37.1 pt   13. 1816 g   15. 0.924 lb   17. 270 min
19. Approx. 0.025 km/sec   21. 16.5 in.   23. 16.1 ft   25. 18 ft 4 in.
27. 55 ft or $18\frac{1}{3}$ yd   29. 25.3 in.   31. 30 ft   33. 62 dm   35. 71.8 ft
37. 106 yd   39. 6.4 m   41. 1,190.25 cm²   43. 252 in.²   45. 3759.1 ft²
47. 14.5 yd   49. 14.62 ft²   51. 91.26 yd²   53. 448.8 in.³   55. 288π cm³

TEASERS, PAGE 253
1. 5984 pt   3. Approx. 0.02 km/sec   5. 252 ft or 84 yd
7. No, the sum of the lengths of any two sides must be greater than the length of the third side; 20.6 cm + 344 mm = 2.06 dm + 3.44 dm = 5.5 dm.
9. 232.5 in.²

CUMULATIVE REVIEW EXERCISES, CHAPTERS 1–5, PAGE 255
1. 69%   3. $6\frac{5}{8}$ 5. 7.5   7. 19   9. 310
11. 3%   13. $\frac{5}{20}$ 15. $\frac{2}{3}$ 17. 93.6 m   19. 56.8 cm
21. 0.123   23. 60,900   25. $1495.72

EXERCISES 6.1, PAGE 263
1. -5 3. $+6$ 5. -10 7. $+19$ 9. 0
11. -27 13. $+93$ 15. $+319$ 17. Negative   19. Negative
21. Positive   23. Negative   25. Negative   27. Negative   29. Positive
31. 3   33. 9   35. 0   37. 101   39. 97
41. 981   43. 6   45. 0   47. 16   49. 7
51. 14   53. 1   55. $<$ 57. $>$ 59. $<$
61. $>$ 63. $<$ 65. $=$ 67. $<$ 69. $=$

EXERCISES 6.2, PAGE 266
1. $\frac{-5}{-7}$ or $\frac{5}{7}$ 3. $\frac{-3}{2}$ or $\frac{3}{-2}$ 5. $\frac{-7}{-2}$ or $\frac{7}{2}$ 7. $\frac{-6}{5}$ or $\frac{6}{-5}$ 9. $\frac{-10}{3}$ or $\frac{10}{-3}$

Answers to Odd-Numbered Exercises   763

11. $\dfrac{-14}{-9}$ or $\dfrac{14}{9}$ **13.** $\dfrac{0}{-3}$ or $\dfrac{0}{3}$ **15.** $\dfrac{0}{13}$ or $\dfrac{0}{-13}$ **17.** Positive   **19.** Positive
21. Negative   **23.** Zero   **25.** Negative   **27.** Negative   **29.** Zero
31. $\dfrac{3}{7}$ **33.** 2.31   **35.** $\dfrac{23}{101}$ **37.** 4.03   **39.** 37.1
41. $\dfrac{401}{38}$ **43.** 0.81   **45.** 1.007   **47.** <   **49.** =
51. >   **53.** >   **55.** <

EXERCISES 6.3, PAGE 273

1. +16   **3.** −12   **5.** 0   **7.** +17   **9.** −4
11. −7   **13.** +47   **15.** +780   **17.** −45   **19.** −238
21. +14   **23.** +6   **25.** −47   **27.** −38   **29.** +94
31. +1   **33.** −19   **35.** +56   **37.** +19.97   **39.** −92.29
41. −102.7   **43.** −16.32   **45.** −7.451   **47.** −69.172   **49.** −0.298
51. −16.921   **53.** −33.722   **55.** +124.291   **57.** −20.271   **59.** +29.58
61. $\dfrac{-1}{63}$ **63.** $\dfrac{-7}{12}$ **65.** $\dfrac{-127}{77}$ or $-1\dfrac{50}{77}$ **67.** $\dfrac{+2}{3}$
69. $\dfrac{-34}{21}$ or $-1\dfrac{13}{21}$ **71.** $\dfrac{-29}{117}$ **73.** $\dfrac{-55}{63}$ **75.** $\dfrac{13}{12}$ or $1\dfrac{1}{12}$
77. $\dfrac{-1}{12}$ **79.** $\dfrac{-13}{60}$ **81.** $8\dfrac{1}{6}$ **83.** $4\dfrac{11}{30}$
85. $-6\dfrac{1}{4}$ **87.** $-6\dfrac{5}{12}$ **89.** $-14\dfrac{23}{180}$ **91.** The signed number
93. 46   **95.** 139 lb   **97.** $6\dfrac{97}{120}$ **99.** $36\dfrac{1}{4}$

EXERCISES 6.4, PAGE 277

1. +2   **3.** +10   **5.** −9   **7.** +5   **9.** +3
11. +30   **13.** −2   **15.** −189   **17.** −184   **19.** −1138
21. −1.78   **23.** −7.92   **25.** +15.78   **27.** +45.78   **29.** +6.53
31. $\dfrac{55}{63}$ **33.** $\dfrac{19}{28}$ **35.** $\dfrac{6}{77}$ **37.** $\dfrac{26}{45}$ **39.** $\dfrac{37}{91}$
41. $-1\dfrac{1}{6}$ **43.** $-9\dfrac{38}{45}$ **45.** $-2\dfrac{17}{20}$ **47.** $35\dfrac{92}{99}$ **49.** $-10\dfrac{29}{65}$
51. 1.9   **53.** 20   **55.** 18.53   **57.** 8.07   **59.** −45
61. −44.089   **63.** −27.373   **65.** $\dfrac{-509}{770}$ **67.** $\dfrac{125}{1008}$ **69.** $15\dfrac{43}{60}$
71. −42   **73.** 77   **75.** $\dfrac{-9}{77}$ **77.** 29   **79.** 302 ft
81. $6\dfrac{11}{12}$ c   **83.** $6452.67   **85.** $1\dfrac{41}{1260}$

EXERCISES 6.5, PAGE 285

1. 24   **3.** 45   **5.** −56   **7.** −14.3   **9.** 0
11. 348   **13.** 1392   **15.** 3.36   **17.** 1320   **19.** 4959
21. −1000   **23.** −30,360   **25.** 420   **27.** 43.7   **29.** 129,168
31. −8.51   **33.** 0.0006   **35.** −24.108   **37.** −35.496   **39.** 180.908
41. −0.000016   **43.** −49,504.73   **45.** 2.21085   **47.** 51.87   **49.** 7.1289
51. $\dfrac{-1}{12}$ **53.** $\dfrac{-8}{21}$ **55.** $\dfrac{20}{117}$ **57.** 0   **59.** $\dfrac{-1}{24}$

61. $\dfrac{-1}{10}$ **63.** $-16\dfrac{1}{3}$ **65.** $-20\dfrac{1}{15}$ **67.** $-27\dfrac{6}{7}$ **69.** $-14\dfrac{7}{24}$

71. 210   **73.** 23.46   **75.** $23\dfrac{23}{24}$

EXERCISES 6.6, PAGE 289

1. -3 **3.** -0.4 **5.** 0   **7.** -0.024 **9.** 14
11. -6.25 **13.** 83   **15.** 3.5625   **17.** -6 **19.** -0.475
21. -2 **23.** -6 **25.** -0.02 **27.** $\dfrac{-4}{3}$ or $-1\dfrac{1}{3}$ **29.** $\dfrac{35}{18}$ or $1\dfrac{17}{18}$
31. 1   **33.** $\dfrac{6}{7}$ **35.** -1 **37.** $\dfrac{8}{27}$ **39.** $\dfrac{-28}{51}$
41. $\dfrac{146}{129}$ or $1\dfrac{17}{129}$ **43.** $\dfrac{-93}{110}$ **45.** 4   **47.** $\dfrac{-10}{63}$ **49.** 6
51. $\dfrac{231}{10}$ or $23\dfrac{1}{10}$ **53.** $\dfrac{-45}{22}$ or $-2\dfrac{1}{22}$ **55.** $\dfrac{-6}{25}$ **57.** 81   **59.** $2\dfrac{2}{3}$
61. $-3\dfrac{3}{14}$ **63.** $5\dfrac{13}{24}$

EXERCISES 6.7, PAGE 292

1. $\dfrac{-4}{9}$ **3.** $\dfrac{-1}{3}$ **5.** $\dfrac{-56}{23}$ **7.** $\dfrac{1}{3}$ **9.** $\dfrac{81}{17}$
11. $\dfrac{-9}{5}$ **13.** 0   **15.** -3 **17.** 0   **19.** $\dfrac{-7}{11}$
21. $\dfrac{-1}{4}$ **23.** $\dfrac{-5}{6}$ **25.** $\dfrac{-32}{25}$ **27.** $\dfrac{-1098}{385}$ **29.** $\dfrac{-1}{75}$

EXERCISES 6.8, PAGE 294

1. -11 **3.** -10 **5.** -1152 **7.** 1   **9.** 0
11. -2.4 **13.** 4.143   **15.** -4.8 **17.** 13   **19.** 11
21. $\dfrac{-11}{24}$ **23.** $\dfrac{4}{105}$ **25.** $\dfrac{40}{441}$ **27.** $\dfrac{9}{196}$ **29.** $\dfrac{5}{6912}$
31. $\dfrac{-343}{8}$ or $-42\dfrac{7}{8}$ **33.** $\dfrac{4913}{27}$ or $181\dfrac{26}{27}$ **35.** $32\dfrac{5}{6}$ **37.** $\dfrac{667}{3136}$ **39.** $-151\dfrac{77}{288}$

EXERCISES 6.9, PAGE 297

1. 34° C   **3.** $37.43   **5.** $378.44   **7.** 33; $5   **9.** $13,138
11. $-$56.50   **13.** -51 **15.** $27,600   **17.** -38 units   **19.** 153.1 m
21. 100.0° F   **23.** $2591.40   **25.** 5000 ft higher   **27.** -11.53 **29.** -1.5

REVIEW EXERCISES, PAGE 299

1. -6 **3.** -12 **5.** -127 **7.** 0   **9.** 1010
11. Negative   **13.** Negative   **15.** Positive   **17.** Negative   **19.** 7
21. 0   **23.** 2   **25.** >   **27.** <   **29.** <
31. $\dfrac{4}{9}$ **33.** $\dfrac{-2}{11}$ **35.** $\dfrac{-4}{13}$ **37.** Negative   **39.** Positive
41. Positive   **43.** $\dfrac{5}{8}$ **45.** $\dfrac{16}{9}$ **47.** $\dfrac{23}{37}$ **49.** >
51. <   **53.** =   **55.** 14   **57.** $\dfrac{-1}{35}$ **59.** -3.16

Answers to Odd-Numbered Exercises

61. $-18\frac{51}{56}$ **63.** 2.36 **65.** 8 **67.** $-17\frac{3}{10}$ **69.** 183
71. $-5\frac{11}{72}$ **73.** -172.3 **75.** -4 **77.** 7.3 **79.** -292
81. 54 **83.** -132 **85.** 33 **87.** 1729 **89.** $-24{,}600$
91. 9 **93.** -3 **95.** 0 **97.** 8 **99.** $1\frac{47}{75}$
101. $\frac{-3}{5}$ **103.** $\frac{1}{2}$ **105.** $\frac{-5}{3}$ **107.** -20 **109.** -6
111. 0 **113.** $\frac{-9}{25}$ **115.** $-2\frac{7}{24}$ **117.** $96\frac{1}{4}$ ft **119.** $431
121. $634

TEASERS, PAGE 301

3. Addition and multiplication
7. No. If either a or b is 0, the product is 0.
9. a. There isn't any. **b.** -1 **c.** 0

CUMULATIVE REVIEW EXERCISES, CHAPTERS 1–6, PAGE 303

1. 19.398 **3.** $\frac{24}{115}$ **5.** -12 **7.** -2 **9.** $(1.2)(-8)$
11. $|-19|$ **13.** 7 **15.** 0.36 **17.** 18.45 ft² or 2.05 yd²
19. $126.62 **21.** Not equivalent **23.** $4\frac{19}{30}$ **25.** $146.25

EXERCISES 7.1, PAGE 310

1. Constants are 3 and -5; variables are x and y.
3. Constants are 2 and -3; variables are u and v.
5. Constants are 2, -3, and 4; variables are x and y.
7. Constants are 2, 3, -4, and 5; variables are p and q.
9. Constant is $\frac{-1}{2}$; variables are A, b, and h.
11. 1; $3x^2y$ **13.** 2; x^2 and $-y^2$ **15.** 1; $(a+b)(a-b)$
17. 2; $3S^2$ and $-4T(3S+R^3)$ **19.** 2; 2 and $-3[x-5(6-5y)]$ **21.** 2
23. $\frac{-1}{6}$ **25.** 81 **27.** 2, -3, and 4
29. -2 **31.** 1, -1, and 1 **33.** 2, -3, and 4

EXERCISES 7.2, PAGE 312

1. 18 **3.** -0.1 **5.** 27 **7.** 0.098 **9.** 66
11. 2.67 **13.** 576 **15.** 4 **17.** -0.5 **19.** 19
21. 11 **23.** -0.462 **25.** -6 **27.** 32 **29.** 16

EXERCISES 7.3, PAGE 316

1. $3x + 2y$ **3.** $6x - 6y$ **5.** $\frac{3}{4}x^2y - \frac{1}{3}xy^2$
7. $5a + 2b - 3c$ **9.** $a^2b + 2ab^2 - 2a^2b^2$ **11.** $2x + 3y - 1$
13. $15.37u - 0.81v - 7.97$ **15.** $-a - 2b$ **17.** $-2.87x - 4.09y + 1.31z$
19. $3ab - 2a + 2b$ **21.** $2x - 10y - 4$ **23.** $\frac{3}{10}p^2q - \frac{5}{3}pq^2$
25. $0.7x^3 - 0.4x^2 - x + 0.9$ **27.** $\frac{-26}{15}x^2y^3 - \frac{1}{4}xy^2 + \frac{1}{9}x^2y$ **29.** $1.6(p^2 - 3q) + 2(p - q^2)$

EXERCISES 7.4, PAGE 321

1. $c + a - b$
3. $p - 2q - 3$
5. $-a + 2b + 1$
7. $-2x^2y - 8xy^2 - 3x^2y^2$
9. $a^2b + 2ab^2 - 2a^2b^2$
11. $\frac{9}{4} - \frac{4}{3}x - \frac{3}{10}y$
13. $6 - 2c + 2d$
15. $-5x - 3y$
17. $10a - 17b - 9c$
19. $\frac{7}{3}r - \frac{3}{2}s - \frac{7}{6}t$
21. $-x^2 + 4y^2 + 35xy + 6x - 12y$
23. $-a$
25. $0.6x - 7.8$
27. $26a - 80b$
29. $-18a + 7b$

EXERCISES 7.5, PAGE 326

1. $a + b$
3. $2(e + f + g)$
5. $2i^3 + 7$
7. m^2n^2
9. $\frac{p}{q}$
11. $r^3 + 7$
13. $(s + t + u)^2$
15. $\frac{vw}{v + w}$
17. $(x + y)^2 - (x - y)^3$
19. $z + (z + 1) + (z + 2)$
21. $2x + 5y$
23. $\frac{1}{3}y - \frac{1}{9}w$
25. $[2(w + 5)]^2$
27. $\frac{p - 4}{9p}$
29. $\frac{2(m + 7)}{3(4 - n)}$
31. Solution
33. Not a solution
35. Solution
37. Solution
39. Not a solution
41. Solution
43. Solution
45. Solution
47. Solution
49. Not a solution

REVIEW EXERCISES, PAGE 327

1. $2; (2a^2b - c), b^3c^4$
3. $1; -3(a^2 - 2ab + 3b^3)$
5. 1
7. -3
9. 18
11. 25
13. 0
15. $-2a^2b + 2ab^2 - 2a^2b^2$
17. $4p - 6q$
19. $10 - 7x - 7y$
21. $-12a - 19b$
23. $4k^2$
25. $3t - 6$

TEASERS, PAGE 328

9. 4

CUMULATIVE REVIEW EXERCISES, CHAPTERS 1–7, PAGE 330

1. 3.9567
3. $\frac{1}{3}$
5. 35
7. $\frac{-12}{17}$
9. $|-3|$
11. 3.9
13. 38.63 cm
15. -16
17. $3a + 4b$
19. $\frac{x + 10}{(2x)(5)}$
21. $\frac{67}{250}$
23. 800 cm^2
25. Not equivalent

EXERCISES 8.1, PAGE 339

1. $x = -1$
3. $y = -6$
5. $w = -10$
7. $p = -1$
9. $y = 9$
11. $x = \frac{11}{12}$
13. $z = 1\frac{1}{6}$
15. $x = \frac{1}{6}$
17. $p = -2.3$
19. $u = 1.9$
21. $z = 1\frac{5}{6}$
23. $s = \frac{-13}{33}$
25. $r = 4.3$
27. $t = -7$
29. $s = 7$
31. $m = -6\frac{1}{2}$
33. $q = \frac{1}{6}$
35. $x = 21.53$
37. $z = 2\frac{5}{57}$
39. $q = 1\frac{7}{18}$

Answers to Odd-Numbered Exercises

EXERCISES 8.2, PAGE 344

1. $x = 18$ **3.** $t = 6$ **5.** $z = -8.4$ **7.** $u = 3.2$ **9.** $x = 3\frac{1}{2}$

11. $y = -2\frac{1}{4}$ **13.** $x = \frac{3}{14}$ **15.** $r = \frac{-3}{5}$ **17.** $w = 0.55$ **19.** $u = \frac{3}{4}$

21. $t = 1\frac{5}{6}$ **23.** $w = -0.3$ **25.** $m = -2\frac{8}{9}$ **27.** $t = -20\frac{10}{23}$ **29.** $y = 14.03$

31. $p = -1\frac{1}{6}$ **33.** $x = \frac{27}{40}$ **35.** $p = 0.638$ **37.** $x = \frac{-2}{9}$ **39.** $t = -3.175$

EXERCISES 8.3, PAGE 348

1. $x = 2$ **3.** $m = \frac{1}{3}$ **5.** $x = 5$ **7.** $t = 5\frac{2}{3}$ **9.** $y = \frac{7}{8}$

11. $p = 10$ **13.** $x = -1\frac{2}{3}$ **15.** $u = -8\frac{1}{2}$ **17.** $w = 3\frac{3}{23}$ **19.** $r = \frac{2}{3}$

21. $x = 3\frac{1}{2}$ **23.** $t = 12\frac{1}{4}$ **25.** $w = \frac{-179}{510}$ **27.** $u = 3\frac{2}{3}$ **29.** $y = 1\frac{3}{10}$

31. $x = 5\frac{7}{9}$ **33.** $w = 1\frac{5}{6}$ **35.** $t = \frac{-7}{11}$ **37.** $u = 35\frac{7}{12}$ **39.** $q = \frac{-1}{3}$

EXERCISES 8.4, PAGE 352

1. $x = 2$ **3.** $x = 3\frac{1}{7}$ **5.** $w = 15$ **7.** $t = \frac{-1}{6}$ **9.** $x = -6\frac{1}{3}$

11. $r = \frac{13}{14}$ **13.** $x = -1\frac{2}{3}$ **15.** $v = \frac{-1}{17}$ **17.** $p = \frac{11}{24}$ **19.** $t = \frac{18}{19}$

21. $x = \frac{5}{6}$ **23.** $x = 4\frac{3}{4}$ **25.** $t = \frac{3}{7}$ **27.** $x = -2\frac{3}{4}$ **29.** $y = 4\frac{2}{3}$

31. $w = \frac{1}{5}$ **33.** $p = -2\frac{7}{30}$ **35.** $u = \frac{9}{34}$ **37.** $w = 1\frac{12}{19}$ **39.** $w = \frac{15}{71}$

41. $y = \frac{-5}{24}$ **43.** No solution **45.** Infinitely many solutions

47. $s = 4\frac{1}{3}$ **49.** $u = \frac{-21}{17}$

EXERCISES 8.5, PAGE 358

1. -7 **3.** 55 and 110 **5.** 30 and 75
7. 8 cm, 10 cm, 12 cm **9.** 11 m, 11 m, 15 m, 27 m, 27 m
11. A, 25°; C, 65° **13.** 30.1 ft
15. $14.52 **17.** 3.5 hr
19. 1350 **21.** $23 for Norayne, $71 for Sarah
23. 37, 38, and 39 **25.** 21, 22, 23, and 24
27. 14 yr and 1 yr **29.** Nancy is 22, John is 50

EXERCISES 8.6, PAGE 367

1. 17 pennies, 12 nickels, 10 dimes, and 5 quarters
3. 28 nickels, 20 dimes, and 10 quarters
5. 26 @ 22¢, 8 @ 39¢, and 13 @ 17¢
7. 1525 @ $1.50 and 3205 @ $.90

9. $5\frac{11}{25}$ sec **11.** $1\frac{4}{5}$ hr after motorist starts

13. 55 mph on interstate, 30 mph on two-lane

15. $2\frac{1}{2}$ hr

17. $7\frac{2}{7}$ ft

19. 8 ft from fulcrum on other side

21. Not balanced

23. $8\frac{14}{15}$ ft from end with Johnny

25. $40\frac{6}{7}$ gal

27. $7000 in savings and $12,000 in CDs
29. $7000
33. 9.5% per year

31. 9.5% per year
35. 400 mi

EXERCISES 8.7, PAGE 374

1. $101\frac{1}{4}$ mi
3. $r = \frac{d}{t}$; 39 mph
5. 15 sec
7. No

9. $C = \frac{5}{9}(F - 32°)$
11. 90
13. $C = \frac{100M}{IQ}$; 10 yr
15. -7.5

17. 214.75 ft
19. 89
21. 104 ft
23. 120 volts

25. $R = \frac{E}{I}$; $7\frac{1}{3}$ ohms
27. $159
29. 5
31. $D = \frac{C(a + 12)}{a}$

33. $8050
35. $C = \frac{V}{1 - rt}$
37. $W = \frac{P - 2L}{2}$; 2.5 ft
39. $r = \frac{C}{2\pi}$; $r = \frac{14}{\pi}$ yd

41. $v = \frac{s - 16t^2}{t}$
43. $b = \frac{2A}{h}$
45. $C = \frac{B}{A}$
47. $d = \frac{L - a}{n - 1}$

49. $b = \frac{a - 3c^2d}{2}$

EXERCISES 8.8, PAGE 383

1. $\frac{9}{22}$
3. $\frac{19}{36}$
5. $\frac{48}{67}$
7. $\frac{69}{7}$
9. $\frac{67 \text{ in.}}{135 \text{ lb}}$

11. 4 to 3
13. 1:4
15. 2 to 17
17. Extremes are 15 and 7, means are 21 and 5.
19. Extremes are $5x$ and 3, means are 9 and $x - 2$.
21. Yes
23. No
25. Yes
27. $n = 4.5$
29. $n = 12.5$

31. $n = 4.2$
33. $n = \frac{22}{9}$
35. $3.36
37. 450
39. 17,150

41. 15
43. 420.8
45. 27.5 ft
47. 2.5
49. 165.2 cm
51. $17,500
53. $10,000 in stock and $12,500 in bonds
55. 1844
57. 273
59. 165
61. 48-oz bottle
63. Bill
65. Bette
67. John
69. bushel
71. $5 rebate
73. Bruce
75. Group A

REVIEW EXERCISES, PAGE 386

1. $x = -5$
3. $u = -3.5$
5. $t = 9$
7. $t = \frac{5}{3}$
9. $z = -15.3$

11. $x = -3\frac{1}{2}$
13. $x = -1.5$
15. $y = 3.5$
17. $p = \frac{11}{3}$
19. $y = -9$

21. $w = -9$
23. $x = \frac{-7}{13}$
25. Dan is 35 and Karen is 29.

27. Fillet-O-Fish®, 435; fries, 220; shake, 352
29. no, $70 \times 5 \neq 85 \times 6$

31. $d = 4A - a - b - c$; 85
33. $D = \frac{C(a + 12)}{a}$; 30

35. 6480 to 5123
37. 3 to 1
39. $n = 12$
41. $n = 1$

Answers to Odd-Numbered Exercises

TEASERS, PAGE 387

1. $s = 1$
3. $p = \dfrac{-95}{8}$
5. Second plan; $24.50
7. 972 ft^2
9. $6000 @ 8%, $8000 @ 6%

CUMULATIVE REVIEW EXERCISES, CHAPTERS 1–8, PAGE 389

1. 2
3. 0.004032
5. 1
7. 0.012
9. 53.6 cm
11. $8u - 13v$
13. $(2xy^2)^3$
15. $r = 10$
17. $p = \dfrac{43}{26}$
19. $w = \dfrac{-21}{8}$
21. 31.9%
23. $3841.25
25. 2,592,000

EXERCISES 9.1, PAGE 400

1. x^6
3. $x^2\ (x \neq 0)$
5. 512
7. u^{11}
9. x^6
11. Not possible; *sum* of powers of *different* bases
13. $-p^6$
15. t^{19}
17. 256
19. -4
21. $(0.6)^9$
23. $\left(\dfrac{1}{3}\right)^9$
25. q^{22}
27. $p^{12} - 3q^{15}$
29. $-6x^3y^4$
31. $\dfrac{-2p}{q^5}\ (p \neq 0, q \neq 0)$
33. $1\ (x \neq 0)$
35. $\dfrac{1}{m^{14}}\ (m \neq 0)$
37. $(-5)^{14}$
39. $(0.2)^{16}$
41. $(a + 2b)^8$
43. $(2p + 3q)^{28}$
45. $(x^2 + y)^9$
47. $\dfrac{51}{5}$
49. $(x + y)^4\ (x + y \neq 0)$

EXERCISES 9.2, PAGE 404

1. x^2y^2
3. $16a^6b^2$
5. u^4v^{12}
7. $x^2y^4z^6$
9. $x^4\ (x \neq 0, y \neq 0)$
11. $a^{10}b^9c^{12}$
13. $\dfrac{-8p^4q^5}{9}\ (p \neq 0, q \neq 0)$
15. $-54a^6b^7$
17. $-36p^8q^{18}$
19. $\dfrac{-x^{17}y}{2}\ (x \neq 0, y \neq 0)$
21. 1,000,000
23. $\dfrac{1}{625}$
25. 121
27. $-29x^6y^9$
29. $(a + b)^9(b - c)^6$
31. $(3u - v)^{12}(2u - 3v)^{15}$
33. $(x^2 + y)^2(x - y^2)^2\ (x - y^2 \neq 0, x^2 + y \neq 0)$
35. **a.** 72   **b.** 46,656
37. **a.** 128   **b.** 262,144
39. **a.** 243   **b.** 27

EXERCISES 9.3, PAGE 409

1. $\dfrac{1}{16}$
3. 1
5. $\dfrac{-1}{2}$
7. $\dfrac{8}{15}$
9. $-16\dfrac{1}{16}$
11. $\dfrac{1}{x^5}\ (x \neq 0)$
13. $x^2\ (x \neq 0)$
15. $x^4\ (x \neq 0)$
17. $\dfrac{1}{x^6}\ (x \neq 0)$
19. $\dfrac{-8}{a^3}\ (a \neq 0)$
21. $\dfrac{-6x}{y}\ (x \neq 0, y \neq 0)$
23. $\dfrac{-4y^2}{x^2}\ (x \neq 0, y \neq 0)$
25. $\dfrac{1}{y}\ (y \neq 0)$
27. $\dfrac{a^2b^3p^3}{c^4m^2}\ (c \neq 0, m \neq 0, p \neq 0)$
29. $\dfrac{-2a^5}{b^4c^3}\ (a \neq 0, b \neq 0, c \neq 0)$
31. $\dfrac{u}{x^2y^2}\ (x \neq 0, y \neq 0, u \neq 0)$
33. $\dfrac{1}{3}\ (x \neq 0)$
35. $\dfrac{4b^6}{c^8}\ (a \neq 0, b \neq 0, c \neq 0)$
37. $\dfrac{1}{3}$
39. $\dfrac{128a^3d^2}{9b^2}\ (b \neq 0, c \neq 0)$
41. $0\ (x \neq 0, y \neq 0)$
43. $a + \dfrac{1}{b}\ (b \neq 0)$
45. $\left(a + \dfrac{1}{b}\right) \div b\ (b \neq 0)$
47. $(a + b)b\ (b \neq 0)$
49. $\left(\dfrac{1}{p + q}\right)\left(\dfrac{1}{p} + \dfrac{1}{q}\right)\ (p \neq 0, q \neq 0, p + q \neq 0)$

EXERCISES 9.4, PAGE 413

1. 3.26×10^1
3. 1.7×10^1
5. 4.18×10^{-3}
7. 3.6×10^7
9. 8.1×10^1
11. 4×10^{-2}
13. 1.47×10^2
15. 6.074×10^{10}
17. 1.234×10^8
19. 1.21×10^2
21. 9×10^{-3}
23. 1×10^{12}
25. 3
27. 2
29. 3
31. 2
33. 2
35. 1
37. 3
39. 4
41. 4
43. 3
45. 1
47. 1
49. 2300
51. 12,340
53. 2,540,000,000
55. 7,146,000,000,000
57. 230,000,000,000,000,000
59. 10,000,000,000
61. 69,070,000,000,000
63. 6.01
65. 0.000000000000000000000000166057
67. 0.000001102331
69. 4,014,490,000
71. 0.00000000000000000000001672
73. 10,540,000,000
75. 38.3
77. 0.0014
79. 20.0
81. 70
83. 0.0044
85. 3.6×10^2
87. 3×10^{-1}
89. 1×10^{17}

REVIEW EXERCISES, PAGE 415

1. x^7
3. Not possible; *sum* of powers of same base
5. $\dfrac{1}{z^2}$ $(z \neq 0)$
7. $\dfrac{-2q^2}{p^2}$ $(p \neq 0, q \neq 0)$
9. $\dfrac{a^5 b}{2}$ $(a \neq 0, b \neq 0)$
11. $-18x^8$
13. $x^{10} y^7 z^{14}$
15. $\dfrac{1}{b^2}$ $(b \neq 0)$
17. $\dfrac{4}{x^6}$ $(x \neq 0)$
19. $\dfrac{-xz^5}{2y^7}$ $(x \neq 0, y \neq 0, z \neq 0)$
21. 2.96×10^1
23. 4.96×10^7
25. 1.302×10^1
27. 360,000
29. 5,670,000
31. 43.6

TEASERS, PAGE 416

1. $3x^0 = 3$ $(x \neq 0)$; $(3x)^0 = 1$ $(x \neq 0)$
5. 1280
7. 1.697×10^4
9. 738.9×10^4

CUMULATIVE REVIEW EXERCISES, CHAPTERS 1–9, PAGE 418

1. 0
3. $\dfrac{-1}{108}$
5. -2.87
7. $16\dfrac{1}{16}$
9. 5.154
11. 0.000316
13. $\dfrac{1}{16}$
15. $\dfrac{x^5}{4y}$ $(x \neq 0, y \neq 0)$
17. 9.9 ft
19. $(y + 2z)^2$
21. $y = \dfrac{2}{5}$
23. 2.692×10^2
25. 0.000601

EXERCISES 10.1, PAGE 425

1. Polynomial in x
3. Polynomial in u
5. Polynomial in s and t
7. Not a polynomial
9. Polynomial in x and y
11. Polynomial in x
13. Polynomial in x and y
15. Polynomial in u and v
17. Not a polynomial
19. Polynomial in p and q
21. 2, 0; 2
23. 1, 4, 3; 4
25. 2, 2, 2; 2
27. 4, 4, 1, 1; 4
29. 5, 3, 4; 5
31. 5, 4, 6; 6
33. 1, 1, 2, 3; 3
35. 4, 3, 6, 0; 6
37. 9, 5; 9
39. 6, 6, 5; 6
41. $x^4 - 3x^2 + 2x - 1$
43. $-5u^6 + 4u^2 - 3u + 2$
45. $2x^4 y^3 + x^3 y - x^2 y^2 - xy^4$
47. $u^4 v^3 + u^3 v^3 + u^3 v^2 - uv$
49. $-(2p^2 q)^3 + p^5 q^2 - p^4 q^3 + pq^4$

EXERCISES 10.2, PAGE 427

1. $5x^4 - 5x^2 + 6$
3. $-0.1u^3 + 0.3u^2 + 1.4u + 1.5$
5. $q^4 - \frac{4}{7}q^3 - \frac{1}{2}q^2 + 8$
7. $6y^4 - 2y^3 - 6y^2 + 8y - 5$
9. $-2z^6 + 8z^2 - 15$
11. $\frac{5}{4}x^2 + \frac{2}{15}xy + \frac{2}{3}y^2$
13. $-1.8x^2y^2 + 1.6x^2y - 0.7xy^2$
15. $3x^3y^2 - x^2y^3 - 2x^2y^2 + xy^3$
17. $\frac{-11}{12}m^2n^2 + \frac{1}{2}m^2n - \frac{1}{6}mn^2 + \frac{1}{10}mn$
19. $-1.7s^3t + 1.7s^2t^3 + 10.7s^2t^2 + 4.4$
21. $3x^3 - x^2 - 2x - 6$
23. $4x^3 - 3x^2 - 5x + 1$
25. $-3x^3y^2 + 4x^2y - xy^3$
27. $\frac{5}{3}x^4 + \frac{1}{2}x^3 - \frac{19}{30}x^2 - \frac{1}{5}x - \frac{33}{5}$
29. $-4u^4 - 5u^2 + 3u$
31. $-2s^4 + 2s^3 + 5s^2 - 3s - 3$
33. $-4y^4 + 3y^3 + 4y^2 - 4y - 6$
35. $3t^6 + 5t^3 - 8t - 5$
37. $\frac{5}{12}u^2v^3 - \frac{1}{6}u^3v + \frac{3}{4}u + \frac{4}{5}v$
39. $-5p^2q^2 + 2p^2q - 3pq^3 + 3q^3$

EXERCISES 10.3, PAGE 430

1. $-3x^4 + 2x^3 + 3x^2 - 5x + 5$
3. $-3.4u^5 - u^4 + 2.3u^3 + 5.3u^2 + 8.5u - 2.2$
5. $q^3 - 2q^2 + 5q - 10$
7. $\frac{-8}{45}t^7 + \frac{3}{5}t^6 - \frac{2}{3}t^5 - \frac{1}{3}t^3 + \frac{21}{20}t^2 - \frac{7}{8}t - \frac{1}{2}$
9. $-9.3x^4 + 7.5x^3 + 4.2x^2 + 2.7x + 2$
11. $-0.1x^3y - 0.6x^2y^3 + 0.3xy^2 + 0.2x + 0.5y$
13. $-p^2q^2 - 10p + 12q + 1$
15. $-2s^4t^2 + 3s^2t^4 + 7s^2t^3 - 13st^4$
17. $5.2y^4 - 2.4y^3 + 1.4y^2 + 19.5y - 19.9$
19. $\frac{-26}{21}p^4 - \frac{2}{5}p^3 - \frac{3}{4}p^2 + \frac{5}{6}p + \frac{151}{99}$
21. $-m^4 + 2m^3 + m^2 - 4m - 1$
23. $-z^3 + z^2 + z + 3$
25. $-2u^3 + 2u^2 - 2u + 2$
27. $3u^3v^2 + 6u^2v^3 + uv^4 - 4u$
29. $-12m^4n^3 - m^3n^3 + 8m^2n$
31. $-3x^3 + 3x^2 - 12x + 8$
33. $u^2 - 6uv + 3v^2$
35. $-5u^3 - 6u^2 + 5u + 6$
37. $15.3y^3 + 7.2y^2 - 8.2y + 5.7$

EXERCISES 10.4, PAGE 435

1. $-10x^3$
3. $-0.03u^7$
5. $-12x^5y^4$
7. $\frac{-1}{6}p^3q^5$
9. $-6p^4q^3r^6$
11. $24x^5y^6$
13. $3x + 6y$
15. $8u^3 - 4u$
17. $5t^5 - 15t$
19. $0.26x^2 + 0.14x$
21. $p^3q^5 - p^4q^7$
23. $-6p^3q + 8p^3q^4$
25. $6x^3 - 30x + 42$
27. $0.69w^5 - 2.3w^2 - 1.61$
29. $8s^3 + 8s^5 - 48s^8$
31. $x^2 - x - 6$
33. $y^2 - 12y + 35$
35. $6u^2 - u - 2$
37. $\frac{4}{3}p^2 - \frac{92}{15}p + \frac{3}{5}$
39. $\frac{-2}{15}t^2 + \frac{7}{5}t - 3$
41. $-14r^2 + 13r - 3$
43. $x^4 + 3x^2 - 4$
45. $x^2y^2 + 2xy - 15$
47. $-p^3q^3 - p^2q^2 + 6pq$
49. $x^3y^4 - 3x^3y^2 + 2x^2y^4 - 6x^2y^2$
51. $6y^3 + 10y^2 - 22y + 6$
53. $3u^3 + 10u^2 + 16u - 16$
55. $2x^4 - 7x^3 + 14x^2 - 2x - 15$
57. $6y^5 - 23y^3 + 12y^2 + 20y - 16$
59. $2p^4 - 7p^3 + 12p^2 - 11p + 4$
61. $2x^4 - 3x^3 - 3x^2 + 17x - 21$
63. $2y^4 - 8y^3 + 19y^2 - 22y + 15$
65. $2w^8 - 3w^6 - 4w^4 + 19w^2 - 24$
67. $4x^4 - 6x^3 - 30x^2 + 46x + 10$
69. $12u^5 + 44u^4 - 33u^3 - 188u^2 - 162u + 21$
71. $8x^4 - 14x^3 + 23x^2 - 25x + 8$

EXERCISES 10.5, PAGE 439

1. $x^2 + 3x + 2$
3. $x^2 - 9$
5. $y^2 + 10y + 21$
7. $0.33u^2 - 0.4u - 5$
9. $2u^2 - 3u - 2$
11. $\frac{8}{35}u^2 - \frac{46}{35}u - 3$

13. $6a^2 - ab - b^2$

15. $8x^2 - 2xy - 15y^2$

17. $\frac{2}{9}x^2 + \frac{2}{35}xy - \frac{4}{35}y^2$

19. $4r^2 + 12rs + 9s^2$

21. $4.37x^2 - 5.2xy - 0.93y^2$

23. $8 - 8p - 6p^2$

25. $21s^2 + 47st + 20t^2$

27. $\frac{9}{28}m^2 + \frac{83}{140}mn + \frac{4}{15}n^2$

29. $x^2 - 4y^2$

31. $7x^2 - 20xy - 3y^2$
33. $4p^2 - 9q^2$
35. $5r^2 - 29rs - 6s^2$
37. $2p^2 - 5pq - 12q^2$
39. $8x^2 + 38xy + 35y^2$
41. $x^2 - 2xy + y^2$
43. $9x^2 - 6xy + y^2$
45. $4m^2 - 12mn + 9n^2$
47. $9p^2 + 12pq + 4q^2$
49. $1 - 4x + 4x^2$
51. $1.44p^2 - 3.12p + 1.69$
53. $0.01 - 0.08q + 0.16q^2$
55. $125.44x^2 - 22.4x + 1$
57. $16a^2 - 24ab + 9b^2$
59. $100x^2 - 500xy + 625y^2$
61. $-5x^2 - 16x - 3$
63. $29p^2 + 10pq + 34q^2$
65. $10x^2 + 11xy - 17y^2$
67. $2.12m^2 + 3.68mn - 1.21n^2$
69. $6x^2 - 4xy + 3y^2$

EXERCISES 10.6, PAGE 447

1. $4x^2 - 2x + 3$

3. $\frac{9}{4}u^4 - 2u^2 + \frac{7}{4}$

5. $\frac{-1}{3}w^7 + \frac{1}{2}w^5 - \frac{2}{3}w^3 + \frac{7}{6}w$

7. $\frac{5}{7}p^5 - \frac{2}{7}p^3 + \frac{3}{7}p - 1$

9. $\frac{3}{2}m^3 - 2m + \frac{1}{m} - \frac{7}{2m^2}$

11. $y^2 - \frac{2}{3}y + \frac{4}{3}$

13. $t^5 - \frac{6}{7}t^2 + \frac{1}{7} - \frac{2}{7t^2} + \frac{1}{7t^3}$

15. $2pq - \frac{5}{2}q + \frac{7}{2pq} - \frac{9}{2p^2}$

17. $\frac{-2a^2}{7b} + \frac{3a}{7b^2} - \frac{a^4b}{7}$

19. $\frac{1}{2t} - \frac{2s}{t} + \frac{5t}{2} - \frac{1}{2t^2}$

21. $x + 2$

23. u R(1)

25. $\left(-2y^3 - \frac{17}{4}y^2 - \frac{177}{16}y - \frac{1539}{64}\right)$ R $\left(\frac{14{,}785}{64}\right)$

27. $\left(-3p^3 - \frac{5}{2}p^2 - \frac{15}{4}p - \frac{33}{8}\right)$ R $\left(\frac{-155}{8}\right)$

29. $(t^4 + 4t^3 + 11t^2 + 33t + 101)$ R(297)

31. $(3x + y)$ R$(9y^2)$

33. $u^2 + uv + v^2$

35. $\left(x^3 + x^2y + \frac{3}{2}xy^2 + \frac{11}{4}y^3\right)$ R $\left(\frac{29}{4}y^4\right)$

37. $2y - 3$

39. $x + 3$

REVIEW EXERCISES, PAGE 449

1. Polynomial in v
3. Polynomial in y
5. Polynomial in u and v
7. Polynomial in w
9. 1, 4, 3; 4
11. 5, 1, 2, 4; 5
13. 4, 5, 6; 6
15. $-u^5 - 6u^4 - 3u^3 + 4u + 7$
17. $6m^3n^2 - 7m^3n + m^2n^3 - 8m^2n^2 + 8mn^4$
19. $-3x^3 + 6x^2 - 13x - 7$
21. $-4u^4 + u^3 + 7u^2 - 3u - 19$
23. $12 - 2q^2 - 3q^3 + 10q^4$
25. $-s^7 - 2s^6 - 2s^4 - 5s^2 - 3$
27. $-4u^5 + 2u^4 + 3u^3 - 5u^2 + 4u + 1$
29. $3s^7 + 2s^6 - 2s^5 + s^4 + 5s^3 - 4s^2 - 7s - 2$
31. $-5v^3 + 9v^2 - 12v + 18$
33. $y^7 - y^6 - y^5 + 6y^4 - y^3 - 3y^2 + 7y - 1$
35. $6x^2 + 2x$
37. $8u^2 - 2u - 15$
39. $26p^2 - 47pq + 12q^2$
41. $4s^3 - 14s^2t + 20st^2 - 12t^3$
43. $u^2 + u - 12$
45. $2 - 11w + 15w^2$
47. $y^2 + 14y + 49$
49. $25p^2 - 40pq + 16q^2$
51. $3u^4 - 2u^2 + 5$
53. $\frac{1}{2}pq - \frac{3}{4}pq^3 + \frac{1}{4}p^2q$
55. $s + 3$

TEASERS, PAGE 450

1. $14x - 10$ yd
3. $10x^2 - 31x - 14$ yd^2
5. 20.408 m
7. $64r^2 + 16r^3s - 415r^2s^2 - 52rs^3 + 676s^4$

CUMULATIVE REVIEW EXERCISES, CHAPTERS 1–10, PAGE 452

1. $\frac{-5}{154}$

3. -2.4732

5. -19.36

7. $\frac{-29}{1024}$

9. 9.1

11. 330,000

13. 4.369×10^{-4}

15. $2x^4 - 3x^3 - 5x^2 + 1$

Answers to Odd-Numbered Exercises

17. $r = \frac{12}{13}$ **19.** 52.2 cm **21.** 30.9% × 14.2 **23.** 0.0000509
25. $\frac{1048}{5}$

EXERCISES 11.1, PAGE 458

1. 1, 3, 9, 27 **3.** 1, 2, 4, 5, 8, 10, 20, 40 **5.** 1, 3, 19, 57
7. 1, 3, 5, 15, 25, 75 **9.** 1, 2, 5, 10, 11, 22, 55, 110
11. 1, 2, 3, 5, 6, 9, 10, 15, 18, 25, 30, 45, 50, 75, 90, 150, 225, 450
13. $2 \times 3 \times 13$ **15.** $2^3 \times 3 \times 5$ **17.** $2^4 \times 3^2$
19. Prime **21.** $2 \times 5 \times 409$ **23.** $2 \times 19 \times 163$
25. Yes **27.** Yes **29.** Yes **31.** Yes **33.** Yes
35. Yes **37.** Yes **39.** No **41.** Yes **43.** Yes
45. No **47.** Yes **49.** No **51.** Yes **53.** No
55. Yes **57.** No **59.** No **61.** Yes **63.** No
65. No **67.** No **69.** Yes **71.** No **73.** No
75. Yes **77.** Yes **79.** No

EXERCISES 11.2, PAGE 464

1. $2(x - 2)$ **3.** $5(u + 12)$ **5.** $7(3a - 7)$ **7.** $3x(x - 17)$
9. $5u^2(u + 4)$ **11.** $2xy(x - 2y)$ **13.** $7p^2q(p^2q^2 - 3)$ **15.** $ab^2(ab - 2)$
17. $27st(s^3 + 3t^3)$ **19.** $15a^2b^3(b^2 + 4a^2)$ **21.** $ab(a - 3b + 4ab)$ **23.** $xy(4x^2 - 5y + 6x^2y)$
25. $11u^3v^3(3uv^3 + 4v^2 - 6u^2)$ **27.** $5(2a^2b - 3a + 5b^3)$ **29.** $8m^4n^4(3n^2 - 4m + 6m^3n^5)$
31. $3a^2b^2c^3(2ac - 3b^3)$ **33.** $2pq^2t(8p^2 - 9q^2t^2)$ **35.** $4xy^2z^2(8xy - 9z^4)$
37. $a^2(2b^3 - 3ac^4 + 4a^3b^5)$ **39.** $12(s^3t^4u^5 + 2s^2t^3u - 3)$ **41.** $2(2a - 6b + 9c - 10d + 12ac)$
43. $3ab(2ab^2 - 3b + 4a^2b^3 - 12a^3)$ **45.** $3r^2s^2(2r - 3r^2 + 8s^3 - 14r^3s^2)$ **47.** $6abc^2(abc - 2b^2 - 3a^2c^2 - 6a^2b^3c^4)$
49. $12a^2b^3(ab - 2 + 4a^2b^2 - 8b^2 + 16a^3b^4)$

EXERCISES 11.3, PAGE 468

1. $(p + q)(p - q)$ **3.** $(m + n)(m - n)$ **5.** $(4 + y)(4 - y)$ **7.** $(2a + b)(2a - b)$
9. $(2a + 3b)(2a - 3b)$ **11.** $(3x + 4)(3x - 4)$ **13.** $2(a + 2b)(a - 2b)$ **15.** $x(x + y)(x - y)$
17. $4t(4 + t)(4 - t)$ **19.** $(xy + uv)(xy - uv)$ **21.** $x^3y(x + y)(x - y)$ **23.** $(x^2 + y^2)(x + y)(x - y)$
25. $(4a^2 + 9b^2)(2a + 3b)(2a - 3b)$ **27.** $3p(p + q)(p - q)$ **29.** $ab(a^2 + b^2)(a + b)(a - b)$
31. $(2x^2y^2 + 1)(2x^2y^2 - 1)$ **33.** $5ab(a^2 + 2b)(a^2 - 2b)$ **35.** $2(7u^3 + 6v^2)(7u^3 - 6v^2)$
37. $(10p^2 + 9q^3)(10p^2 - 9q^3)$ **39.** $(1 + 2abc^2)(1 - 2abc^2)$

EXERCISES 11.4, PAGE 471

1. $(x + 1)(x - 2)$ **3.** $(y - 3)(y - 1)$ **5.** $(u - 3)(u - 4)$ **7.** $(r - 2)(r - 3)$
9. Not factorable **11.** $(x - 6)(x + 1)$ **13.** Not factorable **15.** $(u + 5)(u + 3)$
17. $(s - 3)(s + 2)$ **19.** $(t - 7)(t - 11)$ **21.** $(r - 13)(r + 2)$ **23.** Not factorable
25. $(x - 9)(x + 2)$ **27.** $(y - 10)(y + 3)$ **29.** $(u + 9)(u - 6)$ **31.** $(z + 5)(z + 5)$
33. $(w - 3)(w - 12)$ **35.** $(r + 5)(r + 9)$ **37.** $(m + 5)(m + 10)$ **39.** $(p - 14)(p + 3)$
41. $(x + 3)(x + 7)$ **43.** $(y + 6)(y - 5)$ **45.** $(z + 6)(z - 9)$ **47.** $(p + 11)(p - 9)$
49. $(q - 25)(q + 4)$ **51.** $(u - 17)(u + 4)$ **53.** $(v - 14)(v - 5)$ **55.** $(w + 9)(w + 12)$
57. $(s + 14)(s - 11)$ **59.** $(t + 15)(t + 17)$

EXERCISES 11.5, PAGE 476

1. $(2x + 1)(x - 3)$ **3.** $(3y - 2)(y + 4)$ **5.** $3(2u - 1)(2u - 1)$ **7.** $(2p + 1)(3p + 1)$
9. $(2x - 1)(6x + 1)$ **11.** $(4y - 5)(5y - 1)$ **13.** $2(5m - 2)(m - 3)$ **15.** $(3u + 1)(7u - 2)$
17. $(4s + 1)(4s + 3)$ **19.** $(7t + 1)(7t + 1)$ **21.** $(5x - 1)(5x - 1)$ **23.** $(5y - 3)(5y - 2)$
25. $(8u - 3)(3u + 5)$ **27.** $(5p + 2)(9p - 1)$ **29.** $(11q - 3)(2q + 9)$ **31.** $(3 + 7t)(2 + 5t)$
33. $(6 - 5t)(7 - t)$ **35.** $(5 + w)(3 - 4w)$ **37.** $-3(2 - 3y)(1 - 2y)$ **39.** $3(4 + p)(3 - p)$

EXERCISES 11.6, PAGE 478

1. $(a + b)(m + n)$ **3.** $(x + 2)(y + 3)$ **5.** $(2a + 1)(b + 1)$ **7.** $(x + y)(x - y + 4)$
9. $(p - 1)(q + 1)$ **11.** $(m - n)(x - y)$ **13.** $(2n - 3)(m + 4)$ **15.** $(3t - 4)(s + 2)$

17. $(2m + n)(m - 1)$
19. $(2a - 5)(b + 4)$
21. $(p + q)(p - q + 4)$
23. $3(y - 2)(x + y + 2)$
25. $(x + 4)(x + y - 4)$
27. $(t - 3)(t - 3)(t + 3)$
29. $(w - 5)(w + v + 5)$

EXERCISES 11.7, PAGE 480

1. $3(x + 2y)(x - 2y)$
3. $2(2x + 3y)(2x - 3y)$
5. $(a^2 + b^2)(a + b)(a - b)$
7. $3(x - 7)(x + 3)$
9. $4(2p - 5)(p + 4)$
11. $2(4t - 1)(3t + 2)$
13. $(3a + 4b)(x + y)$
15. $7(p - 1)(q + 1)$
17. $4(b + 5)(3 + b^2)$
19. $(x - y)(9x + 9y - 4)$
21. $(a + 2b + 6)(a + 2b - 6)$
23. $(2p - 3q - 7)(2p - 3q + 7)$
25. $(y + 2)(y - 2)(y + 1)(y - 1)$
27. $3(t + 3)(t - 3)(t + 1)(t - 1)$
29. $-5(w^2 - 5)(w + 2)(w - 2)$

EXERCISES 11.8, PAGE 485

1. $x = 0, 5$
3. $y = \frac{-4}{3}, \frac{1}{2}$
5. $u = \frac{-3}{2}, 0, 2$
7. $a = -1, 2, 3$
9. $b = \frac{-5}{4}, \frac{2}{5}, 2$
11. $z = \frac{-2}{5}, \frac{3}{4}, 1$
13. $x = -1, \frac{10}{11}, \frac{9}{8}$
15. $s = \frac{1}{7}, \frac{2}{9}, 3$
17. $x = -4, -1$
19. $y = -3, -1$
21. $x = \frac{-3}{2}, \frac{3}{2}$
23. $y = -2, 0, 2$
25. $u = \frac{-4}{3}, \frac{5}{2}$
27. $t = \frac{-4}{3}$
29. $x = \frac{-10}{11}, \frac{4}{3}$
31. $y = 0, 1, 6$
33. $u = 0, 1, 15$
35. $p = -3, 0, 16$
37. $z = 0, 2, 4$
39. $m = -6, 0, 1$

EXERCISES 11.9, PAGE 490

1. -12 and -9 or 9 and 12
3. 4 and 11 or $\frac{-11}{2}$ and -8
5. $3, 5,$ and 7
7. -1
9. 12 m by 29 m
11. 16 ft by 23 ft
13. 7 cm
15. 23
17. 27
19. Height = 6 cm, base = 9 cm, Base = 19 cm

REVIEW EXERCISES, PAGE 492

1. 3×7
3. $3^2 \times 11$
5. $2^3 \times 5^3$
7. $10(t - 4)$
9. $11x^2y^2(4x - 11y^2)$
11. $4b(a^2 - 2abc + 3b^2c^4)$
13. $(2x + 3y)(2x - 3y)$
15. $2pq(2p + 3q)(2p - 3q)$
17. $st(s^2 + t^2)(s + t)(s - t)$
19. $(x + 6)(x - 3)$
21. $(u + 3)(u + 7)$
23. $(s - 14)(s + 4)$
25. $(r - 13)(r - 4)$
27. $(2t + 1)(t - 3)$
29. $(8x - 3)(2x + 9)$
31. $7(u + 5)(u - 1)$
33. $2m(3m + 1)(3m + 1)$
35. $(a + b)(c - d)$
37. $(2x + 1)(y - 3)$
39. $(x + y)(y - 3)$
41. $4(x - 3)(x + 2)$
43. $5(a + 2b)(a - b)$
45. $3p(2p + q)(3p - q)$
47. $x = -6, 1$
49. $p = -1, 0, 2$
51. $y = -1, 0, 7$
53. 9 and 16
55. Base = 20 ft, height = 17 ft

TEASERS, PAGE 493

1. Yes. If the number is divisible by 9, the sum of its digits is divisible by 9. Reversing the digits will not change the sum of the digits.
3. 54
5. Divide the expression by $2y - 1$.
7. Yes. 21 is exactly divisible by 1.
9. $(u - 5)(v + 3u)$

CUMULATIVE REVIEW EXERCISES, CHAPTERS 1–11, PAGE 495

1. 11.97
3. 0.05497
5. 7
7. $\frac{-47}{27}$
9. 400,000
11. 3.691×10^{-4}
13. $-a - 15b - 18c$
15. $p = -1, 6$
17. $r = -1, \frac{4}{5}$
19. $n^2 - 3n = \frac{1}{2}(n + 5)$
21. 0.00000109
23. $2x(x + 2)(x - 2)$
25. 118 cm

EXERCISES 12.1, PAGE 503

1. $x = 9$
3. $y = 0$
5. $u = 2$
7. $t = -1, 0, 1$
9. $s = -7, 0, 1$
11. $\frac{1}{2}x^2 \ (x \neq 0)$
13. $\frac{-x}{3y^2} \ (x \neq 0, y \neq 0)$
15. $\frac{-u^2v^2}{3w} \ (v \neq 0, w \neq 0)$
17. $2 \ (x \neq -3)$
19. $\frac{5}{7} \ (u \neq -2)$
21. $-1 \left(s \neq \frac{5}{2}\right)$
23. $u - 2 \ (u \neq -2)$
25. $\frac{x+2}{x-2} \ (x \neq 2)$
27. $\frac{s^2}{2s+3} \left(s \neq \frac{-3}{2}, 0\right)$
29. $\frac{1}{2x} \ (x \neq 0, 5)$
31. $\frac{-(a+b)}{5} \ (a \neq b)$
33. $\frac{2y+1}{3y+2} \left(y \neq \frac{-2}{3}, 1\right)$
35. $\frac{2u+3}{u+7} \left(u \neq -7, \frac{1}{3}\right)$
37. $\frac{3(y-5)}{y+3} \ (y \neq -5, -3)$
39. $\frac{t+2}{3} \ (t \neq -7, 2)$
41. $\frac{t^2(t-2)}{2(t+2)} \ (t \neq -2, 0)$
43. $\frac{x-y}{2x-y} \left(y \neq \frac{-x}{2}, 2x\right)$
45. $\frac{2p-3q}{3p-2q} \left(p \neq \frac{2}{3}q, q\right)$
47. $\frac{s+t}{2s-3t} \left(s \neq \frac{2}{3}t, \frac{3}{2}t\right)$
49. $\frac{r+2}{2r-s} \left(r \neq \frac{2}{3}s, \frac{1}{2}s\right)$

EXERCISES 12.2, PAGE 509

1. $\frac{3}{4} \ (x \neq 0, y \neq 0)$
3. $\frac{2}{t^2} \ (s \neq 0, t \neq 0)$
5. $-b \ (a \neq 0, b \neq 0)$
7. $\frac{1}{3}(x-2)^2 \ (x \neq -2)$
9. $\frac{6x^2}{y(x+y)} \ (y \neq 0, \pm x, x \neq 0)$
11. $\frac{-3(x+y)}{2x^2} \ (z \neq 0, y \neq -x)$
13. $y^3(x-4) \ (y \neq 0, x \neq 4)$
15. $\frac{t+5}{3-2s} \left(s \neq \frac{\pm 3}{2}, t \neq 5\right)$
17. $\frac{x^2}{3-x} \ (x \neq 2, 3)$
19. $-1 \ (s \neq \pm t)$
21. $\frac{5x^2 + 7x + 2}{8x - 12x^2} \left(x \neq 0, \frac{2}{3}, \frac{3}{2}, 2\right)$
23. $\frac{y-4}{y+2} \ (y \neq -6, -5, -2, 4)$
25. $\frac{-1}{2} \ (z \neq x, y, y \neq x)$
27. $\frac{-(x+y)}{2} \ (y \neq -1, 7, x)$
29. $\frac{(a-b)(a-2b)}{(a+b)^3} \left(a \neq -b, \frac{-1}{2}b, 2b\right)$
31. $\frac{xy}{z} \ (y \neq 0, z \neq 0)$
33. $-2(x+y) \ (y \neq x)$
35. $\frac{7}{3}x^2(y-1) \ (x \neq 0, y \neq 1)$
37. $\frac{-1}{x^2(x+2)} \ (x \neq -2, 0, 5)$
39. $\frac{a-3}{a+1} \ (a \neq -1, 2, 3)$
41. $\frac{4s+3}{2s-1} \left(s \neq \frac{-3}{4}, \frac{-1}{2}, \frac{-1}{3}, \frac{1}{2}, 3\right)$
43. $\frac{d+2}{3-d} \ (d \neq -1, 1, 2, 3)$
45. $\frac{x+y}{2x+y} \ (y \neq -2x, x)$
47. $\frac{3t^2 - 2t}{5t^2 - 8t - 4} \left(t \neq -1, \frac{-2}{5}, 0, \frac{3}{2}, 2\right)$
49. $\frac{w+2}{w-4} \ (w \neq -6, -5, -2, 4)$
51. $\frac{x+3}{(2x+1)(x+7)(x+4)} \left(x \neq -7, -4, \frac{-1}{2}, 5\right)$
53. $\frac{1}{6y^2 + 52y - 18} \left(y \neq -9, -3, \frac{1}{3}, \frac{3}{2}\right)$
55. $\frac{1}{3u^3 + 2u^2 - u} \left(u \neq -1, \frac{-4}{5}, 0, \frac{1}{3}, \frac{3}{2}\right)$
57. $\frac{y(y+7)(y-7)^2}{y+1} \left(y \neq -7, -1, \frac{-1}{3}, 7\right)$
59. $6p^2 - 19p + 10 \left(p \neq 1, \frac{-1}{2}, \frac{2}{3}\right)$

EXERCISES 12.3, PAGE 513

1. $\frac{5}{x} \ (x \neq 0)$
3. $\frac{2a+3b}{x} \ (x \neq 0)$
5. $\frac{a-b}{5x} \ (x \neq 0)$
7. $\frac{3}{y} \ (y \neq 0)$
9. $\frac{x-y-2z}{5c} \ (c \neq 0)$
11. $\frac{p - 5q + 7r}{6u} \ (u \neq 0)$

13. $\dfrac{x + 2y - 7}{pq}$ $(p \neq 0, q \neq 0)$
15. $\dfrac{-2}{x + 1}$ $(x \neq -1)$
17. $\dfrac{2a - 3b}{w + 1}$ $(w \neq -1)$
19. $\dfrac{12}{x^2 - 5}$ $(x^2 \neq 5)$
21. $\dfrac{4}{3m + 2}$ $\left(m \neq \dfrac{-2}{3}\right)$
23. $\dfrac{x - 2y - 3z}{p^2 + 2q}$ $\left(q \neq \dfrac{-1}{2}p^2\right)$
25. $\dfrac{x + 4}{a - 1}$ $(a \neq 1)$
27. $\dfrac{5a + 9}{x + 5}$ $(x \neq -5)$
29. $\dfrac{2 - 2a}{2y + 3}$ $\left(y \neq \dfrac{-3}{2}\right)$
31. $\dfrac{2x + 2xy + y^2}{x + y}$ $(y \neq -x)$
33. $\dfrac{4y - x}{x^2 - y^2}$ $(y \neq -x, x)$
35. $\dfrac{6p - 8q}{2p + 1}$ $\left(p \neq \dfrac{-1}{2}\right)$
37. $\dfrac{-4}{q + 7}$ $(q \neq -7)$
39. $\dfrac{1}{m - 2}$ $(m \neq -2, 2)$

EXERCISES 12.4, PAGE 517

1. $6x^2$
3. $30u^3$
5. $15(y - 5)$
7. $10(t - 2)$
9. $6(s - 2)$
11. $7(3p + 4)^2$
13. $a^2 - b^2$
15. $3(a^2 - b^2)$
17. $3(x^2 - 4)$
19. $2(y^2 - 36)$
21. $(m^2 - 9)(m + 4)$
23. $(q - 1)(q - 4)(q + 7)$
25. $(u^2 - 9)(u + 5)$
27. $4y(y - 3)(y + 3)^2$
29. $(s + 1)(s + 2)^2(s + 3)$
31. $(4m^2 - 1)(m - 1)^3$
33. $(2p + q)(2p - q)^2$
35. $2t^2(5 + t)(t - 5)^2$
37. $6(u - 7)(u + 7)^2(u^2 + 8u + 1)$
39. $(y + 3)(y - 6)^2$

EXERCISES 12.5, PAGE 521

1. $\dfrac{4a + 5}{6a^2}$ $(a \neq 0)$
3. $\dfrac{6x^2 + 5x - 10}{5x^3}$ $(x \neq 0)$
5. $\dfrac{3b + 2ab - 5a}{a^2b^2}$ $(a \neq 0, b \neq 0)$
7. $\dfrac{3x - 4}{x^2 - 3x + 2}$ $(x \neq 1, 2)$
9. $\dfrac{15 - y}{y^2 - 9}$ $(y \neq -3, 3)$
11. $\dfrac{8a - 23}{a^2 + a - 20}$ $(a \neq -5, 4)$
13. $\dfrac{26b + 15}{6b^2 + b - 2}$ $\left(b \neq \dfrac{-2}{3}, \dfrac{1}{2}\right)$
15. $\dfrac{23x - 5}{12(x + 1)}$ $(x \neq -1)$
17. $\dfrac{10s^2 + 19st + 12t^2}{5s(2t - s)}$ $(s \neq 0, 2t)$
19. $\dfrac{13y - 6}{3(y^2 - 9)}$ $(y \neq -3, 3)$
21. $\dfrac{2}{x - 1}$ $(x \neq -1, 1)$
23. $\dfrac{x^2 - 5x}{x^2 - 9}$ $(x \neq -3, 3)$
25. $\dfrac{u^2 - 2u + 4}{2(u^2 - 16)}$ $(u \neq -4, 4)$
27. $\dfrac{7x - 1}{(x + 1)^2(x - 1)}$ $(x \neq -1, 1)$
29. $\dfrac{u - 8}{(u - 7)^3}$ $(u \neq 7)$
31. $\dfrac{5 - 2s}{(s + 5)(s^2 - 4)}$ $(s \neq -5, -2, 2)$
33. $\dfrac{2a^2 + 4ab + a + b}{(a^2 - b^2)(a + 2b)}$ $(a \neq -b, b, -2b)$
35. $\dfrac{9t + 28}{2t(4 - t^2)}$ $(t \neq -2, 0, 2)$
37. $\dfrac{2p - 13}{3(2p - 1)(4p^2 - 9)}$ $\left(p \neq \dfrac{-3}{2}, \dfrac{1}{2}, \dfrac{3}{2}\right)$
39. $\dfrac{2t + 17}{(t + 5)(t^2 - 36)}$ $(t \neq -6, -5, 6)$

EXERCISES 12.6, PAGE 525

1. $\dfrac{11}{15}$
3. $\dfrac{21}{17}$
5. $\dfrac{26}{9}$
7. $\dfrac{1}{xy}$ $(x \neq 0, y \neq 0)$
9. $\dfrac{3}{2x^2y}$ $(x \neq 0, y \neq 0)$
11. $\dfrac{1}{b^2 + 1}$ $(b \neq 0)$
13. $\dfrac{x + y}{y - x}$ $(x \neq 0, y \neq 0, y \neq x)$
15. $\dfrac{2}{s^2 + t^2}$ $(s \neq 0, t \neq 0)$
17. $\dfrac{a^2 + b^2}{a^2 - b^2}$ $(a \neq 0, b \neq 0, a \neq -b, a \neq b)$
19. $\dfrac{a + 2}{a^2 - 5}$ $(a \neq 0, a^2 \neq 5)$
21. $\dfrac{32c - 3}{40c + 6}$ $\left(c \neq \dfrac{-3}{20}, 0\right)$
23. $\dfrac{3 + y}{y^2 - 1}$ $(y \neq -1, 0, 1)$
25. $y + 1$ $(x \neq 0, y \neq 0)$
27. $\dfrac{2}{z^2 - 1}$ $(z \neq -1, 0, 1)$
29. $a^2 - b^2$ $(a \neq 0, b \neq 0)$
31. $\dfrac{x^2 - x}{x^2 + 1}$ $(x \neq 0)$
33. $2(x + y)$ $(x \neq -y, y)$
35. $\dfrac{x + y}{y}$ $(x \neq -y, 0, y; y \neq 0)$
37. $\dfrac{3(p + q)}{4(p - q)}$ $(p \neq -q, q)$
39. $\dfrac{xy(2y - 3x)}{5y^2 - x^2}$ $(x \neq 0, y \neq 0, x^2 \neq 5y^2)$

Answers to Odd-Numbered Exercises

EXERCISES 12.7, PAGE 530

1. $x = 11$
3. $x = \frac{1}{10}$
5. $x = -3$
7. $p = \frac{4}{17}$
9. $x = \frac{11}{2}$
11. $s = -60$
13. $r = \frac{-140}{31}$
15. $x = 32$
17. $t = \frac{61}{2}$
19. $s = \frac{33}{16}$
21. $x = \frac{-11}{14}$
23. $s = \frac{14}{5}$
25. $x = \frac{-4}{11}$
27. $r = \frac{1}{3}$
29. $x = \frac{10}{7}$
31. $y = \frac{-28}{13}$
33. $s = 12$
35. $p = \frac{4}{3}$
37. $t = \frac{7}{4}$
39. $x = \frac{17}{3}$

EXERCISES 12.8, PAGE 534

1. 1 and 4
3. 3 and 9
5. 30 and 45
7. $\frac{5}{9}$
9. 13, 15, and 17
11. $27\frac{3}{11}$ min
13. $18\frac{18}{19}$ min
15. 10 yd by 15 yd
17. 12 hr
19. 6 hr

REVIEW EXERCISES, PAGE 535

1. $y = -3$
3. $x = \frac{7}{3}$
5. $s = -1$
7. $\frac{-1}{3xy^3}$ $(x \neq 0, y \neq 0)$
9. 3 $(y \neq -3)$
11. $\frac{t^2}{2rs}$ $(r \neq 0, s \neq 0, r \neq 2s)$
13. $\frac{2p - q}{p - q}$ $(p \neq -2q, q)$
15. $\frac{2y^2}{5x}$ $(x \neq 0, y \neq 0)$
17. $\frac{st + t^2}{6s^2}$ $(s \neq 0, t \neq 0, s \neq t)$
19. $\frac{3}{7a^2(b - 1)}$ $(a \neq 0, b \neq 1)$
21. $\frac{(x + 1)^2}{(x - 5)(2x + 1)}$ $\left(x \neq -1, \frac{-1}{2}, \frac{2}{3}, 1, 5\right)$
23. $\frac{5}{y + 1}$ $(y \neq -1)$
25. $\frac{2a + 3b - 1}{3x - 1}$ $\left(x \neq \frac{1}{3}\right)$
27. $\frac{7a - b}{2x - 3y}$ $(2x \neq 3y)$
29. $18y^4$
31. $10(s - 2)$
33. $(y + 3)^2(y - 3)^3$
35. $3s^5(s + 3)(s - 3)^2$
37. $\frac{5y + 4 - 16y^2}{4y^3}$ $(y \neq 0)$
39. $\frac{13t - 20}{3(t^2 - 16)}$ $(t \neq -4, 4)$
41. $\frac{8s^2 - 27s + 7}{(s - 1)(s + 2)(s - 3)}$ $(s \neq -2, 1, 3)$
43. $\frac{-2}{a^2b}$ $(a \neq 0, b \neq 0)$
45. $\frac{1}{t^2 - 1}$ $(t \neq -1, 0, 1)$
47. $\frac{s^2 + t^2}{s^2 - t^2}$ $(s \neq -t, 0, t, t \neq 0)$
49. $y = \frac{6}{7}$
51. $x = 42$
53. $t = \frac{4}{3}$
55. 2 and 10
57. $3\frac{3}{7}$ hr
59. 7 cm by 15 cm

TEASERS, PAGE 537

7. **a.** No solution, since $x = \frac{5}{2}$ is not an integer.   **b.** $x = \frac{5}{2}$

CUMULATIVE REVIEW EXERCISES, CHAPTERS 1–12, PAGE 539

1. -8
3. 0.08094
5. -3.73
7. $61,000,000$
9. 6.9879×10^8
11. $7y^3 - 2y^2 - 2y - 3$
13. $p = -1, 0, 6$
15. $x = -1, \frac{4}{5}$
17. $\frac{1}{2}(x - 3)(2y - 1)$ $\left(y \neq \frac{-1}{2}, \frac{1}{2}, x \neq -3\right)$
19. 7139.118 cm^3
21. $-\frac{2}{3} \times \frac{3}{7}$
23. $n = -60$
25. $t(2t - 1)(t + 5)$

EXERCISES 13.1, PAGE 549

1. I
3. I
5. Vertical axis
7. IV
9. Horizontal axis
11. Vertical axis
13. II
15. IV
17. II
19. Vertical axis
21. (1, 3)
23. (−7, 0)
25. (8, 0)
27. (12, 5)
29. (0, 0)
31. (−15, −6)
33. (−10, −12)
35. (6, −12)
37. (−6, 6)
39. (−8, −7)

41–59. (odd) Given in the accompanying graph.

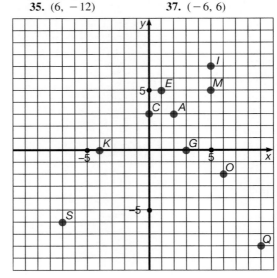

EXERCISES 13.2, PAGE 552

1. $y = 3 - x$
3. $y = \dfrac{-1}{3}x$
5. $y = \dfrac{x - 5}{2}$
7. $y = \dfrac{4x + 1}{3}$
9. $y = \dfrac{-2x - 3}{3}$
11. $y = \dfrac{-5}{3}$
13. $y = \dfrac{-1}{9}$
15. $y = \dfrac{2}{3}$
17. $y = 4$
19. $y = -2.6$
21. $x = \dfrac{4}{3}$
23. $x = 2$
25. $x = \dfrac{13}{3}$
27. a. $x = \dfrac{8}{3}$ b. $y = -13$

EXERCISES 13.3, PAGE 558

1.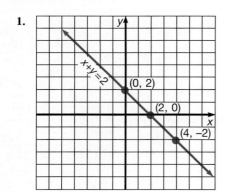

3.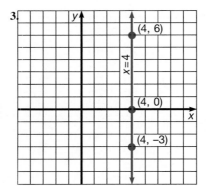

5.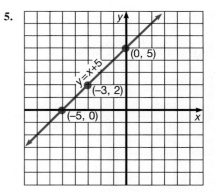

7.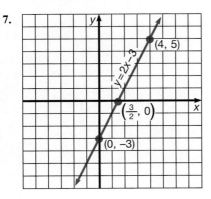

9.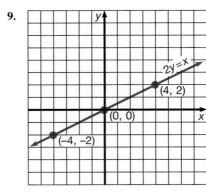

11.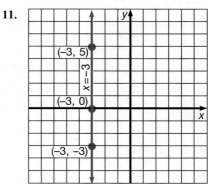

13.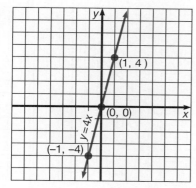
15.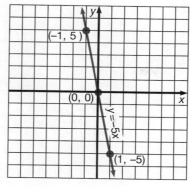
17.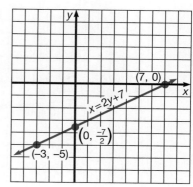
19.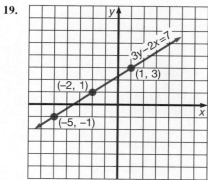

21. x-intercept = 3
y-intercept = 3

23. x-intercept = 0
y-intercept = 0

25. x-intercept = $\frac{5}{2}$
y-intercept = -5

27. x-intercept = 6
y-intercept = -2

29. x-intercept = 4
y-intercept = 2

31.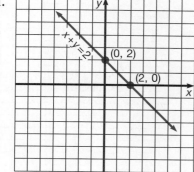
33.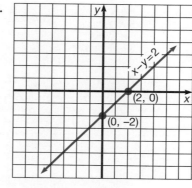
35.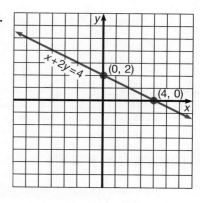
37.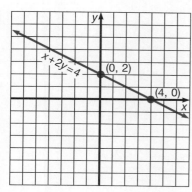
39.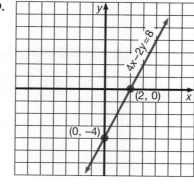
41.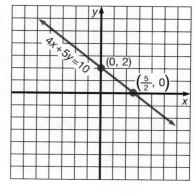

43. 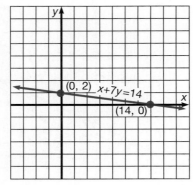   45. 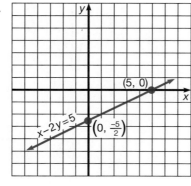   47.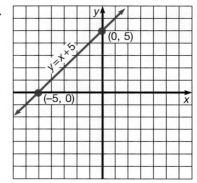

49.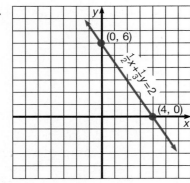

EXERCISES 13.4, PAGE 565

1. 1
3. $\dfrac{-4}{5}$
5. $\dfrac{-3}{2}$
7. $\dfrac{1}{11}$
9. $\dfrac{3}{4}$

11. $\dfrac{-3}{8}$
13. 0
15. $\dfrac{-3}{2}$
17. 0
19. -1

21. -1
23. 0
25. $\dfrac{4}{5}$
27. $\dfrac{1}{6}$
29. 4

31. Undefined
33. $\dfrac{1}{3}$
35. 2
37. 0
39. Undefined

41. Positive   43. Positive   45. Positive   47. Undefined   49. Negative
51. Positive   53. Negative   55. 0   57. Negative   59. Positive

EXERCISES 13.5, PAGE 571

1.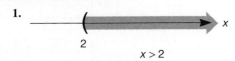
3.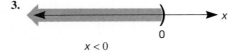

5.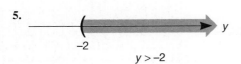
7.

9.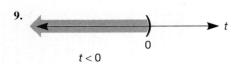

11. $y > 8$ **13.** $u < 3$ **15.** $x < \dfrac{-5}{3}$ **17.** $p > 3$ **19.** $x < 4$

21. $u \le \dfrac{6}{5}$ **23.** $x > \dfrac{23}{9}$ **25.** $t \le \dfrac{47}{20}$ **27.** $x \le \dfrac{3}{4}$ **29.** $t < \dfrac{7}{8}$

31. $y > -8$ **33.** $p < \dfrac{570}{169}$ **35.** $q \le \dfrac{30}{7}$ **37.** $u > \dfrac{83}{177}$ **39.** $x > -3$

EXERCISES 13.6, PAGE 574

1. $y < 3 - x$ **3.** $y \le \dfrac{-1}{3}x$ **5.** $y > \dfrac{x - 5}{2}$ **7.** $y \le \dfrac{4x + 1}{3}$ **9.** $y \le \dfrac{7 - 6x}{5}$

11. $y > \dfrac{4 - x}{2}$ **13.** $y > x + 3$ **15.** $y < \dfrac{5x + 26}{2}$ **17.** $y \ge \dfrac{9x - 1}{5}$ **19.** $y \le \dfrac{3x + 24}{2}$

21. Any values greater than 1   **23.** $\dfrac{-7}{3}$ **25.** No, since $8 + 6 < 0$ is not true.

EXERCISES 13.7, PAGE 581

1.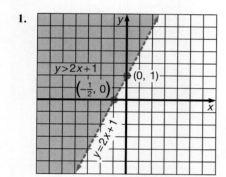

3.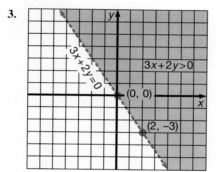

5.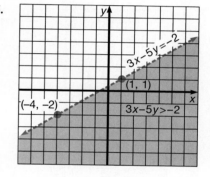

7.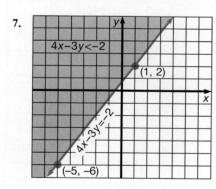

9.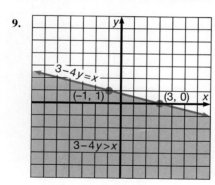

11.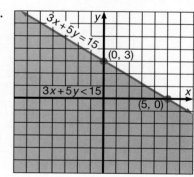

13.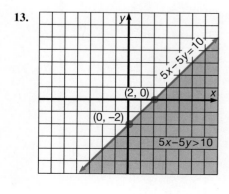

15.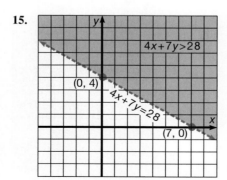

17.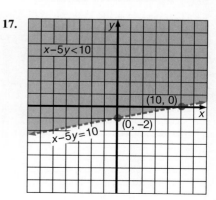

19.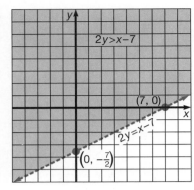

21.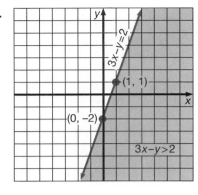

23.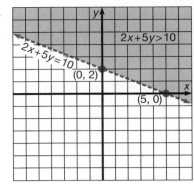

25.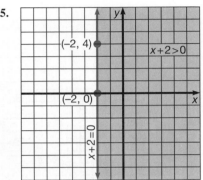

27.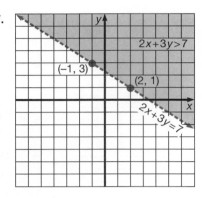

29.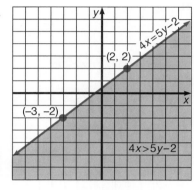

31.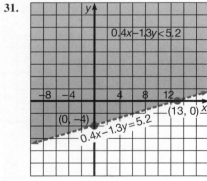

33.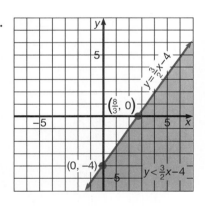

35.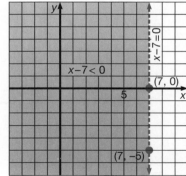

37.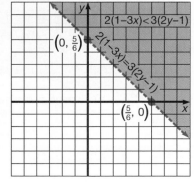

39.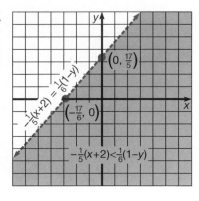

REVIEW EXERCISES, PAGE 582

1. IV **3.** Vertical axis **5.** Horizontal axis

7.–11. (odd) given in accompanying graph.

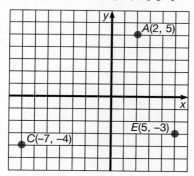

13. $y = \dfrac{4 - 2x}{3}$ **15.** $y = \dfrac{x + 17}{3}$ **17.** $y = \dfrac{-13}{4}$ **19.** Vertical **21.** Neither

23.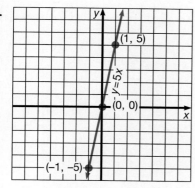

25.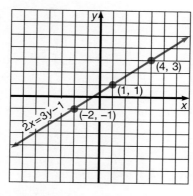

27.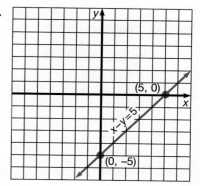

29.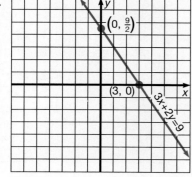

31.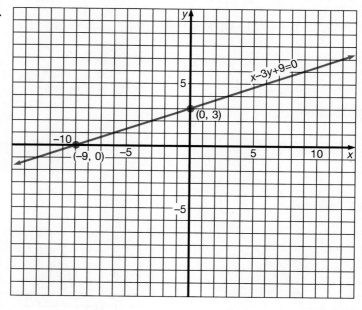

33. $\dfrac{-4}{3}$ **35.** $\dfrac{-3}{4}$ **37.** $\dfrac{-2}{9}$ **39.** Positive

41. Negative **43.** Undefined

45.

47.

49. $x < 7.1$ **51.** $p \leq \frac{12}{7}$ **53.** $y \geq \frac{3-x}{7}$ **55.** $y < \frac{9-4x}{4}$

57. No, since $-4 + 9 < 2$ is not true.

59.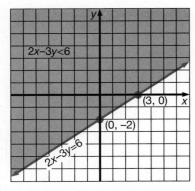

TEASERS, PAGE 584

1. No, they could be coincident.   **5.** $\frac{-5}{3}$ **7.** $y < 2$

9. u is any real number.

CUMULATIVE REVIEW EXERCISES, CHAPTERS 1–13, PAGE 586

1. -68 **3.** 77.44 **5.** 23.06 **7.** 9.286

9. $15y - 32x$ **11.** $3p(2p - 1)(p + 3)$ **13.** $u = 2, 3$ **15.** $t = \frac{-7}{2}, -3, 3$

17. $16\frac{3}{4}$ ft   **19.** No, since $16 + 6 < 7$ is not true.   **21.** $p = -5.1$

23.    **25.** $295.15

EXERCISES 14.1, PAGE 597

1. 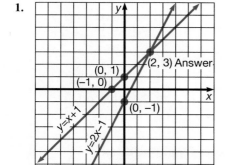   **3.** 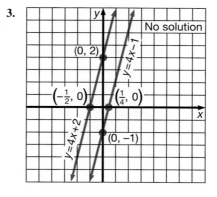   **5.**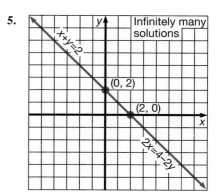

Answers to Odd-Numbered Exercises

7.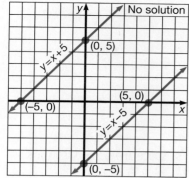

9.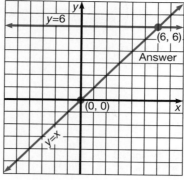

11.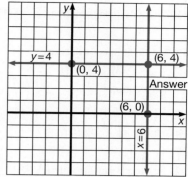

13.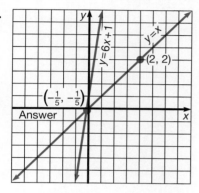

15.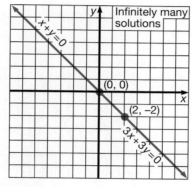

17.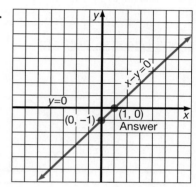

19.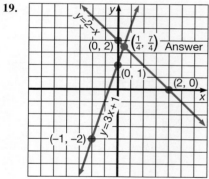

21. $y = -x + 9$ **23.** $y = \dfrac{1}{3}$ **25.** $y = \dfrac{-1}{4}x$ **27.** $y = \dfrac{-1}{5}x + 2$ **29.** $y = 6x + 1$

31. Intersecting   **33.** Intersecting   **35.** Parallel   **37.** Intersecting   **39.** Intersecting

EXERCISES 14.2, PAGE 605

1. $(-2, -3)$ **3.** No solution   **5.** No solution   **7.** $(-3, 4)$ **9.** $\left(\dfrac{1}{8}, \dfrac{1}{8}\right)$

11. $\left(\dfrac{11}{29}, \dfrac{5}{29}\right)$ **13.** $\left(\dfrac{1}{19}, \dfrac{-26}{19}\right)$ **15.** $\left(\dfrac{4}{3}, \dfrac{-1}{3}\right)$ **17.** $\left(\dfrac{1}{5}, \dfrac{2}{5}\right)$ **19.** $\left(\dfrac{19}{7}, \dfrac{3}{7}\right)$

21. $\left(\dfrac{96}{23}, \dfrac{-120}{23}\right)$ **23.** $\left(\dfrac{9}{8}, \dfrac{1}{8}\right)$ **25.** $\left(\dfrac{29}{7}, \dfrac{-20}{7}\right)$ **27.** $\left(\dfrac{11}{12}, \dfrac{-7}{12}\right)$ **29.** $\left(\dfrac{12}{13}, \dfrac{-3}{13}\right)$

EXERCISES 14.3, PAGE 609

1. $\left(\dfrac{1}{3}, \dfrac{5}{3}\right)$ **3.** $\left(\dfrac{9}{11}, \dfrac{-7}{11}\right)$ **5.** $\left(\dfrac{2}{7}, \dfrac{3}{7}\right)$ **7.** $\left(\dfrac{22}{17}, \dfrac{3}{17}\right)$ **9.** $\left(\dfrac{22}{13}, \dfrac{7}{13}\right)$

11. Infinitely many solutions; $y = 2x - 2$ **13.** $\left(\dfrac{16}{15}, \dfrac{-11}{30}\right)$ **15.** $\left(\dfrac{11}{5}, \dfrac{2}{15}\right)$ **17.** $\left(\dfrac{-18}{7}, \dfrac{27}{7}\right)$

19. Infinitely many solutions; $y = \dfrac{5x - 2}{6}$ **21.** (5, 4)   **23.** (−7, 2)   **25.** (3, 4)

27. No solution   **29.** $\left(\dfrac{17}{13}, \dfrac{-7}{13}\right)$

EXERCISES 14.4, PAGE 615

1. 11 and 6   **3.** 19 and 20   **5.** $9000 at 7% and $8000 at 8.5%
7. 59 nickels and 26 dimes   **9.** 59 and 87   **11.** 42 m by 49 m
13. 12 and 14   **15.** 13 and 20   **17.** 239 and 573
19. −3 and −12   **21.** 46   **23.** 26

REVIEW EXERCISES, PAGE 616

1. 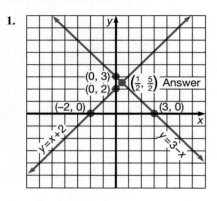   **3.**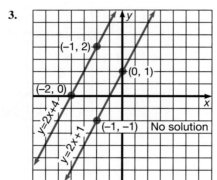

5. (1, −1)   **7.** $\left(\dfrac{28}{17}, \dfrac{13}{17}\right)$ **9.** $\left(\dfrac{2}{7}, \dfrac{1}{7}\right)$ **11.** $\left(\dfrac{16}{17}, \dfrac{-5}{17}\right)$

13. 57 and 96   **15.** $.60 and $1.60   **17.** 26

TEASERS, PAGE 618

1. No   **3.** $\left(\dfrac{31}{47}, \dfrac{-3}{47}\right)$ **5.** $\left(\dfrac{13}{7}, \dfrac{-16}{7}\right)$ **7.** $a = \dfrac{1}{3}, b = \dfrac{4}{3}$

9. $a = \dfrac{x - y}{x^2 + y^2}, b = \dfrac{-(x + y)}{x^2 + y^2}$

CUMULATIVE REVIEW EXERCISES, CHAPTERS 1–14, PAGE 620

1. $\dfrac{81}{140}$ **3.** −301   **5.** −8

7. 210,000   **9.** $\dfrac{-2y^2}{3x^3z^4}$ $(x \neq 0, y \neq 0, z \neq 0)$ **11.** $2p^3 + 3p^2 - 10p - 15$

13. $\dfrac{-2}{5q(q - 2)}$ $(q \neq -2, 0, 2)$ **15.** $y = \dfrac{-24}{29}$ **17.** $p \geq \dfrac{3 + 3q}{5}$

19. $u = -3, \dfrac{-5}{2}, 3$ **21.** (2, −1)   **23.** 396,000,000

25. Not a solution

EXERCISES 15.1, PAGE 626

1. 25   **3.** 100   **5.** 441   **7.** $9y^2$ **9.** $(y - 3)^2$
11. $4(x - 4)^2$ **13.** x^4y^2 **15.** u^6 **17.** 9   **19.** 13

Answers to Odd-Numbered Exercises

21. 15 **23.** a **25.** $9c^2$ **27.** $4(2a+1)$ **29.** $4a(b+1)$
31. 6 **33.** Meaningless; negative radicand **35.** 1 **37.** -1
39. x **41.** Meaningless **43.** Meaningless **45.** Meaningless
47. $7y^2$ **49.** Meaningless

EXERCISES 15.2, PAGE 631

1. $2\sqrt{2}$ **3.** $3\sqrt{2}$ **5.** $2\sqrt{6}$ **7.** $2\sqrt{7}$ **9.** $2\sqrt{10}$
11. $3\sqrt{5}$ **13.** $5\sqrt{2}$ **15.** $3\sqrt{6}$ **17.** $2\sqrt{15}$ **19.** $2\sqrt{17}$
21. $5\sqrt{3}$ **23.** $4\sqrt{5}$ **25.** $2\sqrt{22}$ **27.** $2\sqrt{23}$ **29.** $3\sqrt{11}$
31. $2\sqrt{26}$ **33.** $10\sqrt{2}$ **35.** $5\sqrt{11}$ **37.** $20\sqrt{2}$ **39.** $8\sqrt{15}$
41. $x\sqrt{x}$ **43.** $u^2\sqrt{u}$ **45.** t^4 **47.** $p^5\sqrt{p}$ **49.** x^7
51. $z^8\sqrt{z}$ **53.** s^{10} **55.** $w^{11}\sqrt{w}$ **57.** $2x$ **59.** $4z^3$
61. $3y\sqrt{y}$ **63.** $7t^3\sqrt{t}$ **65.** $2x^2y\sqrt{y}$ **67.** $x^2y^4\sqrt{3}$ **69.** $4n\sqrt{3mn}$
71. $2y\sqrt{2xy}$ **73.** $11x^2y^3\sqrt{x}$ **75.** $5t^3\sqrt{10t}$ **77.** $xy^3\sqrt{2a}$ **79.** $3pq^2\sqrt{10ps}$

EXERCISES 15.3, PAGE 634

1. $\dfrac{2}{5}$ **3.** $\dfrac{6}{5}$ **5.** $\dfrac{5}{8}$ **7.** $\dfrac{9}{4}$ **9.** $\dfrac{11}{13}$
11. $\dfrac{11}{19}$ **13.** $\dfrac{x}{2}$ **15.** $\dfrac{u^3}{7}$ **17.** $\dfrac{2z^2}{3}$ **19.** $\dfrac{2}{y^2}$
21. $\dfrac{x}{y^2}$ **23.** $\dfrac{2x}{y^2}$ **25.** $\dfrac{x^2y^3}{z}$ **27.** $\dfrac{2ab^2}{c^2d}$ **29.** $\dfrac{11u^3}{v^2w^4}$
31. $\dfrac{4m}{5n^2p^3}$ **33.** $\dfrac{p\sqrt{7p}}{st^2}$ **35.** $\dfrac{2a}{y}$ **37.** $\dfrac{st}{4}$ **39.** $\dfrac{y\sqrt{7z}}{x}$
41. $\dfrac{3\sqrt{2}}{2}$ **43.** $\dfrac{8\sqrt{5}}{5}$ **45.** $\dfrac{9\sqrt{7}}{7}$ **47.** $\dfrac{y\sqrt{10}}{10}$ **49.** $\dfrac{6\sqrt{y}}{y}$
51. $\dfrac{\sqrt{3y}}{3y}$ **53.** $\dfrac{\sqrt{6}}{2}$ **55.** $\dfrac{\sqrt{35}}{7}$ **57.** $\dfrac{\sqrt{6x}}{3}$ **59.** $\sqrt{2}$
61. $\dfrac{\sqrt{x}}{x^2}$ **63.** $\dfrac{3\sqrt{z}}{z^4}$ **65.** $\dfrac{\sqrt{xy}}{x}$ **67.** $\dfrac{\sqrt{ab}}{b}$ **69.** $\dfrac{\sqrt{y}}{3}$
71. $\dfrac{7\sqrt{5y}}{5y^4}$ **73.** $\dfrac{x\sqrt{xy}}{2y^2}$ **75.** $\dfrac{\sqrt{10y}}{2y^2}$ **77.** $4x$ **79.** $\dfrac{cd\sqrt{5c}}{3}$

EXERCISES 15.4, PAGE 638

1. $7\sqrt{3}$ **3.** $-\sqrt{5}$ **5.** $2\sqrt{7}$ **7.** $5\sqrt{x}$ **9.** $3\sqrt{2u}$
11. $(a+b)\sqrt{p-q}$ **13.** $8\sqrt{3}$ **15.** $-24\sqrt{5}$ **17.** $-9\sqrt{2}$ **19.** $5\sqrt{3}$
21. $50\sqrt{2}$ **23.** $\dfrac{14\sqrt{6}}{3}$ **25.** $9\sqrt{2x}$ **27.** $-7\sqrt{2b}$ **29.** $13\sqrt{2b}$
31. $(3a-7ab)\sqrt{b}$ **33.** $\dfrac{(a-b-2ab)\sqrt{ab}}{ab}$ **35.** $\dfrac{(7-2a)\sqrt{2a}}{2}$
37. $\dfrac{(a+4y+5abx)\sqrt{2xy}}{a}$ **39.** $(a-bcx+cy)\sqrt{abc}$ **41.** $(6ab-6b^2)\sqrt{2b}$
43. $(4bcx^2y-2axy^2)\sqrt{ac}$ **45.** $(10xy^3-28x^3)\sqrt{2y}$ **47.** $(0.6yz^2-1.2xz)\sqrt{xy}$
49. $\left(12pqx+\dfrac{1}{6}py-\dfrac{2}{15}qz\right)\sqrt{2q}$

EXERCISES 15.5, PAGE 644

1. $\sqrt{6}$ **3.** $2\sqrt{6}$ **5.** 7 **7.** 30 **9.** 100
11. $-6\sqrt{6}$ **13.** -24 **15.** -36 **17.** $\sqrt{6}-\sqrt{10}$ **19.** $84-3\sqrt{14}$

21. $1 + \sqrt{6}$ 23. $4\sqrt{3} - 3\sqrt{6}$ 25. $x\sqrt{6}$ 27. $2\sqrt{2xy}$ 29. $2a - 3\sqrt{ab}$
31. $a - b$ 33. $2x^2y^3\sqrt{6x}$ 35. $2s^2\sqrt{3s}$ 37. $\dfrac{\sqrt{b}}{ac}$
39. $a^3b - ab^3 + (ab^2 - a^2b)\sqrt{ab}$
41. $\dfrac{5}{2}$ 43. $\dfrac{6}{5}$ 45. $\dfrac{15}{2}$ 47. 14   49. $-2\sqrt{5}$
51. 1   53. $\dfrac{1}{b}$ 55. $\dfrac{y\sqrt{2}}{2}$ 57. $\dfrac{q^3\sqrt{p}}{p}$ 59. -7
61. $2\sqrt{2} + 6\sqrt{3}$ 63. $\dfrac{7\sqrt{3}}{2} - \dfrac{9\sqrt{2}}{2}$ 65. $\dfrac{2a}{c} + \dfrac{b\sqrt{6}}{c}$ 67. $\dfrac{x\sqrt{10}}{5} - \dfrac{y\sqrt{15}}{5}$ 69. $\dfrac{-3\sqrt{10}}{10}$

EXERCISES 15.6, PAGE 648

1. $x = 9$ 3. $u = 16$ 5. $x = 26$ 7. $u = \dfrac{1}{2}$ 9. $x = \dfrac{5}{3}$
11. $x = 0$ 13. $u = \dfrac{1}{12}$ 15. $u = \dfrac{-2}{7}$ 17. No solution   19. No solution
21. $a = 102$ 23. $y = \dfrac{-27}{8}$ 25. $x = \dfrac{17}{60}$ 27. $p = \dfrac{57}{46}$ 29. $x = \dfrac{34}{59}$

REVIEW EXERCISES, PAGE 649

1. 7   3. x^3 5. $y + 3$ 7. $(abc)^4$ 9. 9
11. Meaningless   13. $-y^2$ 15. $3\sqrt{2}$ 17. $2\sqrt{70}$ 19. $3y^3$
21. $10a^2b^2\sqrt{b}$ 23. $\dfrac{5}{9}$ 25. $\dfrac{\sqrt{35}}{7}$ 27. $\dfrac{3y\sqrt{x}}{5}$ 29. $\dfrac{\sqrt{6ab}}{3b}$
31. $6\sqrt{5}$ 33. $(2a - 3b + 4c)\sqrt{2}$ 35. $(a^2 - 2ab^2 - 5ab)\sqrt{b}$
37. $-48\sqrt{3}$ 39. $3\sqrt{30} + 30$ 41. 15   43. 0   45. $x = \dfrac{25}{3}$
47. No solution   49. No solution

TEASERS, PAGE 650

1. $a = 0$ or $b = 0$ 3. a. 4   b. $-u$ 5. $a\sqrt{ab}$ 7. $u = \dfrac{-2}{7}$ 9. No solution

CUMULATIVE REVIEW EXERCISES, CHAPTERS 1–15, PAGE 652

1. Distributive property for multiplication over addition   3. Addition and subtraction
5. $2^2 \times 3 \times 41$ 7. 43,200   9. $\dfrac{553}{50}$ 11. $\dfrac{-57}{5}$
13. $(2u - 7)(u + 3)$ 15. $(2t + 3)(t^2 - 2)$ 17. $\dfrac{4a^2b}{3}$ $(a > 0, b > 0)$ 19. $s = -2, 0, \dfrac{3}{2}$
21. 29.0   23. $s = -33$ 25. $y \leq \dfrac{2x - 4}{3}$

EXERCISES 16.1, PAGE 657

1. $3x^2 - 2x - 4 = 0$ 3. $y^2 - y + 4 = 0$ 5. $u^2 + 2u - 3 = 0$ 7. $6r^2 + r - 2 = 0$
9. $t^2 - 4t = 0$ 11. $5p^2 - 6p + 6 = 0$ 13. $6x^2 - 4x + 3 = 0$ 15. $3y^2 + 6y - 10 = 0$
17. $12u^2 + 5u - 23 = 0$ 19. $39s^2 + 14s - 53 = 0$

EXERCISES 16.2, PAGE 661

1. $x = \frac{-3}{2}, 1$
3. $x = -9, 9$
5. $y = \frac{-5}{3}, 0$
7. $p = 0, \frac{20}{3}$
9. $y = 3, 4$
11. $t = \frac{3}{2}, 6$
13. $r = 0, 10$
15. $x = \frac{-5}{2}, 7$
17. $p = 0, 4$
19. $v = \frac{-1}{2}, 0$
21. $y = -2, 2$
23. $u = -3, 3$
25. $t = 0, 9$
27. $t = -2, 2$
29. $p = 0, \frac{6}{7}$

EXERCISES 16.3, PAGE 665

1. $x = \pm 9$
3. $u = \pm 5$
5. $u = \pm 2\sqrt{2}$
7. $x = 0, 6$
9. $v = -4, 14$
11. $p = 0, 1$
13. $t = -2, \frac{-1}{2}$
15. $x = 3 \pm \sqrt{3}$
17. $u = 1 \pm \sqrt{3}$
19. $y = -6 \pm \sqrt{5}$
21. $x = -4, -2$
23. $u = 6, 12$
25. $p = 2 \pm \sqrt{11}$
27. $t = -2, 3$
29. $r = \frac{2 \pm \sqrt{3}}{5}$

EXERCISES 16.4, PAGE 670

1. $x = 1, 5$
3. $w = -2, 3$
5. $x = -1 \pm \sqrt{6}$
7. $u = -12, -2$
9. $y = 2, 8$
11. $u = -1, \frac{3}{8}$
13. $t = \frac{-2}{3}, \frac{3}{2}$
15. $r = \frac{+2}{3}, \frac{-3}{2}$
17. $s = \frac{1}{2}, 3$
19. $t = \frac{-5}{2}$
21. $x = 3 \pm \sqrt{5}$
23. $u = -5 \pm \sqrt{6}$
25. $t = \frac{-5 \pm \sqrt{11}}{3}$
27. $q = \frac{5 \pm \sqrt{11}}{4}$
29. $s = 11 \pm \sqrt{11}$

EXERCISES 16.5, PAGE 675

1. 1, exist
3. -80, not real
5. 8, exist
7. 81, exist
9. -47, not real
11. 320, exist
13. -87, not real
15. 100, exist
17. -135, not real
19. 61, exist
21. $x = 2, 3$
23. $x = \frac{-5 \pm \sqrt{109}}{14}$
25. $y = \frac{-3 \pm \sqrt{3}}{2}$
27. $y = -1, 3$
29. $u = \frac{-2}{3}, 1$
31. $u = 2 \pm \sqrt{3}$
33. $s = -3, \frac{1}{2}$
35. $s = \frac{-3}{4}, \frac{4}{3}$
37. $t = \frac{7 \pm \sqrt{19}}{2}$
39. $t = 1 \pm \sqrt{2}$

EXERCISES 16.6, PAGE 679

1. 17 and 19 or -19 and -17
3. 27 and 29
5. 20 or -2
7. 12 yd by 17 yd
9. Base = 26 in., height = 13 in.
11. 3
13. 13
15. 3.75 sec
17. 6
19. $\frac{1}{2}$
21. 10 ft
23. 2 in
25. 12
27. 2 yd
29. 4

EXERCISES 16.7, PAGE 688

1. $\sqrt{13} \approx 3.6$ ft
3. $\sqrt{73} \approx 8.5$ cm
5. $\sqrt{346} \approx 18.6$ in.
7. $\sqrt{30.77} \approx 5.5$ yd
9. $\sqrt{51} \approx 7.1$ in.
11. $\sqrt{35.84} \approx 6.0$ ft
13. $2\sqrt{22} \approx 9.4$ cm
15. $\sqrt{187.92} \approx 13.7$ ft
17. $\sqrt{153} \approx 12.4$ ft
19. $2\sqrt{165.11} \approx 25.7$ in.
21. Yes
23. Yes
25. No
27. Yes
29. No
31. $\sqrt{287} \approx 17$ ft
33. 13 mi
35. 125 mi
37. $90\sqrt{2} \approx 127$ ft
39. No

REVIEW EXERCISES, PAGE 690

1. $2x^2 - 5x - 4 = 0$ **3.** $2y^2 + 2y + 3 = 0$ **5.** $6u^2 + u + 1 = 0$ **7.** $x = -4, \frac{3}{2}$

9. $y = \frac{-4}{3}, \frac{7}{2}$ **11.** $t = \frac{-4}{3}, \frac{11}{2}$ **13.** $y = -8, 8$ **15.** $x = 1, 7$

17. $w = -8, 2$ **19.** $x = \frac{-5 \pm \sqrt{33}}{2}$ **21.** $u = \frac{-1 \pm \sqrt{65}}{2}$ **23.** $t = \frac{-1 \pm \sqrt{73}}{6}$

25. 153, exist **27.** 277, exist **29.** $x = \frac{-2}{3}, \frac{1}{2}$ **31.** $y = \frac{5 \pm \sqrt{61}}{2}$

33. $u = \frac{-3}{7}, 1$ **35.** 16 and 19 or -19 and -16

37. Height = 11 yd, base = 20 yd **39.** $2\sqrt{34} \approx 11.7$ in.

41. 12 ft **43.** Yes **45.** Yes

TEASERS, PAGE 692

1. 12 and 13, or -13 and -12 **3.** $x = -2; k = -14$ **5.** 12 cm by 15 cm

7. $\frac{1}{14}$ or 7 **9. a.** 15 m **b.** 54 m^2

CUMULATIVE REVIEW EXERCISES, CHAPTERS 1–16, PAGE 694

1. Subtraction and division **3.** 3^2 **5.** 2.53 **7.** $x = -1$

9. $u = -1, \frac{5}{2}$ **11.** $\frac{25}{4}$ **13.** $\frac{-5}{3}$ **15.** $y = \frac{-4}{5}x + \frac{7}{5}$

17. $-4a - 21b$ **19.** $x = -3, -2, \frac{3}{2}$ **21.** No **23.** $11\sqrt{3} - 9\sqrt{2}$

25. No

EXERCISES 17.1, PAGE 702

1. False **3.** False **5.** True **7.** True **9.** False
11. Infinitely many **13.** A line **15.** One **17.** Yes **19.** 9
21. 8 **23.** $\overline{BE}$ and $\overline{DF}$ **25.** 1 **27.** Infinitely many **29.** 4.5

EXERCISES 17.2, PAGE 706

1. $\angle BOF, \angle FOD, \angle AOE, \angle EOC$ **3.** $\angle BOC, \angle AOC, \angle AOD, \angle BOD$ **5.** $\angle HOJ, \angle GOK$
7. $\angle KOJ$ **9.** $\angle GOK, \angle GOJ, \angle GOH$ **11.** 20°
13. 85° **15.** 25° **17.** 65°
19. 40° **21.** 90°

23. 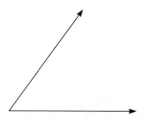   **25.** 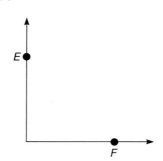   **27.**
<pre>
◄————●————●————●————►
      A   0   B
</pre>

29. 42°

Answers to Odd-Numbered Exercises

EXERCISES 17.3, PAGE 711

1. $\angle AOB$ and $\angle BOC$; $\angle BOC$ and $\angle COD$; $\angle COD$ and $\angle AOD$; $\angle AOD$ and $\angle AOB$
3. $\angle AOB$ and $\angle BOC$; $\angle BOC$ and $\angle COD$; $\angle COD$ and $\angle AOD$; $\angle AOD$ and $\angle AOB$
5. $110°$ 7. $71°$ 9. $82°$ 11. $(90 - 2y)°$ 13. $(89 - 3u)°$
15. $(110 - p)°$ 17. $143°$ 19. $127°$ 21. $(200 - x)°$ 23. $\left(177 - \dfrac{2}{5}p\right)°$
25. $(393 - 5q)°$ 27. $43°$ 29. $90°$ 31. $133°$ 33. $180°$
35. $90°$ 37. $77°$ 39. $70°$

EXERCISES 17.4, PAGE 716

1. Angles 2, 4, 5, 7, 10, 12, 13 and 15
3. Angles 2 and 5
5. Angles 2, 4, 7, 10, 12, 13, and 15
7. Angles 5, 6, 9, and 10
9. Angles 3, 4, 15, and 16
11. Angles 1 and 9, 2 and 10, 5 and 13, 6 and 14
13. Angles 2 and 7, 3 and 6
15. Angles 1 and 6, 2 and 5, 3 and 8, 4 and 7, 9 and 14, 10 and 13, 11 and 16, 12 and 15
17. Angles 1, 3, 6, 8, 9, 11, 14, and 16
19. 50 21. 150 23. 55 25. 65 27. 64
29. $38\dfrac{2}{3}$ 31. $t \perp q$ 33. $90°$ 35. Angles 3 and 6, 4 and 5
37. Angles 1, 2, 3, 4, 6, 7, and 8   39. $p \parallel r$

EXERCISES 17.5, PAGE 724

1. Scalene   3. Isosceles   5. Equilateral   7. Obtuse   9. Obtuse
11. $66°$ 13. $30°$ 15. $84°$ 17. $(180 - 3x)°$ 19. $(179 - 3y)°$
21. $56°$ 23. $38°$ 25. $54°$ 27. $\left(90 - \dfrac{1}{2}x\right)°$ 29. $(91 - 2u)°$
31. $98°$ 33. $122°$ 35. $6°$ 37. $(180 - 2y)°$ 39. $(182 - 8u)°$
41. $63°$ 43. $47°$ 45. $81°$ 47. True   49. True
51. False   53. False   55. $54°, 54°, 72°$ 57. $46°, 67°, 67°$ 59. $55°$ and $61°$

EXERCISES 17.6, PAGE 729

1. False   3. False   5. True   7. True   9. True
11. $x = 78, y = 40$ 13. $x = 33, y = 9$ yd   15. $x = 8$ ft, $y = 15$ ft   17. Not similar   19. Similar

EXERCISES 17.7, PAGE 737

1. $m(\angle C) = 30°$, $AB = 4$ ft, $AC = 4\sqrt{3}$ ft
3. $m(\angle J) = 45°$, $GH = 7$ in., $GJ = 7\sqrt{2}$ in.
5. $(18 + 6\sqrt{3})$ yd
7. $x = 4\sqrt{3}$ ft, $y = 8$ ft
9. $x = 2.5$ in., $y = 2.5\sqrt{2}$ in.
11. $x = 12\sqrt{2}$ ft, $y = 4\sqrt{6}$ ft
13. $x = 4\sqrt{3}$ ft, $y = 8$ ft
15. $x = 16\sqrt{2}$ cm, $y = 16$ cm
17. $12\sqrt{2}$ yd   19. 28 in.   21. 10 cm   23. $135°$
25. $(20 + 10\sqrt{2})$ cm   27. 8 ft   29. $(12 + 4\sqrt{3})$ ft   31. 13 ft
33. $6\sqrt{2}$ ft   35. $90\sqrt{2} \approx 127$ ft   37. 8 ft   39. 18.5 m

EXERCISES 17.8, PAGE 744

1. Yes, SAS   3. Yes, ASA   5. Not enough information given
7. Yes, ASA   9. Yes, SAS   11. Yes, ASA   13. Yes, SSS or SAS
15. $\angle PQR$ and $\angle QRS$, $\angle PRQ$ and $\angle SQR$, $\angle QPR$ and $\angle QSR$, $\angle SPR$ and $\angle PRQ$, $\angle PSQ$ and $\angle SQR$, $\angle PTQ$ and $\angle STR$, $\angle PTS$ and $\angle QTR$, $\angle PQT$ and $\angle SRT$
17. Yes, ASA   19. Yes, SAS   21. ASA
25. Yes, corresponding parts of congruent triangles XYW and UVW
29. Yes, ASA or SSS

REVIEW EXERCISES, PAGE 748

1. $\overline{CE}, \overline{DF}$
3. The point with coordinate 0
5. 2
7. Right
9. Straight
11. 33°
13. 45°
15. $\left(154\frac{2}{7}\right)°$
17. Angles 2, 5, 13
19. Angles 1 and 14, 2 and 13
21. $n \perp j$
23. 60°
25. 39°
27. Yes
29. Not similar
31. Not similar
33. $m(\angle E) = 90°, DE = EF = 8$ cm
35. $20\sqrt{2}$ cm
37. Yes, SAS or ASA

TEASERS, PAGE 750

1. $\left(\dfrac{181 - 2x}{2}\right)°$
3. $(2x + x\sqrt{2})$ cm
5. $795
7. Two half lines and the point
9. No, they could be vertical angles.

CUMULATIVE REVIEW EXERCISES, CHAPTERS 1–17, PAGE 752

1. 2^{-3}
3. 2
5. $p = 2, 3$
7. $y = -2x + \dfrac{5}{3}$
9. $x - 25y$
11. $\dfrac{a^6}{4b^6c^6}$ $(a \neq 0, b \neq 0, c \neq 0)$
13. $a = \dfrac{1}{5}$
15. $9\sqrt{2} - 18\sqrt{3}$
17. 81 yd²
19. $\dfrac{256\pi}{3}$ ft³
21. $m(\angle B) = 90°, m(\overline{BC}) = 9$ yd
23. 90°
25. SSS, SAS, and ASA

Answers to Chapter Pretests

PRETEST FOR CHAPTER 1, PAGE 2

1. Ten thousands
2. Eight million, sixty-two thousand, seven hundred ninety
3. 15,607,891
4. 301
5. 187
6. 89
7. 1169
8. 1288
9. 215
10. 573
11. 2228
12. 4902
13. 45,872
14. 24
15. 45 R32
16. 346,000
17. 4,880,000
18. 5400
19. 23
20. 6
21. 17
22. 0
23. Not possible, division by 0.

PRETEST FOR CHAPTER 2, PAGE 56

1. Eleven-fifteenths
2. $\frac{11}{16}$
3. $7\frac{17}{22}$
4. $\frac{62}{5}$
5. $\frac{74}{90}$
6. Not equivalent
7. $\frac{32}{12}$
8. $\frac{9}{4}$
9. 2^7
10. 240
11. $\frac{4}{9}$
12. $1\frac{1}{29}$
13. $1\frac{2}{105}$
14. $\frac{309}{319}$
15. $\frac{663}{713}$
16. $\frac{388}{693}$
17. $\frac{3}{8}$
18. $\frac{5}{13}$
19. $\frac{23}{42}$
20. $7\frac{19}{42}$
21. $19\frac{37}{55}$
22. $6\frac{5}{14}$
23. $7\frac{34}{45}$
24. $\frac{5}{21}$
25. $\frac{1}{6}$
26. $14\frac{5}{21}$
27. $\frac{13}{17}$
28. $\frac{10}{9}$
29. $1\frac{1}{2}$
30. $\frac{5}{33}$

PRETEST FOR CHAPTER 3, PAGE 122

1. Thousandths
2. One hundred twenty-three and four hundred sixty-nine thousandths
3. 0.23
4. 0.6; 0.62; 0.819; 0.901
5. 2.563
6. 28.136
7. 39.571
8. 62.915
9. 87.37
10. 97
11. 3.6504
12. 23.6462
13. 4.3758
14. 19.4
15. 0.51
16. 2482.9
17. $23\frac{6}{25}$
18. 0.136
19. 6.1
20. 17.7

PRETEST FOR CHAPTER 4, PAGE 160

1. 175%
2. $\frac{41}{20}$
3. $\frac{69}{500}$
4. 37%
5. 700%
6. 0.0196
7. 125.6%
8. 22.568
9. 25%
10. 75%
11. 300%
12. 20%
13. $302.58
14. $2.25
15. $1139.05
16. $154.50
17. $211.05
18. $51

793

PRETEST FOR CHAPTER 5, PAGE 208

1. $9\frac{1}{3}$ yd
2. 1362 g
3. 11,520 sec
4. 270 mm
5. 1.08 kg
6. 18.0 cm
7. 30.9 ft
8. 29.8 dm
9. 24.492 yd
10. 53.6 cm
11. 77.49 ft² or 8.61 yd²
12. $572\frac{11}{14}$ cm²
13. 193.21 dm²
14. 28 ft²
15. 17.5 ft²
16. 184 cm²
17. 12.852 cm³
18. 6.859 yd³
19. 192π cm³
20. $\frac{1372\pi}{3}$ in.³

PRETEST FOR CHAPTER 6, PAGE 258

1. 16
2. 2
3. -9
4. $\frac{23}{-7}$ or $\frac{-23}{7}$
5. $\frac{17}{36}$
6. -1
7. $-1\frac{7}{12}$
8. -2.522
9. -8
10. 5.95
11. $14\frac{11}{40}$
12. 176
13. 10.08
14. 3
15. $-2\frac{11}{102}$
16. $\frac{-8}{3}$
17. -30
18. -10
19. $\frac{13}{35}$
20. $914

PRETEST FOR CHAPTER 7, PAGE 306

1. Terms
2. Product
3. -2
4. x
5. -12
6. -10.8
7. $3x - y$
8. $-3a + 8b$
9. $-4a - 4b$
10. $7 + 3p$
11. $\frac{1}{2}(q + 9)$
12. $2x - 4$
13. $x + y^2$
14. $(3y + 7) - 10$
15. $(x + 2)^2 + (x^3 \div 3)$

PRETEST FOR CHAPTER 8, PAGE 332

1. $x = -7$
2. $x = -0.7$
3. $y = \frac{1}{6}$
4. $w = 3.59$
5. $p = \frac{-17}{3}$
6. $w = -0.48$
7. $y = \frac{-1}{6}$
8. $u = \frac{-4}{21}$
9. $x = \frac{-2}{3}$
10. $u = 6.4$
11. $y = 3$
12. $w = \frac{37}{55}$
13. 17, 18, and 19
14. 51 m by 99 m
15. 2 hr
16. 18 yr and 36 yr
17. $L = \frac{P - 2W}{2}$; 16 m
18. $W = \frac{2R - T}{3S^2}$; 3
19. $\frac{12}{52}$
20. $n = \frac{7}{5}$

PRETEST FOR CHAPTER 9, PAGE 392

1. x^9
2. y^6
3. $\frac{-1}{27}$
4. $\frac{1}{2}p^4$ ($p \neq 0$)
5. $\frac{-3p^2}{q^3}$ ($p \neq 0, q \neq 0$)
6. q^3 ($q \neq 0$)
7. x^6y^9
8. $16p^{12}q^8$
9. $24s^9$
10. $-432s^{13}t^{14}$
11. $\frac{1}{4}x^8y^4$ ($x \neq 0, y \neq 0$)
12. $\frac{5}{8m^3n}$ ($m \neq 0, n \neq 0$)
13. $\frac{1}{a^4}$ ($a \neq 0$)
14. $\frac{1}{p^5}$ ($p \neq 0$)
15. 1 ($x \neq 0, y \neq 0, z \neq 0$)
16. $\frac{2r^6t^3}{3s}$ ($r \neq 0, s \neq 0, t \neq 0$)
17. $\frac{-3v^4}{u^2}$ ($u \neq 0, v \neq 0, w \neq 0$)
18. $\frac{a^4b^6}{36c^{10}}$ ($a \neq 0, b \neq 0, c \neq 0$)

Answers to Chapter Pretests

19. 3.62×10^1 **20.** 9.7×10^{-3} **21.** 2.76×10^0 **22.** 3.962×10^{-1} **23.** 29,000
24. 0.0096 **25.** 0.0001063 **26.** 39.6 **27.** 0.043

PRETEST FOR CHAPTER 10, PAGE 420

1. 7
2. 6
3. $3y^4 - 4y^3 + y^2 - 6y + 5$
4. $p^4q^2 - 4p^3q^2 + 2p^2q - 3pq^2$
5. $-4y^3 + 2y^2 + 6y + 8$
6. $2a^3b^2 - ab^3 - 2a^2b^2 + 6a^2b^3$
7. $-u^3 + 11u^2 + 12u - 15$
8. $x^3 + 9x^2 - 12x - 5$
9. $9p^5 - 4p^4 - 4p^3 - 3p^2 - 7p - 3$
10. $11uv - 4uv^2 + 4u^2v$
11. $-6x^3y^2 + 3x^3y^4$
12. $2u^3 + 3u^2v - 4uv - 6v^2$
13. $6r^7 - 17r^5 - 21r^4 + 17r^3 + 7r^2 - 30r - 42$
14. $2w^2 + 13w - 7$
15. $2x^2 + 5xy - 12y^2$
16. $x^2 + 6xy + 9y^2$
17. $-2x^4 + \frac{3}{2}x^2 - \frac{7}{4}x + 2$
18. $-3xy^3 + y^2 - \frac{11}{3}x^2 - \frac{x^3}{y}$
19. $u^2 - 3u + 2$
20. $2v^3$ R7

PRETEST FOR CHAPTER 11, PAGE 454

1. $2 \times 7 \times 19$
2. Yes, $4 + 6 + 1 + 0 + 7 = 18$, which is divisible by 9
3. No, 07 is not divisible by 4
4. $2(13t - 9)$
5. $6p^2(p - 3)$
6. $27a^2bc^3(c - 3ab)$
7. $5s^2t^3u^2(4s^2 - 3tu^3 + 7st)$
8. $(y + 9)(y - 9)$
9. $2(u + 5)(u - 5)$
10. $(v^2 + w^2)(v + w)(v - w)$
11. $ab(a^2 + b^2)(a + b)(a - b)$
12. $(y - 7)(y + 6)$
13. $(w + 11)(w - 17)$
14. $(2x - 1)(3x + 4)$
15. $(3 - 2y)(1 - 2y)$
16. $7(p - 1)(p + 5)$
17. $3(2t + 1)(t - 3)$
18. $(y - 3)(x - 2)$
19. $(m - 1)(m + 2n)$
20. $2y(x + 2)(x + 3)$
21. $(x + 1)(x - 1)(2x + 3)$
22. $y = -2, \frac{1}{2}$
23. $y = -3, \frac{4}{3}, 3$
24. $p = \frac{1}{2}, \frac{3}{2}$
25. 9 and 23

PRETEST FOR CHAPTER 12, PAGE 498

1. $y = \frac{5}{3}$
2. $t = -3, -2$
3. $\frac{-1}{3xy}$ $(x \neq 0, y \neq 0)$
4. $\frac{p - 4}{2}$ $(p \neq -4)$
5. $r - 3$ $\left(r \neq -3, \frac{1}{2}\right)$
6. $\frac{3a^3b^2}{4}$ $(a \neq 0, b \neq 0)$
7. $\frac{6p^2}{q(p + q)}$ $(p \neq 0, q \neq 0, p \neq \pm q)$
8. $\frac{7}{3}(s - 1)t^2$ $(s \neq 1, t \neq 0)$
9. $\frac{x + 2}{x - 4}$ $(x \neq -6, -5, -2, 4)$
10. $\frac{-4}{2y - 3}$ $\left(y \neq \frac{3}{2}\right)$
11. $\frac{a + 3b}{m + 2}$ $(m \neq -2)$
12. $(y + 1)(y + 2)^2(y + 3)$
13. $\frac{2x^2 - x + 3}{4x^2 - 1}$ $\left(x \neq \frac{\pm 1}{2}\right)$
14. $\frac{5 - 2s}{(s^2 - 4)(s + 5)}$ $(s \neq -5, -2, 2)$
15. $\frac{1}{x^2 + 1}$ $(x \neq 0)$
16. $\frac{2}{a^2 + b^2}$ $(a \neq 0, b \neq 0)$
17. $x = 32$
18. $p = 12$
19. 5 and 10
20. $\frac{1}{3}$

PRETEST FOR CHAPTER 13, PAGE 542

1. IV
2. III
3. $y = \frac{3x - 5}{4}$
4. $y = \frac{4}{7}(x + 3)$
5. $x = \frac{7}{5}$
6. $y = -7$
7. Vertical
8. Neither

9.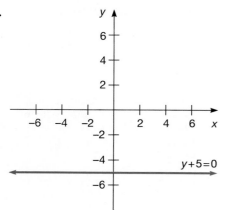

10.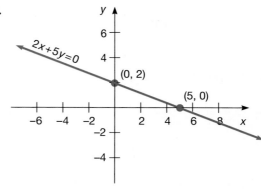

11. $\dfrac{10}{3}$

12. -1

13. Negative

14. Undefined

15.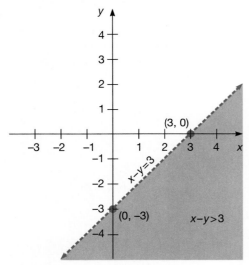

16. $q < \dfrac{2p + 7}{4}$

17. $t < \dfrac{3s + 8}{3}$

18. Any values greater than $\dfrac{-13}{6}$

19.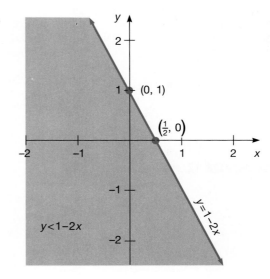

20.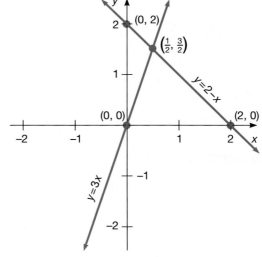

PRETEST FOR CHAPTER 14, PAGE 588

1.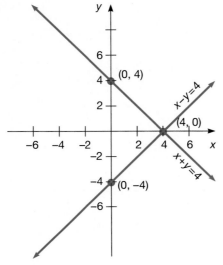

2.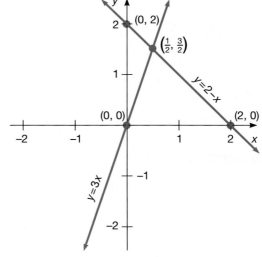

Answers to Chapter Pretests

3. $y = \frac{-2}{3}x + \frac{5}{3}$; $y = \frac{1}{2}x - \frac{1}{2}$; intersecting lines

4. $y = \frac{1}{3}x - \frac{2}{3}$; $y = \frac{1}{3}x - \frac{2}{3}$; coinciding lines

5. $\left(x = \frac{28}{17}, y = \frac{13}{17}\right)$ **6.** $(x = 1, y = -1)$ **7.** $(x = 1, y = -1)$

8. $\left(x = \frac{16}{17}, y = \frac{-5}{17}\right)$ **9.** 18 and 27 **10.** 16

PRETEST FOR CHAPTER 15, PAGE 622

1. 7 **2.** x^3 **3.** $4a^2$ **4.** $9x^2y3$ **5.** $\frac{2x + y}{4}$

6. $(3ab^2c)^4$ **7.** 10 **8.** -15 **9.** a^3 **10.** $-x^5$

11. $4xy^2z^3$ **12.** $10\sqrt{3}$ **13.** $xy^2\sqrt{x}$ **14.** $5abc^2\sqrt{2ac}$ **15.** $\frac{4}{9}$

16. $\frac{\sqrt{21}}{3}$ **17.** $\frac{\sqrt{6ab}}{2b}$ **18.** $3\sqrt{7}$ **19.** $6\sqrt{2} - 7\sqrt{3}$ **20.** $(x^2 - 2xy^2 - 4xy)\sqrt{y}$

21. $2\sqrt{6} - 18$ **22.** $6\sqrt{10} - 10 + 3\sqrt{6} - \sqrt{15}$ **23.** $\frac{3x - 2y}{xy}$ **24.** $t = 8$

25. $p = \frac{-4}{3}$

PRETEST FOR CHAPTER 16, PAGE 654

1. $3y^2 - 6y - 7 = 0$ **2.** $9u^2 - 12u - 4 = 0$ **3.** $x = -7, 3$ **4.** $y = -6, \frac{1}{2}$

5. $u = \frac{-1}{7}, 3$ **6.** $t = \frac{-4}{3}, \frac{7}{2}$ **7.** $x = -9, 9$ **8.** $y = -5, 5$

9. $t = 0, 8$ **10.** $p = -3, 6$ **11.** 16 **12.** $\frac{49}{4}$

13. $x = \frac{-5 \pm \sqrt{33}}{2}$ **14.** $u = -1, \frac{7}{5}$ **15.** 137 **16.** $y = \frac{-2 \pm \sqrt{7}}{3}$

17. $q = \frac{1 \pm \sqrt{22}}{3}$ **18.** 10 cm by 17 cm **19.** $\sqrt{170}$ cm **20.** No, since $14^2 \neq 8^2 + 10^2$

PRETEST FOR CHAPTER 17, PAGE 696

1. Line segment with endpoints C and D
2. Ray from R and in the direction of S
3. Line determined by the points E and F
4. 90° **5.** Acute **6.** 84° **7.** 46° **8.** 3
9. No **10.** $p \parallel r$ **11.** 93° **12.** 98° **13.** Not similar
14. Similar **15.** $10\sqrt{3}$ yd **16.** $8\sqrt{2}$ cm
17. $\angle B \cong \angle E$, $\overline{AB} \cong \overline{DE}$, $\overline{AC} \cong \overline{DF}$, $\overline{BC} \cong \overline{EF}$ **18.** Yes, SSS

INDEX

Abscissa, 545
Absolute value, 262
Acute angle, 704
Acute triangle, 719
Addition
  algebraic, 313–320, 322
  of decimals, 128–129
  of fractions, 78–84
  of mixed numbers, 91–93
  of polynomials, 426–427
  of rational expressions, 511–513, 518–521
  of signed numbers, 267–272
  of square roots, 635–638
  of whole numbers, 7–11
Additive identity, 8, 47
Additive property of 0, 8, 47
Adjacent angles, 708
Algebra, 308–325
Algebraic equations, 324
  solution of, 324, 334, 528, 658–675
Algebraic expressions, 308
  evaluating, 311
  factoring, 456
  simplifying, 313–320
Alternate exterior angles, 713
Alternate interior angles, 713
Angles, 703–710
  acute, 704
  adjacent, 708
  alternate exterior, 713
  alternate interior, 713
  base angles of isosceles triangle, 722
  bisector of, 705
  complementary, 709
  complement of, 709
  congruent, 705
  corresponding, 713
  exterior, 713
  interior, 713
  obtuse, 704
  right, 704
  side of, 703
  straight, 704
  supplementary, 709
  supplement of, 709
  vertex angle of isosceles triangle, 722
  vertex of, 703
  vertical, 708

Area, 230–240
  of circle, 237
  of parallelogram, 233
  of polygons, 24, 230–237
  of rectangle, 230
  of square, 232
  surface area of solid, 239
  of trapezoid, 236
  of triangle, 234
Assessed value, 185
Associative property
  for addition, 8
  for multiplication, 24
Axis, 545
  horizontal, 545
  vertical, 545

Base, 43, 722
Base angles of isosceles triangle, 722
Binomial, 423
  multiplying, 436–439
  squaring, 438
Bisector of angle, 705

Circles, 225–229, 237–238
  area of, 237
  circumference of, 152, 226
  diameter of, 225
  radius of, 225
Circumference of a circle, 152, 226
Coefficient, 309
  leading, 468, 656
Collinear points, 700
Commission, 188
Common factor, 459
Commutative property
  for addition, 8
  for multiplication, 23
Complementary angles, 709
Complement of angle, 709
Complete factorization, 479
Completing the square, 665–670
Complex rational expression, 522
  simplifying, 523
Composite number, 72, 456
  prime factorization of, 73–75, 456
Congruent
  angles, 705

799

Congruent (*Cont.*):
   line segments, 698
   polygons, 247
   triangles, 740
Consecutive integers, 356
Constant, 308
Coordinate, 545
Coplanar points, 700
Corresponding angles, 713
Counting, 4
Cube, 240, 246
   surface area of, 240
   volume of, 246
Cubing a number, 43
Cylinder, 248
   volume of, 249

Decimal fractions, 125
Decimal number, 125
Decimal point, 124
Decimals, 124–152
   addition of, 128–129
   approximating, 143
   division of, 138–143
   multiplication of, 131–134
   ordering of, 126
   repeating, 148
   rounding, 135–136
   subtraction of, 128–130
   terminating, 148
Decimal to fraction conversion, 145
Decimal to percent conversion, 169
Degree
   of a polynomial, 424
   of a term, 424
Denominator, 59
   least common, 82
Descending powers, 425
Diameter of a circle, 225
Difference, 14
Difference of two squares, 465
Digit, 4
   significant digits, 412
Discount, 192
Discriminant, 674
Distributive property
   for multiplication over addition, 41
   for multiplication over subtraction, 41
Dividend, 31
Divisibility tests, 74, 457
Division
   algebraic, 322
   of decimals, 138–143
   of fractions, 108–111
   of mixed numbers, 108–111
   of polynomials, 441–447
   of rational expressions, 507–509
   of signed numbers, 286–289
   of square roots, 642–644
   of whole numbers, 30–35
Division involving 0, 48–49
Divisor, 31

Divisor (*Cont.*):
   proper, 493

Endpoint
   of line segment, 217, 698
   of ray, 700
Equations
   algebraic, 324
   equivalent, 335
   graphing method for solving system of, 590–597
   involving rational expressions, 526–530
   involving square roots, 646–648
   linear in one variable, 334
   linear in two variables, 550
   quadratic in one variable, 656
   solution of, 324, 334, 658–674
   solving by factoring, 481–485, 658–660
   solving quadratic equations by completing the square, 665–670
   solving quadratic equations by quadratic formula, 671–674
   solving quadratic equations using square roots, 661–664
   standard form of a quadratic equation, 656
Equiangular triangle, 719
Equilateral triangle, 218, 719
Equivalent equations, 335
Equivalent fractions, 65
Estimating, 38, 136
Evaluating algebraic expressions, 311–312
Excluded value, 500
Exponent, 43, 394–408
   negative integer, 406–408
   positive integer, 394–404
   zero, 405
Exterior angles, 713
Extremes, 380

Factor, 23, 308, 456
   common, 459
   greatest common, 461
Factoring, 456–480
   algebraic expressions, 456
   common, 459–464
   difference of two squares, 464–467
   factoring completely, 479–480
   by grouping, 476–478
   strategy for, 479
   trinomials of the form $ax^2 + bx + c$, 472–475
   trinomials of the form $x^2 + bx + c$, 468–471
FOIL, 436
Formula, 369
Fractions, 58–111
   addition of, 78–84
   decimal fractions, 125
   denominator of, 59
   division of, 108–111
   equivalent, 65
   Fundamental Principle of, 67

Fractions (*Cont.*):
   improper, 59
   like, 78
   in lowest terms, 69
   multiplication of, 98–103
   numerator of, 59
   ordering of, 86–88
   proper, 59
   reciprocal of, 105
   simplest form of, 69
   subtraction of, 88–89
   unlike, 78
Fraction to decimal conversion, 146
Fraction to percent conversion, 162
Fulcrum, 364
Fundamental Principle of Fractions, 67, 501

Geometry, 217–250, 698–744
Graphing
   half-planes, 575
   linear equations in two variables, 553–558
   linear inequalities in two variables, 574–580
   straight lines, 553–558
   systems of linear equations, 590–609
Greatest common factor, 461
Gross sales, 188

Half-plane, 575
   boundary of, 575
Hexagon, 218
Hypotenuse, 488, 681

Identity property
   for addition, 8
   for multiplication, 23
Improper fraction, 59
Inequality, 565–580
   addition property of, 567
   graphs of linear inequalities in two variables, 574–580
   linear inequalities in one variable, 565–570
   linear inequalities in two variables, 572–580
   multiplication property of, 568
Integers, 260–297
   absolute value of, 262
   addition of, 267–272
   consecutive integers, 356
   division of, 286–289
   multiplication of, 279–285
   negative, 261
   opposite of, 261
   positive, 261
   subtraction of, 275–277
   zero, 261
Intercepts, 555
   x-intercept, 555
   y-intercept, 555
Interest, 195
   simple, 196
Interior angles, 713
Intersecting lines, 592
Is approximately equal to, 143

Is greater than, 7, 261
Is less than, 7, 261
Isosceles trapezoid, 236
Isosceles triangle, 218, 719

Leading coefficient, 468, 656
Least common denominator, 82
Least common multiple, 76, 515
Leg of right triangle, 488, 681
Lever, 364
Lever principle, 364
Like fractions, 78
Like rational expressions, 511
Like square roots, 636
Like terms, 314
Linear binomial, 468
Linear equations, 334–352, 550–558
   graphing system of, 590–597
   in one variable, 334
   solution of, 334
   in two variables, 550
Linear inequalities, 565–580
   in one variable, 565–570
   in two variables, 572–580
Lines, 553–564, 698
   coinciding, 592
   graphing, 553–558
   horizontal, 563
   intersecting, 592
   oblique, 563
   parallel, 592
   perpendicular, 712
   slope-intercept form of equation of, 595
   slope of, 559
   transversal, 713
   vertical, 563
Line segment, 217, 698
   congruent line segments, 698
   endpoints, 217, 698
   length of, 698
   midpoint of, 699

Means, 380
Measurement, 210–250
   of angle, 704
   area, 230–240
   degree, 704
   English system, 210
   metric system, 210
   radian, 704
   units of, 210
   volume, 244–250
Metric conversions, 212–215
Midpoint of line segment, 699
Mixed numbers, 61
   addition of, 91–93
   division of, 108–111
   multiplication of, 104–106
   subtraction of, 94–95
Mixed operations
   with signed numbers, 292–294
   with whole numbers, 40–46

Moment, 364
Monomial, 423
Multiple of whole number, 76
   common multiple, 76
   least common multiple, 76, 515
Multiplication
   algebraic, 322
   of binomials, 436–439
   of decimals, 131–134
   of fractions, 98–103
   of mixed numbers, 104–106
   by one-half, 323
   of polynomials, 431–434
   of rational expressions, 505–506
   of signed numbers, 279–285
   of square roots, 639–642
   by two, 323
   of whole numbers, 22–28
Multiplication property of 0, 48
Multiplicative identity, 23

Natural numbers, 7
Negative integer, 261
Net sales, 188
Number line, 260
Numbers, 4
   addition of, 7–11, 78–84, 91–93, 128–129, 267–272
   composite, 72, 456
   cubing, 43
   decimal, 124–152
   division of, 30–35, 108–111, 138–143, 286–289
   fractions, 58
   integers, 260
   mixed, 61
   multiplication of, 22–28, 98–103, 131–134, 279–285
   natural, 7
   ordering of, 7, 87
   perfect, 493
   prime, 72, 456
   rational, 264
   signed, 264–297
   squaring, 43
   subtraction of, 14–20, 88–89, 128–130, 275–277
   whole, 7
Numerator, 59
Numerical coefficient, 309

Obtuse angle, 704
Obtuse triangle, 719
Opposite of a number, 261
Ordered pair, 545
Ordering
   of decimals, 126
   of fractions, 86–88
   of integers, 261–262
   of whole numbers, 7
Ordinate, 545
Origin, 544

Parallelogram, 219
   area of, 233
   perimeter of, 222
Pentagon, 218
Percent, 162–199
Percent decrease, 178–180
Percent increase, 178–180
Percent to decimal conversion, 168
Percent to fraction conversion, 165
Perfect number, 493
Perfect square, 466, 663
Perimeter, 219–228
   of a parallelogram, 222
   of a polygon, 10, 219
   of a rectangle, 220
   of a square, 221
Perpendicular lines, 712
Pi (π), 226
Place value, 3
Plane, 545, 700
Please Excuse My Dear Aunt Sally, 44
Points
   collinear, 700
   coplanar, 700
Polygons, 10, 217–223, 230–237
   area of, 24, 230–237
   congruent, 247
   perimeter of, 10, 219–223
   sides of, 10
Polynomials, 422–447
   addition of, 426–427
   classification of, 422–425
   degree of, 424
   division of, 441–447
   multiplication of, 431–434
   in one variable, 422
   subtraction of, 428–430
   in two or more variables, 423
Positive integers, 261
Powers, 323
   descending, 425
   of ten, 134
Prime factorization, 73, 456
Prime number, 72, 456
Principal square root, 625
Prism, 247
   volume of, 248
Product, 23, 308
Proper divisor, 493
Proper fraction, 59
Property tax, 184
Proportion, 379
   extremes, 380
   means, 380
Pythagorean Theorem, 682

Quadrants, 545
Quadratic equations, 656–674
   in a single variable, 656
   solving by completing the square, 665–670
   solving by factoring, 658–660
   solving by quadratic formula, 671–674

Index

Quadratic equations (*Cont.*):
  solving by using square roots, 661–664
  standard form of, 656
Quadratic formula, 671
Quadratic trinomial, 468
Quadrilateral, 218
  parallelogram, 219
  rectangle, 219
  square, 219
Quotient, 31

Radian, 704
Radical sign, 624
Radicand, 624
Radius of circle, 225
Ratio, 377
Rational expressions, 500–534
  addition of, 511–513, 518–521
  complex, 522
  division of, 507–509
  like, 511
  multiplication of, 505–506
  simplifying, 501–503, 523
  subtraction of, 511–513, 518–521
  unlike, 518
Rationalizing the denominator of a fraction, 633
Rational numbers, 264
  simplest form of, 291
Ray, 700
  endpoint of, 700
Reciprocal, 105
Rectangle, 219
  area of, 230
  perimeter of, 220
Rectangular coordinate system, 544–548
  abscissa, 545
  coordinate, 545
  horizontal axis, 545
  ordered pair, 545
  ordinate, 545
  origin, 544
  plane, 545
  quadrants, 545
  scale, 544
  vertical axis, 545
Remainder, 33
Removing symbols of grouping, 317
Repeating decimal, 148
Right angle, 704
Right triangle, 681, 719
  hypotenuse of, 488, 681
  legs of, 488, 681
  special right triangles, 731–737
Rounding, 37–38, 135–136

Sales
  gross, 188
  net, 188
Sales tax, 182
Scale, 544
Scalene triangle, 218, 719
Scientific notation, 409–413

Sides of polygons, 10
Signed numbers, 264–297
  absolute value of, 262
  addition of, 267–272
  division of, 286–289
  mixed operations with, 292–294
  multiplication of, 279–285
  subtraction of, 275–277
Significant digits, 412
Similar triangles, 725–728
Simple interest, 196
Simplest form
  of a fraction, 69
  of a rational number, 291
Simplifying
  algebraic expressions, 313–320
  rational expressions, 501–503
  square roots, 627–634
Slope of a line, 559
Slope-intercept form of equation of a line, 595
Solids
  cylinder, 248
  prism, 247
  rectangular, 245
  sphere, 249
Solution
  of algebraic equations, 324, 334
  elimination method, 598–605
  of equations by factoring, 481–485, 658–660
  of equations involving rational expressions, 526–530
  of linear equations, 334, 550–552
  of linear inequalities, 565–580
  of quadratic equations by completing the square, 665–670
  of quadratic equations by using the quadratic formula, 671–674
  of quadratic equations by using square roots, 661–664
  substitution method, 606–609
  of systems of linear equations, 590–609
Sphere, 249
  volume of, 250
Square, 219
  area of, 232
  completing the, 665
  perfect square, 466, 663
  perimeter of, 221
Square roots, 624–648
  addition and subtraction of, 635–638
  equations involving, 646–648
  like, 636
  multiplication and division of, 639–644
  principal square root, 625
  simplifying, 627–634
Squaring a binomial, 438
Squaring a number, 43
Standard form of a quadratic equation, 656
Straight angle, 704
Subtraction
  algebraic, 322
  of decimals, 129–130

Subtraction (*Cont.*):
   of fractions, 88–89
   of mixed numbers, 94–95
   of polynomials, 428–430
   of rational expressions, 511–513, 518–521
   of signed numbers, 275–277
   of square roots, 635–638
   of whole numbers, 14–20
Sum, 8, 308
Superscript, 43
Supplement of angle, 709
Supplementary angles, 709
Surface area of a solid, 239

Term, 308
Terminating decimal, 148
Tests for divisibility, 74, 457
Transversal, 713
Trapezoid, 236
   area of, 236
   bases of, 236
   isosceles, 236
Triangle, 218
   acute, 719
   area of, 234
   congruent, 740
   equiangular, 719
   equilateral, 218, 719
   hypotenuse, 488, 681
   isosceles, 218, 719
   leg of right triangle, 488, 681
   obtuse, 719
   right, 681, 719, 731–737
   scalene, 218, 719
   similar, 725–728

Triangle (*Cont.*):
   special right triangles, 731–737
Trinomial, 423
   factoring, 468–475
   quadratic, 468

Units of measurement, 210–215
Unlike fractions, 78
Unlike rational expressions, 518
Unlike terms, 314

Variable, 308
Vertex
   of angle, 703
   angle of isosceles triangle, 722
Vertical angles, 708
Volume, 244
   of a cube, 246
   of a cylinder, 249
   of a prism, 248
   of a rectangular solid, 245
   of solids, 244–250
   of a sphere, 250

Whole numbers, 7–49
   addition of, 7–11
   division of, 30–35
   multiple of, 76
   multiplication of, 22–28
   ordering of, 7
   subtraction of, 14–20

Zero
   operations involving, 47–49
   properties of, 8, 47